Rabies

Rabies

Second edition

Edited by

Alan C. Jackson
Queen's University, Kingston, ON, Canada

William H. Wunner
The Wistar Institute, Philadelphia, PA, USA

AMSTERDAM • BOSTON • HEIDELBERG • LONDON • NEW YORK • OXFORD
PARIS • SAN DIEGO • SAN FRANCISCO • SINGAPORE • SYDNEY • TOKYO
Academic Press is an imprint of Elsevier

Academic Press is an imprint of Elsevier
84 Theobald's Road, London WC1X 8RR, UK
30 Corporate Drive, Suite 400, Burlington, MA 01803, USA
525 B Street, Suite 1900, San Diego, CA 92101-4495, USA

First edition 2002
Second edition 2007

The following chapters are US government works in the public domain:
Diagnostic Evaluation – Charles V. Trimarchi and Susan A. Nadin-Davis
Rabies in Terrestrial Animals – Cathleen A. Hanlon, Michael Niezgoda, and Charles E. Rupprecht
Bat Rabies – Ivan V. Kuzmin and Charles E. Rupprecht

Notice
No responsibility is assumed by the publisher for any injury and/or damage to persons or property
as a matter of products liability, negligence or otherwise, or from any use or operation of any
methods, products, instructions or ideas contained in the material herein. Because of rapid
advances in the medical sciences, in particular, independent verification of diagnoses and drug
dosages should be made

British Library Cataloguing in Publication Data
A catalogue record for this book is available from the British Library

Library of Congress Cataloguing in Publication Data
A catalogue record for this book is available from the Library of Congress

ISBN: 978-0-12-369366-2

For information on all Academic Press publications
visit our web site at http://books.elsevier.com

Typeset by Charon Tec Ltd (A Macmillan Company), Chennai, India
www.charontec.com

Printed and bound in Great Britain

07 08 09 10 11   10 9 8 7 6 5 4 3 2 1

Contents

Foreword

Since the publication of the first edition of this book, *Rabies*, in 2002, there has been much progress in research on rabies virus and the other lyssaviruses and the disease they cause, i.e. clinical rabies. The reservoir and terminal hosts and their immune defenses, the pathogenesis of the disease in humans and animals, the natural history of the viruses in nature and the public health strategies and actions for prevention and control in developed and developing countries have also been the focus of intense research. The editors, Alan Jackson and Bill Wunner, recognizing this progress and seeing the continuing need to keep the subject 'on the front burner', envisioned this second edition and, we hope, envision further editions far into the future. The exciting content of this edition proves the foresight of the editors. Those of us who care about rabies virus and the other lyssaviruses and the disease they cause are grateful that the editors and the chapter authors have taken on the hard work to bring this edition to fruition.

Progress in research and, indeed, in public health disease prevention and control strategies and actions seems to be made in fits-and-starts – not smoothly across the whole subject, but concentrated in 'hot spots' that are often grounded in emerging technologies, personal interests of investigators and, most importantly, in changing disease threats. Although I am interested in every facet of rabies, from the viruses themselves, to their discovery and natural history, to their molecular biology, and in all aspects of the disease they cause, from its pathology and epidemiology, to public health disease prevention and control strategies and actions, as I read this edition I found myself exercising old biases and focusing on particular favorite 'hot spots'. I am sure that others will see different 'hot spots', but for the moment here are my favorites:

- *Pasteur* – again and always – the father of the infectious disease sciences, who used rabies as the basis for so many lessons. Like so many other virologists and infectious disease scientists, I never get tired of Pasteur's charge: '*Dans les champs de l'observation le hasard ne favorise que les esprits préparés.*' '*In matters of observation chance favors only the prepared mind.*' These words come from a speech Pasteur gave at the age of thirty-one as he was installed as Professor and Dean of the newly created Faculty of Sciences at Lille, well in advance of any of the discoveries for which he is best remembered. As I read this second edition I thought that Pasteur's words are well aimed at the youngest, most forward-looking of rabies researchers.

- *Modern history* – my classmate, George Baer, has provided a comprehensive history chapter emphasizing the classical aspects of rabies history, a subject I never tire of. As I think about this, I like to reflect also on the modern history of the disease and the virus. I say this because as the generations turn over, newcomers to rabies research and public service can learn much from their predecessors, their pre-*PubMed* predecessors. In the 20th century there were many wonderful discoveries and discoverers that advanced our understanding of rabies: for example, Remlinger (1903) and the proof that rabies is caused by a virus; Negri (1903) and discovery of the Negri body; Goodpasture (1925) and the beginnings of rabies experimental pathology; Hoyt and Jungeblut (1930) and the utility of the laboratory mouse; Queiroz-Lima (1934) and the beginnings of the vampire bat rabies story; Kliger and Bernkopf (1938) and the adaptation of rabies virus to grow in embryonating chicken eggs; Johnson and Koprowski and their colleagues, over many decades making many practical discoveries; Constantine (1968) and the ecology of bat rabies in the USA; Kissling (1958) and early cultivation of rabies virus in cell culture; Kissling and Goldwasser (1958) and the application of immunofluorescence to rabies diagnostics; Davies and colleagues and Atanasiu and colleagues (1963) and the first negative contrast electron microscopic images of rabies virus; Matsumoto and Miyamoto (1965) and the first thin-section electron microscopy of rabies virus and the Negri body; Baer, Schneider, and others (1960s) and initial rabies pathogenesis studies; Dierks and Murphy (1968) and the first study of the nature of rabies infection in the salivary gland; Murphy and colleagues (1973) and comprehensive rabies pathogenesis studies in experimental animals; Shope and colleagues (1970) and the first identification of rabies-related viruses (Mokola and Lagos bat virus); Meredith and colleagues (1970) and the discovery of Duvenhage virus; Wiktor, Sokol, Wunner, Clark, Dietzschold, Obijeski, Flamand and others (1970s) and the first rabies molecular virology studies (RNA, proteins, replication); Wiktor, Koprowski, Smith and others (1978) and the development of panels of monoclonal antibodies to type rabies virus antigenic variants; Smith, Rupprecht and colleagues and the use of partial genomic sequencing to characterize rabies virus genotypes and to build the first phylogenetic trees; Schneider, Bourhy, Tordo, Grauballe, Nieuwenhuijs, King, Fooks, Dietzschold, Rupprecht and others (~1977–1993) and the discovery and genetic characterization of the European bat lyssaviruses, EBL1 and EBL2; Fraser, Gould, Hooper and colleagues (1996) and the discovery of Australian bat lyssavirus; and so many other people and discoveries. This kind of list is dangerous in that I am sure to have left out many key discoveries and discoverers and, in so doing, I will have left myself open to justifiable criticism – nevertheless, I think it is important to recognize and remember these pioneers.

- *Molecular epidemiology* – as Susan Nadin-Davis states in her chapter, molecular epidemiology has become a central theme in the study of virus–host relationships and the co-adaptation of particular hosts and particular virus genotypes. Phylogenic analysis is also providing powerful insight into the natural history of each virus genotype, especially now that we have a data-based global perspective. Recent discoveries of important 'spillover events' as particular virus genotypes have successfully jumped from one reservoir animal species to another are very important; appreciation of these events is largely dependent upon molecular epidemiologic approaches. Darwin lives – evolution continues...

- *Taxonomy and strain characterization* – extending from the genomic sequence analyses used in molecular epidemiology (above), there has been a remarkable application of phylogenetic programs that have uncovered the complex genetic structure of the lyssavirus genus. In the dendrograms, now including lyssavirus isolates from around the world, we can see the ancient rather insular econiches of various genotypes, especially the econiches constituted by the many bat species that serve as reservoir hosts. We can also see the evidence of the long-term radiation of particular genotypes into new niches. This is wonderful stuff for anyone interested in natural history of viruses. On a more formal level, that of the International Committee on Taxonomy of Viruses, some of the genomic information being placed in the genus-wide dendrograms may challenge our notions of the fine structure of the genus, but that is the way of taxonomy. In the end, the lyssavirus dendrogram makes sense. I am reminded of the extension of the Henle-Koch postulates, the elements of the proof of causation of specific diseases by specific microorganisms, by Alfred Evans. Evans' extra criteria for the proof of causation were developed to deal with those viruses and viral diseases where the original postulates were found to be inadequate. Evans' postulates add modern technologies to characterize molecularly viruses and also add improved approaches to the understanding of viral pathogenesis. The last of Evans' ten postulates states, *'And the whole thing should make biologic and epidemiologic sense.'* Clearly, the dendrograms outlining the phylogenetic relationships of the genus *Lyssavirus* 'make sense' from the perspective of understanding the natural history of rabies virus and the related viruses. And, we now can see how the lyssaviruses relate to the other member viruses of the family Rhabdoviridae and to the more distantly related member viruses of the order Mononegavirales. Wonderful!

- *Terrestrial rabies in newly discovered reservoir host econiches* – on one hand, here in the 21st century, we might think we should know about all the econiches occupied by lyssavirus genotypes around the world, but the lesson of recent discoveries is that the world still contains geographic and ecologic niches that hold secrets. In the past few years, three such secrets of particular interest have been unraveled:

  1 rabies in non-human primates in Brazil
  2 rabies-related viruses (lyssaviruses) in carnivores and bats in subarctic and arboreal zones of Eurasia
  3 rabies virus in spotted hyenas in the Serengeti.

The latter, as noted by Alan Jackson in this volume, changes our view of the pathogenicity of the lyssavirus genotypes circulating in terrestrial mammals – the distinct genotype found in spotted hyenas in the Serengeti is not associated with fatal disease, but with subclinical infection, chronic viral carriage and long-term shedding. Much more must be learned about this genotype and this econiche.

- *Bat rabies* – Ivan Kuzmin and Charles Rupprecht have, in this volume, advanced in remarkable fashion our general knowledge about the natural history, population biology and epidemiology of bat rabies. At last we have a discerning assessment of the older literature on this subject and, for the first time, in my view, we are able to separate fact from fiction regarding rabies in diverse Chiropteran species. And, this subject is far from maturity: just in the past few years we have seen several additional lyssaviruses from bats added to our listings (e.g. Aravan, Khujand, Irkut and West Caucasian bat viruses). Everything we learn about bat rabies is important in guiding human and domestic animal disease prevention and control programs and in guiding bat conservation strategies. The notion that bats are the true ancestral reservoir hosts of rabies virus and other lyssaviruses, as reviewed so nicely by Ivan Kuzmin and Charles Rupprecht, suggests that the entrenched viral econiches are here to stay and that tangential means of disease control will have to serve the needs of public health while, at the same time, serving the interests of conservation. Perhaps this is a lesson for those dealing with many other zoonotic diseases.

- *Pathogenesis and molecular pathogenesis* – one reminder here that bears on many public health control strategies is that viral 'exposure' does not always lead to productive viral infection and a detectable immune response. As Alan Jackson states in his chapter on rabies pathogenesis, most of what we know about the events that take place during rabies infection has been learned from experimental animal models, too often employing fixed laboratory strains of virus and exquisitely susceptible hosts. We need more research employing models that more closely mimic street virus disease in humans and reservoir host animal species. Perhaps then we will be better able to understand the meaning of the presence of small amounts of viral nucleic acid, as detected by PCR, in feral hosts in the absence of progressive disease. Separately, we still have a long way to go to understand the neuro-anatomical, neuropathological bases for the behavioral changes in animals with rabies. The mystery so nicely described by Richard Johnson in 1971 still faces us: '*It is difficult to harmonize the dramatic clinical signs and lethal outcome of street rabies virus infection with the paucity of pathological changes in neurons. The greater localization to the limbic system with relative sparing of the neocortex provides a fascinating clinicopathologic correlate with the alertness, loss of natural timidity, aberrant sexual behavior and aggressiveness that may occur in clinical rabies. No other virus is so diabolically adapted to selective neuronal populations*

that it can drive the host in fury to transmit the virus to another host animal.'
Clearly, much more research is needed in this area.

- *Diagnostics* – the direct immunofluorescence technique has served as the cornerstone of rabies diagnosis for the past half century. What was in the days of my initial introduction to virology the most avant-garde diagnostic test in all virology is now the oldest standard test in common use. Charles Trimarchi and Susan Nadin-Davis, in this volume, capture very well the pros-and-cons for change. On the one hand, they point out that improved tests mostly employing PCR variations are available that can be performed more rapidly and at lower cost, and have increased sensitivity and specificity. On the other hand, there is much worry that such tests even when made foolproof in proprietary platforms run the risk of sample cross-contamination (false positives) and unforeseen technical glitches (false negatives) in the setting of the usual rabies labs around the world. While research continues, with a test as robust as the 'home pregnancy test' as the goal, it is no wonder that the WHO still recommends direct immunofluorescence as the gold standard post-mortem diagnostic test for rabies.

- *Post-exposure rabies treatment* – potent anti-rabies immunoglobulin (human or humanized) and potent vaccine regimens have a proven track record, yet there is a crucial need for new ideas, innovative approaches, all favoring cost-containment. Fewer than 1% of post-exposure treatments used in developing countries include modern vaccine and appropriately potent anti-rabies immunoglobulin. Limited availability clearly is a consequence of high costs. Stopgap approaches have been tried, but our real need is for revolutionary technological advances that are exportable to all rabies-endemic countries. This goal seems as far in the future as ever, but if research should be prioritized to deal with our most important problems, then here is the most important research target in the rabies public health enterprise.

- *New vaccines* – following the advent of cell-culture-based rabies vaccines in the 1970s, there have been few major conceptual advances, at least in the downstream end of vaccine development, application and use. As Bill Wunner notes in his chapter, new generation rabies vaccines must be developed, which are inexpensive, stable, requiring one or at most two doses and, hopefully, even efficacious by the oral route of administration. Many rabies 'immunogens' have been developed using various molecular technologies, but it is a long way from expressing rabies virus proteins *in vitro* to the construction of a vaccine ready for clinical trials and field application. More needs to be done to expand the downstream vaccine development infrastructure that will yield the kinds of vaccine for use in rabies-endemic countries, as envisioned by Bill Wunner. Further, the differences in the goals for human vaccines versus domestic animal vaccines versus wildlife vaccines are clear enough, but there are common downstream technology and infrastructure development issues that suggest the need for further cross-fertilization.

- *Public health organization and action, locally, nationally and globally* – despite the work of the WHO over many years, there is still a notorious lack of surveillance and disease burden data for some rabies-endemic countries. At the same time dramatic decreases in human rabies cases have been reported in recent years in other countries, particularly in South America. This has followed implementation of programs for improved post-exposure treatment of humans and vaccination of dogs. As Deborah Briggs and B.J. Mahendra note in their chapter, under-estimating the importance of rabies leads decision-makers to perceive rabies as a rare disease, resulting from the bite of an economically unimportant animal (the dog) and therefore not worthy of much prevention and control funding. The fatalistic sense in some parts of the world that 'rabies will always be with us' adds to the perception. As T. Jacob John stated recently when reviewing two human rabies cases in Kerala State, Southern India: *'The adage "familiarity breeds contempt" is apt here. Dogs are common, dog-bites are also common. People often do not take bites seriously. There are several factors involved in such negligent behavior: belief in pre-destiny or fatalism; non-perception of personal risk; lack of authentic information from public health agencies; availability of non-scientific remedies; inadequate health education in schools, etc. Needless to say, such tragic deaths are preventable.'* Surely, with the experiences gained in changing public perception about other diseases, such as AIDS and tuberculosis, social science research can contribute to solving this kind of problem as it impacts on rabies in developing countries. Additionally, in many countries, rabies falls in the crack between departments (ministries) of health and agriculture, adding to the problem of setting proper priorities, increasing public awareness and gaining appropriate political support.

- *Enlisting a new generation of rabies research scientists and public health practitioners* – seemingly the ghost of Pasteur, the nature of rabies as a global health problem and the nature of the research and public service work to be done should draw young would-be scientists and public health practitioners into the field like 'bears to honey'. What a romantic calling to young people who want fulfilling careers! The questions and sense of purpose that have brought inquiring young people into my office over the years should be answered with career advice pertinent to careers dealing with all the zoonoses. In some instances, careers in rabies research and public service have been described in rather insular terms – for too long the rabies unit and its staff has stood alone down at the end of the hallway in many public health institutions. This has been a mistake, in my view: rabies belongs in the same career advice package as descriptions of careers dealing with all zoonoses, including Ebola hemorrhagic fever, West Nile virus encephalitis,

Hantavirus pulmonary syndrome, SARS, avian influenza, and others that are identified as 'new, emerging or re-emerging infectious diseases'. I think that this volume, reflecting the state of the art in rabies research and public service, is full of 'hooks' that should influence career decisions of the next generation of would-be scientists and public health practitioners.

This second edition of *Rabies* not only brings to the reader a wealth of new information representing the major advances in research on rabies virus and the other lyssaviruses and on the disease, rabies, it also provides an updated scientific base for guiding rabies prevention and control programs. Public health officials are on the front lines in global efforts to minimize the human and economic burden presented by the lyssaviruses. The practical application of research breakthroughs, such as improved vaccines and diagnostics, is the key to dealing with rabies as a global health problem. In this edition, we can see many remarkable advances in disease prevention and control programs in many developing countries. Even so, however, given the prevalence of rabies in many of the least developed countries of the world, we must continue with all facets of research and development, always thinking of least-cost, and develop the most practicable intervention strategies. There is still much to be done.

I would, in closing, like to express again my congratulations to the editors, Alan Jackson and Bill Wunner, for their vision in seeing the need for this second edition and for their hard work in bringing their vision to reality. I would also like to congratulate the authors for contributing such excellent chapters. I sense their love of the work and their fidelity to Pasteur's pioneering leadership. As I think of the editors and authors, I conclude that the scientific base for rabies prevention and control, globally, is in good hands.

Frederick A. Murphy
Department of Pathology
University of Texas Medical Branch
Galveston, Texas, USA

Preface

Globally, rabies continues to be an important disease of humans and animals and it is one of the deadliest human diseases. During the past five years since publication of the first edition of *Rabies*, there have been new developments in many areas on the rabies landscape. This edition takes on a more global perspective on rabies. Many new authors, who are leaders in their areas of expertise, offer fresh outlooks on their topics. Rabies has a very rich and lengthy history dating from antiquity. Pasteur's first human rabies vaccination in 1885 followed centuries of terror associated with the disease and ineffective 'treatments' after bites by rabid animals. There were countless superstitions and misconceptions about the disease, many of which exist disquietly to this day in many parts of the world. Rabid dogs continue to be responsible for most human deaths, while wild carnivores, including foxes, skunks and raccoons, maintain sylvatic reservoirs and enzootic cycles of rabies virus infection. A new chapter on bat rabies has been added in this edition because of its importance. Bats are indigenous, quintessential reservoirs for lyssaviruses (rabies and rabies-related viruses) on a global basis and the associated virus variants are responsible for continuing human and veterinary mortalities in both developed and developing countries. Information on vaccines to control rabies virus infection has been expanded into three separate chapters. There have been considerable new developments in epidemiology and molecular epidemiology. New approaches to therapy of human rabies are discussed as well as issues involving organ and tissue transplantation. This edition describes recent successes in controlling wildlife rabies using a variety of innovative methods in the USA, Canada and Europe.

This new edition is again a truly multidisciplinary effort by contributors in the fields of medicine, veterinary medicine, virology, immunology, wildlife biology, epidemiology and geography, and we are very grateful for their efforts. We believe that this edition will again be very useful to medical and veterinary clinicians, public health officials, research scientists and students in their efforts to understand better rabies and to diagnose and prevent the disease worldwide. We are making progress, but much more work remains to be done on this ancient disease.

Alan C. Jackson
William H. Wunner

Contributors

George M. Baer, Laboratorios Baer, S.A. de C.V., Hacienda Sta. Maria Xalostóc, Tlaxco, Tlaxcala 90250, Mexico

Deborah J. Briggs, Rabies Diagnostic Laboratory, Department of Diagnostic Medicine/ Pathobiology, College of Veterinary Medicine, Kansas State University, Manhattan, Kansas 66506, USA

James E. Childs, Department of Epidemiology and Public Health and YIBS Center for Eco-Epidemiology, Yale University School of Medicine, New Haven, Connecticut 06510, USA

Sarah Cleaveland, Centre for Tropical Veterinary Medicine, Royal (Dick) School of Veterinary Studies, University of Edinburgh, Roslin, Midlothian, EH25 9RG, UK

David W. Dreesen, Department of Infectious Diseases, College of Veterinary Medicine, The University of Georgia, Athens, Georgia 30602, USA

Eric Fèvre, Centre for Infectious Diseases, University of Edinburgh, Ashworth Labs, Edinburgh, EH9 3JF, UK

Chandra R. Gordon, Rabies Diagnostic Laboratory, Department of Diagnostic Medicine/ Pathobiology, College of Veterinary Medicine, Kansas State University, Manhattan, Kansas 66506, USA

Cathleen A. Hanlon, Centers for Disease Control and Prevention, Atlanta, Georgia 30333, USA

Alan C. Jackson, Departments of Medicine (Neurology) and of Microbiology and Immunology, Queen's University, Kingston, ON K7L 3N6, Canada; currently Departments of Internal Medicine (Neurology) and of Medical Microbiology, University of Manitoba, Winnipeg, MB R3A IR9, Canada

David H. Johnston, Johnston Biotech, Sarnia, ON N7V 3B5, Canada

Magai Kaare, Centre for Tropical Veterinary Medicine, Royal (Dick) School of Veterinary Studies, University of Edinburgh, Roslin, Midlothian, EH25 9RG, UK

Darryn Knobel, Centre for Tropical Veterinary Medicine, Royal (Dick) School of Veterinary Studies, University of Edinburgh, Roslin, Midlothian, EH25 9RG, UK

Ivan V. Kuzmin, Rabies Section, Centers for Disease Control and Prevention, Atlanta, Georgia 30333, USA

Monique Lafon, Department of Virology, Pasteur Institute, 75724 Paris Cedex 15, France

B.J. Mahendra, Department of Community Medicine, Mandya Institute of Medical Sciences, Mandya, Karnataka 571401, India

Susan M. Moore, Rabies Diagnostic Laboratory, Department of Diagnostic Medicine/Pathobiology, College of Veterinary Medicine, Kansas State University, Manhattan, Kansas 66506, USA

Frederick A. Murphy, University of Texas Medical Branch, Galveston Department of Pathology, Galveston, Texas 77555-0609, USA

Susan A. Nadin-Davis, Centre of Expertise for Rabies, Canadian Food Inspection Agency, Ottawa Laboratory (Fallowfield), Ottawa, ON K2H 8P9, Canada

Michael Niezgoda, Centers for Disease Control and Prevention, Atlanta, Georgia 30333, USA

Leslie A. Real, Department of Biology and Center for Disease Ecology, Emory University, Atlanta, Georgia 30322, USA

Richard C. Rosatte, Ontario Ministry of Natural Resources, Wildlife Research and Development Section, Trent University, Peterborough, ON K9J 7B8, Canada

John P. Rossiter, Department of Pathology and Molecular Medicine, Richardson Laboratory, Queen's University, Kingston, ON K7L 3N6, Canada

Charles E. Rupprecht, Rabies Section, Centers for Disease Control and Prevention, Atlanta, Georgia 30333, USA

Rowland R. Tinline, Queen's GIS Laboratory, Department of Geography, Queen's University, Kingston, ON K7L 3N6, Canada

Charles V. Trimarchi, Wadsworth Center, New York State Department of Health, Albany, New York 12202-0509, USA

William H. Wunner, The Wistar Institute, Philadelphia, Pennsylvania 19104-4268, USA

1

The History of Rabies

GEORGE M. BAER

Laboratorios Baer, S.A. de C.V., Hacienda Sta. Maria Xalostóc, Tlaxco, Tlaxcala 90250, Mexico

1 INTRODUCTION

For centuries man has felt terror after bites by rabid dogs. Rabies is unique: one almost always knows when and where the bite occurred. But one has to take a mighty leap into the past to realize that, until the 19th century, there was no accurate diagnosis of the disease in man or animals, no isolation of the infectious agent, no animal control and no human treatment. The first written record of the disease comes from the Eshnunna code, 23rd century BC: 'If a dog is mad and the authorities have brought the fact to the knowledge of its owner; if he does not keep it in, and it bites a man and causes his death, then the owner shall pay two-thirds of a mine of silver' (Théodorides, 1986). So it is apparent that the connection between the bite and the danger of death was known even then. Some 1300 years later, Homer, in the 9th century BC compares Hector to a rabid dog ('I cannot kill this raging dog') and in Democritus, 5th century BC, one is able to read a description of the disease in the dog. Aristotle, in the 4th century BC, wrote in the 'Natural History of Animals', '…dogs suffer from the madness. This causes them to become very irritable and all animals they bite become diseased…', although he considered that man was not affected by the disease.

Plato (4th century BC) used the word 'Lyssa' to describe erotic passion. Caelius Aurelianus (500 AD) suggests that Homer had the sufferings of a hydrophobic in mind when describing the torment of Tantalus in Hades, with water before him but not able to drink it. Many Greek or Latin classical authors knew of the existence of rabies; Xenophon speaks of it in his 'Anabasis', Virgil in the 'Georgics' and Ovid in his 'Metamorphosis'.

Some early opinions, such as in the Talmud, suggested that rabies could be caused by witches' spells or evil spirits (Rosner, 1974). Hekabe, the widow of Priam, king of Troy, was thought to be transformed into a raging dog. Avicenna, also known as Ibn Sina in the Islamic world, included the suggestion that rabies could arise through changes in ambient temperatures, the same general belief that the Romans held in attributing rabies to the star Sirius (from the Greek 'Seirious' meaning scorching or blazing, since it appeared with the sun in the hottest part of the year). The Romans adapted the Babylonian stellar system and called it the dog star or

1

Rabies, second edition. Edited by Alan C. Jackson and William H. Wunner
ISBN 978-012-369366-2. Copyright Elsevier Inc. 2007

'canis', with the hottest days of the year being 'dies canicularis' or dog days. Among others, Pliny (1st century AD) believed that this was a time when dogs were most susceptible to rabies, an idea which still persists in some parts of the world. Even in the 1st century, the optimism about curing rabies was not shared by Scribonius Largus, who affirms that a rabid patient is never cured, 'Nemo adhuc, corruptum hoc male expeditus est' (No one until now corrupted with this evil is freed therefrom).

Galen, the 2nd century doctor, contributed to the dehydrative cause of rabies, writing about 'the extreme dryness' that arises in the solid parts of the animal; he also believed that only dogs were susceptible and wrote that, 'so great does the corruption of their humours become that their spittle alone, if it falls upon a man, can make him become rabid'.

One widespread myth was that rabies was caused by a small 'worm' at the base of the tongue. A contemporary poet of Ovid (1st century BC), Grattius Falistcus, knew about the mythical origin of the sublingual 'lyssa' of rabid dogs that Pliny popularized; they believed that extracting the worms completely cured the dog. And as a preventive, this worm was also thought to possess magical curative powers in preventing the disease in the person bitten when it was injected, but only after having been carried three times around a fire. Prevention was also thought to be obtained by eating a cock's brain. Other remedies included one coxcomb, some goose fat mixed with some honey, the salted flesh of a rabid dog, some maggots from the carcass of a rabid dog, etc.; and as many remedies existed for the local application of 'preventers', or for mixing them in drinks or food. Moreover, there were numerous talismans 'capable' of diverting rabid dogs from a person, such as a dog's heart or placing a dog's tongue in one's shoes, or arming oneself with a tail of a weasel. But one of the most common 'preventers' of rabies has always been to pray for divine intervention (Figure 1.1), an evocation which many times was successful, for the variety of reasons later known.

2 REMEDIES FOR DOG RABIES FROM THE EARLY AND MIDDLE AGES TO THE 18th CENTURY

The incurability of the disease has led to superstitions for centuries and one of the most celebrated is that of St Hubert, the 8th century evangelizer of the Ardennes and bishop of Liege (Belgium), most known because he was 'patron of the hunters' and protector against rabies (Tricot-Roye, 1926).

One writ from that time states:

Ik kwam alover Sint-Huybrechts zijn graf	I came across St Hubert his grave
Zonder stok of zonder staf	without stick or without stave:
Kwaden Hond, sta stille:	Evil dog, stand still:
Het is Sint-huybrechts wille	It is Saint-Hubert's will

Te damos grasias Virjencita por livrar a mi hijo de morir. al ser mordido por un perro del mal Cuando Jugaba Con el perlo que al ver Tal desgrasia aclamamos aTi Nos isieras el milagro de salvarle la vida por lo que asi fue. 22-Mayo-1927. Votivo de la Soledad † HVH

Figure 1.1 A votive giving thanks to the holy virgin for saving the life of a boy from rabies: 'We give you thanks, little virgin, for keeping my son from dying after he was bitten by a dog with rabies while playing with him, so when we saw that tragedy we called on you to perform the miracle of saving his life, and that's the way it was.' Soledad District, May 22, 1927. This figure is reproduced in the color plate section.

This power was once demonstrated by a celebrated miracle, reported in the 'Cantatorium'. One day, when St Hubert was preaching, a man afflicted with manic rabies entered into the church, his teeth gnashed, a bloody slobber ran from his mouth, his eyes emitted flames, he gave out some frightful roars, he launched himself like a maddened beast into the midst of the faithful, who fled in the most vivid terror. St Hubert faced him without manifesting the least emotion, extended his hands toward the unfortunate man and said: 'May the Lord Jesus heal you'. Calm immediately reappeared on the convulsed features of the patient, his nerves relaxed. The saint wiped away the bloody slobber from the mouth; the sick man smiled and, himself, brought back to the sanctuary a part of those who had just fled from him with great fright. The cult of St Hubert extended to his being the patron against rabies as well as against 'dullighedon' or mental diseases. People bitten by dogs went to priests for the operation of 'cutting'; the penitent presents the forehead to the priest, who then makes a tiny incision and raises a bit of skin in order to insert a small fragment of the stole that belonged to the great saint. This stole had come from heaven while St Hubert prayed on the tomb of St Peter in Rome. The forehead is then bound up with a black headband. It appears that between 500 and 800 cuttings per year were generally carried out. In addition, animals bitten by rabid dogs were 'treated' in the cult of St Hubert: a key that St Hubert received from the hands of the pope was heated to red hot and applied, glowing, to the animal's wounds. This cauterization was supplemented by five to nine days of penitence imposed on the owner and by a diet of 'blessed' oat bread.

As early as the 1st century BC, Cicero called attention to hydrophobia – thirst and fear of water – and described the precautions to be taken as well as immediate cauterization of the wound or its suction (the latter being the hereditary privilege of the families of the 'suckers of poison'). Cicero also stressed the preventive cauterization or extirpation 'en bloc' of the wound, the same that Galen, the 2nd century AD doctor of the gladiators and father of the medicine of humours, recommended in his turn. Celsus recommended, besides, a bathing, especially in sea water, the moment the disease made its appearance; the patient was suddenly thrown into the water, and if he did not know how to swim one should let him sink so that he swallowed some water, then he was taken out in order to be plunged in again. This saturation of water was to suppress the thirst and the hydrophobia. This heroic therapeutic measure remained in repute until the early part of the 19th century. In 1735, Shrewsbury (1952) cited two individuals who advertised their skills as successful 'Dippers of Man and Beast', charging high fees for immersing unfortunate victims of dog bites in the tidal waters of the River Severn. One advertised as follows: 'William Bennet at the George Inn at Whitminster, six miles below Gloucester, Continueth to Dip Man and Beast…in the Salt Water, with Success'. Popular belief in the efficacy of 'dipping' was widespread.

There has been a plethora of herbal remedies used for rabies throughout the ages, from filbert nuts to hellebore, to rose oil, to camomile tea. The variety of these remedies may be seen in Nicholas Culpepper's 'Complete Herbal', first published in England in 1649. In the last 350 years, 41 editions of this famous work have been published. A 1983 edition lists 40 plants, reputedly valuable in relieving the effects of bites and stings in general. Nine of these plants receive particular mention in the treatment of 'mad dogs' (Table 1.1).

During the Middle Ages, epizootics of rabies in dogs were reported in many countries. Not surprisingly, most reports came from Europe since much of the history of the disease comes from that part of the world. The first description of human rabies in China was recorded in the grey record 'Zhou Chuan' 556 BC: 'The native Chinese chased and caught rabid dogs'. Another description of rabies appears in the book 'Priceless Prescription' by Sun Si-Miao, from the Tang Dynasty,

TABLE 1.1

Herbal remedies recommended by Culpepper for the bites of mad dogs

Plant	Part	How used
Angelica (*Angelica archangelica*)	Powdered	Externally as a plaster
Balm (*Melissa officinalis*)	Leaves in wine	Wine drunk, leaves to bite[a]
Carduus Benedictus (*Cnicus benedictus*)	Whole plant infusion	Internal, drink continually
Garlic (*Allium sativum*)	Whole bulb or juice	Internal/external?
Black horehound (*Ballota nigra*)	Leaves beaten with salt	Applied to bite
Hounds tongue (*Cynoglossum officinale*)	Leaves untreated	Applied to bite
Dog lichen (*Pelteriga canina*)	Whole plant + black pepper	In milk: 4 doses internal
Plaintain (Buck's horn) (*Plantago coronopus*)	Decoction of whole plant or leaves?	Internal
Walnut (*Julgans regia*)	Leaves in wine + onions, salt and honey	Internal

[a]Especially recommended for mad dog bites – Avicenna and Dioscorides quoted under balm.
Compiled from lists in *Culpepper's Coloured Herbal* (D. Potterton and E.J. Shellard, eds).
W. Foulsham & Co, 1983.

618 AD: 'Starting from the end of spring to the beginning of the summer, nearly all dogs became rabid; warning must be given to the children and to the weak people to take sticks in hands to defend against attack by rabid dogs'. Still in 1931, there were 272 human cases in Shanghai, all but one transmitted by rabid dogs; canine rabies was eliminated there in 1949 (Lin Fangtao, personal communication, July 21, 1989). Rabies outbreaks in Europe appear to have been rare until the Middle Ages, with most cases caused by single dog bites. One of the first outbreaks was reported in Franconia in 1271, caused by wolves: 30 people died following bites by the rabid wolves (Steele, 1975). By 1500, Spain was reporting many rabid dogs. By 1586, outbreaks of rabies in dogs were reported in Flanders, Austria, Hungary and Turkey. By 1604, rabies was widespread in Paris. In the early 1700s, rabies was common all over Europe, especially in the central countries, in foxes and wolves. The first mention of the disease in Great Britain is in the laws of Howel the Good (Hywed Dda), in 1026, where it is reported that numerous dogs were suffering from 'madness' (King *et al.*, 2004). It authorized any suspect dog to be killed, provided that clinical signs of rabies had been observed in the animal, i.e. a severely inflamed tongue and a frequent tendency to fight with other dogs. In the 1800s, rabies appears to have been widespread in Europe, especially France, Germany and England. An extensive outbreak in foxes occurred in the Jura Alps in eastern France beginning in 1803; this was reported as the largest

outbreak ever seen, with hundreds of dead foxes seen in the surrounding foothills. Many people, dogs, pigs and other animals were bitten. In 1804, the disease appeared in the West German states and, by 1819, spread into the upper Danube and Bavaria; by 1825, it had entered the Black Forest (Steele and Fernandez, 1991). Fox rabies did not spread to northern Germany, unlike its spread from the Oder-Neisse line in the late 1930s (Duwer-Dohrmann, 1950; Starke *et al.* 1961; Seroka, 1977), which eventually covered almost half of France and even extended into northern Italy.

3 RABIES IN THE NEW WORLD

Even though the first reports of rabies in terrestrial animals in the New World came from Mexico in 1709 (Carrada Bravo, 1978), there are numerous descriptions of attacks by vampire bats on the Spanish conquistadores in the Mexican peninsula of Yucatan and south of the Darien strait (now Colombia) two hundred years earlier, where, in 1514, Fernandez de Oviedo wrote (Molina Solis, 1896) that many soldiers died after being bitten by bats. The native Indians discovered that burning the wounds with hot coals eliminated the danger. But the disease was most likely absent in dogs until the Spanish conquest. At the beginning of the 16th century, a 16th century manuscript from Ecuador states that: 'In Quito, nor in any part of Meridional America, plague is to be found…likewise the disease of rabies in dogs is totally unknown….' (America Meridional). In 1591, the viceroy of Mexico, Luís Velasco, published a book on 'The Indies' in which he states, 'And thus the animals of the Indies never get rabies'. Smithcors (1958), too, states that there are no pre-colonial references to rabies in the Americas. Pferd (1987) and Varner and Varner (1983) cited by Childs (2002) mention that importation of dogs from Europe to the New World was known from the time of the second voyage of Columbus. The limited genetic diversity found among canine rabies isolates (Smith, 1989) suggests that the introduction of dog rabies into the New World is due to European colonization with the simultaneous introduction of dogs and their diseases (Smith *et al.*, 1992). Rabies must have appeared in Mexico before 1709, the first date in which it was reported there in terrestrial animals. Later on, many countries in the New World reported the disease. Rabies was common in Virginia in 1753 (Steele, 1975) and 'alarmingly frequent' in Boston by 1768. George Washington, in his diaries, refers to rabies twice: on July 5, 1769 (Fitzpatrick, 1931) he wrote, 'A Dog coming here which I suspected to be Mad I shot him, Several of the Hounds running upon him may have got bit. Note the consequences' and, on Tuesday, August 28, 1786 (Fitzpatrick, 1932) he wrote, 'A Hound bitch which like most of my other hounds appearing to be going Mad and had been shut up, my servant Will, in attempting to get her in again, was snapped at by her at the Arm. The Teeth penetrating through his Coat and Shirt and contused the Flesh, but he says did not penetrate the skin nor draw any blood'.

The first major epizootic in North America was apparently reported between 1768 and 1771, with foxes and dogs transmitting the disease to pigs and other domestic animals. In 1779, rabies was very common in Philadelphia and Maryland. Canine madness was raging all over the colonial states of North America in 1785, continuing until 1789. In 1797, the disease appeared in Rhode Island in an epizootic in dogs and the domestic animals they bit. The disease reappeared in the eastern states of USA in 1810 and in Ohio in dogs, foxes and wolves (Steele, 1975). Records of rabies cases are more complete in the last two centuries and newspaper articles attest to the burden that rabies inflicted and the widespread distribution of the disease; as examples, articles from New York City (Figure 1.2) and Indiana (Figure 1.3) cite human cases from dog bites. The case reported from Brooklyn, in which six persons were bitten and one died, also shows the way people in 1872 still held age-old beliefs about rabies cures: 'When told that there was no efficacy in applying the dog's hair to the wound, or in per-mitting the child to eat the animal's heart and liver raw, he [the father of one of the children bitten…] was much troubled'. As is apparent, rabid dogs were biting people, other dogs and cats and livestock in urban and rural areas all over the USA. And it was not until twenty years later that Negri discovered 'bodies' in ani-mal brains, making the diagnosis of the disease possible (Negri, 1903).

4 THE EARLY ROLE OF WILD ANIMALS IN THE SPREAD OF RABIES

The role of wild animals in transmitting rabies was recognized by Celsus in the 1st century AD. Many of the histories of violent outbreaks in wild animals and the resultant human cases involved wolves. One recent, notorious report of a wolf attack in Iran (Baltazard *et al.*, 1955) is similar to reports of wolf attacks from the Middle Ages – and describes the anguish of an affected city:

> …towards 1:00 in the morning…a large wolf entered the sleeping vil-lage of Sahane, on the international road from Teheran to Baghdad and Damascus…The part of the village along the road included only a few ghavehkanehs (inns), the filling station, the hospital, the police station, and a few houses or shops. Sahane…is a rather frequent stopping place…because of its location at the halfway point on the road, and on this night a dozen trucks and buses had stopped there…with the drivers sleeping inside the inns or outside because of the heat. The wolf came from the mountains bordering on the northeast, through the orchards and vineyards, where no habitation is found, but where, at this time, slept the guards in charge of watching the fruit. Going from one orchard to another towards the southwest, jumping the low dirt walls, or the thorn hedges, the beast attacked in succession 13 persons, the majority

The Horrors of Hydrophobia—Six Children Bitten by a Rabid Dog in Brooklyn—One of them Dying from Rabies—Death of a Horse from the Malady.

On the 9th of February JULIA CONNOLLY, eleven years of age, residing with her parents at No. 45 Walworth-street, Brooklyn, while returning home from a store in the neighborhood, whither she had been sent by her mother, was bitten in the cheek and upper lip by a rabid dog. The girl ran home, and, it becoming known that the animal was rabid and had bitten several other children, the dog was searched for, found at the Flushing-avenue car stables and killed. The wound on JULIA'S face healed, and it was hoped the girl would escape an attack of hydrophobia; but on Monday last she was taken ill, and physicians being called to prescribe for her, her parents were promptly informed that their daughter was suffering from rabies. Since that time she has rejected all food and drink, and even the medicines prescribed by her medical attendants. She continually froths at the mouth and suffers great agony. She seems to be conscious, however, and expresses a constant desire to go to her mother, who is engaged in another room in the sad duty of watching over her babe who is dying from congestion of the lungs. JULIA was yesterday placed in an arm-chair, and so violent were her writhings that it required several persons to hold her down. The physicians say she cannot recover, and that death may ensue at any moment. Yesterday she was visited by PATRICK McGUIGAN, whose daughter, aged five years, was bitten by the dog from which JULIA received her wounds. When told that there was no efficacy in applying the dog's hair to the wound, or in permitting the child to eat the animal's heart and liver raw, he was much troubled. Four other children—making six in all—were bitten by the same dog, but JULIA is the only one who has been seized with hydrophobia.

WILLIAM CODY, the city undertaker of Brooklyn, lost a valuable horse on Wednesday. Death was caused by hydrophobia, the result of a bite from a dog kept in Mr. CODY'S stables. On Wednesday one of Mr. CODY'S workmen found that the horse was covered with foam, and trembling in every limb. A veterinary surgeon was consulted, who at once said that the animal was suffering from hydrophobia. While a messenger was absent in search of a pistol, with which it was proposed to shoot the horse, the animal struggled so violently that he died from strangulation.

Figure 1.2 'The Horrors of Hydrophobia – Six Children Bitten by a Rabid Dog in Brooklyn.' From the New York Times, March 8, 1872.

HIS PET SPANIEL KILLED HIM.

AN INDIANA FARMER THE VICTIM OF
HYDROPHOBIA.

FORT WAYNE, Ind., June 30.—A clearly-defined case of hydrophobia is that of Reuben Drew, a prominent farmer, seventy-six years old, who died in convulsions at his home, eight miles west of Fort Wayne, yesterday, after terrible torture.

About six weeks ago Mr. Drew was bitten on the back of one of his hands by a pet spaniel, the wound being such as to lay open the flesh for an inch or more. It gave Mr. Drew no alarm until the 25th inst., when the first dread symptoms of hydrophobia became manifest by a severe oppression of the chest, making it difficult for the man to breathe. On the 26th his condition grew so much worse that Dr. Wenger, a country physician, was summoned. He formed an opinion at once, but would not permit himself to believe that the patient suffered from hydrophobia. He prescribed something, but on the following day Mr. Drew was so much worse that he was unable to swallow, so that neither nourishment nor medicine could be forced into his stomach. This condition continued until last Saturday, when Dr. Wenger, now thoroughly awakened to the seriousness of the case, summoned Dr. G. W. McCaskey of this city to a consultation.

Dr. McCaskey saw the patient Saturday evening. There were pronounced symptoms of rabies then, among others being a paroxysm of the larynx. The patient's condition Sunday took a sudden and virulent turn for the worse. He became irrational, frothed at the mouth, and became so violent that it was found necessary to bind him to the bed with ropes, but this did not prevent the madman from throwing his head wildly about and snapping his teeth with a savage vigor that made even strong men in the room shudder. All day Sunday and that night these spells continued, until at 7 o'clock Monday morning death removed the unfortunate victim from more suffering.

The dog that inflicted the fatal wound disappeared an hour after biting Mr. Drew, and some anxiety is felt in the matter, for it is feared that others have been bitten. Mr. Drew leaves a wife, but no children. The farmer and his wife were a highly-respected couple, and the terrible fate of the old man has aroused a good deal of sympathy.

Figure 1.3 'His Pet Spaniel Killed Him. An Indiana Farmer The Victim of Hydrophobia.' From the New York Times, July 1, 1891.

bedded down, which explains the number of bites on the head. Some few, awakened, fought with the wolf. One person, awakened with a start by the attack of the beast, was wounded on the cranium; 5 minutes later he was attacked again. This attack in the orchards and vineyards lasted more than two hours; from all sides the people cried and lit lanterns, and the drive was organized. But during this time the wolf had entered the village itself, and in the center of the village it attacked a blind beggar who slept in the street, entered the bazaar and here attacked another person inside a house where the door was left open; from there it jumped into a yard, then over the terrace of the houses where it attacked an old woman. Then it disappeared, and its itinerary became poorly determined...

As already mentioned, the first mention of rabies in the New World came at the start of the 16th century (Tellez Giron, 1977) from the description of men dying after being bitten by (vampire) bats (as early as 1514). Cattle epizootics attributed to vampire bats were reported during the 16th century in Guatemala (Licenciado Palacio al Rey D. Felipe II, 1576), during the 18th century in Ecuador and during the 19th century in Trinidad (de Verteiul, 1858). Although the first mention of the disease in terrestrial animals in North America was in foxes in Virginia in 1753, skunk rabies was also noted as early as the middle 1800s in the Midwest states. It appears to have been a most serious problem in the 1800s, so much that cowboys had to set up 'phobey tents' to keep from being bitten by rabid skunks at night (Charlton et al., 1975).

Centuries before the beginning of vaccine development in the late 19th century, literally dozens of different types of 'treatments' to prevent rabies had been recommended. It appears that Celsus (25 AD) first wrote about the treatment of wounds by searing them with hot irons or 'cupping' them to draw out the poison. In the first centuries AD, cautery became widely adopted (Baer et al., 1996) as part of the immediate treatment of bites by rabid dogs. Galen (131–200 AD) wrote about the need for prompt local treatment and advised that the wound be kept open to prevent 'absorption' of the poison. Early 'treatments', reviewed by Fleming (1872), include eating a cock's brain or a cock's comb pounded and applied to the wound and using grease and honey as a poultice. The flesh of a mad dog was sometimes salted and taken with food as a remedy. In addition, young puppies of the same sex as the biting dog were drowned and the person bitten ate their liver, raw. Other treatments (Pliny, 23–79 AD) included applying the ashes of the dog's head to the wound, or placing a maggot from the carcass of the dead dog on the wound. Maimonides, a 12th century Talmudic scholar and physician, wrote a treatise on 'Poisons and Their Antidotes' at the request of Sultan Al Afdal, including a description of various 'remedies against the bite of mad dogs'. Questions about the 'source' of the disease came from early times. Numerous doctors of the famous school of Alexandria (Egypt) in their writings localize the 'hydrophobia' sometimes in the stomach, sometimes in the brain; they wrote a book called 'Kynolossus' (rabies in the dog). Celsus, a learned man in the time of Augustus Caesar, belongs to the group of encyclopedists. From his complete work 'De Artibus' ('On the Arts'), one chapter is 'de arte medica,' written between 25 and 35 AD. In it he describes rabies in man and affirms that the saliva of a rabid dog is infectious ('autem omnis morsus habet fere quoaddam virus'); 'virus' in that and other Roman writings, was apparently a synonym for poisons of any kind. Many ancient writers refer to the 'virus' that caused the disease and its transmission by the 'spittle' of the dog. From the first century AD comes the 'first recorded attempt at defining the cause of rabies and prescribing treatment' (Neville, 2004). This comes in a poem entitled 'On Hunting' by Grattius Faliscus, a contemporary of Ovid (43 BC – 18 AD).

5 DEVELOPMENT OF THE FIRST-GENERATION RABIES VACCINE

It was not until 1804 that Zinke demonstrated that rabies could be transmitted by saliva (Zinke, 1804). He took the saliva from a rabid dog and brushed it into incisions that he had made on the foreleg of a young dachshund. The dog sickened on the seventh day; on the eighth day it refused food and water and 'crawled into the corner of his cage'. By the tenth day rabid symptoms were obvious. There was no mention of it being a live agent, although Zinke wrote a series of articles on the pathogenesis of the disease and its 'treatment'. Many of the studies carried out in the 19th century centered on ways to reproduce the disease by animal injection; during those early studies many routes were used in trying to achieve 100% mortality, since intramuscular and subcutaneous injections did not reach that level. Some of the routes used were intravenous, intraocular, intranasal and intraneural. As already emphasized by Théodorides, Galtier's work on rabies was a crucial preamble to Pasteur's work on the disease.

Galtier (1879) showed that the disease could be transmitted to rabbits by injection and by bite. Pasteur followed much of the rabbit studies that Galtier had initiated, citing them in detail in series of communications in his 1881 'Sur Une Maladie Nouvelle Provoquee par la Salive d'un Enfant Mort de La Rage' [On a Disease Caused by the Saliva of an Infant that Died of Rabies]. Or, 'un precieux travail de M. Galtier, professeur a l'Ecole veterinaire de Lyon, travail qu'il al'Academie des sciences…nous a appris: 1er que les symptoms de la rage du chien inoculee au lapin, n'apparaissent que de quatre a quarante jours après l'inoculation du virus; 2nd que le lapin mort de la rage ne presente pas de lesions anatomiques de l'ordre de celles ci-dessus indiquees; 3er que le sang des lapins morts de la rage ne peut communiquer la maladie' [In a very precise work by Mr Galtier, professor of the Veterinary School at Lyon, presented in the Academy of Sciences, he has shared with us the following: first, that material from a dog with the symptoms of rabies, inoculated into rabbits, does not appear until the fourth or fifth day after the

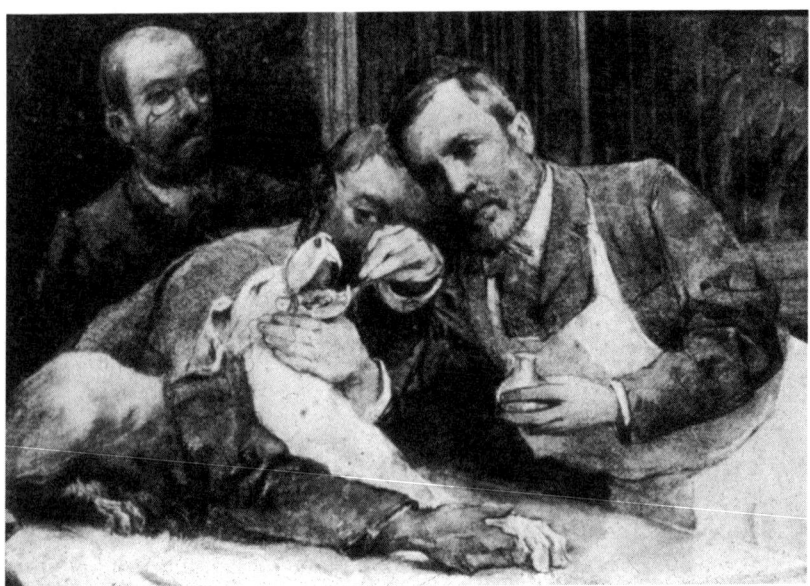

Figure 1.4   Pasteur Removing a Saliva Sample from a Rabid Dog. A painting by Alfonse Marie Mucha. (Reprinted courtesy of the Institut Pasteur, Paris).

inoculation of the virus; second, that the rabbit that died of rabies does not show the lesions in the cells indicated; third, the blood of rabbits that died of rabies cannot transmit the disease]. Pasteur made two landmark discoveries, which eventually led to the famous vaccine that he developed: first, that the brain (and spinal cord) 'housed' the virus. (He had earlier written: 'Le virus rabique, dit ce savant observateur, existe dans la bave, tout le monde le sait. Mais d'ou vient-il? Ou est-il elabore?…' [Rabies virus, says the knowledgeable observer, exists in the spittle, the whole world knows that. But…where does it come from? Where is it produced?]); and, second, that injection of 'virus' material into the brain itself ('…under the dura mater…') ('Je prende un autre exemple portant sur des lapins et par un mode d'inoculation different, celui de la trepanation') invariably resulted in the paralysis and death of the animal from rabies.

  This, of course, is what permitted the 'fixation' of the virus in two ways: 'fixed' in that it always caused the disease in the inoculated animal, and 'fixed' in that the incubation period was a standard 5–7 days after inoculation. Thus, it was not until the landmark studies of Pasteur (Figure 1.4) that the brain was recognized as a crucial organ in the pathogenesis of the disease and that the reliability of the serial intracerebral route for virus passage was recognized (this latter finding was immediately recognized for its importance, as confirmed by an article appearing in the front page of the New York Times a few weeks after Pasteur's discovery (Figure 1.5) (Pasteur, 1881, 1884). After having developed his vaccine and used it on 50 dogs

PASTEUR ON CANINE MADNESS.
Paris Dispatch to the London Standard.

M. Pasteur yesterday made an interesting communication to the Paris Academy of Sciences in relation to canine madness. His experiments had shown him that an injection in the region of the skull of the virus of rabies always produced the malady in an acute form, but that an injection in the veins only occasionally had acute results, being often followed by chronic affection only, without barking or ferocity. If a dog were inoculated with fragments of marrow or of nerve taken from a mad dog, the disease would be communicated. M. Pasteur further stated that he had rendered 20 dogs proof against the disease by inoculating them with other virus than the virus of rabies. Fowls and pigeons injected with the latter became affected, but soon recovered spontaneously.

Figure 1.5 'Pasteur on Canine Madness. Paris Dispatch to the London Standard.' The New York Times, March 16, 1884.

('...jeatais arrive a avoir cinquante chiens...') with much trepidation he began – on July 5, 1885, with the assistance of Vulpian and Grancher – vaccinating young Joseph Meister from the Alsatian town of Meissengott (Pasteur, 1885). The boy had been repeatedly bitten by a dog, the guardian of a grocery store, and was covered in blood; the local physician diagnosed the dog as rabid because it had bloody foam over its mouth; later on the diagnosis was 'confirmed' as rabies by 'pieces of wood in the stomach'.

Meister had his wounds cauterized by phenic acid 12 hours after the bites; Pasteur began injecting Meister the next afternoon with rabbit spinal cords that had been 'treated' in 'dry air' in which the virulence 'slowly disappeared' ('...la virulence dispairait lentement...'). The first cord was from rabbits killed on the 23rd of June (14 days' drying), the second from the 25th of June (12 days' drying), and so on, with reduced drying times for each preparation, until the 16th of July, the 9th day in which cords were used that had only been dried one day. Pasteur had written about the need to dry the cords: 'Si la moelle rabique est mise à l'abri de l'air, dans le gaz acide carbonique, à l'etat humide, la virulence se conserve...sans variation de son intensité rabique...' ['If the rabid spinal cord is kept from the air, and under the carbonic acid gas, away from humidity, the virulence is conserved...without any variation in its rabid intensity...']. Meister survived and the world finally had a treatment for the disease which, for so long, had caused frenzy in people bitten by animals. But the difficulty in administering this treatment can be gleaned by

TABLE 1.2

Treatment for bites on the head by rabid animals by treatment day, 'age' of spinal cord and dose injected (Roux, 1887)

Day		Period (no. of days) spinal cords had been dried	Dose (ml)
1	Morning	14	3
		13	3
	Evening	12	3
		11	3
2	Morning	10	3
		9	3
	Evening	8	3
		7	3
3	Morning	6	2
	Evening	6	2
4	Evening	5	2
5	Evening	5	2
6	Evening	4	2
7	Evening	3	1
8	Evening	4	2
9	Evening	3	1.5
10	Evening	5	2
11	Evening	5	2
12	Evening	4	2
13	Evening	4	2
14	Evening	3	2
15	Evening	3	2
16	Evening	5	2
17	Evening	4	2
18	Evening	3	2
19	Evening	5	2
20	Evening	4	2
21	Evening	3	2

the need for cords dried for different time periods, a Herculean task (Table 1.2). Since that memorable year, 1885, hundreds of thousands of people have been vaccinated for bites by rabid or possible rabid (i.e. escaped) animals. There were modifications to Pasteur's method from the start, due principally to the difficulty in preparing the thousands of rabbit spinal cords needed to immunize exposed persons with 14 or more daily injections (Vodopija and Clark, 1975). Fermi (1908) points out various defects in Pasteur's vaccination regimen. He introduced a new method, treating the virus with carbolic acid. He mentions as advantageous the uniformity of the vaccine so prepared and the simplicity of preparation, compared with the 'complicated and useless' Pasteur method of attenuation. The vaccine could be preserved and sent anywhere so as to be always available.

A major change in Pasteur's method of attenuating vaccine was the inactivation of the virus by phenol, a change initiated by the Englishman Semple, working in India; that was for decades the commonly used type of vaccine worldwide and is still used today in some developing countries (Semple, 1911).

The advent of the growth of virus in tissue culture first reported for polio by Enders *et al.* (1949), inaugurated a new epoch in virology, i.e. the second generation of rabies vaccine development. The first to grow rabies virus in tissue culture was Kissling (1958). The first vaccine was the human diploid cell vaccine (Wiktor and Koprowski, 1965), soon replaced in much of the world by two other vaccines, purified chick embryo cell vaccine (PCEC) (Bijok, 1985) and the VERO vaccine (Montagnon *et al.*, 1985). Instead of the 14 or 21 daily doses of vaccine that the Pasteur regimen required, the number of cell culture vaccines has now been reduced to 5 doses for post-exposure prophylaxis (World Health Organization, 1972). But the impact of Pasteur's discovery in the period immediately following the historic 1885 vaccination schema is shown in Table 1.3 (Kraus *et al.*, 1926), indicating the wildfire spread of his antirabies vaccination to the four corners of the world.

The estimated risk from bites by rabid animals varies greatly, from 75% in persons bitten in the head by rabid wolves to almost nil in persons simply scratched by rabid animals (Babes, 1887). Frascatorius, the celebrated physician, philosopher, poet and Latinist, already in 1546, in his 'Contagions and Contagious Diseases and Their Treatment', wrote, 'whence it happens that not all who are bitten even become rabid but many by reason of their natural constitution either do not contract the contagion at all or if they do, survive'. The 'global' mortality figure often cited in reports from the 1880s and 1890s, prior to Pasteur's finding (Horsley, 1889; Roux, 1891) is 15% in people bitten in the hands or legs (those involved in most cases) by rabid dogs. When people are severely bitten in the head and upper torso, as is the case with wolf bites, the virus dose (or the severity of the bite) is so great that the vaccine treatment often fails. In Iran, wolf bites are common and the mortality after vaccine treatment was considered 'disastrous' and 'no better than no treatment at all' (Baltazard and Ghodssi, 1953). In order to ameliorate this situation, the World Health Organization organized a series of studies in which an adjunct – the infiltration of antirabies serum – was added to Pasteur's vaccination, in order to cover the early period before the full rise of active antibodies that vaccination provides.

Pasteur's preparation of rabbit spinal cords dried under potash for various periods of time was a procedure that changed soon after the decade of the 1880s,

TABLE 1.3

Cities in which rabies vaccination was instituted, by year and number of vaccines

Country/City	Years	Number of Vaccinees
Algiers	1886–1895	4149
	1895–1905	5395
	1910–1924	23 921
Bandung	1895–1924	13 243
Berlin	1898–1925	14 906
Bern	1901–1924	344
Bologna	1889–1923	17 107
Breslau	1907–1924	6127
Budapest	1890–1923	118 258
Buenos Aires	1886–1924	43 500
Bucharest	1888–1905	9250
Cairo	1906–1924	15 105
Celje	–1924	119
Charkow	1887–1924	66 648
Chicago	1910–1915	2352
Columbus	1910–1921	3220
Constantinople	1900–1905	4100
Cracow	1893–1924	22 210
Florence	1899–1923	10 599
Jassy	1891–1924	26 762
Jerusalem	–1924	3685
Kasan	1901–1924	6693
Kasauli	1900–1924	89 797
Klausenburg	1921	3180
Leningrad	1886–1924	34 221
Lille	1895–1924	5491
Lisbon	1893–1920	39 248
Lyon	1900–1924	16 430
Madras	1907–1923	28 898
Madrid	1901–1924	12 251
Marseille	1893–1903	3563
Milan	1889–1903	2942
Minsk	1922–1924	4520
Naples	1886–1908	8446
New York	1890–1901	1608
Nisch	1901–1911	4632
	1919–1924	8032

TABLE 1.3 (continued)

Country/City	Years	Number of Vaccinees
Novi Sad	–1924	6654
Palermo	1887–1895	2221
Paris	1886–1924	46 588
Pernambuco	1889–1903	486
Prague	1919–1923	11 712
Riga	1914–1924	3546
Rome	1889–1902	1940
	1920–1925	5037
Saigon	1886–1895	110
Samara	1886–1924	37 455
Sao Paulo	1903–1914	6502
Sarajevo	–1924	1733
Sasari	1900–1903	1053
	1914–1924	664
Shanghai	–1924	1251
Sofia	1902–1904	1081
	1920–1924	12 796
Tiflis	1888–1924	19 244
Tunis	1894–1924	13 302
Turin	1886–1918	9903
Vienna	1894–1925	15 982
Warsaw	1886–1924	42 189
Zagreb	1919–1924	11 605

After Kraus *et al.*, 1926.

since it was so onerous to prepare the dozens or hundreds of cords needed. As already mentioned, those involved in that preparation began resorting to chemical inactivation for the time periods needed. That difficulty also eliminated any thought of canine vaccination. Nocard, Pasteur's colleague, stated, for instance, that 'For the removal of rabies – would it be feasible to vaccinate all the dogs?' In France alone there are 2 500 000, not even counting the 100 000 in Paris. Since several 'preventive' inoculations are necessary, how many vaccination centers would be needed to hold the many dogs to be vaccinated? If rabbits would have been used to prepare enough vaccine not even all of the rabbits in Australia would have sufficed (Nocard, 1884).

Surprisingly, the first major national program to vaccinate dogs only began 35 years after Pasteur's studies in the 1880s; the Japanese national program in the

1920s was directed by two veterinarians, Umeno and Doi (1921). They used phenolized vaccine in Nagasaki and Tokyo, to start a national urban program in Japan in 1921. Following that, other countries began using phenol and other inactivating agents such as chloroform and ultraviolet rays. But many of the vaccines were of inferior potency and quality and it was the development of a standardized vaccine potency test by Habel in the early 1940s – in mice – that permitted the initiation of large-scale urban programs. These programs were preceded by a series of canine vaccines studies in dogs (Johnson, 1946), which resulted in both national (USA) and international (World Health Organization) recommendations for one or even 3-year durations of immunity in dogs. The use of those vaccines in a major city was first tested in Memphis, Tennessee in 1948. During the first 4 months of that year, canine rabies had reached epidemic levels. During that time the vaccination of approximately 10 000 dogs at private veterinary clinics failed to reduce the numbers of rabid dogs. In order to organize a mass vaccination program, Tierkel (1948) of the US Public Health Service coordinated 71 emergency vaccination clinics in the city and surrounding Shelby County; 23 000 dogs were vaccinated in a two-week period in April, at a nominal charge ($1.00/dog) and the number of cases dropped precipitously, such that no cases were reported after July of that year. Tierkel stressed the importance of vaccinating at least 70% of dogs in short time periods. That campaign gave the impetus to other programs in the USA and other countries. The technology of mass vaccine production and mass canine vaccination resulted in the reply to Nocard's question: 'For the removal of rabies – would it be feasible to vaccinate all the dogs?' Some countries with no wildlife rabies, such as Japan (Shimada, 1971) and Taiwan (Shambaugh, 1960) have been able to eliminate rabies from 1954 and 1955, respectively.

Canine rabies in most industrialized countries has been reduced to almost nil, although the number of reported wildlife rabies cases has increased. The control of wildlife rabies is another matter, with no possibility of vaccinating the hundreds of thousands of animals needed to reduce the disease by needle and syringe. In some cities, however, that method – capture-vaccinate-release – has been successful in reducing rabies in limited urban areas; the best example of that is in Toronto, Canada, where Rosatte and colleagues (Rosatte et al., 1986) have trapped, vaccinated and then released thousands of skunks to combat the disease (see Chapter 18). The method of oral rabies vaccination, applicable to field use, was developed in the USA in 1971 (Baer et al., 1971) and applied in initial field trials in foxes in Switzerland in 1978 (Steck et al., 1982). The oral vaccination method has resulted in the elimination of rabies in many western European countries and eastern Canada, through the use of a variety of attenuated and recombinant rabies vaccines, and is also being used in raccoons and coyotes.

Human rabies cases in the developing world are still almost always transmitted by rabid dogs, but the situation in the developed world has changed radically. Rabies in insectivorous bats was (re)discovered in 1953 in Florida, when the son of a mining engineer was bitten by a yellow (insectivorous) bat; the engineer had worked in Mexico and knew of vampire rabies, quickly realizing the aberrant behavior of the bat, and took it to be diagnosed, with the positive test leading to his son being vaccinated against rabies (Scatterday and Galton, 1954). Since then thousands of rabid insectivorous bats have been diagnosed in the USA, Canada and western Europe, amounting to 5–10% of total human rabies vaccinations. But the situation has changed even more in the last decades, when dozens of human deaths have occurred in which no bite was identified and no treatment was administered. Almost all these cases have resulted from contact with the silver-haired bat (*Lasionycteris noctivagans*). The exact method of exposure is still being investigated. Moreover, bat rabies was reported in bats a few years ago in Australia, a country previously considered free of rabies (Fraser *et al.*, 1996).

We enter the fifth millennium of rabies with the world still suffering tens of thousands of human rabies deaths and millions of persons being vaccinated for bites by rabid animals (or by animals not available for examination) annually. Only a few countries are free of rabies – or only have rabies in bats. What will the future bring?

REFERENCES

Aristotle. In: *Historia Animalum*, Vol. IV, Book VIII, p. 604.

Avicenna (980–1037). *Canon de Medicine*, Book IV, Chap. 6, traité.

Babes, V. (1887). Studien uber die Wutkrankheit. *Virchows Archiv der Pathologie und Anatomie* **110**, 562–601.

Baer, G.M., Abelseth, M.K. and Debbie, J.G. (1971). Oral vaccination of foxes against rabies. *American Journal of Epidemiology* **93**, 487–490.

Baer, G.M., Neville, J. and Turner, G.S. (1996). *Rabbis and Rabies: A Pictorial History of Rabies through the Ages*. Laboratorios Baer, SA, Mexico.

Baltazard, M. and Ghodssi, M. (1953). Prévention de la rage humaine. Traitement des mordus par loups enragés en Iran. *Revue du Immunologie, Therapie e Microbiologie* **17**, 366–371.

Baltazard, M., Bahmanyar, M., Ghodssi, M., Sabeti, A., Gajdusek, C. and Rouzhebi, E. (1955). Essai pratique du serum antirabique chez les mordus par loups enragés. *Bulletin of the World Health Organization* **13**, 747–772.

Bijok, U. (1985). Purified chick embryo cell (PCEC) rabies vaccine. In: *Improvements in Rabies Post-Exposure Treatment* (I. Vodopija, K.G. Nicholson, S. Smerdel and U. Bijok, eds). Zagreb: Zagreb Institute of Public Health.

Carrada Bravo, T. (1978). Investigación documental de la primera epidemia de rabia registrada en la Republica Mexicana en 1709. *Salud Pública de México* **20**, 705–716.

Celsus (ca. 14–37 AD). de Medicina (English Translation, W.G. Spencer, 1938, London).

Centers for Disease Control (2003). Rabies Surveillance Annual Summary, 2002.

Charlton, K.M., Webster, W.A. and Casey, G.A. (1975). Skunk rabies. In: *The Natural History of Rabies* (G.M. Baer, ed.). pp. 307–324. New York: Academic Press.

Childs, J.E. (2002). Epidemiology. In: *Rabies* (A.C. Jackson and W.H. Wunner, eds). pp.113–162. San Diego, CA: Academic Press.

Cicero. Academica i–ii, 46 BC.

Democritus (460–370 BC).

de Verteuil, L.A.A. (1858). Trinidad: Its Geography, Natural Resources and Prospects. (Ward and Lock, eds) quoted by Constantine, 1970.

Duwer-Dohrmann, I. (1950). Tollwut in Fuchse. *Zeitschrift Veterinaermed* **41**.

Enders, J.F., Weller, T.H. and Robbins, F.C. (1949). Cultivation of the Lansing strain of poliomyelitis virus in cultures of various human embryonic tissues. *Science* **109**, 85–87.

Fermi, C. (1908). Immunisierung der muriden mit Wut. *Zeitschrift fur Hygiene und Infectionskrankheiten* **58**, 221.

Fitzpatrick, J.C. (1931). The Writings of George Washington. **179**, 232. Washington, DC: US Government Printing Office.

Fitzpatrick, J.C. (1932). The Writings of George Washington. **181**, 349. Washington, DC: US Government Printing Office.

Fleming, G. (1872). *Rabies and Hydrophobia. Their history, nature, causes, symptoms and prevention.* London: Chapman & Hall.

Frascatorius, H. [translation by William Renwick Riddle] (1924). De Contagione et Contagiosis Morbis et Eorum Curatione, Book II, Chap. X, p. 170.

Fraser, G.C., Hooper, P.T., Lunt, R.A. *et al.* (1996). Encephalitis caused by a lyssavirus in fruit bats in Australia. *Emerging Infectious Diseases* **2**, 327–331.

Galen (131–201 AD). De Arte Medicinae, p. 19.

Galtier, V. (1879). Etudes sur la rage. *Annales Medicin Veterinaire* **28**, 627–639.

Grattius Faliscus. In: *Cynegetica*, pp.383–398, quoted by Neville, J. (2004). In: *Historical Perspective of rabies in Europe* (A.A. King, A.R. Fooks, M. Aubert and A.I. Wandeler, eds). Paris: World Organisation for Animal Health (OIE), p. 3.

Habel, K. (1940). Evaluation of a mouse test for the standardization of the immunizing power of antirabies vaccines. *Public Health Reports* **55**, 1473–1478.

Homer. The Iliad, 8.299.

Horsley, V. (1889). On rabies: its treatment by M. Pasteur. *British Medical Journal* **17**, 342–344.

Johnson, H.N. (1946). Experimental and field studies of canine rabies vaccination. *Proceedings of the 29th Annual Meeting of the US Livestock Sanitary Association,* pp. 99–107.

King, A.A. *et al.* (2004). *Historical perspective of Rabies in Europe and the Mediterranean Basin.* Paris: World Organisation for Animal Health.

Kissling, R.E. (1958). Growth of rabies virus in non-nervous tissue culture. *Proceedings of the Society for Tropical Medicine and Hygiene* **98**, 222–225.

Kraus, R., Gerlanch, F. and Schweinburg, F. (1926). *Lyssa bei Mensch und Tier.* Berlin and Vienna: Urban & Schwarzenberg.

Licenciado Palacio al Rey D. Felipe II (1576). In: *Colección de Documentos Inéditos Relativos Posesiones Españoles en América y Oceanía (1866).* Vol. 6. Madrid: Ministro de Ultramar.

Maimonides, M. (1178). Treatise on Poisons and their Antidotes (1966). In: *The Medical Writings of Moses Maimonides* (S. Munther, ed.). Vol. 2. Philadelphia: Lippincott.

Molina Solis, J.F. (1896). Historia del descubrimiento y conquista de Yucatán., Mérida, p. 911.

Montagnon, B.J., Fournier, P. and Vincent-Falquet, J.C. (1985). Un nouveau vaccin antirabique humaine: rapport preliminaire. In: *Rabies in the Tropics* (E.K. Kuwert, C. Merieux, H. Koprowski and K. Bogel, eds). pp. 138–143. Berlin: Springer Verlag.

Negri, A. (1903). Zur Aetiologie der Tollwuth: Die Diagnose der Tollwuth auf Grund der neueren Befunde. *Zeitschrift fur Hygiene und Infectionskrankheiten* **43**, 507–528.

Neville, J. (2004). Rabies in the Ancient World. In: *Historical Perspective of Rabies in Europe and the Mediterranean Basin* (A.A. King, A.R. Fooks, M. Aubert and A.I. Wandeler, eds). pp.1–12. Paris: World Organisation for Animal Health (OIE).

Nocard (1884). [quoted by Renato Vallery Radot, translation by J. Degiorgi], In: *La Vida de Pasteur*. Barcelona: Juventud.

Ovid. 'Metamorphosis'. [trans. A. Golding] New ed., 1904.

Pasteur, L. (1881). Experiences faites avec la salive d'un enfant mort de la rage. *Bulletin Societé de Ciences Medicin Veterinaire*. **LVIII**, 150–155.

Pasteur, L. (1884). Microbes pathogenes et vaccins. Congrés périodique international des sciences médicales, 8e session, Copenhague, séance de 10 août, 1884.

Pasteur, L. (1885). Méthode pour prévenir la rage après morsure. *Compte Rendue Academie Science (Paris)* **101**, 765–773.

Pferd, W. (1987). *Dogs of the American Indians*. Fairfax: Denlinger's.

Pliny (the Elder, 23–79 AD) *Natural History*, Book. 240, **715**, 29–102.

Rosatte, R.C., Kelly-Ward, P.M. and MacInnes, C.D. (1986). A strategy for controlling rabies in urban skunks and raccoons. In: *Proceedings of the National Symposium on Urban Wildlife*, pp. 16163. Chevy Chase, MD.

Rosner, F. (1974). Rabies in the Talmud. *Medical History* **18**, 198–200.

Roux, E. (1887). Note sur un moyen de conserver les moelles rabiques avec leur virulence. *Annals of the Institut Pasteur* **1**, 87–90.

Roux, E. (1891). Schutzimpfungen gegen Hundswuth. *Wiener Medizinische Presse* **32**, 1327–1330.

Scatterday, J.E. and Galton, M.M. (1954). Bat rabies in Florida. *Veterinary Medicine* **49**, 133–138.

Semple, D. (1911). In: Scientific Memoirs of the Medical and Sanitary Departments of India. No. 44 Calcutta: New Series.

Seroka, D. (1977). The Origin of the Current Fox Rabies Outbreak in Europe. Conference on Surveillance and Control of Rabies, World Health Organization Frankfurt/Main.

Shambaugh. G.E.J. (1960). The last case of rabies in Formosa. *Journal of the Formosan Medical Association* **59**, 775–781.

Shimada, K. (1971). The last rabies outbreak in Japan. In: *Rabies* (Y. Nagano and F.M. Davenport, eds). Baltimore: University Park Press.

Shrewsbury, J.F.N. (1952). Mad dog. *Queens Medical Magazine* **45**, 9–16.

Smith, J.S. (1989). Rabies virus epitopic variation: Use in ecologic studies. *Advances in Virus Research* **36**, 215–253.

Smith, J.S., Orciari, L.A., Yager, P.A., Seidel, H.D. and Warner, C.K. (1992). Epidemiologic and historical relationships among 87 rabies virus isolates as determined by limited sequence analysis. *Journal of Infectious Diseases* **166**, 296–307.

Smithcors, J.F. (1958). The history of some current problems in animal diseases: VII. Rabies. *Veterinary Medicine* **53**, 149–154.

Starke, G., Winkler, C. and Eichwald, C. (1961). 10 Jahre Tollwutprophylaxe. *Monatshefte fur Veterinarmedizin* **16**, 605–609.

Steck, F., Wandeler, A., Bichsel, P., Capt, S. and Schneider, L. (1982). Oral immunization of foxes against rabies. *Comprehensive Immunology and Microbiology of Infectious Diseases* **5**, 165–171.

Steele, J.H. (1975). History of rabies and global aspects. In: *The Natural History of Rabies* (G.M. Baer, ed.). pp. 1–24. New York: Academic Press.

Steele, J.H. and Fernandez, P.J. (1991). History of rabies and global aspects. In: *The Natural History of Rabies*, 2nd edn. Boca Raton: CRC Press.

Tellez-Girón, A. (1977). Apuntes para la Historia de la Rabia en México. *Veterinaria México* **9**, 37–46.

Théodorides, (1986). *Histoire de la Rage*. Paris: Masson.

Tierkel, E.S. (1948). Inauguration of rabies control studies by the US Public Health Service. *Journal of the American Veterinary Medicine Association* **112**, 18–24.

Tricot-Roye (1926). Le Cult de St Hubert en Belgique. *Aesculape*, pp. 97–104.

Umeno, S. and Doi. Y. (1921). A study in the antirabic inoculation of dogs. *Kitisato Archives of Experimental Medicine* **4**, 89–102.

Varner, J.G. and Varner, J.J. (1983). *Dogs of Conquest*. Norman: University of Oklahoma Press.

Virgil. 'Georgics'. In: 'Bude Series', by E. de Saint-Denis, Oxford Classical texts, 1956.

Vodopija, I. and Clark, H.F. (1975). Human rabies vaccination. In: *The Natural History of Rabies*. (G.M. Baer, ed.). pp. 571–595. New York: Academic Press.

Wiktor, T.J. and Koprowski, H. (1965). Successful immunization of primates with rabies vaccine prepared in human diploid cell strain Wi-38. *Proceedings of the Society for Experimental Biology and Medicine* **118**, 1069–1073.

World Health Organization (1972). Expert Committee on Rabies. Geneva: WHO.

Xenophon. 'Anabasis' 5.7.26.

Zinke, G. (1804). Neue Ansichten der Hundswuth, ihre Ursachen und Folgen, nebst einer sichern Behandlungsart der von tollen Tieren gebissenen Menschen. *Jena Review* **16**, 212–218.

2

Rabies Virus

WILLIAM H. WUNNER

The Wistar Institute, Philadelphia, Pennsylvania 19104-4268, USA

1 INTRODUCTION

Rabies virus (RABV) is the prototype virus of the genus *Lyssavirus* (from the Greek *lyssa* meaning 'rage') in the family *Rhabdoviridae* (from the Greek *rhabdos* meaning 'rod'). It has a similar morphology, chemical structure and life cycle to vesicular stomatitis virus (VSV), the prototype virus of the genus *Vesiculovirus*, in the same family (Tordo *et al.*, 2005). RABV is a highly neurotropic virus in the infected mammalian (animal and human) host, invariably causing a fatal encephalomyelitis. RABV and the variants known as the 'rabies-related' lyssaviruses, which share with rabies virus the unique capability to produce a rabies-like encephalomyelitis, along with other RABV characteristics, belong to one of seven lyssavirus genotypes. The prototype RABV is a genotype 1 virus (formerly recognized as serotype 1). Lagos bat virus (LBV, genotype 2/serotype 2), Mokola virus (MOKV, genotype 3/serotype 3), Duvenhage virus (DUVV, genotype 4/serotype 4), European bat lyssavirus type 1 (EBL-1, genotype 5), European bat lyssavirus type 2 (EBL-2, genotype 6) and Australian bat lyssavirus (ABLV, genotype 7) are rabies-related lyssaviruses that reflect the genotypic diversity of the genus *Lyssavirus* (Kuzmin *et al.*, 2005). Any newly isolated lyssaviruses, such as the Irkut and West Caucasian bat viruses recently isolated in Eurasia (Botvinkin *et al.*, 2003) and Aravan and Khujand bat viruses isolated in Central Asia (Arai *et al.*, 2003; Kuzmin *et al.*, 2003) will be assigned to one of the existing genotypes or a new genotype upon careful phylogenetic analysis of their RNA genome sequence (Kuzmin *et al.*, 2005). These viruses, their isolation, genetic typing and phylogenetic relationship to one another, are described in more detail in Chapters 3 and 6. Lyssaviruses share many of the biological and physicochemical features that are associated with other rhabdoviruses. These include the bullet-shaped virus morphology, helical nucleocapsid (NC) or ribonucleoprotein (RNP) core, RNA genome (structure and organization) and viral structural proteins. The five structural proteins of the virus particle (virion) include a nucleocapsid protein (N), phosphoprotein (P), matrix protein (M), glycoprotein (G) and RNA-dependent RNA polymerase or large protein (L). These lyssavirus proteins share many of the biological functions that the same viral proteins have in other rhabdoviruses. Lyssaviruses, like other rhabdoviruses, also

Rabies, second edition. Edited by Alan C. Jackson and William H. Wunner
ISBN 978-012-369366-2. Copyright Elsevier Inc. 2007

use similar mechanisms to enter susceptible cells (albeit, they may use different receptors), express and replicate their genome RNA and release mature virus particles from the plasma membrane of infected cells. Some of the structural proteins of lyssaviruses, on the other hand, can differ dramatically in their antigenic properties and in their post-translational modifications to convey different, often specific properties that distinguish lyssaviruses from other rhabdoviruses. The focus of this chapter will be on the rabies virus structure, molecular composition and morphology, the structure and organization of the RNA genome and the molecular biology of the rabies virus proteins. The chapter also reviews the current knowledge of the virus' life cycle (attachment, penetration, replication, assembly and egress) in infected host cells and mechanisms of virus spread.

2 RABIES VIRUS STRUCTURE

2.1 RNA genome

Lyssaviruses, like other rhabdoviruses, consist mainly of RNA (2–3%), protein (67–74%), lipid (20–26%) and carbohydrate (3%) as integral components (percent of total mass) of their structure (reviewed in Wunner, 1991). The viral RNA genome forms the backbone of the tightly coiled helical RNP (RNA plus protein) core, which extends along the longitudinal axis of the bullet-shaped virus particle. Included in the RNP core are the N, P and L components, which are surrounded by the viral membrane proteins, M and G, and a mixture of lipoprotein components derived from the cell membrane that form the outer envelope or 'membrane matrix' of the virus particle (Figure 2.1). The genome is single-stranded, non-segmented RNA, which has a negative-sense (minus-strand) polarity. This implies that the minus-strand genome RNA (free of protein) is not infectious. The complete sequences of ten lyssavirus RNA genomes have been reported to date (Table 2.1). Nine of these are genomes of genotype 1 strains: Pasteur virus (PV), Street Alabama Dufferin (SAD)-B19, RC-HL, Nishigahara, SRV9, Ni-CE, high egg passage (HEP)-Flury, silver-haired bat rabies virus-18 (SHBRV-18) and RABV. The tenth lyssavirus genome for which the complete sequence has been determined is from the rabies-related MOKV (genotype 3). Its nucleotide sequence is less well conserved than the PV and SAD-B19 genomes in many regions throughout the genome, but particularly in the non-coding regions and in the coding sequences of the P and G genes (Bourhy *et al.*, 1993).

The five structural genes (N, P, M, G and L) of the lyssavirus genome each encode a structural protein of the virus (Figure 2.2). Short (58 and 70 nucleotides) non-coding sequences at the 3' and 5' ends of the genome, called the leader (Le) and trailer (Tr) sequences, respectively, flank the structural genes (Tordo *et al.*, 1986a, 1986b, 1988). The five genes are separated by non-coding intergenic sequences (from the 5' end of one gene to the 3' start of the next gene), one

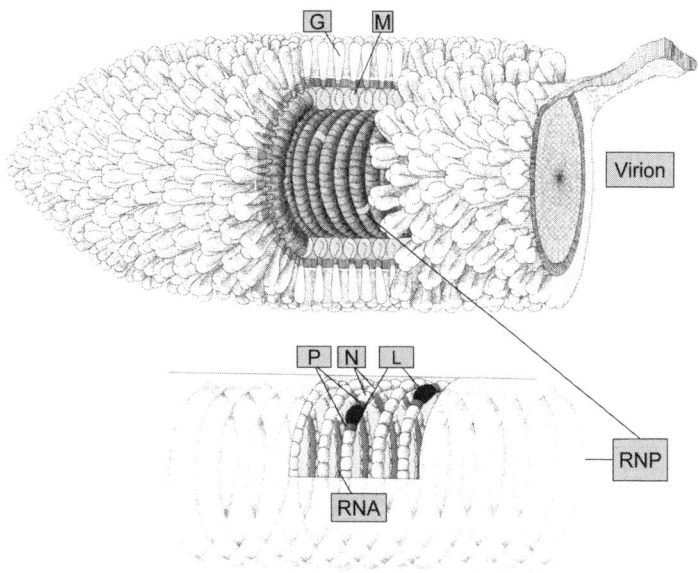

Figure 2.1   Schematic representation of the rabies virion. The drawing shows the internal ribonucleoprotein (RNP) core consisting of the single-strand, negative-sense genome RNA encapsidated with nucleocapsid protein (N), the virion-associated RNA polymerase (L) and polymerase cofactor phosphoprotein (P). The RNP core in association with the matrix protein (M) is condensed into the typical bullet-shape particle that is characteristic of rhabdoviruses. A lipid bilayer envelope (or membrane) in which the surface trimeric glycoprotein (G) spikes are anchored surrounds the RNP-M structure. The membrane 'tail' depicted in the drawing represents the trailing piece of envelope that is frequently observed in the electron microscope attached to the virus as it buds from the plasma membrane of the infected cell. (Reproduced from Wunner, W. H., Larson, J. K., Dietzschold, B. and Smith, C. L. *Review of Infectious Diseases* **10**, Supplement 4, S771–S784, 1988, with permission.) This figure is reproduced in the color plate section.

dinucleotide (N-P), two pentanucleotide (P-M and M-G) and one long 423-nucleotide sequence (G-L), which are unique to lyssaviruses. The long intergenic region between the G and L genes is called a remnant gene or pseudogene (Ψ) in recognition of a sequence of suitable length but lacking an open reading frame (ORF) to code for a detectable protein (Tordo *et al.*, 1986b). The Ψ can vary in length depending on the virus; 423 nucleotides in the RABV genome (Tordo *et al.*, 1986a, 1986b), 504 nucleotides in the MOKV genome (Le Mercier *et al.*, 1997) and 475 nucleotides in the ABLV genome (Gould *et al.*, 1998). The Ψ region represents the most divergent area of the genome (Sacramento *et al.*, 1991).

   The Le sequence at the 3′ end of the genome RNA serves a multifunctional purpose in RABV, as it does in VSV. Within the 3′ terminal non-coding Le sequence, a specific *cis*-acting signal (a specific nucleotide sequence 'acting within' the genome RNA) functions as a signal (or promoter) for template recognition by the viral RNA transcriptase (L alone) or RNA polymerase complex (L plus P). This particular signal initiates genome RNA transcription (Conzelmann and Schnell, 1994; Whelan

TABLE 2.1

Complete lyssavirus genome sequences

Rabies virus (RABV)	Total nucleotides	Database locus[a]	Ref/Submitting author
Pasteur virus (PV)	11 932	NC_001542	Tordo *et al.*, 1986b, 1988
Street Alabama Dufferin (SAD)-B19	11 928	RAVCGA	Conzelmann *et al.*, 1990
Nishigahara	11 926	AB044824	Sakamoto *et al.*, 1994;
			Minamoto, 2000[b]
RC-HL	11 926	AB009663	Minamoto, 1997[b]
Rabies virus SRV9	11 928	AF499686	Hu *et al.*, 2002[b]
Nishigahara (Ni-CE)	11 926	AB128149	Minamoto, 2003[b]
High egg passage (HEP)-Flury	11 615	AB085828	Inoue *et al.*, 2003
Silver-haired bat rabies virus (SHBRV-18)	11 923	AY705373	Prosniak *et al.*, 2004[b]
Rabies virus serotype 1	11 928	AY956319	Pfefferle *et al.*, 2005[b]
Mokola virus (MOKV)	11 939	Y09762	Bourhy *et al*, 1989;
			Le Mercier *et al.*, 1997

[a]Accession numbers; [b] Published only in the database.

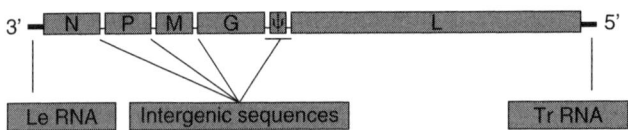

Figure 2.2 Organization of the rabies virus genome. The nucleoprotein (N), phosphoprotein (P), matrix protein (M), glycoprotein (G) and large RNA-polymerase protein (L) genes are separated by intergenic di- and penta-nucleotide sequences and the long pseudogene (Ψ) sequence and are flanked by the leader (Le) RNA and trailer (Tr) RNA sequences at the 3' and 5' ends, respectively.

and Wertz, 1999). Within the first 10 to 20 nucleotides at the 3' and 5' ends of the RABV RNA genome there is a high level of sequence complementarity, including an exact base complementarity between the first and the last 11 nucleotides at the 3' and 5' ends of the genome RNA, respectively. This is strong evidence that the promoter sequences, which are alike in the Le and TrC (3' end of the antigenome RNA that is complementary to the 5' end of the genome) regions provide a common function in transcription and replication (Tordo *et al.*, 1988).

2.2 Viral proteins, lipids and carbohydrate

The five structural proteins of RABV encoded by the lyssavirus comprise the bulk of the virion by weight. The three viral proteins located in the RNP core (see Figure 2.1) are the N, the non-catalytic polymerase-associated P and the catalytic L component of the RNA polymerase. All three proteins are involved in the RNA polymerase activity of the virion. Both the N and P are phosphorylated in RABV, unlike in other rhabdoviruses, including VSV, in which only the P is phosphorylated (Sokol *et al.*, 1974; Gupta *et al.*, 2000). The N in RABV appears to be phosphorylated by a cellular casein kinase II and the P is phosphorylated by a unique cellular protein kinase and specific isomers of protein kinase C (Anzai *et al.*, 1997; Gupta *et al.*, 2000; Wu *et al.*, 2003). The number of molecules of each protein per virion has been estimated in different laboratories with somewhat different results (reviewed in Tordo and Poch, 1988b). Nevertheless, all agree that the most abundant protein in the RNP core is N (1325 or 1800 copies) followed by P (691 or 950 copies) and L (25 or 72 copies) (Madore and England, 1977; Flamand *et al.*, 1993). The stoichiometric relationship that emerges from these independent estimates, however, indicates that the N:P ratio in the RNP complex is 2:1 per virion. This simple relationship, which indicates that two molecules of N interact with 1 molecule of P during nascent protein synthesis and bind to the progeny RNA, is undoubtedly an over-simplification of the dynamics and complexity of the interaction between these two protein molecules. The proteins of the RNP core and the association of N with P and L will be discussed in more detail later.

The remaining two structural proteins of the RABV, G and M, are associated with the lipid-bilayer envelope that surrounds the RNP core. The M is a small-size, non-glycosylated internal protein that lines the viral envelope to form an inner leaflet between the envelope and RNP core. Interestingly, the multifunctional M associates with both the RNP and the viral G, collaborating with G in infected cells, to produce progeny virions in the budding process at the cell membrane (Nakahara *et al.*, 1999). The G is the only glycosylated protein. It produces the trimeric spike-like projections or peplomers on the surface of the viral envelope (Gaudin *et al.*, 1992). The viral G molecule is glycosylated with branched-chain oligosaccharides, which account for 10–12% of the total mass of the protein (reviewed in Wunner, 1991). The structure and function of these two membrane-associated proteins will be discussed in more detail in Section 6. The number of G and M molecules per virion is estimated to be 1205 and 1148 (Flamand *et al.*, 1993), respectively, or 1800 and 1547 (Madore and England, 1977), respectively. This calculates to approximately 450 trimeric G spikes distributed on the outer surface of each virion. The rabies virion envelope contains other host-derived minor protein components such as actin and heat shock proteins of the hsp70 type, CD44 and CD99-related glycoprotein (VAP21) similar to other negative-strand RNA viruses (Naito and Matsumoto, 1978; Sagara and Kawai, 1992; Kawai and Morimoto, 1994; Sagara *et al.*, 1995, 1998). Although the explanation for the presence of cellular proteins in mature virions is not entirely clear, it is possible that the molecular chaperones, such as the heat shock protein calnexin, that associate with the viral proteins during synthesis are incorporated into virions after binding to and assisting in G folding (Gaudin, 1997). Also, a small fraction of calnexin and possibly other chaperone proteins may escape from the endoplasmic reticulum (ER) where they function to ensure proper protein folding before being expressed on the cell surface where virus budding occurs (Okazaki *et al.*, 2000). In a similar manner, cytoskeleton proteins normally expressed on the host cell surface may be incorporated into virions as a consequence of their proximal location and function in virus budding (Sagara *et al.*, 1995, 1998). Other cytoskeletal proteins may ensure intracellular transport of viral RNP (Jacob *et al.*, 2000; Raux *et al.*, 2000). Cellular kinases that activate the transcriptional function of P in RABV may also be packaged into rabies virions (Gupta *et al.*, 2000).

A lipoprotein bilayer forms the viral envelope (or membrane matrix) surrounding the helical RNP core. The lipids, which constitute 20–26% of the viral lipoprotein envelope, are derived entirely from the host cell and, depending on where the virus buds through the cellular membrane, concentrations of certain lipids may be higher in the viral envelope than is represented in the rest of the plasma membrane. In general, the RABV membrane contains a mixture of lipids, including phospholipids (mainly sphingomyelin, phosphatidylethanolamine and phosphatidylcholine), neutral lipids (mainly triglycerides and cholesterol) and glycolipids (reviewed in Wunner, 1991).

2.3 Morphology of standard and defective interfering rabies virus particles

RABV particles are best described as bacilliform or bullet-shaped with one end flattened (planar) and the other rounded (hemispherical), the hallmark of all rhabdoviruses (see Figure 2.1). The physical appearance of this rigid rod-like, bullet-shaped virion was first described in the early 1960s from images seen using the electron microscope (EM) (Matsumoto, 1962, 1963; Davies *et al.*, 1963). The average length of standard-size, infectious virions is 180 nm (130–250 nm) and the average diameter is 75 nm (60–110 nm) (Davies *et al.*, 1963; Hummeler *et al.*, 1967; Sokol, 1975). Beside standard-size virions, other shorter, often cone-shaped 'defective' virions are sometimes co-produced, particularly in cell culture. Defective particles are physically distinguishable from standard-size virions, both in the EM and by size-related physical properties. For example, they contain RNA genomes which are typically shorter than the standard, full-size genomes as a result of internal sequence deletions from the genome (reviewed in Holland, 1987). Cone-shaped defective virions have genomes approximately one-third to one-half the size of the full-length genome of standard-size virions. Other defective virions can have a slightly longer 'thimble-shape' cylindrical dimension, which corresponds to a partially deleted genome that is one-half to two-thirds the size of the standard virion genome.

Defective rabies virions, like defective virions of other rhabdoviruses and other RNA viruses, replicate at the expense of standard 'helper' virus; a phenomenon identified by the Henles in 1943 and studied extensively by von Magnus (see

references in Holland, 1987). Defective virions can grow to become the dominant particle type(s) in infected cells and do so by interfering with production of
standard infectious virus. These are therefore known as defective-interfering (DI)
virions (Wiktor *et al.*, 1977; Clark *et al.*, 1981; Holland, 1987). DI virions of rabies
virus are readily generated in standard cell cultures infected with laboratory-
adapted (fixed) strains of RABV (Wunner and Clark, 1980; Clark *et al.*, 1981). They
have not been described in rabies virus infections *in vivo*, although their role in controlling production of infectious virions *in vivo* has been suggested (see Holland,
1987).

The RNP core of all standard and defective virions is a tightly coiled, yet flexible,
right-handed helical structure that has a periodicity of approximately 7.5 nm per
turn. The length of the tightly coiled RNP core in standard-size infectious virions
measures approximately 165×50 nm in length. The same RNP component when
relaxed and fully extended, like a thread, outside the virion measures between 4.2
and 4.6 mm in length (Sokol *et al.*, 1969; Murphy and Harrison, 1979). During
virus assembly, the condensed RNP core is surrounded by M, to form the 'skeleton'
structure of the virus (Mebatsion *et al.*, 1999). As virus particles mature and bud
through the cellular membrane, the skeleton structure acquires a lipid bilayer
envelope (7.5–10 nm thick) that surrounds the mature virion. Located on the
external surface of the viral envelope are the surface projections that measure
8.3–10 nm in length. Each projection or spike contains three molecules (a trimer)
of the viral G (Gaudin *et al.*, 1992). These have been described when viewed in the
EM as the 'short spikes extending outward with the appearance of hollow knobs
at their distal ends' (Murphy and Harrison, 1979). It is estimated that the height
of the 'hollow knobs' or 'heads' of the spike is about 4.8 nm; the rest of the spike
is made up of the thin 'stalk' on which the head rests (Gaudin *et al.*, 1992).

3 RABIES VIRUS REPLICATION

3.1 Genome transcription and replication

The lyssavirus RNA genome, which is a non-segmented, single-strand RNA molecule with negative-sense polarity cannot be translated directly into protein (i.e. it
is not infectious). The first event in the process of rabies virus replication, therefore,
is to transcribe (copy) the genome RNA to produce complementary (positive-
strand) monocistronic mRNAs representing each of the viral genes or cistrons in
the genome. The monocistronic mRNAs are then translated to produce the viral
proteins. Translation, the process of building a protein from the codons in the RNA
message, involves the cellular ribosome complex, initiation and elongation factors
and transfer RNA charged with individual amino acids. The organization and
specific features of the RABV genome RNA are critical to the transcription and
replication process. The Le sequence at the 3′ end (first 58 nucleotides) of the

11 932-nucleotide genome RNA of rabies virus (PV strain) (see Figure 2.2) is multi-functional in RABV replication, as it is in VSV (Conzelmann and Schnell, 1994; Whelan and Wertz, 1999). Within the 3' terminal Le sequence, a specific *cis*-acting signal functions as a template recognition site for the viral RNA transcriptase (L alone) or RNA polymerase complex (L and P) to bind. This particular signal initiates genome RNA transcription.

Immediately downstream of the Le sequence, in sequential order, are the five structural genes N, P, M, G and L followed by a non-coding trailer (Tr) sequence (last 70 nucleotides) at the 5' end (see Figure 2.2) (Tordo *et al.*, 1986a, 1986b, 1988). The relatively short (dinucleotide or pentanucleotide) sequences between genes (intergenic regions) of the genome (N and P genes, P and M genes and M and G genes) and the long intergenic region between the G and L genes, called the Ψ gene, have the effect of slowing down the transcription rate. These intergenic regions have the effect of causing a progressive slowing of transcription (gene expression) of the non-segmented RNA genome. As a result, the reduction in the rate of transcription is greatest at the distal (5') end of the genome, in particular, causing a severe downregulation of L (polymerase) gene transcription. The severe downregulation of the L gene is correlated with the extraordinary length of the intergenic region provided by the Ψ gene (Finke *et al.*, 2000).

3.1.1 Le RNA transcripts

The first (primary) RNA transcripts produced in the process of transcribing and reproducing the genome RNA are the small 55- to 58-nucleotide-long complementary, positive-strand, non-translated leader RNA (Le$^+$) transcripts. The Le$^+$ transcripts are neither capped nor polyadenylated in contrast to the mRNA transcripts (Colonno and Banerjee, 1978; Leppert *et al.*, 1979) (see Section 3.1.2). In RABV-infected cells, as in VSV-infected cells, the Le$^+$ transcripts interact with the host cellular protein La, which normally associates with cellular RNA polymerase III precursor transcripts (Kurilla *et al.*, 1984). In VSV-infected cells, the association between Le$^+$ transcripts and La protein has been thought to be involved in the shut off of host-cell RNA and protein synthesis. In RABV-infected cells, host macromolecular synthesis is not affected by the association of these two elements (Tuffereau *et al.*, 1985). Le$^+$ transcripts may also interact with conserved *cis*-acting signals that act as transcription promoters (L protein binding sites) at the start of each gene. These transcription promoting Le$^+$ transcripts would appear to initiate mRNA synthesis in a 'stop-start' mechanism of genome transcription as the RNA polymerase moves along the genome (see references in Banerjee and Barik, 1992; Schnell *et al.*, 1996; Yang *et al.*, 1999). Accordingly, a conserved *cis*-acting sequence would be required to signal termination of transcription and polyadenylation of the upstream mRNA, while another *cis*-acting sequence is required to signal reinitiation of the next downstream mRNA transcript (Schnell *et al.*, 1996; Barr *et al.*, 1997; Stillman and Whitt, 1997). Furthermore, another Le termination signal must be required for the switch from primary transcription of mRNAs to secondary transcription of full-length antigenomic RNA (Banerjee *et al.*, 1977; Ball and Wertz, 1981). Precisely how the primary Le$^+$ transcripts function in transcriptional regulation of the mRNAs and switch to virus RNA transcription and replication, i.e. to initiate synthesis of full-length antigenome and progeny genome RNAs, is not entirely clear. It is apparent, however, that after release of the Le$^+$ transcripts, the RNA polymerase complex is fixed in the transcription mode, in which a downstream transcription signal present at the 3′ end of each gene is recognized and transcription proceeds, until the switch from transcription to replication occurs. One way this switch may occur is if the Le$^+$ transcripts become encapsidated by N (see discussion below), preventing them from acting further as initiators of genomic RNA transcription (Yang *et al.*, 1998, 1999).

Just as the extreme 3′ end of the genome RNA provides a specific *cis*-acting signal (promoter) for transcription of the viral genome, the 3′ end of the antigenome RNA (TrC) provides a specific *cis*-acting signal for replication (copy-back) of antigenome RNA. The antigenome RNA is the template for progeny genome RNA synthesis. These signals have been demonstrated for rabies virus (Finke and Conzelmann, 1997) and for VSV (Whelan and Wertz, 1999). In the replication phase (producing progeny genome RNA), the promoter at the 3′ end of the ′replicative intermediate′ (RI) or antigenome RNA is rendered replication competent for a full-length copy-back of the antigenome RNA. Thus, conservation of the extremities, the 3′ and 5′ ends, of the genome (and antigenome) RNA is crucial for successful viral genome RNA transcription and antigenome replication to maintain virus infection (see Section 3.1.3).

Another function of the Le$^+$ transcripts from the 3′ Le sequence in the genomic RNA is to provide specific *cis*-acting signals for RNA encapsidation by soluble N (Yang *et al.*, 1998). Identical *cis*-acting signals are also found in the Le$^-$ (negative-strand) transcripts that are produced from the 3′ TrC sequence of the antigenome RNA, which is complementary to the genome Tr sequence in the genome RNA. The encapsidation signal (N binding site on RNA) in these transcripts is present at or close to the 5′ end of the antigenome RNA (mimicking the location of the Le$^+$ transcripts) and at or near the 5′ end of the progeny genome RNA (mimicking the location of the Le$^-$ transcripts). The encapsidation signal acts as a nucleation site for genome and antigenome RNA encapsidation by N and assembly of RNP complexes. Using synthetic RNA probes that mimic the Le$^+$ transcripts (and sometimes the natural Le$^+$ transcripts) corresponding to the 5′ end of the rabies virus antigenome RNA, researchers have begun to identify the specificity of RNA encapsidation by N (Yang *et al.*, 1998). These RNA probes are also useful in locating the RNA binding site on the RABV N protein (Kouznetzoff *et al.*, 1998). Together the studies indicate that amino acid residues 298 to 352 in a highly conserved region of RABV N bind specifically to a site within nucleotides 20 to 30 of the Le$^+$ transcript, which would be equivalent to nucleotides 20 to 30 of the 5′ end of antigenome RNA.

3.1.2 mRNA transcripts

The RNA transcripts produced immediately after the Le transcripts from the negative-strand genome RNA are the five gene-encoded, positive-strand mono-cistronic mRNAs. The invariant sequence 3'-UUGU-5' in the genome RNA at the 3' end of each gene (5'–AACA–3' at the 5' end of each complementary mRNA) identifies the start of mRNA transcription. During synthesis of the nascent mRNAs, the L protein caps the 5' end of each mRNA by attaching a 7 methyl guanosine (5' m^{7}Gppp-) to the 5' nucleotide of the mRNA (Testa *et al.*, 1980). In the RABV and MOKV genomes, the invariant sequence 3'-AC(U)$_{7-8}$-5' at the 5' end of each gene (upstream of the intergenic sequence) signals both polyadenylation of the mRNA (5'-UGAAAAAAA-3' at the 3' end of each complementary mRNA) (Tordo *et al.*, 1986a; Tordo and Poch, 1988a). In the ABLV genome, the signal to start mRNA transcripts is the same as in the RABV genome, but the signal to terminate and polyadenylate the mRNA is either 3'-AAC(U)$_6$-5' for N-mRNA or 3'-AC(U)$_7$-5' for P-mRNA, M-mRNA and G-mRNA. The latter mRNA includes the ORF sequence for G plus the non-translated Ψ sequence (Gould *et al.*, 1998). When the viral polymerase operating in the transcription mode reaches the string of 7 or 8 Us in the genome, it pauses and begins to stutter or slip as it reiteratively copies the Us. This slows up the transcription process while a polyA tract of variable length (up to ~200 'A's) is produced at the 3' end of the mRNA. The mRNA, which is capped and polyadenylated, is released and the transcriptase, after skipping the next intergenic nucleotides, initiates transcription of the next mRNA.

   In the RABV genome of the PV and Evelyn-Rokitnicki-Abelseth (ERA) strains, two functional polyadenylation (polyA) or transcriptional stop signals exist that terminate the G-mRNA transcript. Only one polyA termination signal exists in the G-mRNA in other virus strains such as the HEP-Flury, CVS, Pitman Moore (PM), SAD-B19 and Nishigahara strains (Tordo *et al.*, 1986b; Morimoto *et al.*, 1989; Conzelmann *et al.*, 1990; Sakamoto *et al.*, 1994) and MOKV (Bourhy *et al.*, 1993). In the genomes where two functional polyA signals exist, the first is located at the end of the G gene and beginning of the Ψ sequence and the second is at the end of the Ψ sequence. As a result of these respective locations, either a short G-mRNA transcript or a long G-mRNA transcript is produced, respectively. Although the signal before the Ψ is functional in the ERA strain, it is leaky and consequently the Ψ region is included in approximately 50% of the G-mRNA transcripts produced (Morimoto *et al.*, 1989). In most of the other RABV strains, MOKV and several street RABV strains, the transcriptional stop signal before the Ψ is either degenerate or absent, leaving only the polyA signal at the end of the Ψ sequence. In such instances, a long G-mRNA transcript is produced that includes the coding sequence for G plus the non-coding Ψ sequence.

   All but one of the monocistronic mRNAs produce a single protein from a single ORF. For these mRNAs, translation starts with the first initiation codon AUG (coding for methionine) from the 5'-end of the mRNA and terminates near the 3'-end with the stop codon UAA or UGA. The one exception is the 3 or 4 proteins derived from the P gene of rabies virus, which are initiated from secondary downstream in-frame AUG initiation codons (Chenik *et al.*, 1995; Takamatsu *et al.*, 1998). These smaller amino-terminal truncated products that are translated from the P-mRNA have been found in purified virions, in infected cells and in cells transfected with a plasmid encoding the complete P sequence. It is thought that a leaky scanning mechanism is responsible for translation of the P-mRNA at several of the internal in-frame AUG initiation sites. The function(s) of these shorter P-related proteins, given the 'anomalous' nature of their derivation and that some of the shorter products end up in the nucleus and the two largest P products remain in the cytoplasm, is unknown. The major ORFs of the five mRNAs encode proteins of approximately 450 (N-mRNA), 297 (P-mRNA), 202 (M-mRNA), 524 (G-mRNA) and 2142 (L-mRNA) amino acids. Some of the lyssavirus mRNAs contain a few more or a few less codons (and hence produce proteins that will contain a few more or a few less amino acids) depending on the genotype of the virus.

3.1.3 Synthesis of antigenome and progeny genome RNA (viral RNA replication)

As rapidly as nascent soluble N protein is produced in the cytoplasm from translation of the primary N-mRNA transcripts, some of the N protein encapsidates the Le^+ RNA transcripts. Protein encapsidation of the Le^+ RNA transcripts either prevents termination of Le^+ RNA transcription at the Le-N gene junction or prevents Le^+ RNA transcripts from initiating the transcription of the individual monocistronic mRNAs. As a result, RNA transcription continues, producing full-length complementary (positive-strand RI) antigenome copies of the parental genome RNA from the infecting virus (Yang *et al.*, 1998, 1999). The RI RNA then becomes the template for progeny (negative-strand) genome RNA, which hereafter in this chapter will be called 'vRNA'. After the RI and vRNAs are synthesized, they, too, are cotranscriptionally encapsidated by soluble N in the cytoplasm (see Section 4.2.1). The 5'-terminal *cis*-acting encapsidation signal in the antigenome and vRNAs acts as a nucleation signal for the nascent soluble N to interact with the viral positive- and negative-strand RNAs. After the specific nucleation signal for encapsidation has been recognized by N, encapsidation proceeds rapidly in the 5' to 3' direction on the RNA, independent of the viral RNA sequence to protect the RNA, particularly from enzymatic degradation in the cell (Banerjee and Barik, 1992).

4 LIFE CYCLE OF RABIES VIRUS INFECTION

The sequence of events in the RABV life cycle, i.e. replication *in vitro* and *in vivo* (in cell culture or animal) can be divided into three phases. The first, or early, phase includes virus attachment to receptors on susceptible host cells, entry via direct virus fusion externally with the plasma membrane and internally with endosomal membranes of the cell and uncoating of virus particles and liberation of the helical RNP in the cytoplasm. The second, or middle, phase includes transcription and replication of the viral genome and viral protein synthesis and the third, or late, phase includes virus assembly and egress from the infected cell. The early phase of the RABV life cycle, often regarded as the most difficult of the events in RABV infection to understand fully, has been studied in many different cell culture systems. These include neuronal and non-neuronal cell lines and primary, dissociated cell cultures derived from dissected pieces of nervous tissue. One caveat that overshadows the use of experimental cell culture systems is that the cells appear to behave differently *ex vivo* in their susceptibility for RABV infection compared with their susceptibility to infection *in vivo*. That is, once cells are removed from their *in vivo* environment, particularly neuronal cells, they lose their natural control over susceptibility (or resistance) to RABV infection. Nevertheless, many studies using *in vitro* cell culture systems describe how virus enters the host cell by direct membrane fusion (Iwasaki *et al.*, 1973; Perrin *et al.*, 1982) or by receptor-mediated endocytosis (Hummeler *et al.*, 1967; Iwasaki *et al.*, 1973; Superti *et al.*, 1984a; Tsiang *et al.*, 1986; Lycke and Tsiang, 1987). No *in vitro* system has yet provided a detailed explanation of how RABV enters muscle cells *in vivo* to support the experimental infections in hamster (Murphy *et al.*, 1973a; Murphy and Bauer, 1974) and skunk (Charlton and Casey, 1979) that show virus replication in striated muscle cells near the site of inoculation.

4.1 Early-phase events (role of the rabies virus receptor)

The life cycle of RABV infection starts with virus attachment to the surface of a target cell and penetration into the cell. Most likely the virus attaches itself to a 'receptor' molecule or cellular receptor unit (CRU) that leads to or permits direct virus entry into susceptible cells in culture (*in vitro*) or specific target cells at the site of inoculation (*in vivo*) (Figure 2.3). Studies using various cell culture systems have implicated various lipids, gangliosides, carbohydrate and protein of the plasma membrane in RABV binding to cells in culture (Perrin *et al.*, 1982; Superti *et al.*, 1984b, 1986; Wunner *et al.*, 1984; Conti *et al.*, 1986, 1988; Broughan and Wunner, 1995), but none have proven to be 'specific' receptors. Others have focused on specific cellular receptor molecules or CRUs *in vivo* that appear to correlate with the defined neurotropism of the virus (reviewed in Lafon, 2005; see also Chapter 8). One of these CRUs is the nicotinic acetylcholine receptor (nAChR) found at neuromuscular junctions, where rabies virus can also be found co-localized *in situ* (Lewis *et al.*, 2000). Not all cell lines infected with rabies virus *in vitro*, however, express the nAChR (Reagan and Wunner, 1985; Tsiang *et al.*, 1986; Tsiang, 1993) and some neuronal cells infected with rabies virus *in vivo* may not express the nAChR (Kucera *et al.*, 1985; Tsiang *et al.*, 1986; Lafay *et al.*, 1991). Two other possibilities are the neural cell adhesion molecule (NCAM) CD56 on the cell surface of RABV-susceptible cell lines (Thoulouze *et al.*, 1998)

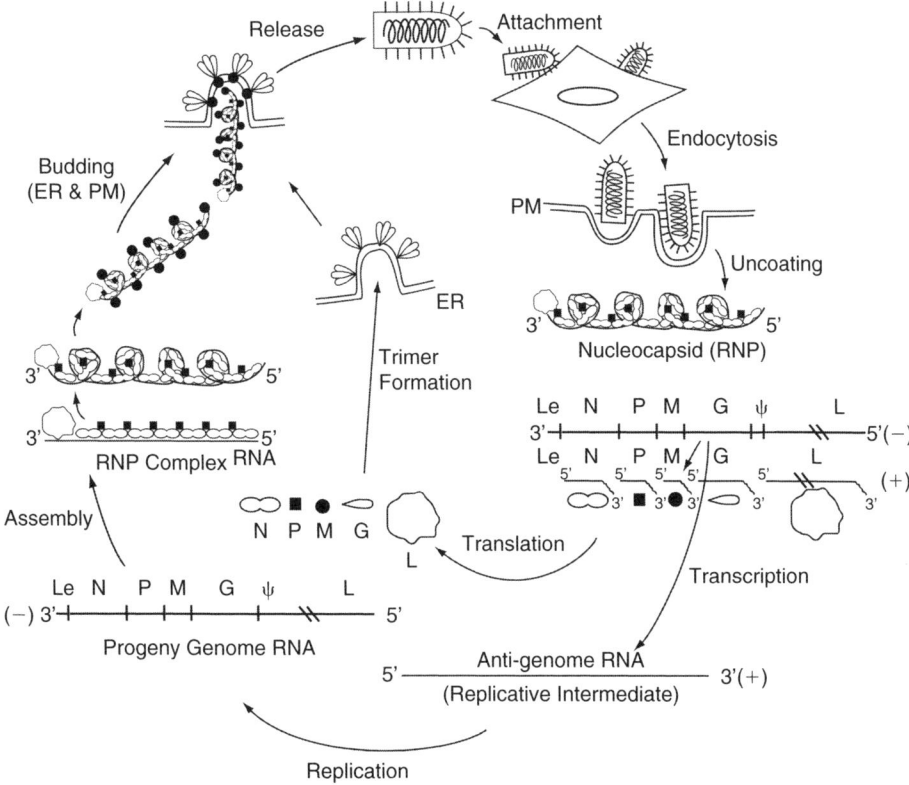

Figure 2.3   Rabies virus life cycle in the cell. Virus enters the cell following attachment through coated pits (viropexis) or via cell surface receptors, mediated by the viral glycoprotein (G) fusing with the cellular membrane (endocytosis). After internalization, the viral G mediates low pH-dependent fusion with the endosomal membrane and the virus is uncoated, releasing the helical nucleocapsid (NC) of the ribonucleoprotein (RNP) core. The five structural genes (N, P, M, G and L) of the genome RNA in the NC are transcribed into five positive (+) strand monocistronic messenger RNAs and a full-length + strand (antigenome) replicative intermediate RNA. The antigenome RNA serves as the template for replication of progeny genome (− strand) RNA. The proteins (N, P, M and L) are synthesized from their respective mRNAs on membrane-free ribosomes in the cytoplasm and the G is synthesized from the G-mRNA on membrane-bound ribosomes (rough endoplasmic reticulum). Some of the N-P molecular complexes produce cytoplasmic inclusion bodies (Negri bodies) *in vivo* and some N-P complexes encapsidate the + strand and − strand viral RNAs. After progeny genome RNA is encapsidated by N-P protein complex and L protein is incorporated to form progeny RNP (both full-length standard and shorter defective) structures, the M protein binds to the RNP and condenses the RNP into the 'skeleton' structures. The skeleton structures interact with the trimeric G protein structures anchored in the plasma membrane and assemble into virus particles that bud from the plasma membrane of the infected cell into adjacent extracellular or interstitial space.

and the low-affinity neurotrophin receptor (p75NTR), a nerve growth factor expressed on the surface of cultured BSR cells (Tuffereau *et al.*, 1998; Langevin *et al.*, 2002). Some doubt has also been cast on the importance of the p75NTR as an obligatory rabies virus receptor since the CVS strain infects p75NTR-deficient mice (Jackson and Park, 1999). Whatever receptor is present on the cell surface, it seems the virus is not limited to choosing only one type of receptor in order to complete the virus life cycle in the infected animal.

After RABV binds to its putative cellular receptor, viral entry (internalization) usually proceeds by fusion of the viral envelope with the cellular membrane (Superti *et al.*, 1984a). RABV, like VSV, may also enter the cell through coated pits and uncoated vesicles (viropexis or pinocytosis), which often incorporate several (two to five) virions per vesicle (Tsiang *et al.*, 1983). After the virus is internalized, whether by receptor-mediated endocytosis (via the endocytic pathway) or through coated pits, the viral G mediates fusion of the viral envelope with the endosomal membrane and releases the viral RNP in endosomal vesicles (Marsh and Helenius, 1989; Whitt *et al.*, 1991; Gaudin *et al.*, 1992). The capacity of RABV to enter the cell via fusion with the endosomal membrane is dependent on the low pH within the endosomal compartment. The threshold pH for fusion activation for RABV is about pH 6.3 and involves a series of specific and discrete conformational changes in RABV G (Gaudin *et al.*, 1993, 1995a, 1995b; Gaudin, 1997). These conformational changes in the G are described in detail below (see Section 6.5). At the same time, changes may occur in the cellular membrane to facilitate virus uptake, which include the formation of membrane fusion pores (Gaudin, 2000).

4.2 Middle-phase events (virus replication)

4.2.1 *Viral RNP release and initiation of genome RNA transcription, replication and protein synthesis*

In the second phase of the RABV life cycle, viral RNA genome transcription is initiated in the cytoplasm of the infected cell after the tightly coiled transcriptionally 'frozen' but potentially active RNP core is released from endosomal vesicles (see Figure 2.3). The tightly coiled RNP structure relaxes to form a loosely coiled helix, conceivably to facilitate the ensuing viral RNA transcription and replication events in the cell described above (see Section 3.1) (Iseni *et al.*, 1998). The transcription process in virus replication is carried out on the infecting genome RNA-N protein complex by the virion-associated RNA polymerase complex (L plus P cofactor) and is independent of host cell functions. The virion-associated polymerase complex either initiates transcription at the 3′ end of the genome RNA or it resumes transcription at the next downstream internal mRNA start site on the viral genome, close to where the polymerase complex was 'frozen' in place during progeny virus assembly in a previously infected cell. This part of the replication process has been termed 'primary transcription' since it takes place on the parental nucleocapsids released into the cytoplasm from the input virus and does not require concomitant host or viral protein synthesis. Each of the five genes produces in sequential order a monocistronic mRNA transcript that is eventually translated into one of the five viral proteins (Flamand and Delagneau, 1978; Holloway and Obejeski, 1980). At each intergenic junction, however, the polymerase pauses before continuing the downstream mRNA transcription process and an estimated 20 to 30% of the polymerase complexes that reach the gene junction dissociate from the nucleocapsid. As a result, fewer polymerase molecules remain associated with the genome RNA-N template after each gene junction to resume the transcription process. Consequently, the number of mRNAs synthesized from the remaining genes downstream in the genome gradually decreases in proportion to the number of polymerase molecules that fall off. This phenomenon of self-regulating viral gene expression is a form of differential transcription attenuation or 'localized' attenuation (Finke *et al.*, 2000 and references therein) (see Section 3.1).

The proteins of the virus are synthesized from the viral mRNAs using the protein synthesis machinery (free and membrane-bound polyribosomes) of the host cell. Four of the viral mRNAs, N-, P-, M- and L-mRNA are translated on free (unbound) ribosomes while the G-mRNA is translated on membrane-bound ribosomes. The nascent G is inserted cotranslationally into the lumen of the ER where disulfide bond formation occurs and the protein-folding enzymes and molecular chaperones are available to assist in the folding of G before the molecule is transported out of the ER (Gaudin, 1997). While in the lumen of the ER, the G monomers undergo modification at specific asparagine (N in single letter code) residues by core glycosylation and N-glycan processing (Shakin-Eshleman *et al.*, 1992) and they form homotrimers (Whitt *et al.*, 1991; Gaudin *et al.*, 1992). The final processing of the N-linked carbohydrate side chains takes place in the Golgi apparatus of the intracellular membrane network. This is accomplished with the sequential removal and then addition of monosaccharides by glycosidases and glycosyltransferases (see references in Gaudin, 1997) before the viral G trimers appear on the plasma membrane surface of the infected cell.

Viral RNA replication (production of progeny genome or vRNA) becomes the dominant event after primary transcription in the mid-phase and late-phase events of virus growth. During this time in the infection, production of vRNA and RI RNA becomes disproportionate. The stoichiometric relationship of the vRNA to RI RNA is 50:1 (Finke and Conzelmann, 1997). The bias for the excessive production of the vRNA over RI RNA in the RABV-infected cell is attributed to the activity of their *cis*-acting sequences (see Section 3.1.3).

4.2.2 Genome instability

Infidelity in lyssavirus genome RNA replication during the life cycle is a consequence of the absence of RNA proofreading/repair and post-replication error correction mechanisms in the cell. As a result, mutations that are introduced at a relatively high frequency, by the 'error-prone' virion-associated RNA polymerase as the enzyme replicates the genome RNA, remain in the vRNA. This produces a population of different progeny genomes that share a common origin, i.e. they are related but distinct (Kissi *et al.*, 1999). Similarly, as progeny genomes are transcribed during secondary transcription and replicated in cells, the error-prone copying of the RNA is magnified. Genome RNA mutations (misincorporation of nucleotides) occur at different rates, in the order of 10^{-4} to 10^{-5} substitutions per nucleotide per cycle, depending on the region of the genome RNA considered (Domingo and Holland, 1997; Bracho *et al.*, 1998; Kissi *et al.*, 1999). Other factors that may also be involved in generating RNA sequence heterogeneity in rabies virus include duration of infection, route of transmission, virus load, host immune response and virus-host protein cooperation (Kissi *et al.*, 1999). Despite these other influences, the infidelity of the RNA polymerase of negative-strand RNA viruses remains the single major factor responsible for the nucleotide misincorporation in the genome RNA. Assuming the distribution of mutations is random in the population of replicating genomes, the variant genomes form a complex 'quasispecies' population that increases very rapidly with successive infection cycles over time (reviewed in Holland *et al.*, 1992; Smith *et al.*, 1997; Domingo and Holland, 1997). As a result of this extreme form of genetic instability, out of a quasispecies population of rabies viral RNA genomes, a true RABV variant may evolve that harbors a specific mutation or set of specific mutations capable of imparting to the virus a distinctive phenotypic or unique virus–host relationship. For example, several mutations have been identified in protein coding regions of the genome of virus variants that correlate with a selective tropism for neurons or for an avirulent phenotype of the virus in a particular animal host (Murphy and Nathanson, 1994; Domingo *et al.*, 1998; Morimoto *et al.*, 1998). Mutations in non-coding regions of the genome may also impact on the balance between replication of standard versus defective genomes and possibly increase or decrease the survival of the infected cell or animal host. They may also influence long-term survival (persistence) of the virus in its host as a result of specific interactions between viral and host determinants (Domingo *et al.*, 1998). The 'quasispecies' model of mixed RNA virus populations provides a plausible explanation for the rapid selection of mutants that fit into any new environmental condition (Morimoto *et al.*, 1998; Dietzschold *et al.*, 2000). This selection process may affect any of the viral genes whose proteins influence the particular structure, function or phenotype of the virus.

4.3 Late-phase events (virus morphogenesis and budding)

Once sufficient pools of vRNA and the viral N, P and L proteins have accumulated in the infected cell, formation of rabies viral RNP complex and assembly of virus particles begins and continues as long as the cells remain metabolically active.

RABV morphogenesis in infected cells is associated with the formation of an intra-cytoplasmic ground substance or 'matrix' commonly found in brain tissue as well as in tissue culture (Hummeler *et al.*, 1968). This filamentous matrix substance constitutes the Negri bodies found in neurons of the infected brain (Matsumoto, 1962; Matsumoto and Miyamoto, 1966), a development that precedes the formation of virus particles (Hummeler *et al.*, 1967) (see Chapter 9). The process of virus assembly really begins with encapsidation of vRNA and RNP formation, i.e. the addition of N to the 5' end of the nascent vRNA, and the formation of vRNA-N-P complexes (Iseni *et al.*, 1998; Mavrakis *et al.*, 2003; Liu *et al.*, 2004). When N binds to the phosphate-sugar backbone (exposing the nucleotide bases), the RNA becomes fully protected from degradation by cellular ribonucleases and possibly innate immune responses directed against the vRNA (Kouznetzoff *et al.*, 1998). As nascent N and P molecules accumulate in the cytoplasm of the rabies virus-infected cell, they form homologous (N-N or P-P) and heterologous (N-P) complexes. N in high concentrations tends to aggregate with itself $(N-N)_n$ to form the large intracytoplasmic 'matrix' bodies found in brain tissue and the P forms homotrimers or tetramers that seem to be necessary for the cooperative polymerase activity of the protein itself (Gigant *et al.*, 2000). As the equilibrium shifts in concentration of soluble N and P monomers and to different levels of P phosphorylation, several unique heterologous complexes may form between the N and P (Chenik *et al.*, 1994; Fu *et al.*, 1994; Kawai *et al.*, 1999; Gigant *et al.*, 2000; Toriumi *et al.*, 2002; Eriguchi *et al.*, 2002; Mavrakis *et al.*, 2003). In some of the N-P complexes the ratio of N:P is 1:1, in some the ratio is 2:1, similar to that found in rabies virions (Fu *et al.*, 1994) and in some the N:P ratio is 1:2 (Mavrakis *et al.*, 2003). When P interacts with N in a 1:1 ratio, it prevents soluble N from self-aggregating and maintains N in a soluble form for efficient RNA encapsidation during replication (Prehaud, *et al.*, 1992; Fu *et al*, 1994). By binding to *de novo*-synthesized N, however, P prevents the immediate phosphorylation of N, enabling N to bind more strongly to RNA (Yang *et al.*, 1999; Liu *et al.*, 2004). N-P complexes, which contain unphosphorylated N, are then responsible for the specificity of viral RNA encapsidation by N and for preventing encapsidation of non-viral RNA species (Liu *et al.*, 2004). Therefore, phosphorylation of N occurs after N-P complex formation and either during or after the process of RNA encapsidation.

Despite current knowledge of the virus assembly process, little is known about how the L is added to the RNP complex. From studies of other negative-strand RNA viruses, it is thought that the P in the RNA-N-P complex mediates L binding to complete the transcriptionally active virion RNP core (Mellon and Emerson, 1978; Buchholz *et al.*, 1994). More recent studies with rabies virus describe the formation of ring structures by recombinant RNA-N complexes, which have biochemical and biophysical properties that appear to be indistinguishable from RABV RNPs (Schoehn *et al.*, 2001; Albertini *et al.*, 2006). These structures define the spatial relationships between the core proteins and RNA in the assembled rabies virus RNP structure (see Section 6.1). Clearly, the crystallization studies on these and other recombinant RNP-like structures will clarify the role of specific amino acid residues or motifs in the molecular mechanisms controlling RNP assembly.

The M is the next viral protein to associate with RNP complexes. M is a multifunctional protein and it plays several key roles in the dynamics of the formation of progeny virus. First, the association of M with newly formed transcriptionally active RNP (vRNA+ N+ P+ L) changes the balance between viral RNA transcription and replication by selectively inhibiting transcription and stimulating replication (Ito *et al.*, 1996; Finke *et al.*, 2003). M imparts this differential effect in RNA synthesis on encapsidated vRNA binding either by the RNP template or the L polymerase in the complex (Finke *et al.*, 2003). It is conceivable, though not proven, that the different regulatory functions of M in RNA synthesis may also be provided by the different conformational forms of the monomer M, Mα and Mβ, recently described (Ameyama *et al.*, 2003) (see Section 6.4). Finally and possibly as an ongoing regulatory role of M, the M alters the structure of the RNP by condensing it into a tightly coiled 'skeleton-like' form in which the polymerase activity is 'frozen'. In this function of M, which may involve M as a dimer, M initiates virus budding as the complexes migrate toward the cellular membranes where they become enveloped by the cellular membrane and enter the virus budding process (Mebatsion *et al.*, 1996, 1999; Nakahara *et al.*, 2003). From the time M enters the virus assembly pathway as a soluble protein in the cytoplasm it is involved in all steps that lead to virus budding. This is based on the observation that assembly and budding of bullet-shaped particles can occur in the absence of the transmembrane G spikes that are normally associated with infectious particles (Mebatsion *et al.*, 1996, 1999; Robison and Whitt, 2000). The M will then localize the RNP coil at the cellular membrane where the nascent viral G is concentrated and where the M is able to interact with G (Mebatsion *et al.*, 1999). In the mature rabies virion that buds from the cell membrane, the M lies between the lipid bilayer envelope (formed by interaction with the host cell membrane) and the helical RNP that it covers (Mebatsion *et al.*, 1999). Thus, the M covering and condensing the helical RNP is thought to play an important role in virion morphogenesis, i.e. giving the particle its bullet-shape morphology. If the M is missing from rabies virus particles that contain G spikes on the outer surface, a morphological variation in budded particles is observed, suggesting that the particles contain uncondensed RNP. M-deficient rabies virus also causes increased cell–cell fusion and enhanced cell death, in contrast to the relatively benign cytopathic effect that is observed with wild-type virus. Also, virus budding is much less efficient in the absence of M, demonstrating again the multifunctional role of M in virus assembly and budding.

In the final stages of RABV assembly, the mature virions acquire their lipid bilayer envelope as the assembled skeleton (RNP + M) structure buds through the host cell plasma membrane. Mature virions that bud through the plasma membrane into the extracellular space are frequently observed in extraneural tissue cells *in vivo* (Murphy *et al.*, 1973a) and in a variety of *in vitro* tissue culture systems (Davies *et al.*, 1963; Hummeler *et al.*, 1967; Matsumoto and Kawai, 1969;

Iwasaki *et al.*, 1973; Matsumoto *et al.*, 1974; Tsiang *et al.*, 1983). Occasionally, virions mature intracellularly by budding through the cytoplasmic ER or Golgi apparatus. Cytoplasmic maturation within infected neuronal cells of brain commonly appears to occur by budding on the intracytoplasmic membranes of the ER; Golgi membranes are occasionally involved (Murphy *et al.*, 1973b; Matsumoto *et al.*, 1974; Matsumoto, 1975; Gosztonyi, 1994). If budding occurs at a site in the cell membrane where the nascent RABV transmembrane G is also targeted, then infectious virions will be produced bearing the G molecules arranged as trimeric spike-like structures tightly packed and anchored in the viral envelope (Whitt *et al.*, 1991). The way the spikes are oriented in the viral membrane, more than 80% of the G molecule (the ectodomain) is exposed on the surface of the virion and the cytoplasmic tail, C-terminal to the transmembrane domain of the G, extends beneath the lipid bilayer envelope. The C-terminal tail of the G molecule is then free to interact with the M of the RNP + M skeleton structure. Interaction of G with M is essential for stabilization of the G in trimers on the virion surface and for efficient budding of RABV (Mebatsion *et al.*, 1996, 1999). This does not preclude the possibility that skeleton structures may bud from membrane regions where no G exists. In this case, budding would be inefficient, producing low levels and the bullet-shaped particles produced would be spikeless (free of G) and non-infectious (Mebatsion *et al.*, 1996). If skeletons bud through ER or Golgi membranes, they bud into the lumen of vesicles produced from these membranes and may be secreted from the cell through the normal secretory pathway.

5 VIRUS CELL-TO-CELL SPREAD – PERPETUATING THE VIRUS LIFE CYCLE

Progeny rabies virions that bud from infected cells are able to spread from cell to cell in cell or tissue culture (*in vitro*), presumably as they do in the animal (*in vivo*). They have the option of spreading to contiguous cells (direct cell-to-cell spread) or to non-contiguous cells, which are surrounded by interstitial space. In the case of direct cell-to-cell spread, RABV spreads despite a continuous presence of serum virus-neutralizing antibody (VNA) (Dietzschold *et al.*, 1985). Alternatively, virus that buds from an infected cell into the surrounding interstitial space must find another cell to infect. In this case, the spread of virus is limited by the presence, *in vitro* and *in vivo*, of VNA that blocks virus attachment to cellular receptors, fusion and subsequent virus entry into a susceptible cell (Dietzschold *et al.*, 1985; Flamand *et al.*, 1993). *In vivo*, rabies virus can travel long distances within a cell, particularly within cells of peripheral nerves and neuronal cells of the CNS, via intra-axonal cytoplasmic transport in a microtubule network-dependent process. Virus that moves intra-axonally can cover great distances, particularly in bipolar neurons, before reaching and crossing the synapse of one dendritic process of one cell into another (Kucera *et al.*, 1985; Gillet *et al.*, 1986; Ceccaldi *et al.*, 1989; Coulon *et al.*,

1989; Lafay *et al.*, 1991) (see Chapter 8). It was also postulated that naked viral nucleocapsids might be transported in the axoplasmic flow along axons and through the synapse into the post-synaptic neuron, particularly as this pathway provides an alternative mechanism of virus spread along neuronal networks when mature virions cannot be detected at synapses (Gosztonyi, 1994). This hypothesis of trans-synaptic transfer of naked nucleocapsids, however, was weakened considerably, if not negated, by recent studies that demonstrate an absolute requirement for the virion G in trans-synaptic spread of RABV, both *in vivo* and *in vitro* (Etessami *et al.*, 2000). Other significant studies of RABV spread also point to the nature of the viral tropism and its relationship to the G of the virion. In particular, viruses that are pathogenic or virulent *in vivo* differ in their ability to invade the CNS and spread within the brain in comparison with the apathogenic or avirulent virus phenotype (Coulon *et al.*, 1989; Lafay *et al.*, 1991). This phenotypic difference, which is determined by the surface G (Coulon *et al.*, 1983; Dietzschold *et al.*, 1983b) is discussed further in Section 6.5.

6 STRUCTURE OF RABIES VIRUS PROTEINS IN RELATION TO FUNCTION

6.1 Nucleocapsid protein (N)

The RABV N (PV strain) contains 450 amino acids (in some viruses 451 amino acid residues), one of which is phosphorylated, and has a molecular weight of ~57 kDa. The amino acid sequence of N is the most conserved of the viral proteins among the seven genotypes. Despite the conserved nature of the N, there is a relatively high degree of genetic diversity within short segments of the N gene between the genotypes (Conzelmann *et al.*, 1990; Bourhy *et al.*, 1993, 1999; Kissi *et al.*, 1995; Kuzmin *et al.*, 2005). An important reason for the high level of amino acid sequence conservation, particularly within specific regions in N, is that it must retain certain key functions. On the other hand, the noted amino acid differences provide unique, genotype-specific epitopes on the N that enable the assignment of viruses to genotypes on the basis of their reactivity patterns (antigenicity) with a panel of anti-N monoclonal antibodies (MAbs) (Flamand *et al.*, 1980a; Dietzschold *et al.*, 1987a; Smith, 1989). This qualitative diversity in the N has also been exploited at the nucleotide level by the polymerase chain reaction (PCR) technology (Sacramento *et al.*, 1991; Bourhy *et al.*, 1993; Kuzmin *et al.*, 2003) (see Chapters 3 and 10).

The binding site on N for its interaction with viral RNA has been localized to a region of N between amino acid residues 298 and 352 (Kouznetzoff *et al.*, 1998). After binding to viral RNA, the N undergoes sufficient conformational change to acquire a number of conformation-dependent epitopes, one of which enables the serine (S) residue at position 389 of N to be phosphorylated (Dietzschold *et al.*,

1987a; Anzai et al., 1997; Kawai et al., 1999; Toriumi and Kawai, 2004). During genome RNA replication in the infected cell, it is thought that the encapsidation of the new viral RNA by N triggers an encapsidation-associated conformational change in the N that makes the S389 residue accessible for phosphorylation (Kawai et al., 1999). Phosphorylation of N then stabilizes the interaction between N and P in the rabies viral RNP complex (Toriumi and Kawai, 2004). It has also been suggested that the phosphorylation of the N after RNA encapsidation is important to regulate viral RNA transcription and replication (Yang et al., 1999; Wu et al., 2002; Liu et al., 2004).

Interesting ring structures were produced with recombinant phosphorylated N and viral RNA to model the viral RNP, which closely resembled the helical structure of the viral RNP core (Iseni et al., 1998; Schoehn et al., 2001; Albertini et al., 2006). The 'bilobed' shape of the model ring structures reveal how the N molecules (protomers) are closely packed in the RNP structure, binding to every nine nucleotides of the vRNA. This tight packing appears to be sufficient to protect the viral RNA against ribonuclease activity (Iseni et al., 1998). Yet, it is necessary that the N protomer dissociate itself from the RNA for the RNA to become a template for the polymerase. One or several N protomers might dissociate locally to provide sufficient space for the RNA polymerase to bind. As a model for authentic viral RNP, it is interesting to see how the P binds to the N protomer in the N-RNA ring. The P, which binds to the unique trypsin-sensitive cleavage site close to the C-terminal end of N, in the region of serine 389 (the phosphorylated site on N) (Kouznetzoff et al., 1998; Albertini et al., 2006), ends up in different positions. In one position, the P bends toward the inside and, in the other position, it bends toward the outside of the ring (Schoehn et al., 2001). In either position, the interaction of N with P causes a further conformational change in N. This is similar to the conformational change that is believed to occur when N first associates with P after N encapsidates the viral RNA (Kawai et al., 1999; Toriumi et al., 2002; Toriumi and Kawai, 2004).

The N is the second most extensively analyzed of the RABV proteins (after the G) with respect to its antigenic and immunogenic structure and function. The immunological interest in N stems from the observation that the RNP of RABV induces protective immunity against a peripheral challenge of lethal rabies virus in animals (Dietzschold et al., 1987b; Tollis et al., 1991). B and T cell-specific epitopes on N were initially defined using MAbs and synthetic peptides in competitive binding assays. These antibodies delineated the topography of functional antigenic sites common to both rabies and rabies-related viruses (Flamand et al., 1980a; Lafon and Wiktor, 1985; Dietzschold et al., 1987a; Ertl et al., 1989). Subsequently, several linear B cell and T cell epitopes were physically mapped on N (reviewed in Fu et al., 1994; Goto et al., 1995, 2000). Three linear antigenic epitopes (antibody binding sites) on N were mapped to amino acids 358 to 367 (antigenic site I) and three linear epitopes (antigenic site IV) were mapped to two independent regions, amino acids 359 to 366 and 375 to 383 (Minamoto et al., 1994; Goto et al., 2000). Although the region between residues 359 and 366 is shared by the two independent antigenic sites, I and IV, the MAbs that recognize epitopes within these sites do not compete with each other for binding to the N antigen. Thus, it would appear that the respective epitopes are detected on different forms of N, one that represents N that is diffusely distributed in the cytoplasm and the other that is associated with cytoplasmic inclusion bodies. The two forms of N might even have different degrees of protein folding or maturation (Goto *et al.*, 2000). The fact that the N associated with inclusion bodies may be N in its mature form, having gained a greater degree of folding, is suggested by another MAb, specific for antigenic site II, which only recognizes a conformation-specific epitope on the inclusion body-associated N antigen. The linear epitope (amino acids 373 to 383) that appears to overlap but may not be identical to another epitope (375 to 383) in site IV, is recognized by a different anti-N MAb (Dietzschold *et al.*, 1987a). Three of the five epitopes from sites I and IV in RABV N, which are also shared by rabies-related viruses, represent cross-reactive determinants (Goto *et al.*, 2000). Conformation-dependent epitopes are present in antigenic sites II and III and also at the phosphorylation site (serine 389) of N, where a phosphorylation-dependent epitope is formed after the newly synthesized N associates with the P in N-P complex formation (Minamoto *et al.*, 1994; Kawai *et al.*, 1999).

The N is also a major target for T helper (Th) cells that cross-react among rabies and rabies-related viruses (Celis *et al.*, 1988a, 1988b; Ertl *et al.*, 1989). Several Th cell epitopes in the RABV N were identified and mapped using a series of overlapping synthetic peptides corresponding to N sequences of approximately 15 amino acids in length (Ertl *et al.*, 1989). The antigenic peptides (bearing a specific epitope for each subset of Th cells) that were capable of stimulating RABV-specific Th cells *in vitro* were selected and subsequently tested for stimulation of RABV-specific Th cells *in vivo* (Ertl *et al.*, 1989, 1991). One such peptide, designated 31D, which corresponds to N residues 404–418, was found to be an immunodominant epitope capable of stimulating production of RABV-specific Th cells, at least *in vitro*. The same peptide also induced a significant Th cell response *in vivo* and an accelerated VNA response after a booster immunization with inactivated RABV vaccine *in vivo* (Ertl *et al.*, 1989). However, neither the peptide-induced Th cell response nor the increase in VNA titer induced by this peptide epitope in a vaccine boost, regardless of the immunodominance, was sufficient to protect against a lethal virus challenge dose in mice.

Finally, certain other properties and immune responses that the rabies viral NC specifically elicits in humans suggest that the RABV N functions as an exogenous superantigen (Lafon *et al.*, 1992). It is perhaps the only viral superantigen that has been identified in humans (Lafon, 1997). Some of the properties and responses found not only in humans but also in mice that are attributable to the RABV N in the role of superantigen include:

1 its potent activation of peripheral blood lymphocytes in human vaccinees
    (Herzog *et al.*, 1992)

2 its ability to produce a more rapid and heightened VNA response upon injection of inactivated rabies vaccines (Dietzschold *et al.*, 1987b; Fu *et al*, 1991)

3 its induction of early T cell activation steps and expansion and mobilization of CD4$^+$ Vβ8 T cells to trigger and support production of VNA (Lafon *et al.*, 1992; Martinez-Arends *et al.*, 1995)

4 its ability to bind to HLA class II antigens expressed on the surface of cells (Lafon *et al.*, 1992).

6.2 Phosphoprotein (P)

The RABV P (PV strain) contains 297 amino acids (303 amino acids in MOKV P) (38–41 kDa) and is highly conserved (>97%) among genotype 1 lyssaviruses. The P of lyssaviruses, like the P in other negative-strand viruses, exists in a variety of phosphorylated forms (Gupta *et al.*, 2000; Toriumi *et al.*, 2004). It is a multifunctional protein interacting with N (N-P) and being a key component and regulatory protein of the virion-associated RNA polymerase (P-L) complex in viral genome replication. The P acts as a chaperone for soluble nascent N preventing its polymerization (self-assembly) and non-specific binding to cellular RNA (Mavrakis *et al.*, 2003) and specifically directs N encapsidation of the viral RNA (Chenik *et al.*, 1994; Fu *et al.*, 1994; Gigant *et al.*, 2000). As a subunit of the RNA polymerase (P-L) complex, the P plays a pivotal role as a non-catalytic cofactor in transcription and replication of the viral genome (Chenik *et al.*, 1994, 1998; Fu *et al.*, 1994). The P serves to stabilize the L protein and place the polymerase complex on the RNA template, which the L protein alone is unable to do.

Two prominent forms of P are present in both rabies virions and in virus-infected cells. One is a major hypophosphorylated 37-kDa form and the other is a minor hyperphosphorylated 40-kDa form (Toriumi *et al*, 2004). Other forms of P, detectable in sodium dodecyl sulfate-polyacrylamide gel electrophoresis (SDS-PAGE), may also exist in RABV-infected cells (Gupta *et al.*, 2000). The smaller nascent cellular forms have fewer amino acids than the P and are products of N-terminal truncations due to translation initiation at internal AUG codons located in-frame within the P mRNA (Conzelmann *et al.*, 1990; Fu *et al.*, 1994; Chenik *et al.*, 1995; Takamatsu *et al.*, 1998).

Studies investigating protein kinases present in cell extract from rat brain have shown that RABV P is phosphorylated in the N-terminal portion by two distinct types of protein kinases, one of which is a unique heparin-sensitive protein kinase (Takamatsu *et al.*, 1998; Gupta *et al.*, 2000). This unique 71-kDa kinase, designated rabies virus protein kinase (RVPK), phosphorylates recombinant P (36 kDa, expressed in *Escherichia coli*) at S63 and S64 (CVS strain) and nascent P (37 kDa) expressed in BHK-21 cells infected with the HEP-Flury strain of RABV. In both cases, hyperphosphorylation alters the mobility of P (36 kDa and 37 kDa) in SDS-PAGE causing it to migrate more slowly, as a protein of 40 kDa (Toriumi *et al.*, 2004). The other phosphorylating enzyme is protein kinase C, which has several isomers (PKCα, β, χ and δ). In contrast to the RABV-PK, phosphorylation of P by the PKC isoforms, dominated by PKCγ activity, did not alter the migration of P in SDS-PAGE (Gupta *et al.*, 2000). Upon analyzing the PK activity in rabies virions for the presence of these two types of enzymes, it was concluded the RABV-PK is selectively packaged in mature rabies virions along with a smaller amount of the predominant PKCγ isoform as the rabies virion-associated PKs.

RABV P, like the P molecules in other negative-strand viruses, is multifunctional and its interactions are multifaceted. It interacts with the N and L via domains of P that are specific for each of these two proteins. In addition, at least two independent N-binding sites have been found on P that allow nascent RABV P to bind to nascent soluble N to maintain N in a competent form for RNA encapsidation and to the C-terminal part of N in the RNA-N complex (Chenik *et al.*, 1994; Fu *et al.*, 1994; Schoehn *et al.*, 2001). One site is located within the C-terminal 30 amino acids (between amino acids 267 and 297) of P and the other is located in the N-terminal portion of the protein between amino acids 69 and 177 (Chenik *et al.*, 1994). Both sites interact with N in a manner that is mutually independent (Fu *et al.*, 1994). The interaction involving the N-terminal binding site of P requires that the P interacts with N soon after the two proteins are synthesized *in vivo*, but that it may compete with another molecule that interacts with N independently. The other interacting molecular species that competes for the binding site on N might be endogenous vRNA. Why should N need two P monomers or a P dimer to prevent its non-specific binding to cellular RNA? Perhaps the interaction between N and P dimers is favored through the energetic stability of the P dimer (Mavrakis *et al.*, 2003).

After the P binds to the RNA-N template in the formation of progeny RNP, the P is required to bind L, the large catalytic subunit of the viral RNA polymerase, to produce a virus-encoded RNA polymerase complex that is fully active. The P subunit in the P-L complex has a major binding site for L protein within the first 19 amino acids of P (Chenik *et al.*, 1998). The P is able to form complexes with itself, though it is not clear whether trimers or tetramers, or both, are formed and which of these oligomers is necessary for binding to L (Gao *et al.*, 1996; Spadafora *et al.*, 1996; Gigant *et al.*, 2000). Oligomerization of RABV P does not require phosphorylation nor is the N-terminal domain (first 52 amino acids) necessary for oligomerization or binding to N-RNA template (Gigant *et al.*, 2000; Mavrakis *et al.*, 2003). This is in contrast to the P of VSV which requires phosphorylation for oligomer formation to be fully active and available for binding both to L and to the RNA template (Gao *et al.*, 1996).

Other protein–protein interactions that involve the RABV P are equally interesting, specifically that P should interact with cellular proteins that influence or help regulate virus tropism and cell-to-cell spread or even inhibit innate immune responses that typically prohibit productive virus replication. Using the yeast two-hybrid approach to identify interactive cellular proteins, the 10 kDa cytoplasmic dynein light chain (LC8), which is involved in the intracellular transport of organelles, was found to interact strongly with the P of RABV and MOKV (Jacob *et al.*, 2000; Raux *et al.*, 2000). In both studies, the P domain that interacts with dynein LC8 was mapped to the N-terminal half of the P; one of them mapped the interactive site to a location between amino acids 138 and 172 in the P (Raux *et al.*, 2000). It is of special interest, with regard to the manner in which RABV spreads along neurons *in vivo*, often over long distances, to know how dynein LC8 might facilitate the transport of virus through neurons. Dynein LC8, as a part of cytoplasmic dynein and myosin V, participates in the myosin V complex, a microtubule-associated motor protein complex that is implicated in the actin-based motor transport of ER vesicles in brain neurons (Jacob *et al.*, 2000). Thus, the retrograde axonal transport of uncoated RNP along axons to the perikaryon and transport of nascent RNP from the perikaryon along dendrites to the next neuron might be mediated by the P-dynein LC8 complex in transporting the viral RNP along the microtubule network (Ceccaldi *et al.*, 1989). RABV P is also responsible for inhibiting the activation of latent interferon regulatory factor (IRF-3), which is the key factor for initiating an interferon (IFN) response. It prevents the phosphorylation of multiple serine residues at the C terminus of the IRF-3 protein, which allows dimers of IRF-3 to form and be recruited to the IFN-β enhancer as part of a larger protein complex that includes transcription factors and coactivators. (Brzózka *et al.*, 2005). Thus, RABV P is necessary and sufficient to prevent a critical IFN response in virus-infected cells.

6.3   Virion-associated RNA polymerase or large protein (L)

The RABV L (PV strain) contains 2142 amino acids (2127 amino acids in SAD-B19 strain and MOKV) (244 kDa) and is encoded in the fifth gene, which comprises more than half (54%) of the coding potential of the RABV genome. The L is the catalytic component of the polymerase complex which, along with the non-catalytic cofactor P, is responsible for the majority of enzymatic activities involved in viral RNA transcription and replication. Many of the activities of this multifunctional enzyme have been identified in genetic and biochemical studies using VSV, the prototype virus and model for studying the virion-associated RNA polymerase of negative-strand RNA viruses (Banerjee and Chattopadhyay, 1990). Like all negative-strand RNA viruses, the virion-associated viral RNA polymerase plays a unique role at the start of infection by initiating the primary transcription of the genome RNA after the RNP core is released into the cytoplasm of the infected cell. The enzymatic steps of transcription include initiation and elongation of the Le$^+$ RNA and mRNA transcripts as well as the cotranscriptional modifications of the mRNAs that include 5'-capping, methylation, and 3'-polyadenylation.

Comparisons of L sequences from different negative-strand RNA viruses have helped to map the functionally homologous and unique sequences in attempts to locate the ascribed enzyme activities (Tordo *et al.*, 1988; Barik *et al.*, 1990; Poch *et al.*, 1990). One of the main features of L that comes out of the sequence comparison is the number of clusters of conserved amino acid residues that appear to be purposefully aligned in blocks, I–VI, along the protein (Poch *et al.*, 1990). Within these blocks, some residues form strongly conserved domains, with a high proportion of amino acids either strictly or conservatively maintained in identical positions, while other domains are more variable, consistent with the multifunctional nature of L (Tordo *et al.*, 1988; Banerjee and Chattopadhyay, 1990; Poch *et al.*, 1990). One of the blocks (block III), the catalytic domain, in the central part of the RABV L, between residues 530 and 1177, contains four motifs, A, B, C and D, that represent regions of highest similarity (Tordo *et al.*, 1988; Poch *et al*, 1989). These motifs, which are thought to constitute the 'polymerase module' of L, maintain the same linear arrangement and location in all viral RNA-dependent RNA and DNA polymerases (Barik *et al.*, 1990; Delarue *et al.*, 1990; Poch *et al.*, 1990). Among the conserved sequences in these four motifs is the tri-amino acid core sequence GDN (glycine, aspartic acid and asparagine) in motif C, which is extensively conserved in all non-segmented negative-strand RNA viruses, suggesting a precise catalytic function in the polymerase activity (Poch *et al.*, 1989). Not only the GDN core sequence but also specific amino acids downstream from the core sequence are crucial for the maintenance of polymerase activity, which catalyzes the polymerization of nucleotides (Schnell and Conzelmann, 1995). In addition, at least two other sequences between amino acid residues 754 to 778 and 1332 to 1351 in the VSV L have been identified as consensus sites for binding and utilization of ATP, similar to those found in cellular kinases (Barik *et al.*, 1990; Canter *et al.*, 1993). Three essential activities encoded by L are involved in the binding and utilization of ATP. These are:

1 the transcriptional activity that requires binding to substrate ribonucleoside triphosphates (rNTPs)
2 polyadenylation
3 protein kinase activity for specific phosphorylation of the P in transcriptional activation (Sanchez *et al.*, 1985; Banerjee and Chattopadhyay, 1990).

Many of the putative functions of this multifunctional protein, including mRNA capping, methylation and polyadenylation, remain to be delineated and mapped within the RABV L. The process of mapping active sites in the RABV L using the mutational and deletion approach will be helped considerably now that it is feasible to apply the powerful technique of reverse genetics (Schnell and Conzelmann, 1995). Using reverse genetics, it is possible to express the RABV L with selected amino acid deletions and point mutations in the cDNA of the L gene to map the locations in L that carry out the various functional activities. Similarly, the role of the cofactor P in the RNA polymerase (L-P) complex can be better defined using reverse genetics. Since the L protein relies exclusively on its interaction with the phosphorylated P to be fully active, the major question is whether the P complements any of the specific enzymatic functions of L or if the P functions solely as a regulatory protein in the RNA transcription and replication process. For example, does the cofactor P unwind the RNA-N complex of the NC to facilitate entry as well as movement of L on the genome template (De and Banerjee, 1985)? The cooperative function of the non-catalytic cofactor P and catalytic L in the polymerase complex is clearly intriguing and of critical importance to warrant further examination.

6.4 Matrix protein (M)

The M of RABV and the other genotype lyssaviruses is the smallest of the virion proteins. The RABV M contains 202 amino acids (25 kDa) (Tordo *et al.*, 1986a; Conzelmann *et al.*, 1990; Bourhy *et al.*, 1993; Hiramatsu, *et al.*, 1993; Gould *et al.*, 1998) and exists in two isoforms. The major (24 kDa) form, Mα, accounts for 70–75% of total M in rabies virions and the minor (23 kDa) form, Mβ, accounts for 25–30% of total M in the virion and only 10–15% of total M in the cell (Ameyama *et al.*, 2003). The Mβ isoform, in particular, is found in the Golgi apparatus of the cell, where it colocalizes with viral G, as well as on the surface of infected cells and the virion. In addition, the M forms dimers (54 kDa; 10–20% of total M in the cell and 20–30% in the virion) that form a strong association with G (Nakahara *et al.*, 2003). M binds to and condenses the nascent NC core into a tightly coiled, helical RNP-M protein complex, forming a sheath around it and producing the bullet-shaped 'skeleton' structure of the virion. Approximately 1200 to 1500 copies of M molecules bind to rabies virus RNP core. At the same time M binds to the RNP structure, it mediates binding of the viral core structure to the host membrane at the marginal region of the cytoplasm where it initiates virus budding from the cell plasma membrane (Mebatsion *et al.*, 1999). A proline-rich motif (PPPY or PY motif) in the M located at residues 35–38 within the highly conserved 14-amino acid sequence near the N-terminus of the RABV M appears to be associated with virus budding from the cell membrane surface (Harty *et al.*, 1999, 2001). Corresponding proline-rich motifs and related core sequences are found in the M of VSV and other viruses (references in Irie *et al.*, 2004). The PY motif, now referred to as a late-budding domain (L-domain), is similar to the L-domains identified in other viral proteins (Wills *et al.*, 1994; references in Irie *et al.*, 2004). These proteins are all involved in a late step of the virus budding process. Although the exocytotic release of virus particles requires the M in RNP-M skeletons, the efficiency of virus budding is greatly enhanced by the interaction of the RNP-M complex with the envelope G (Mebatsion *et al.*, 1996). Increased virion production as a result of direct interaction of the cytoplasmic domain of the transmembrane spike G and the viral RNP-M core suggests that a concerted action of both core and spike proteins is necessary for efficient recovery of virions. However, the interaction of M with the cytoplasmic domain of G does not need to be optimal, i.e. the interaction is sufficient if the G of different viruses are substituted for the homologous G in budding virions (Mebatsion *et al.*, 1995; Morimoto *et al.*, 2000).

6.5 Glycoprotein (G)

The G of all rabies and rabies-related viruses is comprised of four distinct domains, the signal peptide (SP) domain, the ectodomain (ED), the transmembrane (TM) domain and the cytoplasmic domain (CD). The mature G (minus the SP) of all RABV strains for which the amino acid sequence has been determined has 505 amino acids (~65 kDa). The RABV G is translated from a G-mRNA transcript that encodes 524 amino acids, which includes the SP domain (Tordo *et al.*, 1988; Conzelmann *et al.*, 1990). Mature MOKV G has 503 amino acids (522 amino acids are encoded in the MOKV G-mRNA) (Benmansour *et al.*, 1992; Bourhy *et al.*, 1993). The G of RABV and MOKV is a type I membrane glycoprotein, which indicates that the TM (anchoring) domain is located near the C-terminus. The first 19 N-terminal amino acids of the G-mRNA ORF represent the SP and provide the membrane insertion signal, which transports the nascent protein into the rough ER-Golgi-plasma membrane pathway before it is cleaved from the N-terminus of the G molecule in the Golgi apparatus. The resulting transmembrane G is organized into trimers (three monomers of 65 kDa each) in the Golgi apparatus, which later form the G (trimeric) spikes embedded in the plasma membrane and on the virion surface (Whitt *et al.*, 1991; Gaudin *et al.*, 1992). The G spikes in the viral envelope extend 8.3 nm from the virus surface and represent the major surface protein of the virion. Each G is anchored in the plasma membrane and viral lipid envelope by a 22-amino acid TM domain located between residues 439 and 461 of the RABV G. The C-terminal portion (last 44 amino acids) of the G, the CD, extends inward from the plasma membrane into the cytoplasm of the infected cell where it interacts with M of the skeleton particle to complete the virion assembly. The ED of G (residues 1 to 439 of the mature RABV G) in each spike, the portion that extends outward on the plasma membrane and surface of mature virus particles, is the business end of the molecule for a variety of functional virus interactions. The trimeric spike is responsible for interaction with cellular binding sites (receptors) and, therefore, is important in viral pathogenesis by targeting the appropriate cells for infection (Sissoëff *et al.*, 2005). It is also responsible for low pH-induced fusion of the viral envelope with plasma and endosomal membranes in the cell in the early phase of the RABV life cycle (Gaudin *et al.*, 1993). And it is critical for the induction of a host humoral immune response to RABV infection and as the target of VNA as well as for virus-specific helper and cytotoxic T cells (Macfarlan *et al.*, 1986; Celis *et al.*, 1988a).

Appropriate glycosylation of RABV G is important for its proper expression and function. The oligosaccharides that are associated with the G are linked to asparagine (N, in single letter code) residues in the tripeptide sequence (sequon) asparagine-X-serine (NXS) or asparagine-X-threonine (NXT), where 'X' is any amino acid other than proline (Shakin-Eshleman *et al.*, 1992). RABV G molecules can have one, two or three (and sometimes four) sequons or N-linked carbohydrate (N-glycan) sites per molecule depending on the virus strain. Even if an N-glycan site is present, it may be inefficiently glycosylated or not occupied at all depending on the amino acid composition of the sequon (Dietzschold *et al.*, 1983a; Wunner *et al.*, 1985; Shakin-Eshleman *et al.*, 1993; Kasturi *et al.*, 1995). A virus can be heterogeneous with regard to the number of sites that are occupied with N-glycans (macroheterogeneity) and with regard to the types of glycan structures (microheterogeneity) present at each site (Wojczyk *et al.*, 2005). Clearly, various factors can affect the glycosylation of individual sequons in Gs; some influence efficiency of N-glycosylation and others influence processing of N-glycans into a variety of oligosaccharide structures. The sugar residues of the N-glycan are presynthesized in the cell and transferred from a lipid precursor by the enzyme oligosaccharyltransferase as a precursor core-oligosaccharide unit ($Glc_3Man_9GlcNAc_2$) to the specific sequons of the nascent G molecule. Typically, the transfer occurs as the protein is synthesized on the membrane-bound ribosomes, then begins to fold cotranslationally and during translocation to the cytoplasmic ER membrane. After the precursor core-oligosaccharide is transferred, the high-mannose triglucoslylated oligosaccharide is trimmed and processed in the lumen of the rough ER and Golgi stacks to form the 'complex type' N-linked monoglucosylated oligosaccharide of the mature G molecule (see references in Gaudin, 1997). Interestingly, the typical steps in N-glycosylation processing are not always achieved efficiently and uniformly (Wojczyk *et al.*, 2005). The molecular chaperone calnexin recognizes the partially trimmed monoglucosylated glycan and binds to it, assisting the G to fold correctly and completely in order to achieve its biological activity, stability and antigenicity (Shakin-Eshleman *et al.*, 1992; Gaudin, 1997; Okazaki *et al.*, 2000). This dependence of the RABV G on the molecular interaction with calnexin explains why it is critical that at least one of the asparagine residues, i.e. N319, in the RABV G is conserved in all virus genotypes. It has now been demonstrated that N319 glycosylation is essential for correct and complete folding of the nascent RABV G and required for subsequent transport to the cell surface (Wojczyk *et al.*, 2005).

The RABV G is also modified by the addition of palmitic acid (referred to as fatty acid acylation or palmitoylation) at cysteine 461, located in the intra-CD on the C-terminal side of its TM region (Gaudin *et al.*, 1991a). Although the functional significance of palmitoylation is not entirely clear, it is presumed to have a stabilizing effect on the trimeric G spike anchored in the membrane. Palmitoylation may also play a role in the virus budding process by facilitating the interaction between the CD 'tail' of the G and the M in the RNP-M complex at the cell membrane.

The role of RABV G as a fusion protein is to facilitate virus entry into host cells. After targeted binding of virus to its receptor(s) on host cells, what follows to enable the virus to become internalized is mediated by the G trimer spike on the virus particle as it triggers a low pH-dependent fusion with the plasma and endosomal membranes to enter cellular endosomes. In this process, the G goes through significant and critical conformational changes whereby it assumes at least three structurally distinct 'conformational' states (Gaudin *et al.*, 1993, 1999; Gaudin, 2000). Prior to virus binding to the cellular receptor, the G on the virion surface is in its 'native' state. After the virus attaches to the receptor and the virus is internalized, the G is 'activated' to a hydrophobic state enabling it to interact with the hydrophobic endosomal membrane. Upon entering the endosomal compartment and low pH environment of the cellular compartment, the fusion capacity of the G is activated via a major structural change in the G that exposes the fusion domain, which interacts with and destabilizes one or both of the participating membranes (Gaudin *et al.*, 1995a). The low pH-induced fusion domain, which is thought to lie between amino acids 102 and 179 (Gaudin *et al.*, 1995a), is not to be confused with the proposed fusogenic domain (amino acids 360–386) on the RABV G that appears to be involved in pH-independent (neutral pH) cell fusion (Morimoto *et al.*, 1992). The condition that induces endosomal membrane fusion in the viral entry and uncoating process is pH 6.2–6.3. After low-pH fusion, the G assumes a reversible 'fusion-inactive' conformation, which makes the G monomer appear longer than the 'native' conformation and assume selective antigenic distinctions (Gaudin *et al.*, 1993; Kankanamge *et al.*, 2003). The fusion-inactivated G, which is no longer relevant to the fusion process, is highly sensitive to cellular proteases and appears to be in a dynamic equilibrium with the 'native' G that is regulated by lowering and raising the pH (Gaudin *et al.*, 1991b, 1996). Interestingly, the fusion-inactive conformation serves the G in another capacity. During nascent viral protein synthesis, the G assumes an 'inactive state-like' conformation, protecting the G post-translationally from fusing with the acid nature of Golgi vesicles, while it is transported through the Golgi stacks to the cell surface. At the cell surface, the G acquires its 'native' conformation and structure (Gaudin *et al.*, 1995a, 1995b, 1999). The MAbs that recognize specific low pH-sensitive conformational epitopes of the G can identify certain acid-induced conformational changes, as well as detect the various stages of nascent G monomer folding and its association with molecular chaperones like BiP and calnexin (Gaudin, 1997; Maillard and Gaudin, 2002; Kankanamge *et al.*, 2003).

The mechanism of RABV G fusion with the endosomal lipid bilayer membrane can be defined in terms of how the low pH-induced conformational change exposes the G fusion domain allowing it to fuse with the cellular membrane (Gaudin, 2000). The portion of the RABV G, which becomes exposed in the acidified state (below pH 6.7), the so-called low pH fusion domain, lies within a larger peptide segment between amino acid residues 103 and 179 (Durrer *et al.*, 1995).

Interestingly, this fusion domain-containing peptide segment is particularly devoid of any extended hydrophobic amino acids and the degree of homology is low in the region that corresponds to the location of a stretch of hydrophobic residues (102 to 120) in VSV G. The highest degree of homology within the RABV G in relation to other rhabdoviruses is found elsewhere, in the region between residues 144 and 197. The functional state of the RABV G in which it acquires fusogenic activity is correlated with at least one specific conformational epitope. This epitope, which appears to be formed by combining two separate regions, the neurotoxin-like region (residues 189–214) of RABV G and the conformational antigenic site III (residues 330–340) is abrogated when the G is exposed to acidic conditions (Kankanamge *et al.*, 2003; Sakai *et al.*, 2004). To complicate further the mechanism of RABV G-mediated membrane fusion, a second putative RABV G fusogenic domain (residues 360–386) has been identified that is pH-independent and resembles the fusion domain of other virus fusion proteins (Morimoto *et al.*, 1992). It also is not cleaved to generate a terminal fusion peptide as in the hemagglutinin protein of influenza virus. The behavior of both of these putative (low pH-dependent and neutral pH-independent) fusion domains suggests that RABV-induced membrane fusion proceeds via alternative mechanisms. The most plausible fusion mechanism recently postulated for the RABV G is the formation of lipidic (fusion) pores, each generated from a hemifusion diaphragm that is formed by the binding of RABV to the lipid membrane (Gaudin, 2000). This binding mechanism, which is accomplished in a hydrophobic manner, results in inserting the exposed fusion domain on the RABV G into the target membrane. The transition from virus interaction to generating the fusion pore is a high-cost energy step and depends on the integrity and correct folding of the G trimers directly involved in the fusion process. It has been shown that more than one trimer of G is required to build a competent fusion site (Gaudin *et al.*, 1993).

Another important function of the RABV G that is critical in establishing the phenotype of the virus is its role in defining the pathogenicity and determining the neuroinvasive pathway of the virus (Kucera *et al.*, 1985; Ito *et al.*, 2001; Yan *et al.*, 2002). While the neurotropism of a particular RABV strain is primarily a function and major defining characteristic of its G, it is relevant to note that other viral components and attributes are also important in altering the viral pathogenesis of rabies. These might include transcriptional activity of the RABV L, the Tr sequence, Ψ gene and optimal protein–protein (e.g. G-M) interactions (Morimoto *et al.*, 2000; Faber *et al.*, 2004). Whether or not the virus will cause a lethal infection (follow a specific pathway to induce a fatal disease) is determined first by the interaction of the G spike with a specific receptor on neuronal cells *in vivo* (Lafon, 2005; Sissoëff *et al.*, 2005). At least one attempt to map the p75NTR receptor binding site on RABV G has suggested that the receptor binds to a region of the G (within residues 318–352) on both sides of antigenic site III and site 'a' that is not neutralized by anti-G antibody (Langevin and Tuffereau, 2002). Lentiviral vectors pseudotyped with RABV G have demonstrated that the G not only allows entry into the nervous system, upon infection of neurons at distal connected sites within the nervous system; it facilitates and enhances the retrograde axonal transport of the virus to the CNS (Mazarakis *et al.*, 2001). Second, the pathogenic or virulence phenotype of the virus correlates with a single amino acid at a specific position in the RABV G. For example, arginine (R) 333 or lysine (K) 333 of wild-type G determines the virulence phenotype or neuroinvasive pattern of RABV in the CNS. Virus variants that substitute glutamine (Q), isoleucine (I), glycine (G), methionine (M) or serine (S) for R333 in the rabies virus G express a phenotype that is either less pathogenic or totally avirulent compared to the parental wild-type virus in adult immunocompetent mice (Dietzschold *et al.*, 1983b; Seif *et al.*, 1985; Tuffereau *et al.*, 1989). While these mutations in the G protein reduce or abolish neuroinvasiveness, they do not impair the ability of the virus to multiply in cell culture. The same substitutions (e.g. Q333 for R333) can, however, affect the rate of virus spread from cell to cell (Dietzschold *et al.*, 1985) and the ability to infect motor neurons *in vivo* and *in vitro* (Coulon *et al.*, 1998). They can modify the host range spectrum of the virus and determine the choice of neuronal pathways the virus uses to reach the CNS (Kucera *et al.*, 1985; Etessami *et al.*, 2000) as well as the distribution of virus to different areas of the brain (Yan *et al.*, 2002). Still other amino acid substitutions at other positions on the RABV G appear to confer or influence viral pathogenicity. For example, amino acid substitutions located between positions 34 and 42 and at positions 198 and 200 (related to epitopes specific to antigenic site II) of the challenge virus standard (CVS) strain G, reduced the pathogenicity of the CVS strain when inoculated intramuscularly in adult mice (Prehaud *et al.*, 1988). In the pathogenic parental Nishigahara strain and the avirulent variant, RC-HL strain, which is derived from the Nishigahara strain (see Table 2.1), R333, the pathogenic determinant residue that is common among the representative laboratory strains of rabies virus, is present in both. In the parental Nishigahara strain, one or more residues between amino acid residues 164 and 303 define the lethality that is characteristic of this virus (Takayama-Ito *et al.*, 2004).

It is interesting that RABV is able to survive at least the humoral arm of the immune protection mechanisms during infection by spreading from cell to cell without budding into the culture medium or extracellular space where it would be neutralized by anti-rabies VNA (Dietzschold *et al.*, 1985). The precise mechanism of direct cell-to-cell spread of virus *in vitro* or *in vivo* has not been determined. The observation that virus is internalized by a cell without being compromised, i.e. prevented from attaching to the cell surface receptor in cell culture, points to the importance of the pH-dependent and pH-independent fusion functions of the G in virus spread. When cultures of mouse neuroblastoma (NA) cells and baby hamster kidney (BHK) cells were infected with the virulent phenotype of RABV and maintained in the presence of anti-rabies VNA, the virus was able to spread throughout the NA and the BHK cell cultures. Under similar conditions, the avirulent RABV failed to spread from cell to cell in the NA cell culture as if there was some impairment in the G-mediated cell fusion process in the NA cells. The avirulent virus spread cell to cell in the BHK cell culture equally well in comparison with the virulent virus (Dietzschold *et al.*, 1985). In many ways, the pathogenic and avirulent viruses behaved *in vitro* in a manner that reflects their ability to spread *in vivo*, after direct inoculation into the brain of the mouse. The pathogenic virus spread *in vivo* more rapidly in the CNS and infected more neurons than did the avirulent virus. Could it be that the G was not 'activated' to a hydrophobic state due to the amino acid substitution at position 333, enabling it to interact with the hydrophobic endosomal membrane? Others using NA cells and BHK cells, which constitutively express the RABV G in the absence of other viral proteins, have shown that only G with R333 as the determinant for its pathogenic phenotype demonstrated an ability to induce syncytium formation (cell–cell fusion) at neutral pH (pH-independent fusion) in the NA cell culture (Morimoto *et al.*, 1992). Thus, the RABV G with R333, but not with Q333, has the ability to mediate virus spread among neuronal cells in culture. Since some Q333 variants can kill adult immunocompetent mice when infected by stereotaxic inoculation (Yang and Jackson, 1992), it appears that R333 in the G is required for the neuropathogenicity of the virus from a peripheral site of inoculation and for its trans-synaptic and axonal spread *in vivo*. It is also apparent that the pH-independent cell fusion induced by the RABV G may involve the interaction of one or more neuronal cell-specific host cell factors, which are expressed in the NA cells but not in BHK-21 cells.

Another difference between the pathogenic and avirulent RABV G that is determined by R333 and Q333, respectively, is the ability of fixed laboratory virus strains of the pathogenic phenotype to invade the CNS from a peripheral site at different rates (Kucera *et al.*, 1985; Etessami, *et al.*, 2000). By selecting different neuronal pathways to the brain from a peripheral site of inoculation, the pathogenic phenotype virus reaches the brain faster than the avirulent phenotype. Another study showed, as a contraindication of the influence of the amino acid residue at position 333, that a similar rate of spread and pathway of spread to the CNS was observed with both phenotypes after peripheral inoculation (Jackson, 1991). Where differences in rate and pathways of virus spread are apparent, it is conceivable that the pathogenic virus may use different receptor sites on cells that determine neurotropic pathways that the avirulent virus can no longer recognize or uses inefficiently, if at all, to spread from cell to cell. Since it is possible that not all RABV-specific receptors have been identified or meet the biological function criterion, one can only speculate on the reasons why virus of the avirulent phenotype is sometimes unsuccessful or inefficient in spreading to the CNS.

Finally, the RABV G is a potent immunogen and of major importance immunologically for the induction of the host immune response against virus infection. Because it is such an important antigen for the immune system to mount a response against, it is probably the most extensively studied RABV protein in terms of structure in relation to its immunogenicity and antigenicity for VNA, i.e. as a target for VNA. Over the past two decades of study, several major and minor antigenic sites have been defined both functionally using virus neutralization-escape or monoclonal antibody resistant (MAR) mutants and by physical mapping of key VNA epitopes on the RABV G (Lafon *et al.*, 1984; Dietzschold *et al.*, 1988; Benmansour *et al.*, 1991). The G induces VNA that recognize both conformational and linear epitope (antibody binding) sites and stimulates helper as well as cytotoxic T-cell activity. Consequently, studies first to map functionally and then locate physically specific epitopes within antigenic domains for binding antibody and for binding T cells to the RABV G have been an ongoing and extensive process. At least eight antigenic sites (I–VI, 'a' and G1) have been located on the ED (amino acid residues 1–439) of the G of different virus strains (Flamand *et al.*, 1980b; Lafon *et al.*, 1984; Prehaud *et al.*, 1988; Dietzschold *et al.*, 1988, 1990; Benmansour *et al.*, 1991; references in Langevin and Tuffereau, 2002). Sites I, III, VI and 'a' involve the amino acids located at position 231, 330–338, 264 and 342–343, respectively. Site II is a discontinuous antigenic site that involves two separate stretches of amino acids in position 34–42 and 198–200 that are presumably linked by a disulfide bridge (Prehaud *et al.*, 1988). Sites VI and G1 are defined as linear or non-conformational, while the others are conformational and readily destroyed upon denaturation. Epitopes recognized by T cells have been mapped on the G using chemically cleaved and synthetic peptides or T-cell lines and clones derived from individuals immunized with RABV vaccine (Macfarlan *et al.*, 1984, 1986; Celis *et al.*, 1988a, 1988b). These fine-mapping studies have given some limited insight into the structure of the various functional domains on the RABV G. To gain a fuller understanding of the function of the RABV G, it is necessary to determine its overall conformation and, ultimately, its three-dimensional structure. However, crystallization of viral membrane glycoproteins is difficult, due to the TM domain and to oligosaccharide microheterogeneity and attempts at crystallization of the RABV G have so far been unsuccessful.

REFERENCES

Albertini, A.A.V., Wernimont, A.K., Muziol, T. *et al.* (2006). Crystal structure of the rabies virus nucleoprotein-RNA complex. *Science* **313**, 360–363.

Ameyama, S., Toriumi, H., Takahashi, T. *et al.* (2003). Monoclonal antibody 3-9-16 recognizes one of the two isoforms of rabies virus matrix protein that exposes its N-terminus on the virion surface. *Microbiology and Immunology* **47**, 639–651.

Anzai, J., Takamatsu, F., Takeuchi, K. *et al.* (1997). Identification of a phosphatase-sensitive epitope of rabies virus nucleoprotein which is recognized by a monoclonal antibody 5-2-26. *Microbiology and Immunology* **41**, 229–240.

Arai, Y.T., Kuzmin, I.V., Kameoka, Y. and Botvinkin, A.D. (2003). New lyssavirus genotypes from the lesser mouse-eared bat (*Myotis blythi*), Kyrghyzstan. *Emerging Infectious Diseases* **9**, 333–337.

Ball, L.A. and Wertz, G.W. (1981). VSV RNA synthesis: How can you be positive? *Cell* **26**, 143–144.

Banerjee, A.K. and Barik, S. (1992). Gene expression of vesicular stomatitis virus genome RNA. *Virology* **188**, 417–428.

Banerjee, A.K. and Chattopadhyay, D. (1990). Structure and function of the RNA polymerase of vesicular stomatitis virus. *Advances in Virus Research* **38**, 99–124.

Banerjee, A.K., Abraham, G. and Colonno, R.J. (1977). Vesicular stomatitis virus: Mode of transcription. *Journal of General Virology* **34**, 1–8.

Barik, S., Rud, E.W., Luk, D., Banerjee, A.K. and Kang, C.Y. (1990). Nucleotide sequence analysis of the L gene of vesicular stomatitis virus (New Jersey serotype): identification of conserved domains in L proteins of nonsegmented negative-strand RNA viruses. *Virology* **175**, 332–337.

Barr, J.N., Whelan, S.P. and Wertz, G.W. (1997). *cis*-acting signals involved in termination of vesicular stomatitis virus mRNA synthesis include the conserved AUAC and the U7 signal for polyadenylation. *Journal of Virology* **71**, 8718–8725.

Benmansour, A., Brahimi, M., Tuffereau, C., Coulon, P., Lafay, F. and Flamand, A. (1992). Rapid sequence evolution of street rabies glycoprotein is related to the highly heterogeneous nature of the viral population. *Virology* **187**, 33–45.

Benmansour, A., Leblois, H., Coulon, P., Tuffereau, C., Gaudin, Y. and Flamand, A. (1991). Antigenicity of rabies virus glycoprotein. *Journal of Virology* **65**, 4198–4203.

Botvinkin, A.D., Poleschuk, E.M., Kuzmin, I.V. *et al.* (2003). Novel lyssavirus isolated from bats in Russia. *Emerging Infectious Diseases* **9**, 1623–1625.

Bourhy, H., Kissi, B., Audry, L. *et al.* (1999). Ecology and evolution of rabies virus in Europe. *Journal of General Virology* **80**, 2545–2557.

Bourhy, H., Kissi, B. and Tordo, N. (1993). Molecular diversity of the *Lyssavirus* genus. *Virology* **194**, 70–81.

Bourhy, H., Tordo, N., Lafon, M. and Sureau, P. (1989). Complete cloning and molecular organization of a rabies-related virus, Mokola virus. *Journal of General Virology* **70**, 2063–2074.

Bracho, M.A., Moya, A. and Barrio, E. (1998). Contribution of the polymerase-induced errors to the estimation of RNA virus diversity. *Journal of General Virology* **79**, 2921–2928.

Broughan, J.H. and Wunner, W.H. (1995). Characterization of protein involvement in rabies virus binding to BHK-21 cells. *Archives of Virology* **140**, 75–93.

Brzózka, K., Finke, S. and Conzelman, K.-K. (2005). Identification of the rabies virus alpha/beta interferon antagonist: phosphoprotein P interferes with phosphorylation of interferon regulatory factor 3. *Journal of Virology* **79**, 7673–7681.

Buchholz, C.J., Retzler, C., Homann, H.E. and Neubert, W.J. (1994). The carboxy-terminal domain of Sendai virus nucleocapsid protein is involved in complex formation between phosphoprotein and nucleocapsid-like particles. *Virology* **204**, 770–776.

Canter, D.M., Jackson, R.L. and Perrault, J. (1993). Faithful and efficient in vitro reconstitution of vesicular stomatitis virus transcription using plasmid-encoded L and P proteins. *Virology* **194**, 518–529.

Ceccaldi, P.E., Gillet, J.P. and Tsiang, H. (1989). Inhibition of the transport of rabies virus in the central nervous system. *Journal of Neuropathology and Experimental Neurology* **48**, 620–630.

Celis, E., Karr, R.W., Dietzschold, B., Wunner, W.H. and Koprowski, H. (1988a). Genetic restriction and fine specificity of human T cell clones reactive with rabies virus. *Journal of Immunology* **141**, 2721–2728.

Celis, E., Ou, D., Dietzschold, B. and Koprowski, H. (1988b). Recognition of rabies and rabies-related viruses by T cells derived from human vaccine recipients. *Journal of Virology* **62**, 3128–3134.

Charlton, K.M. and Casey, G.A. (1979). Experimental rabies in skunks: Immunofluorescence, light and electron microscopic studies. *Laboratory Investigations* **41**, 36–41.

Chenik, M., Chebli, K. and Blondel, D. (1995). Translation initiation at alternate in-frame AUG codons in the rabies virus phosphoprotein mRNA is mediated by a ribosomal leaky scanning mechanism. *Journal of Virology* **69**, 707–712.

Chenik, M., Chebli, K., Gaudin, Y. and Blondel, D. (1994). In vivo interaction of rabies virus phosphoprotein (P) and nucleoprotein (N), existence of two N binding sites on P protein. *Journal of General Virology* **75**, 2889–2896.

Chenik, M., Schnell, M., Conzelmann, K.K. and Blondel, D. (1998). Mapping the interacting domain between rabies virus polymerase and phosphoprotein. *Journal of Virology* **72**, 1925–1930.

Clark, H.F., Parks, N.F. and Wunner, W.H. (1981). Defective interfering particles of fixed rabies viruses: lack of correlation with attenuation or auto-interference in mice. *Journal of General Virology* **52**, 245–248.

Colonno, R.J. and Banerjee, A.K. (1978). Complete nucleotide sequence of the leader RNA synthesized in vitro by vesicular stomatitis virus. *Cell* **15**, 93–101.

Conti, C., Hauttecoeur, B., Morelec, M.J., Bizzini, B., Orsi, N. and Tsiang, H. (1988). Inhibition of rabies virus infection by soluble membrane fraction from the rat central nervous system. *Archives of Virology* **98**, 73–86.

Conti, C., Superti, R. and Tsiang, H. (1986). Membrane carbohydrate requirement for rabies virus binding to chicken embryo related cells. *Intervirology* **26**, 164–168.

Conzelmann, K.-K. and Schnell, M. (1994). Rescue of synthetic genomic RNA analogs of rabies virus by plasmid-encoded proteins. *Journal of Virology* **68**, 713–719.

Conzelmann, K.-K., Cox, J.H., Schneider, L.G. and Thiel, H.-J. (1990). Molecular cloning and complete nucleotide sequence of the attenuated rabies virus SAD B19. *Virology* **175**, 485–499.

Coulon, P., Derbin, C., Kucera, P., Lafay, F., Prehaud, C. and Flamand, A. (1989). Invasion of the peripheral nervous systems of adult mice by the CVS strain of rabies virus and its avirulent derivative AvO1. *Journal of Virology* **63**, 3550–3554.

Coulon, P., Rollin, P. and Flamand, A. (1983). Molecular basis of rabies virulence. II. Identification of a site on the CVS glycoprotein associated with virulence. *Journal of General Virology* **64**, 693–696.

Coulon, P., Ternaux, J.P., Flamand, A. and Tuffereau, C. (1998). An avirulent mutant of rabies virus is unable to infect motoneurons in vivo and in vitro. *Journal of Virology* **72**, 273–278.

Davies, M.C., Englert, M.E., Sharpless, G.R. and Cabasso, V.J. (1963). The electron microscopy of rabies virus in cultures of chicken embryo tissues. *Virology* **21**, 642–651.

De, B.P. and Banerjee, A.K. (1985). Requirements and functions of vesicular stomatitis virus L and NS proteins in the transcription process *in vitro*. *Biochemical and Biophysical Research Communication* **126**, 40–49.

Delarue, M., Poch, O., Tordo, N., Moras, D. and Argos, P. (1990). An attempt to unify the structure of polymerases. *Protein Engineering* **3**, 461–467.

Dietzschold, B., Gore, M., Marchadier, D. *et al.* (1990). Structural and immunological characterization of a linear virus-neutralizing epitope of the rabies virus glycoprotein and its possible use in a synthetic vaccine. *Journal of Virology* **64**, 3804–3809.

Dietzschold, B., Lafon, M., Wang, H. *et al.* (1987a). Localization and immunological characterization of antigenic domains of the rabies virus internal N and NS proteins. *Virus Research* **8**, 103–125.

Dietzschold, B., Morimoto, K., Hooper, D.G., Smith, J.S., Rupprecht, C.E. and Koprowski, H. (2000). Genotypic and phenotypic diversity of rabies virus variants involved in human rabies: implications for postexposure prophylaxis. *Journal of Human Virology* **3**, 50–57.

Dietzschold, B., Rupprecht, C.E., Tollis, M. *et al.* (1988). Antigenic diversity of the glycoprotein and nucleocapsid proteins of rabies and rabies-related viruses: implications for epidemiology and control of rabies. *Reviews of Infectious Diseases* **10**, S785–S798.

Dietzschold, B., Wang, H., Rupprecht, C.E. *et al.* (1987b). Induction of protective immunity against rabies by immunization with rabies virus nucleoprotein. *Proceedings of the National Academy of Sciences USA* **84**, 9165–9169.

Dietzschold, B., Wiktor, T.J., Trojanowski, J.Q. *et al.* (1985). Differences in cell-to-cell spread of pathogenic and apathogenic rabies virus in vivo and in vitro. *Journal of Virology* **56**, 12–18.

Dietzschold, B., Wiktor, T.J., Wunner, W.H. and Koprowski, H. (1983a). Chemical and immunological analysis of the rabies virus glycoprotein. *Virology* **124**, 330–337.

Dietzschold, B., Wunner, W.H., Wiktor, T. *et al.* (1983b). Characterization of an antigenic determinant of the glycoprotein that correlates with pathogenicity of rabies virus. *Proceedings of the National Academy of Sciences USA* **80**, 70–74.

Domingo, E. and Holland, J. J. (1997). RNA virus mutations and fitness for survival. *Annual Reviews in Microbiology* **5**, 151–178.

Domingo, E., Baranowski, E., Ruiz-Jarabo, C.M., Martin-Hernández, A.M., Sáiz, J.C. and Escarmis C. (1998). Quasispecies structure and persistence of RNA viruses. *Emerging Infectious Diseases* **4**, 521–527.

Durrer, P., Gaudin, Y., Ruigrok, R.W.H., Graf, R. and Brunner, J. (1995). Photolabeling identifies a putative fusion domain in the envelope glycoprotein of rabies and vesicular stomatitis viruses. *Journal of Biological Chemistry* **270**, 17 575–17 581.

Eriguchi, Y., Toriumi, H. and Kawai, A. (2002). Studies on the rabies virus RNA polymerase: 3. Two-dimensional electrophoretic analysis of the multiplicity of non-catalytic subunit (P protein). *Microbiology and Immunology* **46**, 463–474.

Ertl, H.C.J., Dietzschold, B., Gore, M. *et al.* (1989). Induction of rabies virus-specific T-helper cells by synthetic peptides that carry dominant T-helper cell epitopes of the viral ribonucleoprotein. *Journal of Virology* **63**, 2885–2892.

Ertl, H.C.J., Dietzschold, B. and Otvos, L. Jr (1991). T helper cell epitope of rabies virus nucleoprotein defined by tri- and tetrapeptides. *European Journal of Immunology* **21**, 1–10.

Etessami, R., Conzelmann, K.-K., Fadai-Ghotbi, B., Natelson, B., Tsiang, H. and Ceccaldi, P.-E. (2000). Spread and pathogenic characteristics of a G-deficient rabies virus recombinant: an *in vitro* and *in vivo* study. *Journal of General Virology* **81**, 2147–2153.

Faber, M., Pulmanausahakul, R., Nagao, K. *et al.* (2004). Identification of viral genomic elements responsible for rabies virus neuroinvasiveness. *Proceedings of the National Academy of Sciences USA* **101**, 16 328–16 332.

Finke, S. and Conzelmann, K.-K. (1997). Ambisense gene expression from recombinant rabies virus: random packaging of positive- and negative-strand ribonucleoprotein complexes into rabies virions. *Journal of Virology* **71**, 7281–7288.

Finke, S., Cox, J.H. and Conzelmann, K.-K. (2000). Differential transcription attenuation of rabies virus genes by intergenic regions: Generation of recombinant viruses overexpressing the polymerase gene. *Journal of Virology* **74**, 7261–7269.

Finke, S., Mueller-Waldeck, R. and Conzelman, K.-K. (2003). Rabies virus matrix protein regulates the balance of virus transcription and replication. *Journal of General Virology* **84**, 1613–1621.

Flamand, A. and Delagneau, J.F. (1978). Transcriptional mapping of rabies virus in vivo. *Journal of Virology* **28**, 518–523.

Flamand, A., Raux, H., Gaudin, Y. and Ruigrok, R.W.H. (1993). Mechanisms of rabies virus neutralization. *Virology* **194**, 302–313.

Flamand, A., Wiktor, T.J. and Koprowski, H. (1980a). Use of hybridoma monoclonal antibodies in the detection of antigenic differences between rabies and rabies-related virus proteins. I. The nucleocapsid protein. *Journal of General Virology* **48**, 97–104.

Flamand, A., Wiktor, T.J. and Koprowski, H. (1980b). Use of hybridoma monoclonal antibodies in the detection of antigenic differences between rabies and rabies-related virus proteins. II. The glycoprotein. *Journal of General Virology* **48**, 105–109.

Fu, Z.F., Dietzschold, B., Schumacher, C.L., Wunner, W.H., Ertl, H.C.J. and Koprowski, H. (1991). Rabies virus nucleoprotein expressed in and purified from insect cells is efficacious as a vaccine. *Proceedings of the National Academy of Sciences USA* **88**, 2001–2005.

Fu, Z.F., Zheng, Y., Wunner, W.H., Koprowski, H. and Dietzschold, B. (1994). Both the N- and the C-terminal domains of the nominal phosphoprotein of rabies virus are involved in binding to the nucleoprotein. *Virology* **200**, 590–597.

Gao, Y., Greenfield, N.J., Cleverley, D.Z. and Lenard, J. (1996). The transcriptional form of the phosphoprotein of vesicular stomatitis virus is a trimer: structure and stability. *Biochemistry* **35**, 14 569–14 573.

Gaudin, Y. (1997). Folding of rabies virus glycoprotein: epitope acquisition and interaction with endoplasmic reticulum chaperones. *Journal of Virology* **71**, 3742–3750.

Gaudin, Y. (2000). Rabies virus-induced membrane fusion pathway. *Journal of Cell Biology* **150**, 601–611.

Gaudin, Y., Moreira, S., Benejean, J., Blondel, D., Flamand, A. and Tuffereau, C. (1999). Soluble ectodomain of rabies virus glycoprotein expressed in eukaryotic cells folds in a monomeric conformation that is antigenically distinct from the native state of the complete, membrane-anchored glycoprotein. *Journal of General Virology* **80**, 1647–1656.

Gaudin, Y., Raux, H., Flamand, A. and Ruigrok, R.W.H. (1996). Identification of amino acids controlling the low-pH-induced conformational change of rabies virus glycoprotein. *Journal of Virology* **70**, 7371–7378.

Gaudin, Y., Ruigrok, R.W.H. and Brunner, J. (1995a). Low-pH induced conformational changes in viral fusion proteins: implications for the fusion mechanism. *Journal of General Virology* **76**, 1541–1556.

Gaudin, Y., Ruigrok, R.W.H., Knossow, M. and Flamand, A. (1993). Low-pH conformational changes of rabies virus glycoprotein and their role in membrane fusion. *Journal of Virology* **67**, 1365–1372.

Gaudin, Y., Ruigrok, R.W.H., Tuffereau, C., Knossow, M. and Flamand, A. (1992). Rabies virus glycoprotein is a trimer. *Virology* **187**, 627–632.

Gaudin, Y., Tuffereau, C., Benmansour, A. and Flamand, A. (1991a). Fatty acylation of rabies virus proteins. *Virology* **184**, 441–444.

Gaudin, Y., Tuffereau, C., Durrer, P., Flamand, A. and Ruigrok, R.W.H. (1995b). Biological function of the low-pH, fusion-inactive conformation of rabies virus glycoprotein (G): G is transported in a fusion-inactive state-like conformation. *Journal of Virology* **69**, 5528–5534.

Gaudin, Y., Tuffereau, C., Segretain, D., Knossow, M. and Flamand, A. (1991b). Reversible conformational changes and fusion activity of rabies virus glycoprotein. *Journal of Virology* **65**, 4853–4859.

Gigant, B., Iseni, F., Gaudin, Y., Knossow, M. and Blondel, D. (2000). Neither phosphorylation nor the amino-terminal part of rabies virus phosphoprotein is required for its oligomerization. *Journal of General Virology* **81**, 1757–1761.

Gillet, J.P., Derer, P. and Tsiang, H. (1986). Axonal transport of rabies virus in the central nervous system of the rat. *Journal of Neuropathology and Experimental Neurology* **45**, 619–634.

Gosztonyi, G. (1994). Reproduction of lyssaviruses: Ultrastructural composition of lyssavirus and functional aspects of pathogenesis. *Current Topics in Microbiology and Immunology* **187**, 43–68.

Goto, H., Minamoto, H., Ito, H. *et al.* (1995). Expression of the nucleoprotein of rabies virus in *Escherichia coli* and mapping of antigenic sites. *Archives of Virology* **140**, 1061–1074.

Goto, H., Minamoto, N., Ito, H. *et al.* (2000). Mapping of epitopes and structural analysis of antigenic sites in the nucleoprotein of rabies virus. *Journal of General Virology* **81**, 119–127.

Gould, A.R., Hyatt, A.D., Lunt, R., Kattenbelt, J.A., Hengstberger, S. and Blacksell, S.D. (1998). Characterisation of a novel lyssavirus isolated from *Pteropid* bats in Australia. *Virus Research* **54**, 165–187.

Gupta, A., Blondel, D., Choudhary, S. and Banerjee, A. (2000). Phosphoprotein (P) of rabies virus is phosphorylated by a unique cellular protein kinase and specific isomers of protein kinase C. *Journal of Virology* **74**, 91–98.

Harty, R.N., Paragas, J., Sudol, M. and Palese, P. (1999). A proline-rich motif within the matrix protein of vesicular stomatitis virus and rabies virus interacts with WW domains of cellular proteins: implications for viral budding. *Journal of Virology* **73**, 2921–2929.

Harty, R.N., Brown, M.E., McGettigan, J.P. *et al.* (2001). Rhabdoviruses and the cellular ubiquitin-proteosome system: a budding interaction. *Journal of Virology* **75**, 10623–10629.

Herzog, M., Lafage, M., Montano-Hirose, J.A., Fritzell, C., Scott-Algara, D. and Lafon, M. (1992). Nucleocapsid specific T and B cell responses in humans after rabies vaccination. *Virus Research* **24**, 77–89.

Hiramatsu, K., Mannen, K., Mifune, K., Nishizono, A., Takita-Sonoda, Y. (1993). Comparative sequence analysis of the M gene among rabies virus strains and its expression by recombinant vaccinia virus. *Virus Genes* **7**, 83–88.

Holland, J.J. (1987). Defective interfering rhabdoviruses. In: *The Rhabdoviruses* (R.R. Wagner, ed.). pp. 297–360. New York: Plenum Press.

Holland, J.J., De la Torre, J.C., and Steinhauer, D.A. (1992). RNA virus populations as quasispecies. *Current Topics in Microbiology and Immunology* **176**, 1–21.

Holloway, B.P. and Obejeski, J.F. (1980). Rabies virus-induced RNA synthesis in BHK-21 cells. *Journal of General Virology* **49**, 181–195.

Hummeler, K., Koprowski, H. and Wiktor, T.J. (1967). Structure and development of rabies virus in tissue culture. *Journal of Virology* **1**, 152–170.

Hummeler, K., Tomassini, N., Sokol, F., Kuwert, E. and Koprowski, H. (1968). Morphology of the nucleoprotein component of rabies virus. *Journal of Virology* **2**, 1191–1199.

Inoue, K., Shoji, Y., Kurane, I., Iijima, T., Sakai, T. and Morimoto, K. (2003). An improved method for recovering rabies virus from cloned cDNA. *Journal of Virological Methods* **107**, 229–236.

Irie, T., Licata, J.M., McGettigan, J.P., Schnell, M.J. and Harty, R.N. (2004). Budding of PpxY-containing rhabdoviruses is not dependent on host proteins TGS101 and VPS4A. *Journal of Virology* **78**, 2657–2665.

Iseni, F., Barge, A., Baudin, F., Blondel, D. and Ruigrok, R.W.H. (1998). Characterization of rabies virus nucleocapsids and recombinant nucleocapsid-like structures. *Journal of General Virology* **79**, 2909–2919.

Ito, Y., Nishizono, A., Mannen, K., Hiramatsu, K. and Mifune, K. (1996). Rabies virus M protein expressed in *Escherichia coli* and its regulatory role in virion-associated transcriptase activity. *Archives of Virology* **141**, 671–683.

Ito, N., Takayama, M., Yamada, K., Sugiyama, M. and Minamoto, N. (2001). Rescue of rabies virus from cloned cDNA and identification of the pathogenicity-related gene: glycoprotein gene is associated with virulence for adult mice. *Journal of General Virology* **75**, 9121–9128.

Iwasaki, Y., Wiktor, T.J. and Koprowski, H. (1973). Early events of rabies virus replication in tissue cultures: An electron microscopic study. *Laboratory Investigation* **28**, 142–148.

Jackson, A.C. (1991). Biological basis of rabies virus neurovirulence in mice: Comparative pathogenesis study using the immunoperoxidase technique. *Journal of Virology* **65**, 537–540.

Jackson, A.C. and Park, H. (1999). Experimental rabies virus infection of p75 neurotrophin receptor-deficient mice. *Acta Neuropathologica* **98**, 641–644.

Jacob, Y., Badrane, H., Ceccaldi, P.-E. and Tordo, N. (2000). Cytoplasmic dynein LC8 interacts with lyssavirus phosphoprotein. *Journal of Virology* **74**, 10 217–10 222.

Kankanamge, P.J., Irie, T., Mannen, K., Tochikura, T.S. and Kawai, A. (2003). Mapping of the low pH-sensitive conformational epitope of rabies virus glycoprotein recognized by a monoclonal antibody 1-30-44. *Microbiology and Immunology* **47**, 507–519.

Kasturi, L., Eshleman, J.R., Wunner, W.H., and Shakin-Eshleman, S.H. (1995). The hydroxy amino acid in an Asn-X-Ser/Thr sequon can influence the efficiency of N-linked core-glycosylation and the level of expression of a cell surface glycoprotein. *Journal of Biological Chemistry* **270**, 14 756–14 761.

Kawai, A. and Morimoto, K. (1994). Functional aspects of lyssavirus proteins in Lyssaviruses. In: *Current Topics in Microbiology and Immunology*, Vol. 187 (C.R. Rupprecht, B. Dietzschold, and H. Koprowski, eds). pp. 27–42. Berlin and Heidelberg: Springer-Verlag.

Kawai, A., Toriumi, H., Tochikura, T.S., Takahashi, T., Honda, Y. and Morimoto, K. (1999). Nucleocapsid formation and/or subsequent conformational change of rabies virus nucleoprotein (N) is a prerequisite step for acquiring the phosphatase-sensitive epitope of monoclonal antibody 5-2-26. *Virology* **263**, 395–407.

Kissi, B., Badrane, H., Audry, L. *et al.* (1999). Dynamics of rabies virus quasispecies during serial passages in heterologous hosts. *Journal of General Virology* **80**, 2041–2050.

Kissi, B., Tordo, N. and Bourhy, H. (1995). Genetic polymorphism in the rabies virus nucleoprotein gene. *Virology* **209**, 526–537.

Kouznetzoff, A., Buckle, M. and Tordo, N. (1998). Identification of a region of the rabies virus N protein involved in direct binding to the viral RNA. *Journal of General Virology* **79**, 1005–1013.

Kucera, P., Dolivo, M., Coulon, P. and Flamand, A. (1985). Pathways of the early propagation of virulent and avirulent rabies strains from the eye to the brain. *Journal of Virology* **55**, 159–162.

Kurilla, M.G., Cabradilla, C.D., Holloway, B.P. and Keene, J.D. (1984). Nucleotide sequence and host La protein interactions of rabies virus leader RNA. *Journal of Virology* **50**, 773–778.

Kuzmin, I.V., Hughes, G., Botvinkin, A.D., Orciari, L.A. and Rupprecht, C.E. (2005). Phylogenetic relationships of Irkut and West Caucasian bat viruses within the *Lyssavirus* genus and suggested quantitative criteria based on the N gene sequence for lyssavirus genotype definition. *Virus Research* **111**, 28–43.

Kuzmin, I.V., Orciari, L.A., Arai, Y.T. *et al.* (2003). Bat lyssaviruses (Aravan and Khujand) from Central Asia: phylogenetic relationships according to N, P and G gene sequences. *Virus Research* **97**, 65–79.

Lafay, F., Coulon, P., Astic, L. *et al.* (1991). Spread of the CVS strain of rabies virus and of the avirulent mutant AvO1 along the olfactory pathways of the mouse after intranasal inoculation. *Virology* **183**, 320–330.

Lafon, M. (1997). Rabies virus superantigen. In: *Viral Superantigens* (K. Tomonari, ed.). pp. 151–170. Boca Raton: CRC Press.

Lafon, M. (2005). Mini-review –The rabies virus. Rabies virus receptors. *Journal of Neurovirology* **11**, 82–87.

Lafon, M. and Wiktor, T. J. (1985). Antigenic sites on the ERA rabies virus nucleoprotein and nonstructural protein. *Journal of General Virology* **66**, 2125–2133.

Lafon, M., Ideler, J. and Wunner, W.H. (1984). Investigation of antigenic structure of rabies virus glycoprotein by monoclonal antibodies. *Developments in Biological Standardization* **57**, 219–225.

Lafon, M., Lafage, M., Martinez-Arends, A. *et al.* (1992). Evidence for a viral superantigen in humans. *Nature* **358**, 507–510.

Langevin, C., Jaaro, H., Bressanelli, S., Fainzilber, M. and Tuffereau, C. (2002). Rabies virus glycoprotein (RVG) is a trimeric ligand for the N-terminal cysteine-rich domain of the mammalian p75 neurotrophin receptor. *Journal of Biological Chemistry* **277**, 37655–37662.

Langevin, C. and Tuffereau, C. (2002). Mutations conferring resistance to neutralization by a soluble form of the neurotrophin receptor (p75NTR) map outside of the known antigenic sites of the rabies virus glycoprotein. *Journal of Virology* **76**, 10756–10765.

Le Mercier, P., Jacob, Y. and Tordo, N. (1997). The complete Mokola virus genome sequence: Structure of the RNA-dependent RNA polymerase. *Journal of General Virology* **78**, 1571–1576.

Leppert, M., Rittenhouse, L., Perrault, J., Summers, D.F. and Kolakofsky, D. (1979). Plus and minus strand leader RNAs in negative strand virus-infected cells. *Cell* **18**, 735–747.

Lewis, P., Fu, Y. and Lentz, T.L. (2000). Rabies virus entry at the neuromuscular junction in nerve-muscle cocultures. *Muscle Nerve* **23**, 720–730.

Liu, P., Yang, J., Wu, X. and Fu, Z.F. (2004). The interactions amongst rabies virus nucleoprotein, phosphoprotein, and the genomic RNA in virus-infected and transfected cells. *Journal of General Virology* **85**, 3725–3734.

Lycke, E. and Tsiang, H. (1987). Rabies virus infection of cultured rat sensory neurons. *Journal of Virology* **61**, 2733–2741.

Macfarlan, R.I., Dietzschold, B. and Koprowski, H. (1986). Stimulation of cytotoxic T-lymphocyte responses by rabies virus glycoprotein and identification of an immunodominant domain. *Molecular Immunology* **23**, 733–741.

Macfarlan, R.I., Dietzschold, B., Wiktor, T.J. *et al.* (1984). T cell responses to cleaved rabies virus glycoprotein and to synthetic peptides. *Journal of Immunology* **133**, 2748–2752.

Madore, H.P. and England, J.M. (1977). Rabies virus protein synthesis in infected BHK-21 cells. *Journal of Virology* **22**, 102–112.

Maillard, A. P. and Gaudin, Y. (2002). Rabies virus glycoprotein can fold in two alternative, antigenically distinct conformations depending on membrane-anchor type. *Journal of General Virology* **83**, 1465–1476.

Marsh, M. and Helenius, A. (1989). Virus entry into animal cells. *Advances in Virus Research* **36**, 107–151.

Martinez-Arends, A., Astoul, E., Lafage, M. and Lafon, M. (1995). Activation of human tonsil lymphocytes by rabies virus nucleocapsid superantigen. *Clinical Immunology and Immunopathology* **77**, 177–184.

Matsumoto, S. (1962). Electron microscopy of nerve cells infected with street rabies virus. *Virology* **17**, 198–202.

Matsumoto, S. (1963). Electron microscope studies of rabies virus in mouse brain. *Journal of Cell Biology* **19**, 565–591.

Matsumoto, S. (1975). Electron microscopy of central nervous system infection. In: *The Natural History of Rabies* (G.M. Baer, ed.). pp. 217–233. New York: Academic Press.

Matsumoto, S. and Kawai, A. (1969). Comparative studies on development of rabies virus in different host cells. *Virology* **39**, 449–459.

Matsumoto, S. and Miyamoto, K. (1966). Electron-microscopic studies on rabies virus multiplication and the nature of the Negri body. *Symposium Series in Immunobiology Standardization* **1**, 45–54.

Matsumoto, S., Schneider, L.G., Kawai, A. and Yonezawa, T. (1974). Further studies on the replication of rabies and rabies-like viruses in organized cultures of mammalian neural tissues. *Journal of Virology* **14**, 981–996.

Mavrakis, M., Iseni, F., Mazza, C. *et al.* (2003). Isolation and characterization of the rabies virus N° P complex produced in insect cells. *Virology* **305**, 406–414.

Mazarakis, N.D., Azzouz, M., Rohll, J.B. *et al.* (2001). Rabies virus glycoprotein pseudotyping of lentiviral vectors enable retrograde axonal transport and access to the nervous system after peripheral delivery. *Human Molecular Genetics* **10**, 2109–2121.

Mebatsion, T., König, M. and Conzelmann, K. (1996). Budding of rabies virus particles in the absence of the spike glycoprotein. *Cell* **84**, 941–951.

Mebatsion, T., Schnell, M.J. and Conzelmann, K.-K. (1995). Mokola virus glycoprotein and chimeric proteins can replace rabies virus glycoprotein in the rescue of infectious defective rabies virus particles. *Journal of Virology* **69**, 1444–1451.

Mebatsion, T., Weiland, F. and Conzelmann, K.-K. (1999). Matrix protein of rabies virus is responsible for the assembly and budding of bullet-shaped particles and interacts with the transmembrane spike glycoprotein G. *Journal of Virology* **73**, 242–250.

Mellon, M.G. and Emerson, S.U. (1978). Rebinding of transcriptase components (L and NS proteins) to the nucleocapsid template of vesicular stomatitis virus. *Journal of Virology* **27**, 560–567.

Minamoto, N., Tanaka, H., Hishida, M. *et al.* (1994). Linear and conformation-dependent antigenic sites on the nucleoprotein of rabies virus. *Microbiology and Immunology* **38**, 449–455.

Morimoto, K., Foley, H.D., McGettigan, J.P., Schnell, M.J. and Dietzschold, B. (2000). Reinvestigation of the role of the rabies virus glycoprotein in viral pathogenesis using a reverse genetics approach. *Journal of Neurovirology* **6**, 373–381.

Morimoto, K., Hooper, D.C., Carbaugh, H., Fu, Z.F., Koprowski, H. and Dietzschold, B. (1998). Rabies virus quasispecies: implications for pathogenesis. *Proceedings of the National Academy of Sciences USA* **95**, 3152–3156.

Morimoto, K., Ni, Y.-J. and Kawai, A. (1992). Syncytium formation is induced in the murine neuroblastoma cell cultures which produce pathogenic type G proteins of the rabies virus. *Virology* **189**, 203–216.

Morimoto, K, Ohkubo, A. and Kawai, A. (1989). Structure and transcription of the glycoprotein gene of attenuated HEP-Flury strain of rabies virus. *Virology* **173**, 465–477.

Murphy, F.A. and Bauer, S.P. (1974). Early street rabies virus infection in striated muscle and later progression to the central nervous system. *Intervirology* **3**, 256–268.

Murphy, F.A. and Harrison, A. K. (1979). Electron microscopy of the rhabdoviruses of animals. In: *Rhabdoviruses* (D.H.L. Bishop, ed.). Vol. 1, pp. 65–106. Boca Raton: CRC Press.

Murphy, F.A., Bauer, S.P., Harrison, A.K. and Winn, W.C. Jr (1973a). Comparative pathogenesis of rabies and rabies-like viruses: Viral infection and transit from inoculation site to the central nervous system. *Laboratory Investigation* **28**, 361–376.

Murphy, F.A. and Nathanson, N. (1994). The emergence of new virus diseases: an overview. *Seminars in Virology* **5**, 87–102.

Murphy, F.A., Harrison, A.K., Washington, W.C. and Bauer, S.P. (1973b). Comparative pathogenesis of rabies and rabies-like viruses: Infection of the central nervous system and centrifugal spread of virus to peripheral tissues. *Laboratory Investigation* **29**, 1–16.

Naito, S. and Matsumoto, S. (1978). Identification of cellular actin within the rabies virus. *Virology* **91**, 151–163.

Nakahara, K., Ohnuma, H., Sugita, S. *et al.* (1999). Intracellular behaviour of rabies virus matrix protein (M) is determined by the viral glycoprotein (G). *Microbiology and Immunology* **43**, 259–270.

Nakahara, K., Toriumi, H., Irie, T. *et al.* (2003). Characterization of a slow-migrating component of the rabies virus matrix protein strongly associated with the viral glycoprotein. *Microbiology and Immunology* **47**, 977–988.

Okazaki, Y., Ohno, H., Takase, K., Ochiai, T. and Saito, T. (2000). Cell surface expression of calnexin, a molecular chaperone in the endoplasmic reticulum. *Journal of Biological Chemistry* **275**, 35751–35759.

Perrin, P., Portnoi, D. and Sureau, P. (1982). Etude de l'adsorption et de la penetration du virus rabique: Interactions avec les cellules BHK21 et des membranes artificielles. *Annals of Virology (Institut Pasteur)* **133E**, 403–422.

Poch, O., Blumberg, B.M., Bougueleret, L. and Tordo, N. (1990). Sequence comparison of five polymerases (L proteins) of unsegmented negative-strand RNA viruses: theoretical assignment of functional domains. *Journal of General Virology* **71**, 1153–1162.

Poch, O., Sauvaget, I., Delarue, M. and Tordo, N. (1989). Identification of four conserved motifs among the RNA-dependent polymerase encoding elements. *European Molecular Biology Journal* **8**, 3867–3874.

Prehaud, C., Coulon, P., Lafay, F., Thiers, C., and Flamand, A. (1988). Antigenic site II of the rabies virus glycoprotein: Structure and role in viral virulence. *Journal of Virology* **62**, 1–7.

Prehaud, C., Nel, K. and Bishop, D.H.L. (1992). Baculovirus-expressed rabies virus M1 protein is not phosphorylated: It forms multiple complexes with expressed rabies N protein. *Virology* **189**, 766–770.

Raux, H., Flamand, A. and Blondel, D. (2000). Interaction of the rabies virus P protein with the LC8 dynein light chain. *Journal of Virology* **74**, 10 212–10 216.

Reagan, K.J. and Wunner, W.H. (1985). Rabies virus interaction with various cell lines is independent of the acetylcholine receptor. *Archives of Virology* **84**, 277–282.

Robison, C.S. and Whitt, M.A. (2000). The membrane-proximal stem region of vesicular stomatitis G protein confers efficient virus assembly. *Journal of Virology* **74**, 2239–2246.

Sacramento, D., Bourhy, H. and Tordo, N. (1991). PCR techniques as an alternative method for diagnosis and molecular epidemiology of rabies virus. *Molecular and Cellular Probes* **6**, 229–240.

Sagara, J. and Kawai, A. (1992). Identification of heat shock protein 70 in the rabies virion. *Virology* **190**, 845–848.

Sagara, J., Tochikura, T.S., Tanaka, H. *et al.* (1998). The 21-kDa polypeptide (VAP21) in the rabies virion is a C99-related host cell protein. *Microbiology and Immunology* **42**, 289–297.

Sagara, J., Tsukita, S., Shigenobu, Y., Tsukita, S. and Kawai, A. (1995). Cellular actin-binding exrin-radixin-moesin (ERM) family proteins are incorporated into the rabies virion and closely associated with viral envelope proteins in the cell. *Virology* **206**, 485–494.

Sakai, M., Kankanamge, P.J., Shoji, J., Kawata, S., Tochikura, T.S. and Kawai, A. (2004). Studies on the conditions required for structural and functional maturation of rabies virus glycoprotein (G) in G cDNA-transfected cells. *Microbiology and Immunology* **48**, 853–864.

Sakamoto, S.-I., Ide, T., Nakatake, H. *et al.* (1994). Studies on the antigenicity and nucleotide sequence of the rabies virus Nishigahara strain, a current seed strain used for dog vaccine production in Japan. *Virus Genes* **8**, 35–46.

Sanchez, A., De, B.P. and Banerjee, A.K. (1985). In vitro phosphorylation of NS protein by the L protein of vesicular stomatitis virus. *Journal of General Virology* **66**,1025–1036.

Schnell, M. J. and Conzelmann, K.-K. (1995). Polymerase activity of in vitro mutated rabies virus L protein. *Virology* **214**, 522–530.

Schnell, M.J., Buonocore, L., Whitt, M.A. and Rose, J.A. (1996). The minimal conserved transcription stop-start signal promotes stable expression of a foreign gene in vesicular stomatitis virus. *Journal of Virology* **70**, 2318–2323.

Schoehn, G., Iseni, F., Mavrakis, M., Blondel, D. and Ruigrok, R.W.H. (2001). Structure of recombinant rabies virus nucleoprotein-RNA complex and identification of the phosphate binding site. *Journal of Virology* **75**, 490–498.

Seif, I., Coulon, P., Rollin, P.E. and Flamand, A. (1985). Rabies virus virulence: Effect on pathogenicity and sequence characterization of mutations affecting antigenic site III of the glycoprotein. *Journal of Virology* **53**, 926–935.

Shakin-Eshleman, S.H., Remaley, A.T., Eshleman, J.R., Wunner, W.H. and Spitalnik, S.L. (1992). N-linked glycosylation of rabies virus glycoprotein. Individual sequons differ in their efficiencies and influence on cell surface expression. *Journal of Biological Chemistry* **267**, 10 690–10 698.

Shakin-Eshleman, S.H., Wunner, W.H. and Spitalnik, S.L. (1993). Efficiency of N-linked core glycosylation at asparagine-319 of rabies virus glycoprotein is altered by deletions C-terminal to the glycosylation sequon. *Biochemistry* **32**, 9465–9472.

Sissoëff, L., Mousli, M., England, P. and Tuffereau, C. (2005). Stable trimerization of recombinant rabies virus glycoprotein ectodomain is required for interaction with the p75^{NTR} receptor. *Journal of General Virology* **86**, 2543–2552.

Smith, D.B., McAllister, J., Casino, C. and Simmonds, P. (1997). Virus 'quasispecies': making a mountain out of a molehill? *Journal of General Virology* **78**, 1511–1519.

Smith, J.S. (1989). Rabies virus epitopic variation: Use in ecological studies. *Advances in Virus Research* **36**, 215–253.

Sokol, F. (1975). Chemical composition and structure of rabies virus. In: *The Natural History of Rabies* (G.M. Baer, ed.). Vol. I, pp. 79–102. New York: Academic Press.

Sokol, F., Clark, H.F., Wiktor, T.J., McFalls, M.L., Bishop, D.H.L. and Obijeski, J.F. (1974). Structural phosphoproteins associated with ten rhabdoviruses. *Journal of General Virology* **24**, 433–445.

Sokol, F., Schlumberger, H.D., Wiktor, T.J., Koprowski, H. and Hummeler, K. (1969). Biochemical and biophysical studies on the nucleocapsid and on the RNA of rabies virus. *Virology* **38**, 651–665.

Spadafora, D., Canter, D.M., Jackson, R.L. and Perrault, J. (1996). Constitutive phosphorylation of the vesicular stomatitis virus P protein modulates polymerase complex formation but is not essential for transcription or replication. *Journal of Virology* **70**, 4538–4548.

Stillman, E.A. and Whitt, M.A. (1997). Mutational analyses of the intergenic dinucleotide and the transcriptional start sequence of vesicular stomatitis virus (VSV) define sequences required for efficient termination and initiation of VSV transcripts. *Journal of Virology* **71**, 2127–2137.

Superti, F., Derer, M. and Tsiang, H. (1984a). Mechanism of rabies virus entry into CER cells. *Journal of General Virology* **65**, 781–789.

Superti, F., Hauttecoeur, B., Morelec, M.J., Goldoni, P., Bizzini, B. and Tsiang, H. (1986). Involvement of gangiosides in rabies virus infection. *Journal of General Virology* **67**, 47–56.

Superti, F., Seganti, L., Tsiang, H. and Orsi, N. (1984b). Role of phospholipid in rhabdovirus attachment to CER cells. *Archives of Virology* **81**, 321–328.

Takamatsu, F., Asakawa, N., Morimoto, K. *et al.* (1998). Studies on the rabies virus RNA polymerase. 2. Possible relationships between the two forms of the non-catalytic subunit (P protein). *Microbiology and Immunology* **42**, 761–771.

Takayama-Ito, M., Ito, N., Yamada, K., Minamoto, N. and Sugiyama, M. (2004). Region at amino acids 164 to 303 of the rabies virus glycoprotein plays an important role in pathogenicity for adult mice. *Journal of Neurovirology* **10**, 131–135.

Testa, D., Chanda, P.K. and Banerjee, A.K. (1980). Unique mode of transcription in vitro by vesicular stomatitis virus. *Cell* **21**, 267–275.

Thoulouze, M.-I., Lafage, M., Schachner, M., Hartmann, U., Cremer, H. and Lafon, M. (1998). The neural cell adhesion molecule is a receptor for rabies virus. *Journal of Virology* **72**, 7181–7190.

Tollis, M., Dietzshold, B., Volia, C.B. and Koprowski, H. (1991). Immunization of monkeys with rabies ribonucleoprotein (RNP) confers protective immunity against rabies. *Vaccine* **9**, 134–136.

Tordo, N. and Poch, O. (1988a). Strong and weak transcription signals within the rabies genome. *Virus Research* **2** (Supplement), 30.

Tordo N. and Poch, O. (1988b). Structure of rabies virus. In: *Rabies* (J.B. Campbell and K.M. Charlton, eds). pp. 25–45. Boston: Kluwer Academic Publishers.

Tordo, N., Poch, O., Ermine, A. and Keith, G. (1986a). Primary structure of leader RNA and nucleoprotein genes of the rabies genome: segmented homology with VSV. *Nucleic Acid Research* **14**, 2671–2683.

Tordo, N., Poch, O., Ermine, A., Keith, G. and Rougeon, F. (1986b). Walking along the rabies genome: is the large G-L intergenic region a remnant gene? *Proceedings of the National Academy of Sciences USA* **83**, 3914–3918.

Tordo, N., Poch, O., Ermine, A., Keith, G. and Rougeon, F. (1988). Completion of the rabies virus genome sequence determination: highly conserved domains among the L (polymerase) proteins of unsegmented negative-strand RNA viruses. *Virology* **165**, 565–576.

Tordo, N., Benmansour. A., Calisher, C. *et al.* (2005). In: *Virus Taxonomy – Classification and Nomenclature of Viruses.* Eighth Report of the International Committee on the Taxonomy of Viruses (C.M. Fauquet, M.A. Mayo, J. Maniloff, U. Desselberger and L.A. Ball, eds). pp. 623–644. Boston: Elsevier Academic Press.

Toriumi H., Eriguchi Y., Takamatsu, F. and Kawai, A. (2004). Further studies on the hyperphosphorylated form (p40) of the rabies virus nominal phosphoprotein (P). *Microbiology and Immunology* **48**, 865–874.

Toriumi, H., Honda, Y., Morimoto, K. Tochikura, T.S. and Kawai, A. (2002). Structural relationship between nucleocapsid-binding activity of the rabies virus phosphoprotein (P) and exposure of epitope 402-13 located at the C terminus. *Journal of General Virology* **83**, 3035–3043.

Toriumi H. and Kawai, A. (2004). Association of rabies virus nominal phosphoprotein (P) with viral nucleocapsid (NC) is enhanced by phosphorylation of the viral nucleoprotein (N). *Microbiology and Immunology* **48**, 399–409.

Tsiang, H. (1993). Pathophysiology of rabies virus infection of the nervous system. *Advances in Virus Research* **42**, 375–412.

Tsiang, H., De la Porte, S., Ambroise, D.J., Derer, M. and Koenig, J. (1986). Infection of cultured rat myotubes and neurons from the spinal cord by rabies virus. *Journal of Neuropathology and Experimental Neurology* **45**, 28–42.

Tsiang, H., Derer, M. and Taxi, J. (1983). An in vivo and in vitro study of rabies virus infection of the rat superior cervical ganglia. *Archives of Virology* **76**, 231–243.

Tuffereau, C., Benegean, J., Blondel, D., Kieffer, G. and Flamand, A. (1998). Low-affinity nerve-growth factor receptor (P75NTR) can serve as a receptor for rabies virus. *European Molecular Biology Journal* **17**, 7250–7259.

Tuffereau, C., Fischer, S. and Flamand, A. (1985). Phosphorylation of the N and M1 proteins of rabies virus. *Virology* **165**, 565–576.

Tuffereau, C., Leblois, H., Benejean, J., Coulon, P., Lafay, F. and Flamand, A. (1989). Arginine or lysine in position 333 of ERA and CVS glycoprotein is necessary for rabies virulence in adult mice. *Virology* **172**, 206–212.

Whelan, S.P.J. and Wertz, G.W. (1999). Regulation of RNA synthesis by the genomic termini of vesicular stomatitis virus: Identification of distinct sequences essential for transcription but not replication. *Journal of Virology* **73**, 297–306.

Whitt, M.A., Buonocore, L., Prehaud, C. and Rose, J. K. (1991). Membrane fusion activity, oligomerization, and assembly of the rabies virus glycoprotein. *Virology* **185**, 681–688.

Wiktor, T.J., Dietzschold, B., Leamnson, R.N. and Koprowski, H. (1977). Induction and biological properties of defective interfering particles of rabies virus. *Journal of Virology* **21**, 626–635.

Wills, J.W., Cameron, C.E., Wilson, C.B., Xiang, Y., Bennett, R.P. and Leis, J. (1994). An assembly domain of the Rous sarcoma virus Gag protein required late in budding. *Journal of Virology* **68**, 6605–6618.

Wojczyk, B.S., Takahashi, N., Levy, M.T. *et al.* (2005). N-glycosylation at one rabies virus glycoprotein sequon influences N-glycan processing at a distant sequon on the same molecule. *Glycobiology* **15**, 655–666.

Wu, X., Gong, X., Foley, H.D., Schnell, M.J. and Fu, Z.F. (2002). Both viral transcription and replication are reduced when the rabies virus nucleoprotein is not phosphorylated. *Journal of Virology* **76**, 4153–4161.

Wu, X., Lei, X. and Fu, Z.F. (2003). Rabies virus nucleoprotein is phosphorylated by cellular casein kinase II. *Biochemical Biophysical Research Communication* **304**, 333–338.

Wunner, W.H. (1991). The chemical composition and molecular structure of rabies viruses. In: *The Natural History of Rabies* (G.M. Baer, ed.). pp. 31–67. Boca Raton: CRC Press.

Wunner, W.H. and Clark, H F. (1980). Regeneration of DI particles of virulent and attenuated rabies virus: genome characterization and lack of correlation with virulence phenotype. *Journal of General Virology* **51**, 69–82.

Wunner, W.H., Dietzschold, B., Smith, C.L. and Lafon, M. (1985). Antigenic variants of CVS rabies virus with altered glycosylation sites. *Virology* **140**, 1–12.

Wunner, W.H., Reagan, K.J. and Koprowski, H. (1984). Characterization of saturable binding sites for rabies virus. *Journal of Virology* **50**, 691–697.

Yan, X., Mohankumar, P.S., Dietzschold, B., Schnell, M.J. and Fu, Z.F. (2002). The rabies virus glycoprotein determines the distribution of different rabies virus strains in the brain. *Journal of Neurovirology* **8**, 345–352.

Yang, C. and Jackson, A. C. (1992). Basis of neurovirulence of rabies virus variant AvO1 with stereotaxic brain inoculation in mice. *Journal of General Virology* **73**, 895–900.

Yang, J., Hooper, C., Wunner, W.H., Koprowski, H., Dietzschold, B. and Fu, Z.F. (1998). The specificity of rabies virus RNA encapsidation by nucleoprotein. *Virology* **242**, 107–117.

Yang, J., Koprowski, H., Dietzschold, B. and Fu, Z.F. (1999). Phosphorylation of rabies virus nucleoprotein regulates viral RNA transcription and replication by modulating leader RNA encapsidation. *Journal of Virology* **73**, 1661–1664.

3

Molecular Epidemiology

SUSAN A. NADIN-DAVIS

Centre of Expertise for Rabies, Canadian Food Inspection Agency, Ottawa Laboratory (Fallowfield), Ottawa, ONK2H8P9, Canada

1 INTRODUCTION

The discipline of molecular epidemiology, in which patterns of disease transmission are followed using selected markers that distinguish different populations of the disease-causing agent, has become a central theme in the study of infectious diseases, as evidenced by recent publications devoted to this topic (see Thompson, 2000). Technological advancements, in particular those that have simplified the generation of nucleotide sequence data, have played a very significant part in the development of this field. Moreover, the nature of certain infectious disease agents, particularly negative-strand RNA viruses, makes them especially amenable to analysis by molecular epidemiological methods, as reviewed by Moya *et al.* (2004). Such viruses, including the Rhabdoviridae family to which rabies virus belongs, employ error prone RNA polymerases for their replication; the infidelity in genome copying that results from the action of these enzymes ensures that a population of viruses does not exist as a discrete entity but as a collection of genetic variants, a phenomenon frequently referred to as a quasispecies (Domingo *et al.*, 1985). It is now established that even single isolates of a RNA virus exist as a collection of mutant sequences distributed around a central consensus sequence. In a relatively constant and balanced environment, e.g. when a particular viral strain is well adapted to a specific host species, the consensus sequence is relatively stable and exhibits very limited change over time. However, sometimes a virus population undergoes very severe bottlenecks, i.e. virus transmission is severely constrained. The population that successfully emerges from such situations will represent a sub-population, probably with a different consensus sequence from the original viral population and will represent a founder lineage. Sometimes this founder lineage has beneficial properties that allow it to replicate and spread quickly within its new environment. Frequently, however, founder lineages differ from their precursors by a rather small number of neutral mutations only. Nevertheless, if these changes in consensus sequences can be readily and consistently discriminated, they can form the basis of molecular epidemiological investigation. Indeed, the potential importance of the quasispecies nature of rabies virus was documented by relatively

69

early molecular studies on a dog-adapted street strain of the virus that underwent mutations in the glycoprotein (G) gene upon selective pressure, a phenomenon that was clearly related to the heterogeneous nature of the initial viral population (Benmansour *et al.*, 1992).

Several properties of the rabies virus are important to consider when applying molecular epidemiological principles to rabies. First, it is generally established that a given viral lineage is always associated with a particular animal species, known as the reservoir host (Niezgoda *et al.*, 2002); this is due to the relative efficiency with which virus is propagated by individuals of this animal species and transmitted between conspecifics as detailed for fox rabies by Blancou (1988). It follows that the geographical spread of a particular viral population associated with its host is determined by the overall range of the animal species itself as well as aspects of the host animal's biology that determines the extent of interaction between different subpopulations of the host. The maintenance of this virus–host relationship is assumed to involve significant co-adaptation by both parties, though the mechanisms responsible at the level of either the host or the virus are not yet understood. Long-term maintenance of such a virus–host relationship will isolate this virus population from others and lead to the emergence (e.g. via genetic drift) of a virus with distinct markers that can be used to distinguish it from other rabies virus populations. Establishing that a particular rabies viral strain is associated with a specific host is not always straightforward, however, and this requires evidence collected through epidemiology, surveillance and, in some cases, pathogenesis studies.

A 'spillover event' occurs when a rabies virus that is normally maintained in a reservoir host successfully infects another animal species. It is assumed that due to the adaptation of virus to its reservoir host, sustained transmission of virus within the second host is unlikely to occur; spillover transmission is thus regarded as a 'dead-end' infection in the vast majority of situations. However, occasionally such an event can initiate a new virus–host relationship in which sustained propagation and independent transmission of the virus within the new host species occurs such that the new host species becomes a rabies reservoir. Such a 'species jump' will usually be associated with some level of viral adaptation to its new host thereby leading to the emergence of a distinct viral population and hence a new lineage.

Over the years, a number of viral typing methods employing technological innovations of the period have been developed. All of these methods have provided insights into modes of virus transmission. It is the purpose of this chapter to summarize the knowledge gained from such endeavors as it relates to rabies and the rabies-related viruses that constitute the *Lyssavirus* genus.

2 METHODS AND DEFINITION OF TERMS

In this chapter, a viral strain is defined as a viral population that is maintained within a particular host reservoir in a geographically defined area and which can be distinguished between other sympatric or allopatric viral populations. The two main methods for typing of lyssaviruses employ antigenic and genetic methods and the advantages and drawbacks of both techniques will be discussed.

2.1 Antigenic typing

Traditionally, certain lyssaviruses were distinguished on the basis of their reactivity with hyperimmune sera (Shope *et al.*, 1970), although currently, most antigenic typing is conducted using panels of monoclonal antibodies (MAbs). The discriminatory ability of this method relies on the ability of each MAb of the panel to react with a specific epitope of a viral protein. According to the primary sequence of the protein and, occasionally, as a consequence of post-translational modifications to the protein structure, the epitope will be either present or absent and MAb reactivity will be scored as either positive or negative accordingly. Since each MAb theoretically binds to just one epitope, a structure that usually comprises just a small number of amino acids, it follows that the more epitopes that can be targeted with an MAb panel, the greater is the likelihood of improving the discriminatory capability of the panel. Maximizing panel efficiency and discriminatory capability requires not just a significant number of MAbs but ensuring that the individual Mabs that are included within the panel target independent epitopes that are spaced along the length of the protein. While this is theoretically simple, in practice it is not so readily achieved. During the generation of hybridomas from which MAbs are derived it is frequently the case that certain antigenic sites that are immunologically dominant in mice limit the range of reactivities of the hybridoma populations that are selected. Thus, an appropriately discriminatory MAb panel may require selection of MAbs from hybridomas generated from several different fusions. Of course, the nature of the MAb panel must also consider the viral populations to be targeted and regional variations in the nature of the strains circulating in a particular geographical region must be taken into consideration. Thus, a panel developed for use in one country or region will not necessarily be useful for strain typing in another geographic area. A limitation of antigenic typing is that it cannot, of course, discriminate between populations of viruses that do not differ antigenically; thus, if two closely related but genetically distinct viral variants encode similar viral proteins, antigenic typing may not discriminate between them.

The utility of antigenic typing is reflected by the fact that many of our current founding principles of rabies epidemiology were formulated based on studies employing this methodology exclusively (Rupprecht *et al.*, 1991). The viral protein targeted most frequently by this method is the nucleoprotein (N), product of the N gene; this protein is produced in large quantities in infected brain tissue thereby providing an abundant target for MAb binding that is usually assayed by an indirect fluorescent antibody test (FAT) that is readily available in rabies diagnostic facilities. Other proteins that exhibit greater variation in their primary structure,

TABLE 3.1

Staining patterns obtained for four rabies virus strains using a 16 MAb panel

MAb	Eastern Canada (Arctic)	Western skunk	Mid-Atlantic raccoon	ERA vaccine
1C5-5-2[a]	−	−	−	−
5DF12[b]	+	+	+	+
11DD1	−	−	−	+
M1329	−	−	+	−
M1341	+	+	+	+
26BH11	+	+	−	+
20CB11	+	+	+	+
24FF11	−	+	+	+
M993	+	+	+	+
26AF11	+	+	−	+
26BD6	+	+	+	+
32FE10	+	+	+	+
32FF1	+	+	+	+
38FG5	+	+	+	+
M1347	−	−	+	−
7D2-7-4	−	−	+	−

[a]Negative control MAb that binds an adenovirus protein. [b]Broadly cross-reactive MAb used as a positive control.

particularly the G and phosphoprotein (P), represent other potential targets for antigenic typing. Nadin-Davis *et al.* (2000) reported on the development of a panel of anti-P MAbs that could be used in an indirect FAT to discriminate certain lyssaviruses, while a group of anti-G MAbs, applied to formalin-fixed tissues in a histochemical approach, was reported to differentiate between several strains of rabies and other lyssaviruses (Warner *et al.*, 1999a). However, to date by far the most useful epidemiological information has been garnered using MAb panels targeting the N.

To illustrate the utility of antigenic typing, Tables 3.1–3.3 summarize the differential reactivity of selected MAbs against particular rabies virus strains that circulate in the Americas. In Table 3.1, the differential binding of 16 MAbs to four distinct strains is shown. Three of these strains circulate in terrestrial hosts within Canada while the fourth is a live attenuated vaccine strain, Evelyn-Rokitnicki-Abelseth (ERA), that is known to elicit clinical disease in wildlife on rare occasions. Based on comparative studies of the reactivity profile of an isolate with this panel to profiles for reference specimens, strain type can usually be unequivocally assigned. Another MAb panel, shown in Table 3.2, has been used to discriminate between viruses that circulate in several species of insectivorous bats in Canada (Nadin-Davis *et al.*, 2001). It is evident in this table that a significant

TABLE 3.2

The 10 principal rabies virus strains circulating in insectivorous bats within Canada as identified by their reactivity patterns to selected MAbs

					Strain					
MAb #	BBCAN1	BBCAN2	BBCAN3	BBCAN4	BBCAN5	BBCAN6	BBCAN7	MYCAN	LACAN	SHCAN
7AG8	+	+	+	+	+	+	+	Var	−	−
11DD1	−	−	−	−	−	−	−	−	+	−
16DA2	+	+	+	+	+	(+)	+	Var	−	−
1A5	+	+	+	+	−	+	+	Var	(+)	(+)
26BE2	+	+	+	Var	+	+	+	+	−	+
26BH11	+	Var	Var	+	Var	Var	+	Var	Var	+
26FH5	+	−	−	+	Var	+	+	Var	+	+
38AE5	−	+	−	−	−	(+)	+	+	+	+
M319	+	+	+	+	−	+	+	Var	(+)	(+)
M623	−	+	−	+	+	+	+	+	+	+
M862	+	+	+	ND	ND	+	+	+	+	−
M880	−	+	−	ND	ND	+	+	+	+	+
M992	+	+	−	ND	ND	+	+	+	+	−
M996	−	+	−	ND	ND	+	+	+	−	−
M1324	+	+	+	ND	ND	+	+	−	+	+

ND = Not determined. Var = variable reactivity. (+) = very weak and variable + reaction.
Data from Nadin-Davis et al., 2001.

amount of variation in reactivity to certain MAbs, particularly in isolates recovered from species of the *Myotis* genus, can complicate definitive strain typing. Indeed, the wide antigenic variation observed in viruses recovered from these bat species is reflective of the large genetic divergence observed in isolates from this reservoir (see Section 2.2).

Other MAb panels have been used to discriminate additional strains circulating in the USA (Smith, 2002). Indeed, a version of this CDC panel, as depicted in Table 3.3 (Diaz *et al.*, 1994; Delpietro *et al.*, 1997), is now being used extensively to investigate the nature of the viruses circulating throughout Latin America. This panel discriminates between 11 recognized rabies virus types, including several that are associated with two major reservoir hosts of the region – the dog and the vampire bat. It has, however, also significantly contributed to the identification and recognition of additional sylvatic rabies reservoirs. For example, this panel, together with confirmatory data from genetic studies in many cases, has helped to identify the probable role of several species of insectivorous bats as reservoirs in many areas of Latin America (Yung *et al.*, 2002; Cisterna *et al.*, 2005). Additionally, it was used to identify the presence of a fox-associated strain and

TABLE 3.3

Use of the CDC panel of eight monoclonal antibodies to discriminate rabies virus types

RABV antigenic variant	Reactivity pattern with the following MAbs								Reservoir host
	C1	C4	C9	C10	C12	C15	C18	C19	
1	+	+	+	+	+	+	−	+	Dog/mongoose
2	+	+	−	+	+	+	−	+	Dog
3	−	+	+	+	+	−	−	+	D. rotundus
4	−	+	+	+	+	−	−	−	T. brasiliensis
5	−	+	Var	+	+	Var	−	Var	D. rotundus
6	Var	+	+	+	+	−	−	−	Lasiurus spp.
7	+	+	+	−	+	+	−	+	Grey fox
8	−	+	+	+	−	+	+	+	Skunk (south central USA)
9	+	+	+	+	+	−	−	−	T. brasiliensis
10	+	+	+	+	−	+	−	+	Skunk (Baja, California)
11	−	+	+	+	−	−	−	+	D. rotundus
CVS/SAD	+	+	+	+	+	+	+	+	Laboratory strain

All profiles are as reported by Diaz *et al.*, 1994 and Delpietro *et al.*, 1997. Var = variable reactivity.

several distinct skunk-associated strains in Mexico (Velasco-Villa *et al.*, 2002), as well as a distinct strain that appears to be associated with a non-human primate in the Ceara region of Brazil (Favoretto *et al.*, 2001). Where an isolate yields an antigenic pattern that was previously unrecognized, additional isolate evaluation and further isolations of the same viral type will be required to identify unambiguously the reservoir responsible. Moreover, as for any MAb panel, as knowledge of sympatric strains circulating in a specific area improves, panel refinement to meet regional and local needs will be necessary to preserve the value of antigenic typing regimens. Thus, MAbs useful for discriminating between distinct viral populations circulating in dogs and hoary foxes were identified by a rational combination of genetic and antigenic evaluation of a collection of Brazilian isolates (Bernardi *et al.*, 2005).

Antigenic typing relies on MAbs that consistently bind to an antigenic site that is preserved within a strain. Consequently, antigenic data may, in some cases, be usefully applied to mapping of particular epitopes bound by selected MAbs, thereby providing useful information on amino acid residues critical to the maintenance of these structures. In practice, however, only a few residues critical to MAb binding have yet been identified. For example, Smith previously described the process by which amino acids that contribute to the epitopes recognized by three MAbs, CR54, C12 and C6/C20, were located to nucleoprotein residues 112, 36, 128 and 181, respectively (Smith, 2002). However, since most epitopes recognized by the MAbs routinely employed in rabies strain typing are conformational in nature, efforts to identify the precise residues that contribute to epitope structure are clearly complicated. Variations in protein folding patterns due to substitution in other regions of the protein can have subtle effects on binding of a MAb to its epitope. Moreover, since no specific amino acid change in any lyssavirus protein has been correlated with altered host association, extrapolation of antigenic differences between viruses to their evolutionary relationships is not presently possible. Despite the evidence from comparative sequencing data that certain protein-encoding regions of the rabies virus genome (e.g. central region of the P gene and 3'-terminal region of the G gene) are particularly divergent, to date no clear evidence supports the association of certain structural features of viral proteins and their adaptation to particular host species. Further studies may reveal a limited number of genetic changes that contribute to viral adaptation, but presently, the most striking functional requirement of the rabies virus is the requirement for particular amino acids (Arg or Lys) to be retained at position 333 of the mature G in order to maintain a pathogenic phenotype (Tuffereau et al., 1989). Similarly other important structural requirements may reside in single amino acids yet to be identified.

2.2 Genetic typing

MAb panels are often insufficiently discriminating to allow fine epidemiological monitoring of viruses belonging to a single strain and, in such situations, genetic typing is by far the most useful tool. Original studies on the genetic characterization of certain rabies virus strains employed traditional methods of cDNA generation and molecular cloning (Tordo et al., 1986; Conzelmann et al., 1990). The development of polymerase chain reaction (PCR) technology has enormously simplified the process of generating specific segments of nucleic acid and the application of this technique to facilitate molecular epidemiological studies of rabies viruses is now being realized (Tordo et al., 1992). Based upon the published sequences of selected lyssaviruses that are available in publicly accessible databases, primers (synthetic oligonucleotides of defined base sequence) that will direct the amplification of selected portions of the lyssavirus genome can usually be designed readily (see Chapter 10). Use of these primers in a standard reverse-transcription (RT)-PCR allows generation of sufficient quantities of amplicon for characterization by a variety of methods that discriminate small differences (often, single nucleotide polymorphisms or SNPs) between nucleotide sequences. Frequently, the product of the PCR is characterized directly by sequence analysis so as to generate a consensus sequence of the viral population. Alternatively, in some situations the amplicon is cloned into a plasmid vector by standard molecular cloning techniques prior to characterization. It should be recognized that this cloning process generates a collection of clones, each of which is derived from a single molecule of the PCR product; thus each clone will represent a single sequence of the quasispecies population under study. To generate a consensus sequence in such a situation, multiple clones should be sequenced and a consensus derived from combination of these data. In some situations, cloning of PCR products has been undertaken purposely to evaluate the quasispecies nature of particular virus samples (Benmansour *et al.*, 1992; Nadin-Davis *et al.*, 2006a).

Since negative strand RNA viruses exhibit limited recombination capability (Chare *et al.*, 2003), probably due to the nature of their genome replication mechanism, lyssavirus mutation almost always occurs via single base substitutions. Due to the base triplet nature of the genetic code, a deletion/insertion of one or two bases within a protein coding region will result in a change in reading frame and, depending on its location, may result in major alteration to the encoded product. Such a change is highly likely to be deleterious to viral viability and thus, immediately lost from the population. Furthermore, the occurrence of a three base insertion or deletion within a very limited region of the genome, so as to maintain the reading frame, is mechanistically rather unlikely. Regulatory regions of these viruses are small and almost certainly need to maintain primary sequence and/or three-dimensional structures that are dependent on certain sequence motifs. Thus, the only region that is likely to support a significant number of small insertions or deletions is the non-coding G-L intergenic region, for which a function has not yet been identified (Sacramento *et al.*, 1992; Ravkov *et al.*, 1995). Within protein coding regions, substitutions between the four nucleotides (A, C, G and U) that constitute RNA molecules can be either synonymous (not resulting in an amino acid-coding change due to code degeneracy) or non-synonymous (resulting in an amino acid-coding change). Reviews of rabies virus sequence datasets reveal that the majority of mutations are third base changes, predominantly synonymous in nature, although first and second base changes are also observed; first base changes are sometimes synonymous while second base changes are always non-synonymous. Observation of a mutation requires first that a base change be incorporated into a newly synthesized RV genome, but also that the change becomes predominant within the viral population, even if this is only within a single individual. While it is assumed that the initial mutation event occurs by chance at a constant rate along the genome as dictated by the error rate of the viral RNA polymerase, only those mutations that confer a fitness advantage to the virus or those which are neutral in nature and are retained by chance become fixed in the population at least long enough to be observed. Consequently, levels of nucleotide similarity, as observed between two divergent members of the lyssaviruses (rabies Pasteur virus and Mokola virus) are highly variable along the length of the genome (Le Mercier *et al.*, 1997). Functional constraints operate, to different degrees, to limit those mutations that are retained within the lyssavirus's five coding regions; N and L genes are the most conserved while the P and G genes are more variable. Variation in similarity values within each gene further identifies specific protein coding regions or domains that are more or less conserved; e.g. conserved and highly variable domains of the P have been identified (Nadin-Davis *et al.*, 2002) and the central region of the N is known to be less variable than either the N- or C-terminus of N (Kissi *et al.*, 1995). The G gene region that encodes the transmembrane and cytoplasmic domains of the G is the least conserved coding region of the lyssavirus genome (Le Mercier *et al.*, 1997).

Direct nucleotide sequence determination over a predetermined sequence window is the most commonly applied method for genetically comparing a collection of lyssaviruses. Alignments of such sequence data allow direct comparison of isolates to reference sequences and hence allocation to a particular type. An example of such a comparison between three groups of viruses is shown in Figure 3.1. From visual inspection of these aligned nucleotide sequences (Figure 3.1A), it is apparent that within the Arctic fox (AFX) and raccoon strains there is very limited intragroup variation, but a significant difference between groups. The isolates within the bat group exhibit a significant degree of nucleotide substitutions and, indeed, the four isolates examined here represent different bat-associated strains. As shown in Figure 3.1B, only some of the nucleotide variation found within this dataset results in coding differences in the translation products, but consistent intergroup differences are evident. Such a comparison can easily be performed manually for a small sample set. However, where large numbers of isolates are being compared with respect to long sequences of nucleotides, the complexity of the dataset requires that the output of such an analysis be in the form of a phylogenetic tree, a diagram of hierarchical branches that depicts the evolutionary relationships between samples according to their nucleotide substitution patterns. Within this diagram, samples that form a discrete cluster on one branch of the tree are said to form a clade; where there is strong support for this cluster (see below) the samples are said to form a monophyletic clade, indicating that all members originated from a common precursor. A taxon is normally defined as a species or group of species that clearly identifies a specific group of organisms; this term is sometimes used to refer to a particular group of specimens that form a monophyletic clade.

Phylogenetic trees are generated by computer-based phylogeny programs that employ a variety of different algorithms for tree reconstruction. The most commonly used programs include the PHYLIP package available at http://evolution. gs.washington.edu/phylip.html, PAUP (Phylogenetic Analysis Using Parsimony) available from Sinauer Associates and MEGA (Molecular Evolutionary Genetic Analysis) available on-line at http://www.megasoftware.net. A comprehensive listing of packages available for sequence alignments and phylogenetic analysis can be found at the website http://evolution.genetics.washington.edu/phylip/ software.html. Most phylogenetic analysis packages incorporate algorithms that employ distance-based methods: e.g. the neighbor joining (NJ) or unweighted

A.

```
1578T1ON    GAAGACTGGACCAGCTATGGGATCCTGATTGCACGGAAGGAGATAGATCACCCAGATTCTCTGGTGGAGATAAAGCGTACCGGTGTAGAAGGGAATTGGGCT
2756T2ON    ...........................T............................................................................
783T3ON     ....................................................................C...................................
9196T4ON    ....................................................................................T...................
ARCT5CAN    ...........................................C..........................................................

R516NY      ...............A..T.......................C...A..T..A.C.........C.A..C.....A....A.A..............C......
V125FL      ...............A..T.............T.........C...A..T..A.C.........C.A..C.....A....A.A..............C......
3306RON     ...............A..T.............T.........C...A..T..A.C.........C.A..C.....A....A.A..............C......

EF31CAN     ..T..........A...........G.......A.......T.....C.C.A.....AA...G.......C.C...
ML04CAN     ..T..........A...........G.......A.......T..........A.....TCA..G.
LAN12CAN    ..T.....GTT......C.A.....GG..C..A.G......A.....T...GA....TAA..G.
LC01CAN     ..T.....GTT...C......A.....G..C..T..A.....GCA...T...GA....AA...G.
```

BAT-specific primer site   AFX-specific primer site

B.

```
            E D W T S Y G I L I A R K G D K I T P D S L V E I K R T G V E G N W A
1578T1ON    . . . . . . . . . . . . . . . . . . . . . . . . . . . . . . . . . . .
2756T2ON    . . . . . . . . . . . . . . . . . . . . . . . . . . . . . . . . . . .
783T3ON     . . . . . . . . . . . . . . . . . . . . . . . . . . . . . . . . . . .
9196T4ON    . . . . . . . . . . . . . . . . . . . . . . . . . . . . . . . . . . .
ARCT5CAN    . . . . . . . . . . . . . . . . . . . . . . . . . . . . . . . . . . .

R516NY      . . . . . . . . . . . . . . . . . . N . . . D . . . . . D . . . . . . .
V125FL      . . . . . . . . . . . . . . . . . . N . . . D . . . . . D . . . . . . .
3306RON     . . . . . . . . . . . . . . . . . . N . . . D . . . . . D . . . . . . .

EF31CAN     D . . . . . . . . . . . . . . . . . N . . . D . . . . . N . . . H . . .
ML04CAN     D . . . . . . . . . . . . . . . . . N . . . . . . . . . H . . . . . . .
LAN12CAN    D . . V . . . . . . . R . . . . . . N . G . D . R . . . N . . . . . S .
LC01CAN     D . . V . . . . . . . . . . . . . . G T . . D . R . . . N . . . . . . .
```

Figure 3.1 Alignments of 105 bases of internal coding region of the N gene (A) and corresponding amino acid sequences of the encoded nucleoprotein (B) for representative specimens of several viral groups circulating in Canada. The upper group of five sequences represents five distinct variants of the Arctic fox (AFX) strain that circulate in Ontario and northern Canada. The second group of three sequences comprises the North American raccoon strain with representation from Florida to Ontario. The third group consists of four distinct bat strains that are harbored by different species of insectivorous bats. Refer to Table 3.4 for specimen details. Amino acid sequences were predicted from the nucleotide sequences of 1A using the DNAsis package (Hitachi). Alignments were performed using CLUSTALX v1.8 (Thompson *et al.*, 1997) using the Type 1 (T1) Ontario (T1ON) variant of the AFX strain as the reference. For all other specimens, dots represent identity to that reference sequence and only base substitutions/amino acid differences are indicated. Two of the sequences targeted by strain-specific primers, as detailed in Section 2.2.2, are identified below the nucleotide alignment. Since these primers were designed in the negative-sense orientation, their 3′ termini critical for dictating primer specificity would correspond to the 5′ nucleotide in the figure.

pair group method using arithmetic averages (UPGMA), which consider the overall genetic distance between all pairs of sequences rather than the actual sequences themselves. Alternative algorithms include the character-based maximum parsimony (MP) and maximum likelihood (ML) methods that use individual substitutions to determine all the tree constructions that are supported by the data and then identify the optimal tree through tree comparisons. MP identifies the optimal tree by selecting the minimal number of evolutionary steps required to explain the data, while ML identifies the optimal tree as that most likely to have occurred according to the assumed evolutionary model. For a more detailed explanation of the various methods employed for tree reconstruction, see Graur and Li (2000).

The choice of method to be employed for a particular analysis will often depend on the purpose of the analysis. Distance methods are frequently preferred from a practical standpoint due to their relatively rapid execution and their ability to identify groups or clades as well as the other more computationally intensive methods. For the purpose of epidemiological studies, the association of an isolate within a clade, rather than its precise position within this clade, normally identifies the strain or variant responsible for that case thereby providing the key information sought. Thus, for many analyses, distance methods are sufficiently predictive. However, where the data are to be analysed for the purpose of exploring mechanisms of mutation, use of the other algorithms may be advantageous. When independent analysis of a certain dataset by several different algorithms consistently predicts a particular branching pattern and the existence of specific clades, the overall predictive strength of the study is significantly enhanced.

An alternative strategy to explore the robustness of the predictions of phylogenetic analyses is the use of non-parametric bootstrap analysis. Most software programs that predict phylogenetic relationships from nucleotide sequence data can incorporate this statistical method into their analyses and all molecular epidemiological studies should be encouraged to include such methods wherever possible. The method is valuable because even relatively small datasets generate multiple possible branching patterns or trees and statistical methods must be employed to predict the most likely branching pattern, often referred to as the consensus tree. In non-parametric bootstrap analysis, the nucleotide sequence data are resampled randomly with replacement thereby generating pseudoreplicates of the original data; the number of replicates is set by the operator, usually between 100 and 1000, depending on the program to be employed subsequently. Generally, a smaller number of replicates is employed when applying computationally intensive programs such as MP, while an NJ analysis can normally readily incorporate 1000 replicates. Upon analysis by one of these algorithms, the proportion of times that each clade occurs within all trees is calculated and this value is considered as a measure of support for that grouping. In the case of RNA viruses, bootstrap values >90% are generally regarded as providing strong support for a clade, while values >70% are often considered significant (Bauldauf, 2003).

Some consideration must be given to the sequence window to be targeted and this will be determined in part by the goals of an epidemiological study. In general, it has been observed that similar conclusions on the overall epidemiological relationships of particular lyssavirus are obtained regardless of the specific sequence window targeted. Thus, studies on representatives of all the recognized lyssavirus genotypes do generally exhibit similar branching patterns irrespective of whether N, G or P genes were targeted, an observation consistent with the notion that recombination does not occur for members of this genus. Furthermore, although there is a minimum target sequence length below which comparison becomes meaningless (consider a minimum of 250 nucleotides), once such lengths have been reached, additional data often play minimal roles in influencing the overall conclusions of such analyses. A study of Canadian bat rabies viruses that employed sequences for the complete N gene, partial N gene and partial G gene predicted, with some small differences, rather similar associations (Nadin-Davis et al., 2001). However, it should be noted that the robustness of these analyses, as evaluated using bootstrap analysis, varied significantly; the clades predicted using the longer sequence sets were more strongly supported, in some cases to a significant degree. It is self-evident that the longer a sequence window the greater the likelihood of finding one or more differences between samples. As genetic variation increases, the number of informative characters available to a phylogenetic analysis increases and the better the chances of obtaining well supported finely detailed epidemiological details of a sample set. Thus, for studies that seek to monitor variation within a closely related viral population, where differences are small, determination of a long sequence window or a relatively variable region of the genome, or both, is the most effective approach to gain statistically meaningful information. Indeed, even in studies that do not require a high level of detail, use of such targets maximizes the possibility of uncovering strongly supported associations and relationships. The read length typically obtained with the use of older methods of nucleotide sequencing, that relied on the use of radioisotopes and autoradiography, was limited to 200–300 bases per sample loading, while current automated systems can routinely generate 500–1000 bases of sequence from a single loading. As a consequence, the sequence comparison of complete genes is increasingly employed for epidemiological studies. For simple identification of an isolate to a particular variant or strain, very long read lengths are probably unnecessary, although the availability of such data will be useful for documenting differences in protein sequences that may exist between viral strains and for evaluation of the role of specific amino acid residues in host adaptation (see prior section). Indeed, as sequencing technologies continue to develop in the future, the complete characterization of viral genomes may become routine; the resulting information would greatly benefit studies exploring aspects of viral evolution and the molecular mechanisms of viral adaptation.

While non-coding segments of the genome may be considered to be suitable for detailed epidemiological study, the increased likelihood of insertions/deletions in such regions can complicate tree reconstruction. It is assumed by all phylogenetic reconstruction programs that all the nucleotides aligned to a particular position within a sequence window represent homologous sites. However, alignment of non-coding regions differing in total length can result in ambiguous alignments that, in turn, generate phylogenies that are inaccurate or poorly supported.

An additional advantage of targeting coding regions for molecular epidemiological studies is that the protein translation products encoded by such regions can be readily predicted. The resulting amino acid sequences can themselves be used to infer phylogenetic relationships using modified algorithms based on distance and parsimony methods that are frequently included in the more commonly used phylogenetic inference packages. Comparison of trees generated using nucleotide and amino acid sequence data of the N locus of rabies viruses showed rather similar topologies, although bootstrap values tended to be higher for nucleotide sequence data (Kissi *et al.*, 1995), an observation that may simply reflect the fact that nucleotide sequences comprise three times more data points than amino acid sequences.

Another important aspect of any molecular epidemiological study is the selection of samples to be included for comparison. In general, as broad a selection of samples as is appropriate to the questions posed by the phylogenetic study should be included. Compliance with this criterion may not be simple. Most collections of rabies viruses are drawn from passive surveillance systems that can introduce significant bias. Areas of low human population density, lack of human contact with rabies reservoir species or an inadequate system for dealing with suspect cases all result in limited numbers of samples being submitted to rabies testing facilities and thus a lack of truly representative viral isolates for study. In North America, such factors may contribute significantly to limit knowledge of the *Chiroptera* species that maintain rabies. Although certain species, e.g. big brown bats (*Eptesicus fuscus*), are well established as rabies reservoirs due to their frequent contact with humans, the role of other less common bats that are infrequently observed and rarely tested for rabies remains unclear.

To illustrate the complexity of the situation with North American bats and to demonstrate some of the difficulties and challenges that can be encountered in molecular epidemiological investigations, the tree depicted in Figure 3.2 is the result of a study attempting to identify the source of infection of a North American domestic cat. Initial studies with a monoclonal antibody panel failed to identify the viral strain responsible, but preliminary phylogenetic studies strongly suggested the animal had been infected with a bat strain (data not shown). The detailed study presented in Figure 3.2 includes representatives (see Table 3.4 for details) of all of the major viral strains known to be associated with bat hosts in the Americas (Smith *et al.*, 1995; Nadin-Davis *et al.*, 2001). It also includes as outgroups two terrestrial strains, AFX and raccoon, which represent the extent of

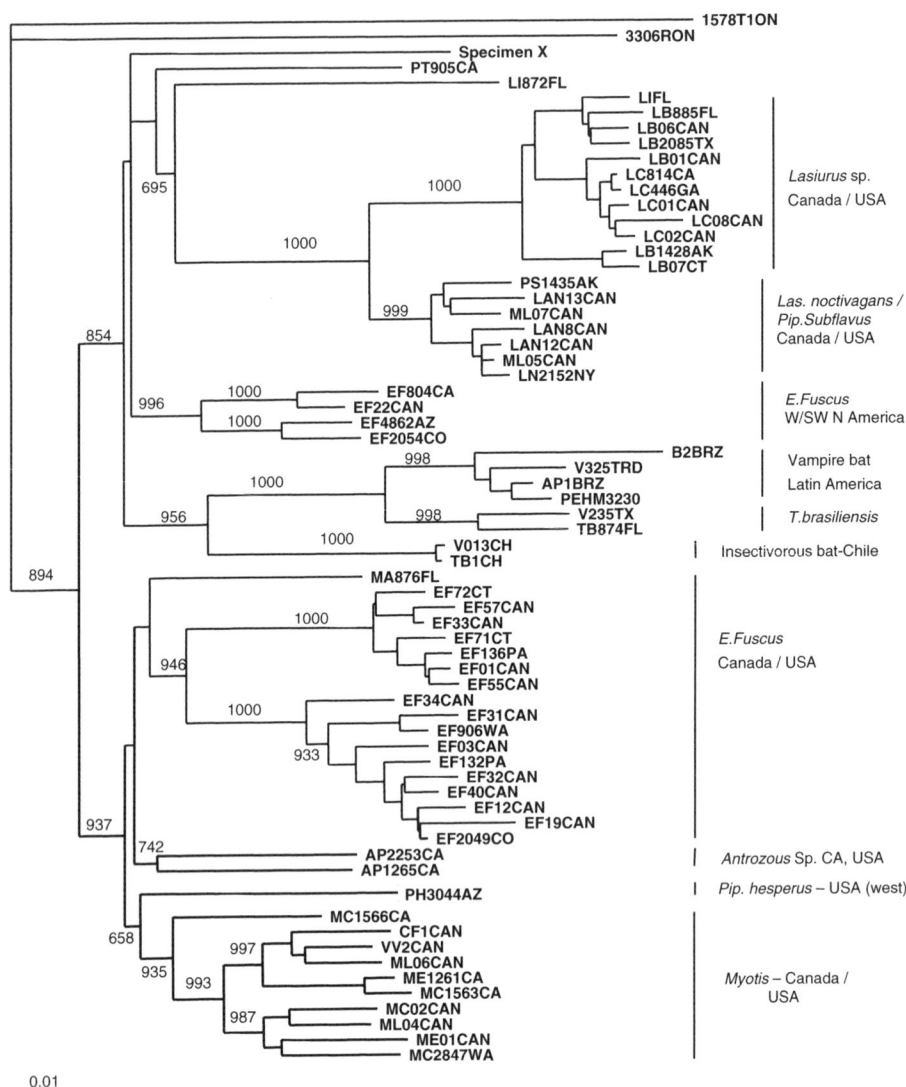

Figure 3.2 A phylogenetic analysis to explore the chiropteran source of exposure for a rabies case in a terrestrial mammal in North America. Using nucleotide sequence data from 64 rabies viruses, representative of the strains circulating in bats of the Americas, together with the sequence under investigation (specimen X from a cat) and two terrestrial rabies isolates of the AFX and raccoon strains as outgroups, an alignment of 600 bp of N gene sequences was generated using the CLUSTAL X program. This alignment was subjected to an NJ analysis using the PHYLIP 3.61 package essentially as described previously (Nadin-Davis *et al.*, 2001). The output treefile was converted into a graphical format using the TREEVIEW package (Page, 1996). Horizontal lines represent genetic distance according to the scale at bottom left. Bootstrap values (out of 1000 replicates) are indicated on the branch to the left of the corresponding clade. The group descriptions to the right of the tree identify the strain represented by each clade and indicate their geographical location.

TABLE 3.4

Source of sequence files referred to in this work

Country	Specimen name	Year	Species of isolation	Reservoir species	Reference or source	GenBank accession # N gene	GenBank accession # P gene
Genotype 1							
Argentina	A1ARG	1985	Insectivorous bat	Insectivorous bat	Nadin-Davis et al., 2002		AF369361
Botswana	V037BWA	1991	Mongoose	Mongoose	Nadin-Davis et al., 2002		AF369300
Botswana	V039BWA	1992	Genet	Genet or mongoose spp.	Nadin-Davis et al., 2002		AF369304
Brazil	AP1BRZ	–	Artibeus planirostris	Vampire bat	Shoji et al., 2004	AB117972	
Brazil	B1BRZ	1993	Dog	Dog	Nadin-Davis et al., 2002		AF369314
Brazil	B2BRZ (DR.braz)	1993	Bovine	Vampire bat	Nadin-Davis et al., 2001, 2002	AF351847	AF369365
Brazil	V904BRZ	1989	Dog	Dog	Bernardi et al., 2005		AY962049
Brazil	V908BRZ	1997	Histiotus velatus	Insectivorous bat	Bernardi et al., 2005		AY962053
Brazil	V913BRZ	2002	Hoary fox	Hoary fox	Bernardi et al., 2005		AY962057
Brazil	V947BRZ	2000	Bovine	Vampire bat	Bernardi et al., 2005		AY962066
Brazil	V951BRZ	1991	Bovine	Vampire bat	Bernardi et al., 2005		AY962068
Brazil	V970BRZ	–	Dog	Dog	Bernardi et al., 2005		AY962077
Brazil	V990BRZ	2002	Molossus molossus	Insectivorous bat	Bernardi et al., 2005		AY962088
Brazil	V1002BRZ	2002	Hoary fox	Hoary fox	Bernardi et al., 2005		AY962094
Canada, ON	1578T1ON	1991	Striped skunk	Red fox	Nadin-Davis et al., 1993, 2002	L20673	AF369265
Canada, ON	2756T2ON	1991	Striped skunk	Red fox	Nadin-Davis et al., 1993	L20674	
Canada, ON	3306RON	1999	Raccoon	Raccoon	Nadin-Davis et al., 2001, 2005	AF351826	AY854603

(Continued)

TABLE 3.4 (Continued)

Country	Specimen name	Year	Species of isolation	Reservoir species	Reference or source	GenBank accession # N gene	GenBank accession # P gene
Canada, ON	783T3ON	1993	Red fox	Red fox	Nadin-Davis et al., 1993, 2002	L20675	AF369267
Canada	867WSKCAN	1992	Striped skunk	Striped skunk	Nadin-Davis et al., 1997		AF369285
Canada, ON	9196T4ON	1990	Red fox	Red fox	Nadin-Davis et al., 1993	L20676	
Canada	ARCT5CAN	1993	Dog	Arctic fox	Nadin-Davis et al., 1994, 2002	U03769	AF369270
Canada	CF1CAN	1998	Dog	Myotis spp.	Nadin-Davis et al., 2001	AF351848	
Canada	EF01CAN	1993	Big brown bat	Big brown bat	Nadin-Davis et al., 2001	AF351827	
Canada	EF03CAN	1972	Big brown bat	Big brown bat	Nadin-Davis et al., 2001	AF351861	
Canada	EF12CAN	1992	Big brown bat	Big brown bat	Nadin-Davis et al., 2001	AF351833	
Canada	EF19CAN	1991	Big brown bat	Big brown bat	Nadin-Davis et al., 2001	AF351855	
Canada	EF22CAN (V078.BBB)	1988	Big brown bat	Big brown bat	Nadin-Davis et al., 2001, 2002	AF351830	AF369341
Canada	EF31CAN	1989	Big brown bat	Big brown bat	Nadin-Davis et al., 2001	AF351831	
Canada	EF32CAN (2994.BBB)	1993	Big brown bat	Big brown bat	Nadin-Davis et al., 2002	AF351828	AF369339
Canada	EF33CAN (058.BBB)	1993	Big brown bat	Big brown bat	Nadin-Davis et al., 2002	AF351829	AF369338
Canada	EF34CAN	1972	Big brown bat	Big brown bat	Nadin-Davis et al., 2001	AF351832	
Canada	EF40CAN	1995	Big brown bat	Big brown bat	Nadin-Davis et al., 2001	AF351862	
Canada	EF55CAN	1997	Big brown bat	Big brown bat	Nadin-Davis et al., 2001	AF351853	
Canada	EF57CAN	1997	Big brown bat	Big brown bat	Nadin-Davis et al., 2001	AF351859	
Canada	LAN08CAN	1988	Silver-haired bat	Silver-haired bat	Nadin-Davis et al., 2001	AF351842	
Canada	LAN12CAN (V077.SHB)	1988	Silver-haired bat	Silver-haired bat	Nadin-Davis et al., 2001, 2002	AF351840	AF369346

Country	Strain	Year	Host	Host	Reference	GenBank	GenBank
Canada	LAN13CAN (4398.SHB)	1980	Silver-haired bat	Silver-haired bat	Nadin-Davis et al., 2002	AF351841	AF369345
Canada	LB1CAN	1994	Red bat	Red bat	Nadin-Davis et al., 2001	AF351844	
Canada	LB6CAN	1991	Red bat	Red bat	Nadin-Davis et al., 2001	AF351856	
Canada	LC01CAN (V103.HB)	1992	Hoary bat	Hoary bat	Nadin-Davis et al., 2001	AF351845	AF369347
Canada	LC02CAN	1993	Hoary bat	Hoary bat	Nadin-Davis et al., 2001	AF351846	
Canada	LC08CAN	1996	Hoary bat	Hoary bat	Nadin-Davis et al., 2001	AF351858	
Canada	MC02CAN (V089)	1992	Myotis californicus	Myotis californicus	Nadin-Davis et al., 2001	AF351836	
Canada	ME01CAN	1992	Myotis evotis	Myotis evotis	Nadin-Davis et al., 2001	AF351835	
Canada	ML04CAN (V089)	1992	Little brown bat	Little brown bat	Nadin-Davis et al., 2001, 2002	AF351839	AF369343
Canada	ML05CAN	1992	Little brown bat	Silver-haired bat	Nadin-Davis et al., 2001	AF351834	
Canada	ML06CAN (4887.LBB)	1994	Little brown bat	Little brown bat	Nadin-Davis et al., 2001, 2002	AF351838	AF369344
Canada	ML07CAN	1979	Little brown bat	Silver-haired bat	Nadin-Davis et al., 2001	AF351837	
Canada	VV2CAN	1993	Red fox	Myotis spp.	Nadin-Davis et al., 2001	AF351851	
Chad	V825CHD	–	–	Canid	This report		DQ275559
Chad	V828CHD	–	–	Canid	This report		DQ275561
Chile	TB1CH	–	Free-tailed bat	Free-tailed bat	Warner et al., 1999b	AF070450	
Chile	VO13CH (IB.Ch)	1988	Insectivorous bat	Insectivorous bat	Nadin-Davis et al., 2001, 2002	AF351850	AF369360
CSFR	V285CSFR	1992	Red fox	Red fox	Nadin-Davis et al., 2002		AF369273
Cuba	V1039CU	2000	Cat	Mongoose	Nadin-Davis et al., 2006c		AY854575
Cuba	V1044CU	2000	Dog	Mongoose	Nadin-Davis et al., 2006c		AY854504
Cuba	V1060CU	2000	Ovine	Mongoose	Nadin-Davis et al., 2006c		AY854537
Cuba	V1069CU	2000	Caprine	Mongoose	Nadin-Davis et al., 2006c		AY854552

(Continued)

TABLE 3.4 (Continued)

Country	Specimen name	Year	Species of isolation	Reservoir species	Reference or source	GenBank accession # N gene	GenBank accession # P gene
Cuba	V1117CU	2002	Ovine	Mongoose	Nadin-Davis et al., 2006c		AY854563
Ethiopia	V667ETH	–	Dog	Dog	Nadin-Davis et al., 2002		AF369331
Ethiopia	V676ETH	–	Jackal	Jackal or other canids	Nadin-Davis et al., 2002		AF369336
France	9147FRA	1991	Red fox	Red fox	Kissi et al., 1995	U22474	
France	V280FRA	1976	Red fox	Red fox	Nadin-Davis et al., 2002		AF369278
Greenland	V864GLD	1990	Fox	Fox	This report		DQ275551
Greenland	V888GLD	2001	Fox	Fox	This report		DQ275553
India	I15IND	1995	Dog	Dog	Nadin-Davis et al., 2002		AF369309
India	I19IND	1995	Dog	Dog			AF369310
India	V458IND	1991	Bovine	Dog	Nadin-Davis et al., 2002		AF369330
Iran	Ir5IRN	1993	Dog	Dog	Nadin-Davis et al., 2002		AF369311
Iran	Ir14IRN	1993	Human	Dog	Nadin-Davis et al., 2002		AF369312
Israel	V661IRL	–	Fox	Fox	Nadin-Davis et al., 2002		AF369281
Israel	V662IRL	–	Fox	Fox	Nadin-Davis et al., 2002		AF369282
Kazakhstan	RU7	1988	Fox	Fox	Nadin-Davis et al., 2002		AF369274
Korea	V739KOR	–	Dog	Dog &/or raccoon dog	This report		DQ275562
Laboratory strain	CVS	–	Challenge virus standard		–		D42112
Laboratory strain	NISH		Nishigahara				AB044824
Laboratory strain	PV	–	Pasteur virus		–		NC_001542

Laboratory strain	SAD$_{B19}$		Street Alabama Dufferin				M31046
Mexico	M4MX	1991	Bovine	Vampire bat	Nadin-Davis et al., 2002		AF369366
Mexico	M29MX	1991	Cat	Dog	Nadin-Davis et al., 2002		AF369313
Mexico	V229MX	1994	Bovine	Vampire bat	Nadin-Davis et al., 2002		AF369362
Mexico	V587MX	1996	Vampire bat	Vampire bat	Loza-Rubio et al., 2005	AY854587	
Mexico	V588MX	1995	Insectivorous bat	Free-tailed bat	Loza-Rubio et al., 2005; Nadin-Davis & Loza-Rubio, 2006	AY854588	AY998246
Mexico	V591MX	1998	Bobcat	Grey fox?	Nadin-Davis & Loza-Rubio, 2006		AY998249
Mexico	V684MX	1992	Skunk	Skunk	Nadin-Davis & Loza-Rubio, 2006		AY998258
Mexico	V854MX	2002	Skunk	Skunk	Nadin-Davis & Loza-Rubio, 2006		AY998275
Namibia	V243NAM	1980	Kudu	?	Nadin-Davis et al., 2002		AF369324
Nepal	V121NEP	1989	Dog	Dog	Nadin-Davis et al., 2002		AF369317
Nepal	V123NEP	1989	Dog	Dog	Nadin-Davis et al., 2002		AF369318
Nigeria	V461NIG	1996	Dog	Dog	Nadin-Davis et al., 2002		AF369326
Nigeria	V464NIG	1996	Dog	Dog	Nadin-Davis et al., 2002		AF369328
Paraguay	P1PGY	1994	Dog	Dog	Nadin-Davis et al., 2002		AF369315
Paraguay	P10PGY	1994	Bovine	Vampire bat	Nadin-Davis et al., 2002		AF369363
Peru	PEHM3230	–	Human	Vampire bat	Warner et al., 1999b	AF045166	
Peru	V001PER	1985	Dog	Dog	Nadin-Davis et al., 2002		AF369323
Philippines	V1371PHIL	2002	Cat	Dog	This report		DQ275546
Philippines	V1374PHIL	2004	Dog	Dog	This report		DQ275547
Philippines	V1375PHIL	2004	Dog	Dog	This report		DQ275548
Puerto Rico	90002PR	1990	Human	Mongoose	Nadin-Davis et al., 2002		AF369303
Puerto Rico	V469PR	1997	Mongoose	Mongoose	This report		DQ275564
Puerto Rico	V470PR	1997	Mongoose	Mongoose	This report		DQ275565
Rep. S. Africa	1500AFS	1987	Mongoose	Mongoose	Kissi et al., 1995	U22628	

(Continued)

TABLE 3.4 (Continued)

Country	Specimen name	Year	Species of isolation	Reservoir species	Reference or source	GenBank accession # N gene	GenBank accession # P gene
Rep. S. Africa	V046AFS	1990	Yellow mongoose	Yellow mongoose	Nadin-Davis et al., 2002		AF369298
Rep. S. Africa	V050AFS	1990	Yellow mongoose	Yellow mongoose	Nadin-Davis et al., 2002		AF369301
Rep. S. Africa	V250AFS	1994	Canid	Canid	Nadin-Davis et al., 2002		AF369297
Rep. S. Africa	V264AFS	1995	Mongoose	Mongoose	Nadin-Davis et al., 2002		AF369302
Russia	RU9	1988	Raccoon dog	Raccoon dog	Nadin-Davis et al., 2002		AF369284
Sri Lanka	1077SRL	1996	Mongoose	?	Arai et al., 2001	AB041967	
Sri Lanka	V113SRL	1986	Dog	Dog	Nadin-Davis et al., 2002		AF369320
Switzerland	V034SWZ	1987	Red fox	Red fox	Nadin-Davis et al., 2002		AF369272
Tanzania	V040TAN	1992	Lycaon pictus	Canid	Nadin-Davis et al., 2002		AF369275
Tanzania	V271TAN	–	Bat-eared fox	Bat-eared fox	Nadin-Davis et al., 2002		AF369277
Thailand	V1148THA	2001	Dog	Dog	This report		DQ275541
Thailand	V1149THA	2001	Dog	Dog	This report		DQ275542
Thailand	V1152THA	2002	Dog	Dog	This report		DQ275545
Trinidad	V325TRD (DR.Td2)	1995	Bovine	Vampire bat	Nadin-Davis et al., 2001, 2002	AF351852	AF369368
Tunisia	V027TUN	1988	Dog	Dog	Nadin-Davis et al., 2002		AF369322
USA. AK	PS1435AK	1991	Eastern pipistrelle bat	Eastern pipistrelle bat		AF394881	
USA. AK	LB1428AK	1991	Red bat	Red bat		AY039224	
USA. AZ	EF4862AZ	1999	Big brown bat	Big brown bat		AY170397	
USA. AZ	PH3044AZ	1993	Western pipistrelle bat	Western pipistrelle bat		AF394870	

Location	Sample	Year	Species	Variant	Reference	GenBank	
USA, CA	AP1265CA	1991	*Antrozous pallidus*	?		AF394868	
USA, CA	AP2253CA	1993	*Antrozous pallidus*	?		AF394869	
USA, CA	EF804CA	1987	Big brown bat	Big brown bat		AF394887	
USA, CA	LC814CA	1986	Hoary bat	Hoary bat		AF394883	
USA, CA	MC1563CA	1986	*Myotis californicus*	*Myotis* spp.		AF394873	
USA, CA	MC1566CA	1987	*Myotis californicus*	*Myotis* spp.		AF394871	
USA, CA	ME1261CA	1990	*Myotis evotis*	*Myotis* spp.		AF394874	
USA, CA	PT905CA	1989	*Plecotus townsendii*	?		AF394877	
USA, CA	V650CA	1997	Skunk	Skunk			DQ275556
USA, CA	V652CA	1997	Skunk	Skunk			DQ275557
USA, CO	EF2049CO	1985	Big brown bat	Big brown bat		AY039228	
USA, CO	EF2054CO	—	Big brown bat	Big brown bat		AF394888	
USA, CT	EF71CT	1998	Big brown bat	Big brown bat	Nadin-Davis *et al.*, 2001	AF351860	
USA, CT	EF72CT	1998	Big brown bat	Big brown bat	Nadin-Davis *et al.*, 2001	AF351854	
USA, CT	LB7CT	1998	Red bat	Red bat	Nadin-Davis *et al.*, 2001	AF351857	
USA, FL	LB885FL	1988	Red bat	Red bat		AF394885	
USA, FL	LI872FL	1988	Yellow bat	Yellow bat		AF394878	
USA, FL	LIFL	1987	Yellow bat	*Lasiurus* spp.		AF351843	
USA, FL	MA876FL	1988	*Myotis austroriparius*	?	Nadin-Davis *et al.*, 2001	AY039225	
USA, FL	TB874FL	1988	Free-tailed bat	Free-tailed bat		AF394876	
USA, GA	LC446GA	1982	Hoary bat	Hoary bat		AF394884	
USA, NY	LN2152NY	1984	Silver-haired bat	Silver-haired bat		AF394880	
USA, NY	R516NY	1992	Raccoon	Raccoon	Nadin-Davis *et al.*, 1997, 2002	U27218	AF369293
USA, KY	KY2877USA	1995	Dog	Skunk	Nadin-Davis *et al.*, 2002		AF369292
USA, FL	V125FL	1987	Raccoon	Raccoon	Nadin-Davis *et al.*, 1997	U27220	AF369294
USA, PA	EF132PA	1984	Big brown bat	Big brown bat		AY039229	
USA, PA	EF136PA	1984	Big brown bat	Big brown bat		AY039226	

(Continued)

TABLE 3.4 (Continued)

Country	Specimen name	Year	Species of isolation	Reservoir species	Reference or source	GenBank accession # N gene	GenBank accession # P gene
USA, TX	LB2085TX	1986	Red bat	Red bat		AF394886	
USA, TX	V211TX	1994	Skunk	Skunk	Nadin-Davis et al., 2002		AF369287
USA, TX	V212TX	1994	Skunk	Skunk	Nadin-Davis et al., 2002		AF369288
USA, TX	V217TX	1994	Coyote	Coyote/dog	Nadin-Davis et al., 2002		AF369337
USA, TX	V224TX	1994	Bovine	Grey fox	Nadin-Davis et al., 2002		AF369271
USA, TX	V230TX	1994	Big brown bat	Big brown bat	Nadin-Davis et al., 2002		AF369342
USA, TX	V231TX	1994	Red bat	Red bat	Nadin-Davis et al., 2002		AF369348
USA, TX	V235TX (TB01)	1994	Free-tailed bat	Free-tailed bat	Nadin-Davis et al., 2001, 2002	AF351849	AF369359
USA, WA	EF906WA	1987	Big brown bat	Big brown bat	Nadin-Davis et al., 2002	AY039227	
USA, WA	MC2847WA	1995	*Myotis californicus*	*Myotis* spp.	Nadin-Davis et al., 2002	AF394872	
Zimbabwe	V269ZM	–	Civet	Civet or mongoose	Nadin-Davis et al., 2002		AF369295
Zimbabwe	V284ZM	–	Dog	Dog	Nadin-Davis et al., 2002		AF369325
Genotypes 2–7 and unclassified							
Ethiopia	LBV.ETH	1990/1	Dog	Bat	Kuzmin et al., 2005	AY333110	
Nigeria	LBV.8619	1958	*Eidolon helvum*	Bat	Bourhy et al., 1993	U22842	
Ethiopia	MOKV.ETH	1990/1	Cat	Unknown	Kuzmin et al., 2005	AY333111	
Zimbabwe	MOKV	1981	Cat	Unknown	Bourhy et al., 1993	U22843	
S. Africa	DUVV.86132	1986	Human	Bat	Bourhy et al., 1993	U22848	
S. Africa	DUVV.94286	1981	*Minopterus* spp.	Bat	Davis et al., 2005	AY996324	
France	EBLV1.8918	1989	*E. serotinus*	*E. serotinus*	Bourhy et al., 1992	U22845	
Poland	EBLV1.8615	1985	*E. serotinus*	*E. serotinus*	Bourhy et al., 1992	U22844	
Finland	EBLV2.9007	1986	Human	*M. daubentoni*	Bourhy et al., 1992	U22846	

The Netherlands	EBLV2.9018	1986	Myotis sp.	M. daubentoni	Bourhy et al., 1992	U22847	
Australia	ABLV.FF	1996	Pteropus alecto	Flying fox	Gould et al., 1998	AF006497	
Australia	ABLV.IB	1996	Saccolaimus flaviventris	Insectivorous bat	Gould et al., 2002	AF081020	
Australia	V474.ABL	1996	Pteropus alecto	Flying fox	Nadin-Davis et al., 2002		AF369369
Kyrgizstan	ARAV	1991	Myotis blythi	Bat species	Kuzmin et al., 2003	AY262023	
Tajikistan	KHUV	2001	Myotis daubentoni	Bat species	Kuzmin et al., 2003	AY262024	
Russia	IRKV	2002	Murina leucogaster	Bat species	Kuzmin et al., 2005	AY333112	
Russia	WCBV	2002	Miniopterus schreibersi	Bat species	Kuzmin et al., 2005	AY333113	

Names in parentheses refer to alternate designations of these same specimens which have appeared in the literature.

divergence of the viruses circulating in terrestrial mammals on the continent. Based on the branching structure of this tree and the bootstrap support for many of these clades, it is evident that distinct viral strains associate with the following species of bats: members of the *Lasiurus* genus, including *Lasiurus cinereus* (hoary bat), *Lasiurus borealis* (red bat) and *Lasiurus intermedius* (yellow bat); *Lasionycteris noctivagans* (silver-haired bat) and *Pipistrellus subflavus* (eastern pipistrelle bat); several distinct lineages of viruses are associated with *Eptesicus fuscus*; two very divergent groups of viruses (based on the branch lengths of the isolates within the clades) are associated with *Myotis* species, particularly *Myotis lucifugus* (little brown bat) and *Myotis californicus*; in the state of California, USA, another strain associated with *Pipistrellus hesperus* (western pipistrelle) bats was recently described (Franka *et al.*, 2004); in Latin America other viral lineages are associated with insectivorous bats, particularly *Tadarida brasiliensis* (free-tailed bat), as well as the vampire bat (*Desmodus rotundus*). Some of the branches of this tree represent just single isolates in which case the host reservoir of these strains remains uncertain until confirmed by additional isolations. Indeed, the cat isolate under investigation (specimen X) itself forms a branch that is not strongly associated with any other specimen of the analysis. Although its position within one of the main chiropteran-associated clades of the tree is well supported with a bootstrap value of 854 out of 1000 replicates, indicating that this virus probably was from a chiropteran host, the specific reservoir responsible is apparently not represented within this sample set. Consequently, no conclusions can be made regarding the source of this infection and, moreover, it is evident that viral lineages yet to be assigned to specific bat species circulate in North America.

Physical isolation of a viral population, due to geographical features that limit host species movements and interactions, for example, can result in localized viral variation that may complicate or preclude unequivocal identification of an unknown if the entire range of the viral strain is not represented in a sample set. Furthermore, the inclusion of outgroups is a useful tool, as represented by two strains of terrestrial rabies in Figure 3.2. An outgroup is a representative of a taxon that is known or assumed to be less closely related to all other taxa of the analysis than these taxa are to each other. Thus, the outgroup essentially provides a root to the tree and determines the direction of character change. An unexpected association of an unknown sample with a particular outgroup will normally imply that the original assumptions regarding the origins of the unknown were incorrect and a different sample set should be employed for an alternate phylogenetic prediction.

Because of the need to include good representation of all appropriate taxa in a phylogenetic analysis and because comparison of an unidentified specimen with representatives of any database require that the same sequence window be employed, choice of the genome target may be strongly influenced by the availability of sequence data generated by prior studies. The majority of comparative rabies virus studies, which include viruses recovered in many parts of the world, have employed the N gene, or parts thereof, and consequently a large publicly accessible database for this gene is now widely available. While databases for other genes (G and P) are now increasing in scope, the N gene repository is currently by far the most extensive.

While improvements in automated nucleotide sequencing make it the optimal method of genetically characterizing PCR products generated from rabies virus collections, other methods can provide useful epidemiological information. Some of the more useful are described here.

2.2.1 Restriction fragment length polymorphism (RFLP) analysis using restriction endonucleases

Restriction endonucleases are enzymes, used extensively in molecular biology, that cleave double-stranded DNA at specific sequences, usually encompassing a 4–6 bp motif. Large numbers of these enzymes that target a wide range of sequence motifs are commercially available. Due to their absolute sequence specificity for cleavage they can discriminate between SNPs if such a change results in the acquisition or loss of a cutting site. Visualization of the products of PCRs treated independently with several of these endonucleases is readily achieved by standard DNA electrophoresis, which is technically much simpler than the denaturing gel electrophoresis employed for nucleotide sequencing. From a collection of specimens, the association of certain viral types with specific digestion patterns can be established; subsequently RFLP analysis of unidentified samples can assign them to a particular type by comparison to a series of reference isolates. The more enzymes employed for targeting sites distributed along the length of a genome segment, the greater the discriminatory capabilities of this method.

This strategy was first applied to the identification of the variants responsible for three cases of unexplained human rabies in the USA (Smith *et al.*, 1991). It was determined that in each case the patient had been exposed to rabies in their country of origin prior to travel to the USA. Subsequently, an extensive study incorporating RFLP analysis allowed the identification of several AFX strain variants that circulate in geographically restricted areas of Ontario and in northern Canada (Nadin-Davis *et al.*, 1993, 1994, 1999), a finding that has facilitated tracking of the movements of specific outbreaks. In addition, a study of the genetic diversity of Iranian rabies viruses permitted generation of a panel of restriction endonucleases capable of discriminating between several of the different viral lineages circulating within the country (Nadin-Davis *et al.*, 2003b). Other examples include the use of RFLP to map distinct rabies virus variants in Europe (Bourhy *et al.*, 1999), to distinguish street and vaccine stains in Estonia (Kulonen and Boldina, 1993), to discriminate different strains associated with distinct host reservoirs in Mexico (Loza-Rubio *et al.*, 1999) and to distinguish various bat-associated strains in Brazil (Schaefer *et al.*, 2005).

2.2.2 Strain-specific PCRs

Another strategy has employed PCR primers that target variable sequences; by adjusting annealing conditions so that these primers hybridize selectively only to their matched target sequence, a PCR product or amplicon that is type specific can be developed. Such an approach was used to develop an assay that discriminated between the mid-Atlantic raccoon strain and established rabies lineages maintained in Ontario wildlife in Canada. This assay employed a multiplex RT-PCR that generated amplicons of different sizes according to whether the target present in the assay represented fox, raccoon or a bat-related viral lineage (Nadin-Davis et al., 1996); the assay also incorporated a second PCR, that generated a larger product from flanking sequences for all the taxa and thus acted as an internal positive control. Identification of the strain within a sample merely required analysis of the sizes of the RT-PCR products by standard DNA electrophoresis. Generation of the larger non-discriminatory PCR product but no specific smaller band would be indicative of the presence of an alternate taxon and, indeed, such a situation was observed in 1998 during the investigation of a rabies-positive cow that was later determined to be infected by the ERA vaccine strain (Nadin-Davis, unpublished data). Rohde et al. (1997) also employed variant specific primers to discriminate between rabies strains circulating in Texas, USA, and similar approaches were applied to distinguish between two distinct biotypes of rabies that circulate within S. Africa (Nel et al., 1998) and to discriminate several distinct viral types circulating in Brazil (Sato et al., 2005).

Methods employing genotype-specific PCR primers used in combination with oligonucleotide probes and Southern blotting techniques have allowed the discrimination of genotype 1 rabies viruses and other lyssavirus genotypes, especially the European bat lyssaviruses (Black et al., 2000; Picard-Meyer et al., 2004b). In some of these methods a PCR-ELISA methodology was used to allow relatively rapid genotype discrimination (Black et al., 2000), although even more rapid real-time PCR methods may be preferred in the future (see below).

2.2.3 Heteroduplex mobility assay (HMA)

A novel method for discriminating between PCR amplicons generated from a collection of rabies viruses recovered in Turkey was described by Johnson et al. (2003a). This technique compares two PCR products generated from two different specimens by annealing them and then assessing the change in mobility of the resulting products by standard gel electrophoresis. The method relies on the fact that the greater the genetic difference between the two products, the greater is the degree of retardation of the heteroduplex compared with the homoduplex during electrophoresis. The authors found that the HMA approach was able to discriminate the three main viral lineages identified by standard sequence analysis and they proposed this alternative methodology as a cost-effective method of comparing large numbers of PCR products.

2.2.4 Discriminatory probes and quantitative PCR

Until recently, few strain-discriminatory probes have been applied to rabies molecular epidemiology or rabies studies in general due to the difficulty in identifying genomic regions that are sufficiently divergent as to be practically useful. One such region is the highly variable central region of the P gene. A method designed for application, specifically to formalin-fixed brain tissues, and which employs strain-specific probes and *in situ* hybridization methods, can discriminate between most of the terrestrial and chiropteran rabies virus types that circulate in Canada (Nadin-Davis *et al.*, 2003a). Although this method is utilized in only exceptional circumstances, it is currently the only one available for typing of Canadian variants when formalin-fixed tissue is the sole material available.

Developments in PCR technology have provided fluorescence-based methods that link increasing production of the amplicon to increased fluorescent signal intensity in real-time mode. Utilization of this technology to discriminate between lyssaviruses of different genotypes has been reported (Black *et al.*, 2002; Wakeley *et al.*, 2005) and increased use of this technology for both rabies diagnosis and epidemiological investigations would appear likely in the future (see Chapter 10).

2.3 Antigenic versus genetic typing methods

The major advantage of genetic typing in which sequence determination is employed is the capability of using such data for phylogenetic investigation and gaining a totally objective strain identification as well as evolutionary information that is inherent to this type of analysis. However, the costs associated with the performance of this type of variant identification, in terms of technical expertise as well as the acquisition of equipment and reagents, are substantial. Indeed, no rabies diagnostic facility that handles significant numbers of rabies cases has the resources to characterize, by sequence analysis, all rabies-positive cases and such an undertaking is probably not worthwhile.

In contrast, antigenic typing methods are less technically demanding and less costly and can thus be applied routinely to large numbers of cases. However, discrimination by antigenic methods does depend on having a suitable panel of MAbs that are capable of discrimination of all possible taxa; in practice this may be difficult to achieve and in those situations where the appropriate selectivity in binding of the available MAbs is insufficient, development of new MAbs may be required to address fully the deficiency. While typing of all rabies-positive cases by antigenic methods is feasible, even in developed countries this is rarely undertaken. It is usually considered sufficient to characterize a selection of cases that are deemed to be representative of the geographical range of each particular taxon. By its nature, antigenic typing, together with the alternative genetic methods discussed above that do not involve sequence determination, cannot provide information on the evolutionary relationships of viral collections; however, they do allow isolates to be assigned to a particular group and this is frequently all that is required from an epidemiological perspective.

Thus, the final choice of which method(s) to apply to a situation will depend on the information sought. Ideally, a combination of these methods, in which a wide selection of isolates is characterized antigenically or by other simple genetic typing regimens, supplemented with nucleotide sequence determination on representative specimens, is the preferred and most cost effective means of generating comprehensive epidemiological data.

3 VIRAL TAXONOMY

3.1 Lyssavirus species

One issue that continues to elicit controversy in virology is the concept of viral species and the issue of designating species names to viruses (van Regenmortel *et al.*, 2000). This difficulty certainly applies to the viruses designated as belonging to the genus *Lyssavirus*, a grouping that became necessary upon the discovery of two rabies-related viruses originating from Africa that were able to elicit the clinical disease of rabies. These two viruses, Mokola virus (MOKV) and Lagos Bat virus (LBV), were sufficiently similar to classical rabies virus such that antibodies directed to the rabies ribonucleoprotein antigen could detect them (Schneider *et al.*, 1973) while, in contrast, cross-neutralization tests readily distinguished them from classical rabies virus (Shope *et al.*, 1970). It was argued that, despite their evident taxonomic relatedness, these African viruses are sufficiently phenotypically distinct from rabies virus that they represent different viral species. This has subsequently been borne out by genetic studies of these viruses as discussed below. Moreover, since human rabies immune globulin exhibits little neutralizing activity against LBV and MOKV, this has important public health implications. However, many other rabies-related viruses that exhibit variable levels of distinction from rabies virus have since been identified and the issue of species assignment has become far less clear. As was stated by van Regenmortel (1998): 'Virus species are fuzzy sets with hazy boundaries....'

As detailed in Table 3.5, seven distinct lyssavirus genotypes (GTs) are currently recognized as species by the International Committee on Taxonomy of Viruses (Tordo *et al.*, 2004). GTs 4–7, which all appear to circulate in various species of bats, are all neutralized by anti-rabies virus antibodies and phenotypic differences between them and classical rabies are not clearly apparent. In addition, four presently unclassified viruses, recovered from bats of Eurasia and which appear to represent several new lyssavirus genotypes, have recently been described (Table 3.6, see also Chapter 6). Calculations of pairwise genetic distances between members of all of these groups, as illustrated in Table 3.7, demonstrate that intragenotypic nucleotide distances tend to be ≤0.2 while

TABLE 3.5

Currently recognized species in the *Lyssavirus* Genus[a,b]

	Rabies virus (RABV)	Lagos bat virus (LBV)	Mokola virus (MOKV)	Duvenhage virus (DUVV)	European bat lyssavirus 1 (EBLV-1)	European bat lyssavirus 2 (EBLV-2)	Australian bat lyssavirus (ABLV)
Genotype	1	2	3	4	5	6	7
Distribution	Worldwide with the exception of Antarctica and some islands	Several African countries, including Central African Republic, Ethiopia, Nigeria, Senegal, South Africa, Zimbabwe	Several African countries, including Cameroon, Central African Republic, Ethiopia, Nigeria, South Africa, Zimbabwe	African nations, including Guinea, South Africa, Zimbabwe	Europe, including Denmark, France, Germany, The Netherlands, Poland, Russia, Spain, Ukraine	Several countries of western Europe, particularly The Netherlands, Switzerland, Finland and the UK	Australia and possibly areas of SE Asia
Reservoir	*Canivora*: domestic and wild canid species, mongoose, raccoons, skunks *Microchiroptera*: (Americas only) several species of insectivorous bats, including members of the *Eptesicus*, *Myotis* and *Lasiurus* genera, vampire bats	*Megachiroptera*: Cases reported from *Eidolon helvum*, *Micropterus pusillus* and *Epomophorus wahlbergi*	Possible bat reservoir but data inconclusive – single cases in shrews (*Crocidura* sp.) and a small rodent (*Lophyromys sikapusi*) with most reported cases in domestic cats and dogs	*Microchiroptera*: Single cases assigned to *Miniopterus schreibersii*, *Nycteris gambiensis* and *N. thebaica*	*Microchiroptera*: Almost all cases in *Eptesicus serotinus*	*Microchiroptera*: Almost all cases in *Myotis* species (*M. dasycneme* and *M. daubentonii*)	*Megachiroptera*: Pteropid species *Microchiroptera*: the insectivorous yellow-bellied sheathtail bat (*Saccolaimus flaviventris*)

[a] Species are listed in order of increasing divergence from the classical rabies virus group. [b]Data are summarized from several articles, including Mebatsion *et al.* (1992), Swanepoel *et al.* (1993) and Mackenzie (1999) and several other papers referenced in this chapter.

TABLE 3.6

Unclassified lyssaviruses

	Aravan virus (ARAV)	Khujand virus (KHUV)	Irkut virus (IRKV)	West Caucasian bat virus (WCBV)
Distribution and reservoir	Single case attributed to a bat (*Myotis blythi*) from Kyrgizstan	Single case attributed to a bat (*Myotis daubentonii*) from Tajikistan	Single case attributed to a bat (*Murina leucogaster*) from Irkutsk province, Russia	Single case attributed to a bat (*Miniopteris schreibersi*) from the Krasnodar region of Russia

TABLE 3.7

Pairwise distance values calculated for N gene and nucleoprotein sequences of

	1077 SRL	783T3 ON	1500 AFS	9147 FRA	V587 MX	ABLV .FF	ABLV .IB	EBLV2 .9007	EBLV2 .9018	ARAV
1077SRL	–	0.0268	0.0335	0.0155	0.0377	0.0700	0.0934	0.1268	0.1318	0.0958
783T3ON	0.1424	–	0.0472	0.0312	0.0518	0.0875	0.1085	0.1366	0.1416	0.1104
1500AFS	0.1577	0.1443	–	0.0405	0.0540	0.0865	0.1060	0.1367	0.1417	0.1134
9147FRA	0.1368	0.1209	0.1286	–	0.0470	0.0840	0.1053	0.1368	0.1418	0.1082
V587MX	0.1675	0.1724	0.1931	0.1813	–	0.0790	0.0988	0.1385	0.1435	0.1126
ABLV.FF	0.2683	0.2563	0.2700	0.2742	0.2729	–	0.0443	0.1272	0.1393	0.0825
ABLV.IB	0.2827	0.2971	0.3008	0.2856	0.2815	0.1806	–	0.1303	0.1426	0.0918
EBLV2.9007	0.3176	0.3306	0.3190	0.3196	0.3175	0.2796	0.3058	–	0.0219	0.1168
EBLV2.9018	0.3199	0.3263	0.3071	0.3142	0.3089	0.2924	0.3136	0.0409	–	0.1242
ARAV	0.3076	0.3032	0.2977	0.3216	0.2856	0.2835	0.2872	0.2766	0.2837	–
KHUV	0.2776	0.2796	0.2930	0.3112	0.2833	0.2676	0.2979	0.2475	0.2483	0.2501
EBLV1.8615	0.3138	0.3091	0.3154	0.3035	0.3136	0.2776	0.2802	0.2800	0.2961	0.2673
EBLV1.8918	0.3102	0.3107	0.3132	0.3072	0.3181	0.2866	0.2874	0.2666	0.2824	0.2692
DUVV.86132	0.3367	0.3321	0.3370	0.3196	0.3280	0.2869	0.2967	0.3003	0.3024	0.2642
DUVV.94286	0.3399	0.3297	0.3345	0.3127	0.3289	0.2881	0.2980	0.3038	0.3026	0.2685
IRKV	0.3282	0.3134	0.3232	0.3169	0.3244	0.3017	0.3102	0.2898	0.2853	0.2928
LBV.8619	0.3311	0.3599	0.3461	0.3391	0.3318	0.3569	0.3501	0.3572	0.3411	0.3306
LBV.ETH	0.3476	0.3494	0.3518	0.3515	0.3452	0.3151	0.3423	0.3585	0.3680	0.3168
MOKV	0.3993	0.3898	0.4007	0.4142	0.3967	0.3916	0.4129	0.4234	0.4262	0.3654
MOKV.ETH	0.3623	0.3472	0.3421	0.3431	0.3650	0.3626	0.3673	0.3685	0.3647	0.3347
WCBV	0.3569	0.3585	0.3621	0.3658	0.3656	0.3504	0.3492	0.3695	0.3669	0.3498

Values, rounded to 4 decimal places, were calculated using the DNADIST and PROTDIST sequences while those in the upper right refer to nucleoprotein comparisons.

intergenotype distance values tend to be ≥0.23, values which substantiate the proposed cut-off value of 80–82% nucleotide identity for discriminating between genotypes (Kuzmin *et al.*, 2005). At the protein level, a distance value of 0.065 appears to provide a convenient cut-off, whereby viruses that differ by more than this value are consistently placed in separate genotypes; this cut-off also supports the distinctive nature of the newly described Eurasian bat lyssaviruses.

The N gene sequences used to calculate these distance values were also employed with various algorithms to construct phylogenetic trees illustrating these species groupings (Figure 3.3A–D). These resulting diagrams illustrate how the same sequence set can be analysed and depicted in rather different ways; some of these methods are concerned primarily in evaluation of distance

representative lyssaviruses

KHUV	EBLV1 .8615	EBLV1 .8918	DUVV .86132	DUVV .94286	IRKV	LBV .8619	LBV .ETH	MOKV	MOKV .ETH	WCBV
0.0917	0.1208	0.1179	0.1194	0.1169	0.1371	0.1900	0.1623	0.2582	0.2045	0.1985
0.0949	0.1333	0.1253	0.1315	0.1291	0.1402	0.2000	0.1705	0.2690	0.2142	0.2131
0.1013	0.1338	0.1308	0.1337	0.1313	0.1455	0.2083	0.1804	0.2672	0.2132	0.2154
0.1018	0.1308	0.1279	0.1313	0.1288	0.1448	0.1972	0.1694	0.2662	0.2119	0.2149
0.0981	0.1332	0.1302	0.1269	0.1244	0.1435	0.1884	0.1571	0.2534	0.1950	0.2153
0.0790	0.1100	0.1071	0.1059	0.1083	0.1299	0.1999	0.1634	0.2386	0.1886	0.1935
0.0981	0.1260	0.1241	0.1173	0.1197	0.1296	0.2073	0.1762	0.2486	0.1987	0.2099
0.0998	0.1339	0.1268	0.1467	0.1491	0.1484	0.2339	0.2028	0.2815	0.2338	0.2242
0.1023	0.1441	0.1321	0.1521	0.1546	0.1490	0.2342	0.2133	0.2846	0.2367	0.2231
0.0766	0.0882	0.0831	0.0853	0.0877	0.0980	0.1688	0.1311	0.2216	0.1713	0.1825
–	0.1153	0.1034	0.1124	0.1148	0.1287	0.2077	0.1662	0.2544	0.2091	0.2118
0.2799	–	0.0132	0.0749	0.0772	0.0848	0.1864	0.1546	0.2483	0.1931	0.2053
0.2842	0.0443	–	0.0680	0.0703	0.0757	0.1785	0.1519	0.2457	0.1902	0.1982
0.3014	0.2478	0.2398	–	0.0022	0.0994	0.1546	0.1355	0.2227	0.1761	0.1696
0.3049	0.2541	0.2439	0.0097	–	0.1017	0.1571	0.1379	0.2254	0.1736	0.1721
0.2978	0.2576	0.2626	0.2670	0.2713	–	0.1728	0.1541	0.2276	0.1745	0.1875
0.3666	0.3241	0.3260	0.3457	0.3471	0.3149	–	0.0562	0.1726	0.1166	0.1943
0.3394	0.3251	0.3333	0.3143	0.3145	0.3221	0.1970	–	0.1583	0.1039	0.1822
0.4394	0.4032	0.4039	0.3776	0.3790	0.3861	0.3200	0.2996	–	0.0631	0.2509
0.3885	0.3486	0.3482	0.3659	0.3657	0.3322	0.2929	0.2856	0.1546	–	0.2020
0.3932	0.3656	0.3617	0.3371	0.3432	0.3664	0.3321	0.3293	0.3936	0.3525	–

programmes of PHYLIP v. 3.61; those in the lower left of the table were generated from N gene

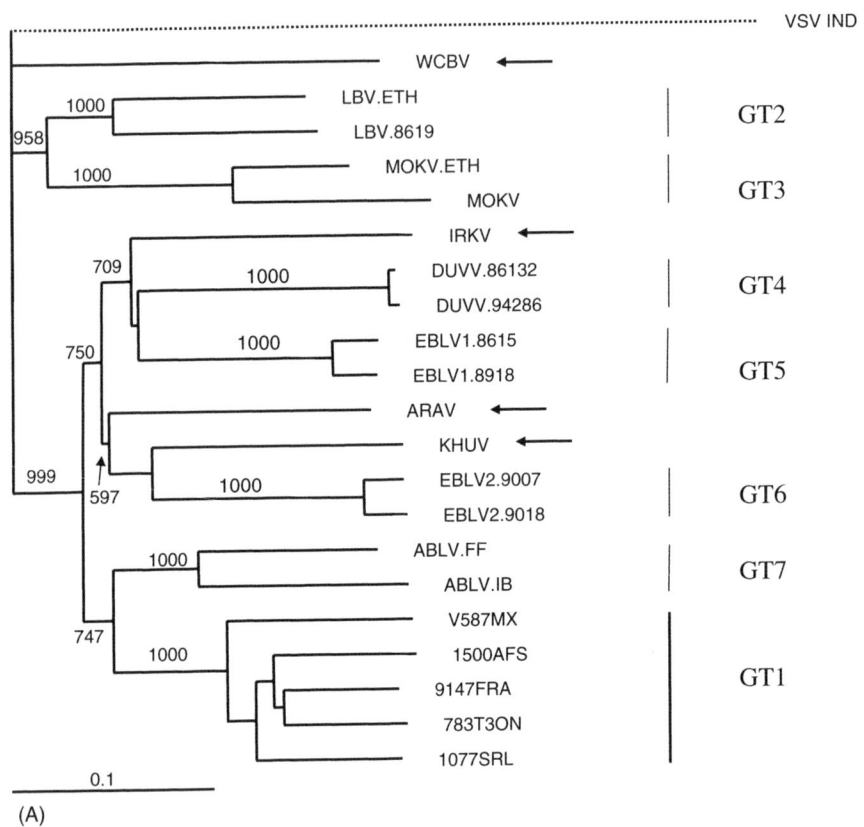

Figure 3.3 Phylogenetic analysis of 21 representative members of the lyssavirus genus. An alignment of 21 complete N gene sequences from selected lyssaviruses, together with the N gene of vesicular stomatitis virus (Indiana strain, GenBank accession number U12967) from the *Vesiculovirus* genus used as the outgroup, was generated using CLUSTALX. The PHYLIP 3.61 package of programmes used this alignment to generate the following phylogenetic trees: (A) Phylogram of a UPGMA analysis. A distance scale, shown at bottom left, was re-applied to the consensus tree using the FITCH package with distance values generated from the DNADIST programme. Thus, the lengths of horizontal lines separating isolates accurately represent the genetic distances between them. Note that the branch of the outgroup (VSV IND), shown as a dashed line, is not to scale so as to allow the relative expansion of the lyssavirus branches. Clades representing the seven established genotypes are indicated; arrows identify the unclassified Eurasian bat isolates. Bootstrap values out of 1000 replicates are shown to the left of the major clades.

between specimens with prediction in branching order being a secondary priority (Figure 3.3A), while other methods are concerned primarily in finding the correct branching order (Figure 3.3B–D). It is apparent from these trees that members of each of the seven established genotypes do indeed form well supported clades and that the GT2 and GT3 viruses are the more genetically divergent of the genus. Indeed, Badrane *et al.* (2001) proposed, based upon genetic,

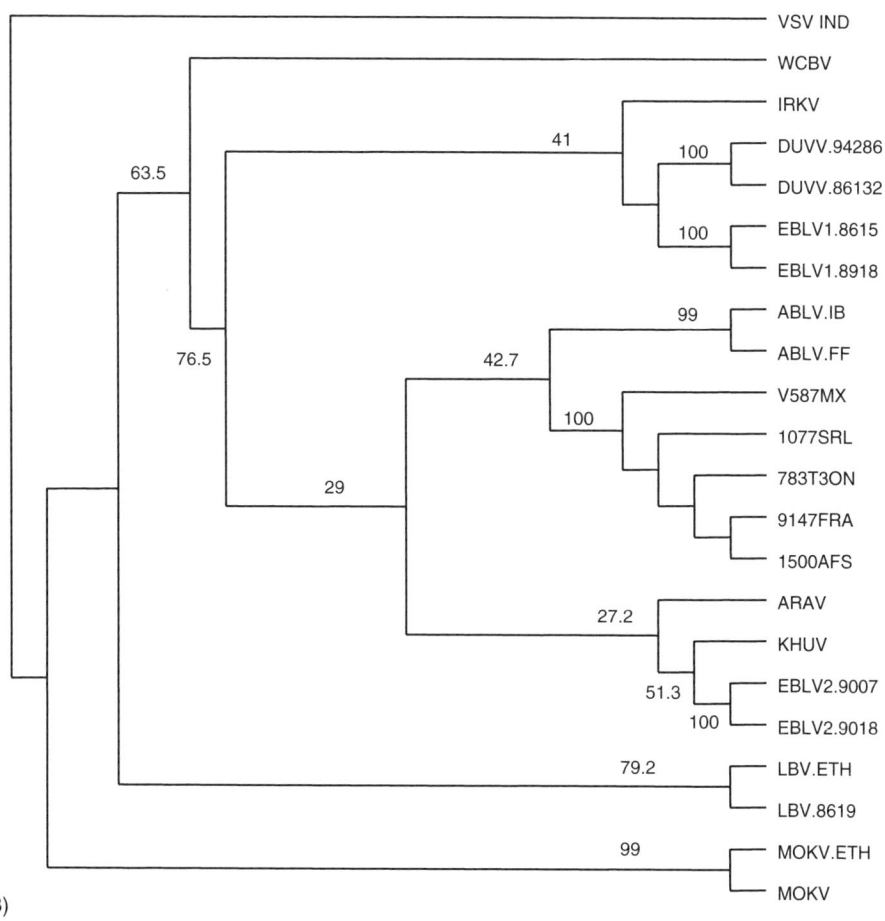

(B)

Figure 3.3 (Continued) (B) Rectangular cladogram of an MP analysis using 100 bootstrap replicates of the data. The lengths of the horizontal lines in this diagram are not representative of the degree of difference between isolates.

immunologic and pathologic characteristics of lyssaviruses, that the distinctive MOKV and LBV viruses should be classified as a subgeneric group of lyssaviruses. Consequently, the concept of phylogroups was developed whereby all seven lyssavirus genotypes are divided into two phylogroups, 1 and 2 (Figure 3.3D). These trees also include the four unclassified lyssaviruses that, according to their placements, significantly complicate current lyssavirus taxonomy. Phylogenetic studies targeting N, P and G genes supported the placement of Aravan and Khujand viruses on distinct branches segregating them from the European bat lyssavirus (EBLV) lineages, though Khujand virus often clustered more closely with, yet still distinctly from, the EBLV-2 branch (Arai *et al.*, 2003; Kuzmin *et al.*,

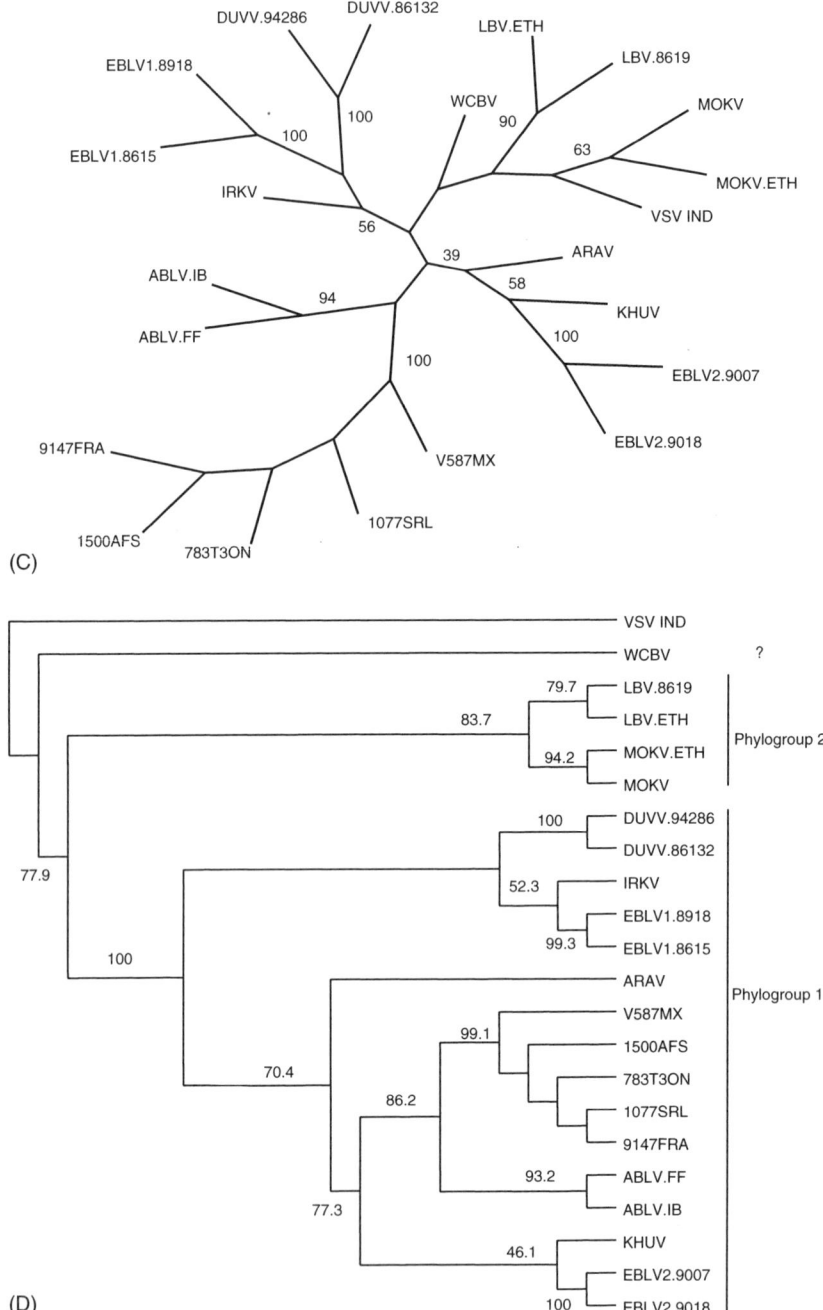

(C)

(D)

Figure 3.3 (Continued) (C) Unrooted radial tree of an ML analysis using 100 bootstrap replicates of the data; the lengths of the branches are not indicative of the difference between isolates. (D) Rectangular cladogram of an MP analysis of an alignment of nucleoprotein sequences (450 residues). These sequences had been predicted from the N gene sequences used to construct trees of (A)–(C). Bootstrap values, generated from 100 replicates of the data, are shown on branches as for the other trees. The division of all isolates into two phylogroups is illustrated to the right of this tree; the WCBV isolate appears, by this analysis, to lie outside of both groups.

2003). Subsequent studies that included the more recently isolated Irkut virus and the West Caucasian bat virus (WCBV) concluded that N gene comparisons provided the most unequivocal separation of all lyssaviruses into genotypes (Kuzmin *et al.*, 2005). Irkut virus formed an outlier to the clade comprising both GT 4 and 5 viruses, while the WCBV clustered as an outlier with members of phylogroup 2 and, indeed, may represent the most divergent of all lyssaviruses yet characterized. These relationships are illustrated in the N gene phylogenies depicted in Figure 3.3A–D. The geographical range of these Eurasian bat lyssaviruses remains to be established though serological evidence has suggested that these or closely related viruses may circulate in certain bat populations of Thailand (Lumlertdacha *et al.*, 2005).

3.2 Strain definition

At a finer level of detail, distinct viral lineages are readily identified within many of the established lyssavirus genotypes and many issues of nomenclature and taxonomy for these lineages have yet to be fully addressed. For ease of description, a strain is often designated according to its host and geographical location. Such a system works well for those strains that are clearly defined, e.g. European red fox rabies, and attempts to name certain well-defined lineages have been published (Kissi *et al.*, 1995; Nadin-Davis *et al.*, 2002). However, in some situations, the major host species is in dispute (e.g. for certain lineages of American bat rabies viruses and in middle-eastern regions where the relative importance of dogs and wildlife vectors is unclear), thereby complicating the nomenclature and necessitating use of 'region' or 'country' complemented with an alphanumeric system where multiple but distinct lineages of virus co-circulate. Moreover, localized variations in nomenclature sometimes can cause confusion when a single population is given two different names.

Within a viral population representing a strain, small genetic differences (that may or may not be identifiable by antigenic methods of discrimination) will arise through continued mutation, thereby generating viral variants that may become established in the population with time. If these genetic variations are not associated with a change in host or major jumps in geographical distribution, they are not normally considered to represent a new strain. However, due to the many levels of bifurcation that can occur in phylogenetic trees, the definition of what constitutes a strain and a variant can become hazy in the same way that the definition of viral species is problematic. It has been proposed that genetic distances between isolates together with strong support for a lineage using bootstrap values may prove to be good indicators of what constitutes a strain (Kissi *et al.*, 1995; Nadin-Davis *et al.*, 2002), but such parameters must take into account the genomic interval examined and may not be universally applicable in all situations.

3.3 Molecular epidemiology of GT 1 lyssaviruses

To illustrate most of the main rabies virus strains currently recognized, a dendrogram, produced by an NJ analysis of partial P gene sequences of 101 rabies virus isolates together with a single Australian bat lyssavirus (ABLV) isolate as outgroup, is shown in Figure 3.4. The following description of the global status of rabies virus molecular epidemiology uses this tree as a guide, but also refers to the extensive literature that is now published on this topic. Space permits only major themes to be described here and for more detailed information on the viruses of particular regions the reader should refer to the many papers referenced in this section.

The uppermost clade identified in this tree represents many different strains of the ARCTIC lineage. While originally thought to be restricted to fox populations of arctic and temperate regions of North America and Russia, this lineage is now known to be also widely distributed in several countries of Asia including Pakistan, Korea, Iran, Nepal and India (Nadin-Davis *et al.*, 1993, 2002; Kuzmin *et al.*, 2004; Mansfield *et al.*, 2006; Park *et al.*, 2005) where other species, including dogs and raccoon dogs, seem to act as the viral reservoirs. Indeed, this lineage has been found to predominate in dogs of southern India as far south as the city of Bangalore (Nadin-Davis, personal communication). The well-supported dichotomy of this clade differentiates old and new world viruses, previously referred to as Asian and American variants, respectively (Nadin-Davis *et al.*, 2002).

Another distinct clade, previously referred to as AFRICA 3 (Kissi *et al.*, 1995) or AFRICA:HP 1 (Nadin-Davis *et al.*, 2002), represents viruses circulating in mongooses in southern countries of Africa including South Africa, Botswana and Zimbabwe (von Teichman *et al.*, 1995; Johnson *et al.*, 2004a). Based upon the distinctness of these viruses from other lineages found in Africa and their relatively high level of genetic heterogeneity, von Teichman *et al.* (1995) have suggested that these viruses represent an ancient lineage that was introduced into mongoose populations of the South African central plateau region quite independently of other African viruses harbored by various canid species (see below). While the yellow mongoose is apparently the principal host for these viruses in South Africa, the slender mongoose may be the main reservoir host in Zimbabwe. Furthermore, these viruses are not infrequently found in various viverrid species (e.g. genets, civets) and other mongoose species that often live in very close proximity to the mongoose reservoir populations. Strong geographical partitioning of variants of this viral group has been reported and the possibility that this viral lineage extends into other parts of Africa requires further investigation (Nel, 2005).

Geographically, the most widely distributed group of rabies viruses is undoubtedly those belonging to the cosmopolitan lineage. As reviewed in detail previously (Nadin-Davis and Bingham, 2004), the phylogeny of this lineage supports the concept that progenitors of the strains represented within this lineage were widely distributed to many parts of the world during colonial times and that viruses

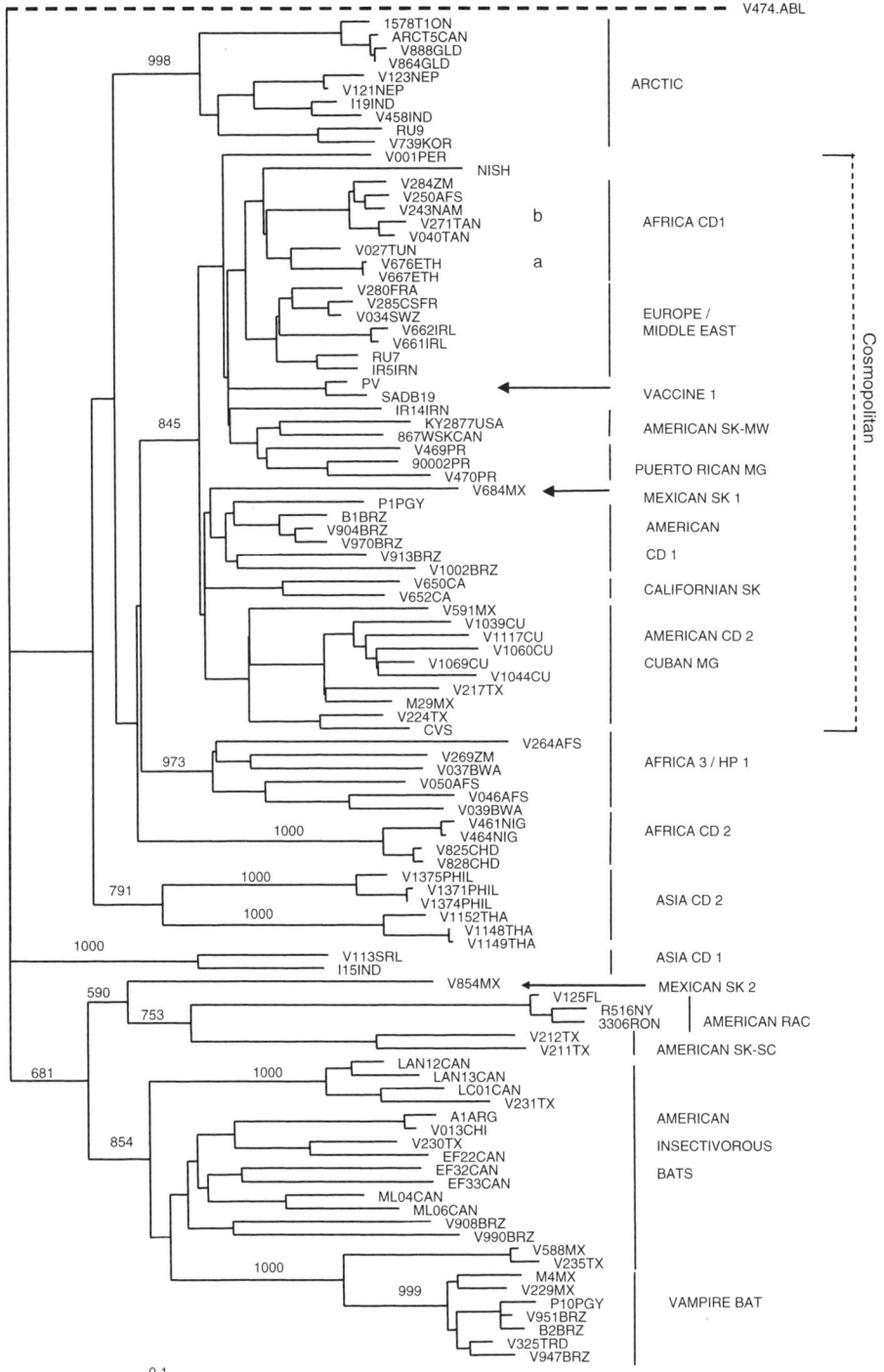

Figure 3.4 A detailed phylogeny of 101 isolates of GT 1 lyssaviruses using partial P gene sequences. This phylogram was generated by the NJ algorithm with re-application of distance data to the consensus tree as indicated for Figure 3.3A. The branch connecting the ABLV isolate used as the outgroup (dashed) is not to scale with the rest of the tree. The groupings discussed in the text are identified to the right of the tree.

circulating in dogs in Europe during the 14th century or later served as the source for these current enzootics. This explains the observation, evident in Figure 3.4, that many isolates recovered from geographically diverse regions are genetically very closely related. Notwithstanding this limited genetic heterogeneity, several distinct clades are readily identified within this lineage. The viruses that until very recently circulated widely in Europe, and which still persist in certain European countries, form one well supported clade. It is well documented that as dog rabies declined in Europe in the early 20th century, the virus adapted to the red fox host and, after World War II, an epizootic in foxes spread westwards from the Russian–Polish border to infect foxes and other species in most of western Europe by the 1960s (see Bourhy et al., 1999). Closely related viruses also apparently spread eastwards and currently circulate throughout the Middle-East where all rabies viruses that have been examined fall within the European/Middle-eastern clade or in closely related branches of the cosmopolitan lineage (see Israeli and Iranian isolates in Figure 3.4) (Kissi et al., 1995). In Israel, for example, jackals and, more recently, foxes have served as reservoirs for rabies viruses closely related to the European variants (David et al., 2000). A detailed study of the evolution of rabies in Europe identified several geographically restricted variants that circulated in western, central, eastern and north-eastern Europe; in the latter region, the rac-coon dog has become a major reservoir apparently as a result of adaptation of the fox virus to this host (Bourhy et al., 1999).

Another sublineage of the cosmopolitan cluster, referred to originally as AFRICA 1 (Kissi et al., 1995) or AFRICA CD1 (Nadin-Davis et al., 2002) and which is adapted to dogs and wild canids, is widely distributed throughout Africa. This lineage apparently underwent an early subdivision that clearly discriminates the viruses of northern and southern countries of the continent; thus viruses from Algeria, Ethiopia, Gabon, Madagascar, Morocco and Tunisia, as well as those from Sudan (Johnson et al., 2004b), cluster within clade AFRICA 1a, while isolates from the Central African Republic, Kenya, Mozambique, Namibia, Tanzania, Zaire, Zambia and some from Zimbabwe and South Africa constitute AFRICA 1b (Kissi et al., 1995; Sabeta et al., 2003). While dogs are major reservoirs for these viruses, wildlife may also play some role in their maintenance.

The cosmopolitan lineage also includes strains from the American continent. Two viruses associated with herpestid hosts: the mongoose on the Caribbean island of Puerto Rico (PUERTO RICO MG) and the skunk in western Canada and the mid-western states of the USA (AMERICAN SK-MW), are closely related geneti-cally despite their geographical separation and different host species (Smith et al., 1995; Nadin-Davis et al., 1997); this cluster in particular provides strong evidence for iatrogenic translocation of this disease to newly discovered lands during colo-nization. Additional clades within the cosmopolitan lineage include a rather het-erogeneous Californian skunk group (Crawford-Miksza et al., 1999), VACCINE 1, represented by two laboratory strains PV and SADB19, and a cluster (AMERICAN CD 2/Cuban MG) comprising the laboratory CVS strain, a strain associated with the grey fox that circulates in Texas (Rohde et al., 1997) and Arizona, several viruses of Mexican origin, including strains recovered from bobcats, but which may circulate predominantly in foxes of Northern Mexico (de Mattos *et al.*, 1999; Velasco-Villa *et al.*, 2002, 2005; Nadin-Davis and Loza-Rubio, 2006) as well as the Mexican dog type (Velasco-Villa *et al.*, 2005), which gave rise to a coyote epizootic in Texas in the 1990s (Clark *et al.*, 1994), and which is also epidemiologically closely linked to mongoose rabies on the island of Cuba (Nadin-Davis *et al.*, 2006c). Another clade (AMERICAN CD 1) included within the cosmopolitan lineage comprises dog isolates representative of urban rabies from many parts of Latin America, represented here by isolates from Brazil (also see Ito *et al.*, 2001) and Paraguay, but this cluster also includes a strain that appears to be maintained in the Brazilian hoary fox, a species recently recognized as a reservoir host (Bernardi *et al.*, 2005). A single isolate of a Mexican skunk strain (MEXICAN SK 1) forms an outlier to this clade. For some countries, specimens representing two or more distinct branches within the cosmopolitan lineage can be identified (e.g. IR5IRN and IR14IRN in Iran). This suggests the independent incursion of variants of this lineage into the country on at least two occasions.

Other distinctive rabies viruses have also been documented. In western Africa, for example, a group of dog-associated viruses form a distinct clade (AFRICA 2 or AFRICA CD 2), represented here by isolates from Nigeria and Chad, but also recovered from Benin, Burkina Faso, Cameroon, Guinea, Ivory Coast, Mauritania, Niger, Senegal and Somaliland (Kissi *et al.*, 1995). Even more distinctive viruses are found in Asia as originally reported by Smith *et al.* (1992). In this tree, isolates from Sri Lanka and parts of India form a very distinct lineage, ASIA CD 1 (Nadin-Davis *et al.*, 2002) as reported elsewhere (Arai *et al.*, 2001), while viruses from the Philippines and Thailand clustered as separate branches of another Asian lineage, designated here as ASIA CD 2, although distinct designations were originally ascribed to viruses that presumably represent these lineages (Smith *et al.*, 1992). Detailed epidemiological studies of 27 viruses recovered from Thailand confirmed their monophyletic nature (Ito *et al.*, 1999) and subsequent studies indicated extensive circulation of multiple variants of this strain in both cats and dogs within the city of Bangkok (Kasempimolporn *et al.*, 2004). Detailed genetic studies of isolates from the Philippines also confirmed the distinctive nature of these viruses and identified two subtypes that appeared to be sequestered differentially within the Philippine archipelago (Nishizono *et al.*, 2002).

A second major group of rabies viruses (previously referred to as sub-genotype 1–2 due to its distinctive place within genotype 1 (Nadin-Davis *et al.*, 2002)) comprises viruses recovered exclusively from the Americas. Hence these viruses, which are found in members of both the *Carnivora* and *Chiroptera*, have been referred to as American indigenous strains in recognition of their very distinctive evolutionary origins compared with other American viruses belonging to the cosmopolitan lineage. In many parts of the Americas, viruses that have evolved from these two separate origins are sympatric. Certain viruses of this cluster are associated with terrestrial reservoir species including skunks of the south central USA and Mexico (Smith *et al.*, 1995; de Mattos *et al.*, 1999; Nadin-Davis and

Loza-Rubio, 2006) as well as raccoons. The raccoon strain, which emerged in Florida in the 1940s, exhibits rather limited sequence variation despite its now extensive range throughout the eastern seaboard of North America (Nadin-Davis *et al.*, 2006a), an observation that supports the relatively recent adaptation of its progenitor to this host. A third terrestrial strain, not shown here, is a variant, apparently adapted to marmosets of Ceará state in Brazil (Favoretto *et al.*, 2001), isolates of which formed a monophyletic branch lying in between the terrestrial- and bat-associated strains of the Americas.

The highly divergent viruses associated with many different chiropteran species are represented by several representative strains that form the remainder of the American indigenous clade of Figure 3.4; the Americas are currently considered unique by having genotype 1 viruses associated with chiropteran hosts. As described above (Section 2.2B) and illustrated in more detail in Figure 3.2, distinctive rabies virus strains are harbored by several insectivorous bats of North America and there is also evidence to support the circulation of viruses associated with *Lasiurus* and *Myotis* species in countries of South America (de Mattos *et al.*, 2000; Sheeler-Gordon and Smith, 2001; Yung *et al.*, 2002). Another chiropteran species, *Tadarida brasiliensis*, is a known rabies reservoir in countries as far apart as the USA (Smith *et al.*, 1995), Mexico (Velasco-Villa *et al.*, 2002; Nadin-Davis and Loza-Rubio, 2006) and Chile (de Mattos *et al.*, 2000), though the complete geographical range and extent of genetic variation of viruses associated with this host remain to be fully determined. Other insectivorous bats, including members of the *Histiotus* and *Molossus* genera have been found infected with distinct viral strains (see isolates V908BRZ and V990BRZ), but it remains to be confirmed whether these bats act as the reservoir host (Yung *et al.*, 2002; Páez *et al.*, 2003; Bernardi *et al.*, 2005). Another important strain, both from an economical standpoint and increasingly as a significant public health factor, is vampire bat rabies that is broadly distributed throughout Latin America and parts of the Caribbean (see de Mattos *et al.*, 1996; Ito *et al.*, 2001; Wright *et al.*, 2002; Favi *et al.*, 2003). In Figure 3.4, this rather homogeneous lineage is represented by specimens from Mexico, Paraguay, Brazil and Trinidad.

3.4 Molecular epidemiology of lyssavirus of GTs 2–7

Due to limitations in the numbers of isolates collected for members of the other lyssavirus genotypes, rather less is known about their molecular epidemiology, although a few papers report some interesting findings.

3.4.1 Genotype 2

About a dozen isolates of LBV, recovered principally from bats from countries scattered throughout the African continent, have been characterized variously at the N, P and G loci. Despite the limited sampling, significant variation within this genotype is clearly evident with the trend that the greater the physical separation between samples the greater is their genetic diversity (Bourhy *et al.*, 1993; Johnson *et al.*, 2002; Nadin-Davis *et al.*, 2002).

3.4.2 Genotype 3

Although MOKVs have been recovered from several African countries, not all isolates have been studied genetically. Characterization of the N locus of representative isolates from southern Africa identified three distinct viral groups, two from geographically distant areas of South Africa and the third represented by a single Zimbabwean isolate (Nel *et al.*, 2000). Indeed, P gene characterization of several Mokola virus isolates also identified significant genetic divergence within this group that correlated with the geographical origins of these isolates; viruses from South Africa and Zimbabwe segregated to different branches and were very distinct from a Nigerian isolate (Nadin-Davis *et al.*, 2002).

3.4.3 Genotype 4

Duvenhage virus (DUVV) isolates are extremely limited; probably no more than half a dozen specimens have been recovered from human and bat cases in South Africa and Zimbabwe. All viruses characterized at the N, P or G loci are relatively homogeneous, possibly reflecting the limited geographical range over which samples have been retrieved (Amengual *et al.*, 1997; Badrane *et al.*, 2001; Johnson *et al.*, 2002; Nadin-Davis *et al.*, 2002). The geographic range and genetic diversity of this genotype remain to be further explored.

3.4.4 Genotypes 5 and 6

Although cases of rabies in European bats were known since 1954, reports were relatively rare until 1985 when several isolations of rabies-related EBLVs were made from certain bat species in Denmark and from a bat biologist from Finland, events that triggered significantly increased surveillance of European bat species (Kappeler, 1989). Subsequent characterization of many rabies-like viruses from bats revealed the circulation of two viral biotypes, EBLV-1 and EBLV-2, that circulate, respectively, in *Eptesicus serotinus* and species of *Myotis*, particularly *M. daubentonii* and *M. dasycneme* (Bourhy *et al.*, 1992); these two biotypes were later recognized as two distinct lyssavirus genotypes (Bourhy *et al.*, 1993). From a detailed phylogenetic study of these two viral groups, Amengual *et al.* (1997) reported that EBLV-1 was widely distributed in Germany, France, Denmark, Holland and parts of eastern Europe and that isolates formed two genetically distinguishable lineages (1a and 1b) that now appear to represent two independent introductions of this biotype into Europe (Davis *et al.*, 2005). Support for such a hypothesis comes from a study of EBLV-1 isolates made in France in which spatial clustering of these two types was evident (Picard-Meyer *et al.*, 2004a). A smaller sampling of EBLV-2 isolates from Holland, Finland and Switzerland also appeared to segregate into two distinct lineages according to their geographical location (Amengual *et al.*, 1997). Moreover, the recent isolation of EBLV-2 from *M. daubentonii* bats in the UK (Johnson *et al.*, 2003b) and the indigenously acquired case of human rabies reported by Fooks *et al.* (2003), strongly indicate that this genotype circulates in indigenous bat populations of the UK.

3.4.5 *Genotype 7*

A study of the G gene of 24 ABLV isolates readily classified these viruses as a distinct monophyletic group that can readily be differentiated into two biotypes associated with *Pteropus* (flying fox) species and an insectivorous bat (*Saccolaimus flaviventris*), respectively (Guyatt *et al.*, 2003). The geographical range of the Pteropid-associated viruses appears to correlate with that of the host. Indeed, the range of ABLVs may extend outside of Australia given some serological evidence for the circulation of these, or closely related viruses, in bat populations of the Philippines (Arguin *et al.*, 2002) and Cambodia (Reynes *et al.*, 2004).

4 ASPECTS OF RABIES PATHOGENESIS AND EVOLUTION REVEALED BY MOLECULAR EPIDEMIOLOGY

4.1 Long incubation period

It has long been known that the incubation period between infection with a lyssavirus and development of clinical signs by the host is variable and can extend to several months (Charlton, 1994). However, investigation of certain human cases has greatly extended the length of incubation periods now known to be possible. Through molecular epidemiological studies on the source of infection for three immigrants to the USA, Smith *et al.* (1991) demonstrated that, in each case, the infection had originated in the country of origin. Based on the travel histories of the three individuals, it was concluded that the clinical manifestations of rabies occurred after periods of 11 months, 4 years and 6 years in these patients. Similarly, McColl *et al.* (1993) determined that a case of rabies diagnosed in a young Vietnamese girl in Australia was the result of infection obtained in her home country at least 5 years prior to the emergence of symptoms. The longevity of these incubation periods could impact on decisions regarding the need for post-exposure treatment for individuals in situations where significant delays in treatment after the exposure have occurred.

4.2 Role of bat reservoirs in human rabies

Another aspect of human rabies is the significant role played by specific strains of bat rabies in indigenously acquired cases in North America. In a review of 57 human rabies cases reported in the USA from 1958 to 2000, Messenger *et al.* (2002) noted that a high proportion of indigenously acquired cases, especially cryptic cases in which no bite had been reported, was the consequence of exposure to the strain of rabies associated with the silver-haired/eastern pipistrelle bat reservoir. This is notable given the relatively few cases of rabies that have been reported over this time period in these two bat species. Moreover, in Canada, where human cases are very rarely reported, the same strain was responsible for the death of a young boy in Quebec in 2000 (Elmgren *et al.*, 2002). To explain the USA findings, Messenger *et al.* (2002) considered the role that might be played by the seasonal distribution of these bats, especially young bats that constitute a high percentage of positive cases. Alternatively, this strain of rabies virus has phenotypic characteristics that may allow it to propagate successfully even after superficial contact. Such information provides a basis for formulating recommendations for appropriate post-exposure prophylaxis in situations where humans have known or potential contact with bats.

Another notable case was that of an indigenously acquired case of human rabies in the UK due to exposure to EBLV-2 (Fooks *et al.*, 2003), a situation which has identified the need for bat handlers in that country, and indeed throughout Europe, to receive rabies pre- and or post-exposure treatments (Office Internationale des Epizooties, 2005).

4.3 Vaccine coverage

The consequence of the genetic and immunological distinctiveness between the two lyssavirus phylogroups is that traditional rabies vaccines, that were developed based on classical rabies virus isolates of GT 1, as well as biologicals used to provide passive immunity, are ineffective in protecting against infection with the more divergent lyssaviruses LBV and MOKV (Badrane *et al.*, 2001; Hanlon *et al.*, 2001). Moreover, in animal models in which many of the current anti-rabies vaccines and biologicals were tested, reduced protection against the newly described Eurasian bat lyssaviruses was observed for both pre- and post-exposure treatment regimens (Hanlon *et al.*, 2005). Since the emergence of such divergent lyssaviruses is a real possibility, strategies to combat this deficiency in vaccine coverage are being explored (Nel, 2005). To date, efforts have focused on the development of novel genetic vaccines in which the glycoprotein genes of GT 1 and 3 viruses have been combined so as to encode hybrid proteins that can elicit the production of antibodies capable of neutralizing antigenic sites of the viruses representative of both phylogroups (Jallet *et al.*, 1999). In addition, to increase the spectrum of passive protection provided by rabies immune globulin, as well as to increase global supplies, alternate sources of antibodies, e.g. humanized forms of broadly reactive MAbs, are being investigated (Hanlon *et al.*, 2001).

4.4 Molecular clocks and spatial dynamics

While phylogenetic trees suggest the evolutionary relationships between isolates and strains, most studies are not usually able to deduce the timeframe over which these evolutionary processes have occurred. Compared to the timeframe of the outbreaks under investigation, most virus collections are represented by a very limited time period. Consequently, trying to assign a time scale to a certain level of mutation from such a collection and then extrapolating back over decades or centuries to the origins of a strain using a linear regression analysis can result in significant errors and wide variations in estimates. However, the ability to assign timeframes to trees would be valuable, not just for improved understanding of the historical spread of lyssaviruses, but to provide insight into the evolutionary origins of this genus and its original host species. In an early study that explored the timeframe of lyssavirus evolution using a collection of G gene sequences, Badrane and Tordo (2001) reported similar rates of evolution were apparent for all lyssavirus lineages and thus employed distance data with phylogenetic tree reconstruction to estimate timeframes for pivotal stages of lyssavirus evolution. They inferred that chiropteran hosts were the initial lyssavirus reservoirs and that carnivoran rabies emerged from chiropteran lyssaviruses between 888 and 1459 years ago, while the emergence of the cosmopolitan lineage was dated to between 284 and 504 years ago.

Recent developments in coalescent theory that provide advantages over linear regression methods have been described that may be advantageous for estimation of nucleotide substitution rates using dated isolates. Using both ML and Markov chain Monte Carlo (MCMC) methods, rates of molecular evolution can be estimated from molecular sequence data and thereby employed to estimate times of the most recent common progenitor of a phylogeny; moreover, models of nucleotide substitution can be tested in this process. For a thorough review of the role of molecular clocks in RNA virus studies, see Holmes (2003).

Application of such methodologies to the phylogenetic reconstruction of canine rabies in Colombia suggested that an older lineage, GV2, that had circulated in the country since the 1960s gave rise to a second lineage, GV1, sometime between 1983 and 1988 (Hughes *et al.*, 2004). Although these two lineages circulated in the country over most of the study period, population size estimates of the two lineages in more recent years indicated that GV2 was by far the more predominant type, possibly having replaced GV1 entirely.

Another study, employing similar methods to investigate the evolution of American bat rabies viruses, proposed that the most recent common ancestor for all these viruses was dated to the mid 1600s (Hughes *et al.*, 2005). Moreover, those variants presently associated with *Desmodus* and *Tadarida* species in Latin America were the earliest to evolve with subsequent emergence of the viruses associated with other insectivorous bat species of North America. If these estimates are indeed correct, then this could bring into question the hypothesis that all terrestrial rabies viruses emerged from chiropteran strains (Badrane and Tordo, 2001). Furthermore, since the vampire bat strain is not, based on current analyses, particularly genetically divergent compared with some of the other bat strains studied (see Figures 3.2 and 3.4), it would clearly indicate that evolutionary rates are highly host dependent and that levels of genetic variation observed within strains are not good indicators of the evolutionary time period of association with a particular host. Such a feature of rabies viruses was not previously recognized.

Similar conclusions were drawn from a study of the molecular evolution of the European bat lyssaviruses in which current genetic diversity of the EBLV-1 group was proposed to have emerged between 500 and 750 years ago (Davis *et al.*, 2005). This group of viruses exhibited an especially low rate of nucleotide substitution compared to many other RNA viruses. This finding prompted speculation that adaptation of the virus to this particular bat reservoir, and the emergence of a distinctive epidemiology, in which many apparently healthy bats exhibit evidence of long-term EBLV-1 infection, has resulted in a very high level of constraint that severely limits non-synonymous changes due to purifying selection. Moreover, differences in the evolution and dispersal patterns of subgroups 1a and 1b were evident (Davis *et al.*, 2005).

More extensive application of the ML and MCMC methods employed in the papers described above are likely to shed new light on many aspects of lyssavirus evolution.

Recent trends combining the statistical analysis of molecular data together with the geographical distribution of rabies virus variants may also provide useful insights into the factors that influence viral spread and evolution. It had been reported some years ago that distinct variants of the AFX strain were segregated to discrete regions within the outbreak area of Ontario (Nadin-Davis *et al.*, 1993, 1994, 1999). It had been proposed that the geographical clustering of sequence types might be due to ecological substructuring of the virus population into ecotypes associated with distinct habitats. However, a re-evaluation of these sequences suggested that an isolation by distance population structure likely accounts for at least 90% of the observed genetic variation in this viral population; in other words, the spatial organization of sequence variants probably arose due to neutral evolution linked to limited local dispersal (Real *et al.*, 2005). The data on sequence variation and geographical distance supported the existence of two groups of viruses corresponding to two distinct arms of the outbreak that moved from northern Ontario into southern and eastern regions of the province in the late 1950s to the early 1960s, in agreement with case surveillance reports of the period. Through such studies, a better understanding of how viral evolutionary change affects the spatial spread and ecological dynamics of rabies may be achieved. Such knowledge would provide a framework for greatly improved understanding of disease emergence and may ultimately suggest improved intervention strategies.

5 CONCLUSIONS AND FUTURE TRENDS

Knowledge of the antigenic and genetic diversity of rabies viruses has increased enormously over the last 20 years and it may be anticipated that this knowledge will continue to accumulate in the near future. In particular, the combined application of methodical surveillance practices and current tools of molecular phylogenetics in countries of the developing world, particularly in Africa and Asia, is likely to uncover an increasingly complex structure for the *Lyssavirus* genus. Such information may increasingly challenge our notions of the classification of these viruses based on current genotypic designations.

The information garnered from such investigations should provide information helpful in the control of this disease by clearly identifying the species involved in maintaining particular rabies virus variants and thereby allowing the most cost-effective strategies to be applied in rabies control. While dogs are by far the most problematic reservoir in terms of human exposure in developing countries, one might anticipate that as dog rabies comes under increasing control through improved dog vaccination and control, the role of wildlife, particularly chiropteran species, is deserving of further scrutiny in light of the extensive role played by wildlife in maintaining rabies, especially on the American continent.

Studies to explore the reservoir host's population structure and its role in influencing the spread of specific rabies virus strains are becoming increasingly popular. In particular, this is being investigated for the raccoon rabies virus strain that has successfully spread throughout the eastern seaboard of North America in a relatively short period and which now threatens to move westwards across the USA. Attempts to correlate virus phylogeny with host population substructure, analysed using mitochondrial DNA markers, are in progress and promise to provide interesting and pertinent information about rabies spread (Cullingham *et al.*, 2005; Reeder *et al.*, 2005).

It is becoming clear that adaptation of rabies viruses to new hosts is a phenomenon that is far more frequent than originally supposed as evidenced by the emergence of sylvatic rabies from dog reservoirs (Bourhy *et al.*, 1999; Johnson *et al.*, 2003a) and the emergence of rabies in terrestrial hosts after spillover from chiropteran reservoirs (Badrane and Tordo, 2001; Smith, 2001). It thus follows that even if all present terrestrial reservoirs of this disease were eliminated, re-emergence of rabies in terrestrial species may occur if the disease persists in chiropteran hosts. Indeed, even the EBLV-1 virus, which is apparently well adapted to its chiropteran host (Davis *et al.*, 2005), can on occasion infect terrestrial species (see Vos *et al.*,

2004) and, in the long term, a successful adaptation to a new host is possible. Development of vaccination strategies to eliminate lyssaviruses in mammals with an aerial lifestyle will thus be the only means of permanently eradicating rabies.

To assist in such long-term goals, better understanding of the evolutionary forces acting on these viruses during the adaptive process to a new host is greatly needed. Since we currently have no indications of what features of the virus facilitate its adaptation to a specific host, understanding of this adaptive process will require that complete viral genomes are characterized, a technical feat that is becoming more realistic as improved, rapid sequencing technologies evolve. In situations where such host adaptation can be followed almost as it occurs, as in the situation in Flagstaff, Arizona, where a bat-associated virus adapted to transmission in skunks (Smith, 2001), complete genomic analysis will be invaluable for gaining an appreciation of the evolutionary forces operating during this process. Incorporation of information on the rate and nature of genetic change that occurs during host adaptation into spatial models may allow prediction of how new rabies epizootics will develop; the models currently in use assume an equilibrium between host and virus. Improved predictive capabilities will ultimately facilitate cost-effective control strategies.

Molecular epidemiological investigations may also help re-shape our understanding of how rabies is maintained in specific hosts, or indeed whether in some cases secondary hosts may have some impact on viral persistence within a region. A role for skunks in maintenance of the AFX strain of rabies virus has been proposed (Nadin-Davis *et al.*, 2006b) while the exclusive role of *E. serotinus* bats in maintaining and dispersing EBLV-1 viruses has been questioned due to the presence of this viral genotype in a number of other insectivorous bat species in Europe, including certain migratory species of *Myotis* (Serra-Cobo *et al.*, 2002).

Finally, more fundamental aspects of lyssavirus evolution are now being studied with the aim of eventually defining how the original lyssavirus progenitor evolved and in which host species this occurred. As developments in the analytical tools applicable to molecular sequence data continue to unfold, so will our understanding of this important class of viruses.

ACKNOWLEDGMENTS

I am most grateful to Louis Nel (University of Pretoria, South Africa) and Lorraine McElhinney (Veterinary Laboratories Agency, UK) for their reading of the first draft of this work and for their thoughtful comments and suggestions that have been incorporated in this revised version.

REFERENCES

Amengual, B., Whitby, J.E., King, A., Cobo, J.S. and Bourhy, H. (1997). Evolution of European bat lyssaviruses. *Journal of General Virology* **78**, 2319–2328.

Arai, Y.T., Takahashi, H., Kameoka, Y., Shiino, T., Wimalaratne, O. and Lodmell, D.L. (2001). Characterization of Sri Lanka rabies virus isolates using nucleotide sequence analysis of nucleoprotein gene. *Acta Virologica* **45**, 321–333.

Arai, Y.T., Kuzmin, I.V., Kameoka, Y. and Botvinkin, A.D. (2003). New lyssavirus genotype from the Lesser mouse-eared bat (*Myotis blythi*), Kyrghyzstan. *Emerging Infectious Diseases* **9**, 333–337.

Arguin, P.M., Murray-Lillibridge, K., Miranda, M.E.G., Smith, J.S., Calaor, A.B. and Rupprecht, C.E. (2002). Serologic evidence of lyssavirus infections among bats, the Philippines. *Emerging Infectious Diseases* **8**, 258–262.

Badrane, H. and Tordo, N. (2001). Host switching in *Lyssavirus* history from the Chiroptera to the Carnivora orders. *Journal of Virology* **75**, 8096–8104.

Badrane, H., Bahloul, C., Perrin, P. and Tordo, N. (2001). Evidence of two *Lyssavirus* phylogroups with distinct pathogenicity and immunogenicity. *Journal of Virology* **75**, 3268–3276.

Bauldauf, S.L. (2003). Phylogeny for the faint of heart: a tutorial. *Trends in Genetics* **19**, 345–351.

Benmansour, A., Brahimi, M., Tuffereau, C., Coulon, P., Lafay, F. and Flamand, A. (1992). Rapid sequence evolution of street rabies glycoprotein is related to the highly heterogeneous nature of the viral population. *Virology* **187**, 33–45.

Bernardi, F., Nadin-Davis, S.A., Wandeler, A.I. *et al.* (2005). Antigenic and genetic characterization of rabies viruses isolated from domestic and wild animals of Brazil identifies the hoary fox as a rabies reservoir. *Journal of General Virology* **86**, 3153–3162.

Black, E.M., McElhinney, L.M., Lowings, J.P., Smith, J., Johnstone, P. and Heaton, P.R. (2000). Molecular methods to distinguish between classical rabies and the rabies-related European bat lyssaviruses. *Journal of Virological Methods.* **87**, 123–131.

Black, E.M., Lowings, J.P., Smith, J., Heaton, P.R. and McElhinney, L.M. (2002). A rapid RT-PCR method to differentiate six established genotypes of rabies and rabies related viruses using TaqMan™ technology. *Journal of Virological Methods* **105**, 25–35.

Blancou, J. (1988). Ecology and epidemiology of fox rabies. *Review of Infectious Diseases* **10**, S606–S609.

Bourhy, H., Kissi, B., Lafon, M., Sacramento, D. and Tordo, N. (1992). Antigenic and molecular characterization of bat rabies virus in Europe. *Journal of Clinical Microbiology* **30**, 2419–2426.

Bourhy, H., Kissi, B. and Tordo, N. (1993). Molecular diversity of the *Lyssavirus* genus. *Virology* **194**, 70–81.

Bourhy, H., Kissi, B., Audry, L. *et al.* (1999). Ecology and evolution of rabies in Europe. *Journal of General Virology* **80**, 2545–2557.

Chare, E.R., Gould, E.A. and Holmes, E.C. (2003). Phylogenetic analysis reveals a low rate of homologous recombination in negative-sense RNA viruses. *Journal of General Virology* **84**, 2691–2703.

Charlton, K.M. (1994). The pathogenesis of rabies and other lyssaviral infections: recent studies. In: *Lyssaviruses* (C.E. Rupprecht, B. Dietzschold and H. Koprowski, eds). pp. 95–120. Berlin: Springer-Verlag KG.

Cisterna, D., Bonaventura, R., Caillou, S. *et al.* (2005). Antigenic and molecular characterization of rabies virus in Argentina. *Virus Research* **109**, 139–147.

Clark, K.A., Neill, S.U., Smith, J.S., Wilson, P.J., Whadford V.W. and Mckirahan, G.W. (1994). Epizootic canine rabies transmitted by coyotes in south Texas. *Journal of the American Veterinary Medical Association* **204**, 536–540.

Conzelmann, K.-K., Cox, J.H., Schneider, L.G. and Thiel, H.-J. (1990). Molecular cloning and complete nucleotide sequence of the attenuated rabies virus SAD B19. *Virology* **175**, 485–499.

Crawford-Miksza, L.K., Wadford, D.A. and Schnurr, D.P. (1999). Molecular epidemiology of enzootic rabies in California. *Journal of Clinical Microbiology* **14**, 207–219.

Cullingham, C.I., Kyle, C.J. and White, B.N. (2005). An investigation of raccoon subspecies designations and their association with differences in population response to the rabies virus. *Presented at the XVI International Rabies in the Americas conference*, Ottawa, Canada, October 16–21, 2005.

David, D., Yakobson, B., Smith, J.S. and Stram, Y. (2000). Molecular epidemiology of rabies virus isolates from Israel and other middle- and near-Eastern Countries. *Journal of Clinical Microbiology* **38**, 755–762.

Davis, P.L., Holmes, E.C., Larrous, F. *et al.* (2005). Phylogeography, population dynamics, and molecular evolution of European bat lyssaviruses. *Journal of Virology* **79**, 10 487–10 497.

de Mattos, C.A., de Mattos, C.C., Smith, J.S. *et al.* (1996). Genetic characterization of rabies field isolates from Venezuela. *Journal of Clinical Microbiology* **34**, 1553–1558.

de Mattos, C.C., de Mattos, C.A., Loza-Rubio, E., Aguilar-Setién, A., Orciari, L.A. and Smith, J.S. (1999). Molecular characterization of rabies virus isolates from Mexico: Implications for transmission dynamics and human risk. *American Journal of Tropical Medicine and Hygiene* **61**, 587–597.

de Mattos, C.A., Favi, M. Yung, V., Pavletic, C. and de Mattos, C.C. (2000). Bat rabies in urban centers in Chile. *Journal of Wildlife Diseases* **36**, 231–240.

Delpietro, H.A., Gury-Dhomen, F., Larghi, O.P., Mena-Segura, C. and Abramo, L. (1997). Monoclonal antibody characterization of rabies virus strains isolated in the river plate basin. *Journal of Veterinary Medicine* **44**, 477–483.

Diaz, A.-M., Papo, S., Rodriguez, A. and Smith, J.S. (1994). Antigenic analysis of rabies-virus isolates from Latin America and the Caribbean. *Journal of Veterinary Medicine* **41**, 153–160.

Domingo, E., Martinez-Salas, E., Sobrino, F. *et al.* (1985). The quasispecies (extremely heterogeneous) nature of viral RNA genome populations: biological relevance – a review. *Gene* **40**, 1–8.

Elmgren, L.D., Nadin-Davis, S.A., Muldoon, F.T. and Wandeler, A.I. (2002). Diagnosis and analysis of a recent case of human rabies in Canada. *Canadian Journal of Infectious Diseases* **13**, 129–133.

Favi, M., Nina, A., Yung, V. and Fernández, J. (2003). Characterization of rabies virus isolates in Bolivia. *Virus Research* **97**, 135–140.

Favoretto, S.R., de Mattos, C.C., Morais, N.B., Alves Araújo, F.A. and de Mattos, C.A. (2001) Rabies in marmosets (*Callithrix jacchus*), Ceará, Brazil. *Emerging Infectious Diseases* **7**, 1062–1065.

Fooks, A.R., McElhinney, L.M., Pounder, D.J. *et al.* (2003). Case report: isolation of a European bat lyssavirus type 2a from a fatal human case of rabies encephalitis. *Journal of Medical Virology* **71**, 281–289.

Franka, R., Constantine, D.G., Kuzmin, I., Velasco Villa, A. and Rupprecht, C.E. (2004). Molecular analysis of rabies virus variants associated with bats in California. Presented at the XV International Meeting on Rabies in the Americas, October 31–November 4, 2004, Santo Domingo, Dominican Republic.

Gould, A.R., Hyatt, A.D., Lunt, R., Kattenbelt, J.A., Hengstberger, S. and Blacksell, S.D. (1998). Characterisation of a novel lyssavirus isolated from *Pteropid* bats in Australia. *Virus Research* **54**, 165–187.

Gould, A.R., Kattenbelt, J.A., Gumley, S.G. and Lunt, R.A. (2002). Characterisation of an Australian bat lyssavirus variant isolated from an insectivorous bat. *Virus Research* **89**, 1–28.

Graur, D. and Li, W.-H. (2000). Molecular phylogenetics. In: *Fundamentals of Molecular Evolution*. pp. 165–247. Sunderland: Sinauer Associates, Inc.

Guyatt, K.J., Twin, J., Davis, P. *et al.* (2003). A molecular epidemiological study of Australian bat lyssavirus. *Journal of General Virology* **84**, 485–496.

Hanlon, C.A., DeMattos C.A., DeMattos, C.C. *et al.* (2001). Experimental utility of rabies virus-neutralizing human monoclonal antibodies in post-exposure prophylaxis. *Vaccine* **19**, 3834–3842.

Hanlon, C.A., Kuzmin, I.V., Blanton, J.D., Weldon, W.C., Manangan, J.S. and Rupprecht, C.E. (2005). Efficacy of rabies biologics against new lyssaviruses from Eurasia. *Virus Research* **111**, 44–54.

Holmes, E.C. (2003). Molecular clocks and the puzzle of RNA virus origins. *Journal of Virology* **77**, 3893–3897.

Hughes, G.J., Páez, A., Bóshell, J. and Rupprecht, C.E. (2004). A phylogenetic reconstruction of the epidemiological history of canine rabies virus variants in Colombia. *Infections, Genetics and Evolution* **4**, 45–51.

Hughes, G.J., Orciari, L.A. and Rupprecht, C.E. (2005). Evolutionary timescale of rabies virus adaptation to North American bats inferred from the substitution rate of the nucleoprotein gene. *Journal of General Virology* **86**, 1467–1474.

Ito, N., Sugiyama, M., Oraveerakul, K. *et al.* (1999). Molecular epidemiology of rabies in Thailand. *Microbiology and Immunology* **43**, 551–559.

Ito, M., Arai, Y.T., Itou, T. *et al.* (2001). Genetic characterization and geographic distribution of rabies virus isolates in Brazil: identification of two reservoirs, dog and vampire bats. *Virology* **284**, 214–222.

Jallet, C., Jacob, Y., Bahloul, C. *et al.* (1999). Chimeric lyssavirus glycoproteins with increased immunological potential. *Journal of Virology* **73**, 225–233.

Johnson, N., McElhinney, L.M., Smith, J., Lowings, P. and Fooks, A.R. (2002). Phylogenetic comparison of the genus *Lyssavirus* using distal coding sequences of the glycoprotein and nucleoprotein genes. *Archives of Virology* **147**, 2111–2123.

Johnson, N., Black, C., Smith, J. *et al.* (2003a). Rabies emergence among foxes in Turkey. *Journal of Wildlife Diseases* **39**, 262–270.

Johnson, N., Selden, D., Parsons, G. *et al.* (2003b). Isolation of a European bat lyssavirus type 2 from a Daubenton's bat in the United Kingdom. *Veterinary Record* **152**, 383–387.

Johnson, N., Letshwenyo, M., Baipoledi, E.K., Thobokwe, G. and Fooks, A.R. (2004a). Molecular epidemiology of rabies in Botswana: a comparison between antibody typing and nucleotide sequence phylogeny. *Veterinary Microbiology* **101**, 31–38.

Johnson, N., McElhinney, L.M., Ali, Y.H., Saeed, I.K. and Fooks, A.R. (2004b). Molecular epidemiology of canid rabies in Sudan: evidence for a common origin of rabies with Ethiopia. *Virus Research* **104**, 201–205.

Kappeler, A. (1989). Bat rabies surveillance in Europe. *Rabies Bulletin of Europe* **13**, 12–13.

Kasempimolporn, S., Saengseesom, W., Tirawatnapong, T., Puempumpanich, S. and Sitprija, V. (2004). Genetic typing of feline rabies virus isolated in greater Bangkok, Thailand. *Microbiology and Immunology* **48**, 307–311.

Kissi, B., Tordo, N. and Bourhy, H. (1995). Genetic polymorphism in the rabies virus nucleoprotein gene. *Virology* **209**, 526–537.

Kulonen, K. and Boldina, I. (1993). Differentiation of two rabies strains in Estonia with reference to recent Finnish isolates. *Journal of Wildlife Diseases* **29**, 209–213.

Kuzmin, I.V., Orciari, L.A., Arai, Y.T. *et al.* (2003). Bat lyssaviruses (Aravan and Khujand) from Central Asia: phylogenetic relationships according to N, P and G gene sequences. *Virus Research* **97**, 65–79.

Kuzmin, I.V., Botvinkin, A.D., McElhinney, L.M. *et al.* (2004). Molecular epidemiology of terrestrial rabies in the former Soviet Union. *Journal of Wildlife Diseases* **40**, 617–631.

Kuzmin, I.V., Hughes, G.J., Botvinkin, A.D., Orciari, L.A., and Rupprecht, C.E. (2005). Phylogenetic relationships of Irkut and West Caucasian bat viruses within the Lyssavirus genus and suggested quantitative criteria based on the N gene sequence for lyssavirus genotype definition. *Virus Research* **111**, 28–43.

Le Mercier, P., Jacob, Y. and Tordo, N. (1997). The complete Mokola virus genome sequence: structure of the RNA-dependent RNA polymerase. *Journal of General Virology* **78**, 1571–1576.

Loza-Rubio, E., Aguilar-Setien, A., Bahloul, C., Brochier, B., Pastoret, P.-P. and Tordo, N. (1999). Discrimination between epidemiological cycles of rabies in Mexico. *Archives of Medical Research* **30**, 144–149.

Loza-Rubio, E., Rojas-Anaya, E., Banda-Ruiz, V.M., Nadin-Davis, S.A. and Cortez-Garcia, B. (2005). Detection of multiple strains of rabies virus RNA using primers designed to target Mexican vampire bat variants. *Epidemiology and Infection* **133**, 927–934.

Lumlertdacha, B., Boongird, K., Wanghongsa, S. *et al.* (2005). Survey for bat lyssaviruses, Thailand. *Emerging Infectious Diseases* **11**, 232–236.

Mackenzie, J.S. (1999). Emerging viral diseases: an Australian perspective. *Emerging and Infectious Diseases* **5**, 1–8.

Mansfield, K.L., Racloz, V., McElhinney, L.M. *et al.* (2006). Molecular epidemiological study of Arctic rabies virus isolates from Greenland and comparison with isolates from throughout the Arctic and Baltic regions. *Virus Research* **116**, 1–10.

McColl, K.A., Gould, A.R., Selleck, P.W., Hooper, P.T., Westbury, H.A. and Smith, J.S. (1993). Polymerase chain reaction and other laboratory techniques in the diagnosis of long incubation rabies in Australia. *Australian Veterinary Journal* **70**, 84–89.

Mebatsion, T., Cox, J.H. and Frost, J.W. (1992). Isolation and characterization of 115 street rabies virus isolates from Ethiopia by using monoclonal antibodies: Identification of 2 isolates as Mokola and Lagos bat viruses. *Journal of Infectious Diseases* **166**, 972–977.

Messenger, S.L., Smith, J.S. and Rupprecht, C.E. (2002). Emerging epidemiology of bat-associated cryptic cases of rabies in humans in the United States. *Emerging Infectious Diseases* **35**, 738–747.

Moya, A., Holmes, E.C. and González-Candelas, F. (2004). The population genetics and evolutionary epidemiology of RNA viruses. *Nature Reviews* **2**, 1–10.

Nadin-Davis, S.A. and Bingham, J. (2004). Europe as a source of rabies for the rest of the world. In: *Historical perspective of rabies in Europe and the Mediterranean basin* (A.A. King, A.R. Fooks, M. Aubert and A.I. Wandeler, eds). pp. 259–280. Paris: OIE.

Nadin-Davis, S.A. and Loza-Rubio, E. (2006). The molecular epidemiology of rabies associated with chiropteran hosts in Mexico. *Virus Research* **117**, 215–226.

Nadin-Davis, S.A., Casey, G.A. and Wandeler, A. (1993). Identification of regional variants of the rabies virus within the Canadian province of Ontario. *Journal of General Virology* **74**, 829–837.

Nadin-Davis, S.A., Casey, G.A. and Wandeler, A.I. (1994). A molecular epidemiological study of rabies virus in central Ontario and western Quebec. *Journal of General Virology* **75**, 2575–2583.

Nadin-Davis, S.A., Huang, W. and Wandeler, A.I. (1996). The design of strain-specific polymerase chain reactions for discrimination of the raccoon rabies virus strain from indigenous rabies viruses of Ontario. *Journal of Virological Methods* **57**, 141–156.

Nadin-Davis, S.A., Huang, W. and Wandeler, A.I. (1997). Polymorphism of rabies viruses within the phosphoprotein and matrix protein genes. *Archives of Virology* **142**, 979–992.

Nadin-Davis, S.A., Sampath, M.I., Casey, G.A., Tinline, R.R. and Wandeler A.I. (1999). Phylogeographic patterns exhibited by Ontario rabies virus variants. *Epidemiology and Infection* **123**, 325–336.

Nadin-Davis, S.A., Sheen, M., Abdel-Malik, M., Elmgren, L., Armstrong, J. and Wandeler A.I. (2000). A panel of monoclonal antibodies targeting the rabies virus phosphoprotein identifies a highly variable epitope of value for sensitive strain discrimination. *Journal of Clinical Microbiology* **38**, 1397–1403.

Nadin-Davis, S.A., Huang, W., Armstrong, J. *et al.* (2001). Antigenic and genetic divergence of rabies viruses from bat species indigenous to Canada. *Virus Research* **74**, 139–156.

Nadin-Davis, S.A., Abdel-Malik, M., Armstrong, J. and Wandeler, A.I. (2002). Lyssavirus P gene characterisation provides insights into the phylogeny of the gene and identifies structural similarities and diversity within the encoded phosphoprotein. *Virology* **298**, 286–305.

Nadin-Davis, S.A., Sheen, M. and Wandeler, A.I. (2003a). Use of discriminatory probes for strain typing of formalin-fixed, rabies virus-infected tissues by *in situ* hybridization. *Journal of Clinical Microbiology* **41**, 4343–4352.

Nadin-Davis, S.A., Simani, S., Armstrong, J., Fayaz, A. and Wandeler, A.I. (2003b). Molecular and antigenic characterization of rabies viruses from Iran identifies variants with distinct epidemiological origins. *Epidemiology and Infection* **131**, 777–790.

Nadin-Davis, S.A., Muldoon, F. and Wandeler, A.I. (2006a). A molecular epidemiological analysis of the incursion of the raccoon strain of rabies virus into Canada. *Epidemiology and Infection* **134**, 534–547.

Nadin-Davis, S.A., Muldoon, F. and Wandeler, A.I. (2006b). Persistence of genetic variants of the arctic fox strain of Rabies virus in Southern Ontario. *Canadian Journal of Veterinary Research* **70**, 11–19.

Nadin-Davis, S.A., Torres, G., de los Angeles Ribas, M. *et al.* (2006c). A molecular epidemiological study of Rabies in Cuba. *Epidemiology and Infection* In press.[**3.6]

Nel, L.H. (2005). Vaccines for lyssaviruses other than rabies. *Expert Reviews in Vaccines* **4**, 533–540.

Nel, L.H., Bingham, J., Jacobs, J.A. and Jaftha, J.B. (1998). A nucleotide-specific polymerase chain reaction assay to differentiate rabies virus biotypes in South Africa. *Onderstepoort Journal of Veterinary Research* **65**, 297–303.

Nel, L., Jacobs, J., Jaftha, J., von Teichman, B. and Bingham, J. (2000). New cases of Mokola virus infection in South Africa: a genotypic comparison of southern African virus isolates. *Virus Genes* **20**, 103–106.

Nel, L.H., Sabeta, C.T., von Teichman, B., Jaftha, J.B., Rupprecht, C.E. and Bingham, J. (2005). Mongoose rabies in southern Africa: a re-evaluation based on molecular epidemiology. *Virus Research* **109**, 165–173.

Niezgoda, M., Hanlon, C.A. and Rupprecht, C.E. (2002). Animal rabies. In: *Rabies* (A.C. Jackson and W.H. Wunner, eds). pp. 163–218. San Diego: Academic Press.

Nishizono, A., Mannen, K., Elio-Villa, L.P. *et al.* (2002). Genetic analysis of rabies virus isolates in the Philippines. *Microbiology and Immunology* **46**, 413–417.

Office Internationale des Epizooties (OIE) (2005). http://www.oie.int/downld/Rabies_Conclusions_Recom_Kiev.pdf

Páez, A., Nůñez, C., Garcia, C. and Bóshell, J. (2003). Molecular epidemiology of rabies epizootics in Colombia: evidence for human and dog rabies associated with bats. *Journal of General Virology* **84**, 795–802.

Page, R.D.M. (1996). TREEVIEW: An application to display phylogenetic trees on personal computers. *Computational and Applied Biosciences* **12**, 357–358.

Park, Y.J., Shin, M.K. and Kwon, H.M. (2005). Genetic characterization of rabies virus isolates in Korea. *Virus Genes* **30**, 341–347.

Picard-Meyer, E., Barrat, J., Tissot, E., Barrat, M.J., Bruyεre, V. and Cliquet, F. (2004a). Genetic analysis of European bat lyssavirus type 1 isolates from France. *Veterinary Record* **154**, 589–595.

Picard-Meyer, E., Bruyεre, V., Barrat, J., Tissot, E., Barrat, M.J. and Cliquet, F. (2004b). Development of a hemi-nested RT-PCR method for the specific determination of European bat Lyssavirus 1. Comparison with other rabies diagnostic methods. *Vaccine* **22**, 1921–1929.

Ravkov, E.V., Smith, J.S. and Nichol, S.T. (1995). Rabies virus glycoprotein contains a long 3' noncoding region which lacks pseudogene properties. *Virology* **206**, 718–723.

Real, L.A., Henderson, J.C., Biek, R. *et al.* (2005). Unifying the spatial population dynamics and molecular evolution of epidemic rabies virus. *Proceedings of the National Academy of Sciences USA* **102**, 12 107–12 111.

Reeder, S.A., Hanlon, C.A., Rupprecht, C.E. and Real, L.A. (2005). Phylogeography and population genetics of raccoons (*Procyon lotor*) in the Southeastern United States. Presented at the XVI International Rabies in the Americas conference, Ottawa, Canada, October 16–21, 2005.

Reynes, J.-M., Molia, S., Audry, L. *et al.* (2004). Serologic evidence of lyssavirus infection in bats, Cambodia. *Emerging Infectious Diseases* **10**, 2231–2234.

Rohde, R.E., Neill, S.U., Clark, K.A. and Smith, J.S. (1997). Molecular epidemiology of rabies epizootics in Texas. *Clinical and Diagnostic Virology* **8**, 209–217.

Rupprecht, C.E., Dietzschold, B., Wunner, W.H. and Koprowski, H. (1991). Antigenic relationships of lyssaviruses. In: *The Natural History of Rabies* (G.M. Baer, ed.). pp. 69–100. Boca Raton: CRC Press.

Sabeta, C.T., Bingham, J. and Nel, L.H. (2003) Molecular epidemiology of canid rabies in Zimbabwe and South Africa. *Virus Research* **91**, 203–211.

Sacramento, D., Badrane, H., Bourhy, H. and Tordo, N. (1992). Molecular epidemiology of rabies virus in France: comparison with vaccine strains. *Journal of General Virology* **73**, 1149–1158.

Sato, G., Tanabe, H., Shoji, Y. *et al.* (2005). Rapid discrimination of rabies viruses isolated from various host species in Brazil by multiplex reverse transcription-polymerase chain reaction. *Journal of General Virology* **33**, 267–273.

Schaefer, R., Batista, H.B.R., Franco, A.C., Rijsewijk, F.A.M. and Roehe, P.M. (2005). Studies on antigenic and genomic properties of Brazilian rabies virus isolates. *Veterinary Microbiology* **107**, 161–170.

Schneider, L.G., Dietzschold, B., Dierks, R.E., Matthaaeus, W., Enzmann, P.J. and Strohmaier, K. (1973). Rabies group-specific ribonucleoprotein antigen and a test system for grouping and typing of rhabdoviruses. *Journal of Virology* **11**, 748–755.

Serra-Cobo, J., Amengual, B., Abellán, C. and Bourhy, H. (2002). European bat *Lyssavirus* infection in Spanish bat populations. *Emerging Infectious Diseases* **8**, 413–420.

Sheeler-Gordon, L.L. and Smith, J.S. (2001). Survey of bat populations from Mexico and Paraguay for rabies. *Journal of Wildlife Diseases* **37**, 582–593.

Shoji, Y., Kobayashi, Y., Sato, G. *et al.* (2004). Genetic characterization of rabies viruses isolated from frugivorous bat (Artibeus spp.) in Brazil. *Journal of Veterinary and Medical Science* **66**, 1271–1273.

Shope, R.E., Murphy, F.A., Harrison, A.K. *et al.* (1970). Two African viruses serologically and morphologically related to rabies virus. *Journal of Virology* **6**, 690–692.

Smith, J.S. (2001). Molecular evidence for sustained transmission of a bat variant of rabies virus in skunks in Arizona. Presented at the12th International Meeting on Advances in Rabies Research and Control in the Americas, Nov 12–16, 2001, Peterborough, ON, Canada.

Smith, J.S. (2002). Molecular epidemiology. In: *Rabies* (A.C. Jackson and W.H. Wunner, eds). pp. 79–111. San Diego: Academic Press.

Smith, J.S., Fishbein, D.B., Rupprecht, C.E. and Clark, K. (1991). Unexplained rabies in three immigrants in the United States. A virologic investigation. *New England Journal of Medicine* **324**, 205–211.

Smith, J.S., Orciari, L.A., Yager, P.A., Seidel, H.D. and Warner, C.K. (1992). Epidemiologic and historical relationships among 87 rabies virus isolates as determined by limited sequence analysis. *Journal of Infectious Diseases* **166**, 296–307.

Smith, J.S., Orciari, L.A. and Yager, P.A. (1995). Molecular epidemiology of rabies in the United States. *Seminars in Virology* **6**, 387–400.

Swanepoel, R., Barnard, B.J.H., Meredith, C.D. *et al.* (1993). Rabies in southern Africa. *Onderstepoort Journal of Veterinary Research* **60**, 325–346.

Thompson, R.C.A. (2000). *Molecular Epidemiology of Infectious Diseases*. London: Arnold Press.

Thompson, J.D., Gibson, T.J., Plewniak, F., Jeanmougin, F. and Higgins, D.G. (1997). The ClustalX windows interface: flexible strategies for multiple sequence alignment aided by quality analysis tools. *Nucleic Acids Research* **25**, 4876–4882.

Tordo, N., Poch, O., Ermine, A., Keith, G. and Rougeon, F. (1986). Walking along the rabies genome: Is the large G-L intergenic region a remnant gene? *Proceedings of the National Academy of Sciences USA* **83**, 3914–3918.

Tordo, N., Bourhy, H. and Sacramento, D. (1992). Polymerase chain reaction technology for rabies virus. In: *Diagnosis of human viruses by polymerase chain reaction technology*, Frontiers in Virology, Vol. 1 (Y. Becker and G. Darai, eds). pp. 389–405. Berlin: Springer-Velag Press.

Tordo, N., Benmansour, A., Calisher, C. *et al.* (2004). Rhabdoviridae. In: *Virus Taxonomy, VIIIth Report of the ICTV (International Committee on Taxonomy of Viruses)* (C.M. Fauquet, M.A. Mayo, J. Maniloff, U. Desselberger and L.A. Ball, eds). pp. 623–644. London: Elsevier/Academic Press.

Tuffereau, C., LeBlois, H., Benejean, J., Coulon, P., Lafay, F. and Flamand, A. (1989). Arginine or lysine in position 333 of ERA and CVS glycoprotein is necessary for rabies virulence in adult mice. *Virology* **172**, 206–212.

van Regenmortel, M.H. (1998). From absolute to exquisite specificity: Reflections on the fuzzy nature of species, specificity and antigenic sites. *Journal of Immunological Methods* **216**, 37–48.

van Regenmortel, M.H.V., Mayo, M.A., Fauquet, C.M. and Maniloff, J. (2000). Virus nomenclature: Consensus versus chaos. *Archives of Virology* **145**, 2227–2232.

Velasco-Villa, A., Gómez-Sierra, M., Hernández-Rodríguez, G. *et al.* (2002). Antigenic diversity and distribution of rabies virus in Mexico. *Journal of Clinical Microbiology* **40**, 951–958.

Velasco-Villa, A., Orciari, L.A., Souza, V. *et al.* (2005). Molecular epizootiology of rabies associated with terrestrial carnivores in Mexico. *Virus Research* **111**, 13–27.

von Teichman, B.F., Thomson, G.R., Meredith, C.D. and Nel, L.H. (1995). Molecular epidemiology of rabies virus in South Africa: evidence for two distinct virus groups. *Journal of General Virology* **76**, 73–82.

Vos, A., Müller, T., Neubert, L. *et al.* (2004). Rabies in red foxes (*Vulpes vulpes*) experimentally infected with European bat lyssavirus type 1. *Journal of Veterinary Medicine Series B* **51**, 327–332.

Wakeley, P.R., Johnson, N., McElhinney, L.M., Marston, D., Sawyer, J. and Fooks, A.R. (2005). Development of a real-time, TaqMan reverse transcription-PCR assay for detection and differentiation of lyssavirus genotypes 1, 5 and 6. *Journal of Clinical Microbiology* **43**, 2786–2792.

Warner, C., Fekadu, M., Whitfield, S. and Shaddock, J. (1999a). Use of anti-glycoprotein monoclonal antibodies to characterize rabies virus in tissues. *Journal of Virological Methods* **77**, 69–74.

Warner, C.K., Zaki, S.R., Shieh, W.J. *et al.* (1999b). Laboratory investigation of human deaths from vampire bat rabies in Peru. *American Journal of Tropical Medicine and Hygiene* **60**, 502–507.

Wright, A., Rampersad, J., Ryan, J. and Ammons, D. (2002). Molecular characterization of rabies virus isolates from Trinidad. *Veterinary Microbiology* **87**, 95–102.

Yung, V., Favi, M. and Fernández. (2002). Genetic and antigenic typing of rabies virus in Chile. *Archives of Virology* **147**, 2197–2205.

4

Epidemiology

JAMES E. CHILDS[1] AND LESLIE A. REAL[2]

Department of Epidemiology and Public Health and YIBS Center for Eco-Epidemiology, Yale University School of Medicine, New Haven, Connecticut 06510[1]; Department of Biology and Center for Disease Ecology, Emory University, Atlanta, Georgia 30322[2]

1 INTRODUCTION TO CONCEPTS

Rabies is a zoonotic disease. The diagnosis of rabies among humans and animals has traditionally been restricted to the symptoms of acute fatal encephalomyelitis caused by rabies virus, serotype 1 (ST 1)/genotype 1 (GT 1) of the genus *Lyssavirus*, family Rhabdoviridae. However, with the discovery of additional lyssaviruses, most notably Australian bat lyssavirus (ABLV) in 1996 (Speare *et al.*, 1997), capable of causing a fatal human disease indistinguishable from classic rabies (Samaratunga *et al.*, 1998; Hanna *et al.*, 2000), the disease rabies has been ascribed to any of the fatal diseases resulting from infection by any of the viruses in the genus *Lyssavirus* (Hanlon *et al.*, 2005). As of 2005, the proposed number of genotypes within the genus *Lyssavirus* has grown from seven in 2002 (Childs, 2002) to possibly 11; since 2003, four novel lyssaviruses have been recovered and characterized from bats originating from Eurasia (Table 4.1) (Hanlon *et al.*, 2005) and their confirmation as new genotypes appears certain (World Health Organization, 2004; Hanlon *et al.*, 2005).

Human deaths are an unfortunate consequence of pathogenic processes in affected animals which serve to enhance rabies virus transmission (e.g. behavioral changes leading to increased aggression or biting, virus present in salivary glands and saliva) in maintenance cycles involving diverse mammalian species within the mammalian Orders Carnivora and Chiroptera (Figure 4.1). As humans are but rarely implicated in the non-iatrogenic transmission of rabies virus from human to human (Fekadu *et al.*, 1996), humans do not contribute to the maintenance of rabies virus or other lyssaviruses and are considered 'dead-end' hosts.

The reservoir for rabies is the animal pool that circulates different rabies virus variants, which collectively define ST 1/GT 1 lyssaviruses, with only occasional spillover of rabies virus into humans. Although rabies virus is often regarded or referred to as a single viral entity with multiple mammalian hosts (Cleaveland *et al.*, 2001; Dobson and Foufopoulos, 2001), rabies virus more resembles a metapopulation in which genotypically and phenotypically distinguishable rabies virus variants are adapted to and maintained by a single or a few mammalian

123

Rabies, second edition. Edited by Alan C. Jackson and William H. Wunner
ISBN 978-012-369366-2. Copyright Elsevier Inc. 2007

TABLE 4.1

Recognized or proposed members of the genus *Lyssavirus*, family Rhabdoviridae

Serotype/genotype ICTV abbreviation[a]	Species implicated in maintenance	Distribution	Annual human deaths	Reference
Rabies (ST 1/GT 1) RABV	Dogs, wild carnivores, bats	Worldwide (with exception of Australia, Antarctica, and designated rabies-free countries)	~55 000	Shope, 1982
Lagos bat (ST 2/GT 2) LBV	Bats – Megachiroptera; *Eidolon helvum*, *Micropterus pusillus*, *Epomophorus wahlbergi*	Africa: Central African Republic, Ethiopia, Nigeria, Senegal, South Africa	not reported	Bougler and Porterfield, 1958
Mokola (ST 3/GT 3) MOKV	Shrew – Insectivora: *Crocidura* spp.; Rodentia; *Lopyhromys sikapusi*	Africa: Cameroon, Central African Republic, Ethiopia, Nigeria, South Africa, Zimbabwe	occasional	Shope *et al.*, 1970
Duvenhage (ST 4/G 4) DUVV	Bats – Microchiroptera; *Miniopterus schreibersii, Nycteris gambiensis, N. thebaica*	Africa: South Africa, Guinea, Zimbabwe	occasional	Meredith *et al.*, 1971
European bat Lyssavirus 1 (GT 5) EBLV-1	Bats – Microchiroptera; *Myotis dasycneme, M. daubentonii*	Europe	occasional	Bourhy *et al.*, 1993
European bat Lyssavirus 2 GT 6) EBLV-2	Bats – Microhiroptera: *Eptesicus serotinus*	Europe	occasional	Bourhy *et al.*, 1993
Australian bat Lyssavirus (GT 7) ABLV	Bats – Megachiroptera; *Pteropus alecto, P. scapulatus*	Australia, 1996; possibly SE Asia mainland	occasional	Speare *et al.*, 1997
Aravan virus[b] ARAV	Bats – Microhiroptera: *Myotis blythi*	Kyrgyzstan, 1991	not reported	Arai *et al.*, 2003
Khujand virus[b] KHUV	Bats – Microhiroptera: *Myotis mystacinus*	Tajikistan, 2001	not reported	Kuzmin *et al.*, 2003
Irkut virus[b] IRKV	Bats – Microhiroptera: *Murina leucogaster*	Eastern Siberia, 2002	not reported	Kuzmin *et al.*, 2003
Wet Caucasian bat virus[b] WCBV	Bats – Microhiroptera; *Miniopterus schreibersi*	Caucasus Mountains, 2003	not reported	Botvinkin, 2003

[a] ICTV=International Committee on Taxonomy of Viruses. [b] As yet unclassified new lyssaviruses (WHO, 2004).

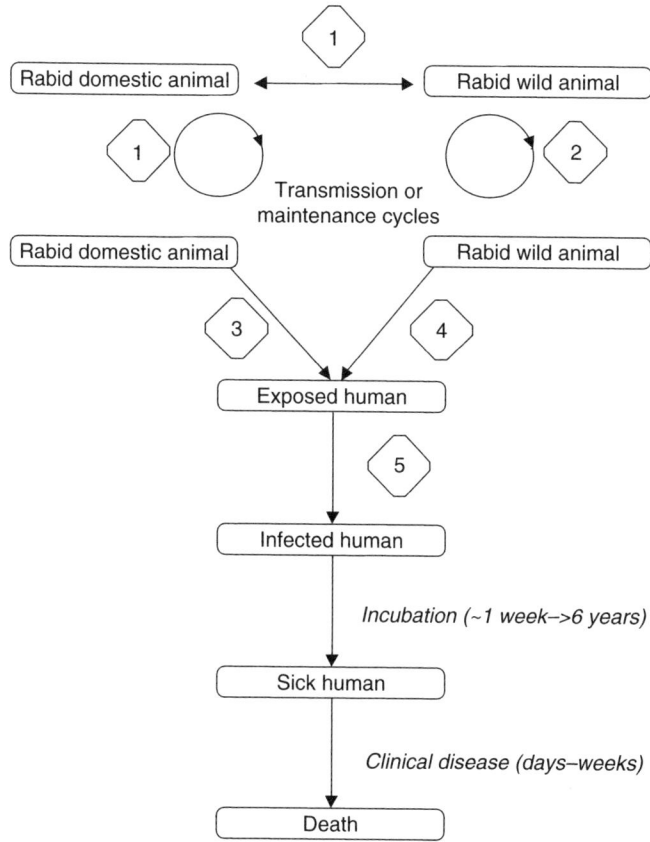

Figure 4.1 The natural history of rabies annotated with opportunities for prevention, treatment and control. The circular symbols indicate natural maintenance of rabies virus transmission among domestic dogs and wild animals. (1) Prevention and control of wildlife transmission of rabies virus or dog-to-dog maintenance of transmission by vaccination and control of domestic animals; (2) prevention and control of wildlife transmission of rabies virus through vaccination of natural reservoir species; (3) prevention of human exposure through avoidance and population control of rabid domestic animals; (4) prevention of human exposure through avoidance of rabid wild animals; (5) human infection by rabies virus can be prevented either by pre-exposure immunization (and appropriate post-exposure treatment) or by post-exposure treatment with vaccine and rabies immunoglobulin. Once a person develops clinical signs of disease, rabies is nearly always fatal. (Figure modified from Hattwick, 1974, Fishbein, 1991 and Childs, 2002.)

species which, *in toto*, constitute the GT 1/ST 1 of genus *Lyssavirus*. The classic definition of a metapopulation is a population of subpopulations that are linked by occasional movements (Levins, 1969). In the case of rabies virus, movements are represented by the occasional cross-species transmission (i.e. spillover) of rabies virus variants from a reservoir host to a secondary species in which the original variant subsequently becomes adapted to sustained transmission within the secondary species, coincident with the genetic and phenotypic changes indicative of its emergence as a novel and unique subvariant of rabies. Although the generation of novel subvariants is difficult to capture in historical time, all rabies virus variants of terrestrial carnivores are believed to have their origin from variants of rabies virus associated with bats (Badrane and Tordo, 2001).

Rabies virus variants are differentiated by their antigenic makeup, as disclosed by differential binding patterns of panels of monoclonal antibodies (Rupprecht *et al.*, 1987; Smith, 1988) and by characteristic patterns of nucleotide substitutions in their single stranded negative-sense RNA genome (Sacramento *et al.*, 1991; Nadin-Davis and Casey, 1994; Smith *et al.*, 1995). This molecular variation has permitted identification of the primary reservoir hosts for different rabies virus variants, detailed descriptions of the geographic distribution of virus variants circulating among reservoir hosts and identification of specific virus variants which, through spillover, cause rabies in humans and animals other than the specific reservoir host (Smith *et al.*, 1995). Significantly, the identification of specific virus metapopulations within readily identifiable mammalian host species has served as a basis for species-specific design of control measures for delivering rabies vaccines by parenteral or oral routes for immunizing sylvatic reservoir hosts (Creekmore *et al.*, 1994; Linhart *et al.*, 1997, 2002) and domestic dogs (Rupprecht *et al.*, 2004, 2005).

Given the dependency of rabies virus on maintenance cycles or transmission cycles involving specific mammalian hosts (Figure 4.1), the geographic distribution of genetically distinguishable variants of rabies virus can be mapped with a fair degree of accuracy where reasonable levels of national or regional surveillance for animal and human rabies cases and laboratory typing of rabies virus variants are available. For example, in the USA, mapping of the co-localization of rabies virus variants circulating in some portion of a reservoir host's natural geographic range (Figure 4.2) provides information on where to target specific hosts within delimited areas for delivery of oral rabies vaccine (ORV).

Currently, the ORV most used is a vaccinia-rabies glycoprotein recombinant virus (V-RG) vaccine expressing the ERA (Evelyn-Rokitnicki-Abelseth) rabies virus glycoprotein (Rupprecht *et al.*, 1986, 1988); ORV was the first live-recombinant vaccine to be released in the field (Hanlon *et al.*, 1998). The vaccine has been distributed either in naked sachets or, most commonly, in sachets covered with a polymer containing additives designed preferentially to attract the target reservoir host species (Linhart *et al.*, 1997, 2002); non-target species find these vaccine-laden baits attractive (Olson and Werner, 1999).

Rabies control programs using ORV to target rabies variants associated with raccoons, coyotes and gray foxes, are ongoing in the USA (Slate *et al.*, 2005) (see Chapter 17). Similar information on the circulation of rabies virus variants among sylvatic reservoirs is becoming available from Mexico (de Mattos *et al.*, 1999; Velasco-Villa *et al.*, 2002), South America nations, such as Brazil (Sato *et al.*, 2004;

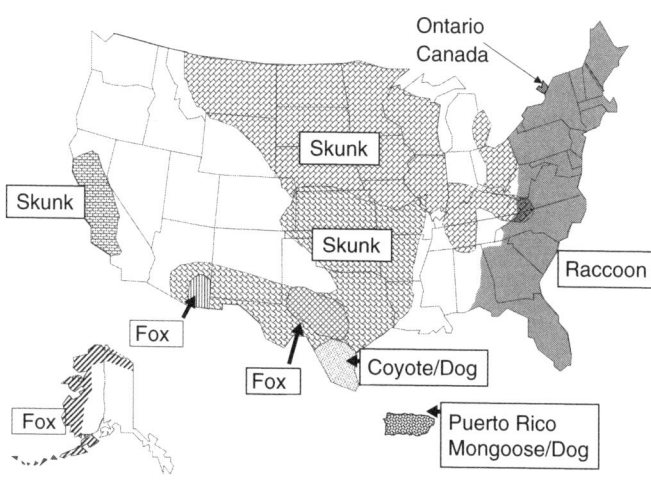

Figure 4.2 The distribution of rabies virus variants in the USA delineated by the primary host species or the common name of the group of terrestrial carnivore species serving as reservoir hosts. Overlaying this distribution of rabies virus variants is the reservoir present in species of bats which includes the entire continental USA.

Bordignon *et al.*, 2005; Schaefer *et al.*, 2005), countries in southern Africa (Sabeta *et al.*, 2003; Johnson *et al.*, 2004; Nel *et al.*, 2005) and in Eurasia and south-eastern Asia (Kuzmin *et al.*, 2004; Hyun *et al.*, 2005; Kim *et al.*, 2005; Park *et al.*, 2005). Although targeted vaccination of sylvatic reservoir host species by ORV is largely restricted to North America and Europe, application of ORV in South Korea is targeting an Arctic strain of rabies virus circulating among raccoon dogs (*Nyctereutes procyonoides koreensis*) in several districts adjacent to the border with North Korea (Hyun *et al.*, 2005).

Finer discrimination of genetic lineages of rabies virus at the subvariant level is also being used to target specific metapopulations of rabies virus circulating within a common mammalian host species associated with a definable geographic area within a country. Regional dog populations within Thailand have been linked to the spread of specific rabies virus subvariants along networks of roads linking commercial urban centers with rural sites, presumably through movements of dogs accompanying their human owners (Denduangboripant *et al.*, 2005). Subvariant structure of rabies virus circulating among skunks in California has been described (Crawford-Miksza *et al.*, 1999) and, recently, the raccoon variant of rabies virus invading Canada was demonstrated to have three independent origins based on molecular biologic methods (Nadin-Davis *et al.*, 2005).

The epidemiology of human and animal rabies almost invariably reflects the regional terrestrial virus variants maintained by carnivores, the specific animal reservoirs for virus variants and the opportunity for human–animal interaction (Noah *et al.*, 1998; McQuiston *et al.*, 2001). Exceptions to this rule are the occasional spillover of bat-associated rabies virus variants to domestic animals and humans (McQuiston *et al.*, 2001; Messenger *et al.*, 2002), which are more difficult to map with accuracy due to the movements and migration of the bat species involved, and the instances of 'imported' cases of rabies where a human or animal infected with rabies in a foreign country succumbs to rabies in a different country. Recent examples include human rabies diagnosed in the USA where patients were infected by dog bites received in Ghana (Krebs *et al.*, 2004) and Haiti (Centers for Disease Control and Prevention, 2005b), a rabies case in the UK where the patient was infected by a dog bite received in the Philippines (Smith *et al.*, 2003), an Austrian tourist bitten by a dog while traveling in Morocco (Krause *et al.*, 2005) and multiple cases of human rabies diagnosed in France where the origin of infection was Africa (Peigue-Lafeuille *et al.*, 2004).

Animals are also subject to intentional or accidental long distance movement through human endeavor. Translocations of sylvatic hosts for rabies virus have been responsible for outbreaks of coyote/dog variant rabies among coyotes in Florida, where the virus was introduced with translocated coyotes from Texas (Centers for Disease Control and Prevention, 1995) and for seeding the epidemic of raccoon-associated rabies in the mid-Atlantic states following translocation of infected raccoons from the southeastern focus of raccoon rabies (Nettles *et al.*, 1979). Instances of animals being diagnosed with rabies following their introduction into a foreign country include a horse diagnosed with rabies in Zimbabwe having been infected in South Africa (Sabeta and Randles, 2005) and several instances involving bats (Constantine, 2003). On one occasion, a rabid bat-stowaway on a ship was detected in rabies-free Hawaii after passage from the mainland USA (Sasaki *et al.*, 1992).

An understanding of patterns of rabies virus maintenance within animal host populations and the degree to which rabies virus has co-adapted with particular animal hosts has only been appreciated for the last few decades. The genetic basis for host specificity and the accompanying changes in virus phenotype are still not understood, although the members of the genus *Lyssavirus* can be broadly classified within two phylogroups of different virulence (Badrane *et al.*, 2001). An important feature of the ST 1/GT 1 grouping of lyssaviruses of overriding public health importance is that all human infections caused by rabies virus, ST 1/GT 1, are preventable by pre- or post-exposure treatment (PET) regimens using the various tissue-culture-derived rabies virus vaccines and immunoglobulin products, irrespective of host affiliations or subtle genetic variation (Centers for Disease Control and Prevention, 1999a, 2005a).

Details of the maintenance cycles for lyssaviruses other than rabies virus, such as Duvenhage (DUVV, ST 2/GT 2), Lagos bat (LBV, ST 3/GT 3) and Mokola (MOKV ST 4/GT 4) viruses (see Table 4.1) are unclear (Nel and Rupprecht, 2006). However, like rabies virus, their perpetuation is assumed to involve introduction of virus into a susceptible individual of the reservoir host species, almost invariably by bite transmission involving another conspecific, then infection, dissemination and multiplication of the virus within the individual prior to reintroduction of viral progeny into a new susceptible animal. Other species than the presumed bat reservoir host for DUVV and LBV and the insectivore reservoir host for MOKV have been affected by spillover of these rarely identified lyssaviruses, including a human with DUVV (Meredith *et al.*, 1971; Swanepoel *et al.*, 1993), domestic cats and a dog with LBV (Foggin, 1988; King and Crick, 1988; Mebatsion *et al.*, 1992) and humans, domestic cats and dogs with MOKV (Familusi and Moore, 1972; Familusi *et al.*, 1972; Foggin, 1983; Nel *et al.*, 2000; Bingham *et al.*, 2001).

Rabies virus maintenance in animal reservoirs has served as a model system for illustrating many important concepts in infectious disease epidemiology and the theoretical modeling of the population biology of a virus and its host. Examples of these efforts include the notion of threshold population density for virus transmission to occur among individuals of the reservoir host species (Anderson *et al.*, 1981; Cruickshank *et al.*, 1999); the relevance and critical threshold level of herd immunity necessary to achieve rabies control by interfering with virus transmission (Coleman and Dye, 1996); statistical models of rabies spread (Moore, 1999); deterministic dynamics models of rabies infection (Anderson *et al.*, 1981; Murray *et al.*, 1986; Coyne *et al.*, 1989); the development of stochastic simulators of the spatial spread of rabies (Smith *et al.*, 2002; Russell *et al.*, 2004); and model validation of the population and the temporal structure of rabies epizootics among reservoir species (Childs *et al.*, 2000). Predictions from model outcomes have been used to identify temporal linkages in the risk of rabies virus spillover from sylvatic reservoirs to domestic species (Gordon *et al.*, 2004) and other wildlife (Guerra *et al.*, 2003), the design of rational intervention schemes based on geographic simulators projecting rabies spread following breaches in ORV immune barriers (Russell *et al.*, 2005) and the development of analytic methods to inform economic models with finer scale resolution of cost structures associated with different temporal stages of rabies epizootics, providing improved estimates of the savings potentially accrued through active interventions (Gordon *et al.*, 2005).

2 THE EPIDEMIOLOGY OF HUMAN RABIES

2.1 Rabies mortality in Africa and Asia

In most areas of the world, especially in tropical zones where the majority of human rabies deaths occur, accurate estimates of human rabies deaths are impossible to obtain as surveillance systems and regional laboratories are inadequate or non-existent for the systematic detection and laboratory confirmation of human or animal rabies cases. The proportion of rabies cases detected and reported to the World Health Organization (WHO) in 1999 was estimated to represent but 3% of the total global rabies mortality (Knobel *et al.*, 2005). As most human rabies deaths in developing countries result from infection by rabies virus variants maintained by domestic dogs and transmitted by dog bite (Knobel *et al.*, 2005), novel methods of estimating human rabies deaths have focused on extrapolations from studies estimating domestic dog population densities in different regions in Africa and Asia (e.g. Childs *et al.*, 1998; Pal, 2001), or directly from the incidence of dog bite in African countries such as Uganda and Tanzania (Cleaveland *et al.*, 2002a; Fevre *et al.*, 2005). Risk models based on the likelihood of clinical rabies developing after being bitten by a rabid dog suggest some 55 000 (90% confidence interval = 24 000–93 000) human rabies deaths occur annually in Africa and Asia (Knobel *et al.*, 2005). This estimate is greater than the WHO's estimate of between 35 000 and 50 000 human deaths worldwide annually (World Health Organization, 1999), largely as a consequence of earlier underestimation of rabies deaths occurring in China (Knobel *et al.*, 2005).

Recent estimates of the annual incidence of rabies deaths per 100 000 persons in India (0.37 in urban locations and 2.49 in rural locations; Table 4.2) (Knobel *et al.*, 2005) are, on average, lower than the total estimated annual incidence of 28.8 human rabies deaths per 10^6 persons reported for 1986 (Bögel and Motschwiller, 1986), although the rural rate was virtually unchanged from the earlier calculation. The current overall estimated incidence of human rabies in urban and rural areas of Africa are 2.0 and 3.6 persons per 10^5 persons, respectively (Knobel *et al.*, 2005). These values exceed earlier estimates, published in 1986, by as much as an order of magnitude; the estimated incidence of rabies per 10^6 persons in 1986 was 1.1 for Sudan, 2.2 for Zimbabwe, 2.4 for Algeria and Morocco, 2.9 for Tunisia, 6.3 for Botswana and 12.6 for Ethiopia (Bögel and Motschwiller, 1986). The reasons for such discrepancies are always uncertain. Although dog-associated rabies has increased throughout most of sub-Saharan Africa during the last 70 years (Nel and Rupprecht, 2006), it remains difficult to assess the magnitude of rabies deaths in countries without adequate surveillance and laboratory facilities in the region; in 1997, laboratory confirmation of rabies was available for less than 0.5% of the estimated human rabies cases (World Health Organization, 1999). Sensitive, specific and, perhaps most importantly, widely available laboratory testing is an essential element for surveillance for rabies, as it is with any infectious disease.

2.2 The burden and cost of rabies in Africa and Asia

An additional health burden to that directly resulting from rabies mortality is the high percentage of adverse reactions to PET involving vaccines of nervous tissue origin (VNTO). An estimated 200 000 persons in Africa and 7 500 000 persons in Asia (India, China and other southeastern Asian nations) annually receive PET for rabies exposures (Knobel *et al.*, 2005); approximately 10% of those persons in Africa and 33% of those in Asia receive a VNTO. Of those individuals receiving Semple type vaccines, derived from phenol-treated sheep- or goat-brain tissue (Swaddiwuthipong *et al.*, 1988), or those receiving vaccines

TABLE 4.2

Incidence or estimates of human rabies deaths from selected countries in the Americas, 1993–2002 and for Asia and Africa[a]

Region and/or country	Range/estimate of annual incidence of human rabies	Annual estimated incidence of rabies per 100 000 persons
Latin America		
Argentina	0–1	0.00
Bolivia	2–16	0.02
Brazil	10–50	0.01
Chile	0–1	0.00
Columbia	0–9	0.02
Ecuador	0–65	0.00
Paraguay	0–9	0.09
Peru	1–41	0.00
Venezuela	0–5	0.00
Central America		
Costa Rica	0–2	0.00
El Salvador	0–15	0.10
Guatemala	0–20	0.00
Honduras	0–2	0.00
Nicaragua	0–2	0.00
Panama	0–2	0.07
Mexico	3–31	0.00
Latin Caribbean		
Cuba	0–1	0.00
Haiti	0–9	0.06
Dominican Republic	0–3	0.02
Surinam	0–1	0.00
North America		
Canada	0–1	0.00
USA	0–5	0.00
India[b]		
Urban	1058	0.37
Rural	18 201	2.49
China[b]		
Urban	1324	0.29
Rural	1275	0.15

(*Continued*)

TABLE 4.2 (Continued)

Region and/or country	Range/estimate of annual incidence of human rabies	Annual estimated incidence of rabies per 100 000 persons
Other Asian countries		
Urban	853	0.29
Rural	8135	1.55
Africa[b]		
Urban	5886	2.00
Rural	17 937	3.60

[a]Adapted from Belotto et al., 2005; [b]Estimates from Knobel et al., 2005.

derived from suckling-mouse brain tissue (Held and Adaros, 1972), an estimated 360 (CI = 142–586) and 44 525 (CI = 17 585–72 575) disability-adjusted life years (DALY) are lost annually due to adverse reactions among individuals treated in Africa and Asia, respectively (Knobel et al., 2005).

The proportion of neuroparalytic complications among persons receiving VNTO has been estimated at between 0.3 and 0.8 adverse reactions per 1000 vaccinees (Meslin et al., 1994; World Health Organization, 2004); one estimate of the incidence is as high as 4.5 per 1000 vaccinees (Abdussalem and Bögel, 1971). The duration of incurred disability following an adverse reaction has been estimated at between 4.9 and 6.6 months (World Health Organization, 2004). However, mortality among persons experiencing severe neuroparalytic reactions to VNTO has been estimated to be as high as 15% (Abdussalem and Bögel, 1971) and up to 25% of patients suffer permanent disability or death (Meslin et al., 1994).

The global public health cost of rabies is far in excess of metrics limited to the loss of human life. Estimates of the annual burden of canine rabies in Africa and Asia, based on direct medical expenses and costs incurred by patients seeking treatment, amount to 20.5 million US$ (CI = 19.3–21.8) and 563 million US$ (CI = 520–605.8), respectively (Knobel et al., 2005). These values vary with location and a large degree of uncertainty is warranted for estimates from Asia which include China (Knobel et al., 2005). In local studies, such as in Thailand, an estimated 200 000 persons received PET with tissue-culture-derived vaccines in 1997 at a cost of approximately 10 million US$ (Knobel et al., 2005).

2.3 Human demographics and the epidemiology of human rabies in Africa and Asia

From the public health perspective, rabies remains a major threat only among regions of Asia and Africa and, to a lesser degree Latin America (see below).

Approximately 99% of all rabies deaths occur in the developing world and throughout these regions the domestic dog is still the major reservoir host for virus variants associated with most human deaths (World Health Organization, 2004). For example, approximately 96% of rabies deaths in India are directly attributable to dog bite (Dutta, 1999) and from 95% to 100% of human rabies deaths in northern Nigeria were related to dogs (Nawathe, 1980; Ezeokoli and Umoh, 1987). Within the broad category of developing countries, the geographic area of the tropics has accounted for more than 99% of human deaths and ~90% of PET for rabies (Acha and Arambulo, 1985). The trend toward increasing global urbanization has resulted in approximately one-half the world's human population living in cities in 2002, as compared with one-third in the 1970s (Lederberg et al., 1992). In India alone, the percentage of persons living in urbanized areas rose from 17.3% (44 million) in 1966–1967 to 28.3% (217 million) by 1997; human population density in India grew from 77 persons/km^2 in 1901 to 273 persons/km^2 in 1991 (Gupte et al., 2001). Approximately 36% of India's population is aged 1–14 years, which is the highest risk group for rabies (see below). Diseases such as rabies that are prone to transmission in crowded urban centers with inadequate public health infrastructure remain a constant or increasing threat in much of the world.

In general, most rabies deaths are reported from urbanized areas where dog and human populations reach their highest population densities (Beran, 1991). However, examination of the rates of rabies in urban and rural areas of India, other Asian countries (China excluded; see Table 4.2) and Africa (see Table 4.2), shows risk of rabies infection by dog bite is greater in rural locations than in urban settings. Rural or 'sylvatic' rabies (Turner, 1971; Acha and Arambulo, 1985) involving rabies virus variants maintained and transmitted from indigenous wildlife to humans is rarely reported from Africa and Asia where canine rabies is an overwhelming problem.

In countries where canine rabies persists, the incidence of rabies in humans closely mirrors the incidence of rabies among dogs (Figure 4.3). The incidence of rabies is highest among human males and among individuals <20 years of age (Fekadu, 1982; Lakhanpal and Sharma, 1985; Bhatia et al., 1988; Cleaveland et al., 2002a). The age and sex distribution of human rabies deaths mirrors the age distribution of dog-bite victims with the highest risk among persons under the age of 15 years (Fekadu, 1982; Cleaveland et al., 2002a; Kayali et al., 2003a) (Figure 4.4).

The rate at which children are bitten by dogs and potentially exposed to rabies virus is underestimated in developed (Beck and Jones, 1985) and developing countries (Eng et al., 1993). Where measured, the community impact of dog bite is substantial, as exemplified by the city of Hermosillo, Mexico, where approximately 2.5% of the resident population are bitten by dogs annually (Eng et al., 1993). In India, as many as 1.9% of the population (i.e. 10 million persons) may be bitten annually (cited in Meslin, 2005). In Bangkok, 5.3% of injuries seen at an emergency room associated with a teaching hospital were due to dog bite (Bhanganada et al., 1993). Overall estimates of the annual incidence of bites

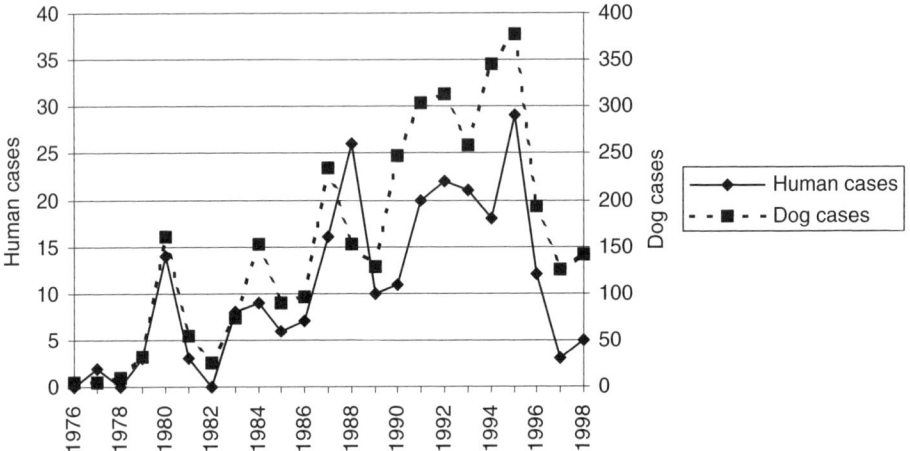

Figure 4.3 Human and dog rabies deaths reported from Kwazulu/Natal Province in South Africa from 1976 to 1998; data from unpublished observations from Bishop and Swanepoel and figure from Nel and Rupprecht (2006). Enzootic canine rabies drives human rabies mortality in high endemicity locations where rabies virus variants are maintained by dogs and some wildlife species. Reproduced with permission from L H Nel, based on unpublished work by G Bishop.

received from suspect rabid dogs are 100 per 100 000 persons in urban and rural settings in Africa and 120 and 100, respectively, for persons in urban and rural settings in Asia (Cleaveland *et al.*, 2003; Knobel *et al.*, 2005; Meslin, 2005).

Rabies in Africa and Asia disproportionately afflicts residents in areas of lower socio-economic status (SES) in rural and urban settings (Fagbami *et al.*, 1981; Knobel *et al.*, 2005). The estimated direct and indirect costs of rabies PET per treated individual is US $39.57 in Africa and US $49.41 in Asia, which represents approximately 5.8% and 3.4% of the annual income for the average person in Africa and Asia, respectively (Knobel *et al.*, 2005). Rabies epizootics among dogs in cities may continue for longer periods in areas of lower SES (Eng *et al.*, 1993), in part due to the large populations of free-roaming dogs, lack of adequate vaccination coverage of owned or neighborhood dogs and the lack of resources available for timely control of epizootics (World Health Organization, 1987; Wilde *et al.*, 2005). Even within the USA, studies of the ecology of free-ranging dogs indicate that more severe problems associated with higher densities of free-ranging dogs occur in urban neighborhoods of lower SES (Beck, 1973).

The actual risk of human exposure to rabies virus in developing countries is difficult to estimate, given the inadequate infrastructure to conduct surveillance in high-endemicity areas of Asia and Africa (Meslin, 2005). Surveys based on the experiences of foreign citizens from developed nations traveling or working overseas or native residents receiving bites from suspect rabid dogs provide a range of values for rabies virus exposures or PETs for Africa of 0.8 to 2.5 persons per 1000 persons per month and 0.6 to 2.8 for Asia (Table 4.3).

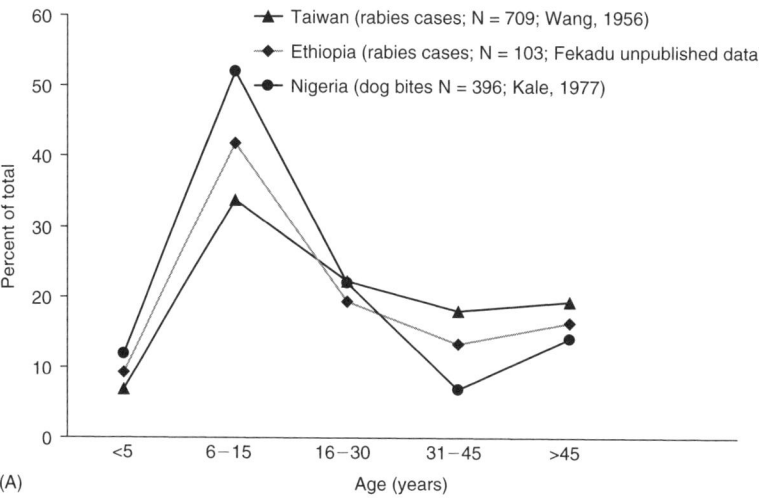

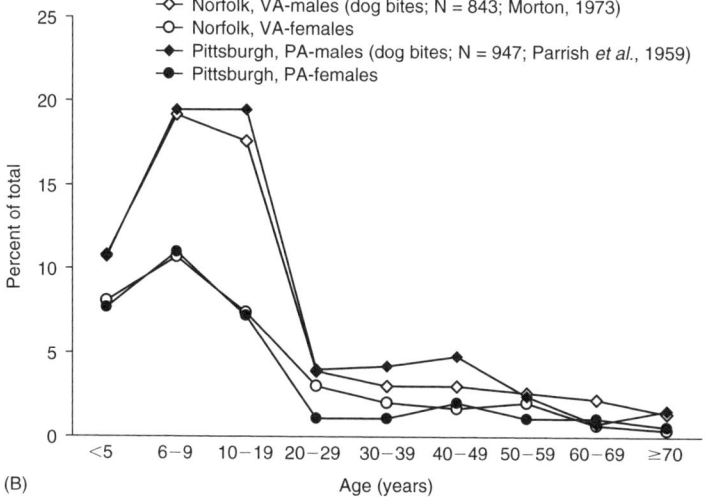

Figure 4.4 (A) The young-age distribution of rabies cases in developing countries (illustrated by recent data from Ethiopia and past data from Taiwan) is mirrored by the age distribution of dog-bite cases (illustrated with data from Nigeria). (B) In the developed world, dog bite is also a pediatric problem; however, the reduction of rabies in dogs and the availability of excellent biologics for the post-exposure prophylaxis for rabies have reduced human disease to a sporadic occurrence.

2.4 Epidemiology of rabies in Latin America: a region in transition

Reports of human rabies deaths in Latin America (South and Central America plus the Latin Caribbean nations) are substantially lower than reports in Asia and Africa.

TABLE 4.3

Estimated rates of exposure to rabid animals based on the experiences of citizens from the USA or other developed nations who are stationed or living overseas and for local residents of urban and rural locations in Africa and Asia

Continent/country or number of countries	Population	Rate of exposure to a suspect rabid animal (or dog) per 1000 persons per month or rate of postexposure treatment per 1000 persons per month	Source
Africa/several	Missionaries	1.0	Arguin et al., 2000
Africa/21	Peace Corps volunteers	2.5	Bernard and Fishbein, 1991
Asia/several	Missionaries	0.6	Arguin et al., 2000
Asia/4	Peace Corps volunteers	2.8	Bernard and Fishbein, 1991
Asia/Nepal	Travelers	0.4	Shlim et al., 1991
Latin America/several	Missionaries	0.2	Arguin et al., 2000
Latin America/6	Peace Corps volunteers	6.2	Bernard and Fishbein, 1991
Worldwide/(several African countries, India, Bangladesh, Ecuador, Bolivia)	Missionaries	1.3	Bjorvatn and Gundersen, 1980
Africa/urban settings[a]	Resident population	0.8	Knobel et al., 2005
Africa/rural settings[a]	Resident population	0.8	Knobel et al., 2005
Asia/urban settings[a]	Resident population	1.0	Knobel et al., 2005
Asia/rural settings[a]	Resident population	0.8	Knobel et al., 2005

[a]Data extracted and modified from estimated suspect dog bites per 100 000 persons per year.

Regional reports of human rabies ranged from approximately 216 in 1993 to 36 in 2002 (see Table 4.2; Figure 4.5A). The annual incidence of human rabies in Latin America per 10^5 persons now ranges between zero and 0.09 in South America, zero and 0.10 in Central America and zero and 0.06 in Latin Caribbean islands (see Table 4.2) (Belotto *et al.*, 2005). In large part, the lower rates in the developing nations of the Americas reflect the highly effective Rabies Elimination Regional Program started in the 1980s and directed at urban dog rabies control in Latin America and the Latin Caribbean nations (Organización Panamericana de la Salud (OPS), 1983). The regional success in human rabies prevention throughout Latin-speaking counties in the Americas is testimony to the effectiveness of traditional control efforts focused on dog population management and mass vaccination campaigns targeting dogs in reducing human exposures to rabies virus, as indicated by the relatively few rabid dogs reported in 1993 from the region (Figure 4.5B).

Irrespective of the gains in urban and rural dog population management and vaccination coverage, dog-associated rabies remains the principal source of human rabies throughout Latin America. During the interval between 1993 and 2002, 65.2% of the 1147 human deaths recorded were associated with dogs (Belotto *et al.*, 2005). In Mexico, where vampire bat and dog rabies co-occur, dog exposures account for ~81% of human rabies deaths (mainly urban) and vampire bats account for ~11% of cases (mostly rural) (de Mattos *et al.*, 1999). As in Mexico, rabies occurring among domestic animals and humans in Brazil is either of vampire bat origin or associated with dogs (Ito *et al.*, 2001), irrespective of the presence of other variants circulating among other species of bats (Kobayashi *et al.*, 2005). On islands within the Latin Caribbean, dog variants of rabies viruses have become established within introduced populations of mongooses and mongooses are responsible for sporadic cases of human rabies such as reported from Puerto Rico (Krebs *et al.*, 2004).

Surveys based on the experiences of foreign citizens from developed nations traveling or working in Latin America or native residents of Latin America receiving bites from suspect rabid dogs provide a range of values for the incidence of rabies exposure by dog bite, or among persons receiving PET, of 0.2 to 6.2 per 1000 persons per month (see Table 4.3). Given the decline in canine rabies and reported human rabies mortality in Latin Central and South America (Belotto *et al.*, 2005), newer estimates, such as those provided by Arguin *et al.* (2000), may provide a more accurate estimate of risk than the older values provided by Bernard and Fishbein (1991) on US Peace Corps volunteers.

Detailed information on rabies virus variants circulating among sylvatic animal reservoirs is becoming available from Mexico (de Mattos *et al.*, 1999; Velasco-Villa *et al.*, 2002, 2005) and several South American countries, such as Bolivia (Favi *et al.*, 2003), Brazil (Favi *et al.*, 2003; Shoji *et al.*, 2004; Bordignon *et al.*, 2005; Schaefer *et al.*, 2005), Chile (de Mattos *et al.*, 2000; Favi *et al.*, 2002; Yung *et al.*, 2002) and Columbia (Páez *et al.*, 2003, 2005), as the need to concentrate solely on canine

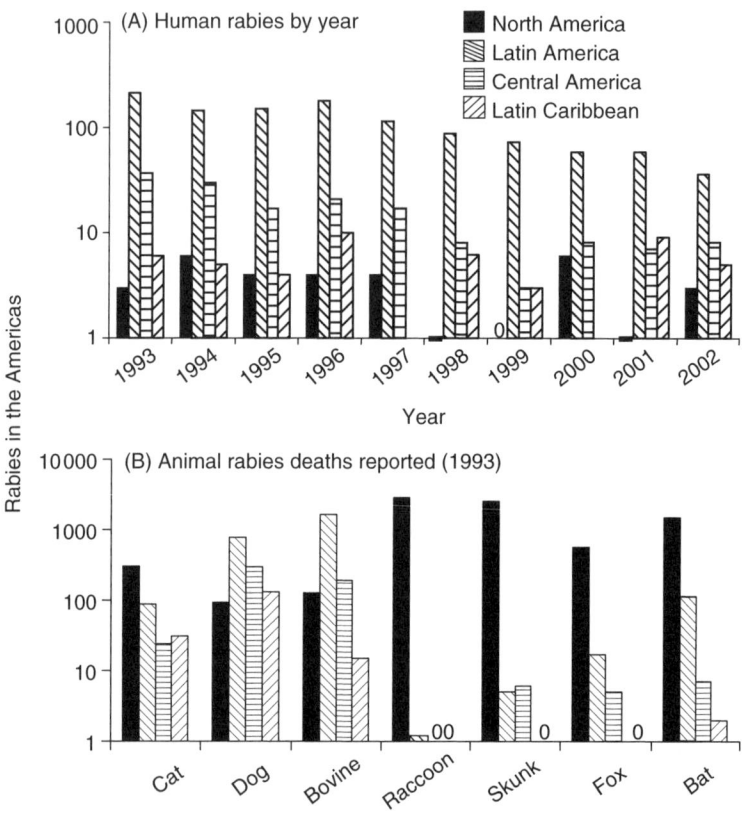

Figure 4.5 Human and animal rabies from countries in North, South and Central Americas and Caribbean nations. (A) Human rabies has steadily declined throughout the Americas from 1993 to 2004. (B) Animal rabies cases reported from the Americas in 1993 indicates how canine cases have fallen below 1000 reports per year. Data from Belotto *et al.* (2005).

reservoirs is lessening. Molecular epidemiologic data on sylvatic rabies virus variants maintained by non-canine reservoirs offer insight into the diverse maintenance cycles of rabies virus variants, as demonstrated for the USA (see Figure 4.2).

In Brazil, molecular sequence data based on the rabies virus N gene have identified four genetic lineages of rabies virus clustering with four bat genera: one lineage is associated with the common vampire bat, *Desmodus rotundus*; the other three segregate with three families/genera of insectivorous bats, *Eptesicus*, *Molossus* and *Nyctinomops* (Kobayashi *et al.*, 2005). A novel variant of rabies isolated from marmosets (*Callithix jacchus*) in Brazil was linked to a human rabies case (Favoretto *et al.*, 2001). Rare human rabies deaths in Chile, Columbia and Brazil have been attributed to rabies virus variants circulating in insectivorous bats (Favi *et al.*, 2002; Páez *et al.*, 2003, 2005). In Columbia, variants of rabies virus detected from a case of human rabies and three cases of dog rabies indicated infection by bat-associated variants of rabies virus circulating in two species/ families of bats, *Eptesicus brasiliensis* and *Molossus molossus* (Páez *et al.*, 2003). The domestic dog variant of rabies virus has also successfully jumped species to establish an enzootic cycle of rabies virus maintenance among gray foxes (*Urocyon cineroargenteus*) in Columbia (Páez *et al.*, 2005).

In Central America, where human rabies is associated with vampire bat and canine rabies virus variants, a newly identified rabies virus variant present among skunks (species not identified) and distinguishable from those rabies virus variants circulating among skunks in North America, has been identified in Mexico (de Mattos *et al.*, 1999). In addition, a focus, or an extension of the existing focus of Arizona Fox (*U. cineroargenteus*) rabies variant in the USA, was identified from bobcats (*Felix rufus*) obtained from the provinces of Chihuahua and Sonora, Mexico (de Mattos *et al.*, 1999).

2.5 Epidemiology of human rabies in North America and Europe

Historically, the pattern of human rabies occurring in countries and regions which have since controlled canine-associated rabies, mirrored the present situation in Africa and Asia. When the annual numbers of human rabies deaths numbered between 10 and 50 in the USA and Europe (Figures 4.6 and 4.7), most were attributable to dog-bite exposure or secondary transmission from an animal infected by a canine rabies virus variant. Most cases of dog bite and rabies occurred among young males (see Figure 4.4B). From 1946, the year the Communicable Disease Center established its national rabies control program, to 1965, 236 cases of human rabies were reported from the USA, of which 70% of the cases were male, 51.3% were ⩽15 years of age and approximately 82% were attributed to dog exposures (Held *et al.*, 1967). The age distribution of persons bitten by dogs in the USA was indistinguishable from that occurring in Asia and Africa (see Figure 4.4A) (Wang, 1956; Parrish *et al.*, 1959; Morton, 1973; Kale, 1977).

Human rabies in North America and Europe is now a very rare disease (Figures 4.6 and 4. 7). Over the decade from 1994 to 2003, the number of rabies deaths in the USA has ranged between 0 and 6, with tentatively 7 cases reported in 2004 (Figure 4.6) and the number in Europe (including Turkey and the Russian Federation) has ranged between 0 and 21 (Figure 4.7). The combination of dog vaccination and free-ranging dog management (Centers for Disease Control and Prevention, 2005a), coupled with the universal availability of highly effective tissue culture-derived vaccines and human rabies immunoglobulin (HRIG) for PET (Advisory Committee on Immunization Practices, 1999), have resulted in

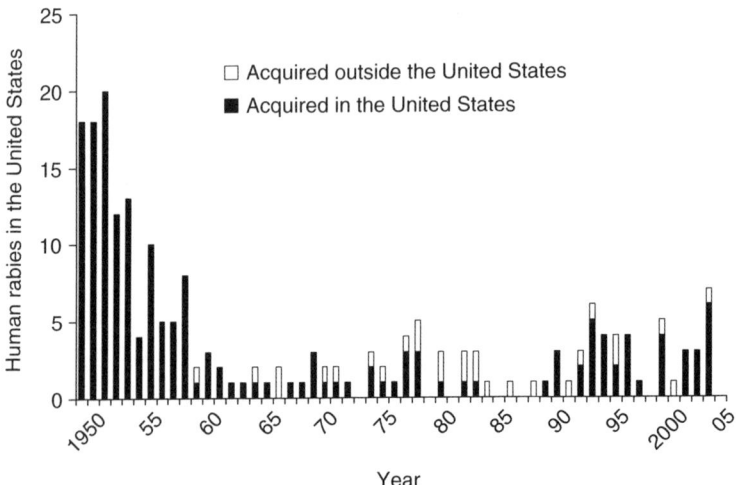

Figure 4.6 Human rabies in the USA, 1950 to 2004. The number has declined dramatically as a result of rabies control among dogs and because of the availability of excellent biologics for the post-exposure treatment for rabies exposures. Data through 1990 taken from Hattwick (1974) and Fishbein (1991); where data disagreed, numbers from Fishbein were used.

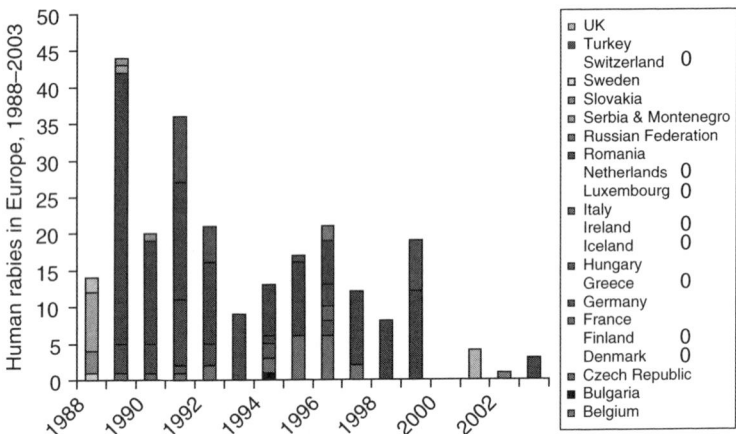

Figure 4.7 Human rabies in Europe, 1988–2003. The number has declined dramatically as a result of rabies control among dogs and because of the availability of excellent biologics for the post-exposure treatment for rabies exposures. Data from Rabnet. Numbers from Rabnet may vary from numbers published by the WHO in annual reports of rabies from a given year.

the significant decline in reports of human rabies. In addition, Europe has dramatically decreased the incidence of red fox-associated rabies through an aggressive multination effort to immunize foxes using live-attenuated rabies strains in addition to V-RG (Cliquet and Aubert, 2004).

The history and molecular typing of rabies virus obtained from human rabies cases over the past several decades in the USA and Europe show that many individuals dying of rabies received their exposures to the virus in another country (Noah *et al.*, 1998). These 'imported' cases have occurred both among citizens or residents of the USA (Centers for Disease Control and Prevention, 1982, 1983; Noah *et al.*, 1998) and among immigrants or visitors to the USA (Smith *et al.*, 1991; Krebs *et al.*, 2001, 2004). Similarly, in Europe, imported cases of rabies have been documented among European tourists, as well as in immigrants or foreign visitors exposed to dog bites in countries such as Nigeria, Morrocco and Thailand, where canine rabies is endemic (Hojer *et al.*, 2001; Johnson *et al.*, 2002; Krause *et al.*, 2005). The recent cases of human rabies in France from 1995 to1997 (Figure 4.7) were all related to exposures received outside of Europe (World Health Organization, 1997, 1998, 1999). In each case of imported rabies, molecular biologic analyses of human brain material has implicated canine-associated rabies virus variants from the country in which the dog bite was received; imported human rabies in the USA and Europe will continue to occur until canine rabies control in Asia and Africa is achieved.

Whereas the decline in human rabies in Europe appears to be continuing, irrespective of imported cases, the USA has seen a resurgence of human rabies deaths since 1990. In the USA and, to a lesser degree, in Canada (Krebs *et al.*, 2003a; Public Health Agency of Canada, 2003), most of the indigenously acquired human rabies cases over the last two decades have been due to 'cryptic' human exposure to variants of rabies virus maintained by native North American species of insectivorous bats (Gibbons, 2002; Messenger *et al.*, 2002, 2003a, 2003b). Europe has no ST 1/GT 1 lyssavirus circulating among its insectivorous bats, although sporadic human deaths are caused by two other lyssaviruses, European bat lyssavirus type 1 (GT 5) and European bat lyssavirus type 2 (GT 6; EBLs). Since 1977, four human deaths have been attributed to EBLs (Fooks *et al.*, 2003); a recent case of fatal EBL2 infection in a Scottish bat conservationist was the first indigenously acquired case of rabies in the UK in 100 years (Nathwani *et al.*, 2003).

It should be noted that some human rabies cases within developed nations remain undiagnosed as dramatically illustrated by rabies deaths among three recipients of organ-transplants and one recipient of a vascular segment in 2004 in the USA, in which rabies was not diagnosed in the donor (Centers for Disease Control and Prevention, 2004b; Srinivasan *et al.*, 2005). Three rabies deaths among recipients of organ transplants in Germany were reported in 2005 (Eursurveillance Weekly, 2005; Homola, 2005). Numerous cases of human rabies have been detected only through post-mortem examination of prepared brain tissue (Noah *et al.*, 1998) or, less frequently, analysis of serum for rising antibody titers to rabies virus (Krebs *et al.*, 2002). A 2001 case of imported rabies in a California resident was diagnosed retrospectively on the basis of detection of rabies virus-specific antibody as part of an ongoing surveillance effort for unexplained encephalitis (Krebs *et al.*, 2003a).

Human rabies deaths in the USA occur throughout the year. Historically, the majority of human exposures occurred from spring through fall, following the peak in rabies reports among terrestrial carnivores that occurred in the winter and spring (Held *et al.*, 1967). In recent years, with the emergence of bat-associated rabies virus variants as a major contributor to human disease, a marked trend for late summer occurrence of human cases can be observed (Figure 4.8). This peak coincides with the peak reporting of rabies among bats in the fall (Childs *et al.*, 1994; Mondul *et al.*, 2003).

A disturbing recent trend in human rabies in the USA has been the inability to elicit a history of animal bite from the victim or close family members (Noah *et al.*, 1998). Public health experts believe that a bat bite is still the most plausible explanation for these cases, as documented instances of non-bite routes of infection are exceedingly rare and generally have occurred under exceptional situations (see below) which are not comparable to those described for recent cases (Gibbons, 2002). The most common explanations for the lack of history of bat bite obtained from recent rabies cases or their close personal contacts are that cases failed to report an event perceived as insignificant (Gibbons, 2002) or that the bat bite went unnoticed (Feder *et al.*, 1997). The phenotype of rabies virus variants associated with the bat species *Pipistrellus subflavus* and *Lasionycteris noctivagans* (P.s./L.n. variant), the most frequent variant associated with human rabies due to bat variants in North America (Messenger *et al.*, 2003b), may promote transmission via superficial wounds inflicted on peripheral body sites (Morimoto *et al.*, 1996; Messenger *et al.*, 2003b; Dietzschold *et al.*, 2005). The indication that unnoticed or trivial contact with bats may result in rabies transmission to humans has resulted in the additional recommendation to treat individuals

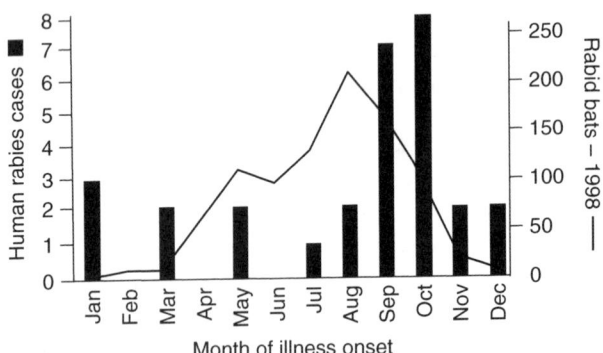

Figure 4.8   Cases of human rabies (January 1990–November, 2003) caused by variants of rabies virus associated with bats show a peak date of death in the fall. National surveillance data on reports of rabid bats also peak in the fall (data from Krebs *et al.*, 2004). The four transplant-related cases of rabies in humans are omitted, although the index case, who died in May 2004, is included. There is a marked increase in reports of rabid bats in the fall.

when bat bite cannot be ruled out (Centers for Disease Control and Prevention, 1999a).

3 ROUTES OF RABIES VIRUS TRANSMISSION TO HUMANS

3.1 Natural (non-iatrogenic) routes of transmission

The two natural (non-iatrogenic) routes of rabies virus entry occur through animal bite in which rabies virus in saliva is transdermally inoculated and through direct contact of rabies virus with mucosal membranes (World Health Organization, 2004). The most important route of rabies virus transmission to humans remains animal bite and little has changed since a review of human rabies cases from 1927 to1946, reported from Pasteur Institutes around the world, documented an animal bite in 99.8% of the 3920 cases (McKendrick, 1941). Although in 1999, 494 of the 1866 human rabies cases were characterized as exposure unknown (World Health Organization, 2002), many of these unknown cases presumably reflect difficulties in obtaining accurate histories from critically ill patients. Sporadic reports exist of human rabies following transmission by licks to mucous membranes (Leach and Johnson, 1940), transdermal scratches contaminated with infectious material and even improperly inactivated rabies vaccines (Para, 1965).

Two cases of natural transmission of rabies virus from human-to-human through either human bite or infectious saliva exposure to mucous membranes have been described from Ethiopia (Fekadu et al., 1996). Although human transplacental transmission of rabies virus has been reported in a single case (Sipahioglu and Alpaut, 1985), infants have survived delivery from mothers infected with rabies, when the child received PET (Lumbiganon and Wasi, 1990).

3.2 Iatrogenic routes of transmission

Although rare, iatrogenic human-to-human transmission of rabies has been well documented for recipients of transplanted human tissues. In 2004, four recipients of kidneys, a liver and an arterial segment from a common donor were diagnosed with rabies in the USA (Srinivasan et al., 2005). The donor had presented to two emergency rooms prior to his hospital admission and brain death was diagnosed due to a subarachnoid hemorrhage; his symptoms had included changes in mental status and autonomic instability, consistent with a diagnosis of rabies (Srinivasan et al., 2005). Rabies virus was recovered from CSF, 10% clarified homogenates of brain tissue, spinal cord and kidney suspensions of the 2004 transplant recipients (Srinivasan et al., 2005). In 2005, three cases of rabies in transplant patients from Germany were reported (Eursurveillance Weekly, 2005; Homola, 2005).

Prior to the recent organ transplant cases, eight human rabies deaths had been reported among corneal transplant recipients from France (Centers for Disease Control and Prevention, 1980), the USA (Houff *et al.*, 1979), Thailand (Centers for Disease Control and Prevention, 1981), India (Gode and Bhide, 1988) and Iran (World Health Organization, 1994). Administration of PET has been credited with preventing rabies in one recipient of a corneal transplant (Sureau *et al.*, 1981).

Under exceptional circumstances, such as use of blenders to homogenize rabies virus-infected brain material (Conomy *et al.*, 1977), infections of humans with rabies virus by air-borne droplets can occur. Rabies infection possibly acquired by droplets or aerosolized virus has been described in two persons visiting Frio cave in Texas (Irons *et al.*, 1957). Millions of Mexican free-tailed bats (*Tadarida brasiliensis*) congregate in this cave and rabies virus is present within the bat population (Humphrey *et al.*, 1960), and there is evidence of prenatal transmission among Mexican free-tailed bats (Steece and Calisher, 1989). Experimental studies with animals and with an electrostatic precipitation device suggest that air-borne transmission of rabies virus can occur under these exceptional circumstances (Constantine, 1962; Winkler, 1968).

There is little documentation for natural rabies transmission by simple contact with virus-infected tissue, although isolated reports suggest infection following butchering of infected carcasses (Tariq *et al.*, 1991). In the USA, ingestion of unpasteurized milk from rabid cows has been treated as a possible exposure to virus (Centers for Disease Control and Prevention, 1999b). Scratches received from a rabid animal could potentially be contaminated with saliva containing rabies virus and, in the USA, such exposures are treated with PET, although documented cases of this injury leading to rabies are rare (Babes, 1912).

It has been previously noted that for many of the recent human rabies cases in the USA, no history of animal bite could be solicited from the patient, relatives, or close companions (Noah *et al.*, 1998; Rupprecht and Gibbons, 2004). The reasons, as noted above, could include lack of awareness of the minor wound likely derived from a bite inflicted by small bats.

4 RISK AND PREVENTION OF RABIES FOLLOWING AN EXPOSURE

The risk of developing rabies depends on the anatomical site and severity of the bite, the species inflicting the wound and, presumably, the rabies virus variant. Data published by Babes and others (Babes, 1912; Baltazard and Ghodssi, 1954; Shah and Jaswal, 1976) and reviewed by Hattwick (1974), indicate the risk of developing clinical rabies in unvaccinated persons was 50–80% following multiple, severe head bites, 15–40% following multiple, severe finger, hand, or arm bites and 3–10% following multiple, severe leg bites inflicted by large terrestrial carnivores, such as wolves or bears. In general, exposure to most body fluids or blood from a rabid animal or a rabid human, with the notable exception of saliva and tears (Anderson *et al.*, 1984a, 1984b; Helmick *et al.*, 1987), is not regarded as reason for treatment (Centers for Disease Control and Prevention, 1999a). However, in certain circumstances, laboratory technicians reporting a definite and significant exposure (e.g. a technician cut by a broken specimen container from a rabies patient) to CSF or urine from a human rabies case have been given PET (Anderson *et al.*, 1984b).

In general, modern cell culture-derived vaccines, when properly administered with anti-rabies immunoglobulin (RIG), are virtually 100% effective in preventing rabies after an exposure has occurred (Centers for Disease Control and Prevention, 1999a). The need for administration of both vaccine and RIG for the prevention of rabies following exposure has been appreciated for decades (Baltazard and Bahmanyar, 1955; Hemachudha *et al.*, 1999). Vaccine failures have been reported when RIG was not infiltrated around the bite site (Centers for Disease Control and Prevention, 1987; Wilde *et al.*, 1996) or when RIG was omitted in treatment (Gacouin *et al.*, 1999). Potential vaccine failures and failure to seroconvert to rabies tissue and cell culture-derived vaccines among persons concurrently taking chloroquine for malaria have been documented (Pappaionou *et al.*, 1986). In India, PET with PVRC protected 17 of 19 victims bitten by a confirmed rabid fox (Matha and Salunke, 2005); the two patients who died had not received equine immunoglobulin (Matha and Salunke, 2005). Pregnancy, infancy, old age or concurrent infections are not contraindications for rabies PET (World Health Organization, 2004).

4.1 PET for non-rabies lyssaviruses

Rabies vaccines and RIG have also been used to treat exposures to other lyssaviruses (Nel, 2005), notably ABLV (Fielding and Nayda, 2005). As ABLV is an ST 1/GT 7 lyssavirus, rabies PET appears to be an acceptable alternative until a specific vaccine is developed. In cross-neutralization studies using human sera from persons vaccinated with human diploid-cell vaccine (HDCV), human antibodies protected against challenge with ABLV and ELBV types 1 and 2 in a modified fluorescent antibody virus neutralization assay at titers ⩾0.5 IU/ml (Fielding and Nayda, 2005). In experimental trials using PET to treat Syrian hamsters and domestic ferrets, rabies biologicals produced reduced protection against the four newly described Eurasian bat lyssaviruses (see Table 4.1; ARAV, KHUV, IRKV, and WCBV) (Hanlon *et al.*, 2005). The effectiveness of vaccine and RIG in animal models was described as '...inversely related to the genetic distance between the new isolates and traditional rabies virus', providing no significant protection against WCBV, the most divergent of the four new bat lyssaviruses from rabies virus (Hanlon *et al.*, 2005).

4.2 WHO and the Advisory Committee on Immunization Practices (ACIP) recommendations for pre-and post-exposure treatment (PET)

In the USA, all potential rabies virus exposures are treated with intramuscular vaccination and HRIG (Centers for Disease Control and Prevention, 1999a). Three vaccines are licensed in the USA for PET or pre-exposure immunization (Rupprecht and Gibbons, 2004). The WHO recommends PET treatment according to categorical grades of potential rabies virus contact and exposure (World Health Organization, 2004) and lists nine suitable vaccines for production for human and animal use. Grade II exposures (potential contact with rabies virus through nibbling of uncovered skin or minor scratches or abrasions with bleeding) require vaccination, unless or until the animal is determined to be negative for rabies virus infection. Grade III exposures (potential contact with rabies virus by single or multiple transdermal bites or scratches, licks, or broken skin, or contamination of mucous membrane with saliva [i.e. licks], or exposure to bats) require vaccination and RIG (World Health Organization, 2004). An additional difference from ACIP recommendations in the USA is that WHO recommends the use of intradermal vaccination for PET and pre-exposure vaccination with two commercially available vaccines (Goswami *et al.*, 2005), which permits cost savings of up to 84% (World Health Organization, 2004). The efficacy of this regimen using PCEC in protecting persons exposed to bites from animals with laboratory confirmed rabies has been well demonstrated (Quiambao *et al.*, 2005).

In developed countries, a bite from an animal is generally sufficient to ensure assessment of rabies risk and initiation of PET when required. As many as 40 000 persons annually may receive PET in the USA (Krebs *et al.*, 1998), although frequently treatment may be recommended in situations where it is unwarranted (Noah *et al.*, 1996). Increasingly, mass human exposures to rabid animals have taxed the local availability of rabies biologicals in the USA and resulted in expensive prevention programs (Rotz *et al.*, 1998; Robbins *et al.*, 2005). Of some surprise to public health officials, it appears that when recommendations for PET are not adhered to in an emergency room setting in the USA, most often PET is withheld when it should be recommended (Moran *et al.*, 2000). These findings suggest that closer adherence to recommended policies for deciding when to recommend PET may increase rather than decrease rabies biological use in the USA.

4.3 Rabies with no exposure history

Unfortunately, in the USA, the absence of a clear history of an exposing animal bite has been increasingly associated with the rise in bat-associated rabies cases (Messenger *et al.*, 2003b). The number of human rabies deaths where the exposing animal was unknown, although molecular biologic analyses may have implicated an ultimate reservoir host for the infecting rabies virus variant, has increased from 12% (N=16 cases) in the decade 1966–1975, to 55% (N=20) from 1976 to1985, to 82% and 71% (N=17 and 21, respectively) for the decade 1986–1995, and 8-year period from 1996 to 2003 (Rupprecht and Gibbons, 2004). It appears certain that rabies cases will continue to be identified from persons unaware that an exposure to rabies virus has occurred.

5   EPIDEMIOLOGY OF RABIES IN MAMMALIAN POPULATIONS

Rabies disproportionately affects populations of the species of terrestrial carnivore serving as the reservoir host. Transmission of rabies virus is primarily among conspecifics with cross-species transmission to other mammals, usually depending upon direct interactions with infected individuals of the reservoir host species (McQuiston et al., 2001; Krebs et al., 2003a, 2003b). Although virus spillover can ultimately lead to viral evolution and adaptation to a new species, which then serves as a true reservoir host, such events have been difficult to monitor on historical time scales and are generally inferred by molecular epidemiologic reconstruction (Smith, 2002; Hughes et al., 2005; Nel et al., 2005). By molecular phylogenetic analyses of rabies virus, a bat ancestry is hypothesized as the origin of rabies virus variants affecting terrestrial carnivores (Badrane and Tordo, 2001). Bat variants of rabies virus spill over sporadically to infect wild and domestic animals (McQuiston et al., 2001). In certain instances, as with red foxes on Prince Edward Island and striped skunks in Arizona (Daoust et al., 1996; Engeman et al., 2003), sustained transmission of bat variants of rabies virus within terrestrial carnivore populations has succeeded until natural extinction, or by vaccination control programs as occurred with skunk rabies in Arizona.

Rabies epidemics among wildlife reservoir host populations within defined regions frequently follow a distinct course. Intervals of increased disease activity (epidemics) are separated by intervals (inter-epidemics) in which rabies may seem to disappear or reach undetectable levels within a local mammalian community (Wandeler et al., 1974; Steck and Wandeler, 1980; Anderson et al., 1981; Childs et al., 2000). Following an initial epidemic of rabies, which is typically the largest of a possible series of epidemics that may emerge over time as wildlife rabies enters into a new region to infect previously naïve populations, a series of successively smaller epidemics may occur at increasing frequency (Anderson et al., 1981; Smith, 1985; Coyne et al., 1989); over time, the periodic epidemic structure of rabies epidemics may become indistinguishable against a background level of sporadic disease. A pattern of diminishing epidemic size but increasing frequency of epidemics has occurred with red fox rabies in regions of Europe (Lloyd, 1976; Toma and Andral, 1977; Macdonald and Voigt, 1985) and appears to be occurring in regions of the USA invaded by raccoon-associated rabies over the past few decades (Childs et al., 2000; Gordon et al., 2004, 2005). This pattern is best

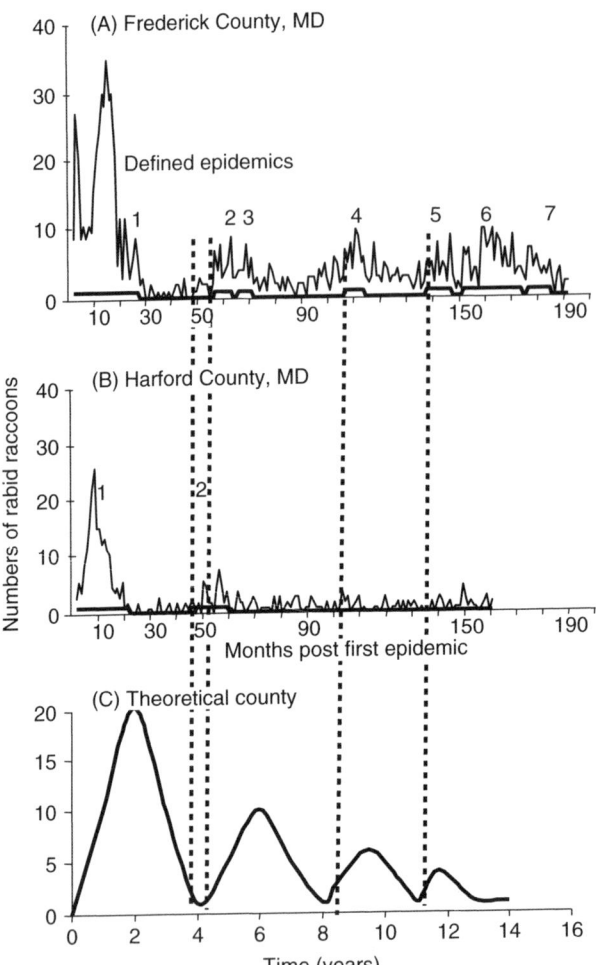

Figure 4.9 The temporal dynamics of rabies in raccoon populations is illustrated by actual counts of rabid animals from two counties in Maryland (A) and (B) and by the damped oscillations in the infected/rabid compartment of raccoons in a theoretical population (for details see Childs *et al.*, 2000). (A) The start time, end time and duration of successive epidemics of raccoon rabies are shown for Frederick County, Maryland, in the USA, as the bold line varying between 0 (no epidemic) and 1 (epidemic); the time scale is plotted relative to the starting time of the first epidemic in Frederick County. The numbered epidemics were identified through an epidemic algorithm that determined epidemic patterns for each individual county based on time series data for the number of rabid raccoons reported by each county through national surveillance data collected at the CDC. (B) An additional time series with epidemic intervals depicted and numbered is shown for Harford County, Maryland, which experienced raccoon rabies at a later date than did Frederick County. Note the close correspondence between epidemic outbreaks with the predicted periodicity of the damped oscillations of rabies predicted from a mathematical model of the population dynamics of the disease in a raccoon reservoir population (for details see Childs *et al.*, 2000). The analysis made no effort to combine what appear to be falsely divided epidemics (such as epidemics 2 and 3 in Panel A) into a single epidemic, as the results produced strong modes at the mathematically predicted time intervals without any modification.

described mathematically as a damped oscillation (Figure 4.9A, B, C) and over broad ranges in wildlife host population densities, such oscillatory behavior appears to be a robust pattern with wildlife rabies.

Rabies-induced temporal changes in the population density, or K, for a wildlife species serving as the sole reservoir host for a specific rabies virus variant in a region was first explored for the red fox in western Europe (Anderson *et al.*, 1981). Similar models, but introducing an additional immune compartment for animals exposed to rabies virus but surviving with immunity, were developed for raccoon rabies in the USA (Coyne *et al.*, 1989). These models and many others (e.g. Smith, 1985; Artois *et al.*, 1997; Shigesada and Kawasaki, 1997; Evans and Pritchard, 2001) have identified and explored the interplay of critical elements underlying the fluctuating behavior of rabies epidemics among populations of terrestrial reservoir hosts, including the influence of population density at varying environmental carrying capacity (K) on the number and magnitude (counts of rabid individuals of the reservoir host species) of successive epidemics; the significance and estimation of critical threshold densities (K_T) required to sustain rabies virus transmission with a basic reproductive number R_0 (the expected number of new infectious hosts that one infectious host will produce during its period of infectiousness in a large population of completely susceptible individuals [Halloran, 1998]) exceeding unity; the interplay between culling, vaccination and fertility control as potential interventions to control rabies in a terrestrial reservoir host (Suppo *et al.*, 2000; Smith and Wilkinson, 2003); and the influence of recovery with immunity to rabies infection (Coyne *et al.*, 1989; Murray and Seward, 1992).

In addition to models examining the population dynamics of rabies virus within reservoir host populations, a large number of spatial models have examined the characteristic of the traveling wave of epidemic rabies in a wildlife reservoir species and identified the impact of spatially heterogeneous landscapes in advancing or impeding the rate of epidemic expansion (Kallen *et al.*, 1985; Murray *et al.*, 1986; Gardner *et al.*, 1990; Jeltsch *et al.*, 1997; Evans and Pritchard 2001; Lucey *et al.*, 2002; Russell *et al.*, 2004). Examples of specific uses of models and validation of model outcomes are discussed below. A review of models developed for red fox rabies is available (Harris and White, 2004).

5.1 Epidemic cycles of terrestrial wildlife rabies and critical population parameters

5.1.1 Red fox rabies in Europe and North America

Estimates of the epidemic period for red fox rabies derived from surveillance data collected in Europe, the USA and Canada range from 2 to 7 years (Toma and Andral 1977; Steck and Wandeler, 1980; Carey, 1985; Tinline and Macinnes,

2004) in keeping with model predictions of 3–5 years (Anderson *et al.*, 1981; Smith, 1985). Prolonged cycles of rabies epidemics and increased duration of the damped oscillatory phase are associated with areas supporting relatively high fox population densities ($\geqslant 2$ per km^2) before rabies outbreaks (Steck and Wandeler, 1980; Smith, 1985); periodic oscillations of about a 5-year cycle, with a serrated profile, have been obtained when fox population densities are set to very high densities, such as $K = 10$ foxes per km^2 (Smith, 1985). Allowing for a high percentage of animals to become immune and recovered after rabies virus exposure can also dramatically alter the pattern of epidemic rabies among wildlife, such as red foxes and raccoons (Coyne *et al.*, 1989; Murray and Seward 1992), although immunity is not thought to play a significant role in the epidemiology of rabies among terrestrial carnivores (Shigesada and Kawasaki, 1997 and see following section on raccoon rabies). Areas of low fox density may fail to support rabies virus transmission or experience only sporadic disease (Steck and Wandeler, 1980); rabies epidemics are an unlikely occurrence in red fox populations where $K < K_T$, (Figure 4.10A) (Steck and Wandeler, 1980).

The available estimates from field data of the hunter indicator of fox population density (HIPD; foxes shot per km^2 per year) (Steck and Wandeler, 1980) suggest K_T is on the order of one red fox per km^2 (Figure 4.10A) (Anderson *et al.*, 1981). Actual estimates of red fox population densities indicate the typical range is between 0.1 and 4.0 foxes per km^2, although substantially higher population densities have been documented in suburban locations in the UK (Macdonald, 1980).

Areas of high quality habitat supporting high population densities of red foxes suffer the greatest population depression due to rabies and the highest incidence of disease during the initial epidemic, as based on HIPD estimates of red fox population size (Steck and Wandeler 1980). Based on HIPD, rabies reduced populations of red foxes in numbers to 50–60% below pre-epidemic levels (Bögel *et al.*, 1974). Population densities of red foxes rarely recover to population levels predating rabies, supporting mathematical model outcomes and indicating host populations may be regulated by rabies around K_T (Müller and Breitenmoser, 2004). Estimates of the time in years to recovery of red fox populations to prerabies densities, as a function of the percentage of the population surviving a rabies epidemic (Figure 4.10B), illustrate how high reproductive rates among these carnivores can rapidly increase densities above the minimum values required to sustain periodic re-emergence of disease outbreaks (Bögel *et al.*, 1974; Macdonald, 1980).

Prevalence is a difficult attribute to estimate for a wildlife disease such as rabies, as the required denominator (the population at a specific time or average population size during an interval of time) is almost never known. However, theoretical estimates and reported estimates from HIPD of the equilibrium prevalence of rabies suggest the value remains fairly constant at 3–7% during outbreaks (Bögel *et al.*, 1974; Anderson *et al.*, 1981).

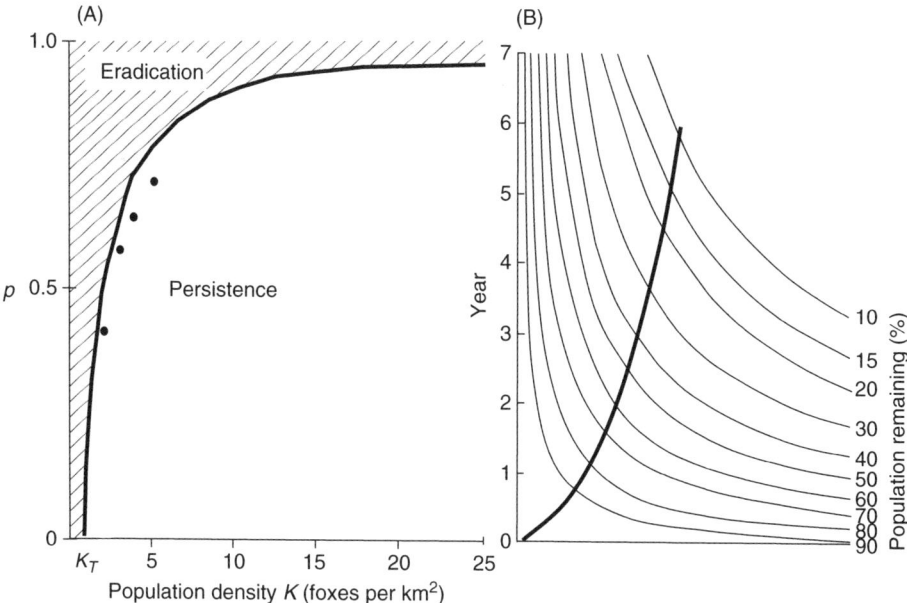

Figure 4.10 (A) The relationship between the proportion of a red fox population protected against rabies by vaccination and the disease-free carrying capacity of the fox habitat, K. The threshold density required for sustained transmission of rabies virus within a fox population, K_T, is taken to be $K_T = 1.0$ fox per km^2. The shaded region represents values for which rabies cannot persist, as determined from equation 11 in the original paper (Anderson *et al.*, 1981). The four points are taken from a detailed model developed by Berger (1976) and indicate the predicted boundary levels of vaccination which result in rabies persistence or eradication. Panel A is reproduced (with slight modification) from Figure 5 in Anderson *et al.* (1981), with permission from Macmillan Publishers Ltd. (B) A rule of thumb diagram for best estimating the speed of recovery of red fox populations decimated by a rabies epidemic. The intersection of the bold line with the number lines indicates the number of years required for the population to recover, depending on the proportion of the population escaping infection and death. Reproduced (with slight modification), with permission, from Figure 5.14 in Macdonald (1980). The original figure was first published by Bögel *et al.* (1974).

5.1.2 Raccoon rabies in the USA

Raccoons have been the wild animal species most frequently diagnosed as rabid since the early 1990s (Figure 4.11). The process of epidemic rabies among raccoons in the USA, as defined by time series data collected through national surveillance efforts (for a review of the strengths and weaknesses of the animal-based national surveillance program for rabies in the USA, see Childs *et al.*, 2002) is similar in most respects to that described for red foxes in Europe. The median predicted first epidemic period (interval from the start date of the initial epidemic as raccoon rabies enters a previously unaffected area and the start date of the second epidemic as defined by a statistical algorithm [Childs *et al.*, 2000]) obtained from

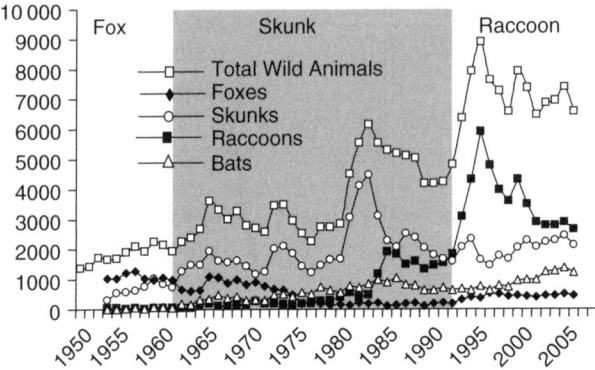

Figure 4.11 Numbers of reported cases of rabies among wildlife in the USA, 1955–2005, as reported through national surveillance (data obtained from annual CDC surveillance reports). The different zones of gray and white overlayed on the figures indicate the dominant wildlife species reported rabid during a defined interval.

time series data from the eastern USA was 48 months. This value was in excellent agreement with theoretical predictions (Coyne *et al.*, 1989; Childs *et al.*, 2000) when the immune compartment of rabid raccoons resulting from survival after rabies virus exposure was set to be ≤5%. The original model value for the percentage of raccoons surviving rabies to become immune was set at 20% (Coyne *et al.*, 1989). This value was based on serological survey data which may have included a large number of false positive results due to cytotoxic substances present in some wildlife sera, notably among raccoons and striped skunks (Barton and Campbell, 1988). These substances can interfere with rabies serological tests based on detecting neutralizing antibody. There are no recent data to suggest that many infected raccoons survive infection by the raccoon variant of rabies virus (Seidensticker *et al.*, 1988; Torrence *et al.*, 1992). The low expectation of raccoon survival following rabies virus exposure is also in keeping with the experiences with red fox rabies in Europe (Torrence *et al.*, 1992).

Successive epidemics of raccoon rabies were smaller than the initial epidemic and occurred at increasing frequency, with a decline in each successive epidemic period of approximately 5% (Childs *et al.*, 2000, 2001). The characteristics of the damped oscillations of raccoon rabies epidemics were concordant with expectations from solutions obtained to mathematical models for the infected or rabid class of raccoons subjected to the same statistical algorithm used on national surveillance data (Coyne *et al.*, 1989; Childs *et al.*, 2000)

There are no published estimates for the critical threshold value of the raccoon population density required to support rabies transmission in areas of the USA. However, direct and indirect estimates of raccoon population size or density indicate significant declines in population size following epidemics of raccoon rabies (Anthony *et al.*, 1990; Riley *et al.*, 1998). For reasons which are unclear, but may be related to surveillance methods in addition to biological factors, epidemics of raccoon rabies at the county level occur less often and at diminished size as one progresses from north to south in the USA (Childs *et al.*,

2001). The magnitude of recorded epidemics, as well as the number of animals tested for possible rabies infection, is clearly correlated with the human population size or density at the level of township and county (Wilson *et al.*, 1997; Gordon *et al.*, 2005). Consequently, assessments of epidemic behavior must be adjusted in analyses based on surveillance data, as they are inherently dependent upon numerous human–animal interactions (Childs *et al.*, 2006).

5.1.3 Skunk rabies in North America

Rabies among skunks has a long history in North America. The earliest reports of rabies, from California in 1826, incriminate spotted skunks (genus *Spilogale*) as the source of human disease (Parker, 1975). Altogether seven species of skunks in three genera (*Mephites*, the striped skunk and *Conepatus*, the hog-nosed skunks, in addition to *Spilogale*) exist in North America, however, the striped skunk is by far the most common (Parker, 1975).

Currently, two rabies virus biotypes, the South-Central and North-Central biotypes, circulate among skunks in the central USA (see Figure 4.2), with the North-Central skunk biotype extending into central and western Canada; a third biotype of rabies virus circulates among skunks in California (Crawford-Miksza *et al.*, 1999). Beginning in the 1960s and continuing until 1990 (see Figure 4.11), skunks were the group of terrestrial mammals most frequently reported rabid in the USA.

Analyses of the temporal patterning of epidemics among skunks in North America reveal major outbreaks with a period of 6 to 8 years (Gremillion-Smith and Woolf, 1988), although shorter periods of 4 years have been reported from western Canada (Pybus, 1988). In general, the percentage of skunk submissions testing positive for rabies virus antigen is strongly bimodal, with peaks in the spring and fall (Gremillion-Smith and Woolf, 1988). However, in Texas, Illinois and Arkansas, the spring peak predominates (Heidt *et al.*, 1982; Pool and Hacker, 1982).

In the USA and Canada, skunk rabies is common in prairie habitats (Pool and Hacker, 1982; Greenwood *et al.*, 1997), where periodically epidemic disease has been postulated to be a major factor in driving the cyclic variations in skunk population numbers (Pybus, 1988). The potential impact of rabies on skunk populations was amply demonstrated by Greenwood *et al.* (1997), who followed a population of radio-collared animals during an epidemic in South Dakota. Estimated densities of skunks fell from 0.85 skunks per km^2 during April to June, 1991, to 0.17 skunks per km^2 in April to July, 1992, during the rabies epidemic.

5.1.4 Vampire bat rabies and epidemic cycles of bovine rabies

Cycles of rabies outbreaks have only been determined for the common vampire bat (*D. rotundus*). Although cycles in rabies outbreaks among insectivorous bats have not yet been described, a well-defined seasonality in the reporting of rabid bats in the late-summer/early fall is well known from national and state surveillance data gathered from temperate zones of the USA (Childs *et al.*, 1994; Mondul *et al.*, 2003).

Vampire bats have been associated with rabies epidemics among cattle in South and Central America since this association was first noted in 1910 in Brazil (Carini, 1911). The first human deaths attributed to bites received from vampire bats were documented on the Island of Trinidad (Hurst and Pawan, 1931; Pawan, 1936), where vampire bat-transmitted rabies continues to be a sporadic disease of cattle (Wright *et al.*, 2002). Outbreaks of human rabies due to vampire bats continue to be reported from many South and Central American countries and rabies virus variants originating from vampire bats continue to be isolated from cattle and other species in Argentina (Cisterna *et al.*, 2005), Bolivia (Favi *et al.*, 2003), Brazil (Batista-da-Costa *et al.*, 1993; Ito *et al.*, 2001; Kobayashi *et al.*, 2005), Peru (Lopez *et al.*, 1992), Venezuela (Caraballo, 1996), Costa Rica (after a 31-year interval without any reports of human rabies [Badilla *et al.*, 2003]) and Mexico (Martinez-Burnes *et al.*, 1997; de Mattos *et al.*, 1999).

Vampire bats feed preferentially on livestock and most of the economic burden they pose is through bovine paralytic rabies (Baer, 1991; Delpietro and Russo, 1996). Estimates from the 1960s placed the loss of cattle at 100 000 to 500 000 head per year in Latin America (Acha, 1967), whereas estimates from 1985 place the figure at 100 000 head at a cost of 30 million US$ (Acha and Arambulo, 1985). Because of this economic burden, vampires have been the target of major control efforts. Anticoagulants, applied by topical treatment to the back of captured bats, subsequently released to return to roosting sites (Linhart *et al.*, 1972), or through systemic treatment of cattle, have been used to achieve reductions in vampire bat biting rates on cattle of 85 to 96% (Flores-Crespo and Arellano-Sota, 1991). Unfortunately, vampire bat control also leads to the death of many non-target species of bats (Mayen, 2003), such that an ORV based on V-RG, has been developed for topical application to vampires and may some day replace the use of anticoagulants (Almeida *et al.*, 2005).

Vampire bat colonies appear to be naturally decimated by rabies virus infections and discrete colonies may fall to population densities incapable of maintaining transmission. The so-called 'migration' of vampire bat-associated rabies described from South America (for detailed discussion of the epidemiology of vampire bat rabies, see Brass, 1994) is compatible with a traveling wavefront of rabies propagating among neighboring vampire bat metapopulations, whereby populations are decimated in the trough behind the traveling wavefront and require time to recover above some critical population density to support rabies virus transmission. Control of population densities below the K_T required to maintain and transmit rabies virus has been recommended as a control strategy for vampire bats (Delpietro and Russo, 1996).

The migratory wave of vampire bat rabies has been estimated to travel at 40–50 km per year (Brass, 1994), remarkably similar to the rates (30–60 km per year) established for the spread of red fox and raccoon rabies epidemics (Lucey *et al.*, 2002; Wandeler, 2004). Cyclic changes in vampire bat populations could drive cyclic and periodic epidemics of cattle rabies caused by vampire bat transmitted rabies. In regions of Central and South America, areas affected by vampire rabies experience outbreaks every 2 to 3 years (Ruiz-Martínez, 1963). Rabies virus prevalence increases or is only present among vampire bats captured just before or after an outbreak of cattle rabies (Fornes *et al.*, 1974; Lord *et al.*, 1975; Delpietro and Konolsaisen, 1991), although there may be also a seasonal increase in bovine rabies cases and possibly the percentage of vampire bats testing positive for rabies during the rainy season (Turner, 1975).

6 DISSEMINATION OF CANINE RABIES AND EVIDENCE FOR PERIODIC CYCLES

In contrast to rabies virus variants found in wildlife, limited antigenic and genetic diversity is found among rabies virus isolates from dogs from many locations in the world (Smith, 1989; Nadin-Davis and Bingham, 2004), allowing reconstruction of the probable introduction of rabies into South America, Africa and parts of Asia due to translocation of dogs along with early European colonialists (Smith *et al.*, 1992; Smith and Seidel, 1993; Nadin-Davis and Bingham, 2004). The similarity of canine rabies isolates from Latin America, Africa, Asia and Eastern Europe, reflect a global reservoir of rabies in dogs that arose from a common source and the isolate has been termed 'the cosmopolitan strain' (Nel and Rupprecht, 2006). The cosmopolitan strain of rabies is believed to have its origins in the Palearctic region, which includes Europe, the Middle East and northern Africa (Badrane *et al.*, 2001; Nadin-Davis and Bingham, 2004).

Rabies in sub-Saharan Africa, West Africa, East Africa and southern Africa is dominated by the cosmopolitan strain of rabies virus, which in many regions only became a significant problem in the 1900s (Badrane *et al.*, 2001; Johnson *et al.*, 2004; Bingham, 2005). The domestic dog remains the major reservoir species for rabies virus throughout Africa, although other rabies virus strains may have circulated in West Africa since antiquity (Nel and Rupprecht, 2006) and Africa serves as the ancestral home of several other lyssaviruses (DUVV, MOKV, LGBV) (Calisher *et al.*, 1989).

Importation of dogs from Europe to the New World was known from the time of the second voyage of Columbus (Varner and Varner, 1983) and, within a few centuries, European breeds had essentially replaced native dogs in the Americas. Dog rabies was first recognized in the Greater Antilles in the 18th century during the time of Spanish dominion in this area and in Mexico as early as 1709 (Smithcors, 1958; Steele and Fernandez, 1991). The first outbreaks of dog rabies in South America were recorded in 1803 in Peru and in 1806 in LaPlata, Argentina, among sporting dogs belonging to British officers (Steele and Fernandez, 1991).

However, through mutation of rabies virus within canine metapopulations there is sufficient sequence variation among variants of rabies solely associated with dogs to detect differences within canine populations. This diversity permits epidemiological tracking of how dog dispersal, along with their human hosts, influences national patterns of rabies in countries such as Thailand (Denduangboripant *et al.*, 2005).

The strongest evidence of epidemic cycles of dog rabies comes from studies conducted in South America, often before the initiation of large-scale mass vaccination campaigns. In Bolivia, a 5- to 6-year cycle in canine rabies was evident during the interval between 1972 and 1997, as was a 5-year cycle in Chile between 1950 and 1960 (Ernst and Fabrega, 1989; Widdowson *et al.*, 2002). As the incidence of canine rabies declined in Chile during the 1960s and 1970s, the cyclic pattern to canine rabies disappeared (Ernst and Fabrega, 1989). Detailed epidemiologic data on the characteristics, such as comparative magnitude, of successive epidemics among dogs are lacking. However, young dogs, frequently <1 year of age, are a major demographic class contributing to rabies outbreaks (Malaga *et al.*, 1979; Eng *et al.*, 1993; Mitmoonpitak *et al.*, 1998). Young dogs are less likely to have completed a full course of vaccination and may be inherently more susceptible to rabies virus infection (Beran, 1982; Centers for Disease Control and Prevention 2005a).

Although significant numbers of feral, stray or neighborhood dogs contribute to free-ranging populations in developed and developing countries (Beck, 1973; Flores-Ibarra and Estrella-Valenzuela, 2004), the demographic features of dog populations are strongly influenced by humans; cyclic behavior driven by natural rates or reproduction and population recovery are readily masked (Perry, 1993; Kitala *et al.*, 2001; Flores-Ibarra and Estrella-Valenzuela 2004). Human behaviors with regard to recruitment of companion animals, in addition to their interactions with stray or neighborhood dogs (Beck, 1973, 2000; Matter and Daniels, 2000), can rapidly alter the demographic features of free-ranging dog populations.

There is no clear consensus as to the degree of seasonality to rabies outbreaks among dog populations or varying risk of human rabies exposure through bites inflicted by stray dogs. In Africa, the incidence of dog-associated rabies appears to be largely non-seasonal and relatively consistent throughout the year in Tanzania and Zimbabwe (Bingham *et al.*, 1999a; Kayali *et al.*, 2003a). However, in Namibia, dog rabies incidence is lowest from January through June and there is a cyclic pattern to dog rabies incidence, with a 24-month lag between cycles (Courtin *et al.*, 2000). Clustering of canine rabies cases within two months (April and September) has also been reported from Nigeria (Ezeokoli and Umoh, 1987) and Ghana (July–September and January–March) (Addy, 1985).

Peaks in canine rabies reports associated with a spring breeding season for dogs, measured by household surveys or observation of the proportion of pups in the canine population, have been demonstrated from several Southern Hemisphere nations, including Nigeria, Peru and Chile (Malaga *et al.*, 1979; Ezeokoli and Umoh 1987; Ernst and Fabrega 1989). In Chile, there is some evidence of an increase in canine rabies during the spring/summer months of November and December; however, these data were acquired prior to the initiation of canine rabies control (Ernst and Fabrega, 1989).

In the USA, a peak in reports of rabid dogs submitted through national surveillance efforts was discernable in the spring (March and April) when counts of rabid dogs numbered in the 1000s (Held *et al.*, 1967). A modest surplus in counts of rabid dogs during the spring can still be noted in some annual surveillance reports obtained over the last three decades (e.g. Krebs *et al.*, 1996, 2003a).

7 CROSS-SPECIES TRANSMISSION (SPILLOVER) OF RABIES VIRUS

7.1 Bats as ancestral hosts of rabies virus

Evidence has accumulated that all rabies virus variants affecting terrestrial carnivores originated from cross-species transmission of bat-associated variants of rabies virus (Badrane and Tordo, 2001). In North America, a molecular clock model suggests the date of divergence of extant bat-associated rabies virus from the most recent common ancestor occurred about 1651–1660 (Hughes *et al.*, 2005). The bat rabies virus variants found in Latin America among the common vampire bat (*D. rotundus*) and species of the genus *Tadarida*, family Mollosidae or free-tailed bats, are closest to the earliest common ancestor. Adaptation of rabies virus variants to colonial species of bats, genera *Eptesicus* and *Myotis*, occurred more rapidly and earlier in time than did adaptation to the more solitary genera, *Lasionycteris*, *Pipistrellus*, and *Lasiuris* (Hughes *et al.*, 2005).

7.2 Spillover of bat-associated rabies virus variants to humans in North America

The role of insectivorous bat rabies in causing human rabies death in North America has been described above and reviewed by others (Gibbons 2002; Messenger *et al.*, 2002, 2003a). The bat variants most commonly associated with human rabies in Canada and the USA are those associated with the species *P. subflavus* and *L. noctivagans* (Krebs *et al.*, 2001; Messenger *et al.*, 2003b), followed in frequency by variants associated with *T. brasiliensis*, with the rare case of other species involvement as occurred in two human rabies cases from Washington State with *Eptesicus fuscus*- and *Myotis* species-associated variants (Krebs *et al.*, 1996, 2001).

7.3 Spillover of wild carnivore rabies to humans in North America

Although the raccoon rabies epidemic has been one of the most intensive and extensive wildlife epidemics ever recorded, involving hundreds of thousands of animals affected by spillover of this variant (Childs *et al.*, 2000; Gordon *et al.*, 2004, 2005), only a single human rabies death has been attributed to this particular variant (Centers for Disease Control and Prevention, 2003). Since the 1950s, nine cases of human rabies associated with skunk exposures have been diagnosed in the USA and two in Canada, the last occurring in 1970 in Arizona (Tabel *et al.*, 1974; Anderson *et al.*, 1984a). The last human case attributable to exposure to a rabid fox was in Kentucky in 1961 (Anderson *et al.*, 1984a).

7.4 Rabies epidemics and human exposure as assessed by PET

Most studies documenting an increase in human PETs have only spanned the interval of the first rabies epidemic as raccoon rabies enters new regions (Uhaa *et al.*, 1992; Hanlon and Rupprecht, 1998; Moore *et al.*, 2000). In New York State, the number of individuals receiving post-exposure prophylaxis increased from 84 in 1989, prior to the introduction of raccoon rabies, to 1125 in 1992 and 2905 in 1993 (Centers for Disease Control and Prevention, 1997); similar dramatic increases have been reported from other states (Centers for Disease Control and Prevention, 1996).

There are few data regarding the long-term need for PET following the initial epidemic stage of raccoon rabies. Data on PET use during successive epidemics and into the endemic phase of raccoon-associated rabies are required for understanding the long-term risk of human and animal exposure to rabies virus as reports of rabies within principal reservoir host populations wax and wane (Gordon *et al.*, 2004, 2005). Such data are also essential for better estimates of the cost and benefits potentially accrued by active interventions against wildlife rabies through ORV (Meltzer, 1996; Foroutan *et al.*, 2002; Kemere *et al.*, 2002; Gordon *et al.*, 2005).

In a pattern analogous to persistently elevated risk of rabies spillover, the number of human PETs can remain at elevated levels in regions of the USA where counts of rabid raccoons have greatly diminished in post-epidemic temporal stages (Chang *et al.*, 2002). The number and costs associated with testing animals suspected of being rabid do not decline even as the number of rabid raccoons reported from an area decline, but remain elevated well above pre-epidemic levels (Gordon *et al.*, 2005).

7.5 Epidemiology and spillover of dog-associated rabies virus variants

Descriptive studies of human rabies associated with epidemics or outbreaks of domestic dog rabies provide some of the clearest demonstrations of how human disease, acquired through spillover transmission, is driven by incidence or reporting levels of rabies in a reservoir host species.

Quantitative estimates of human mortality due to rabies have been frequently based on dog-bite injuries and estimates of the densities of dog populations (Perry *et al.*, 1988; Brooks, 1990; Robinson *et al.*, 1996), human population density and human-to-dog ratios (Knobel *et al.*, 2005) and knowledge of the incidence of bite injuries and risk of canine rabies within different populations of dogs (Cleaveland *et al.*, 2002a, 2003; Fevre *et al.*, 2005). The close relationship between reports of rabid dogs and human rabies deaths, within countries where the cosmopolitan variant of rabies virus circulates within canines, has been demonstrated in Africa (see Figure 4.3) (Bingham, 2005; Nel and Rupprecht, 2006), Asia (Denduangboripant *et al.*, 2005) and South America (Ernst and Fabrega 1989; Belotto *et al.*, 2005); this association was widely appreciated within developed countries prior to canine rabies control (Tierkel *et al.*, 1950). As dog rabies becomes controlled, cycles of rabies among wildlife reservoirs can become increasingly apparent with increased spillover of rabies associated with distinct increases in the incidence or reporting of the disease among the wildlife reservoir host species.

Variants of rabies virus introduced by domestic dogs have been implicated as the recent source of rabies virus independently circulating among several populations of wild carnivore. In South Africa and Zimbabwe, rabies virus variants recovered from the side-striped jackal (*Canis adustus*), the black-headed jackal (*Canis mesomelas*) and the bat-eared fox (*Otocyon megalotis*) fall within the broad phylogenetic group of the cosmopolitan type of rabies virus associated with domestic dogs (Bingham *et al.*, 1999b; Sabeta *et al.*, 2003; Bingham, 2005). Dogs, however, remain the primary rabies threat for humans and animals in Africa. During an outbreak of rabies in Zimbabwe from 1980 to 1983, the majority of documented animal cases occurred in jackals (74.3% of 404 cases; *C. mesomelus* and *C. adustus*), but domestic dogs were responsible for most human rabies (Kennedy, 1988).

A similar situation to that in Zimbabwe may be developing in northern South America. In Columbia, rabies virus variants believed to be circulating independently among a gray fox (*U. cinereoargenteus*) reservoir are most closely related to, and group phylogenetically with, domestic dog variants (Páez *et al.*, 2003).

It must be noted and emphasized that spillover of rabies virus variants from domestic dogs is not only a public health and veterinary problem, but a major concern to conservation efforts to protect species of endangered carnivores in Africa. In East Africa, rabies spillover from domestic dogs has been confirmed as the cause of death in wild dogs (*Lycaon pictus*) and has been incriminated in the disappearance of more than eight separate packs of these highly social canids in Kenya and Tanzania (Gascoyne *et al.*, 1993; Kat *et al.*, 1995). In 1991–1992, rabies was implicated in the rapid disappearance of 75 individuals of the 111 known Ethiopian wolves (*Canis simensis*) living in five packs within a national park in Ethiopia (Sillero-Zubiri *et al.*, 1996). Rabies virus was confirmed as the cause of death in samples taken from carcasses and decomposed brain tissues (Whitby *et al.*, 1997).

7.6 Epidemiology of mongoose-associated rabies; dog-associated and native biotype

Dog-associated variants of rabies virus have been implicated in the inception of novel maintenance cycles in wildlife within the past few centuries. The Asian yellow mongoose (*Herpestes javanicus*; formerly designated as *Cynicus penicillata*; family Herpestidae)- associated rabies in the Caribbean has been historically linked to the timing of this species' introduction from Asia into the Caribbean and the mongooses' subsequent infection with dog-associated rabies virus variants; inter-island variation among rabies virus variants from mongooses is substantial and suggests multiple cross-species introductions of virus from dogs (Smith *et al.*, 1992).

All mongooses present today in the Caribbean are descendants of animals brought from India to Jamaica in the 1870s for rodent control on sugar cane plantations (Everard and Everard, 1985). Although it is possible that rabies virus was introduced with these animals, official reports of rabies in mongooses in the Caribbean were not made until 1950 in Puerto Rico (Tierkel *et al.*, 1952). Within a few years, rabies was diagnosed in mongooses in Cuba, Grenada and the Dominican Republic (Everard and Everard, 1985). Mongooses continue to be a source of human rabies in the Caribbean where the virus is referred to as the dog/mongoose variant of rabies virus (see Figure 4.2) (Krebs *et al.*, 2004).

In sub-Saharan Africa, yellow mongoose (*C. penicillata*) rabies appears to involve rabies virus variants distinct from those associated with domestic dogs. Data generated from recent molecular epidemiologic studies indicate an extended history of evolutionary adaptation of rabies virus within yellow mongoose and slender mongoose (*Galerella sanguinea*) populations in South Africa and Zimbabwe, respectively, and suggest that the endemic area affected by these variants has a heretofore unappreciated geographic range (Nel *et al.*, 2005). The mongoose 'biotype' of rabies virus has been identified from other mammals in Zimbabwe and South Africa (King *et al.*, 1993; Nel *et al.*, 1997), indicating the potential for cross-species transmission of mongoose rabies. However, such occurrences are rarely documented due to the poor surveillance coverage for rabies and other lyssaviral infections in this region (Chaparro and Esterhuysen, 1993).

7.7 Spillover of wild carnivore and insectivorous bat rabies to wildlife

Although the population and temporal dynamics of rabies among wildlife reservoir species has been described for single reservoir host populations, such as red foxes and raccoons (Anderson *et al.*, 1981; Coyne *et al.*, 1989; Artois *et al.*, 1997; Ireland *et al.*, 2004), there are no formal mathematical models addressing the phenomenon of how the temporal dynamics of rabies within an animal reservoir relate to patterns of rabies spillover to humans and other mammalian species.

During epidemics of rabies in a wildlife reservoir host species, the reported number of cases among other domestic and wildlife species also increases. Specific examples include an increase in rabid stone martins and other species associated with epidemic red fox rabies in Europe (Wandeler *et al.*, 1974; Toma and Andral 1977), an increase in rabid skunks and raccoons associated with increasing incidence of fox-associated rabies in New York State in the 1940s and 1950s (Carey, 1985) and an increase in rabid skunks, cats and other species associated with the epidemic occurrence of raccoon-associated rabies in the eastern USA (Guerra *et al.*, 2003; Gordon *et al.*, 2004, 2005). Initially, as rabies counts among the principal reservoir host species decline, reports of rabies among other species may also decline precipitously. Wandeler *et al.* (1974) summarize the temporal effects of red fox variant-associated rabies virus and its spillover to other species in the following sentence (p. 743): 'It is important to note that rabies cases among other wildlife species (notably stone martens, badgers, and roedeer) or domestic animals are proportional to fox rabies and have disappeared in parallel or shortly after the disappearance of fox rabies'. However, studies from the USA have documented a persistent and longstanding effect due to spillover of raccoon-associated rabies virus, even after rabies cases in the reservoir host species have waned (Gordon *et al.*, 2005).

As noted by Carey (1985), a major deficiency in rabies studies among wildlife has been the failure to investigate patterns of endemic disease. While the correlation between increasing reports of rabies among a wildlife reservoir host and all other species during an initial epidemic of rabies appears clear enough, much less is known about rabies persistence and re-emergence in a sylvatic reservoir host and spillover. Although numerous descriptive studies have demonstrated that cases of spillover increase with the initial epidemic of raccoon rabies (Jenkins and Winkler, 1987; Fischman *et al.*, 1992; Wilson *et al.*, 1997), in many instances, the long-term correlation between counts of rabies among raccoons and cases of spillover in various species is less clear and may appear to become 'uncoupled' (Fischman *et al.*, 1992). In Maryland, an unexpected excess reporting of cat rabies persisted years after the initial wavefront of an epidemic of raccoon variant-associated rabies had passed (Fischman *et al.*, 1992).

In various areas of Ontario, Canada, persistence of the Arctic fox variant of rabies was frequently identified in animals other than red foxes, the principal reservoir host, from sites monitored by quarter year intervals over a 13-year span (Macinnes *et al.*, 2001). Skunks are the dominant wildlife species affected by spillover of the Artic fox variant of rabies virus in Canada (Tinline and Macinnes, 2004), as is also the situation in areas of the USA affected by the raccoon variant of rabies virus (Guerra *et al.*, 2003). In the USA, the number of reported cases of spillover of raccoon variant rabies virus to skunks can be predicted by the number of rabid raccoons with a 1-month lag (Figure 4.12). Of special interest was the observation that the biannual, late winter–early spring and fall peaks in skunk rabies (Figure 4.12), typical of midwestern regions

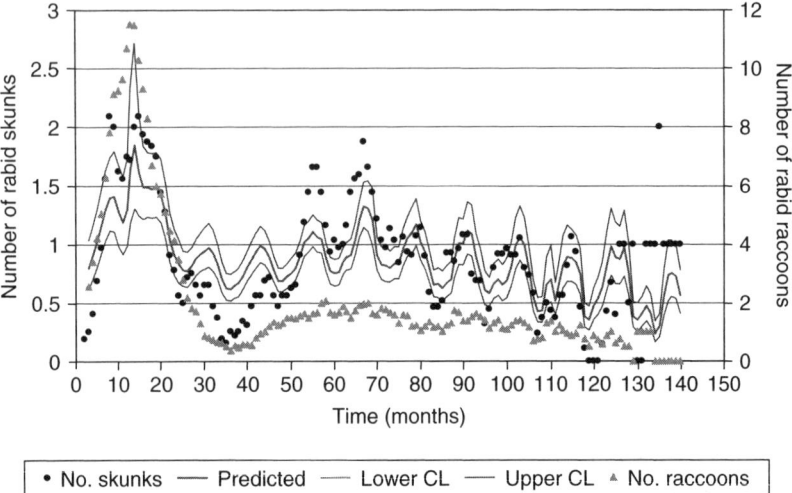

Figure 4.12 The temporal dynamics of the raccoon-associated variant of rabies virus among raccoons and spillover of this variant to skunks are indicated relative to the first month of the first epidemic of raccoon rabies, averaged from individual determinations from 32 counties in 11 eastern states. The green triangles (rabid raccoons) and black points (rabid skunks) are actual average values obtained from individual counties from 11 eastern states. The fitted lines predict the temporal dynamics of rabies in skunks and is best modeled with a lag of one-month following the onset of an epidemic among raccoons. The epidemic pattern of rabies among skunks maintains a distinct bimodal annual pattern that is also present among rabid skunks in the mid-western USA where skunks are the primary reservoir of skunk-adapted rabies virus variants (see Figure 4.2 and text for more details). Figure from Guerra *et al.* (2003). This figure is reproduced in the color plate section.

where skunk-associated variants of rabies virus drive epidemics among skunks (Bartlett and Martin, 1982; Gremillion-Smith and Woolf 1988), were preserved in epidemics involving skunks driven by a different variant of rabies virus maintained in different reservoir hosts (Guerra *et al.*, 2003).

7.8 Spillover of wild carnivore and insectivorous bat rabies to domestic animals

In the USA, virtually all the rabies cases occurring among wild and domestic animals within any geographic region are caused by the variant of rabies virus circulating in the species of terrestrial carnivore serving as the regional reservoir host. Of the 308 rabies virus variants typed from brain samples obtained from domestic dogs and cats in the USA in 1999, 307 were of the expected terrestrial rabies virus variant circulating among wildlife in the region from which the domestic animal sample originated (McQuiston *et al.*, 2001). National surveillance for animal rabies, coupled with genetic typing of viruses sampled from different geographic regions, provides a reasonable indicator of the geographic extent of endemic areas for different rabies virus variants and can be used to predict accurately the likely source of most animal rabies in a given locale (see Figure 4.2) (Childs *et al.*, 2002).

The obvious and most significant exception to the general rule concerning terrestrial carnivores as the source of rabies virus infection is the complex overlay of insectivorous-bat associated rabies virus variants across the USA and Canada. As previously described, these viral variants have been the source of most human cases of rabies acquired in the USA over the past decade and a half (Gibbons, 2002; Messenger *et al.*, 2003a). Among a survey of rabies virus infections among domestic animals a single bat-associated variant was identified from all 308 samples examined (McQuiston *et al.*, 2001).

7.8.1 Rabies virus spillover to domestic dogs

With the advent of effective dog management programs, which includes high levels of vaccination in Europe and North America, spillover of wildlife variants of rabies virus to dogs has declined in these locations to levels below that reported for cats. Over the past several decades in Western Europe, dog rabies has been reported at approximately half the annual rate of the figure for cat rabies, with the average number of annual reports well below 50 to 100 for most nations (for comparative numbers see section on cats below) (Aubert *et al.*, 2004; Müller *et al.*, 2004; Mutinelli *et al.*, 2004).

In North America, dog rabies has become increasingly rare. From 2000 through 2003, Canada reported an annual average of 12.8 rabid dogs; however, the range of values was 4 to 23, with the highest number in 2000 and the lowest in 2003 (Krebs *et al.*, 2001, 2003a, 2004). In the USA, the annual number of dog rabies cases, primarily due to spillover from rabies virus variants circulating among terrestrial wildlife, has hovered around 100 for the past decade; in 2003, 117 dogs were reported rabid (Figure 4.13) (Krebs *et al.*, 2004). In southern Texas, a number of domestic dogs were infected with a dog/coyote variant of rabies virus in the 1990s; however, an aggressive ORV campaign has successfully controlled this variant (Clark *et al.*, 1994; Farry *et al.*, 1998; Sidwa *et al.*, 2005).

7.8.2 Rabies virus spillover to domestic cats

For public and veterinary health practitioners, the problem of rabies spillover is of foremost importance. In areas of the world affected by epidemics of wildlife rabies, such as Europe and North America, cats have replaced dogs as the companion animal species most commonly reported rabid.

In Western Europe, where red fox rabies is the dominant form of endemic rabies and domestic dog rabies has long been controlled, cat rabies remains a

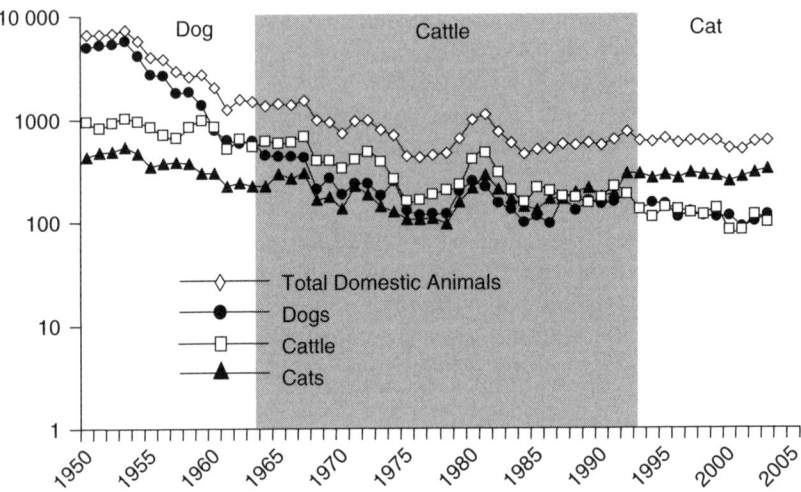

Figure 4.13 Numbers of reported cases of rabies among domestic animals in the USA, 1955–2003, as reported through national surveillance (data obtained from annual CDC surveillance reports). The different zones of gray and white overlayed on the figure indicate the dominant domestic species reported rabid during a defined interval.

public health problem. Between 1977 and 2000, Germany reported rabies from 4302 cats, as compared to 2307 dogs and, during the same interval, Austria reported 471 rabid cats and 108 rabid dogs (Müller *et al.*, 2004) and Italy reported 14 rabid cats and no rabid dogs (Mutinelli *et al.*, 2004). Between 1979 and 2000, France reported rabies among 1256 cats and 694 dogs and between 1966 and 2000, Belgium reported 295 rabid cats and 64 rabid dogs (Aubert *et al.*, 2004).

Reports from European countries from the 1950s through the 1970s indicated that cats were more frequently associated with human exposures to rabies virus than were dogs (reviewed in Vaughn, 1975). Among 1104 persons receiving PET in France in 1988, following exposure to an animal proven to be rabid by laboratory testing, 88 (8%) had been bitten by a rabid dog and 285 (26%) by a rabid cat; almost half (522; 47%) were treated for contact with rabid herbivores (Sureau, 1990).

In a pattern similar to that in Europe, since 1992, the cat has been the domestic animal most commonly reported rabid in the USA (see Figure 4.13). In 2003, 321 cats were reported rabid (Krebs *et al.*, 2004). The majority of rabid cats were reported from the eastern USA, where the raccoon-adapted variant of rabies virus is endemic. The large and disproportionate number of rabid cats being identified in the USA presumably reflects poorer vaccine coverage in this animal than is achieved for dogs (Fischman *et al.*, 1992; Nelson *et al.*, 1998). Required vaccination for cats is still not legally mandated in some states or counties.

In 1996, a survey of the 50 states, the District of Columbia, and three of five territories revealed that 74% of these political units required dog vaccination compared with 52% requiring cat vaccination (Johnston and Walden, 1996). The large number of stray and unvaccinated cats in rural environments has contributed greatly to the increase in rabies in this species.

In the late 1980s, a national survey identified rabid cats as being more commonly identified than dogs as the source of potential human exposure to rabies virus in the USA (Eng and Fishbein, 1990). However, since that time, local studies in rural areas, such as in Pennsylvania (Moore *et al.*, 2000), and studies conducted in urban emergency rooms have found that more PET is still occasioned by dog bite, although a higher percentage of humans reporting cat bite receive PET (Moran *et al.*, 2000). Cats have been the cause of several large-scale exposures of humans to potentially rabid animals (Rotz *et al.*, 1998), including one situation involving over 600 PETs (Noah *et al.*, 1996).

In a study assessing risk of cat rabies associated with different temporal stages of the raccoon rabies epidemics, there was a significant urban to rural trend in the increased risk of a cat testing positive for rabies. The risk of a cat testing positive for rabies (odds ratio [OR]) in rural counties in the lowest quartile for human population density (<61.6 inhabitants per mi^2) was 2.7-fold above the referent value (OR = 1) for counties in the highest quartile of human population density (>420.2 inhabitants per mi^2); counties with intermediate human population densities, in the second and third quartiles, also showed increasing risk for cat rabies above counties with the highest human density, with ORs of 1.7 and 2.0 above the referent value, respectively (Gordon *et al.*, 2004).

Attempts to model the risk of rabies spillover to domestic cats and other species during different temporal stages of the epidemic transition of raccoon rabies from epidemic to endemic disease states indicate that risk of rabies and the numbers of animals tested for rabies can persist at several multiples of pre-epidemic levels, even as reports of rabies among the principal reservoir species show significant declines (Baker, 2002; Gordon *et al.*, 2005). The risk (OR) of cat rabies in the northeastern USA during the initial epidemic of raccoon rabies was elevated >12-fold above the referent value, which was defined as risk of cat rabies during the pre-epidemic stage of raccoon rabies when only sporadic numbers of rabid raccoons were being recorded. The risk of cat rabies declined to ~6-fold pre-epidemic levels during the first inter-epidemic interval, before increasing to 7.5-fold over baseline levels during the time interval subsequent to the end of the first epidemic; this final interval contained several smaller epidemics which were insignificantly larger than the background level of sporadic rabies. The number of skunks testing positive for raccoon variant rabies also was associated with increasing risk of cat rabies, suggesting their potential role as bridging or secondary reservoirs, capable of transient maintenance of the raccoon-associated variant of rabies virus (Guerra *et al.*, 2003).

7.8.3 Rabies virus spillover from terrestrial carnivores to cattle

Spillover of red fox rabies to cattle in Western Europe has exceeded the levels of dog and cat rabies over the past several decades. Between 1977 and 2000, 6047 cases of cattle rabies were reported from Germany, 681 cases were reported from Austria (Müller *et al.*, 2004) and five cases were reported from Italy (Mutinelli *et al.*, 2004). Between 1979 and 2000, 2153 cases of cattle rabies were reported from France and between 1966 and 2000, 1629 cases of cattle rabies were reported from Belgium (Aubert *et al.*, 2004).

In North America, annual reports of cattle rabies have exceeded the numbers reported for domestic dogs and cats in Canada over the past few decades. From 2000 through 2003, Canada reported an annual average of 24.5 rabid cattle; however, the range of values was 11 to 43, with the highest number in 2000 and the lowest in 2003 (Krebs *et al.*, 2001, 2003a, 2004). In the USA, cattle rabies has declined in a similar manner to that of dog rabies (see Figure 4.13); in 2003, 98 cattle were reported rabid (Krebs *et al.*, 2004).

8 SPATIAL SPREAD AND EPIDEMIOLOGY OF WILDLIFE RABIES

Rabies virus can cause sensational epidemics among wildlife reservoirs. Epidemics driven by rabies virus variants associated with red foxes in Europe and North America and raccoons in North America have been chronicled since the 1940s and 1950s, and have involved infections of hundreds of thousands of wildlife and domestic animals (Figure 4.14A and B) and have spread to cover extensive regions of entire continents.

8.1 Red fox rabies in Europe

Beginning in the 1940s, an epidemic of red fox rabies began spreading from Russia and Poland towards western Europe, eventually affecting much of the continent (Figure 4.14A) (Wandeler *et al.*, 1974; Steck and Wandeler 1980). The origin of the European red fox variant of rabies virus is hypothesized to be the domestic dog (Bourhy *et al.*, 1999). Physical barriers have contributed to the localized evolution of several genetic lineages of viruses found in red foxes and raccoon dogs in Europe (Bourhy *et al.*, 1999).

The epidemic front of red fox rabies in Europe advanced in an irregular wave-like fashion at an estimated 25–60 km per year (Wandeler, 2004). Beginning in 1978, the first large-scale experimentation with oral rabies vaccination was initiated in Europe (for review, see Pastoret *et al.*, 2004 and Chapter 17 in this volume). Initially, live modified vaccines (South Alabama Dufferin (SAD) varieties and others) and later, in 1984, the V-RG vaccine contained in flavored baits,

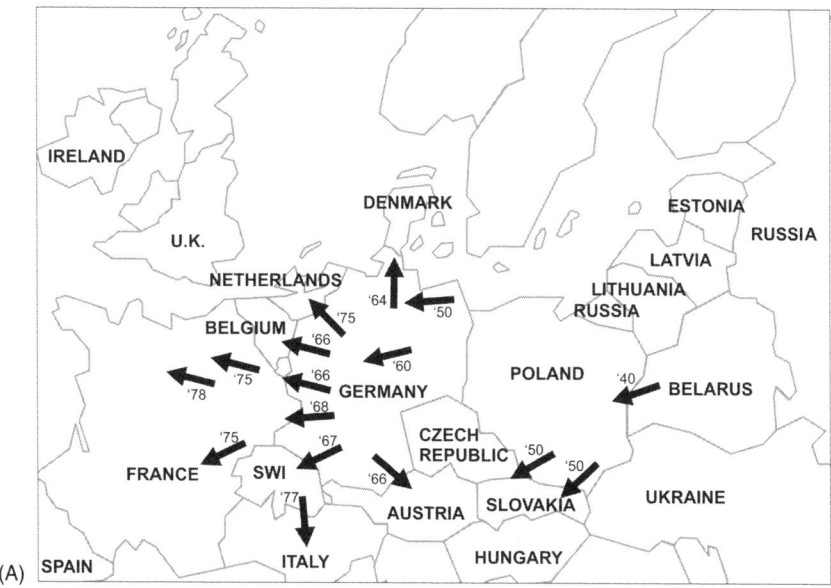

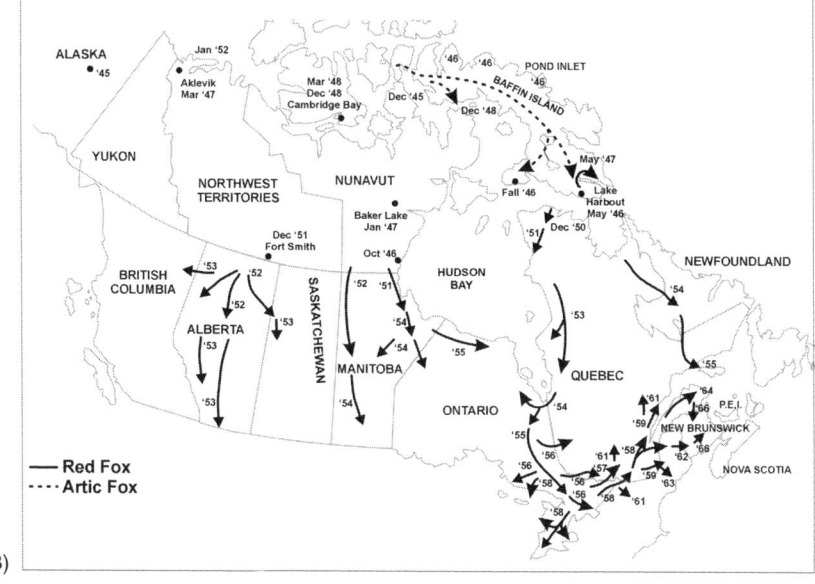

Figure 4.14 Example of the spread of fox-associated rabies epidemics in Europe (A) and North America (B). Both epidemics are traced from their beginning in the 1940s to the approximate maximum extent of their spread. (A) Modified from Macdonald (1980) and reproduced with permission by Academic Press. (B) Reproduced (with slight modification) from Figure 3 in Tabel *et al.* (1974), with permission from the Canadian Veterinary Medical Association.

were distributed in the field (Schneider *et al.*, 1988; Brochier *et al.*, 1991). Between 1978 and 1999, over 151 million vaccine-laden baits were distributed in 18 European countries, which has greatly reduced and even eliminated fox rabies from many previously affected regions (Pastoret *et al.*, 2004).

8.2 Fox rabies in North America

In North America, a major epidemic of the Artic fox variant of rabies virus, involving red and Arctic foxes (*Alopex lagopus*), began in northern Canada in the 1940s (Tabel *et al.*, 1974) (Figure 4.14B). Artic fox rabies virus variants have a near circumpolar distribution (Smith, 1995) and remain an occasional source of human rabies in Asia (Kuzmin, 1999), and have successfully spilled over to initiate independent cycling of rabies virus among raccoon dogs in Europe and Asia (Bourhy *et al.*, 1999; Hyun *et al.*, 2005).

In the early 1960s, the epidemic of red fox rabies expanded from Ontario into the northeastern states of New York, New Hampshire, Vermont and Maine (Tabel *et al.*, 1974; Blancou *et al.*, 1991). Red fox rabies and spillover to domestic and wild animals in the USA occurred until the mid-1990s (Gordon *et al.*, 2004), when effective control efforts, initiated by the Canadians, brought fox rabies under control in neighboring Ontario (Macinnes *et al.*, 2001) and coincidently resulted in the disappearance of red fox-associated rabies in the northeastern USA.

The first recorded case of rabies in foxes in the USA occurred in gray foxes (*U. cinereoargenteus*) in Georgia in 1940 (Macinnes *et al.*, 2001). Within years, rabies was endemic among gray foxes in Alabama, Florida and Tennessee and, from 1940 to 1960, gray and red foxes were the wild carnivore most commonly reported rabid in the USA (see Figure 4.11).

Since the 1940s, the endemic area affected by the gray fox-associated variant of rabies virus has diminished in size such that endemic gray fox rabies in the USA is presently known only from relatively small areas in Texas and Arizona (see Figure 4.2) (Rohde *et al.*, 1997); in Texas this focus is the target of an extensive ORV campaign using V-RG (Sidwa *et al.*, 2005). However, a phylogenetically related virus has been recovered from bobcats in Mexico (de Mattos *et al.*, 1999), suggesting this particular rabies virus variant may enjoy a more extensive range than previously appreciated.

8.3 Raccoon rabies in North America

The epidemic associated with raccoons in the eastern USA is believed to have been initiated in the mid-Atlantic region by the inter-state translocation of raccoons incubating rabies from an established focus of raccoon rabies in the southeastern USA for the purpose of restocking dwindling local populations (Nettles *et al.*, 1979). Since the mid-1970s, this raccoon-adapted variant of rabies virus has spread north to Maine and Ontario, Canada, and west to Ohio, causing one of the most intensive outbreaks of animal rabies ever recorded (Figure 4.15) (Childs *et al.*, 2000). The magnitude of this epidemic was enhanced by the spread of virus through naïve raccoon populations of very high density, often in states that had not experienced terrestrial rabies for decades (Hanlon and Rupprecht, 1998).

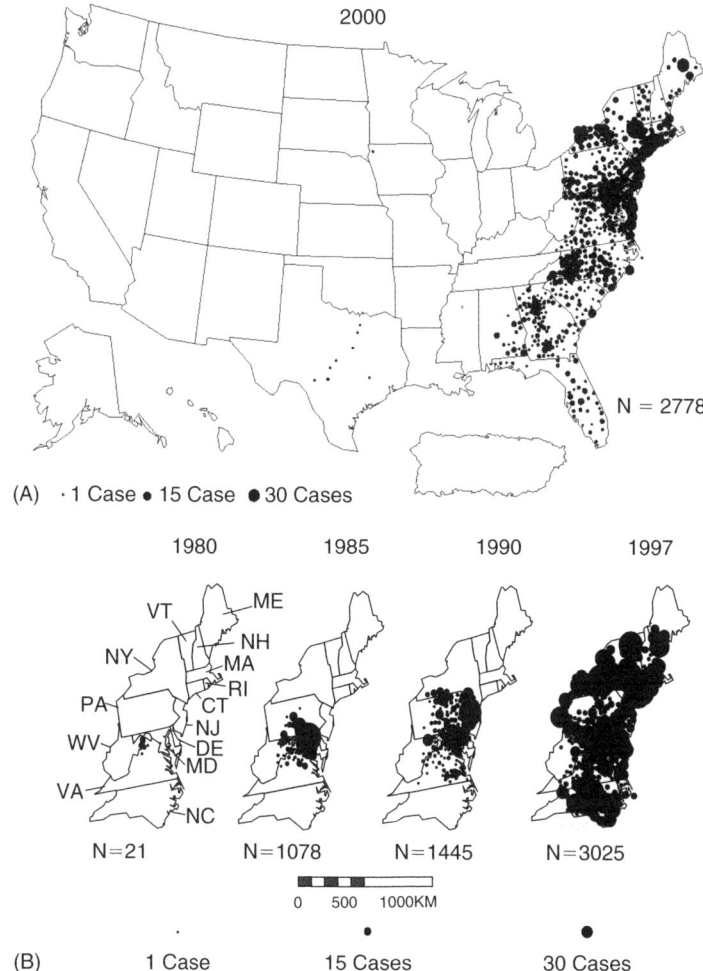

Figure 4.15   (A) The distribution of the raccoon-associated rabies virus in 2000, as indicated by the number of rabid raccoons reported through national surveillance for that year (Krebs *et al.*, 2001). The size of each dot is proportional to the number of cases reported by an individual county in that year. (B) The early expansion of raccoon rabies from a new focus seeded in the mid-Atlantic region of the USA for selected years, 1980–1997. Images from USDA website, www.aphis.usda.gov.

As with the expansion of red fox rabies in Europe, the raccoon rabies epidemic has spread in an irregular wave in a northeasterly direction from the introduction focus in the mid-Atlantic region of the USA. Similar to the rate of rabies spread estimated for red foxes in Europe, the average velocity of spread has been estimated to be between 30 and 48 km per year (Wilson *et al.*, 1997; Moore, 1999; Lucey *et al.*, 2002).

The local rate of disease propagation is significantly affected by local environmental heterogeneities, such as those posed by major rivers running orthogonal to the major direction of raccoon rabies spread. In Connecticut, models using a stochastic simulator determined the reduction in local transmission of raccoon rabies to be sevenfold for townships separated by a major river compared to townships without such a physical barrier (Smith *et al.*, 2002).

The impact of habitat variation affecting the equilibrium raccoon population density appears to have an influence on the rate of spread of raccoon rabies as estimated by empirical data on the date of first appearance of rabies in townships of New York State and simulations of the expected rate of rabies spread (Russell *et al.*, 2004). In New York State, the rate of rabies spread in townships in the northwestern portion of the state was approximately 30% lower than the rate of spread in the southern portion of the state. The raccoon epidemic slowed approximately 4 years after entering New York State from the south in 1991. The slowing was coincident with the rabies wavefront colliding with the Adirondack Mountains. The coniferous forests of the Adirondack Mountains are not a preferred habitat for raccoons (Merriam, 1886) and the population density of raccoons in this region is extremely low (Godin, 1977). In addition to the impasse posed by the Adirondack Mountains, ORV delivered by the New York State Department of Health and Cornell University at two sites at the northeastern and northwestern edges of the Adirondack Mountain range in front of the advancing epidemic, may have played a role in the rate of spread of the raccoon wavefront of rabies (Hanlon and Rupprecht, 1998). However, the Adirondack region of New York continued to be a formidable barrier to raccoon rabies through 2003 (Krebs *et al.*, 2004) and raccoon rabies has gone around rather than through this region to threaten Canada.

Raccoon rabies was first detected in Ontario, Canada in 1999, from across the St Lawrence River border with the USA (Wandeler and Salsberg, 1999; Centers for Disease Control and Prevention, 2000). Emergency vaccination of raccoons in the Ontario area, where the outbreak was first detected, may have halted spread of this initial epidemic (Rosatte *et al.*, 2001). However, molecular epidemiologic analyses of raccoon variants from Canada indicate that three independent incursions of raccoon rabies have occurred, two in 1999 in Ontario and one in 2000 in New Brunswick (Nadin-Davis *et al.*, 2005). The threat of raccoon rabies incursion into Canada and western regions of the USA will exist so long as this rabies virus variant remains endemic.

8.4 Common features to rabies epidemiology in Europe and the USA

In addition to the similar estimated rates of wavefront movement for red fox rabies in Europe and raccoon rabies in the USA (roughly 30–60 km per year; Lucey *et al.*, 2002; Wandeler, 2004), other factors relating to the impact of spatial heterogeneity on the spread and dissemination of wildlife rabies bear mentioning.

The impact of rivers and mountainous terrain on the spread of red fox rabies in Europe has also been noted (Wandeler *et al.*, 1988), although quantitative estimates of the effect of such barriers are not available for comparison with values obtained from raccoon rabies. In Switzerland, mountains and major rivers clearly impede or prevent the flow of red fox rabies into adjacent valleys where the disease persists or wanes independently (Müller and Breitenmoser, 2004).

Rivers or other physical barriers have been hypothesized as playing a role in the emergence of subvariants of red fox rabies virus in both Europe and North America. Genome sequence variation present in isolates of the red fox rabies virus variant obtained from red foxes and raccoon dogs separated by the Vistula River in Poland suggest this barrier presented an obstacle to viral gene flow (Bourhy *et al.*, 1999).

In northern Ontario, the Arctic fox rabies virus variant descended as an irregular wave with two arms invading into southern Ontario over the 1980s and 1990s, with evidence of ecological partitioning of subvariants based on both molecular epidemiologic and epidemiologic analyses (Nadin-Davis *et al.*, 1993; Tinline and Macinnes, 2004). However, additional analyses examining correlations between genetic and geographic distance of red fox subvariants suggest an isolation-by-distance population structure for the virus. The divergence among viral lineages since the most recent common ancestor correlates with position along the advancing wavefront with more divergent lineages near the origin of the epidemic (Real *et al.*, 2005).

8.5 Long-distance translocations

One of the most obvious epidemiological features of the raccoon rabies epidemic in the USA was the critical role long-distance translocations (LDT) played in the emergence of raccoon rabies in the mid-Atlantic region (Nettles *et al.*, 1979). On numerous occasions isolated foci of rabies in raccoons have appeared in states, such as New York (C. Trimarchi, personal communication), Pennsylvania (Moore, 1999) and Connecticut (Smith *et al.*, 2005). In Connecticut, although local transmission between adjacent townships accounted for most of the transmission of raccoon rabies, of the 159 townships not on the western border where rabies entered the state, twenty-one townships (13%) recorded their first case of raccoon rabies when none of the adjacent townships was infected. At least two putative LDT events were identified in the small State of Connecticut which were deemed to have sparked new foci of infection with raccoon rabies spreading to adjacent townships well removed in space and time from the advancing raccoon rabies wavefront (Smith *et al.*, 2005). The importance of identifying and understanding the role of LDT cannot be overstated where control programs aimed at preventing the spread of raccoon rabies depend upon construction of a vaccine barrier zone or cordon sanitaire, such as the coordinated effort between the US Department of Agriculture (USDA), individual states and other federal agencies (Figure 4.16) (Kemere *et al.*, 2002; Slate *et al.*, 2005).

9 CONTROL OF RABIES DIRECTED TOWARD MAMMALIAN RESERVOIR HOSTS

As the topic of rabies control in domestic dogs and wild carnivores is extensively reviewed in Chapter 17 and Chapter 18, this section will only briefly review the topic within the context.

9.1 Culling

The ultimate prevention strategy for zoonotic agents affecting humans is to abrogate or greatly reduce cross-species transmission by disrupting transmission and maintenance cycles of zoonotic viruses within the reservoir host. The most widespread and most commonly practised control mechanism for interrupting zoonotic pathogen transmission is to reduce directly the total number of individuals in the reservoir host or secondary host population. However, the effectiveness of culling of wildlife reservoir host populations for some pathogens, including rabies virus variants, has varied (Denduangboripant *et al.*, 2005).

Targeted reduction of red fox populations has been employed in Europe and North America (Muller, 1971; Bögel *et al.*, 1974; Debbie, 1991) and, in North America, skunk populations have been targeted for population reduction by culling (Rosatte *et al.*, 1986; Debbie, 1991). Efforts are ongoing in Central and South America to reduce vampire bat populations to curtail the enormous economic losses sustained from transmission of vampire bat rabies to cattle. Anticoagulants applied topically to the back of bats, or systemically by injection into livestock, are the major methods of vampire bat control (Thompson *et al.*, 1972; Fornes *et al.*, 1974; Crespo *et al.*, 1979); unfortunately other bat populations may also be affected by topical applications (Martinez-Burnes *et al.*, 1997; Mayen, 2003).

With the development of ORV, wildlife culling has been largely relegated to a supplementary control measure to be used in conjunction with vaccination efforts (Brochier *et al.*, 1995; Aubert, 1999; Macinnes *et al.*, 2001; Centers for

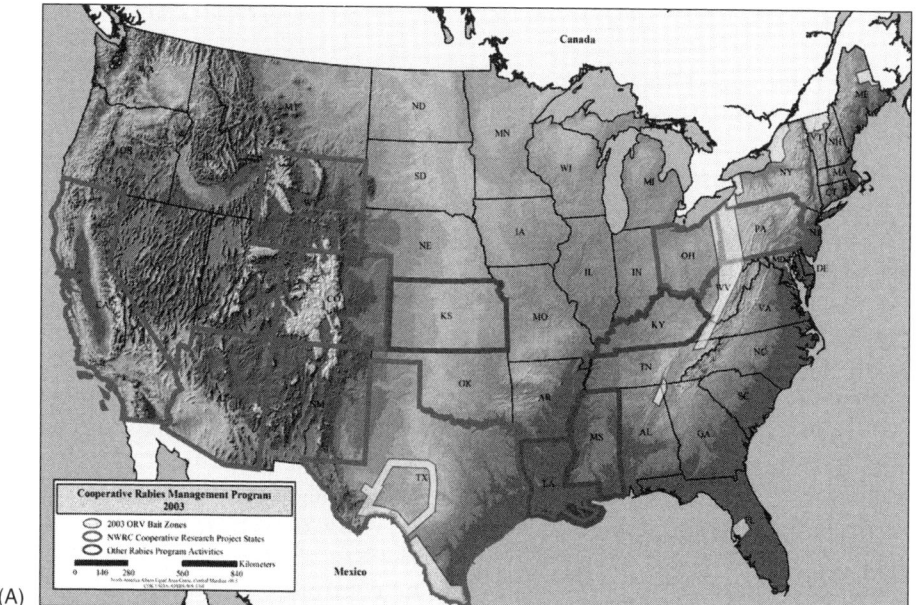

(A)

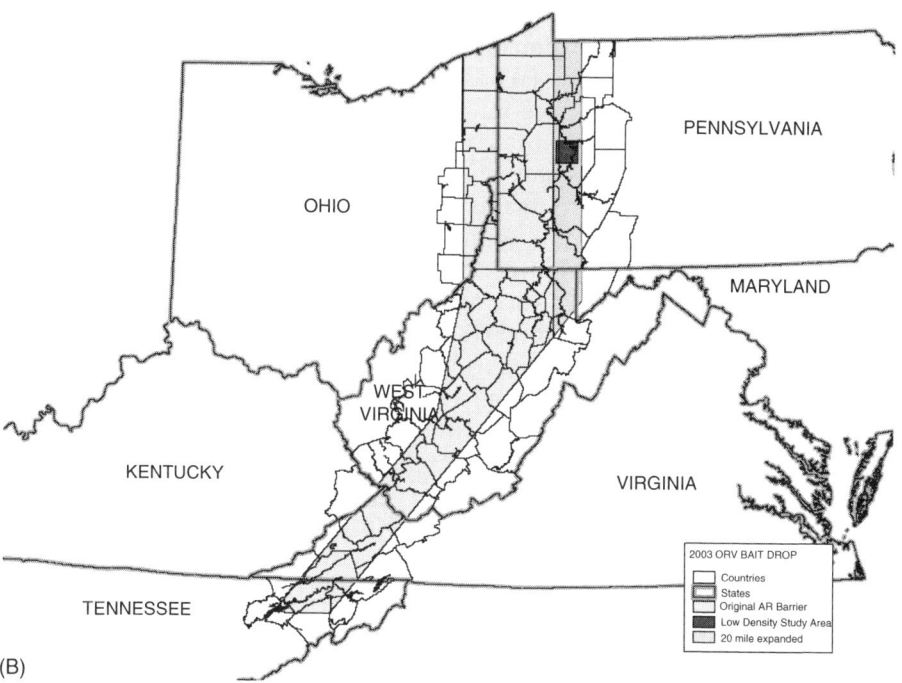

(B)

Figure 4.16 (A) The extent of a national effort to control the westward expansion of the raccoon variant of rabies virus into the Ohio Valley and into western Alabama coordinated by the USDA with individual states. (B) Shows a detail from (A), illustrating the extent and breadth of the ORV barrier established in Ohio, Pennsylvania, West Virginia and Virginia. This figure is reproduced in the color plate section.

Disease Control and Prevention, 2004a; Chapter17). Most mathematical models identify a combination of vaccination and targeted culling as the optimal strategy for red fox rabies control in Europe and the USA (Anderson *et al.*, 1981; Smith and Wilkinson 2003; Chapter 18).

9.2 Management of canine rabies

Canine rabies control (vaccination, reproduction control, movement restrictions and habitat modification) not only reduces the incidence of human rabies (Figure 4.17), but also provides a cost-effective intervention for reducing the need of human PET (Bögel and Meslin, 1990). In the USA, compulsory rabies vaccination laws have been credited with reducing the incidence of canine rabies (Bech-Nielsen *et al.*, 1979) and the WHO endorses national legislation in countries where such a program is affordable and enforceable (World Health Organization, 1987, 2004).

Endemic or epidemic canine rabies can be controlled or eliminated through comprehensive programs of parenteral vaccination and animal management. Removal or culling of free-ranging dogs is not recommended as a primary means of dog population reduction or even as a supplementary measure to mass vaccination of dog populations (World Health Organization, 2004). However, the shooting and poisoning of free-ranging dogs is still widely practised in many parts of the world as a part of routine rabies control (Akkoca *et al.*, 2004).

Mass vaccination campaigns have effectively controlled canine rabies throughout much of Latin America (Organización Panamericana de la Salud, 1983; Chomel *et al.*, 1988) and in some countries in Asia, such as Malaya, Japan and Taiwan (Bögel *et al.*, 1982). Mass vaccination campaigns in Africa indicate that a sufficiently high percentage of vaccine coverage can be achieved by the parenteral route in some urban settings (Kayali *et al.*, 2003b). However, the effectiveness of dog vaccination campaigns when confronted by steadily increasing populations of free-ranging dog populations, as is occurring in urban centers in Thailand, has been questioned (Hemachudha, 2005; Wilde *et al.*, 2005).

Immune coverage to rabies virus of 70% (percent coverage or *pc*) among free-ranging dog (stray, feral and neighborhood dogs) populations has been a traditional target promoted as sufficient to establish herd immunity (World Health Organization, 2004) and drive R_0 below unity. At $R_0 <1$, the average canine rabies case fails to replace itself causing cessation of rabies virus transmission (World Health Organization, 2004). The 70% target coverage level has been demonstrated to be sufficient in interrupting the transmission of rabies virus among dogs (Matter *et al.*, 2000; World Health Organization, 2002); however, a lower level of immunity may suffice. Theoretical estimates of *pc* suggest herd immunity levels between 39 and 57%, with an upper 95% confidence interval of 55 and 71%, respectively, may prove sufficient to end epidemic transmission

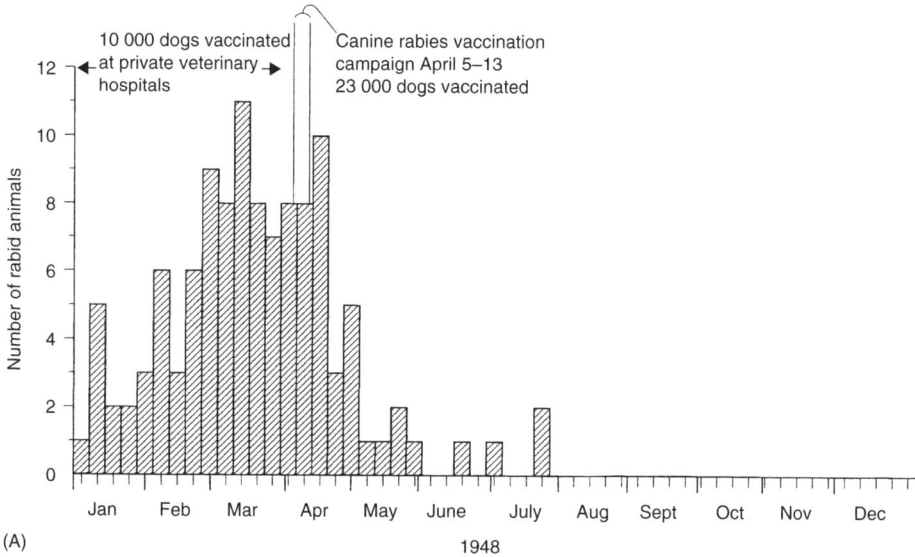

(A)

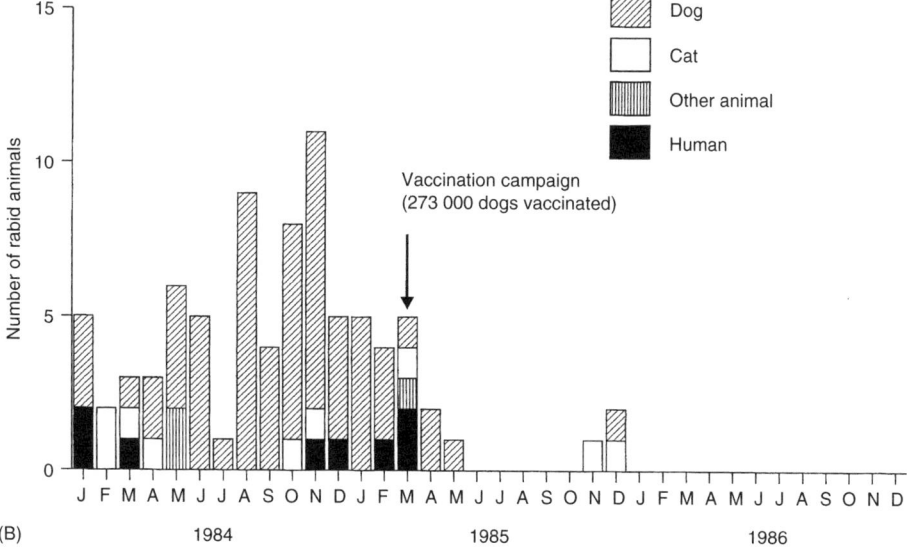

(B)

Figure 4.17 Vaccination campaigns aimed at the domestic dog have been shown to be effective in the USA (A) (reproduced and modified from Tierkel, 1959, with permission by Academic Press) as well as in developing nations (Peru [B]) in preventing dog-to-dog transmission and dog-to-human transmission of rabies virus. (Reproduced with permission from Chomel *et al.*, 1988, from *Reviews of Infectious Diseases*, published by the University of Chicago Press, © 1988 University of Chicago.)

(Coleman and Dye, 1996). Vaccination coverage of 20 to 40% among populations of the endangered Ethiopian Wolf proved sufficient to eliminate the largest epidemics among this species (Cleaveland *et al.*, 2002b), but in this situation virtually all rabies was being transmitted to wolves by domestic dogs as there is no evidence that wolves can sustain transmission among themselves, possibly due to the small population remaining (Nel and Rupprecht, 2006).

9.3 Vaccination of dogs by ORV

The development and experimental testing of oral vaccines for dogs by live attenuated or recombinant vaccines have given mixed results (Vos *et al.*, 2001; Rupprecht *et al.*, 2005; Chapter 18). Field testing of a mixed delivery strategy for dog vaccination in the Philippines indicated that ORV by SAD B-19 was an important complement to vaccination by the parenteral route when confronted with a large population of free-ranging or unrestrained dogs (Estrada *et al.*, 2001). Additional vaccines for dogs, including a DNA vaccine (Lodmell *et al.*, 2003, 2005; Rath *et al.*, 2005), continue to be developed to bring down the costs associated with rabies control and to offer effective means to reach animals in challenging settings such as low SES urban settings.

9.4 Other methods for control of canine rabies

The regulation and control of dog movements within and between countries remains an important strategy for preventing rabies. Rabies-free locations within a country, such as Hawaii in the USA (Fishbein *et al.*, 1990) and rabies-free nations, such as the UK (rabies-free since 1922) and Japan, have traditionally enforced strict laws requiring 6 months of quarantine (Fishbein *et al.*, 1990). The UK initiated a new Pet Travel Scheme in 2000 requiring imported animals to have an implanted microchip identification tag and documented proof of current vaccination, as evidenced by the presence of antibody, within a 6-month period prior to presenting proof of procedure at an official UK port (Fooks *et al.*, 2002; Jones *et al.*, 2005). A quantitative, comparative assessment of the risk of rabies importation into Great Britain identified a greater risk of rabies introduction by the old 6-month quarantine regimen (annual probability of rabies introduction = 1.01×10^{-5}) than through the Pet Travel Scheme (7.22×10^{-6}) (Jones *et al.*, 2005).

The persistence and growth of canine populations in urban and rural environments of Asia, even after decades of canine rabies control, indicate that additional control methodologies, other than mass vaccination, are necessary to achieve control (Hemachudha, 2005; Wilde *et al.*, 2005). Immunocontraception targeting dogs and other reservoir host species for rabies, such as red foxes (Reubel *et al.*, 2005), is an area of active research (Ferro, 2002; Ferro *et al.*, 2004).

Efforts to isolate and quarantine dogs, cats and domestic ferrets exposed to a known rabid animal remain a cornerstone of rabies control in the USA (Centers for Disease Control and Prevention, 2005a). In addition to isolation of pet animals following potential or actual exposures to rabies virus, legislation banning the interstate translocation of carnivore hosts for rabies virus has been implemented following instances of coyote variant rabies introduction into Florida from Texas (Centers for Disease Control and Prevention, 1995b) and the translocation of rabid raccoons in shipments from Florida to the mid-Atlantic states to re-establish local populations diminished by hunting (Nettles *et al.*, 1979).

9.5 Wildlife vaccination by oral rabies vaccines (ORV)

In Western Europe, vaccination of red foxes serving as the principal host of rabies virus has eliminated or significantly reduced rabies in many countries (Steck *et al.*, 1982; Aubert 1999). Models of rabies epidemics among rapidly growing red fox populations suggest the level of vaccination coverage required to control transmission is 75% (Artois *et al.*, 1997), greater than the 70% recommended for dog rabies vaccination coverage (World Health Organization, 2004).

Dog rabies had been controlled in South Korea since 1985 by vaccination campaigns but, in 1993, sylvatic rabies caused by a rabies strain closely related to the Arctic fox variant appeared among raccoon dogs in northern provinces bordering North Korea and resulted in spillover affecting both domestic and wildlife species. In response, South Korea began ORV distribution in 2002 to control rabies maintained by a raccoon dog reservoir (Hyun *et al.*, 2005).

In North America, Canada was the first country to initiate ORV in 1989 to eliminate the Arctic fox variant of rabies which had become epidemic among red foxes (Macinnes *et al.*, 2001; Chapter 18). Control of red fox rabies in Canada also terminated the sporadic cases of red fox rabies which had plagued counties in Vermont, New Hampshire, New York and Maine (Gordon *et al.*, 2004). The Canadians complemented parenteral vaccination of trapped raccoons with ORV distribution in an emergency response to the first cases of raccoon variant rabies where rabid raccoons had crossed the St Lawrence River (Rosatte *et al.*, 2001).

In the USA, ORV campaigns targeting raccoons in several northeastern states (Bigler, 1997; Robbins *et al.*, 1998), Ohio in the midwestern USA (Smith *et al.*, 1999) and a small area in Florida in the southern USA (Olson *et al.*, 2000) and coyotes and gray foxes in Texas began in the 1990s (Fearneyhough *et al.*, 1998; Steelman *et al.*, 2000). The USDA has coordinated an aggressive ORV campaign in collaboration with individual states in an effort to prevent the raccoon-associated variant from extending its range into the Ohio River Valley and further to the west (Slate *et al.*, 2005). An extensive, but discontinuous, immune barrier has been created in parts of Ohio, Pennsylvania, West Virginia, Virginia, Tennessee and Alabama by delivery of ORV (see Figure 4.16) (Slate *et al.*, 2005). The cost benefits of such an ORV barrier appear to outweigh the significant expense associated with maintaining the cordon sanitaire (Kemere *et al.*, 2002; Gordon *et al.*, 2005).

Recent experiments involving topical application of concentrated V-RG mixed with Vaseline to the backs of common vampire bats has shown promise as a tool to vaccinate these reservoir hosts and may provide an alternative to culling. Between 42.8 and 71.4% of bats in groups into which a single vaseline-V-RG-coated individual was introduced survived rabies virus challenge (Almeida *et al.*, 2005). An experimental oral vaccination of Arctic foxes with SAG2 rabies vaccine has also provided positive results that could result in ORV campaigns directed against this pan-arctic variant of rabies virus (Follmann *et al.*, 2004).

9.6 Potential threats and challenges for sustained ORV use

In addition to the costs associated with sustained applications of ORV, there are both public health threats and epidemiological challenges to applying ORV and maintaining vaccine barriers.

Although the risk of human contact with V-RG is small, the risk is finite. As of 2003, almost 50 million doses of V-RG have been distributed across highly heterogeneous landscapes (Hanlon and Rupprecht, 1998; Aubert, 1999; Slate *et al.*, 2005). Relatively few instances of human exposure to the vaccine content have been reported (Slate *et al.*, 2005). In the USA, a single case of systemic vaccinia occurred in a pregnant women after she was bitten by her pet dog while trying to remove a vaccine sachet from the dog's mouth (Rupprecht *et al.*, 2001).

Finally, the LDT of rabid animals, both by intentional and unintended movements of animals by humans, in addition to possible long-distance dispersal events, has played such an extensive role in the emergence of rabies, and the phenomenon is so common on both greater (>100 km) and lesser scales (20–100 km), that it must be considered in conjunction with any control strategy (Nettles *et al.*, 1979; Centers for Disease Control and Prevention, 1995; Constantine 2003; Smith *et al.*, 2005). Breaches in immune barriers are certain to occur and contingency plans for remedial vaccination must be executed rapidly. Some models have identified areas for ring vaccination around outbreak loci, as proposed for Ohio, based on a stochastic simulator predicting likely trajectories of raccoon rabies spread from a breach in the ORV barrier detected in Leroy Township, 11 km west of the vaccine zone (Figure 4.18) (Anonymous, 2004; Russell *et al.*, 2005). Strategic planning for ring vaccination must take into account spatial heterogeneities in the environment, delays in realtime rabies detection through routine surveillance and the incubation period of rabies within the target host species (Childs *et al.*, 2002; Tinline *et al.*, 2002); these factors were considered when proposing a remedial intervention for raccoon rabies in Ohio (Russell *et al.*, 2005; Smith *et al.*, 2005). When resources are limited, a theoretical analysis

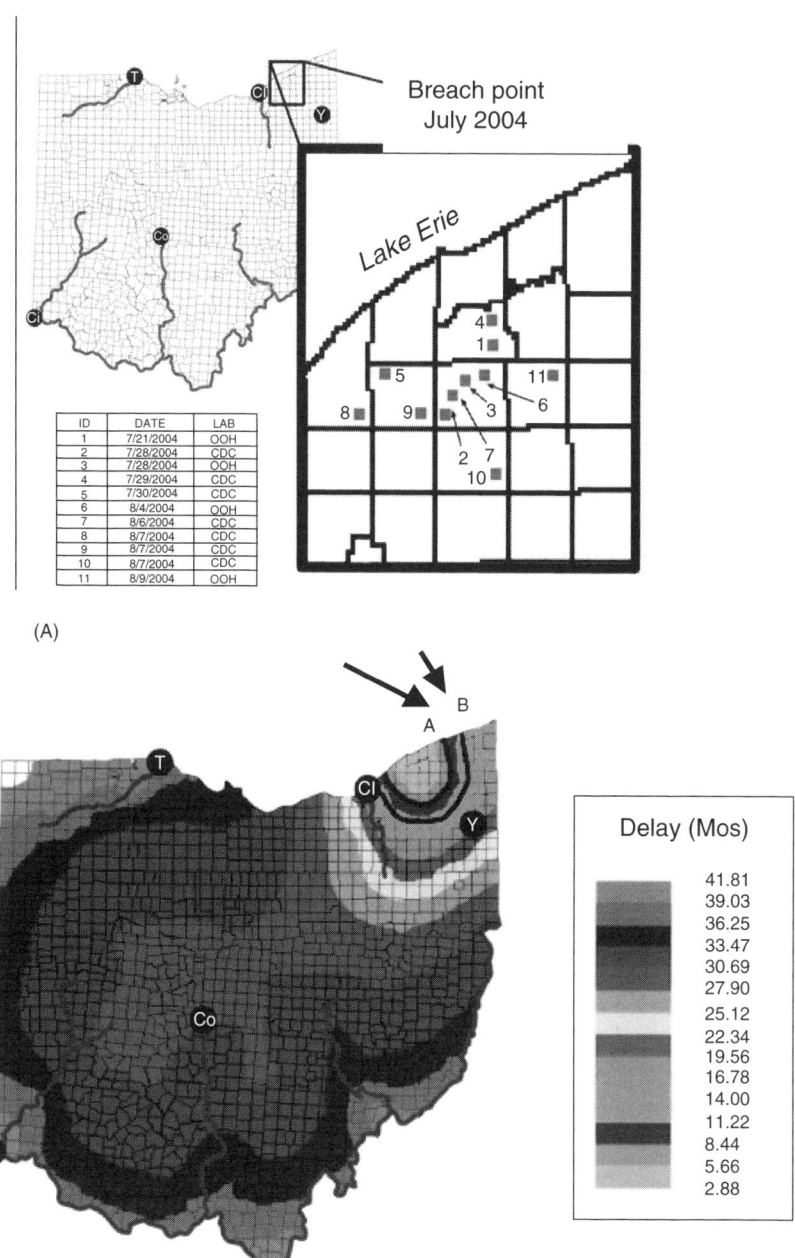

ID	DATE	LAB
1	7/21/2004	OOH
2	7/28/2004	CDC
3	7/28/2004	OOH
4	7/29/2004	CDC
5	7/30/2004	CDC
6	8/4/2004	OOH
7	8/6/2004	CDC
8	8/7/2004	CDC
9	8/7/2004	CDC
10	8/7/2004	CDC
11	8/9/2004	OOH

(A)

Breach point
July 2004

Lake Erie

(B)

Delay (Mos)

41.81
39.03
36.25
33.47
30.69
27.90
25.12
22.34
19.56
16.78
14.00
11.22
8.44
5.66
2.88

Figure 4.18 (A) A potential breach in the ORV barrier established by the USDA in cooperation with the State of Ohio occurred in Leroy Township in July 2004 (Anonymous, 2004). (B) The use of a stochastic simulator model to predict the potential trajectory of raccoon rabies spread from the nascent foci permitted identification of two boundaries for remedial ORV control by ring vaccination to contain raccoon rabies spread (Russell *et al.*, 2005). Arrows indicate the inner boundary in (A) and outer boundary in (B). The boundary lines were estimated by delays anticipated in passive surveillance based on detection of raccoon rabies and the estimated incubation period for rabies among raccoons (Tinline *et al.*, 2002). Ci = Cincinatti; Co = Columbus; Cl = Cleveland; T = Toledo; Y = Youngstown, all part of Ohio. This figure is reproduced in the color plate section.

suggests that an intact circular coverage of ORV, rather than a ring application, might result in better remedial intervention (Eisinger *et al.*, 2005). The comparative efficacy of these approaches to contain breaches in immune barriers may never be tested, as the circular approach subsumes the ring vaccination approach and is a more conservative approach.

REFERENCES

Abdussalem, M. and Bögel, K. (1971). The problem of antirabies vaccination. *International Conference on Applications for Vaccines in Viral, Rickettsial, and Bacterial Diseases in Man.* **No. 226**, 1–54. Washington DC: Pan American Health Organization.

Acha, P.N. (1967). Epidemiology of paralytic bovine rabies and bat rabies. *Bulletin de l' Office International des Epizooties* **67**, 343–382.

Acha, P.N. and Arambulo, P.V. III (1985). Rabies in the Tropics – history and current status. In: *Rabies in the Tropics* (E. Kuwert, C. Merieux, H. Koprowski and K. Bögel, eds). pp. 343–359. Berlin: Springer-Verlag.

Addy, P.A.K. (1985). Epidemiology of rabies in Ghana. In: *Rabies in the Tropics* (E. Kuwert, C. Merieux, H. Koprowski and K. Bögel, eds). pp. 497–519. Berlin: Springer-Verlag.

Advisory Committee on Immunization Practices. (1999). Human rabies prevention – United States, 1999: Recommendations of the Advisory Committee on Immunization Practices (ACIP). *Morbidity and Mortality Weekly Report* **48**, 1–21.

Akkoca, N., Economides, P., Maksoud, G. and Mestom, M. (2004). Rabies in Turkey, Cyprus, Syria and Lebanon. In: *Historical Perspective of Rabies in Europe and the Mediterranean Basin* (A.A.King, A.R.Fooks, M.Aubert and A.Wandeler, eds). pp. 157–169. Paris: OIE.

Almeida, M.F., Martorelli, L.F., Aires, C.C., Sallum, P.C. and Massad, E. (2005). Indirect oral immunization of captive vampires, *Desmodus rotundus*. *Virus Research* **111**, 77–82.

Anderson, L.J., Nicholson, K.G., Tauxe, R.V. and Winkler, W.G. (1984a). Human rabies in the United States, 1960 to 1979: epidemiology, diagnosis, and prevention. *Annals of Internal Medicine* **100**, 728–735.

Anderson, L.J., Williams, L.P. Jr, Layde, J.B., Dixon, F.R. and Winkler, W.G. (1984b). Nosocomial rabies: investigation of contacts of human rabies cases associated with a corneal transplant. *American Journal of Public Health* **74**, 370–372.

Anderson, R.M., Jackson, H.C., May, R.M. and Smith, A.M. (1981). Population dynamics of fox rabies in Europe. *Nature* **289**, 765–771.

Anonymous (2004). Rabid raccoon found in Leroy Township. *Lake County General Health District News Release*, pp. 1–2.

Anthony, J.A., Childs, J.E., Glass, G.E., Korch, G.W., Ross, L. and Grigor, J.K. (1990). Land use associations and changes in population indices of urban raccoons during a rabies epizootic. *Journal of Wildlife Diseases* **26**, 170–179.

Arai, Y.T., Kuzmin, I.V., Kameoka, Y. and Botvinkin, A.D. (2003). New lyssavirus genotype from the Lesser Mouse-eared Bat (*Myotis blythi*), Kyrghyzstan. *Emerging Infectious Diseases* **9**, 333–337.

Arguin, P.M., Krebs, J.W., Mandel, E., Guzi, T. and Childs, J.E. (2000). Survey of rabies preexposure and postexposure prophylaxis among missionary personnel stationed outside the United States. *Journal of Travel Medicine* **7**, 10–14.

Artois, M., Langlais, M. and Suppo, C. (1997). Simulation of rabies control within an increasing fox population. *Ecological Modeling* **97**, 23–34.

Aubert, M.F.A. (1999). Costs and benefits of rabies control in wildlife in France. *Revue Scietifique et Technique des Epizooties* **18**, 533–543.

Aubert, M.F., Cliquet, F., Smak, J.A., Brochier, B., Schon, J. and Kappeler, A. (2004). Rabies in France, The Netherlands, Belgium, Luxembourg and Switzerland. In: *Historical Perspective of Rabies in Europe and the Mediterranean Basin* (A.A. King, A.R. Fooks, M. Aubert and A. Wandeler, eds). pp. 129–145. Paris: OIE.

Babes, V. (1912). *Traité de la Rage*. Paris: J. B. Baillière et fils.

Badilla, X., Perez-Herra, V., Quiros, L. *et al.* (2003). Human rabies: a reemerging disease in Costa Rica? *Emerging Infectious Diseases* **9**, 721–723.

Badrane, H., Bahloul, C., Perrin, P. and Tordo, N. (2001). Evidence of two Lyssavirus phylogroups with distinct pathogenicity and immunogenicity. *Journal of Virology* **75**, 3268–3276.

Badrane, H. and Tordo, N. (2001). Host switching in Lyssavirus history from the Chiroptera to the Carnivora orders. *Journal of Virology* **75**, 8096–8104.

Baer, G.M. (1991). Vampire bat and bovine paralytic rabies. In: *The Natural History of Rabies* (G.M. Baer, ed.). pp. 389–403. Boca Raton: CRC Press.

Baker, S. (2002). *Assessment of the Relationship between the Raccoon Rabies Epizootic in New England and Rabies in Livestock.* pp. 1–48. MPH, Rollins School of Public Health, Emory University.

Baltazard, M. and Bahmanyar, M. (1955). Essai pratique du serum antirabique chez les mordus par loups enrages. *Bulletin of the World Health Organization* **13**, 747–772.

Baltazard, M. and Ghodssi, M. (1954). Prevention of human rabies. Treatment of persons bitten by rabid wolves in Iran. *Bulletin of the World Health Organization* **10**, 797–803.

Bartlett, P.C. and Martin, R.J. (1982). Skunk rabies in Illinois. *Journal of the American Veterinary Medical Association* **180**, 1448–1450.

Barton, L.D. and Campbell, J.B. (1988). Measurement of rabies-specific antibodies in carnivores by an enzyme-linked immunosorbent assay. *Journal of Wildlife Diseases* **24**, 246–258.

Batista-da-Costa, M., Bonito, R.F. and Nishioka, S.A. (1993). An outbreak of vampire bat bite in a Brazilian village. *Tropical Medicine of Parasites* **44**, 219–220.

Bech-Nielsen, S., Hagstad, H.V. and Hubbert, W.T. (1979). Vaccination against dog rabies in the United States. *Journal of the American Medical Association* **174**, 695–699.

Beck, A.M. (1973). *The Ecology of Stray Dogs*. Baltimore: York Press.

Beck, A.M. (2000). The human-dog relationship: a tale of two species. In: *Dogs, Zoonoses and Public Health* (C.N.L. Macpherson, F.X. Meslin and A.I. Wandeler, eds). pp. 1–16. New York: CABI Publishing.

Beck, A.M. and Jones, B.A. (1985). Unreported dog bites in children. *Public Health Reports* **100**, 315–321.

Belotto, A., Leanes, L.F., Schneider, M.C., Tamayo, H. and Correa, E. (2005). Overview of rabies in the Americas. *Virus Research* **111**, 5–12.

Beran, G.W. (1982). Ecology of dogs in the Central Philippines in relation to rabies control efforts. *Comparative Immunology and Microbiology of Infectious Diseases* **5**, 265–270.

Beran, G.W. (1991). Urban rabies. In: *The Natural History of Rabies* (G.M. Baer, ed.). pp. 427–443. Boca Raton: CRC Press.

Berger, J. (1976). Models of rabies control. *Lecture Notes in Biomathematics* **11**, 74–88.

Bernard, K.W. and Fishbein, D.B. (1991). Pre-exposure rabies prophylaxis for travellers: are the benefits worth the cost? *Vaccine* **9**, 833–836.

Bhanganada, K., Wilde, H., Sakolsataydorn, P. and Oonsombat, P. (1993). Dog-bite injuries at a Bangkok teaching hospital. *Acta Tropica (Basel)* **55**, 249–255.

Bhatia, R., Bhardwaj, M. and Sehgal, S. (1988). Canine rabies in and around Delhi – a 16 years study. *Journal of Communicable Diseases* **20**, 104–110.

Bigler, L. (1997). Oral rabies vaccination of raccoons in the St Lawrence, Niagara, and Erie regions of New York State and the Champaign region of Vermont. *Proceedings of the 8th Annual International Meeting on Research Advances in Rabies Control in the Americas*, Peterborough, Ontario, Canada.

Bingham, J. (2005). Canine rabies ecology in southern Africa. *Emerging Infectious Diseases* **11**, 1337–1342.

Bingham, J., Foggin, C.M., Wandeler, A.I. and Hill, F.W. (1999a). The epidemiology of rabies in Zimbabwe. 1. Rabies in dogs (*Canis familiaris*). *Onderstepoort Journal of Veterinary Research* **66**, 1–10.

Bingham, J., Foggin, C.M., Wandeler, A.I. and Hill, F.W. (1999b). The epidemiology of rabies in Zimbabwe. 2. Rabies in jackals (*Canis adustus* and *Canis mesomelas*). *Onderstepoort Journal of Veterinary Research* **66**, 11–23.

Bingham, J., Javangwe, S., Sabeta, C.T., Wandeler, A.I. and Nel, L.H. (2001). Report of isolations of unusual lyssaviruses (rabies and Mokola virus) identified retrospectively from Zimbabwe. *Journal of the South African Veterinary Association* **72**, 92–94.

Bjorvatn, B. and Gundersen, S.G. (1980). Rabies exposure among Norwegian missionaries working abroad. *Scandinavian Journal of Infectious Diseases* **12**, 257–264.

Blancou, J., Aubert, M.F.A. and Artois, M. (1991). Fox rabies. In: *The Natural History of Rabies* (G.M. Baer, ed.). pp. 257–290, Boca Raton: CRC Press.

Bögel, K. and Motschwiller, E. (1986). Incidence of rabies and post-exposure treatment in developing countries. *Bulletin of the World Health Organization* **64**, 883–887.

Bögel, K. and Meslin, F.X. (1990). Economics of human and canine rabies elimination: guidelines for programme orientation. *Bulletin of the World Health Organization* **68**, 281–291.

Bögel, K., Arata, A.A., Moegle, H. and Knorpp, F. (1974). Recovery of reduced fox populations in rabies control. *Zentralblatt fur Veterinarmedizin – Reihe B* **21**, 401–412.

Bögel, K., Andral, L., Beran, G., Schneider, L.G. and Wandeler, A. (1982). Dog rabies elimination. *International Journal of Zoonoses* **9**, 97–112.

Bordignon, J., Brasil-Dos-Anjos, G., Bueno, C.R. *et al.* (2005). Detection and characterization of rabies virus in Southern Brazil by PCR amplification and sequencing of the nucleoprotein gene. *Archives of Virology* **150**, 695–708.

Botvinkin, A.D. (2003). Novel lyssaviruses isolated from bats in Russia. *Emerging Infectious Diseases* **9**, 1623–1625.

Bougler, L.R. and Porterfield, J.S. (1958). Isolation of virus from Nigerian fruit bats. *Transcriptions of the Royal Society of Tropical Medicine and Hygiene* **52**, 421–424.

Bourhy, H., Kissi, B. and Tordo, N. (1993). Taxonomy and evolutionary studies on Lyssaviruses with special reference to Africa. *Onderstepoort Journal of Veterinary Research* **60**, 277–282.

Bourhy, H., Kissi, B., Audry, L. *et al.* (1999). Ecology and evolution of rabies virus in Europe. *Journal of General Virology* **80**, 2545–2557.

Brass, D.A. (1994). *Rabies in Bats*. Ridgefield: Livia Press.

Brochier, B., Costy, F. and Pastoret, P.P. (1995). Elimination of fox rabies from Belgium using a recombinant vaccinia-rabies vaccine: An update. *Veterinary Microbiology* **46**, 269–279.

Brochier, B., Kieny, M.P., Costy, F. *et al.* (1991). Large-scale eradication of rabies using recombinant vaccinia rabies vaccine. *Nature* **354**, 520–522.

Brooks, R. (1990). Survey of the dog population of Zimbabwe and its level of rabies vaccination. *Veterinary Record* **127**, 592–596.

Calisher, C.H., Karabatsos, N., Zeller, H. *et al.* (1989). Antigenic relationships among rhabdoviruses from vertebrates and haematophagous arthropods. *Intervirology* **49**, 241–257.

Caraballo, A.J. (1996). Outbreak of vampire bat biting in a Venezuelan village. *Revista de Saude Publica* **30**, 483–484.

Carey, A.B. (1985). Multispecies rabies in the eastern United States. In: *Population Dynamics of Rabies in Wildlife* (P.J. Bacon, ed.). pp. 23–41. London: Academic Press.

Carini, A. (1911). Sur une grande épizootie de rage. *Annals of the Institute Pasteur (Paris)* **25**, 843–846.

Centers for Disease Control and Prevention (1980). Human-to-human transmission of rabies via a corneal transplant – France. *Mortality and Morbidity Weekly Report* **29**, 25–26.

Centers for Disease Control and Prevention (1981). Human-to-human transmission of rabies via corneal transplant – Thailand. *Morbidity and Mortality Weekly Report* **30**, 473–474.

Centers for Disease Control and Prevention (1982). Human rabies – Rwanda. *Morbidity and Mortality Weekly Report* **31**, 135.

Centers for Disease Control and Prevention (1983). Human rabies – Kenya. *Morbidity and Mortality Weekly Report* **32**, 494–495.

Centers for Disease Control and Prevention (1987). Human rabies despite treatment with rabies immune globulin and human diploid cell rabies vaccine – Thailand. *Morbidity and Mortality Weekly Report* **36**, 759–760, 765.

Centers for Disease Control and Prevention (1995). Translocation of coyote rabies – Florida, 1994. *Morbidity and Mortality Weekly Report* **44**, 580–587.

Centers for Disease Control and Prevention (1996). Rabies postexposure prophylaxis – Connecticut, 1990–1994. *Mortality and Morbidity Weekly Report* **45**, 232–234.

Centers for Disease Control and Prevention (1997). Update: Raccoon rabies epizootic: United States, 1996. *Morbidity and Mortality Weekly Report* **45**, 1117–1120.

Centers for Disease Control and Prevention (1999a). Human rabies prevention – United States, 1999: Recommendations of the Advisory Committee on Immunization Practices (ACIP). *Morbidity and Mortality Weekly Report* **48**, 1–21.

Centers for Disease Control and Prevention (1999b). Mass treatment of humans who drank unpasteurized milk from rabid cows – Massachusetts, 1996–1998. *Morbidity and Mortality Weekly Report* **48**, 228–229.

Centers for Disease Control and Prevention (2000). Update: raccoon rabies epizootic – United States and Canada, 1999. *Morbidity and Mortality Weekly Report* **49**, 31–35.

Centers for Disease Control and Prevention (2003). First human death associated with raccoon rabies – Virginia, 2003. *Morbidity and Mortality Weekly Report* **52**, 1102–1103.

Centers for Disease Control and Prevention (2004a). Compendium of Animal Rabies Prevention and Control, 2004; National Association of State Public Health Veterinarians, Inc (NASPHV). *Mortality and Morbidity Weekly Report* **No. RR-9**, 1–6.

Centers for Disease Control and Prevention (2004b). Investigation of rabies infections in organ donor and transplant recipients – Alabama, Arkansas, Oklahoma, and Texas, 2004. *Mortality and Morbidity Weekly Report* **53**, 586–589.

Centers for Disease Control and Prevention (2005a). Compendium of animal rabies prevention and control, 2005; National Association of State Public Health Veterinarians, Inc. (NASPHV). *Morbidity and Mortality Weekly Report* **54**, 1–8.

Centers for Disease Control and Prevention (2005b). Human rabies – Florida, 2004. *Morbidity and Mortality Weekly Report* **54**, 767–769.

Chang, H.G., Eidson, M., Noonan-Toly, C. *et al.* (2002). Public health impact of reemergence of rabies, New York. *Emerging Infectious Diseases* **8**, 909–913.

Chaparro, F. and Esterhuysen, J.J. (1993). The role of the yellow mongoose (*Cynictis penicillata*) in the epidemiology of rabies in South Africa – preliminary results. *Onderstepoort Journal of Veterinary Research* **60**, 373–377.

Childs, J.E. (2002). Epidemiology. In: *Rabies* (A.C. Jackson and W.H. Wunner, eds). pp. 114–162. New York: Academic Press.

Childs, J.E., Trimarchi, C.V. and Krebs, J.W. (1994). The epidemiology of bat rabies in New York State, 1988–1992. *Epidemiology and Infection* **113**, 501–511.

Childs, J.E., Robinson, L.E., Sadek, R., Madden, A., Miranda, M.E. and Miranda, N.L. (1998). Density estimates of rural dog populations and an assessment of marking methods during a rabies vaccination campaign in the Philippines. *Preventive Veterinary Medicine* **33**, 207–218.

Childs, J.E., Curns, A.T., Dey, M.E. *et al.* (2000). Predicting the local dynamics of epizootic rabies among raccoons in the United States. *Proceedings of the National Academy of Sciences USA* **97**, 13 666–13 671.

Childs, J.E., Curns, A.T., Dey, M.E., Real, A.L., Rupprecht, C.E. and Krebs, J.W. (2001). Rabies epizootics among raccoons vary along a North-South gradient in the Eastern United States. *Vector Borne and Zoonotic Diseases* **1**, 253–267.

Childs, J.E., Krebs, J.W. and Smith, J.S. (2002). Public health surveillance and the molecular epidemiology of rabies. In: *The Molecular Epidemiology of Human Viruses* (T. Leitner, ed.). pp. 273–312. Dordrecht: Kluwer Academic.

Childs, J.E., Krebs, J.W., Real, L.A. and Gordon, E.R. (2006). Animal-based national surveillance for zoonotic disease: quality, limitations, and implications of a model system for monitoring rabies. *Preventive Veterinary Medicine* **78**, 246–261.

Chomel, B., Chappuis, G., Bullon, F., Cardenas, E., de Beublain, T.D. and Lombard, M. (1988). Mass vaccination campaign against rabies: are dogs correctly protected? *Review of Infectious Diseases* **10**, S697–S702.

Cisterna, D., Bonaventura, R., Caillou, S. *et al.* (2005). Antigenic and molecular characterization of rabies virus in Argentina. *Virus Research* **109**, 139–147.

Clark, K.A., Neill, S.U., Smith, J.S., Wilson, P.J., Whadford, V.W. and McKirahan, G.W. (1994). Epizootic canine rabies transmitted by coyotes in south Texas. *Journal of the American Veterinary Medical Association* **204**, 536–540.

Cleaveland, S., Laurenson, M.K. and Taylor, L.H. (2001). Diseases of humans and their domestic mammals: pathogen characteristics, host range and the risk of emergence. *Philosophical Transactions of the Royal Society of London B Biological Sciences* **356**, 991–999.

Cleaveland, S., Fevre, E.M., Kaare, M. and Coleman, P.G. (2002a). Estimating human rabies mortality in the United Republic of Tanzania from dog bite injuries. *Bulletin of the World Health Organization* **80**, 304–310.

Cleaveland, S., Hess, G.R., Dobson, A.P., McCallum, H.I., Roberts, M.G. and Woodroffe, R. (2002b). The role of pathogens in biological conservation. In: *The Ecology of Wildlife Diseases* (P.J. Hudson, A. Rizzoli, B.T. Grenfell, H. Heesterbeekand and A.P. Dobson, eds). pp. 139–150. Oxford: Oxford University Press.

Cleaveland, S., Kaare, M., Tiringa, P., Mlengeya, T. and Barrat, J. (2003). A dog rabies vaccination campaign in rural Africa: impact on the incidence of dog rabies and human dog-bite injuries. *Vaccine* **21**, 1974–1982.

Cliquet, F. and Aubert, M. (2004). Elimination of terrestrial rabies in Western European countries. *Developmental Biology (Basel)* **119**, 185–204.

Coleman, P.G. and Dye, C. (1996). Immunization coverage required to prevent outbreaks of dog rabies. *Vaccine* **14**, 185–186.

Conomy, J.P., Leibovitz, A., McCombs, W. and Stinson, J. (1977). Airborne rabies encephalitis: demonstration of rabies virus in the human central nervous system. *Neurology* **27**, 67–69.

Constantine, D.G. (1962). Rabies transmission by nonbite route. *Public Health Reports* **77**, 287–289.

Constantine, D.G. (2003). Geographic translocation of bats: known and potential problems. *Emerging Infectious Diseases* **9**, 17–21.

Courtin, F., Carpenter, T.E., Paskin, R.D. and Chomel, B.B. (2000). Temporal patterns of domestic and wildlife rabies in central Namibia stock-ranching area, 1986–1996. *Preventive Veterinary Medicine* **43**, 13–28.

Coyne, M.J., Smith, G. and McAllister, F.E. (1989). Mathematic model for the population biology of rabies in raccoons in the mid-Atlantic states. *American Journal of Veterinary Research* **50**, 2148–2154.

Crawford-Miksza, L.K., Wadford, D.A. and Schnurr, D.P. (1999). Molecular epidemiology of enzootic rabies in California. *Journal of Clinical Virology* **14**, 207–219.

Creekmore, T.E., Linhart, S.B., Corn, J.L., Whitney, M.D., Snyder, B.D. and Nettles, V.F. (1994). Field evaluation of baits and baiting strategies for delivering oral vaccine to mongooses in Antigua, West Indies. *Journal of Wildlife Diseases* **30**, 497–505.

Crespo, R.F., Fernandez, S.S., De Anda, L., Velarde, F.I. and Anaya, R.M. (1979). Intramuscular inoculation of cattle with warfarin: a new technique for control of vampire bats. *Bulletin of the Pan American Health Organization* **13**, 147–161.

Cruickshank, I., Gurney, W.S.C. and Veitch, A.R. (1999). The characteristics of epidemics and invasions with thresholds. *Theoretical Population Biology* **56**, 279–292.

Daoust, P.Y., Wandeler, A.I. and Casey, G.A. (1996). Cluster of rabies cases of probable bat origin among red foxes in Prince Edward Island, Canada. *Journal of Wildlife Diseases* **32**, 403–406.

de Mattos, C.C., de Mattos, C.A., Loza-Rubio, E., Aguilar-Setien, A., Orciari, L.A. and Smith, J.S. (1999). Molecular characterization of rabies virus isolates from Mexico: implications for transmission dynamics and human risk. *American Journal of Tropical Hygiene* **61**, 587–597.

de Mattos, C.A., Favi, M., Yung, V., Pavletic, C. and de Mattos, C.C. (2000). Bat rabies in urban centers in Chile. *Journal of Wildlife Diseases* **36**, 231–240.

Debbie, J.G. (1991). Rabies control of terrestrial wildlife by population reduction. In: *The Natural History of Rabies* (G.M. Baer, ed.). pp. 477–484. Boca Raton: CRC Press.

Delpietro, H.A. and Konolsaisen, J.F. (1991). Dinamica da raiva em uma populacao de morcehos hematofagos (*Desmodus rotundus*) no nordeste Argentino, e sua relacao com a raiva paralitica dos herbivoros. *Arquivos de Biologia e Tecnologia (Curitiba)* **31**, 411.

Delpietro, H.A. and Russo, R.G. (1996). Ecological and epidemiological aspects of attacks by vampire bats in relation to paralytic rabies in Argentina, and an analysis of proposals for control. *Revue Scientifique et Technique* **15**, 971–984.

Denduangboripant, J., Wacharapluesadee, S., Lumlertdacha, B. *et al.* (2005). Transmission dynamics of rabies virus in Thailand: Implications for disease control. *BMC Infectious Diseases* **5**, 52–63.

Dietzschold, B., Schnell, M. and Koprowski, H. (2005). Pathogenesis of rabies. *Current Topics in Microbiology and Immunology* **292**, 45–56.

Dobson, A. and Foufopoulos, J. (2001). Emerging infectious pathogens of wildlife. *Philosophical Transactions of the Royal Society of London B Biological Sciences* **356**, 1001–1012.

Dutta, J.K. (1999). Human rabies in India: epidemiological features, management and current methods of prevention. *Tropical Doctor* **29**, 196–201.

Eisinger, D., Thulke, H.H., Selhorst, T. and Muller, T. (2005). Emergency vaccination of rabies under limited resources – combating or containing? *BMC Infectious Diseases* **5**, 10.

Eng, T.R. and Fishbein, D.B. (1990). Epidemiologic factors, clinical findings, and vaccination status of rabies in cats and dogs in the United States in 1988. National Study Group on Rabies. *Journal of the American Veterinary Medical Association* **197**, 201–209.

Eng, T.R., Fishbein, D.B., Talamante, H.E. *et al.* (1993). Urban epizootic of rabies in Mexico: epidemiology and impact of animal bite injuries. *Bulletin of the World Health Organization* **71**, 615–624.

Engeman, R.M., Christensen, K.L., Pipas, M.J. and Bergman, D.L. (2003). Population monitoring in support of a rabies vaccination program for skunks in Arizona. *Journal of Wildlife Diseases* **39**, 746–750.

Ernst, S.N. and Fabrega, F. (1989). A time series analysis of the rabies control programme in Chile. *Epidemiology and Infection* **103**, 651–657.

Estrada, R., Vos, A., De Leon, R. and Mueller, T. (2001). Field trial with oral vaccination of dogs against rabies in the Philippines. *BMC Infectious Diseases* **1**, 23.

Eursurveillance Weekly (2005). **10**(7), 17 February at www.eurosurveillance.org.

Evans, N.D. and Pritchard, A.J. (2001). A control theoretic approach to containing the spread of rabies. *IMA Journal of Mathematics Applied in Medicine & Biology* **18**, 1–23.

Everard, C.O.R. and Everard, J.D. (1985). Mongoose rabies in Grenada. In: *Population Dynamics of Rabies in Wildlife* (P.J. Bacon, ed.). pp. 43–67. London: Academic Press, Inc.

Ezeokoli, C.D. and Umoh, J.U. (1987). Epidemiology of rabies in northern Nigeria. *Transactions of the Royal Society of Tropical Medicine and Hygiene* **81**, 268–272.

Fagbami, A.H., Anosa, V.O. and Ezebuiro, E.O. (1981). Hospital records of human rabies and antirabies prophylaxis in Nigeria 1969–78. *Transactions of the Royal Society of Tropical Medicine and Hygiene* **75**, 872–876.

Familusi, J.B. and Moore, D.L. (1972). Isolation of a rabies related virus from the cerebrospinal fluid of a child with 'aseptic meningitis'. *African Journal of Medical Science* **3**, 93–96.

Familusi, J.B., Osuhkoya, B.O., Moore, D.L., Kemp, G.E. and Fabiyi, A. (1972). A fatal human infection with Mokola virus. *American Journal of Tropical Medicine and Hygiene* **21**, 959–963.

Farry, S.C., Henke, S.E., Beasom, S.L. and Fearneyhough, M.G. (1998). Efficacy of bait distributional strategies to deliver canine rabies vaccines to coyotes in southern Texas. *Journal of Wildlife Diseases* **34**, 23–32.

Favi, M., de Mattos, C.A., Yung, V., Chala, E., Lopez, L.R. and de Mattos, C.C. (2002). First case of human rabies in chile caused by an insectivorous bat virus variant. *Emerging Infectious Diseases* **8**, 79–81.

Favi, M., Nina, A., Yung, V. and Fernandez, J. (2003). Characterization of rabies virus isolates in Bolivia. *Virus Research* **97**, 135–140.

Favoretto, S.R., de Mattos, C.C., Morais, N.B., Alves Araujo, F.A. and de Mattos, C.A. (2001). Rabies in marmosets (*Callithrix jacchus*), Ceará, Brazil. *Emerging Infectious Diseases* **7**, 1062–1065.

Fearneyhough, M.G., Wilson, P.J., Clark, K.A. *et al.* (1998). Results of an oral rabies vaccination program for coyotes. *Journal of the American Medical Association* **212**, 498–502.

Feder, H.M. Jr, Nelson, R. and Reiher, H.W. (1997). Bat bite? *Lancet* **350**, 1300.

Fekadu, M. (1982). Rabies in Ethiopia. *American Journal of Epidemiology* **115**, 266–273.

Fekadu, M., Endeshaw, T., Alemu, W., Bogale, Y., Teshager, T. and Olson, J.G. (1996). Possible human-to-human transmission of rabies in Ethiopia. *Ethiopian Medical Journal* **34**, 123–127.

Ferro, V.A. (2002). Current advances in antifertility vaccines for fertility control and noncontraceptive applications. *Expert Reviews in Vaccines* **1**, 443–452.

Ferro, V.A., Khan, M.A., McAdam, D. *et al.* (2004). Efficacy of an anti-fertility vaccine based on mammalian gonadotrophin releasing hormone (GnRH-I) – a histological comparison in male animals. *Veterinary Immunology and Immunopathology* **101**, 73–86.

Fevre, E.M., Kaboyo, R.W., Persson, V., Edelsten, M., Coleman, P.G. and Cleaveland, S. (2005). The epidemiology of animal bite injuries in Uganda and projections of the burden of rabies. *Tropical Medicine and International Health* **10**, 790–798.

Fielding, J.E. and Nayda, C.L. (2005). Postexposure prophylaxis for Australian bat lyssavirus in South Australia, 1996 to 2003. *Australian Veterinary Journal* **83**, 233–234.

Fischman, H.R., Grigor, J.K., Horman, J.T. and Israel, E. (1992). Epizootic of rabies in raccoons in Maryland from 1981 to 1987. *Journal of the American Veterinary Medical Association* **201**, 1883–1886.

Fishbein, D.B. (1991). Rabies in humans. In: *The Natural History of Rabies* (G.M. Baer, ed.). pp. 519–549. Boca Raton: CRC Press.

Fishbein, D.B., Corboy, J.M. and Sasaki, D.M. (1990). Rabies prevention in Hawaii. *Hawaii Medical Journal* **49**, 98–101.

Flores-Crespo, R. and Arellano-Sota, C. (1991). Biology and control of the vampire bat. In: *The Natural History of Rabies* (G.M. Baer, ed). pp. 461–476. Boca Raton: CRC Press.

Flores-Ibarra, M. and Estrella-Valenzuela, G. (2004). Canine ecology and socioeconomic factors associated with dogs unvaccinated against rabies in a Mexican city across the US-Mexico border. *Preventive Veterinary Medicine* **62**, 79–87.

Foggin, C.M. (1983). Mokola virus infection in cats and a dog in Zimbabwe. *Veterinary Record* **113**, 115.

Foggin, C. M. (1988). *Rabies and rabies-related viruses in Zimbabwe: Historical, virological and ecological aspects.* PhD thesis, University of Zimbabwe.

Follmann, E.H., Ritter, D.G. and Donald, W.H. (2004). Oral vaccination of captive arctic foxes with lyophilized SAG2 rabies vaccine. *Journal of Veterinary Diseases* **40**, 328–334.

Fooks, A.R., McElhinney, L.M., Brookes, S.M. *et al.* (2002). Rabies antibody testing and the UK Pet Travel Scheme. *Veterinary Record* **150**, 428–430.

Fooks, A.R., Brookes, S.M., Johnson, N., McElhinney, L.M. and Hutson, A.M. (2003). European bat lyssaviruses: an emerging zoonosis. *Epidemiology and Infection* **131**, 1029–1039.

Fornes, A., Lord, R.D., Kuns, M.L., Larghi, O.P., Fuenzalida, E. and Lazara, L. (1974). Control of bovine rabies through vampire bat control. *Journal of Wildlife Diseases* **10**, 310–316.

Foroutan, P., Meltzer, M.I. and Smith, K.A. (2002). Cost of distributing oral raccoon-variant rabies vaccine in Ohio: 1997–2000. *Journal of the American Veterinary Association* **220**, 27–32.

Gacouin, A., Bourhy, H., Renaud, J.C., Camus, C., Suprin, E. and Thomas, R. (1999). Human rabies despite postexposure vaccination. *European Journal of Clinical Microbiology and Infectious Diseases* **18**, 233–235.

Gardner, G.A., Gardner, L.R. and Cunningham, J. (1990). Simulations of a fox-rabies epidemic on an island using space-time finite elements. *Zeitschrift für Naturforschung* **45**, 1230–1240.

Gascoyne, S.C., King, A.A., Laurenson, M.K., Borner, M., Schildger, B. and Barrat, J. (1993). Aspects of rabies infection and control in the conservation of the African wild dog (*Lycaon pictus*) in the Serengeti region, Tanzania. *Onderstepoort Journal of Veterinary Research* **60**, 415–420.

Gibbons, R.V. (2002). Cryptogenic rabies, bats, and the question of aerosol transmission. *Annals of Emergency Medicine* **39**, 528–536.

Gode, G.R. and Bhide, N.K. (1988). Two rabies deaths after corneal grafts from one donor. *Lancet* **2**, 791.

Godin, A.J. (1977). *Wild Mammals of New England.* Baltimore: The Johns Hopkins University Press.

Gordon, E.R., Curns, A.T., Krebs, J.W., Rupprecht, C.E., Real, L.A. and Childs, J.E. (2004). Temporal dynamics of rabies in a wildlife host and the risk of cross-species transmission. *Epidemiology and Infection* **132**, 515–524.

Gordon, E.R., Krebs, J.W., Rupprecht, C.E., Real, L.A. and Childs, J.E. (2005). Persistence of elevated rabies prevention costs following post-epizootic declines in rates of rabies among raccoons (*Procyon lotor*). *Preventive Veterinary Medicine* **68**, 195–222.

Goswami, A., Plun-Favreau, J., Nicoloyannis, N., Sampath, G., Siddiqui, M.N. and Zinsou, J.A. (2005). The real cost of rabies post-exposure treatments. *Vaccine* **23**, 2970–2976.

Greenwood, R.J., Newton, W.E., Pearson, G.L. and Schamber, G.J. (1997). Population and movement characteristics of radio-collared striped skunks in North Dakota during an epizootic of rabies. *Journal of Wildlife Diseases* **33**, 226–241.

Gremillion-Smith, C. and Woolf, A. (1988). Epizootiology of skunk rabies in North America. *Journal of Wildlife Diseases* **24**, 620–626.

Guerra, M.A., Curns, A.T., Rupprecht, C.E., Hanlon, C.A., Krebs, J.W. and Childs, J.E. (2003). Skunk and raccoon rabies in the eastern United States: temporal and spatial analysis. *Emerging Infectious Diseases* **9**, 1143–1150.

Gupte, M.D., Ramachandran, V. and Mutatkar, R.K. (2001). Epidemiological profile of India: historical and contemporary perspectives. *Journal of Bioscience* **26**, 437–464.

Halloran, M.E. (1998). Concepts of infectious disease epidemiology. In: *Modern Epidemiology* (K.J. Rothman and S. Greenland, eds). pp. 529–554. Philadelphia: Lippincott Williams & Wilkins.

Hanlon, C.A. and Rupprecht, C.E. (1998). The reemergence of rabies. In: *Emerging infections* (W.M. Scheld, D. Armstrong and J.M. Hughes, eds). pp. 59–80. Washington, DC: ASM Press.

Hanlon, C.A., Niezgoda, M., Hamir, A.N., Schumacher, C., Koprowski, H. and Rupprecht, C.E. (1998). First North American field release of a vaccinia-rabies glycoprotein recombinant virus. *Journal of Wildlife Diseases* **34**, 228–239.

Hanlon, C.A., Kuzmin, I.V., Blanton, J.D., Weldon, W.C., Manangan, J.S. and Rupprecht, C.E. (2005). Efficacy of rabies biologics against new lyssaviruses from Eurasia. *Virus Research* **111**, 44–54.

Hanna, J.N., Carney, I.K., Smith, G.A. *et al.* (2000). Australian bat lyssavirus infection: a second human case, with a long incubation period. *Medical Journal of Australia* **172**, 597–599.

Harris, S. and White, P.C.L. (2004). Epidemiological models. In: *Historical Perspective of Rabies in Europe and the Mediterranean Basin* (A.A. King, A.R. Fooks, M. Aubert and A. Wandeler, eds). pp. 293–309. Paris: OIE.

Hattwick, M.A.W. (1974). Human rabies. *Public Health Reviews* **3**, 229–274.

Heidt, G.A., Ferguson, D.V. and Lammers, J. (1982). A profile of reported skunk rabies in Arkansas: 1977–1979. *Journal of Wildlife Diseases* **18**, 269–277.

Held, J.R. and Adaros, H.L. (1972). Neurological disease in man following administration of suckling mouse brain antirabies vaccine. *Bulletin of the World Health Organization* **46**, 321–327.

Held, J.R., Tierkel, E.S. and Steele, J.H. (1967). Rabies in man and animals in the United States, 1946–1965. *Public Health Reports* **82**, 1009–1018.

Helmick, C.G., Tauxe, R.V. and Vernon, A.A. (1987). Is there a risk to contacts of patients with rabies? *Review of Infectious Diseases* **9**, 511–518.

Hemachudha, T. (2005). Rabies and dog population control in Thailand: success or failure? *Journal of the Medical Association of Thailand* **88**, 120–123.

Hemachudha, T., Mitrabhakdi, E., Wilde, H., Vejabhuti, A., Siripataravanit, S., and Kingnate, D. (1999). Additional reports of failure to respond to treatment after rabies exposure in Thailand. *Clinical Infectious Diseases* **28**, 143–144.

Hojer, J., Sjoblom, E., Berglund, O., Hammarin, A. L. and Grandien, M. (2001). The first case of rabies in Sweden in 26 years. Inform travellers abroad about risks and treatment following suspected infection. *Lakartidningen* **98**, 1216–1220.

Homola, V. (2005). World Briefing Europe: Germany: Rabies after transplants. *The New York Times*. February 18, 2005 Friday Late Edition – Final, 6. New York: New York Times Inc.

Houff, S.A., Burton, R.C., Wilson, R.W. *et al.* (1979). Human-to-human transmission of rabies virus by corneal transplant. *New England Journal of Medicine* **300**, 603–604.

Hughes, G.J., Orciari, L.A. and Rupprecht, C.E. (2005). Evolutionary timescale of rabies virus adaptation to North American bats inferred from the substitution rate of the nucleoprotein gene. *Journal of General Virology* **86**, 1467–1474.

Humphrey, G.L., Kemp, G.E. and Wood, E.G. (1960). A fatal case of rabies in a woman bitten by an insectivorous bat. *Public Health Reports* **75**, 317–326.

Hurst, E.W. and Pawan, J.L. (1931). An outbreak of rabies in Trinidad without history of bites and with the symptoms of acute ascending paralysis. *Lancet* **ii**, 622–625.

Hyun, B.H., Lee, K.K., Kim, I.J. *et al.* (2005). Molecular epidemiology of rabies virus isolates from South Korea. *Virus Research* **114**, 113–125.

Ireland, J.M., Norman, R.A. and Greenman, J.V. (2004). The effect of seasonal host birth rates on population dynamics: the importance of resonance. *Journal of Theoretical Biology* **231**, 229–238.

Irons, J.V., Eads, R.B., Grimes, J.E. and Conklin, A. (1957). The public health importance of bats. *Texas Reports of Biology and Medicine* **15**, 292–298.

Ito, M., Arai, Y.T., Itou, T. *et al.* (2001). Genetic characterization and geographic distribution of rabies virus isolates in Brazil: identification of two reservoirs, dogs and vampire bats. *Virology* **284**, 214–222.

Jeltsch, F., Mueller, M.S., Grimm, V., Wissel, C. and Brandl, R. (1997). Pattern formation triggered by rare events: Lessons from the spread of rabies. *Proceedings of the Royal Society of London B Biological Sciences* **264**, 495–503.

Jenkins, S.R. and Winkler, W.G. (1987). Descriptive epidemiology from an epizootic of raccoon rabies in the Middle Atlantic States, 1982–1983. *American Journal of Epidemiology* **126**, 429–437.

Johnson, N., Lipscomb, D.W., Stott, R. *et al.* (2002). Investigation of a human case of rabies in the United Kingdom. *Journal of Clinical Virology* **25**, 351–356.

Johnson, N., McElhinney, L.M., Ali, Y.H., Saeed, I.K. and Fooks, A.R. (2004). Molecular epidemiology of canid rabies in Sudan: evidence for a common origin of rabies with Ethiopia. *Virus Research* **104**, 201–205.

Johnston, W.B. and Walden, M.B. (1996). Results of a national survey of rabies control procedures. *Journal of the American Veterinary Medical Association* **208**, 1667–1672.

Jones, R.D., Kelly, L., Fooks, A.R. and Wooldridge, M. (2005). Quantitative risk assessment of rabies entering Great Britain from North America via cats and dogs. *Risk Analysis* **25**, 533–542.

Kale, O.O. (1977). Epidemiology and treatment of dog bites in Ibadan: a 12-year retrospective study of cases seen at the University College Hospital Ibadan (1962–1973). *African Journal of Medical Science* **6**, 133–140.

Kallen, A., Arcuri, P. and Murray, J.D. (1985). A simple model for the spatial spread and control of rabies. *Journal of Theoretical Biology* **116**, 377–393.

Kat, P.W., Alexander, K.A., Smith, J.S. and Munson, L. (1995). Rabies and African wild dogs in Kenya. *Proceedings of the Royal Society of London B Biological Sciences* **262**, 229–233.

Kayali, U., Mindekem, R., Yemadji, N. *et al.* (2003a). Incidence of canine rabies in N'Djamena, Chad. *Preventive Veterinary Medicine* **61**, 227–233.

Kayali, U., Mindekem, R., Yemadji, N. *et al.* (2003b). Coverage of pilot parenteral vaccination campaign against canine rabies in N'Djamena, Chad. *Bulletin of the World Health Organization* **81**, 739–744.

Kemere, P.K., Liddel, M.K., Evangelou, P., Slate, D. and Osmek, S. (2002). Economic analysis of a large scale oral vaccination program to control raccoon rabies. *Proceedings of the Third NWRC Special Symposium*, 109–116. Ft Collins, Colorado, National Wildlife Research Center, USDA. Human Conflicts with Wildlife: Economic Considerations.

Kennedy, D.J. (1988). An outbreak of rabies in north-western Zimbabwe 1980 to 1983. *Veterinary Record* **122**, 129–133.

Kim, J.H., Hwang, E.K., Sohn, H.J., Kim, D.Y., So, B.J. and Jean, Y.H. (2005). Epidemiological characteristics of rabies in South Korea from 1993 to 2001. *Veterinary Record* **157**, 53–56.

King, A. and Crick, J. (1988). Rabies-related viruses. In: *Rabies* (J.B. Campbell, ed.). pp. 177–199. Boston: Kluwer Academic Publishers.

King, A.A., Meredith, C.D. and Thomson, G.R. (1993). Canid and viverrid rabies viruses in South Africa. *Onderstepoort Journal of Veterinary Research* **60**, 295–299.

Kitala, P., McDermott, J., Kyule, M., Gathuma, J., Perry, B. and Wandeler, A. (2001). Dog ecology and demography information to support the planning of rabies control in Machakos District, Kenya. *Acta Tropica* **78**, 217–230.

Knobel, D.L., Cleaveland, S., Coleman, P.G. *et al.* (2005). Re-evaluating the burden of rabies in Africa and Asia. *Bulletin of the World Health Organization* **83**, 360–368.

Kobayashi, Y., Sato, G., Shoji, Y. *et al.* (2005). Molecular epidemiological analysis of bat rabies viruses in Brazil. *Journal of Veterinary Medical Science* **67**, 647–652.

Krause, R., Bago, Z., Revilla-Fernandez, S. *et al.* (2005). Travel-associated rabies in Austrian man. *Emerging Infectious Diseases* **11**, 719–721.

Krebs, J.W., Strine, T.W., Smith, J.S., Noah, D.L., Rupprecht, C.E. and Childs, J.E. (1996). Rabies surveillance in the United States during 1995. *Journal of the American Veterinary Medical Association* **209**, 2031–2044.

Krebs, J.W., Long-Marin, S.C. and Childs, J. E. (1998). Causes, costs, and estimates of rabies postexposure prophylaxis treatments in the United States. *Journal of Public Health Management Practice* **4**, 57–63.

Krebs, J.W., Mondul, A.M., Rupprecht, C.E. and Childs, J.E. (2001). Rabies surveillance in the United States during 2000. *Journal of the American Veterinary Medical Association* **219**, 1687–1699.

Krebs, J.W., Noll, H.R., Rupprecht, C.E. and Childs, J.E. (2002). Rabies surveillance in the United States during 2001. *Journal of the American Veterinary Medical Association* **221**, 1690–1701.

Krebs, J.W., Wheeling, J.T. and Childs, J.E. (2003a). Rabies surveillance in the United States during 2002. *Journal of the American Veterinary Medical Association* **223**, 1736–1748.

Krebs, J.W., Williams, S.M., Smith, J.S., Rupprecht, C.E. and Childs, J.E. (2003b). Rabies among infrequently reported mammalian carnivores in the United States, 1960–2000. *Journal of Wildlife Diseases* **39**, 253–261.

Krebs, J.W., Mandel, E.J., Swerdlow, D.L. and Rupprecht, C.E. (2004). Rabies surveillance in the United States during 2003. *Journal of the American Veterinary Medical Association* **225**, 1837–1849.

Kuzmin, I.V. (1999). An arctic fox rabies virus strain as the cause of human rabies in Russian Siberia. *Archives of Virology* **144**, 627–629.

Kuzmin, I.V., Orciari, L.A., Arai, Y.T. *et al.* (2003). Bat lyssaviruses (Aravan and Khujand) from Central Asia: phylogenetic relationships according to N, P and G gene sequences. *Virus Research* **97**, 65–79.

Kuzmin, I.V., Botvinkin, A.D., McElhinney, L.M. *et al.* (2004). Molecular epidemiology of terrestrial rabies in the former Soviet Union. *Journal of Wildlife Diseases* **40**, 617–631.

Lakhanpal, U. and Sharma, R.C. (1985). An epidemiological study of 177 cases of human rabies. *International Journal of Epidemiology* **14**, 614–617.

Leach, C.N. and Johnson, H.N. (1940). Human rabies, with special reference to virus distribution and titer. *American Journal of Tropical Medicine and Hygiene* **20**, 335–340.

Lederberg, J., Shope, R.E. and Oaks, S.C. Jr (1992). *Emerging Infections*. Washington, DC: National Academy Press.

Levins, R. (1969). Some demographic and genetic consequences of environmental heterogeneity for biological control. *Bulletin of the Entomological Society of America* **15**, 237–240.

Linhart, S.B., Flores Crespo, R. and Mitchell, G.C. (1972). Control of vampire bats by means of an anticoagulant. *Boletin de la Oficina Sanitaria Panamericana* **73**, 100–109.

Linhart, S.B., King, R., Zamir, S., Naveh, U., Davidson, M. and Perl, S. (1997). Oral rabies vaccination of red foxes and golden jackals in Israel: preliminary bait evaluation. *Reviews in Science and Technology* **16**, 874–880.

Linhart, S.B., Wlodkowski, J.C., Kavanaugh, D.M. *et al.* (2002). A new flavor-coated sachet bait for delivering oral rabies vaccine to raccoons and coyotes. *Journal of Wildlife Diseases* **38**, 363–377.

Lloyd, H.G. (1976). Wildlife rabies in Europe and the British situation. *Transactions of the Royal Society of Tropical Medicine and Hygiene* **70**, 179–187.

Lodmell, D.L., Parnell, M.J., Weyrich, J.T. and Ewalt, L.C. (2003). Canine rabies DNA vaccination: a single-dose intradermal injection into ear pinnae elicits elevated and persistent levels of neutralizing antibody. *Vaccine* **21**, 3998–4002.

Lodmell, D.L., Ewalt, L.C., Parnell, M.J., Rupprecht, C.E. and Hanlon, C.A. (2005). One-time intradermal DNA vaccination in ear pinnae one year prior to infection protects dogs against rabies virus. *Vaccine* **24**, 412–416.

Lopez, A., Miranda, P., Tejada, E. and Fishbein, D.B. (1992). Outbreak of human rabies in the Peruvian jungle. *Lancet* **339**, 408–411.

Lord, R.D., Fuenzalida, E., Delpietro, H., Larghi, O.P., de Diaz, A.M. and Lazaro, L. (1975). Observations on the epizootiology of vampire bat rabies. *Bulletin of the Pan American Health Organization* **9**, 189–195.

Lucey, B.T., Russell, C.A., Smith, D. *et al.* (2002). Spatiotemporal analysis of epizootic raccoon rabies propagation in Connecticut, 1991–1995. *Vector Borne and Zoonotic Disease* **2**, 77–86.

Lumbiganon, P. and Wasi, C. (1990). Survival after rabies immunisation in newborn infant of affected mother. *Lancet* **336**, 319.

Macdonald, D.W. (1980). *Rabies and Wildlife A Biologists's Perspective.* Oxford: Oxford University Press.

Macdonald, D.W. and Voigt, D.R. (1985). The biological basis of rabies models. In: *Population Dynamics of Rabies in Wildlife* (P.J. Bacon, ed.). pp.71–108. London: Academic Press.

Macinnes, C.D., Smith, S.M., Tinline, R.R. *et al.* (2001). Elimination of rabies from red foxes in eastern Ontario. *Journal of Wildlife Diseases* **37**, 119–132.

Malaga, H., Lopez Nieto, E. and Gambirazio, C. (1979). Canine rabies seasonality. *International Journal of Epidemiology* **8**, 243–245.

Martinez-Burnes, J., Lopez, A., Medellin, J., Haines, D., Loza, E. and Martinez, M. (1997). An outbreak of vampire bat-transmitted rabies in cattle in northeastern Mexico. *Canadian Veterinary Journal* **38**, 175–177.

Matha, I.S. and Salunke, S.R. (2005). Immunogenicity of purified vero cell rabies vaccine used in the treatment of fox-bite victims in India. *Clinical Infectious Diseases* **40**, 611–613.

Matter, H.C. and Daniels, T.J. (2000). Dog ecology and population biology. In: *Dogs, Zoonoses and Public Health* (C.N.L. Macpherson, F.X. Meslin and A.I. Wandeler, eds). pp. 17–62. Trowbridge: CABI Publishing.

Matter, H.C., Wandeler, A.I., Neuenschwander, B.E., Harischandra, L.P. and Meslin, F.X. (2000). Study of the dog population and the rabies control activities in the Mirigama area of Sri Lanka. *Acta Tropica* **75**, 95–108.

Mayen, F. (2003). Haematophagous bats in Brazil, their role in rabies transmission, impact on public health, livestock industry and alternatives to an indiscriminate reduction of bat population. *Journal of Veterinary Medicine, B, Infectious Diseases and Veterinary Public Health* **50**, 469–472.

McKendrick, A.G. (1941). A ninth analytical review of reports from Pasteur Institutes. *Bulletin of the World Health Organization* **9**, 31–78.

McQuiston, J.H., Yager, P.A., Smith, J.S. and Rupprecht, C.E. (2001). Epidemiologic characteristics of rabies virus variants in dogs and cats in the United States, 1999. *Journal of the American Veterinary Medical Association* **218**, 1939–1942.

Mebatsion, T., Cox, J.H. and Frost, J.W. (1992). Isolation and characterization of 115 street rabies isolates from Ethiopia by using monoclonal antibodies. *Journal of Infectious Diseases* **166**, 972–977.

Meltzer, M.I. (1996). Assessing the costs and benefits of an oral vaccine for raccoon rabies: a possible model. *Emerging Infectious Diseases* **2**, 343–349.

Meredith, C.D., Rossouw, A.P. and Van Pragg Koch, H. (1971). An unusual case of human rabies thought to be of chiropteran origin. *South African Medical Journal* **45**, 767–769.

Merriam, C.H. (1886). *The Mammals of the Adirondack Region; Northeastern New York*. New York: Clinton Holt and Company.

Meslin, F.-X. (2005). Rabies as a traveler's risk, especially in high-endemicity areas. *Journal of Travel Medicine* **12**, S30–S40.

Meslin, F.-X., Fishbein, D.B. and Matter, H.C. (1994). Rationale and prospects for rabies elimination in developing countries. *Current Topics in Microbiology and Immunology* **187**, 1–26.

Messenger, S.L., Smith, J.S. and Rupprecht, C.E. (2002). Emerging epidemiology of bat-associated cryptic cases of rabies in humans in the United States. *Clinical Infectious Diseases* **35**, 738–747.

Messenger, S.L., Rupprecht, C.E. and Smith, J.S. (2003a). Bats, emerging virus infections, and the rabies paradigm. In: *Bat Ecology* (T.H. Kunz and M.B. Fenton, eds). pp. 622–679. Chicago: University of Chicago Press.

Messenger, S.L., Smith, J.S., Orciari, L.A., Yager, P.A. and Rupprecht, C.E. (2003b). Emerging pattern of rabies deaths and increased viral infectivity. *Emerging Infectious Diseases* **9**, 151–154.

Mitmoonpitak, C., Tepsumethanon, V. and Wilde, H. (1998). Rabies in Thailand. *Epidemiology and Infection* **120**, 165–169.

Mondul, A.M., Krebs, J.W. and Childs, J.E. (2003). Trends in national surveillance for rabies among bats in the United States (1993–2000). *Journal of the American Veterinary Medical Association* **222**, 633–639.

Moore, D.A. (1999). Spatial diffusion of raccoon rabies in Pennsylvania, USA. *Preventive Veterinary Medicine* **40**, 19–32.

Moore, D.A., Sischo, W.M., Hunter, A. and Miles, T. (2000). Animal bite epidemiology and surveillance for rabies postexposure prophylaxis. *Journal of the American Veterinary Medical Association* **217**, 190–194.

Moran, G.J., Talan, D.A., Mower, W. *et al.* (2000). Appropriateness of rabies postexposure prophylaxis treatment for animal exposures. *Journal of the American Medical Association* **284**, 1001–1007.

Morimoto, K., Patel, M., Corisdeo, S. *et al.* (1996). Characterization of a unique variant of bat rabies virus responsible for newly emerging human cases in North America. *Proceedings of the National Academy of Sciences USA* **93**, 5653–5658.

Morton, C. (1973). Dog bites in Norfolk, VA. *Health Services Reports* **88**, 59–64.

Muller, J. (1971). The effect of fox reduction on the occurrence of rabies. Observations from two outbreaks of rabies in Denmark. *Bulletin de l'Office International des Epizooties* **75**, 763–776.

Müller, U. and Breitenmoser, U. (2004). Computer analysis of the fox rabies epidemic. In: *Historical Perspective of Rabies in Europe and the Mediterranean Basin* (A.A. King, A.R. Fooks, M. Aubert, and A.I. Wandeler, eds). pp. 281–291. Paris: OIE.

Müller, W., Cox, J. and Müller, T. (2004). Rabies in Germany, Denmark, and Austria. In: *Historical Perspective of Rabies in Europe and the Mediterranean Basin* (A.A. King, A.R. Fooks, M. Aubert, and A.I. Wandeler, eds). pp. 79–92. Paris: OIE.

Murray, J.D. and Seward, W.L. (1992). On the spatial spread of rabies among foxes with immunity. *Journal of Theoretical Biology* **156**, 327–348.

Murray, J.D., Stanley, E.A. and Brown, D.L. (1986). On the spatial spread of rabies among foxes. *Proceedings of the Royal Society of London Biology* **229**, 111–150.

Mutinelli, F., Stankov, M., Hristovski, M., Seimenis, A., Theoharakou, H. and Vodopija, I. (2004). Rabies in Italy, Yugoslavia, Croatia, Bosnia, Slovenia, Macedonia, Albania & Greece. In: *Historical Perspective of Rabies in Europe and the Mediterranean Basin* (A.A. King, A.R. Fooks, M. Aubert, and A.I. Wandeler, eds). pp. 93–118. Paris: OIE.

Nadin-Davis, S.A., Casey, G.A. and Wandeler, A. (1993). Identification of regional variants of the rabies virus within the Canadian Province of Ontario. *Journal of General Virology* **74**, 829–837.

Nadin-Davis, S.A. and Casey, G.A. (1994). A molecular epidemiological study of rabies virus in central Ontario and western Quebec. *Journal of General Virology* **75**, 2575–2583.

Nadin-Davis, S.A., Sampath, M.I., Casey, G.A., Tinline, R.R. and Wandeler, A.I. (1999). Phylogeographic patterns exhibited by Ontario rabies virus variants. *Epidemiology and Infection* **123**, 325–336.

Nadin-Davis, S.A. and Bingham, J. (2004). Europe as a source of rabies for the rest of the world. In: *Historical Perspective of Rabies in Europe and the Mediterranean Basin* (A.A. King, A.R. Fooks, M. Aubert, and A.I. Wandeler, eds). pp. 259–292. Paris: OIE.

Nadin-Davis, S.A., Muldoon, F. and Wandeler, A.I. (2005). A molecular epidemiological analysis of the incursion of the raccoon strain of rabies virus into Canada. *Epidemiology and Infection* **134**, 534–547.

Nathwani, D., McIntyre, P.G., White, K. *et al.* (2003). Fatal human rabies caused by European bat Lyssavirus type 2a infection in Scotland. *Clinical Infectious Diseases* **37**, 598–601.

Nawathe, D.R. (1980). Rabies control in Nigeria. *Bulletin de Office International des Epizooties* **92**, 129–139.

Nel, L.H. (2005). Vaccines for lyssaviruses other than rabies. *Expert Reviews of Vaccines* **4**, 533–540.

Nel, L., Jacobs, J., Jaftha, J. and Meredith, C. (1997). Natural spillover of a distinctly Canidae-associated biotype of rabies virus into an expanded wildlife host range in southern Africa. *Virus Genes* **15**, 79–82.

Nel, L., Jacobs, J., Jaftha, J., von Teichman, B. and Bingham, J. (2000). New cases of Mokola virus infection in South Africa: a genotypic comparison of Southern African virus isolates. *Virus Genes* **20**, 103–106.

Nel, L.H., Sabeta, C.T., von Teichman B., Jaftha, J.B., Rupprecht, C.E. and Bingham, J. (2005). Mongoose rabies in southern Africa: a re-evaluation based on molecular epidemiology. *Virus Research* **109**, 165–173.

Nel, L.H. and Rupprecht, C.E. (2006). Emergence of lyssaviruses in the Old World: The case of Africa. In: *Wildlife and Emerging Zoonotic Diseases: The Biology, Circumstances, and Consequences of Cross-species Transmission.* (J.E. Childs, J.A. Richt, and J.S. Mackenzie, eds).

Nelson, R.S., Mshar, P.A., Cartter, M.L., Adams, M.L. and Hadler, J.L. (1998). Public awareness of rabies and compliance with pet vaccination laws in Connecticut, 1993. *Journal of the American Veterinary Medical Association* **212**, 1552–1555.

Nettles, V.F., Shaddock, J.H., Sikes, R.K. and Reyes, C.R. (1979). Rabies in translocated raccoons. *American Journal of Public Health* **69**, 601–602.

Noah, D.L., Smith, G.M., Gotthardt, J.C., Krebs, J.W., Green, D. and Childs, J.E. (1996). Mass human exposure to rabies in New Hampshire: assessment of exposures and adverse reactions. *American Journal of Public Health* **86**, 1149–1151.

Noah, D.L., Drenzek, C.L., Smith, J.S. *et al.* (1998). Epidemiology of human rabies in the United States, 1980 to 1996. *Annals of Internal Medicine* **128**, 922–930.

Olson, C.A., Mitchell, K.D. and Werner, P.A. (2000). Bait ingestion by free-ranging raccoons and nontarget species in an oral rabies vaccine field trial in Florida. *Journal of Wildlife Diseases* **36**, 734–743.

Olson, C.A. and Werner, P.A. (1999). Oral rabies vaccine contact by raccoons and nontarget species in a field trial in Florida. *Journal of Wildlife Diseases* **35**, 687–695.

Organización Panamericana de la Salud (OPS) (1983). *Estragia y Plan de Acción para le Eliminación de la Rabia Urbana en América Latina para et final de la década de 1980.* Guayaquil: Ecuador.

Páez, A., Nunez, C., Garcia, C. and Boshell, J. (2003). Molecular epidemiology of rabies epizootics in Colombia: evidence for human and dog rabies associated with bats. *Journal of General Virology* **84**, 795–802.

Páez, A., Saad, C., Nunez, C. and Boshell, J. (2005). Molecular epidemiology of rabies in northern Colombia 1994–2003. Evidence for human and fox rabies associated with dogs. *Epidemiology and Infection* **133**, 529–536.

Pal, S.K. (2001). Population ecology of free-ranging dogs in West Bengal, India. *Acta Theriologica* **46**, 69–78.

Pappaionou, M., Fishbein, D.B., Dreesen, D.W. *et al.* (1986). Antibody response to preexposure human diploid-cell rabies vaccine given concurrently with chloroquine. *New England Journal of Medicine* **314**, 280–284.

Para, M. (1965). An outbreak of post-vaccinal rabies (rage de laboratoire) in Fortaleza, Brazil, in 1960. *Bulletin of the World Health Organization* **33**, 177–182.

Park, Y.J., Shin, M.K. and Kwon, H.M. (2005). Genetic characterization of rabies virus isolates in Korea. *Virus Genes* **30**, 341–347.

Parker, R.L. (1975). Rabies in skunks. In: *The Natural History of Rabies* (G.M. Baer, ed.). pp. 41–51. New York: Academic Press.

Parrish, H.M., Clack, F.B., Brobst, D. and Mock, J.F. (1959). Epidemiology of dog bites. *Public Health Reports* **74**, 891–903.

Pastoret, P.P., Kappeler, A. and Aubert, M. (2004). European rabies control and its history. In: *Historical Perspective of Rabies in Europe and the Mediterranean Basin* (A.A. King, A.R. Fooks, M. Aubert, and A.I. Wandeler, eds). pp. 337–350. Paris: OIE.

Pawan, J.L. (1936). The transmission of paralytic rabies in Trinidad by the vampire bat. *Annals of Tropical Medicine and Parasitology* **30**, 101–129.

Peigue-Lafeuille, H., Bourhy, H., Abiteboul, D. *et al.* (2004). Human rabies in France in 2004: update and management. *Medecine et Maladies Infectieuses* **34**, 551–560.

Perry, B.D. (1993). Dog ecology in eastern and southern Africa: implications for rabies control. *Onderstepoort Journal of Veterinary Research* **60**, 429–436.

Perry, B.D., Brooks, R., Foggin, C.M., Bleakley, J., Johnston, D.H. and Hill, F.W. (1988). A baiting system suitable for the delivery of oral rabies vaccine to dog populations in Zimbabwe. *Veterinary Record* **123**, 76–79.

Pool, G.E. and Hacker, C.S. (1982). Geographic and seasonal distribution of rabies in skunks, foxes and bats in Texas. *Journal of Wildlife Diseases* **18**, 405–418.

Public Health Agency of Canada. (2003). Human rabies, British Columbia – January 2003. www.phac-aspc.gc.ca/publicat/ccdr-rmtc/03vol29/dr2916ea.html.

Pybus, M.J. (1988). Rabies and rabies control in striped skunks (*Mephitis mephitis*) in three prairie regions of western North America. *Journal of Wildlife Diseases* **24**, 434–449.

Quiambao, B.P., Dimaano, E.M., Ambas, C., Davis, R., Banzhoff, A. and Malerczyk, C. (2005). Reducing the cost of post-exposure rabies prophylaxis: efficacy of 0.1 ml PCEC rabies vaccine administered intradermally using the Thai Red Cross post-exposure regimen in patients severely exposed to laboratory-confirmed rabid animals. *Vaccine* **23**, 1709–1714.

Rath, A., Choudhury, S., Batra, D., Kapre, S.V., Rupprecht, C.E. and Gupta, S.K. (2005). DNA vaccine for rabies: Relevance of the trans-membrane domain of the glycoprotein in generating an antibody response. *Virus Research.* **113**,143–152.

Real, L.A., Henderson, J.C., Biek, R. *et al.* (2005). Unifying the spatial population dynamics and molecular evolution of epidemic rabies virus. *Proceedings of the National Academy of Sciences USA* **102**, 12 107–12 111.

Reubel, G.H., Beaton, S., Venables, D. *et al.* (2005). Experimental inoculation of European red foxes with recombinant vaccinia virus expressing zona pellucida C proteins. *Vaccine* **23**, 4417–4426.

Riley, S.P.D., Hadidian, J. and Manski, D.A. (1998). Population density, survival, and rabies in raccoons in an urban national park. *Canadian Journal of Zoology* **76**, 1153–1164.

Robbins, A., Eidson, M., Keegan, M., Sackett, D. and Laniewicz, B. (2005). Bat incidents at children's camps, New York State, 1998–2002. *Emerging Infectious Diseases* **11**, 302–305.

Robbins, A.H., Borden, M.D., Windmiller, B.S. *et al.* (1998). Prevention of the spread of rabies to wildlife by oral vaccination of raccoons in Massachusetts. *Journal of the American Veterinary Medical Association* **213**, 1407–1412.

Robinson, L.E., Miranda, M.E., Miranda, N.L. and Childs, J.E. (1996). Evaluation of a canine rabies vaccination campaign and characterization of owned-dog populations in the Philippines. *Southeast Asian Journal of Tropical Medicine and Public Health* **27**, 250–256.

Rohde, R.E., Neill, S.U., Clark, K.A. and Smith, J.S. (1997). Molecular epidemiology of rabies epizootics in Texas. *Clinical and Diagnostic Virology* **8**, 209–217.

Rosatte, R.C., Pybus, M.J. and Gunson, J.R. (1986). Population reduction as a factor in the control of skunk rabies in Alberta. *Journal of Wildlife Diseases* **22**, 459–467.

Rosatte, R., Donovan, D., Allan, M. *et al.* (2001). Emergency response to raccoon rabies introduction into Ontario. *Journal of Wildlife Diseases* **37**, 265–279.

Rotz, L.D., Hensley, J.A., Rupprecht, C.E. and Childs, J.E. (1998). Large-scale human exposures to rabid or presumed rabid animals in the United States: 22 cases (1990–1996). *Journal of the American Veterinary Medical Association* **212**, 1198–1200.

Ruiz-Martínez, C. (1963). Epizootologia y profilaxis regional de la rabia paralitica en las Americas. *Revista Veteinaria.Venazolana* **14**, 71–173.

Rupprecht, C.E., Wiktor, T.J., Johnston, D.H. *et al.* (1986). Oral immunization and protection of raccoons (*Procyon lotor*) with a vaccinia-rabies glycoprotein recombinant virus vaccine. *Proceedings of the National Academy of Sciences USA* **83**, 7947–7950.

Rupprecht, C.E., Glickman, L.T., Spencer, P.A. and Wiktor, T.J. (1987). Epidemiology of rabies virus variants. Differentiation using monoclonal antibodies and discriminant analysis. *American Journal of Epidemiology* **126**, 298–309.

Rupprecht, C.E., Hamir, A.N., Johnston, D.H. and Koprowski, H. (1988). Efficacy of a vaccinia-rabies glycoprotein recombinant virus vaccine in raccoons (*Procyon lotor*). *Reviews of Infectious Diseases* **10** (Suppl 4), S803–S809.

Rupprecht, C.E., Blass, L., Smith, K. *et al.* (2001). Human infection due to recombinant vaccinia-rabies glycoprotein virus. *New England Journal of Medicine* **345**, 582–586.

Rupprecht, C.E. and Gibbons, R.V. (2004). Clinical practice. Prophylaxis against rabies. *New England Journal of Medicine* **351**, 2626–2635.

Rupprecht, C.E., Hanlon, C.A. and Slate, D. (2004). Oral vaccination of wildlife against rabies: opportunities and challenges in prevention and control. *Developmental Biology (Basel)* **119**, 173–184.

Rupprecht, C.E., Hanlon, C.A., Blanton, J. *et al.* (2005). Oral vaccination of dogs with recombinant rabies virus vaccines. *Virus Research* **111**, 101–105.

Russell, C.A., Smith, D.L., Waller, L.A., Childs, J.E. and Real, L.A. (2004). A priori prediction of disease invasion dynamics in a novel environment. *Proceedings of the Royal Society of London B Biological Sciences* **271**, 21–25.

Russell, C.A., Smith, D.L., Childs, J.E. and Real, L.A. (2005). Predictive spatial dynamics and strategic planning for raccoon rabies emergence in Ohio. *PLoS Biology* **3**, 1–7.

Sabeta, C.T., Bingham, J. and Nel, L.H. (2003). Molecular epidemiology of canid rabies in Zimbabwe and South Africa. *Virus Research* **91**, 203–211.

Sabeta, C.T. and Randles, J.L. (2005). Importation of canid rabies in a horse relocated from Zimbabwe to South Africa. *Onderstepoort Journal of Veterinary Research* **72**, 95–100.

Sacramento, D., Bourhy, H. and Tordo, N. (1991). PCR technique as an alternative method for diagnosis and molecular epidemiology of rabies virus. *Molecular and Cellular Probes* **5**, 229–240.

Samaratunga, H., Searle, J.W. and Hudson, N. (1998). Non-rabies Lyssavirus human encephalitis from fruit bats: Australian bat Lyssavirus (pteropid Lyssavirus) infection. *Neuropathology and Applied Neurobiology* **24**, 331–335.

Sasaki, D.M., Middleton, T.R., Sawa, T.R., Christensen, C.C. and Kobyashi, G.Y. (1992). Rabid bat diagnosed in Hawaii. *Hawaii Medical Journal* **51**, 181–185.

Sato, G., Itou, T., Shoji, Y. *et al.* (2004). Genetic and phylogenetic analysis of glycoprotein of rabies virus isolated from several species in Brazil. *Journal of Veterinary Medicine and Science* **66**, 747–753.

Schaefer, R., Batista, H.B., Franco, A.C., Rijsewijk, F.A. and Roehe, P.M. (2005). Studies on antigenic and genomic properties of Brazilian rabies virus isolates. *Veterinary Microbiology* **107**, 161–170.

Schneider, L.G., Cox, J.H., Muller, W.W. and Hohnsbeen, K.P. (1988). Current oral rabies vaccination in Europe: an interim balance. *Reviews of Infectious Diseases* **10** (Suppl 4), S654–S659.

Seidensticker, J., Johnsingh, A.J.T., Ross, R., Sanders, G. and Webb, M.B. (1988). Raccoons and rabies in Appalachian Mountain hollows. *National Geographic Research* **4**, 359–370.

Shah, U. and Jaswal, G.S. (1976). Victims of a rabid wolf in India: effect of severity and location of bites on development of rabies. *Journal of Infectious Diseases* **134**, 25–29.

Shigesada, N. and Kawasaki, K. (1997). *Biological Invasions: Theory and Practice*. Oxford: Oxford University Press.

Shlim, D.R., Schwartz, E. and Houston, R. (1991). Rabies immunoprophylaxis strategy in travelers. *Journal of Wilderness Medicine* **2**, 15–21.

Shoji, Y., Kobayashi, Y., Sato, G. *et al.* (2004). Genetic characterization of rabies viruses isolated from frugivorous bat (*Artibeus* spp.) in Brazil. *Journal of Veterinary Medical Science* **66**, 1271–1273.

Shope, R.E. (1982). Rabies-related viruses. *Yale Journal of Biology and Medicine* **55**, 271–275.

Shope, R.E., Murphy, F.A., Harrison, A.K. *et al.* (1970). Two African viruses serologically and morphologically related to rabies virus. *Journal of Virology* **6**, 690–692.

Sidwa, T.J., Wilson, P.J., Moore, G.M. *et al.* (2005). Evaluation of oral rabies vaccination programs for control of rabies epizootics in coyotes and gray foxes: 1995–2003. *Journal of the American Veterinary Medical Association* **227**, 785–792.

Sillero-Zubiri, C., King, A.A. and Macdonald, D.W. (1996). Rabies and mortality in Ethiopian wolves (*Canis simensis*). *Journal of Wildlife Diseases* **32**, 80–86.

Sipahioglu, U. and Alpaut, S. (1985). Transplacental rabies in humans. *Mikrobiyologi Bulteni* **19**, 95–99.

Slate, D., Rupprecht, C.E., Rooney, J.A., Donovan, D., Lein, D.H. and Chipman, R.B. (2005). Status of oral rabies vaccination in wild carnivores in the United States. *Virus Research* **111**, 68–76.

Smith, A.D.M. (1985). A continuous time deterministic model of temporal rabies. In: *Population Dynamics of Rabies in Wildlife* (P.J. Bacon, ed.). pp. 131–146. New York: Academic Press.

Smith, D.L., Lucey, B., Waller, L.A., Childs, J.E. and Real, L.A. (2002). Predicting the spatial dynamics of rabies epidemics on heterogeneous landscapes. *Proceedings of the National Academy of Sciences USA* **99**, 3668–3672.

Smith, D.L., Waller, L.A., Russell, C.A., Childs, J.E. and Real, L.A. (2005). Assessing the role of long-distance translocation and spatial heterogeneity in the raccoon rabies epidemic in Connecticut. *Preventive Veterinary Medicine* **71**, 225–240.

Smith, G.C. and Wilkinson, D. (2003). Modeling control of rabies outbreaks in red fox populations to evaluate culling, vaccination, and vaccination combined with fertility control. *Journal of Wildlife Diseases* **39**, 278–286.

Smith, J., McElhinney, L., Parsons, G. *et al.* (2003). Case report: rapid ante-mortem diagnosis of a human case of rabies imported into the UK from the Philippines. *Journal of Medical Virology* **69**, 150–155.

Smith, J.S. (1988). Monoclonal antibody studies of rabies in insectivorous bats of the United States. *Reviews of Infectious Diseases* **10** (Suppl 4), S637–S643.

Smith, J.S. (1989). Rabies virus epitopic variation: use in ecologic studies. *Advances in Virus Research* **36**, 215–253.

Smith, J.S. (1995). Rabies virus. In: *Manual of Clinical Microbiology*, 6th edn (R.R. Murray, E.J. Baron, M.A. Pfaller, F.C. Tenover and R.H. Yolken, eds). pp. 997–1003. Washington DC: ASM Press.

Smith, J.S. (2002). Molecular epidemiology. In: *Rabies* (A.C Jackson, and W.H.Wunner, eds). pp. 79–111. New York: Academic Press.

Smith, J.S. and Seidel, H.D. (1993). Rabies: a new look at an old disease. *Progress in Medical Virology* **40**, 82–106.

Smith, J.S., Fishbein, D.B., Rupprecht, C.E. and Clark, K. (1991). Unexplained rabies in three immigrants in the United States. A virologic investigation. *New England Journal of Medicine* **324**, 205–211.

Smith, J.S., Orciari, L.A., Yager, P.A., Seidel, H.D. and Warner, C.K. (1992). Epidemiologic and historical relationships among 87 rabies virus isolates as determined by limited sequence analysis. *Journal of Infectious Diseases* **166**, 296–307.

Smith, J.S., Orciari, L.A. and Yager, P.A. (1995). Molecular epidemiology of rabies in the United States. *Seminars in Virology* **6**, 387–400.

Smith, K., Krogwold, A., Smith, R., Hale, F., Collart, R. and Craig, M. (1999). The Ohio ORV program. *Proceedings of the 10th Annual International Meeting on Research Advances in Rabies Control in the Americas*, San Diego, CA.

Smithcors, J.F. (1958). The history of some current problems in animal diseases VII. Rabies. *Veterinary Medicine* **53**, 149–154.

Speare, R., Skerratt, L., Foster, R. *et al.* (1997). Australian bat lyssavirus infection in three fruit bats from north Queensland. *Communicable Diseases Intelligence* **21**, 117–120.

Srinivasan, A., Burton, E.C., Kuehnert, M.J. *et al.* (2005). Transmission of rabies virus from an organ donor to four transplant recipients. *New England Journal of Medicine* **352**, 1103–1111.

Steck, F. and Wandeler, A. (1980). The epidemiology of fox rabies in Europe. *Epidemiological Reviews* **2**, 71–96.

Steck, F., Wandeler, A., Bichsel, P., Capt, S. and Schneider, L. (1982). Oral immunization of foxes against rabies. A field study. *Zentralblatt für Veterinarmedizin – Reihe B* **29**, 377–396.

Steece, R.S. and Calisher, C.H. (1989). Evidence for prenatal transfer of rabies virus in the Mexican free-tailed bat (*Tadarida brasiliensis Mexicana*). *Journal of Wildlife Diseases* **25**, 329–334.

Steele, J.H. and Fernandez, P.J. (1991). History of rabies and global aspects. In: *The Natural History of Rabies*, 2nd edn (G.M. Baer, ed.). pp. 1–24. Boca Raton: CRC Press.

Steelman, H.G., Henke, S.E. and Moore, G.M. (2000). Bait delivery for oral rabies vaccine to gray foxes. *Journal of Wildlife Diseases* **36**, 744–751.

Suppo, C., Naulin, J.M., Langlais, M. and Artois, M. (2000). A modeling approach to vaccination and contraception programmes for rabies control in fox populations. *Proceedings of the Royal Society London B Biological Sciences.* **267**, 1575–1582.

Sureau, P. (1990). Recent data on the epidemiology and prophylaxis of human rabies in France. *Comparative Immunology, Microbiology & Infectious Diseases* **13**, 107–110.

Sureau, P., Portnoi, D., Rollin, P., Lapresle, C. and Chaouni-Berbich, A. (1981). Prevention of interhuman rabies transmission after corneal graft. *Comptes Rendus des Seances de l'Academie de Sciences Serie III, Sciences de la Vie* **293**, 689–692.

Swaddiwuthipong, W., Weniger, B.G., Wattanasri, S. and Warrell, M.J. (1988). A high rate of neurological complications following Semple anti-rabies vaccine. *Transactions of the Royal Society of Tropical Medicine and Hygiene* **82**, 472–475.

Swanepoel, R., Barnard, B.J., Meredith, C.D. *et al.* (1993). Rabies in southern Africa. *Onderstepoort Journal of Veterinary Research* **60**, 325–346.

Tabel, H., Corner, A.H., Webster, W.A. and Casey, C.A. (1974). History and epizootiology of rabies in Canada. *Canadian Veterinary Journal* **15**, 271–281.

Tariq, W.U., Shafi, M.S., Jamal, S. and Ahmad, M. (1991). Rabies in man handling infected calf. *Lancet* **337**, 1224.

Thompson, R.D., Mitchell, G.C., and Burns, R.J. (1972). Vampire bat control by systemic treatment of livestock with an anticoagulant. *Science* **177**, 806–808.

Tierkel, E.S. (1959). Rabies. *Advances in Veterinary Science* **5**, 183–226.

Tierkel, E.S., Graves, L.M. and Wadley, S.L. (1950). Effective control of an outbreak of rabies in Memphis and Shelby County, Tennessee. *American Journal of Public Health* **40**, 1084–1088.

Tierkel, E.S., Arbona, G., Rivera, A. and de Juan, A. (1952). Mongoose rabies in Puerto Rico. *Public Health Reports* **67**, 274–278.

Tinline, R.R. and Macinnes, C.D. (2004). Ecogeographic patterns of rabies in southern Ontario based on time series analysis. *Journal of Wildlife Diseases* **40**, 212–221.

Tinline, R., Rosatte, R. and Macinnes, C. (2002). Estimating the incubation period of raccoon rabies: a time-space clustering approach. *Preventive Veterinary Medicine* **56**, 89–103.

Toma, B. and Andral, L. (1977). Epidemiology of fox rabies. *Advances in Virus Research* **21**, 1–36.

Torrence, M.E., Jenkins, S.R. and Glickman, L.T. (1992). Epidemiology of raccoon rabies in Virginia, 1984 to 1989. *Journal of Wildlife Diseases* **28**, 369–376.

Turner, D.C. (1975). *The Vampire Bat*. Baltimore: The Johns Hopkins University Press.

Turner, G.S. (1971). Rural rabies. *Rural Medicine* **2**, 108–112.

Uhaa, I.J., Dato, V.M., Sorhage, F.E. *et al.* (1992). Benefits and costs of using an orally absorbed vaccine to control rabies in raccoons. *Journal of the American Veterinary Medical Association* **201**, 1873–1882.

Varner, J.G. and Varner, J.J. (1983*). Dogs of Conquest*. Norman: University of Oklahoma Press.

Vaughn, J.B. (1975). Cat rabies. In: *The Natural History of Rabies* (G.M. Baer, ed.). pp. 139–154. New York: Academic Press.

Velasco-Villa, A., Gomez-Sierra, M., Hernandez-Rodriguez, G. *et al.* (2002). Antigenic diversity and distribution of rabies virus in Mexico. *Journal of Clinical Microbiology* **40**, 951–958.

Velasco-Villa, A., Orciari, L.A., Souza, V. *et al.* (2005). Molecular epizootiology of rabies associated with terrestrial carnivores in Mexico. *Virus Research* **111**, 13–27.

Vos, A., Neubert, A., Pommerening, E. *et al.* (2001). Immunogenicity of an E1-deleted recombinant human adenovirus against rabies by different routes of administration. *Journal of General Virology* **82**, 2191–2197.

Wandeler, A. (2004). Epidemiology and ecology of fox rabies in Europe. In: *Historical Perspective of Rabies in Europe and the Mediterranean Basin* (A.A King, A.R. Fooks, M. Aubert and A.I. Wandeler, eds). pp. 201–214. Paris: OIE.

Wandeler, A.I. and Salsberg, E.B. (1999). Raccoon rabies in eastern Ontario. *Canadian Veterinary Journal* **40**, 731.

Wandeler, A., Wachendorfer, G., Forster, U. *et al.* (1974). Rabies in wild carnivores in central Europe. I. Epidemiological studies. *Zentralblatt fur Veterinarmedizin – Reihe B* **21**, 735–756.

Wandeler, A.I., Capt, S., Gerber, H., Kappeler, A. and Kipfer, R. (1988). Rabies epidemiology, natural barriers and fox vaccination. *Parasitologia* **30**, 53–57.

Wang, San P. (1956). Statistical studies of human rabies in Taiwan. *Journal of the Formosan Medical Association* **55**, 548–554.

Whitby, J.E., Johnstone, P. and Sillero-Zubiri, C. (1997). Rabies virus in the decomposed brain of an Ethiopian wolf detected by nested reverse transcription-polymerase chain reaction. *Journal of Wildlife Diseases* **33**, 912–915.

Widdowson, M.A., Morales, G.J., Chaves, S. and McGrane, J. (2002). Epidemiology of urban canine rabies, Santa Cruz, Bolivia, 1972–1997. *Emerging Infectious Diseases* **8**, 458–461.

Wilde, H., Sirikawin, S., Sabcharoen, A. *et al.* (1996). Failure of postexposure treatment of rabies in children. *Clinical Infectious Diseases* **22**, 228–232.

Wilde, H., Khawplod, P., Khamoltham, T. *et al.* (2005). Rabies control in South and Southeast Asia. *Vaccine* **23**, 2284–2289.

Wilson, M.L., Bretsky, P.M., Cooper, G.H. Jr, Egbertson, S.H., Van Kruiningen, H.J. and Cartter, M.L. (1997). Emergence of raccoon rabies in Connecticut, 1991–1994: spatial and temporal characteristics of animal infection and human contact. *American Journal of Tropical Medicine and Hygiene* **57**, 457–463.

Winkler, W.G. (1968). Airborne rabies virus isolation. *Bulletin of the Wildlife Disease Association*. **4**, 37–40.

World Health Organization (1987). Guidelines for Dog Rabies Control. **VPH/83.43**, 1–21. Geneva: World Health Organization.

World Health Organization (1994). Two rabies cases following corneal transplantation. *Weekly Epidemiologic Record* **44**, 330.

World Health Organization (1997). World Survey of Rabies No.31: For the year 1995. 1–29. Geneva: World Health Organization.

World Health Organization (1998). World Survey of Rabies No.32: For the year 1996. 1–27. Geneva: World Health Organization.

World Health Organization (1999). World Survey of Rabies No.33: For the year 1997. 1–29. Geneva: World Health Organization.

World Health Organization (2002). WHO strategies for the control and elimination of rabies in Asia: Report of a WHO interregional consultation. **WHO/CDS/CSR/EPH/2002.8**. Geneva: World Health Organization.

World Health Organization Expert Consultation on Rabies. WHO Expert Consultation on Rabies First Report (2004). WHO Technical Report Series 931, 1–121. Geneva: World Health Organization.

Wright, A., Rampersad, J., Ryan, J. and Ammons, D. (2002). Molecular characterization of rabies virus isolates from Trinidad. *Veterinary Microbiology* **87**, 95–102.

Yung, V., Favi, M. and Fernandez, J. (2002). Genetic and antigenic typing of rabies virus in Chile. Brief report. *Archives of Virology* **147**, 2197–2205.

5

Rabies in Terrestrial Animals

CATHLEEN A. HANLON, MICHAEL NIEZGODA, AND
CHARLES E. RUPPRECHT

Centers for Disease Control and Prevention, Atlanta, Georgia 30333, USA

1 INTRODUCTION

Rabies is the most important viral zoonosis and an occupational hazard to all those exposed to particular animals, intentionally or accidentally. Preventive public health practices of vaccination, leash restraint, confinement, observation, quarantine, etc., relate directly to animals. Pure, potent, safe and effective rabies vaccines are developed through evaluation in experimental trials in animals. Elimination of the source of rabies in animals is a desirable goal. However, elimination of animal populations as a means to control rabies, because a particular species is a reservoir (i.e. the population is a continuing source of rabies infection), is neither achievable nor desirable. In the 21st century, with highly effective methods for the prevention of human rabies, including effective tools for the elimination of rabies in dogs, there has been successful extinction of many unique rabies virus variants (Windiyaningsih *et al.*, 2004). In particular, significant advances have been made in the past 25 years in the control and elimination of rabies in major wildlife sources through oral vaccination. With rabies, it all begins, and ends, with animals (Niezgoda *et al.*, 2002).

In addition to the efforts of dedicated public health professionals at the local, state and federal level, veterinary and human medical professionals have an important role to play in educating the general population on the risks of rabies to their pets and to themselves. Veterinarians in private practice strive to protect and enhance the health of companion and food producing animals. Insofar as zoonotic diseases threaten the client as well as the pet, the veterinarian participates in protection of the owners' health by educating them about the risks of such diseases. Although inarguably the most important zoonosis, many public misconceptions persist about rabies. A stereotypical view of rabies is the 'mad' dog, frothing at the mouth. However, cats have been the leading domestic animal with rabies in the USA for nearly three decades (Krebs *et al.*, 2005).

A recent clinical case, provided by James B. Lawhead, VMD, illustrates the critical role of veterinary practitioners in protecting both animal and human health.

A client found a stray kitten and presented it due to its listlessness. The veterinarian discovered a bite wound at the tail base of the 8-week-old kitten. In view of the history and of the wound, which was compatible with a bite, the veterinarian discussed the potential risk of rabies developing in this kitten. Despite this cautionary guidance, the owner elected to treat the animal's bite wound because she already had substantial emotional attachment to the animal. The owner was quite pleased when the kitten responded well to antibiotics with resolution of the bite-wound infection. However, nearly two months later the owner returned to the veterinary practice with the same kitten. The kitten was ataxic with hind limb paresis. It had also become quite agitated and had bitten the owner when she was handling it. The veterinarian was now very concerned about the risk of rabies. The owner still wanted to try empirical treatment, hoping that the illness was not due to rabies. However, the worsening of neurological signs necessitated euthanasia of the kitten the following day. The kitten was diagnosed as rabid and the owner underwent post-exposure prophylaxis. Such cases of rabies in cats are not uncommon and occur, on average, nearly 300 times each year in the USA, often with large numbers of potentially exposed humans and domestic animals (Noah *et al.*, 1996).

For most people, the word 'rabies' still invokes visions of an unexpected attack from a strange-behaving dog. While reflective of the global situation and historic trends in many countries, rabies in dogs is caused by distinctive rabies virus variants, many of which have been eliminated throughout islands, large geographic areas and even continents, when a majority of the dog population has been vaccinated against rabies (World Health Organization (WHO), 2005). Nonetheless, the sheer magnitude of human and domestic dog populations, such as in Asia and Africa, where vaccination of dogs and stray dog control have not been adequately implemented to effect elimination of this variant, constitutes the main global burden of rabies. Perhaps more importantly, the economic, political and socio-religious impediments to comprehensive dog vaccination and rigorous population control need to be understood and overcome. Global elimination of canine rabies, although often difficult logistically and politically, is a biologically feasible goal, which would result in an exponential reduction of human rabies exposures and deaths. Beyond the problem of rabies in its major reservoir, the dog, there exists a diversity of other competent reservoirs affected by other specific rabies virus variants. The complexity of wildlife species serving as rabies reservoirs, some potentially yet to be discovered, will not lend itself easily to control measures that have been successfully applied towards dog rabies control and elimination. The objectives of this chapter are to review some of the basic principles of animal rabies related to host range, susceptibility, clinical signs, differential diagnosis and transmission, to provide salient examples of these topics in selected domestic and wildlife species and to illuminate certain research issues in need of further inspection.

2 HOST RANGE AND SUSCEPTIBILITY

The infectious spectrum of rabies viruses *in vitro* (under experimental conditions) may be broad and even birds may be experimentally infected in captivity (Yamaoka, 1962; Schneider and Burtscher, 1967; Jorgenson and Gough, 1976; Shannon *et al.*, 1988; Seganti *et al.*, 1990). Nevertheless, fundamentally, rabies is a disease of mammals, caused by viral representatives in the genus *Lyssavirus*, family Rhabdoviridae. All mammals are susceptible, but to varying degrees, and are regarded as the ultimate sources of all rabies infections. However, the infection of a competent reservoir species with a variant specifically adapted to that species is particularly efficient in driving behavior that enhances virus transmission, as well as providing the physiologic environment that maximizes viral shedding in the saliva. Beyond biological susceptibility, ecological characteristics of various mammalian populations affect the probability of exposure to a clinically rabid animal belonging to the reservoir species. How is the infection promulgated in nature? Practically, the existence of rabies virions is limited to an intracellular environment until they are released, often in impressive amounts, in the saliva of clinically rabid animals. How is this possible within immunocompetent hosts? Rabies viruses are exquisitely adept at avoiding immune detection through neurotropism. In addition to trans-synaptic transfer of fully formed virions, neuron-to-neuron spread may be facilitated through the transfer of intact rabies virus or rabies virus ribonucleoprotein complexes at the neuronal synapse. Moreover, rabies viruses are capable of prolonged incubation periods during which the infection is not detectable, nor does it induce an apparent immune response. Preservation of the integrity of a functional neuronal network is critical to both viral survival and an ultimate subversive strategy to minimize immunological detection during the phases of a productive infection (Lafon, 2005).

What makes a successful rabies virus reservoir? Trends in species susceptibility are greatly influenced by species attributes and, to a lesser extent, by individual host factors, such as immunologic status and age. Although generalities about species susceptibility may be inferred, specific species attributes have not been exhaustively characterized, due in part to confounding variables such as virus variant, ecological characteristics of the hosts, which affect exposure risk and outcome, and surveillance bias, which limits the source and number of specimens examined. From an observational standpoint, capable rabies reservoir species are found among members of the order Carnivora in the families Canidae (i.e. dogs, foxes, coyotes, jackals), Herpestidae (e.g. mongoose species), Procyonidae (i.e. the common raccoon and its relatives), Mephitidae (i.e. the skunks) and bats in the families Vespertilionidae (a multitude of rabies virus variants are associated with particular members of this family, including *Lasionycteris noctivagans/Pipistrellus subflavus*, *Eptesicus fuscus*, *Myotis* spp., *Lasiurus borealis*, etc.), Molossidae (the free-tailed bats), Phyllostomidae, particularly the subfamily Desmodontinae

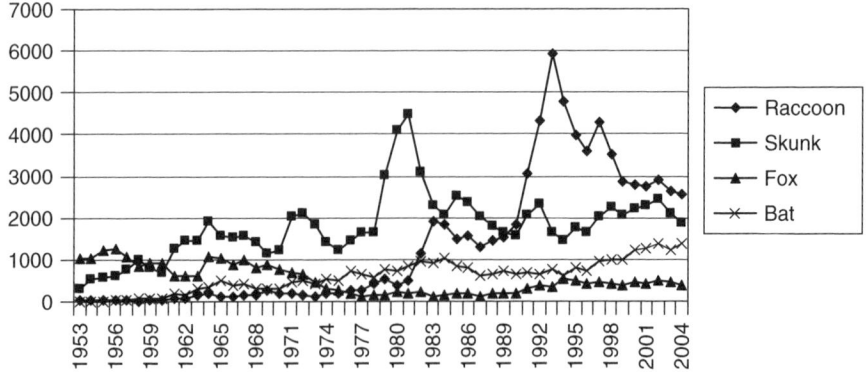

Figure 5.1   Number of rabies cases among rabies reservoir species in the USA.

(i.e. vampire bats) and Pteropodidae (i.e. flying foxes and their allies) (Figure 5.1). The restriction of rabies virus perpetuation to these limited families stimulates introspection in view of class Mammalia consisting of 29 orders of which rabies reservoirs are found only among two: orders Carnivora and Chiroptera. Among the order Chiroptera, only four of 18 families currently encompass species that may serve as rabies reservoirs. Among the suborder Feliformia of the order Carnivora, only one in five families, the Herpestidae consisting of mongooses, may serve as a rabies reservoir species. Comparatively, the suborder Caniformia has nine families among which three are found rabies reservoir species (family Canidae, Mephitidae, and Procyonidae).

Why don't rats and squirrels and other rodents 'carry' rabies? As clarification, no mammals are true carriers and, despite their ubiquity in distribution and abundance and their diversity as the largest mammalian order, documented cases of rabies in rodents are uncommon (Kulonen and Boldina, 1996; Childs *et al.*, 1997). To date, there are no known rodent reservoirs, or documented human rabies cases attributable to contact with rabid rodents. Nevertheless, critical public health resources are devoted to the assessment of bites by small mammals (Nair *et al.*, 1978; Wimalaratne, 1997). Detection of rabies in rodents depends, in part, on several variables, including: the probability of rodent contact with a reservoir species; the ability of a rodent to survive an initial, presumably traumatic, encounter with a naturally infected animal, typically a carnivore; the likelihood that the infected rodent can avoid terminal predation, particularly during the debilitating stages of encephalomyelitis; and the opportunity for human interaction leading to successful submission and testing at a diagnostic laboratory (Moro *et al.*, 1991). Thus, it is not surprising that when rabies is diagnosed, large-bodied (>1 kg) rodents, such as woodchucks, predominate and significant long-term spatial-temporal patterns may gradually appear when particular hosts or viral variants emerge, especially in areas of high human population density. For example, during the last two decades

TABLE 5.1

Number of cases of rabies in rodents compared to rabies reservoir species

	1985–1994	1995–2004
Rabies reservoir species		
Raccoon	27 278	31 868
Skunk	19 443	20 724
Bat	2706	4435
Wild rodents		
Woodchuck	317	459
Beaver	12	21
Squirrel	12	7
Rat	2	2
Muskrat	2	2
Chipmunk	1	1
Porcupine	1	0
Nutria	1	0
Mouse	0	0
Pet rodents		
Rabbit	17	21
Guinea pig	0	2
Chinchilla	0	1
Prairie dog	1	0

of a raccoon rabies epizootic in the USA, there have been nearly 60 000 diagnosed cases in raccoons. In association with this outbreak, there were 776 cases in wood-chucks (*Marmota monax*) (Table 5.1). Similarly, from 1985 through 2004, 33 rabid beavers were recorded (Table 5.1), all within the eastern raccoon rabies enzootic. Little is known concerning the pathogenesis and epizootiology of such cases in beavers or other wild rodents. The issue of body size may be especially relevant, not only as to the opportunity for survival after rabies exposure, but also as to the chance for a person to observe an unusual event in an otherwise wary species, such as a woodchuck or beaver. Multiple samples from rabid beavers were exam-ined for rabies virus antigen. In at least one animal, immunofluorescent and immunohistochemical tests for rabies virus antigen were positive for all tissues examined, including skin, tonsil, tongue, lymph nodes, cranial nerves and salivary glands (CDC, unpublished data). In this case, spillover infection resulted from the rabies virus variant associated with raccoons in the eastern USA. As the largest rodent in North America, with prominent mandibles and incisors, beavers may inflict severe bites (Centers for Disease Control and Prevention, 2002). On the basis of these very preliminary results suggesting vector competence, laboratory submission and testing of suspect beavers and other wild rodents is reasonable and human rabies post-exposure prophylaxis (PEP) should be administered based on test results or continue to be considered on a case-by-case basis where the animal is not available for testing, particularly in situations involving an unprovoked human or domestic animal exposure by an ill animal. Given the diversity of rodent species, global abundance, variety of ecological niches and proven competence as significant hosts for other viral diseases, one would anticipate that at least one salient example may eventually be found to meet the minimum attributes of a rabies virus reservoir in this group.

Species at moderate to low risk for natural rabies infection consist of other members of the order Carnivora, such as Felidae, Viverridae, Hyaenidae, Mustelidae, Ursidae and others. Members of the orders Perrisodactyla (equids, tapirs and rhinoceros), Artiodactyla (10 families including pigs, camels, hippopotamus, cervids, antelopes, giraffes and bovids), Rodentia (33 families) and Primates (15 families) appear to be at moderate to low risk as well. Species with a very low probability of natural rabies virus infection comprise monotremes, marsupials, insectivores, cetaceans and others. The only North American marsupial, the opossum (*Didelphis virginianus*), appears relatively resistant to experimental infection (Beamer *et al.*, 1960), also reflected by consistently low numbers of naturally occurring cases. One of the greatest sources of infection for the opossum is the raccoon, most likely due to ecological overlap of the two species in the suburban environment. The virtual absence of substantial reports, despite the diversity of marsupials widely and abundantly distributed throughout Australia and South America, argues in part for fundamental, taxonomically based, differences in viral–host response. Nonetheless, any individual mammalian species is still considered susceptible to rabies, even marine representatives, as demonstrated by a case report of rabies in a ringed seal (order Carnivora, family Phocidae) from Norway (Odegaard and Krogsrud, 1981). This animal was wounded and appeared confused. It deteriorated over the course of five days and became aggressive. Rabies was confirmed by immunofluorescent testing of the brain (Dean *et al.*, 1996) and the case was presumed to be due to an epizootic of rabies among Arctic foxes in the area. Similarly, rare occurrences of rabies in elephants (order Proboscidae, family Elephantidae) have been reported from Sri Lanka (Wimalaratne and Kodikara, 1999; Nanayakkara *et al.*, 2003).

It is unclear how mammalian reservoirs and associated viral variants may have co-evolved. The greatest genetic differences exist between virus variants of terrestrial species and those found in bats. The capacity to distinguish among variants has become more precise in the past 30 years due to the advent of monoclonal antibodies for antigenic characterization (Wiktor and Koprowski, 1978) and, later, the development of the reverse transcriptase polymerase chain reaction (PCR) assay and genetic sequencing to elucidate differences (Smith, 1996; Tordo, 1996). This level of distinction has led to a clearer understanding of the likely sources of rabies in different animals. Reservoirs are those mammals

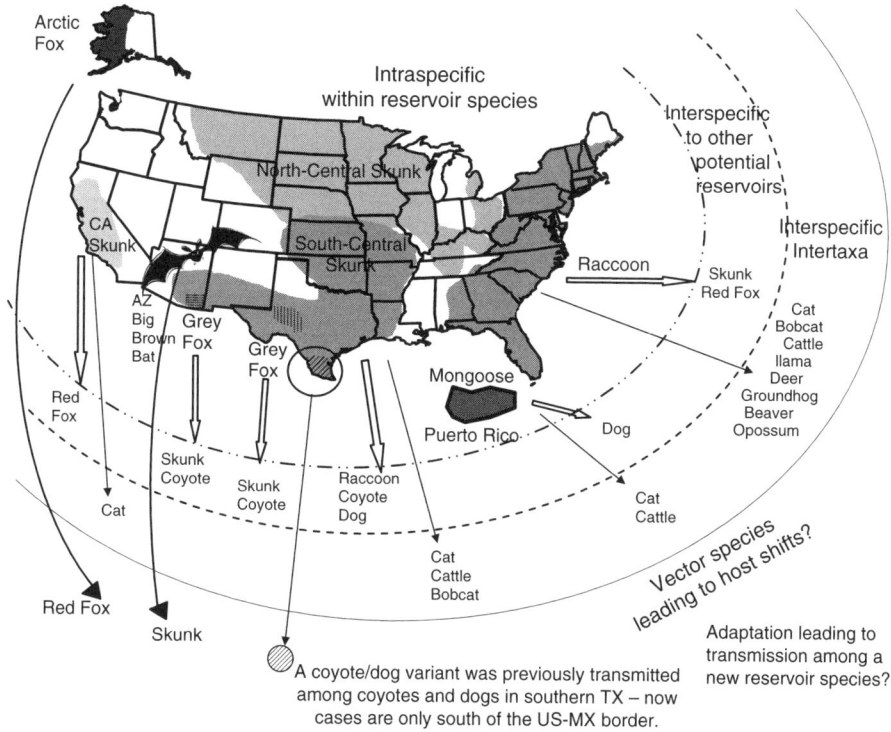

Figure 5.2   Transmission patterns of rabies in the USA. This figure is reproduced in the color plate section.

capable of sustained intraspecies maintenance of a virus variant within a geographic area (Figure 5.2). Critical components of the reservoir in the virus–host interaction would necessarily include distribution, abundance, population density and contact rate. In addition, virus maintenance may also take advantage of potentially competent coexisting vector species (Bell, 1980; Childs *et al.*, 1994). Vector species may closely overlap with the primary reservoir in ecological and behavioral characteristics, as well as in temporal and spatial patterns in a rabies-enzootic area. A recent example is the description of a canid viral variant, maintained and perpetuated in domestic dogs at the Texas–Mexico border. In the 1980s, the variant demonstrated emergent properties, a possible host shift, through its dramatic spread through a newly recognized reservoir – the coyote (Clark *et al.*, 1994; Krebs *et al.*, 2000). Equally interesting was the apparent retention of its capacity for dog-to-dog transmission, evident in border communities where rabies vaccination of domestic dogs was low. Control efforts directed toward containment and elimination of this particular rabies virus variant have included novel, traditional and perhaps natural environmental factors including enhanced parenteral rabies vaccination of owned domestic dogs, control of stray dogs, the distribution of rabies vaccine-laden baits to immunize orally coyotes against rabies and drought conditions perhaps decreasing the abundance of susceptible coyotes. At present, this variant no longer circulates in the USA but it may be found in coyotes and dogs in the neighboring states of Mexico (Figure 5.2).

In addition to the host species, equally important facets in basic measurement of susceptibility include specific attributes of the rabies virus variant, its origin and passage history, the concentration of virus, the type of inoculum and the route of exposure. Transmission does not occur through intact skin. Such trivial contact is not considered an exposure to rabies. Are other activities, such as predation, scavenging or repeated exposure to fomites, important to viral maintenance? As to a specific niche, lyssaviruses replicate predominantly in mammalian neural tissue and do not persist outside of animals. The synergy of ultraviolet radiation, pH extremes, organic solvents, desiccation, excessive heat and putrefaction lead to a relatively rapid diminution of viral load, usually on the order of hours to days (Lewis and Thacker, 1974). Thus, the abiotic environment is not implicated in transmission.

The virus variant profoundly affects host susceptibility to infection and seems uniquely adapted to each reservoir species. Virus variants may possess characteristics that can directly influence tropism, host behavior or clinical outcome (Baer et al., 1977; Flamand et al., 1993; Wilbur and Aubert, 1996). Thus, susceptibility of a species to a virus, for which the species is not a natural reservoir, will vary. Moreover, viral pathogenesis within a 'spillover' or 'victim' species may differ substantially from that found in the reservoir species. For example, a 'canine' virus variant in a raccoon may produce a hyper-acute encephalitis, sometimes resulting in rather sudden death prior to substantial viral shedding in the saliva (Wandeler et al., 1994; Hamir et al., 1996). In contrast, a 'raccoon' variant in a dog may behave differently, with selection eventually enhancing mechanisms that could favor survival and transmission in a new environment.

2.1 A Question of immunity?

Immunity to rabies can involve both anatomical defenses, such as intact, heavily furred, or cornified skin that aids in protection against bites (or licks) and non-specific inflammatory responses to foreign substances, as well as specifically induced responses to viral antigens (Lodmell, 1983; Xiang et al., 1995). Intact virus can induce high lymphokine secretion, but induction of protective antiviral immunity is primarily based on response to the rabies virus glycoprotein (G) found within the envelope of the virus. The G is the only lyssavirus antigen known to induce virus-neutralizing antibody (VNA) and plays a critical role in eliciting immunity (Foley et al., 2000). Nevertheless, there is not always a clear relationship between level of VNA and resistance to rabies, suggesting that other antigens and immune effector mechanisms are likely involved in protection against lethal infection. Besides the G, internal lyssavirus antigens may possess the capacity to enhance immune responsiveness, characterized either as superantigens (Lafon, 1994), or as powerful adjuvants.

Contrary to common perception, viral exposure may not always lead to a productive viral infection and may or may not result in detectable immune responses (Niezgoda *et al.*, 1997). Demonstration of VNA in serum only indicates exposure to viral antigen. The outcome following exposure depends in part upon a complex interplay of viral and host factors (Nathanson and Gonzalez-Scarano, 1991). During the centripetal transport of limited numbers of virions to the brain, especially during long incubation periods, antigenic mass may be so minimal as to avoid immune detection. Alternatively, highly neuroinvasive strains might be quite immunogenic, but the host response could be slow or inadequate to prevent or clear central nervous system (CNS) replication in time to avert profound dysfunction and death, given the rather immunologically privileged location of replication. If detected at all, rabies virus-specific antibodies may be found at the onset of illness, but are more often detected during the terminal stages of disease. The presence of rabies VNA may interfere with salivary excretion of infectious virus (Charlton *et al.*, 1987). Antibody detection in the cerebrospinal fluid (CSF) of clinically ill, rabies-suspected human or animals is considered a reliable indication of current CNS infection. If VNAs are present in the CSF of a clinically normal individual, they may reflect past infection and possible recovery (Fekadu, 1991a, 1991b). The finding of a lymphocytic pleocytosis without rabies virus-specific antibody in the CSF in some animals that were protected from infection has not been adequately explained (Hanlon *et al.*, 1989).

Natural immunity to rabies may vary (Rosatte and Gunson, 1984; Black and Wiktor, 1986; Orr *et al.*, 1988; Follmann *et al.*, 1994), depending in part upon the species and the virus variant (Niezgoda *et al.*, 1998). For example, European red foxes appear seldom to have significant VNA, usually less than a fraction of a percent of surveyed populations (Blancou *et al.*, 1991). A seroprevalence of 2–10% has been reported in some insectivorous bats (Trimarchi and Debbie, 1977), but could be considerably higher in some highly colonial species. Among raccoons, the presence of VNA varied from 1 to 3% in epizootic areas of the mid-Atlantic states (Winkler and Jenkins, 1991) to >20% of those sampled in Florida (Bigler *et al.*, 1983). In mongooses from Grenada, the presence of rabies VNA ranged from 9% to upwards of 55% and was inversely proportional to the number of reported rabies cases. Upon rabies vaccination, animals with pre-existing VNA responded with higher titers (Everard and Everard, 1988). The presence and level of rabies VNA in populations may be an adjunct surveillance tool for rabies activity when direct surveillance is practically limited through the testing of brain material from dead or intentionally collected animals, for example among wildlife populations (Almeida *et al.*, 2001).

On average, productive infection in a single animal is adequate to maintain a chain of transmission. With regard to dose, inoculation of a high viral load may result in a relatively short incubation period with less opportunity for shedding of virus in the saliva. Conversely, infection with lower amounts of virus may result in longer incubation and clinical periods and greater viral shedding over time.

3 TRANSMISSION

A bite is the most reliable route of exposure leading to infection (Centers for Disease Control and Prevention, 1999). Oral exposure may result in infection, but with relatively low efficiency. Consumption of infected carcasses by carnivores may be the most relevant example of this scenario. For example, transmission could be enhanced through the penetration of oral or esophageal mucosa by bone fragments contaminated with highly infectious material, such as brain and salivary gland tissue. This mechanism has been hypothesized to contribute to the maintenance of Arctic fox rabies, where contact between potential hosts is minimal but where transmission may be facilitated due to the preservation of virus in carcasses under polar conditions (Crandell, 1991). Infectious virus may be recovered months later in a frozen fox carcass during winter, but could be inactivated within hours in the decomposing tissues of a road-killed raccoon under summer conditions in the mid-Atlantic USA. Similarly, contamination of other mucous membranes, such as the eyes and nose, is considered a potential exposure to rabies. However, this is largely based upon human infection following corneal or organ transplantation (Anderson et al., 1984; Centers for Disease Control and Prevention, 1999; Hellenbrand et al., 2005; Srinivasan et al., 2005) and a few case histories in which mucous membranes may have been contaminated, but the possibility of a bite could not be completely excluded. In context, during limited experimental studies in Syrian hamsters (with a demonstrated moderate susceptibility to infection), ocular instillation by drops of a virulent canine isolate of virus, which results in rabies when inoculated intramuscularly, did not result in rabies, despite months of observation (C. Hanlon, unpublished data). It is possible that multiple direct applications to the corneal surface using actions mimicking grooming by a rasping tongue and resulting in minor abrasions might have produced a different outcome. Depending on dose, oral exposure to rabies virus may not elicit a fatal infection. In a majority of animals exposed in this fashion, there may be no detectable response. Rather, a small proportion of animals may become immune while another smaller proportion may develop clinical rabies. Intranasal exposure may result in infection more readily than by the oral route (Charlton and Casey, 1979), but is clearly less efficient in the establishment of an infection than intramuscular inoculation. Aerosol transmission of rabies can occur under some field conditions, but these involve unusual circumstances (Centers for Disease Control and Prevention, 1999). Aerosol infection has been implicated as the route of infection during laboratory accidents (Centers for Disease Control, 1972, 1977), in an unusual cave setting involving millions of Mexican free-tailed bats (Constantine, 1962) and in experimental studies with a unique bat rabies virus variant (Winkler *et al.*, 1972). Direct contamination of an open wound with infectious material (i.e. saliva and CNS tissue) is considered an exposure to rabies and may result in infection.

4 CLINICAL COURSE

There are no known definitive or species–specific clinical signs of rabies beyond acute behavioral alterations (Blancou *et al.*, 1991; Charlton *et al.*, 1991; Winkler and Jenkins, 1991; Brass, 1994). To paraphrase, in rabies, '...the abnormal becomes typical...' Severity and variation of signs may be related to the specific site(s) of the primary CNS lesion(s) or to viral strain, dose and route of infection (Smart and Charlton, 1992; Hamir *et al.*, 1996). The prodromal period usually follows a bite by several weeks. However, periods of less than 10 days to several months are well documented (Charlton, 1994). Severe and multiple bites to the head and neck and bites to highly innervated areas may result in shorter incubation periods. At the end of the incubation period, the disease progresses through a short prodromal stage to encephalopathy and death, usually within days.

Initial signs of rabies are non-specific and may include anorexia, lethargy, fever, dysphagia, vomiting, stranguria, straining to defecate and diarrhea. A gradual and subtle alteration in behavior, such as increased aggressiveness in a normally even-tempered animal or vice versa and a qualitative change in phonation, as well as increased vocalization, may be noted fairly early in the clinical course. In addition to behavioral abnormalities, cranial nerve manifestations may develop including facial asymmetry, trismus, choking, a lolling tongue, drooling, drooping of the lower jaw, prolapse of the third eyelid (nictitating membrane) and anisocoria.

An acute neurologic period usually follows the brief prodromal period by 1 to 2 days. A generalized excitative increase in neurologic activity is observed, associated with hyperesthesia to auditory, visual or tactile stimuli and sudden and seemingly unprovoked agitation and extreme aggressive behavior (furious rabies) toward animate or inanimate objects. Wildlife may lose their apparent wariness of humans. Domestic species and other animals alter their activity cycles either by seeking solitude or becoming more gregarious. Head tilt, head pressing or butting and 'star gazing' may be observed. In addition, ataxia and paresis may develop and progress rapidly over a few days.

Similar to paresthesias reported by infected humans, animals may be observed grooming or scratching at the known or presumed site of virus exposure, most likely in response to altered sensation, probably due to viral excitation of sensory ganglia. This abnormal behavior may progress to self-mutilation and even self-consumption of body parts, particularly appendages. In males, aberrant grooming of the penis may lead to gross trauma. During the clinical period, animals

Figure 5.3 A rabid fox with multiple embedded porcupine quills as evidence of abnormal behavior. Photograph courtesy of New York State Department of Health, Rabies Laboratory. This figure is reproduced in the color plate section.

may attack inanimate objects (cages, sticks, moving vehicles) or other animals with no apparent response to pain (Figure 5.3), including cessation of the activity, following tooth or bone fracture or extensive tissue trauma, particularly to the tongue and facial structures.

The clinical presentation of rabies is often generalized as being either furious or dumb. Furious rabies is a clinical presentation that consists predominantly of profound agitation and aggression. With dumb rabies, aggression may be completely lacking but severe paralysis may be paramount. Clinical manifestations may progressively include anorexia, depression, cranial nerve signs and increased salivation, sometimes profuse (Figure 5.4). Paresis may occur and progress to severe paralysis and death. These two extreme clinical presentations may likely reflect differences in viral infection within specific areas of the CNS. An individual animal may alternatively manifest both of these generalized forms, with the furious signs first and then the dumb signs or vice versa, as the clinical course progresses. Profound characteristic neurological manifestations of human rabies (Hemachudha, 1994), namely true hydrophobia and aerophobia, have not been documented in other animals.

The disease has a rapidly declining clinical course with progressive worsening. Most clinical periods are less than 1 week to 10 days, although exceptions have been reported, particularly in reservoir species, such as illness as long as 17 days in a fox with shedding of virus 30 days prior to death (Blancou *et al.*, 1991) or 18 days of viral shedding prior to death in a skunk (Parker and Wilsnack, 1966).

Figure 5.4   Salivation in a clinically rabid dog. Photograph by Ivan Kuzmin. This figure is reproduced in the color plate section.

While recovery from rabies may occur, it is quite rare (Baer and Olson, 1972). Recovery may be complete or it may involve neurological sequelae.

As the CNS infection progresses, signs may include extreme tremors, paresis, ataxia and paralysis. There may be additional CNS excitation with pronounced hyperactivity, disorientation, confusion, photophobia, pica, incoordination and convulsive seizures. Autonomic systemic excitation may also result in labile hypertension, hyperventilation, muscle tremors, priapism, altered libido, hypothermia or hyperthermia. Ascending, flaccid, symmetrical or asymmetrical paralysis leads eventually to respiratory and cardiac failure. Ultimately, rabies most often culminates in coma and generalized multiorgan failure with death in 1–10 days (or rarely more) after primary signs, or it may result in acute death with no premonitory suspicion.

Curiously, reported signs in rabies from different species range the gamut from the subtle neurological illness to maniacal frenzy, the stuff of literature and legend. In contrast to these dramatic scenarios, common mammalian behaviors may also contribute to transmission of virus, especially of a rabies virus variant associated with a particular reservoir species. For example, many mammalian males often mount and stabilize the female by holding at the nape of the neck with the teeth during copulation, combining the potential excretion of virus in the saliva prior to overt clinical signs with an obvious need for mechanical manipulation during procreation, even among less social species. Teeth and the oral cavity (with its associated secretions) are used daily for multiple and critical mammalian behaviors. Beyond obvious foraging activities such as predation and aggressive encounters during resource defense, these can include stimulation of the urogenital reflex of the neonate by the mother, carrying of infants, food begging, food sharing, play, grooming, etc. Viral excretion during the prodromal phase, when the individual may appear normal but is gradually lapsing into non-specific episodes of illness, may be a more important opportunity for transmission than later mania. Exhibition of more extreme forms of behavior do occur, but may not be needed to the extent believed, particularly if the encounters are initiated by the unaffected animal, rather than by the obviously diseased conspecific. No doubt, rabies virus infection changes host behavior and virtually ensures its own transmission in the process, but exploitation of routine social activities may also facilitate transmission, particularly among reservoir species.

5 DIFFERENTIAL DIAGNOSIS

Rabies should be strongly considered in the differential diagnosis of any mammal presenting with signs compatible with an acute encephalitis, particularly in high risk taxa such as the Carnivora and Chiroptera. Depending upon the species in question, signs of rabies may appear clinically similar to a number of viral, bacterial, mycotic, protozoal, helminth or other acquired conditions. For example, for the canid species, both wild and domestic, canine distemper virus infection may be common and clinically indistinguishable from rabies. Any acute infection resulting in a meningitis or encephalitis may result in a clinical presentation that is suggestive of rabies. Moreover, rabies may be mimicked by a multitude of non-infectious diseases, such as acute trauma or toxicity.

6 VIRAL EXCRETION AND PUBLIC HEALTH IMPLICATIONS

Unlike any other common zoonosis, routine concentration upon individual animal health directly determines the public health management of humans after a suspected rabies exposure. All decisions regarding prophylaxis depend upon the basic awareness that the animal may be rabid. Recent data on rabid animal survival, in a report from Thailand, underscore professional attitudes and management options concerning rabies, as concerns the practice of observing the suspect rabid animal until death (Tepsumethanon *et al.*, 2004). Rabid animals die within a very short period after clinical signs appear. In a study of naturally infected animals, Tepsumethanon *et al.* (2004) confirmed the very widely held conviction that rabid dogs and cats succumb in 10 days or less after illness onset. If a suspect animal remains well, prophylaxis is not necessary. However, the rationale for animal observation, and its relationship to rabies infection, evolved as virological knowledge, microscopic insights, pathogenetic revelations and diagnostic techniques improved over the past hundred years.

By the early 20th century, after laboratory recognition of typical histological changes and appreciation of the inclusion (Negri) bodies associated with rabies, the importance of animal observation was renewed. Investigators observed that the longer the animal was allowed to live, the better the opportunity for observing microscopic lesions. The length of clinical illness was believed to be directly related to the subsequent presence, size, relative abundance and development of Negri bodies. Basic reliance upon observation of the animal suspected of rabies for as long as possible enhanced pathological detection. These public health practices were critical during the application of post-exposure neural tissue vaccines to humans, which were often associated with high rates of serious adverse events in recipients, because a negative diagnostic finding, or having a healthy dog, after 10 days meant that human vaccination could cease.

During the late 1950s, diagnostic improvements allowed direct intervention and observation of brain material from the suspect animal without the need to wait for the development of histopathologically observable Negri bodies. Application of the direct fluorescent antibody (DFA) test according to the current national standard protocol, is rapid, sensitive, specific and inexpensive and eliminates the need for animal confinement to obtain reliable diagnostic results.

Currently, behavioral observations of the suspect rabid animal are linked not with lesion development *per se*, but rather directly with important background data on basic virus excretion. Historically, many investigators such as Zinke, Galtier, Pasteur, Roux and others (see Chapter 1) inferred the critical importance of saliva as the vehicle for virus infection. Without primary involvement of the CNS first, virus transit to the salivary glands does not occur. Typically, rabid dogs shed virus concomitant with illness, or a few days before signs begin. A dog that remains healthy for 10 days after a human exposure virtually assures a lack of concern about rabies. Thus, the '10-day rule' of dog observation evolved not with the reliance upon death as an end point, but rather as an appreciation of the probability of infection at the time of exposure, based upon pathogenesis of the disease and transmission dynamics. These same tenets may apply to other mammals that potentially expose humans or other domestic animals but have only been experimentally evaluated, such as cats and domestic pet ferrets.

Observation alone of suspect rabid animals until death, without euthanasia, conflicts with many moral, ethical, medical and public health principles related to modern rabies management of animals. After human exposure, prophylaxis begins whenever rabies is strongly suspected. Given the frequency of bites by healthy dogs, prophylaxis does not need to begin after exposure in regions without enzootic canine rabies, unless a confined dog dies, or sickens and is euthanized and proves rabid by immediate diagnosis. In canine rabies endemic areas, human PEP is reasonable after the bite of a suspect dog (Denduangboripant *et al.*, 2005). However, if minimum observation periods encompass 10 to 14 days before discontinuation of prophylaxis, the usefulness of starting PEP, while the animal is available for observation, is highly questionable. Halting prophylaxis at this point is of little consolation, after application of first aid, administration of immune globulin and the majority of vaccine doses. Human rabies vaccination is a critical but expensive tool in public health. Reliance upon human prophylaxis alone, in the absence of other basic zoonotic disease tools and techniques, such as stray animal control, dog leash laws, application of herd immunity principles, etc., does little to alter the substantial health economics burden of rabies (Coleman *et al.*, 2004; Fevre *et al.*, 2005; Knobel *et al.*, 2005).

7 RESERVOIRS AND OTHER LYSSAVIRUSES

Rabies virus is the representative type species and most significant member of the genus *Lyssavirus* in terms of relative distribution, abundance and public health and veterinary impact. Not surprisingly, most scientific focus has concentrated upon this classical agent and the principal attributes of rabies virus are felt to be largely interchangeable with the rabies-related lyssaviruses in regard to features such as incubation period, clinical signs, etc.

Although not currently recognized as major zoonotic threats, the non-rabies virus lyssaviruses (Lagos bat, Mokola, Duvenhage, European bat, Australian bat and Eurasia bat viruses) (see Chapter 6) may pose future public health problems because of the opportunity for international translocation and local establishment. This latter concern could be compounded because traditional rabies virus vaccines do not always provide adequate protection (Dietzschold *et al.*, 1988; Kuzmin *et al.*, 2003; Hanlon *et al.*, 2005). Two rhabdoviruses, Obodhiang and Kotonkan, isolated from invertebrates in Africa by intracerebral inoculation of mice, were seemingly aligned with the rabies-like viruses on the basis of initial serology at one time, but do not produce rabies in the traditional sense and await further genetic characterization (Shope, 1975).

Within each lyssavirus serotype/genotype, there are numerous viral variants that exist as quasispecies, heterogeneous mutant viral populations, subject to Darwinian evolution and punctuated equilibrium (Eigen and Biebricher, 1988; Nichol *et al.*, 1993). Theses factors may further complicate any in-depth studies of overt susceptibility, incubation periods, associated clinical signs and related outcome in a relevant host.

8 RABIES IN DOMESTIC ANIMALS

Trends in animal rabies change with adaptations of housing practices and annual at-risk populations. For example, increasingly intensive husbandry of food production animals has reduced animal rabies cases in pigs because the majority are housed in confinement where potential exposure to rabies is virtually eliminated. From 1985 to 2004, cats led domestic animal rabies cases with a total of 4913. Comparatively, rabies occurred in 2855 cattle, 2554 dogs, 992 horses and mules, 215 sheep and goats and 33 pigs.

8.1 Dog

The global burden of rabies still arises from sustenance of transmission among domestic dogs, mainly in Asia and Africa. Historically, forceful efforts directed towards stray dog control and vaccination has eliminated the canine rabies virus variant throughout island nations, such as Japan and Great Britain, and broad geographic areas such as North America and, more recently, Latin America. In a subsequent juxtaposition, dogs are no longer important as a primary reservoir in the USA. The annual numbers of rabid dogs is similar to cases in cattle, for example, around 100 each for the past five years and between 100 and 150 cases in the past decade (Figure 5.5). Notably, cases in cats are about three times higher. Vaccination of dogs in the USA remains important, not as a means for rabies control among a reservoir, but as a prevention measure against spillover of rabies from the now predominant wildlife reservoirs (i.e. raccoons, skunks, foxes, bats) to individual dogs (see Figure 5.2). More importantly, maintaining the vaccination of dog populations, even with moderate rates of compliance, is an important safeguard against re-introduction of a canine rabies virus variant through translocation. Importation of domestic pets occurs on a routine and regular basis. Although health examination and certification is required for importation, this practice is insufficient to prevent potential introduction of the canine

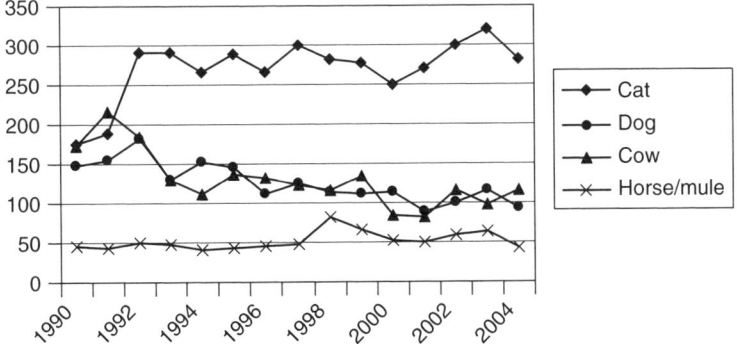

Figure 5.5   Annual rabies cases in major domestic species in the USA.

rabies virus variant due to naturally variable and potentially long incubation periods, as well as exemptions for vaccination of young animals due to possible interference with vaccination by maternal antibodies. The threat of introduction is demonstrated by the importation of a puppy from Thailand that became sick during travel and died of rabies within days of arriving in the USA (personal communication, Dr Ben Sun), as well as the importation of another young dog from Puerto Rico that became rabid shortly after arrival in the USA. The importation of dogs from Puerto Rico remains ongoing as part of an organized dog rescue effort. Other international incidents include the importation into France of a young dog that developed rabies associated with its country of origin.

On a global basis, the domestic dog remains the most important reservoir of rabies in overall case numbers and with regard to transmission to humans. In regions where strict control of free-ranging dogs and mandatory parenteral rabies vaccination were enforced, canine rabies has been successfully eliminated. Examples of such areas include Great Britain, Japan, Canada and the USA. Recently, numerous countries in Latin America have made tremendous progress in the application of stray dog control and mass vaccination, with consequent control of canine rabies and markedly reduced human rabies over large geographic areas. In fact, some canine rabies virus variants have been extirpated through successful dog vaccination programs (Hughes et al., 2004). Although canine rabies has been eliminated in most European countries, canine rabies remains largely uncontrolled throughout most of Asia and Africa. In many developing countries, besides the detrimental effects of poverty upon the delivery of basic health services, the devastating effects of sociopolitical upheaval and intersectorial violence exacerbate local deficits, with consequent increases in human rabies mortality due to lack of resources for and access to human post-exposure biologics and as a direct result of unencumbered rabies transmission among free-ranging, unvaccinated dogs (Hatch et al., 2004).

An additional aspect of canine rabies is its impact upon other animals. True wild dogs, Lycaon pictus, are among the most endangered carnivores in Africa (Gascoyne et al., 1993). In 1990, an adult Lycaon was found dead due to rabies in the Serengeti region of Tanzania. Rabies was also suspected but not confirmed in one adult and six pups of the same pack. In response, two Lycaon packs in the Serengeti National Park were given inactivated rabies vaccine either by dart or by parenteral inoculation following sedation, demonstrating the need for veterinary intervention in wildlife conservation, directly related to dog rabies.

The incubation period of rabies in dogs may be as short as 10 days or as long as several months. A commonly imposed quarantine period for an exposed dog (and often cat, ferret, horse, cow and other animals) is 6 months. This period is based on epizootiologic and experimental evidence that if an exposed dog will develop rabies from an exposure, there is a strong likelihood that it will occur within 6 months of the exposure. Although a combination of clinical signs may be useful in the management of the suspect rabid dog, none is absolute (Tepsumethanon *et al.*, 2005). At the beginning of the clinical course, non-specific signs may include anorexia, fever, dysphagia and a change in behavior, such as more reclusive than normal or restlessness. The animal may be more easily startled by noise, light or touch. As each day passes, other clinical abnormalities may appear and the initial signs may become more pronounced. Substantial injury may occur due to aggression towards objects or other animals. Self-mutilation may occur. Salivation may appear profuse due to the inability to swallow and potentially increased saliva production as well (see Figure 5.4). Dehydration and anorexia secondary to dysphagia may result in acute weight loss. Tremors, ataxia, paresis, paralysis and generalized seizures may develop. Alterations in CSF towards a lymphocytic pleocytosis may be compatible with a non-specific viral meningoencephalomyelitis (Barnes *et al.*, 2003). The actual physiologic mechanism resulting in death can vary but may result from profound autonomic dysfunction resulting in alterations in respiration, body temperature, blood pressure, cardiac rhythm, kidney function, etc.

When a dog has potentially exposed a person to rabies through a bite or other exposure route, the dog is commonly subjected to a 10-day confinement and observation period. The 10-day time period is derived from experimental observations of clinical shedding of rabies virus (Vaughn *et al.*, 1965), as well as supportive field data. If a dog is capable of shedding virus (which has necessarily developed after CNS infection) in its saliva at the time that it bit someone, the dog may reliably be expected to develop clinical signs of rabies within the 10-day confinement and observation period. If the dog remains alive and well, there are no substantive epizootiologic data to contradict the presumption that there was no virus in the dog's saliva at the time of the bite.

There are occasional reports in the literature of apparent exceptions to the widely practiced public health rabies management protocol in North America of a 10-day confinement and observation period for a biting dog. In one report, four beagles were inoculated with mouse brain passaged-saliva from an apparently healthy dog from Ethiopia that had been reported to excrete virus intermittently (Fekadu, 1972; Fekadu *et al.*, 1981, 1983). Two inoculated dogs remained clinically normal. Two dogs developed signs of rabies but apparently recovered; one later died of a bacterial pneumonia. Salivary swabs from the surviving dog were routinely collected and inoculated into mice. Virus was reported to have been present in samples taken on days 42, 169 and 305 days post-inoculation. The amount of virus was extremely low, less than 10 infectious particles. In a related study, 39 dogs were injected intramuscularly with either an Ethiopian strain or a Mexican strain of rabies virus. Virus was recovered from the submaxillary salivary glands of 9 of 17 dogs that died following inoculation with the Ethiopian virus variant. Four of these dogs were reported to have virus in the saliva up to 13 days before overt signs of rabies were observed. Virus was recovered from the submaxillary salivary glands of 16 of 22 dogs that died following inoculation with the Mexican virus variant. Eight of these dogs also excreted virus in the saliva up to 7 days before definitive signs of rabies were observed (Fekadu *et al.*, 1982, 1983). Observations in Asia suggest that, on rare occasions, a dog under a 10-day observation period can still be alive and apparently healthy, while the bitten human may develop signs and symptoms of rabies during that period (Somayajulu and Reddy, 1989). The veracity of these reports is unknown. It is possible that a different dog may have bitten a patient previously several weeks to several months prior to the temporally associated situation. However, if these observations have merit, two potential interpretations may be possible. The dogs may be remaining healthy and shedding rabies virus for longer than the expected 10-day period prior to clinical development of rabies. Alternatively, these dogs may reflect the possibility of recovery from rabies with the potential for intermittent shedding. In the past 20 years, virus detection and identification techniques have greatly advanced in sophistication and sensitivity, yet there has been a conspicuous lack of additional reports of similar findings. No such field observations have come from North or South America and a recent field investigation in China failed to confirm rabies virus in the saliva of brain-negative dogs. A primary explanation for the recent resurgence of rabies in China is the circulation of a virulent canine rabies virus among a rapidly growing population of unvaccinated dogs (Tang *et al.*, 2005). It is unknown if these occasional observations simply reflected rare events or if they have more profound implications as to the difficulty of canine rabies control in Asia, Africa and possibly elsewhere.

Recently, a purported ante-mortem latex agglutination (LA) test for rabies virus antigen has been developed for evaluation of dog saliva (Kasempimolporn *et al.*, 2000). Rabies virus antigen is detected by agglutination on a glass slide using latex particles coated with gamma globulin. In the study of paired saliva and brain specimens from 238 dogs, the LA test using saliva was 99% specific and 95% sensitive compared to the DFA test on brain impressions. It is still unfortunate that, for most species, besides dogs, cats and ferrets (Centers for Disease Control and Prevention, 1999), an animal must be euthanized to rule-out rabies definitively. The obvious disadvantage of a potential ante-mortem test is the biological potential for intermittent shedding of rabies virus in the saliva of an infected animal, even during a typically short clinical course (Vaughn *et al.*, 1965). Moreover, the test itself would require rigorous controlled evaluation with acceptable levels at no less than virtually 100% sensitivity, if it were ever to be considered as a primary diagnostic tool to facilitate human exposure determinations. A false-negative test would have severe implications if human rabies PEP decisions were to be based upon such a test. Certainly, it would seem prudent to conduct further research on the pathogenesis of prominent canine rabies virus variants employing traditional and novel viral detection methods. The findings of such studies would either support the current practices of dog (as well as cat and ferret) management in the face of human exposure or provide enhanced experimental data on which to frame optimal human rabies prevention.

8.2 Cat

Although no unique rabies virus variants are associated with cats and no feline reservoirs have been defined, cats can be significant disease vectors. In the USA, the leading domestic animal diagnosed with rabies is the cat with a total of 2834 positive animals (range 249–321 per year) from 1995 to 2004 (Krebs *et al.*, 2005) (see Figure 5.5). The impact of rabies in cats is reflected in the relatively large number of humans exposed, typically involving caregivers of the cat and veterinary hospital staff (Gordon *et al.*, 2004). One of the most dramatic cases occurred when a kitten was offered for sale in a pet store where customers could freely pet the animals, after which at least 665 persons received PEP (Noah *et al.*, 1996). Although most states and localities require registration of dogs coupled with proof of rabies vaccination, few have extended similar regulations to cats. Thus, cats are less commonly vaccinated against rabies. Also, when cats are allowed outdoors by owners, cats are more commonly unrestricted and unsupervised in comparison to dogs and roam frequently at night. In many communities, free-ranging cats are more likely to be tolerated than free-ranging dogs. Recent controversy has erupted regarding the maintenance of feral cat colonies by humans through provision of shelter, food and limited veterinary care that may include surgical neutering to limit colony size with the ultimate intention of colony extinction, although this may be negated by the addition of new members.

Cases of rabies in developing countries are usually related directly to infection from rabid dogs (Kasempimolporn *et al.*, 2004). In developed countries, rabid cats are affected by the predominant circulating wildlife rabies virus (McQuiston *et al.*, 2001). For example, within the USA, cats along the east coast from Florida to Maine are most commonly infected with the raccoon rabies virus variant. In the Midwest and California, cats are likely to be infected with the predominant skunk rabies virus variant. In suburban and urban settings, raccoons and skunks are often tolerant of human presence and, along with dogs and cats, forage on food offered intentionally for domestic animals and sometimes offered for wild animals, as well as from unintended garbage sources. This may enhance opportunities for conflict that may directly contribute to rabid raccoons and skunks infecting cats and dogs.

Another source of rabies in cats is from bats. Since insectivorous bats and their associated rabies virus variants are widely distributed throughout North America, this provides a constant potential source of infection for cats. Encounters between rabid bats and cats are not surprising given the nocturnal activity patterns of bats and cats, as well as a cat's propensity to investigate and capture small animals, such as bats. Cats are susceptible to rabies virus of bat origin and can shed the virus in their saliva (Trimarchi *et al.*, 1986). In light of these findings, it is not surprising that rabid cats have outnumbered dogs by a margin of two-to-one in the USA for the past 8-year period. Moreover, among 209 rabies-positive cat samples provided to the Centers for Disease Control and Prevention from various state health departments within the USA, 11% (4/47) were bat-associated variants (J.S. Smith, personal communication), a statistically significant higher rate than that of bat-associated variants found in dogs (3/157, 2%; $P < 0.0001$). Notably, cats may be implicated in the transmission of bat-associated rabies virus variants to humans (Badilla *et al.*, 2003).

Based upon a seminal study published over 40 years ago (Vaughn *et al.*, 1963), a 10-day confinement and observation period is considered an option to euthanasia and testing of a cat to rule out rabies exposure when a cat has bitten a human (Centers for Disease Control and Prevention, 1999). Two subsequent studies raised interesting questions about potential chronic and recrudescent rabies in the cat (Perl *et al.*, 1977; Murphy *et al.*, 1980). Since these original studies were conducted, virus detection and identification techniques have greatly advanced in sophistication and sensitivity. Similar to the need for rabies pathogenesis studies in dogs, it would seem prudent to conduct further research on the pathogenesis of important rabies variants employing traditional and novel viral detection methods. Again, the findings of such studies would either support the current practices of cat management in the face of human exposure or provide enhanced experimental data on which to frame optimal human rabies prevention.

8.3 Livestock

Most reports of rabies in livestock involve cattle, and occasionally horses, but other domestic species have been reported (Baer and Olson, 1972; Bergeron *et al.*, 1981). The greatest economic and public health impact of rabies in cattle occurs in Latin America (Lord, 1992; Delpietro and Russo, 1996; Martinez-Burnes *et al.*, 1997). The major source of infection is from vampire bats, predominantly the common vampire bat (*Desmodus rotundus*). Cattle may also be infected from dogs, foxes or jackals, in areas where such rabies is endemic. In North America, predominant sources of infection are from wild carnivore species such as from skunks in the Midwest and raccoons in the East (see Figure 5.2). In addition, cattle may infrequently become infected with insectivorous bat rabies virus variants. Among 47 rabies-positive bovine samples provided to the Centers for Disease Control and Prevention from various state health departments within the USA, 8% (4/47) were bat-associated variants (J.S. Smith, personal communication).

Hudson and colleagues (1996a) provided a recent description of clinical signs of rabies in cattle and sheep. In the diseased cattle (N = 20), the average incubation period was 15 days and the average morbidity period was nearly 4 days. Major clinical signs included excessive salivation (100%), behavioral change (100%), muzzle tremors (80%), vocalization (bellowing; 70%), aggression, hyperesthesia and/or hyperexcitability (70%) and pharyngeal paresis/paralysis (60%). The furious form of rabies was seen in 70% of the cattle and in 80% of sheep (N = 5). In many cases, rabies may be considered only in retrospection (Stoltenow *et al.*,

1999). Often, the animal exhibits signs of choking and well-intentioned individuals may insert a hand in the mouth to search for a foreign body.

In a similar study among 21 experimentally infected horses, the average incubation period was 12 days and average morbidity was nearly 6 days (Hudson *et al.*, 1996b). Naïve animals had significantly shorter incubation and morbidity periods ($P <0.05$) than did test animals vaccinated with products under development. Tremor of the muzzle was most frequently observed (81%) and the most common initial sign. Other common signs were pharyngeal spasm or pharyngeal paresis (71%), ataxia or paresis (71%), lethargy or somnolence (71%). Although some initial presentations began as the dumb form, ultimately 43% of horses developed furious rabies. Clearly, horses should be considered for pre-exposure vaccination, not only due to their typical expense and agrarian utility, but because of their potential involvement in human exposures. Not surprisingly, besides relative health and nutrition, age may be a factor in adequate response to vaccination (Muirhead *et al.*, 2006).

An animal's clinical response to supportive therapy, as well as results of specific diagnostic tests, may confirm a diagnosis other than rabies. With rabies, the clinical course will worsen intractably and is invariably fatal. Despite reports of survivorship among other animals, such as humans, dogs, cats, ferrets and raccoons, there have been no such reports in cattle and horses, probably due in part to limitations in intensive clinical management and also increased physiologic complications due to ponderous size (e.g. rhabdomyolysis). In equids, ataxia may be pronounced after involvement of the brainstem and cerebellum, due in no small part to the substantial balancing act these animals perform upon a single toe. Owing to these morphological complexities, the importance of laboratory examination of appropriate areas of the brain including brainstem, for rabies and other differential diagnoses, should be obvious.

8.4 Ferrets

Ferrets are popular pets in the USA. During 1990, an inactivated rabies vaccine was licensed for use in these domestic carnivores (Rupprecht *et al.*, 1990). However, the potential rabies virus shedding period of ferrets remained uncharacterized, even though a preliminary study using a fox virus found no evidence for excretion in the saliva (Blancou *et al.*, 1982). This situation presented a dilemma for public health officials for the management of those occasions when a person was bitten by a pet ferret. There was no scientifically defined confinement period for which a biting ferret, regardless of vaccination status, could be observed for signs of rabies. Due to the lack of experimental data on ferret response to rabies virus infection, recommendations were to euthanize the ferret and test the brain for rabies virus, even in low risk situations, such as if the ferret was vaccinated, appeared clinically normal and the bite was provoked. Clearly, this was unlike the management of the biting dog and cat in which an observation and confinement period of at least 10 days was recommended. Before recommendations could be made regarding an observation period for biting ferrets, basic parameters of rabies pathogenesis in ferrets needed to be defined (Anonymous, 1990).

To answer a number of basic questions about ferret response to rabies virus infection, studies were designed to determine ferret response using rabies virus variants obtained from important reservoir species in the USA, including primary isolates from naturally infected skunks, raccoons and coyotes, among others. Susceptibility, incubation and morbidity period, clinical signs, serologic response and viral shedding were investigated by routine procedures (Niezgoda $et\ al.$, 1997).

The pathogenesis of rabies in domestic ferrets in regard to overt susceptibility, aggressive behavior, spread of virus to the salivary glands and excretion, was found to vary significantly dependent upon the rabies virus variant. For example, six of 12 (50%) ferrets given a raccoon rabies virus variant at a viral dose of $10^{5.8}$ MICLD$_{50}$ (mouse intracerebral lethal dose 50) succumbed to infection (Niezgoda $et\ al.$, 1998), whereas 10 of 10 (100%) ferrets that received a North Central skunk rabies virus variant at $10^{5.5}$ MICLD$_{50}$ succumbed (Niezgoda $et\ al.$, 1997). In contrast, 12 of 20 (60%) ferrets that received a North Central skunk rabies virus variant at doses that ranged from $10^{3.5}$ to $10^{2.5}$ MICLD$_{50}$ succumbed to infection, whereas 12 of 12 (100%) ferrets that received a canine rabies virus variant at doses that ranged from $10^{3.3}$ to $10^{2.3}$ MICLD$_{50}$ died. At a higher dose, only two of 10 (20%) ferrets that received a raccoon rabies virus variant at $10^{4.0}$ MICLD$_{50}$ succumbed. Overall clinical signs were consistent with an acute viral infection of the CNS and included ataxia, paresis, paraparesis, paralysis, lethargy, tremors, bladder atony, fever and weight loss. Morbidity periods were also similar at 4 to 5 days, regardless of virus variant, but aggressive behavior in ferrets varied significantly with different viruses. For example, none of 33 rabid ferrets that received a North Central skunk rabies virus variant showed signs of aggressive behavior. Only two of 19 (11%) rabid ferrets that received a raccoon rabies virus isolate were aggressive. In contrast, 10 of 10 (100%) rabid ferrets that received a canine rabies virus variant were aggressive. In addition, viral excretion and salivary gland infection results varied significantly depending upon the rabies virus variant. No ferrets succumbing to a North Central skunk rabies virus variant had detectable virus in their saliva and only one of 33 (3%) had detectable virus in its salivary glands. In contrast, in ferrets that received a raccoon rabies virus isolate, 12 of 19 (63%) rabid ferrets had detectable virus in their salivary glands and nine (47%) shed virus in their saliva. Shedding of virus in saliva was largely concomitant with clinical signs of rabies. The antibody response of ferrets also varied significantly depending upon the virus variant. In ferrets inoculated with the North Central skunk rabies virus variant, 14 of 33 (42%) rabid ferrets had rabies VNA and five of 17 (29%) survivors seroconverted. In ferrets given the raccoon rabies virus isolate, two of 19 (11%) rabid ferrets had rabies VNA and only one of 32 (3%) survivors seroconverted. Interestingly, surviving ferrets tended to seroconvert at about the same time that infected members of the same experimental cohort died. Obviously, in the same species, multiple parameters of pathobiology will change, even at the same relative dose and route, as the virus variant in question varies. In addition, when the identical virus variant is used in different species, the clinical response may vary significantly (see Figure 5.3).

Based upon the data generated in these published studies and a subsequent unpublished study using a number of bat variants, the Compendium of Animal Rabies Control (Centers for Disease Control and Prevention, 1998) supported the option of a 10-day confinement and observation period for a ferret that bites a person, as an alternative to euthanasia and diagnostic testing to rule out rabies.

8.5 Wolf hybrids

As with ferrets, the keeping of wolf–dog hybrids as pets has been challenging, not only with regard to rabies prevention, but also for many regulatory agencies responsible for public interests and protection (Jay *et al.*, 1994; Johnson, 1995; Ballard and Krausman, 1997; Overall, 2000). Whether or not these animals are suitable pets will continue to be fiercely debated for the foreseeable future. For example, a 50-year-old woman from Pennsylvania was recently mauled by her pet wolf hybrids and died in the outdoor pen where the nine 70–100 lb (32–45 kg) animals were housed. Nonetheless, at this time, there is no objective scientific method to determine the difference between the genetic make-up of a dog versus a wolf, or of combinations in between. Due to this lack of capacity for distinction, it is illogical to assume that they would substantially differ in their immunological capacity to respond to parenteral inoculation of rabies vaccines with demonstrated efficacy in dogs. Therefore, the US Department of Agriculture, Animal and Plant Health Inspection Service, Veterinary Biologics, has deliberated on this generalization on the approval of the use of licensed vaccines for dogs in offspring of wolves and dogs (Centers for Disease Control and Prevention, 2000a). However, there remains a complete void with regard to rabies pathogenesis and viral shedding studies in wolves. Wolves have been of historical importance in human exposures and present some of the highest case/fatalities per occurrence. In regard to rabies control and prevention, whether an animal is considered a wolf, dog–wolf hybrid, or wolf-appearing dog will continue to be problematic. Some sources now regard them as a single species (*Canis lupus*). Since rabies vaccines are licensed by species, unless wolf or wolf-hybrid ownership is specifically illegal in a given state, owners may request that their 'pets' (i.e. *C. lupus*) be vaccinated and managed in a rabies-risk situation the same as a vaccinated domestic dog. This would be the case despite the lack of scientific data with regard to viral shedding that is needed to establish safe confinement and observation periods for the times when animals, vaccinated or not, bite people, similar to the situation in 1990 with ferrets.

8.6 Other domestic species

When rabies occurs in wild or exotic hoofed stock, such as llamas, these cases tend to occur as isolated events. However, surveillance is less than ideal over high altitude portions of their range in South America and rabies is considered the most notable naturally occurring viral infection in New World camelids (Thedford and Johnson, 1989). With their increasing popularity as pack or pasture animals in North America and their propensity to bite, rabies should be included in the differential diagnosis when acute encephalitis is suspected. In 2002, three rabid llamas were diagnosed in Arizona (Krebs *et al.*, 2003b) and a gray fox was found on the premises (M. Leslie, personal communication). Antigenic typing confirmed the variant as associated with gray foxes, but no virus was isolated from salivary glands of the rabid llamas. In 2004, a rabid llama occurred in an enclosure in Georgia; its pen-mate survived without illness. A 6-month quarantine was invoked due to the risk of potential exposure from the rabid llama, as well as to the unobserved potential exposure at the same time as the rabid llama. In addition, approximately 50 rabies cases occur in horses and mules every year and about 100 rabid cows occur every year, sometimes succumbing to rabies virus variants from their original location rather than rabies virus variants present at the site of their final growth production site. Rabies in small pet mammals occurs at a very low rate considering the size of the population potentially at risk, with 38 cases in domestic rabbits, two rabid guinea pigs (Eidson *et al.*, 2005), one pet chinchilla with rabies and one rabid pet prairie dog from 1985 to 2004.

9 WILDLIFE RESERVOIRS

The potential role of wildlife in rabies maintenance has often been obscured by ubiquitous urban dog rabies and the higher prominence placed upon domestic animals due to the human–pet bond and the economic value of livestock (Smith and Seidel, 1993; WHO, 2005). For example, despite the plethora of mammals found throughout Africa, recognition of the complexity of the disease among wildlife has been very slow to emerge. Given deficiencies in general surveillance of wildlife diseases, recognition may require notoriety related to perceived threats to endangered species and dramatic losses, such as among wild dogs, the Ethiopian wolf and kudu (Rottcher and Sawchuk, 1978; Sawchuk and Rottcher, 1978; Gascoyne *et al.*, 1993; Sillero-Zubiri *et al.*, 1996; Creel *et al.*, 1997; Whitby *et al.*, 1997; Mansfield *et al.*, 2006a). In contrast, in southern Africa, many indigenous inhabitants long believed the bite of certain wild mammals would lead to a fatal disease, principally from viverrids (e.g. mongoose, civet, genet). Recognized today, the yellow mongoose, *Cynictis penicillata*, is a common, widely distributed rabies host, involved in maintaining endemic disease throughout the region (King *et al.*, 1994; Nel *et al.*, 2005).

In developed countries, epizootiologic patterns of rabies cases among reservoir and spillover species are comparatively well described and consist of some of the most common wildlife mammals (Krebs *et al.*, 2003a). In contrast, basic rabies surveillance is a limiting factor in understanding primary wildlife reservoirs in many areas of Africa and Asia. Surveillance would be greatly enhanced through development of the capacity to conduct reliable diagnostic testing on-site in localities, rather than through the current practicality of central diagnostic laboratories due to the expense and expertise demanded by the DFA test and the requisite fluorescent microscope. A direct, rapid, immunohistochemical test (DRIT) has been developed and is currently undergoing field testing in the USA, Tanzania and several other locations (Lembo *et al.*, 2006) (Figure 5.6). The DRIT is conducted on brain impressions, with rapid-fixation in formalin, staining with a combination of biotinylated anti-rabies murine monoclonal antibodies and developing via an immunoperoxidase technique. Rabies virus antigen may be visualized by light microscopy. It is anticipated that the availability of this light

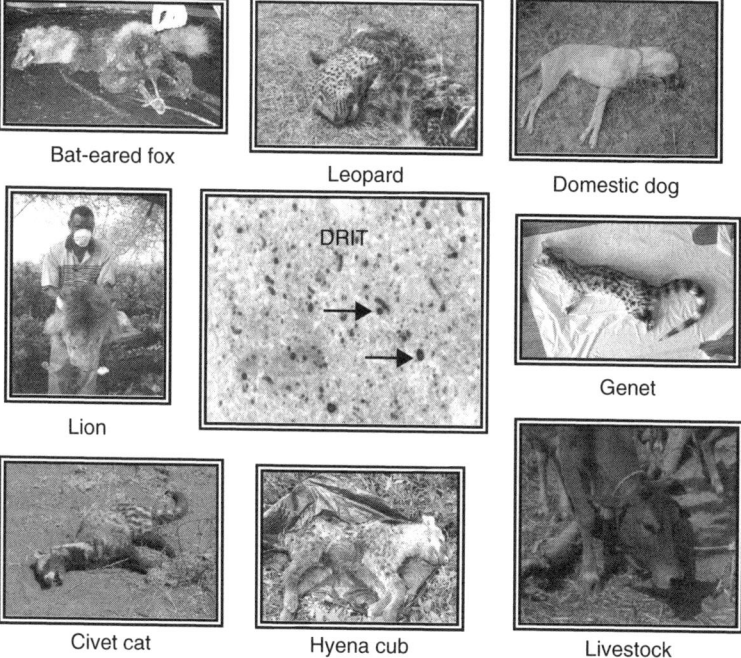

Figure 5.6   Enhanced rabies surveillance using the direct rapid immunohistochemical test (DRIT) in Tanzania. Arrows: identification of rabies virus antigen on a brain impression using a cocktail of monoclonal antibodies directed against the virus nucleoprotein. (Animal photographs courtesy of the Serengeti Carnivore Disease Project (Tiziana Lembo); leopard photograph provided by Sarah Durant; DRIT photograph by Michael Niezgoda.) This figure is reproduced in the color plate section.

microscope technique for the diagnosis of rabies may greatly enhance rabies surveillance and, thus, track well-established epizootiologic patterns and provide a powerful tool for the timely illumination of emergent ones.

9.1 Fox

With its occurrence in Eurasia, North America and northern Africa and introduction to Australia, the red fox (*Vulpes vulpes*) is one of the most widely distributed and abundant wild carnivores in the world. Perhaps more is known about rabies in this fox than in any other wild mammal, not only related to its significance as a major reservoir, but due to the influence of directed research during the past 30 years, coordinated by the World Health Organization (Wandeler, 1980).

Descriptions of rabies among foxes (and other canids) were not uncommon in Western and Central Europe throughout the Middle Ages (Steck, 1968; Wilkinson, 1988; Steele and Fernandez, 1991). Possible confusion with other diseases, such as canine distemper and the possibility of species misidentification, are both dilemmas for all such historical accounts. Nevertheless, by the 19th century, rabies in the red fox seemed prevalent throughout Europe. Oddly, fox rabies largely disappeared during the first decades of the 20th century, but re-emerged during the 1930s and spread throughout much of mainland Europe, until the relatively recent advent of oral vaccination (Wandeler *et al.*, 1974; Bögel *et al.*, 1976; Toma and Andral, 1977; Steck and Wandeler, 1980; Blancou *et al.*, 1991; Steele and Fernandez, 1991). Within some parts of Europe, rabies may re-emerge among foxes due to apparent viral spillover via infected dogs (Johnson *et al.*, 2003). In North America during the mid- to late 1700s, reports of rabid foxes and dogs were common throughout the mid-Atlantic British colonies, probably exacerbated by a cultural predilection for fox hunting and hence European fox introductions and widespread translocation. One of the first descriptions of the disease in Canada was in 1819, when the Governor General was bitten by a pet fox and died (Steele and Fernandez, 1991; Jackson, 1994). The disease may have been present in the New World for centuries (Crandell, 1991). Description of a major rabies outbreak among Arctic (*Alopex lagopus*) and red foxes did not occur until well into the mid-20th century, spreading into southern Canada and the USA by the 1950s (Johnston and Beauregard, 1969; Tabel *et al.*, 1974). Rabies in polar regions has been maintained for decades by the Arctic fox, an important reservoir for Eurasia, North America and Greenland (Ballard *et al.*, 2001; Kuzmin *et al.*, 2004; Mork and Prestrud, 2004; Mansfield *et al.*, 2006b; Nadin-Davis *et al.*, 2006b). If trends in global warming continue, with consequent effects upon population densities of local flora and fauna, rabies outbreaks may increase in both frequency and intensity (Parkinson and Butler, 2005).

Foxes are quite susceptible to experimental infection. The amount of virus needed productively to infect half of red foxes in captivity varied from less than or equal to 1–5 $MICLD_{50}$ in the masseter, neck or cervical muscles (and intradermally in the ear), to 16 $MICLD_{50}$ or more in the gluteal muscles (Sikes, 1962; Parker and Wilsnack, 1966; Black and Lawson, 1970; Winkler, 1975; Wandeler, 1980; Blancou *et al.*, 1991). Foxes were more resistant to rabies when infected by isolates of canine, bat or raccoon dog origin, among others (Blancou *et al.*, 1991). By other routes, it took approximately 5 more logs of rabies virus to infect foxes orally than parenterally (Wandeler, 1980). Minimum incubation periods ranged from as short as 4 days to longer than 15 months. Most appeared between 2 weeks and 3 months, inversely related to dose, as was the proportion of foxes that have virus in the salivary glands and the relative quantity of virus recovered (Wandeler, 1980; Blancou *et al.*, 1991). Thus, in theory, the greater the infective dose, the more foxes that can develop rabies, but the fewer that may transmit to conspecifics because of the decreased probability of salivary shedding. Conversely, small inocula may result in fewer productive infections, but a greater proportion of animals that could shed virus in the saliva. Considering that most foxes tend to excrete on the average between 3 and 4 logs of virus, this appears to be more than adequate to insure infectious chains of transmission, given the extreme susceptibility of the species to selected viral variants. Short morbidity periods of 2–3 days are the usual rule, regardless of dose, but have ranged from less than 1 to over 14 days (Wandeler, 1980; Aubert, 1992). Signs are variable but commonly include anorexia, restlessness, hyperactivity, ataxia, and aggression (George *et al.*, 1980). Radio-tracked rabid foxes seem to experience abnormal behaviors, such as spatial-temporal alterations and frequent prostration and may acquire wounds from normal foxes responding to territorial incursion (Artois and Aubert, 1985). Virus may be excreted in the saliva concomitant with, or as long as a month before, obvious clinical signs (Aubert, 1992). Unlike the case for some other mammals, herd immunity is not believed to be important in red foxes. Insufficient explanations have appeared as to why fox rabies halted mid-stream in France (Blancou *et al.*, 1991), or seemingly disappeared in the USA during the 1970s (Winkler, 1975).

Approximately 20 extant 'fox' species are presumed to exist throughout the world (Nowak, 1991). Although many possess solitary habits and occupy only geographical remnants of a former domain, others are widespread and seem to meet basic social criteria for a successful rabies reservoir. Nevertheless, with the exception of the Arctic and gray foxes (*Urocyon cinereoargenteus*) that have distinct associated rabies virus variants and are regionally important in the epidemiology of the polar and western American regions, related information on other foxes is scanty in this regard. Recent information does suggest that the gray fox may be an important reservoir and vector, not only in the USA, but also in Mexico, portions of Central America and into South America, such as Colombia (Páez *et al.*, 2005; Velasco-Villa *et al.*, 2005). Some biological limitations appear obvious, most related to direct human depredation or encroachment and subsequent habitat loss. Such a tale holds for many representatives among the genus *Vulpes* in the Old World, including the quite social Eurasian corsac fox of the steppe and desert zones, Blanford's fox, the Bengal fox in India, Pakistan and Nepal and the Tibetan sand fox. An apparent exception to these may be the common sand fox, *V. rueppelli*, reported from the desert zones of Morocco and Niger to Afghanistan and Somalia, in which recent reports to the World Health Organization support the notion of a possible rabies reservoir on the Arabian Peninsula. Alternatively, the Fennec fox, *Fennecus zerda*, seems confined to the north African desert biome and the Cape fox has been severely persecuted in its restricted occurrence in dry areas of southern Africa. Little is known of the otherwise rather gregarious pale fox, found throughout Senegal to the Sudan. The bat eared fox, *Otocyon megalotis*, has been reported from Ethiopia to southern Africa and is considered to be occasionally important in the local epidemiology of rabies, but it too declines near human habitation and suffers destruction via domestic dogs. The New World Channel Islands fox, *Urocyon littoralis*, is restricted to offshore sites of southwestern California and the kit and swift foxes are endangered species with extremely fragmented distributions, having disappeared over much of their former ranges. The hoary fox, *Lycalopex*, is confined to Brazil, where it is often killed for presumed predation upon fowl and is suggested as an emerging reservoir (Bernardi *et al.*, 2005; Carnieli *et al.*, 2006). The Culpeo fox, *Pseudalopex* (and its allies), extends over a fairly wide swathe of South America from the equator to Argentina, but is a habitat specialist and may be limited in population density, while restricted to sanctuaries among park reserves. One possible candidate deserving of further study is the crab-eating fox, *Cerdocyon thous*. Populations of this 3–8 kg canid occur in a variety of savanna to woodland habitats from Columbia to Argentina, are omnivorous and locally abundant and could prove to be an important species of concern in Latin America, given adequate attention to rabies surveillance.

9.2 Raccoon

Raccoons remain the sole major terrestrial carnivore reservoir in the eastern USA. Over the past three decades, raccoon rabies in the northeastern states has caused the deaths of tens of thousands of raccoons. Persuasive explanations for the original nidus in Florida during the 1940s remain speculative. From the 1950s to the 1970s, the disease in the southeastern USA among raccoons gradually spread northward. Due to their importance in the fur trade and related recreational hunting, it was not unusual for raccoons to be moved between regions. Rabid raccoons were discovered in some of these shipments (Nettles *et al.*, 1979). In the late 1970s, a new focus of rabid raccoons appeared in the mid-Atlantic region of the USA. By 1991, the number of reported rabid raccoons had outnumbered rabid skunks. At present, the affected area stretches from its southern extent in Florida, east of the Appalachian Mountains, north into Canada in Ontario and New Brunswick, and westward into Tennessee, Alabama and Ohio, making the raccoon the single most important rabies reservoir in the USA (Centers for Disease Control and Prevention, 2000b; Krebs *et al.*, 2005). The large geographical range exploited by raccoons in North America is coupled with their adaptability, tendencies for communal foraging in suitable habitats and potential for high density populations in both suburban and urban environments (Totton *et al.*, 2002). Together with the existence of a persistent viral variant (Smith *et al.*, 1984), raccoon rabies is a particularly salient testament to re-emerging disease (Hanlon and Rupprecht, 1998).

One mathematical model, based on county-level surveillance data, predicted an initial epizootic period of approximately 5 years, with progressive dampening of magnitude and progressive decrease in periodicity (Childs *et al.*, 2000). Despite what was suggested from the southeastern USA prior to the mid-Atlantic translocation in the late 1970s, the best current quantitative agreement between collected case data and the generated model assumed fairly low levels of herd immunity, on the order of 1–5% within raccoon populations, suggesting little or no immunity in naïve animals, in agreement with field surveillance within the region.

Despite its relevance as a reservoir, rabies pathogenesis in raccoons is neither well understood nor well described experimentally. Most published responses of raccoons to rabies virus infection have been conducted using a number of other isolates, including a fox (Sikes and Tierkel, 1961), a bat (Constantine, 1966a), a skunk (Hill and Beran, 1992; Hill *et al.*, 1993) or a dog (Rupprecht *et al.*, 1988). Experimental studies investigating the pathogenesis of raccoon rabies virus in raccoons are limited (McLean, 1975; Winkler *et al.*, 1985; Winkler and Jenkins, 1991), hampered in no small way by the lack of a means definitively to characterize such variants prior to the advent of monoclonal antibodies in 1978 (Wiktor and Koprowski, 1978).

Using data extrapolated from a temporal clustering of raccoon rabies cases in Ontario, the incubation period was estimated at a mode of approximately 5 weeks (Tinline *et al.*, 2002). In a recent study of experimental rabies in raccoons, animals were inoculated with rabies virus several times over one year and their responses were monitored (Niezgoda *et al.*, 1991). The incubation period was independent of viral dose and was ~50 days (range 23–92 days) and the morbidity period was ~4–5 days (range 2–10 days). Not all raccoons succumbed or seroconverted. All raccoons euthanized with clinical signs of rabies contained rabies virus antigen in their CNS, whereas all survivors were free of antigen at final necropsy.

Not unexpectedly, raccoons may not develop clinical rabies following exposure to a rabid animal (McLean, 1975) for a variety of reasons. Rabies virus may not be shed from the infected animal during the exposure event, or the concentration of excreted virus may be minimal, or the severity of the exposure could be limiting. Likely, raccoons may be exposed, and develop a detectable response and apparent immunity. In contrast, no antibodies may be detected following multiple rabies exposures, but with protective immunity nonetheless ensuing.

In an experiment conducted for the titration of challenge virus used in oral rabies vaccination studies, 24 raccoons in three experimental groups (Tables 5.2

TABLE 5.2

Raccoon response to experimental rabies virus infection

Group[a] MICLD$_{50}$	Raccoon No./sex	Died/ survived	Incubation period (days)	Morbidity period (days)	Virus isolation[b] SG	Saliva	Terminal serum[d] rabies VNA
$10^{4.3}$	764 M	D	18	2	+	+	26
	763 M	D	20	<1	+	+[c]	45
	765 M	D	15	4	+	+	135
	766 F	D	20	2	+	+[c]	78
	7510 M	D	24	2	+	−	ND
	223 M	D	16	3	+	+	45
	9553 M	D	17	2	+	+	234
	9034 F	D	16	1	+	−	135
$10^{4.0}$	687 F	D	18	<1	+	+	15
	768 M	S					<5
	456 F	D	15	<1	+	+	405
	221 M	D	20	5	+	+	45
	703 M	D	17	3	+	+	15
	773 F	D	17	5	+	+	405
	704 F[e]	D	16	3	+	+	45
	788 M[e]	D	23	2	−	−	ND
	701 M[e]	D	15	4	+	+	6316
	783 M[e]	D	17	2	+	+	15
	774 M[e]	D	16	1	+	+	135
	702 M[e]	D	18	5	+	+	405
$10^{3.7}$	767 M	D	29	1	+	+	1215
	699 F	D	25	2	+	+	ND
	698 M	D	22	4	+	+	1215
	697 M	S					78

[a]All raccoons received virus in the right masseter muscle with a salivary gland suspension from naturally infected PA raccoons.

[b]Virus isolation from salivary glands (submandibular) and saliva was done by cell culture using baby hamster kidney (BHK) cells.

[c]Virus was isolated in the saliva from two raccoons 3 days prior to observable clinical signs.

[d]Rabies virus neutralizing antibodies (VNA) were determined by the rapid fluorescent focus inhibition test (RFFIT).

[e]Six of 12 raccoons in the group $10^{4.0}$ MICLD$_{50}$ received 1.0 ml of VRG at 10^9 TCID$_{50}$/ml orally 7 days after virus infection.

TABLE 5.3

Raccoon response to experimental rabies virus infection

Inoculum MICLD$_{50}$	Mortality(%)	Incubation period (days) range (med)	Morbidity period (days) range (med)	Virus isolation	
				Salivary glands (%)	Saliva (%)
$10^{4.3}$	8/8 (100)	15 to 24 (18.2)	<1 to 4 (2)	7/8 (88)	6/8 (75)(91)
$10^{4.0}$	11/12 (92)	15 to 23 (17.4)	<1 to 5 (2.9)	11/11 (100)	10/11 (91)
$10^{3.7}$	3/4 (75)	22 to 29 (25.3)	1 to 4 (2.3)	3/3 (100)	3/3 (100)
Total	**22/24 (92)**	**15 to 29 (18.8)**	**<1 to 5 (2.4)**	**21/22 (96)**	**19/22 (86)**

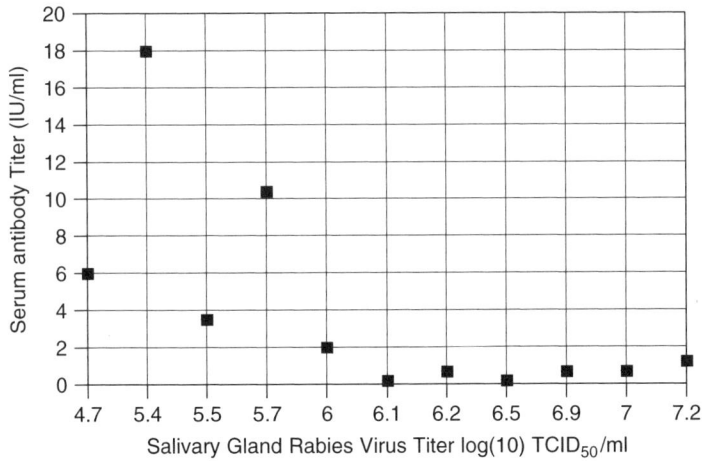

Figure 5.7   Rabies VNA in serum and virus titer in submandibular salivary glands from experimentally infected rabid raccoons.

and 5.3) received a salivary gland suspension obtained from naturally infected raccoons from Pennsylvania and typed as raccoon rabies virus variant at $10^{4.3}$, $10^{4.0}$ and $10^{3.7}$ MICLD$_{50}$ in the masseter muscle. Twenty-two of the 24 (92%) raccoons succumbed to rabies virus inoculation. Incubation periods ranged from 15 to 29 days. Morbidity periods ranged from <1 to 5 days. Rabies virus was isolated from the submandibular salivary glands of 21 of 22 (96%) rabid raccoons and rabies virus was detected in the saliva of 19 (86%) raccoons. Virus titers in the submandibular salivary glands from rabid raccoons were inversely proportional to rabies VNA detected in serum. Raccoons with the higher virus titers in submandibular salivary glands showed the lowest VNA in their serum (Figure 5.7). Saliva collected from one raccoon with clinical illness yielded a virus titer of $10^{4.2}$

tissue culture infectious dose $(TCID_{50})$/ml. To test the response of raccoons incubating rabies to the consumption of an oral rabies vaccine bait, six of 12 raccoons in the group that received raccoon virus variant at $10^{4.0}$ $MICLD_{50}$ received 1.0 ml of a vaccinia-rabies glycoprotein recombinant vaccine at 10^9 $TCID_{50}$ orally 7 days after the exposure to rabies virus. Vaccination at 7 days post-virus exposure had no protective effect; all six animals succumbed to rabies. There was no difference found in overt mortality, incubation period, virus isolation from submandibular salivary glands and shedding of virus in saliva in post-exposure vaccinated raccoons and routinely challenged raccoons.

Based on both laboratory and field observations, several theoretical groups may exist in free-ranging raccoons, dependent upon local rabies epidemiology. For example, in a study conducted in four different areas of Pennsylvania, the lowest geometric mean titer (GMT) of rabies VNA was found in a 'rabies-free' area. A higher GMT was found at the perceived front to two areas and the highest within a rabies epizootic area (Winkler and Jenkins, 1991). All of the antibody positive raccoons were clinically normal when originally live-trapped. Additionally, in raccoons we live-trapped during a raccoon rabies epizootic in Philadelphia's Fairmount Park during 1990, only one of 14 clinically rabid animals had detectable rabies antibody and all were later confirmed rabid by laboratory diagnosis. Hence, detection of rabies antibody in a free-ranging raccoon shows only that exposure to rabies virus antigen has occurred and a detectable response may be reflective of immunity, rather than indicative of a viral incubation phase or ensuing illness *per se*.

Similarly, rabid raccoons that succumb to a productive infection can present either as antibody positive or negative. Experimental data suggest that raccoons in terminal stages of disease will probably be seropositive, whereas during earlier stages it would not be unusual for an animal to be antibody negative. Also, if seroconversion does occur at all, the GMT may be higher in raccoons that succumb to infection, in comparison to those that are exposed, and seroconvert, without detectable illness. Under common field scenarios, rabid raccoons may succumb much earlier than under laboratory situations with supportive veterinary care, due to a variety of adverse environmental conditions and probably would still be antibody negative.

In areas believed to be free of terrestrial rabies based on surveillance (such as much of the mid-Atlantic and northeastern USA immediately prior to the late 1970s), a majority of raccoons are presumed to be largely naïve, non-exposed, antibody-negative animals. Occurrence of minimal seroprevalence (i.e. in the range of 0 to 3%) in low to non-enzootic areas may be coincident to rare rabies exposures, such as a raccoon consuming an infected bat. Supposed detection of some low 'titers' may be artifactual, due to non-specific viral inhibitors within serum (Hill and Beran, 1992; Hill *et al.*, 1992, 1993). Alternatively, intentional human intervention through parenteral immunization with inactivated rabies vaccine via wildlife rehabilitation, purposeful trap-vaccinate-release programs or, more recently, programs evaluating the effect of oral rabies vaccination upon wildlife, could also account for antibody.

Categorically, to include a complete theoretical outcome following exposure, a certain small fragment of a subcompartment of raccoons could be exposed to rabies, develop disease, yet recover with or without obvious sequelae. Nevertheless, documented recovery from clinical rabies is quite rare. A single report exists of an experimentally infected raccoon that showed compatible clinical signs but recovered and it was previously vaccinated (Rupprecht *et al.*, 1988). Moreover, it is difficult to comprehend how viral clearance could spontaneously occur in rabies without some residual neurological damage. The probability for survival in the field for such an obviously debilitated animal would be assuredly low. To date, there is no discrete experimental evidence or epizootiologic suggestion that a chronic carrier state occurs in raccoons, i.e. infected animals that shed virus in the saliva, but remain clinically normal. Rather, the accepted dogma of exposure leading to clinical rabies and death, or the induction of protective immunity, seems a much better fit with current experimental and field data.

9.3 Skunk

Although tales of rabies in foxes were detailed in the New World as early as the 18th century, it took another hundred years for the same case to be made for skunks during Western explorations. These small to moderate sized carnivores with specialized anal glands are widespread in parts of North America, consisting of at least four different species and several subspecies. By the mid-1800s, rabies in skunk was reported from the American prairies and the disease spread north into Canada by the next century. Rabid skunks were noted to be tenacious biters and, at times, rabies seemed so prevalent that some proposed the name 'Rabies mephitica' to describe the disease in spotted skunks (Steele and Fernandez, 1991). Today, in North America, reports of rabies in skunks occur mainly in four geographic regions:

1. the eastern USA
2. the north central USA and the Canadian provinces of Manitoba, Saskatchewan and Alberta
3. California
4. the south central USA and Mexico (Gremillion-Smith and Woolf, 1988; Aranda and Lopez-de, 1999; Crawford-Miksza *et al.*, 1999; de Mattos *et al.*, 1999).

Rabies in these areas (in skunks and, to a large extent, in other terrestrial mammals) is caused mainly by four different street virus variants. In the eastern USA, cases are primarily related to spillover from infected raccoons, whereas in the other three areus, the viruses are adapted to skunks as the primary reservoir. Earlier work suggested that basic differences in host susceptibility would help to explain geographic patterns of maintenance, in that red foxes could succumb relatively quickly to large doses of virus inoculated by rabid skunks (Sikes, 1962). This observation was not found to be universally true and pre-dated the discovery of antigenic variants of rabies virus. Moreover, the role of several independently maintained rabies virus variants and different skunk species is not precisely understood. Other experimental studies do suggest that species specificity of enzootic rabies is due, at least partly, to host immune response and differences in the route, dose and pathogenicity of these variants of rabies virus (Charlton et al., 1987, 1988, 1996).

Besides route of inoculation and virus dose, the rabies virus variant may also produce different outcomes in skunks. For example, groups of skunks were inoculated intramuscularly with different dilutions of virus from salivary glands of naturally infected skunks in Canada and the USA, collected in areas reflective of either red fox or raccoon rabies outbreak. While there was no significant difference in basic susceptibility, incubation period or the spectrum of clinical signs, skunks infected with the rabies virus isolate from the USA exhibited a morbidity period and shed virus in the saliva over 2–3 days. However, skunks infected with the Canadian isolate were clinically rabid and shed virus in the saliva for an average of one week, suggestive of significant differences in the infective potential of the two different variants (Charlton et al., 1988, 1991; Hill et al., 1993).

Other research using skunks suggests a major role for muscle tissue during the incubation period of rabies (Charlton et al., 1997). Two months post-inoculation, muscle at the inoculation site contained viral RNA, even though other relevant tissues on the route of viral migration and early entrance into the CNS were negative. The location of viral antigen was striated muscle fibers and fibrocytes, confirming earlier observations, which indicated that virus may replicate in local tissues prior to invasion of the nervous system.

Rabies virus strains may utilize different cell populations in the CNS of skunks, even by the same route of infection. For example, examination of the cerebellum of skunks experimentally infected with either a skunk isolate of street rabies virus or the fixed virus strain, CVS, revealed a differential response based on viral inoculum (Jackson et al., 2000). The skunk rabies virus variant displayed prominent infection of glia cells, with a relatively small amount of antigen in the perikarya of Purkinje cells. In contrast, the highly adapted CVS strain showed many intensely labeled Purkinje cells and relatively few infected glia cells. Previously, it was suggested that the relative accumulations and effects of different rabies viruses in the CNS could account for either furious or paralytic phases of the disease (Smart and Charlton, 1992). Thus, while lyssaviruses are highly neurotropic, different viruses can show a predilection for various neural elements and locations.

Experimental studies have shed some light on the pathobiology of rabies in skunks, but fewer observations have been made from naturally infected animals. Rabies virus may be detected in a number of glands, particularly in the end stages of disease, but the submandibular salivary glands seem to be one of the sites most consistently infected, with high concentrations of virus (Howard, 1981a,b; Charlton *et al.*, 1984). Other incidental findings have been reported. During necropsy of one pregnant skunk, rabies virus was discovered in the CNS of a single fetus, but the significance of this observation is unclear (Howard, 1981c). The behavior of normal and rabid striped skunks was recently reported unexpectedly, during an unrelated radio telemetry study in North Dakota (Greenwood *et al.*, 1997). In 1991, only one of 23 skunks under study was found to be rabid, whereas during 1992, 35 of 50 (70%) were diagnosed as rabid. The estimated survival rate of skunks was 0.85 during the spring of 1991, but dropped to 0.17 during the same season in 1992. Nearly a third of rabid skunks were located below ground. No differences were observed between healthy and rabid skunks in estimated mean rate of travel per hour, distance traveled per night or home range. Among rabid skunks, mean rate of travel and distance traveled per night tended to decrease with the onset of illness. Mean home ranges of male skunks were greater than for females before illness, but not after the demonstration of clinical signs. Home range of females did not differ when compared before or after signs of rabies. Rabid animals were more spatially clumped than expected, but no relationship was detected between locations of rabid skunks and dates of death. This research not only provided a glimpse of how rabies may affect an animal's use of space, but provided an estimate of the relative impact of rabies upon a local population. It also exemplifies how many rabies cases can typically escape routine surveillance due to the circumstances surrounding an animal's demise, such as predation or carcass loss.

As with any wild animal, clinical signs of rabies are difficult to discern and skunks should not be kept as pets. Illustrative of this point, an apparent outbreak of rabies was reported within a colony of captive striped skunks. One of the animals had been infected in the wild and developed clinical illness approximately 7 weeks after capture. This female transmitted the virus to three of her five offspring and to one other adult. The disease spread when the suckling skunks were adopted by other lactating females. Although the animals were in close contact with each other, the infection spread quite slowly. Furious rabies was not detected. Usually, rabid skunks were found dead without obvious prior clinical signs.

In the raccoon rabies enzootic area of the USA, skunks are the most frequently reported species after raccoons to be diagnosed with rabies virus (Krebs *et al.*, 2005). To date, the infection in all skunks examined has been associated with the raccoon rabies virus variant and appears to be related to infection from raccoons, based on spatial and temporal submission patterns (Guerra *et al.*, 2003). However, on a few occasions (e.g. in Massachusetts, Rhode Island), rabid skunks have outnumbered rabid raccoons. Whether this variant will evolve towards independent rabies transmission in skunks remains to be seen. Such a phenomenon would substantially hamper oral vaccination plans for raccoons in the region, due to the absence of an effective rabies vaccine for skunks (Hanlon *et al.*, 2002).

Recently, the emergence of a bat variant and its apparent adaptation to striped skunks in the Flagstaff area of Arizona resulted in the largest number of geographically related cases of bat rabies virus infection in terrestrial mammals (Leslie *et al.*, 2006). Investigation of the outbreak demonstrated novel evolution and emergence of a rabies virus variant that was successfully adapted from chiroptera to carnivora. Accurate species identification is another unappreciated but necessary fundamental to understanding in rabies epizootiology (Dragoo *et al.*, 2004). A control effort through live-trapping, parenteral rabies vaccination and release, in combination with application of an experimental oral rabies vaccine, is underway in an effort to contain and potentially eliminate the emergence of this novel transmission pattern (Engeman *et al.*, 2003).

9.4 Coyote

The coyote, *Canis latrans*, is a highly adaptable and behaviorally variable 'mesocarnivore', which is actually quite omnivorous, exploiting both western rangelands and eastern suburbs. It occurs widely throughout North America, from Alaska towards Panama. With the near extirpation of the gray wolf, the coyote appears to be in a range expansion and more numerous than in the past (Vila *et al.*, 1999). Flexibility in social organization prevails, from solitary individuals with transient home ranges in excess of 50 sq km, to monogamous pair bonds and small packs.

Provided limitations imposed by predation and competition from wolf populations, it is doubtful if the coyote doubled as an adequate reservoir for rabies in times past. Few accounts in the New World suggest any major problem, prior to reports from North America in the 20th century. For example, during 1952–54, a large rabies outbreak in Alberta, Canada, involved foxes, wolves and coyotes, but the primary role of the latter species is questionable. In the USA, during 1915–1917, coyotes were involved in an extensive epizootic that extended over portions of California, Oregon, Nevada and Utah (Humphrey, 1971). From California alone in this period, records from the state public health department confirmed infection via laboratory examination in at least 94 coyotes, 64 cattle, 31 dogs, eight sheep, six horses, three bobcats, one cat, one goat and one human. These records only serve to underscore the magnitude of the outbreak, considering the hundreds of miles from the field sites to the laboratory. In the same period, at least 192 rabid coyotes were diagnosed in Nevada. Trapping and poisoning campaigns ensued, resulting in the destruction of thousands of coyotes and dogs and hundreds of other species representatives, and the epizootic eventually abated, even if the enzootic focus did not.

Were coyotes the reservoir that infected dogs and other species, or were rabid dogs the instigation that eventually spilled over to coyotes? Some believe that rabies in coyotes was present in Oregon as early as 1910, but this does not explain adequately how or why (Mallory, 1915). Others insinuate '...the disease gradually spread, traveling northward through California and being introduced into Oregon in 1912 by a sheep dog taken across the mountains from Redding, California, to Wallowa County in that state, where this infected dog in a fight with a coyote, first introduced the disease...' (Records, 1932). As with many rabies tales, the chicken or egg origin to this outbreak (like others) cannot be resolved easily, but does point out the intrinsic historical relationship between poorly supervised, unvaccinated dogs and wildlife disease. Similarly, coyotes in northern Baja were believed to act as vehicles behind the persistent infections that started during 1958 along the California and Mexico borders, a thought related in part to long-distance dispersal (Humphrey, 1971). Neither has absolute numbers of laboratory confirmed cases in coyotes throughout the USA, nor geographical spread ever again reached the extent exemplified by the 1915 western states outbreak. Nevertheless, its message should have prepared public health professionals for a repeat lesson more than 70 years later.

Rabies cases in coyotes were quite few and only sporadically reported in the USA from 1960 through the mid-1980s. For example, a Sonora canine rabies virus variant would occasionally be detected in animals along the west Texas border with Mexico (Rhode et al., 1997). This situation began to change slowly at a focus near the south Texas–Mexico border, associated with another rabies virus variant known, at least, from the region since 1978, in coyotes and domestic dogs (Clark et al., 1994). During 1988, a south Texas county reported six confirmed cases of rabies in coyotes and two cases in dogs. At the same time, an adjacent county reported nine cases of rabid dogs. During 1989–90, seven rabid coyotes and 65 dogs were reported in these areas. By 1991, the outbreak expanded approximately 160 km north, with a total of 42 rabid coyotes and 25 dogs over 10 counties. In 1992, it rose to 70 rabid coyotes and 41 dogs from a 12-county area and, by 1993, 71 of the 74 total cases in coyotes and 42 of 130 total cases in dogs reported from the entire USA were from south Texas. By comparison, that year no other state reported more than seven cases in dogs. The risk of artificial spread to other areas was realized during 1993, by identification of the coyote rabies virus variant from a dog infected on a compound in Alabama, where imported coyotes from Texas were released for hunting purposes (Krebs et al., 1994). Over some 18 counties in 1994, the number of coyote rabies cases reached 77, with 32 cases in dogs, and peaked at 80 rabid coyotes; there were 36 cases in dogs in 20 counties during 1995, when an oral vaccination program began to halt the progression of the disease (Fearneyhough et al., 1998). Unfortunately, as in Alabama previously, translocation of coyote rabies happened again, this time from Texas to Florida (Centers for Disease Control and Prevention, 1995). During November and December 1994, rabies was diagnosed in five dogs from two associated kennels in Florida. In addition, two other dogs at one of the kennels died with suspected, but unconfirmed, rabies. The rabies virus recovered from these dogs was identified as a rabies virus variant not previously found in Florida, but rather the same virus that was enzootic among coyotes in south Texas. The suspected source of infection was translocation of infected coyotes from Texas to Florida, also used in hunting enclosures. Luckily, cases of rabies in coyotes at the Texas nidus continued to decline each year from 1996 to 1999, with 19, four, four and two reports, respectively. With the exception of this variant identified recently in a rabid dog, most likely translocated into the USA from Mexico, there have been no further isolations of the variant in the USA. With ongoing oral vaccination efforts in southern Texas, by all accounts this particular variant has been extirpated from coyotes in the USA and now resides in a nidus in Mexico (Sidwa *et al.*, 2005; Velasco-Villa *et al.*, 2005).

At least two human cases were associated with the Texas coyote rabies outbreak, in 1991 and 1994, but the history surrounding each exposure is unclear. With the elimination of dog-to-dog transmission in Canada and the USA, this recent coyote rabies saga and the subsequent re-emergence of canine rabies, should once again impart the sense of wildlife's role in jeopardizing this rather fragile public health success story. Yet, besides the data gained from historical surveillance reports, few research studies concentrate on rabies in coyotes, beyond a demonstration of their basic susceptibility to bat rabies virus, aerosol infection or virological curiosities (Constantine, 1966a, 1966b, 1966c; Behymer *et al.*, 1974). Explanations as to the dearth of knowledge surrounding rabies in coyotes may be best summed up by the following past opinion: '...although they are a potential hazard as a reservoir or vector of rabies, they do not appear to be of major epidemiologic significance...' (Sikes and Tierkel, 1966).

During the 1990s, limited studies were initiated at the Centers for Disease Control and Prevention to elucidate the potential role of coyotes in rabies epizootiology and in response to the need for the development of an oral vaccine in coyotes as part of a potential control method. In the search for an appropriate rabies virus challenge of vaccinated animals, the salivary glands from 43 naturally infected coyotes from Texas were individually homogenized and the concentrations of rabies virus in each were determined. Most glands contained more than 5 logs of rabies virus, at a minimum, despite potential viral deterioration from the time of death in the field until the period of harvest in the laboratory. Adult coyotes of both sexes were captured and maintained in captivity. They were then inoculated with four serial 10-fold dilutions of a homogenized salivary gland from one of the naturally infected rabid coyotes. The virus isolated was representative of the south Texas canine rabies virus variant. At the higher concentrations, all exposed coyotes succumbed to rabies; 80% of animals developed fatal illness when exposed to at least 3.3 logs of rabies virus. Incubation periods ranged from 10 to 26 days, with a suggestion of an inverse relationship to infectious dose. Frozen sections of salivary glands obtained at necropsy of infected coyotes were examined by immunofluorescent microscopy. Whereas all five sections obtained from animals inoculated with at least 4.3 logs of virus contained evidence of rabies virus antigen, only two of five and two of four samples were positive from coyotes inoculated with a higher (5.3) or lower (3.3) concentration of virus, respectively. Clinical signs were characteristic of the paralytic form of the disease and included altered appetite, depression, confusion, anisocoria, excessive salivation, ataxia and paresis. Only a single animal exhibited aggressive signs and charged its cage at the sight of animal handlers. Morbidity periods were typically 3–4 days. Based on these limited findings, coyotes appear quite susceptible to this particular rabies virus variant (as do domestic dogs). Such data suffer the limitations of all experimental studies and field outcomes are dependent, in part, upon the quantity of virus excreted in the saliva over time and the manner in which coyotes actually infect one another. On a minor note, experimental utilization of this canid rabies virus in coyotes at the Centers for Disease Control and Prevention did lead to an unanticipated and unprecedented case of non-bite transmission to a laboratory beagle (Rupprecht *et al.*, 1994). This event re-emphasized the volatile mix especially implicit with certain lyssaviruses, hosts and environmental situations and the danger inherent in cavalier attempts to predict the future when surrounded by profound unknowns.

9.5 Mongoose

As elsewhere in the world, among African Carnivora, the dog figures prominently as the major rabies reservoir to humans and other animals (Bingham, 2005). Besides domestic dogs, other important canines include jackals and bat-eared foxes (von Teichman *et al.*, 1995). Still, one other major type of rabies virus in sub-Saharan Africa is specifically adapted to a variety of mongooses, belonging to the family Herpestidae. For example, Bingham *et al.* (2001) reported typical rabies viruses isolated in Zimbabwe from honey badgers, civets and unidentified mongoose species. Such cases were considered representative of an infrequently reported variant, otherwise maintained by the slender mongoose (*Galerella sanguinea*). In South Africa, the yellow mongoose, *Cynictis penicillata*, is a major reservoir (von Teichman *et al.*, 1995). Not unexpectedly, canine and mongoose viruses may have different effects upon their primary hosts. During one laboratory study, a significantly higher proportion of *Cynictis* inoculated with yellow mongoose rabies virus died, compared with those inoculated with a dog isolate (Chaparro and Esterhuysen, 1993). Moreover, the levels of rabies virus in the saliva and salivary glands were high in all clinically affected animals infected with the mongoose isolate, but only one of two mongooses, which died following inoculation of the dog isolate, contained detectable levels of virus in the salivary glands. In both Zimbabwe and South Africa, data arising from molecular epidemiological studies of mongoose rabies indicate a history of extended evolutionary adaptation of this lineage quite distinct from canine rabies (Nel *et al.*, 2005). Cycles of rabies among African mongoose are distinct from those found in parts of Asia. For example, in Sri Lanka, brown and ruddy mongooses are believed to be reservoirs (Patabendige and Wimalaratne, 2003). Viruses associated with Asian mongoose, translocated to the Caribbean during the 20th century, are phylogenetically distant from viruses circulating in mongooses in other parts of the world (Nadin-Davis *et al.*, 2006a).

Given the diversity of African wildlife, it seems surprising that dogs and mongooses may be the major drivers of regional epidemiology. However, recently, East *et al.* (2001) reported a 37.0% frequency of exposure to rabies virus in spotted hyenas from Tanzania, with intermittent detection of virus in saliva and, unlike other reports from the region, rabies virus isolates differed significantly from those found in other Serengeti carnivores. Reports of a high seroprevalence of rabies VNA, few fatal cases, suspicions of a 'carrier' state and maintenance of novel, multiple strains of virus in hyena, have not been corroborated to date. Regardless of the regional sources, rabies spillover infections to endangered species, such as African wild dogs or the highly threatened Ethiopian wolf, can result in dramatic consequences for populations already decimated by human encroachment (Randall *et al.*, 2004).

10 OTHER ANIMALS AS RESERVOIRS OR SIMPLY SPILLOVER HOSTS?

Considering the existence of several robust rabies reservoirs among multiple carnivore examples (e.g. fox, raccoon, skunk, mongoose), the absence of other reservoirs may at first appear puzzling. Yet, evolutionary bottlenecks and recent constraints imposed by humans may limit the ability of various species to serve in an ideal host capacity as animal reservoirs. For example, the canids are one of the oldest extant taxa in the group, widely distributed and usually small to moderate in size. Many are fairly solitary, but even more social examples (e.g. side-striped jackal, African wild dog, bush dog, dhole [a species of wild dog from southern Asia] and small-eared dog) have experienced a recent lowering of population densities and extreme habitat fragmentation (Gascoyne *et al.*, 1993). Despite current debate over their role as reservoirs or maintenance hosts, jackals are important vectors in many places throughout the Middle East and southern Africa (David *et al.*, 2000; Sabeta *et al.*, 2003). The introduction of rabies into this already complicated scene can have further deleterious effects on threatened and endangered species, as has been observed with both the gray and the Ethiopian wolf (Chapman, 1978; Theberge *et al.*, 1994; Weiler *et al.*, 1995; Sillero-Zubiri *et al.*, 1996; Ballard and Krausman, 1997). No doubt, occasionally, these can act as effective vectors, as seems to be the case for blacked-backed and golden jackals in parts of Africa and Asia, almost akin to the New World coyote, but not often as stable reservoirs (Rhodes *et al.*, 1998). Indeed, historically, throughout temperate portions of the Old World, humans attacked by rabid wolves represent some of the highest fatality rates per event on record (Shah and Jaswal, 1976; Bahmanyar *et al.*, 1976; Selimov *et al.*, 1978; Ianshin *et al.*, 1979; Butzeck, 1987). Nevertheless, it is quite unlikely that these large canids can serve as adequate, long-term reservoirs. No unique rabies virus variants have been acquired from wolves and those that have been typed appear to be of fox origin.

Rabies has been reported extensively from raccoon dogs (*Nyctereutes pro-cynonoides*), particularly in Eastern Europe and the Baltic region. They are highly adaptable canids that originate from eastern Asia, but were raised for fur and released in the former Soviet Union between 1928 and 1955. Individuals gradually dispersed westwards, after their original purposeful human translocation (Cherkasskiy, 1988). Little is known of the direct role they may play in perpetuation of rabies, independent of the red fox. It is questionable if they could sustain rabies for an indefinite period. Recently, characterization of rabies virus isolates from Brazil revealed the likelihood of the hoary fox as a rabies reservoir as only one part of a complexity of wildlife reservoirs (Bernardi *et al.*, 2005, Carnieli *et al.*, 2006).

For most hoofed stock, wild or domestic, rabies usually affects individual animals and results in a dead-end infection. One exotic bovid species may be an exception to this rule. The kudu (*Tragelaphus strepsiceros*) is a rather spectacular antelope that occurs widely throughout southern Africa. Because it is desirable to obtain a trophy animal with large, spiral horns, millions of dollars of revenue are generated each year from the kudu hunting industry. In response to this marketability, kudu may commonly be reared on game ranches under unnaturally dense conditions. An outbreak of rabies in kudu was reported in central Namibia in the late 1970s, apparently involving oral spread of infection between individual kudus. It peaked in 1980 and eventually subsided in 1985, by which time it was suggested that rabies caused a loss of approximately 50 000 animals or 20% per cent of the estimated population. Such incidents appear to be rather sporadic over time. For example, during 2002–03, there was another substantial outbreak in which an estimated 2500 animals died on scores of affected farms. Clinical signs among infected kudu included unusual tameness and, sometimes, wandering into human dwellings; excess salivation; loss of appetite; and frequent urination. Thereafter, rabid animals became aggressive or paralyzed. The unusual spread of rabies may be related to kudu social behavior and artificially high numbers of animals on the game ranches with subsequent overcrowding at foraging areas and watering holes. Some investigators of these outbreaks suggested that oral lesions produced from browsing upon thorn bushes may have been a potentiating factor in animal-to-animal transmission. It has been postulated that thorn bushes may have been readily contaminated with fresh saliva, possibly containing infectious doses of rabies virus, from clinically infected herd mates. Nearly immediate browsing of these contaminated areas by an uninfected herd mate while sustaining oral lesions from the thorns may have resulted in horizontal spread between animals, perhaps providing an example of non-bite transmission among non-reservoir conspecifics. Recent studies of rabies virus isolates from rabid kudu originating in Namibia during 1980–2003 suggested that these variants were all associated with canine rabies of southern Africa, closely associated with jackal, bat-eared fox and domestic dog viruses (Mansfield *et al.*, 2006a).

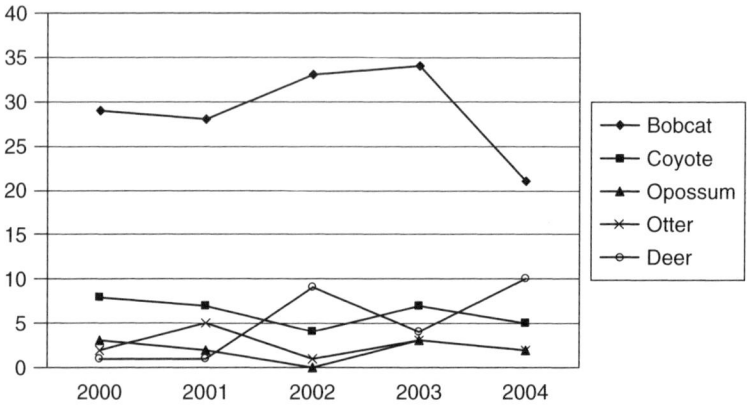

Figure 5.8 Rabies cases in selected wildlife species in the USA.

As with the restrictions on the Canidae, the same could be said for many of the felids, the quintessential predators and one of the most specialized groups in the Carnivora. With few exceptions, they exploit a rather solitary existence. Acquisition of a large body size in some groups (e.g. tiger, panther, jaguar, leopard) has added utility in hunting hoofed stock, but tends to limit absolute numbers. Effective depredations upon humans and their domestic animals have resulted in severe population reductions of most wild cat taxa. Nevertheless, rabid felids can be effective vectors. For example, after raccoons, skunks and foxes, the bobcat is the most common wild carnivore found with rabies in the USA (Figure 5.8) and human cases have resulted from exposure to these wild felids. To date, no reservoirs have been described for any feline species. Similar comments could be made about the bears, hyenas and others (Krebs *et al.*, 2003b). An entirely aquatic existence offered by the cetaceans and other marine mammals such as the dugong and manatee, fashion a rather implausible risk of likely rabies exposure. Many pinnipeds (i.e. seals, sea lions and their relatives) form huge social aggregations during the reproductive season and are thus at risk for exposure from both infected terrestrial carnivores (such as foxes and skunks) and some bats (such as the vampires), but relatively few rabid specimens have been reported thus far. Besides inadequate veterinary surveillance at remote rookeries, the insulation offered by blubber, potential frequent bathing of lesions in seawater and deaths in the marine environment from debilitation or active predation, are educated guesses as to their absence in rabies testing and statistics.

When one scans the published global literature and surveillance reports from Eurasia, Africa and the Americas, one is impressed with the zoological array of cases that have been reported, diverse enough to fill a child's alphabet book. For example, these involve the aardwolf, armadillo, baboon, badger, bison, camel, caracal, chipmunk, civet, duiker, elephant, fox squirrel, genet, honey badger, hyena, hyrax, *Ictonyx*, javelina, kudu, llama, lion, marmoset, *Nasua*, ocelot, opossum, otter, polecat, pig, rabbit, rat, reindeer, roe deer, seal, springbok, suricate, *Taurotragus*, ursids, vole, warthog, weasel, wildcat, *Xerus*, yak, zebra and a host of others (Frye and Cucuel, 1968; Cappucci *et al.*, 1972; Rausch, 1975; Dieterich and Ritter, 1982; Dowda and DiSalvo, 1984; Leffingwell and Neill, 1989; Centers for Disease Control and Prevention, 1990; Taylor *et al.*, 1991; Berry, 1993; Swanepoel *et al.*, 1993; Walroth *et al.*, 1996; Karp *et al.*, 1999; Wimalaratne and Kodidara, 1999; Batista-Morais *et al.*, 2000; Stoltenow *et al.*, 2000). After speculating on the question of diagnostic fidelity and the potential for cross-contamination, bona fide cases do support the contention that practically all mammals are susceptible to rabies within the limitations of niche availability to the virus and adequate human sampling. Almost all occur as single, incidental observations. However, multiple occurrences in a herd, pack, etc., can occur with dramatic finality as illustrated by the thousands of cases reported from kudu (Barnard *et al.*, 1982; Hubschle, 1988), probably enhanced by artificial propagation in game ranching. The tally sheet, however, is simply a record of victims, dead-end infections exposed by those individuals with a combination of evolutionary and ecological attributes that serve as unfortunate host to one of the oldest infectious diseases. Will this always be the case and will new reservoirs arise? As suggested by the bats and their multitude of lyssavirus variants, are there new examples yet to be discovered among other small mammals (or non-mammals)?

11 CONCLUSIONS

Many questions remain unanswered in the search for an advanced understanding of animal rabies. Key facets of basic mammalian susceptibility to rabies still require definition. Adequate explanations of potential long incubation periods in animals do not yet exist. The site and manner of virus sequestration during this period – within or outside the nervous system – await description. Factors that may predispose a neurotropic virus to non-neural sites and the conditions, albeit very rarely, that may drive towards viremia, need identification (Lodmell *et al.*, 2006). A convincing neurochemical mechanism for altered behavior during rabies has not been made. The importance of the non-bite route, including transplacental or aerosol transmission and body fluids, such as milk, is unknown. Coherent algorithms should be able to predict when a productive fatal infection is likely to result versus no evidence of infection but instead acquired immunity from an exposure. A melding of applied ecology and virology may permit the elucidation of common features that are important in the definition of an ideal rabies host. The influence of what appear to be minor genetic changes on phenotypic alterations should be explored, particularly as related to significant biological outcomes. Use of more realistic disease models could allow rabies' re-emergence to be anticipated and, perhaps, even prevented, a priori.

Clearly, a variety of factors influence the biological outcome of rabies exposure to an individual animal, towards the opportunity for primary establishment, spread and persistence in a population and its spillover effects on other populations. The ultimate response is dependent in part upon the virus and mammal in question, as well as individual variables, such as age, nutritional state, underlying disease burden, prior exposure and genetic make-up. Greater progress has been made in the last two decades than ever before in understanding primary rabies pathogenesis, with advancement in major areas that use molecular techniques, such as PCR, *in situ* hybridization and immunohistochemistry. Undoubtedly, additional features of animal rabies pathobiology also may be more fully elucidated, using lyssavirus street isolates of public health relevance, more natural routes of infection, variable exposures to differing concentrations of virus and a greater diversity of natural host species. Ultimately, such endeavors will yield a better understanding of rabies epizootiology such that advances may be made towards renewed control, prevention and, perhaps, continued progress towards extirpation of select rabies virus variants.

REFERENCES

Almeida, M.F., Massad, E., Aguiar, E.A., Martorelli, L.F. and Joppert, A.M. (2001). Neutralizing antirabies antibodies in urban terrestrial wildlife in Brazil. *Journal of Wildlife Diseases* **37**, 394–398.

Anderson, L.J., Williams, L.P. Jr, Layde, J.B., Dixon, F.R. and Winkler, W.G. (1984). Nosocomial rabies: investigation of contacts of human rabies cases associated with a corneal transplant. *American Journal of Public Health* **74**, 370–372.

Anonymous (1990). Compendium of animal rabies control. *Journal of the American Veterinary Medical Association* **196**, 36–39.

Aranda, M. and Lopez-de, B.L. (1999). Rabies in skunks from Mexico. *Journal of Wildlife Diseases* **35**, 574–577.

Artois, M. and Aubert, M.F.A. (1985). Behavior of rabid foxes. *Revue d'Ecologie* **4**, 171–176.

Aubert, M.F.A. (1992). Epidemiology of fox rabies. In: *Wildlife Rabies Control* (K. Bogel, F.-X. Meslin, M. Kaplan, eds). pp. 9–18. Kent: Wells Medical Ltd.

Badilla, X., Perez-Herra, V., Quiros, L. *et al.* (2003) Human rabies: a reemerging disease in Costa Rica? *Emerging Infectious Diseases* **9**, 21–23.

Baer, G.M. and Olson, H.R. (1972). Recovery of pigs from rabies. *Journal of the American Veterinary Medical Association* **160**, 1127–1128.

Baer, G.M., Cleary, W.F., Diaz, A.M. and Perl, D.F. (1977). Characteristics of 11 rabies virus isolates in mice: titers and relative invasiveness of virus, incubation period of infection, and survival of mice with sequelae. *Journal of Infectious Diseases* **136**, 336–345.

Bahmanyar, M., Fayaz, A., Nour-Salehi, S., Mohammadi, M. and Koprowski, H. (1976). Successful protection of humans exposed to rabies infection. Postexposure treatment with the new human diploid cell rabies vaccine and antirabies serum. *Journal of the American Medical Association* **236**, 2751–2754.

Ballard, W.B., Follmann, E.H., Ritter, D.G., Robards, M.D. and Cronin, M.A. (2001). Rabies and canine distemper in an arctic fox population in Alaska. *Journal of Wildlife Diseases* **37**, 133–137.

Ballard, W.B. and Krausman, P.R. (1997). Occurrence of rabies in wolves of Alaska. *Journal of Wildlife Diseases* **33**, 242–245.

Barnard, B.J., Hassel, R.H., Geyer, H.J. and De Koker, W.C. (1982). Non-bite transmission of rabies in kudu (*Tragelaphus strepsiceros*). *Onderstepoort Journal of Veterinary Research* **49**, 191–192.

Barnes, H.L., Chrisman, C.L., Farina, L. and Detrisac, C.J. (2003). Clinical evaluation of rabies virus meningoencephalomyelitis in a dog. *Journal of American Animal Hospital Association* **39**, 547–550.

Batista-Morais, N., Neilson-Rolim, B., Matos-Chaves, H.H., Brito-Neto, J. and Maria-da-Silva, L. (2000). Rabies in tamarins (*Callithrix jacchus*) in the State of Ceara, Brazil, a distinct viral variant? *Memorias do Instituto Oswaldo Cruz* **95**, 609–610.

Beamer, R.D., Mohr, C.O. and Barr, T.R.B. (1960). Resistance of the opossum to rabies virus. *American Journal of Veterinary Research* **21**, 507–510.

Behymer, D.E., Frye, F.L., Riemann, H.P., Franti, C.E. and Enright, J.B. (1974). Observations on the pathogenesis of rabies: experimental infection with a virus of coyote origin. *Journal of Wildlife Diseases* **10**, 197–203.

Bell, G.P. (1980). A possible case of interspecific transmission of rabies in insectivorous bats. *Journal of Mammology* **61**, 528–530.

Bergeron, J.A., Quinn, W.J. and Stackhouse, L.L. (1981). Rabies in a ewe and a 6-week-old lamb. *Modern Veterinary Practice* **62**, 784–785.

Bernardi, F., Nadin-Davis, S.A., Wandeler, A.I. *et al.* (2005). Antigenic and genetic characterization of rabies viruses isolated from domestic and wild animals of Brazil identifies the hoary fox as a rabies reservoir. *Journal of General Virology* **86**(Pt 11), 3153–3162.

Berry, H.H. (1993). Surveillance and control of anthrax and rabies in wild herbivores and carnivores in Namibia. *Revue Scientifique et Technique* **12**, 137–146.

Bigler, W.J., Hoff, G.L., Smith, J.S., McLean, R.G., Trevino, H.A. and Ingwersen, J. (1983). Persistence of rabies antibody in free-ranging raccoons. *Journal of Infectious Diseases* **148**, 610.

Bingham, J. (2005). Canine rabies ecology in southern Africa. *Emerging Infectious Diseases* **11**, 1337–1342.

Bingham, J., Javangwe, S., Sabeta, C.T., Wandeler, A.I. and Nel, L.H. (2001). Report of isolations of unusual lyssaviruses (rabies and Mokola virus) identified retrospectively from Zimbabwe. *J S Afr Vet Assoc* **72**, 92–4.

Black, D. and Wiktor, T.J. (1986). Survey of raccoon hunters for rabies antibody titers: pilot study. *Journal of the Florida Medical Association* **73**, 517–520.

Black, J.G. and Lawson, K.F. (1970). Sylvatic rabies studies in the silver fox (*Vulpes vulpes*). Susceptibility and immune response. *Canadian Journal of Comparative Medicine* **34**, 309–311.

Blancou, J.A., Aubert, M.F.A. and Artois, M. (1982). Rage experimentale du furet (*Mustela putorius furo*). *Revue de Medecine Veterinaire* **133**, 553–557.

Blancou, J., Aubert, M.F.A. and Artois, M. (1991). Fox rabies. In: *The Natural History of Rabies* (G.M. Baer, ed.). pp. 257–290. Boca Raton: CRC Press.

Bögel, K., Moegle, H., Knorpp, F., Arata, A., Dietz, K. and Diethelm, P. (1976). Characteristics of the spread of a wildlife rabies epidemic in Europe. *Bulletin of the World Health Organization* **54**, 433–447.

Brass, D. (1994). *Rabies in Bats*. p. 335. Ridgefield, Connecticut: Livia Press.

Butzeck, S. (1987). The wolf, *Canis lupus L.*, as a rabies vector in the 16th and 17th centuries. *Zeitschrift fur die gesamte Hygiene* **33**, 666–669.

Cappucci, D.T. Jr, Emmons, R.W. and Sampson, W.W. (1972). Rabies in an Eastern fox squirrel. *Journal of Wildlife Diseases* **8**, 340–342.

Carnieli, P. Jr, Brandao, P.E., Carrieri, M.L. *et al.* (2006). Molecular epidemiology of rabies virus strains isolated from wild canids in Northeastern Brazil. *Virus Research* **120**, 113–120.

Centers for Disease Control (1972). Rabies in a laboratory worker – Texas. *Morbidity Mortality Weekly Report* **21**, 113–124.

Centers for Disease Control (1977). Rabies in a laboratory worker – New York. *Morbidity Mortality Weekly Report* **26**, 183–184.

Centers for Disease Control and Prevention (1990). Rabies in a llama – Oklahoma. *Morbidity Mortality Weekly Report* **39**, 203–204.

Centers for Disease Control and Prevention (1995). Translocation of coyote rabies – Florida, 1994. *Morbidity Mortality Weekly Report* **44**, 580–587.

Centers for Disease Control and Prevention (1998). Compendium of Animal Rabies Control, 1998. National Association of State Public Health Veterinarians, Inc. *Morbidity Mortality Weekly Report* **47**(RR-9), 1–9.

Centers for Disease Control and Prevention (1999). Human rabies prevention – United States, 1999 recommendations of the Advisory Committee on Immunization Practices (ACIP). *Morbidity Mortality Weekly Report* **48**(RR-1), 1–21.

Centers for Disease Control and Prevention (2000a). Compendium of Animal Rabies Prevention and Control, 2000: National Association of State Public Health Veterinarians, Inc. *Morbidity Mortality Weekly Report* **49**, 21–30.

Centers for Disease Control and Prevention (2000b). Update: Raccoon rabies epizootic – United States and Canada, 1999. *Morbidity Mortality Weekly Report* **49**, 31–35.

Centers for Disease Control and Prevention (2002) Rabies in a beaver – Florida, 2001. *Morbidity Mortality Weekly Report* **51**, 481–482.

Chaparro, F., Esterhuysen, J.J. (1993) The role of the yellow mongoose (Cynictis penicillata) in the epidemiology of rabies in South Africa – preliminary results. *Onderstepoort Journal of Veterinary Research* **60**, 373–377.

Chapman, R.C. (1978). Rabies: decimation of a wolf pack in arctic Alaska. *Science* **201**, 365–367.

Charlton, K.M. and Casey, G.A. (1979). Experimental rabies in skunks: oral, nasal, tracheal and intestinal exposure. *Canadian Journal of Comparative Medicine* **43**, 168–172.

Charlton, K.M., Casey, G.A. and Webster, W.A. (1984). Rabies virus in the salivary glands and nasal mucosa of naturally infected skunks. *Canadian Journal of Comparative Medicine* **48**, 338–339.

Charlton, K.M., Casey, G.A. and Campbell, J.B. (1987). Experimental rabies in skunks: immune response and salivary gland infection. *Comparative Immunology, Microbiology, and Infectious Diseases* **10**, 227–235.

Charlton, K.M., Webster, W.A., Casey, G.A. and Rupprecht, C.E. (1988). Skunk rabies. *Reviews of Infectious Diseases* **10** (Suppl 4), S626–S628.

Charlton, K.M., Webster, W.A. and Casey, G.A. (1991). Skunk rabies. In: *The Natural History of Rabies* (G.M. Baer, ed.). pp. 307–324. Boca Raton: CRC Press.

Charlton, K.M. (1994). The pathogenesis of rabies and other lyssaviral infections; recent studies. *Current Topics in Microbiology and Immunology* **187**, 95–119.

Charlton, K.M., Casey, G.A., Wandeler, A.I. and Nadin-Davis, S. (1996). Early events in rabies virus infection of the central nervous system in skunks (Mephitis mephitis). *Acta Neuropathologica* **91**, 89–98.

Charlton, K.M., Nadin-Davis, S., Casey, G.A. and Wandeler, A.I. (1997). The long incubation period in rabies: Delayed progression of infection in muscle at the site of exposure. *Acta Neuropathologica* **94**, 73–77.

Cherkasskiy, B.L. (1988). Roles of the wolf and the raccoon dog in the ecology and epidemiology of rabies in the USSR. *Reviews of Infectious Diseases* **10** (Suppl 4), S634–S636.

Childs, J.E., Trimarchi, C.V. and Krebs, J. (1994). The epidemiology of bat rabies in New York State, 1988–92. *Epidemiology and Infection* **113**, 501–511.

Childs, J.E., Curns, A.T., Dey, M.E. *et al.* (2000) Predicting the local dynamics of epizootic rabies among raccoons in the United States. *Proceedings of the National Academy of Sciences USA* **97**, 13666–13671.

Childs, J.E., Colby, L., Krebs, J.W. *et al.* (1997). Surveillance and spatiotemporal associations of rabies in rodents and lagomorphs in the United States 1985–1994. *Journal of Wildlife Diseases* **33**, 20–27.

Clark, K.A., Neill, S.U., Smith, J.S., Wilson, P.J., Whadford, V.W. and McKirahan, G.W. (1994). Epizootic canine rabies transmitted by coyotes in south Texas. *Journal of the American Veterinary Medical Association* **204**, 536–540.

Coleman, P.G., Fevre, E.M. and Cleaveland, S. (2004) Estimating the public health impact of rabies. *Emerging Infectious Diseases* **10**, 140–142.

Constantine, D.G. (1962). Rabies transmission by the non-bite route. *Public Health Reports* **77**, 287–289.

Constantine, D.G. (1966a). Transmission experiments with bat rabies isolates: reaction of certain Carnivora, opossum, and bats to intramuscular inoculations of rabies virus isolated from free-tailed bats. *American Journal of Veterinary Research* **27**, 16–19.

Constantine, D.G. (1966b). Transmission experiments with bat rabies isolates: bite transmission of rabies to foxes and coyote by free-tailed bats. *American Journal of Veterinary Research* **27**, 20–23.

Constantine, D.G. (1966c). Transmission experiments with bat rabies isolates: responses of certain Carnivora to rabies virus isolated from animals infected by nonbite route. *American Journal of Veterinary Research* **27**, 13–15.

Crandell, R.A. (1991). Arctic fox rabies. In: *The Natural History of Rabies* (G.M. Baer, ed.). pp. 291–306. Boca Raton: CRC Press.

Crawford-Miksza, L.K., Wadford, D.A. and Schnurr, D.P. (1999). Molecular epidemiology of enzootic rabies in California. *Journal of Clinical Virology* **14**, 207–219.

Creel, S., Creel, N.M., Munson, L., Sanderlin, D. and Appel, M.J.G. (1997). Serosurvey for selected viral diseases and demography of African wild dogs in Tanzania. *Journal of Wildlife Diseases* **33**, 823–832.

David, D., Yakobson, B., Smith, J.S., Stram, Y. (2000). Molecular epidemiology of rabies virus isolates from Israel and other middle- and Near-Eastern countries. *Journal of Clinical Microbiology* **38**, 755–762.

Dean, D.J., Abelseth, M.K. and Atanasiu, D.P. (1996). The fluorescent antibody test. In: *Laboratory Techniques in Rabies* (F.X. Meslin, M.M. Kaplan and H. Koprowski, eds). pp. 88–95. Geneva: World Health Organization.

de Mattos, C., de Mattos, C., Loza-Rubio, E., Aguilar-Setien, A., Orciari, L. A. and Smith, J.S. (1999). Molecular characterization of rabies virus isolates from Mexico: Implications for transmission dynamics and human risk. *American Journal of Tropical Medicine and Hygiene* **61**, 587–597.

Delpietro, H.A. and Russo, R.G. (1996). Ecological and epidemiologic aspects of the attacks by vampire bats and paralytic rabies in Argentina and analysis of the proposals carried out for their control. *Revue Scientifique et Technique* **15**, 971–984.

Denduangboripant, J., Wacharapluesadee, S., Lumlertdacha, B. *et al.* (2005) Transmission dynamics of rabies virus in Thailand: implications for disease control. *BMC Infectious Diseases* **5**, 52.

Dieterich, R.A. and Ritter, D.G. (1982). Rabies in Alaskan reindeer. *Journal of the American Veterinary Medical Association* **181**, 1416.

Dietzschold, B., Rupprecht, C. E., Tollis, M. *et al.* (1988). Antigenic diversity of the glycoprotein and nucleocapsid proteins of rabies and rabies-related viruses: implications for epidemiology and control of rabies. *Reviews of Infectious Diseases* **10** (Suppl 4), S785–S798.

Dowda, H. and DiSalvo, A.F. (1984). Naturally acquired rabies in an eastern chipmunk (*Tamias striatus*). *Journal of Clinical Microbiology* **19**, 281–282.

Dragoo, J.W., Matthes, D.K., Aragon, A., Hass, C.C. and Terry, L.Y. (2004) Identification of skunk species submitted for rabies testing in the desert southwest. *Journal of Wildlife Diseases* **40**, 371–376.

East, M.L., Hofer, H., Cox, J.H., Wulle, U., Wiik, H. and Pitra, C. (2001) Regular exposure to rabies virus and lack of symptomatic disease in Serengeti spotted hyenas. *Proceedings of the National Academy of Sciences USA* **98**, 15 026–15 031.

Eidson, M., Matthews, S.D., Willsey, A.L., Cherry, B., Rudd, R.J. and Trimarchi C.V. (2005). Rabies virus infection in a pet guinea pig and seven pet rabbits. *Journal of the American Veterinary Medical Association* **227**, 932–935.

Eigen, M. and Biebricher, C. K. (1988). Sequence space and quasispecies distribution. In: *RNA Genetics, Vol. 3, Variability of RNA Genomes* (E. Domingo, J.J. Holland and P. Ahlquist, eds). pp. 211–245. Boca Raton: CRC Press.

Engeman, R.M., Christensen, K.L., Pipas, M.J. and Bergman, D. (2003). Population monitoring in support of a rabies vaccination program for skunks in Arizona. *Journal of Wildlife Diseases* **39**, 746–750.

Everard, C.O. and Everard, J.D. (1988). Mongoose rabies. *Reviews of Infectious Diseases* **10** (Suppl 4), S610–S614.

Fearneyhough, M.G., Wilson, P.J., Clark, K.A. *et al.* (1998). Results of an oral rabies vaccination program for coyotes. *Journal of the American Veterinary Medical Association* **212**, 498–502.

Fekadu, M. (1972). Atypical rabies in dogs in Ethiopia. *Ethiopian Medical Journal* **10**, 79–86.

Fekadu, M. (1991a). Canine rabies. In: *The Natural History of Rabies* (G.M. Baer, ed.). pp. 367–378. Boca Raton: CRC Press.

Fekadu, M. (1991b). Latency and aborted rabies. In: *The Natural History of Rabies* (G.M. Baer, ed.). pp. 191–198. Boca Raton: CRC Press.

Fekadu, M., Shaddock, J.H. and Baer, G.M. (1981). Intermittent excretion of rabies virus in the saliva of a dog two and six months after it had recovered from experimental rabies. *American Journal of Tropical Medicine and Hygiene* **30**, 1113–1115.

Fekadu, M., Shaddock, J.H. and Baer, G.M. (1982). Excretion of rabies virus in the saliva of dogs. *Journal of Infectious Diseases* **145**, 715–719.

Fekadu, M., Shaddock, J.H., Chandler, F.W. and Baer, G.M. (1983). Rabies virus in the tonsils of a carrier dog. *Archives of Virology* **78**, 37–47.

Fevre, E.M., Kaboyo, R.W., Persson, V., Edelsten, M., Coleman, P.G. and Cleaveland, S. (2005). The epidemiology of animal bite injuries in Uganda and projections of the burden of rabies. *Tropical Medicine and International Health* **10**, 790–798.

Flamand, A., Coulon, P., Lafay, F. and Tuffereau, C. (1993). Avirulent mutants of rabies virus and their use as live vaccine. *Trends in Microbiology* **1**, 317–320.

Foley, H.D., McGettigan, J.P., Siler, C.A., Dietzschold, B. and Schnell, M.J. (2000). A recombinant rabies virus expressing vesicular stomatitis virus glycoprotein fails to protect against rabies virus infection. *Proceedings of the National Academy of Sciences USA* **97**, 14 680–14 685.

Follmann, E.H., Ritter, D.G. and Beller, M. (1994). Survey of trappers in northern Alaska for rabies antibody. *Epidemiology and Infection* **113**, 137–141.

Frye, F.L. and Cucuel, J.P. (1968). Rabies in an ocelot. *Journal of the American Veterinary Medical Association* **153**, 789–790.

Gascoyne, S.C., King, A.A., Laurenson, M.K., Borner, M., Schildger, B. and Barrat, J. (1993). Aspects of rabies infection and control in the conservation of the African wild dog (Lycaon pictus) in the Serengeti region, Tanzania. *Onderstepoort Journal of Veterinary Research* **60**, 415–420.

George, J.P., George, J., Blancou, J. and Aubert, M.F.A. (1980) Description clinique de la rage du renard. *Revue de Medecine Veterinaire* **131**, 153–160.

Gordon, E.R., Curns, A.T., Krebs, J.W., Rupprecht, C.E., Real, L.A. and Childs, J.E. (2004). Temporal dynamics of rabies in a wildlife host and the risk of cross-species transmission. *Epidemiology and Infection* **132**, 515–524.

Greenwood, R.J., Newton, W.E., Pearson, G.L. and Schamber, G.J. (1997). Population and movement characteristics of radio-collared striped skunks in North Dakota during an epizootic of rabies. *Journal of Wildlife Diseases* **33**, 226–241.

Gremillion-Smith, C. and Woolf, A. (1988). Epizootiology of skunk rabies in North America. *Journal of Wildlife Diseases* **24**, 620–626.

Guerra, M.A., Curns, A.T., Rupprecht, C.E., Hanlon, C.A., Krebs, J.W. and Childs, J.E. (2003). Skunk and raccoon rabies in the eastern United States: temporal and spatial analysis. *Emerging Infectious Diseases* 9, 1143–1150.

Hamir, A.N., Moser, G. and Rupprecht, C.E., (1996). Clinicopathologic variation in raccoons infected with different street rabies virus isolates. *Journal of Veterinary Diagnostic Investigation* **8**, 31–37.

Hanlon, C.A. and Rupprecht, C.E. (1998) The reemergence of rabies. In: *Emerging Infections I.* pp. 59–80. Washington, DC: ASM Press.

Hanlon, C.A., Ziemer, E.L., Hamir, A.N. and Rupprecht, C.E. (1989). Cerebrospinal fluid analysis of rabid and vaccinia-rabies glycoprotein recombinant, orally vaccinated raccoons (*Procyon lotor*). *American Journal of Veterinary Research* **50**, 364–367.

Hanlon, C.A., Niezgoda, M., Morrill, P. and Rupprecht, C.E. (2002). Oral efficacy of an attenuated rabies virus vaccine in skunks and raccoons. *Journal of Wildlife Diseases* **38**, 420–427.

Hanlon, C.A., Kuzmin, I.V., Blanton, J.D., Weldon, W.C., Manangan, J.S. and Rupprecht, C.E. (2005). Efficacy of rabies biologics against new lyssaviruses from Eurasia. *Virus Research* **111**, 44–54.

Hatch, C., Sneddon, J. and Jalloh, G. (2004) A descriptive study of urban rabies during the civil war in Sierra Leone: 1995–2001. *Tropical Animal Health and Production* **36**, 321–334.

Hemachudha, T. (1994). Human rabies: clinical aspects, pathogenesis, and potential therapy. In: *Lyssaviruses* (C.E. Rupprecht, B. Dietzschold and H. Koprowski, eds). pp. 121–144. Berlin: Springer-Verlag.

Hellenbrand, W., Meyer, C., Rasch, G., Steffens, I. And Ammon, A. (2005). Cases of rabies in Germany following organ transplantation. *Euro Surveillance* **10**, E050224.6.

Hill, R.E. Jr and Beran, G.W. (1992). Experimental inoculation of raccoons (*Procyon lotor*) with rabies virus of skunk origin. *Journal of Wildlife Diseases* **28**, 51–56.

Hill, R.E. Jr, Beran, G.W. and Clark, W.R. (1992). Demonstration of rabies virus-specific antibody in the sera of free-ranging Iowa raccoons (*Procyon lotor*). *Journal of Wildlife Diseases* **28**, 377–385.

Hill, R.E. Jr, Smith, K.E., Beran, G.W. and Beard, P.D. (1993). Further studies on the susceptibility of raccoons (*Procyon lotor*) to a rabies virus of skunk origin and comparative susceptibility of striped skunks (Mephitis mephitis). *Journal of Wildlife Diseases* **29**, 475–477.

Howard, D.R. (1981a). Rabies virus tropism in naturally infected skunks (*Mephitis mephitis*). *American Journal of Veterinary Research* **42**, 2187–2190.

Howard, D.R. (1981b). Rabies virus titer from tissues of naturally infected skunks (*Mephitis mephitis*). *American Journal of Veterinary Research* **42**, 1595–1597.

Howard, D.R. (1981c). Transplacental transmission of rabies virus from a naturally infected skunk. *American Journal of Veterinary Research* **42**, 691–692.

Hubschle, O.J. (1988). Rabies in the kudu antelope (Tragelaphus strepsiceros). *Reviews of Infectious Diseases* **10** (Suppl 4), S629–S633.

Hudson, L.C., Weinstock, D., Jordan, T. and Bold-Fletcher, N.O. (1996a). Clinical features of experimentally induced rabies in cattle and sheep. *Journal of Veterinary Medicine Series B* **43**, 85–95.

Hudson, L.C., Weinstock, D., Jordan, T. and Bold-Fletcher, N.O. (1996b). Clinical presentation of experimentally induced rabies in horses. *Journal of Veterinary Medicine Series B* **43**, 277–285.

Hughes, G.J., Paez, A., Boshell, J. and Rupprecht, C.E. (2004) A phylogenetic reconstruction of the epidemiological history of canine rabies virus variants in Colombia. *Infection, Genetics and Evolution* **4**, 45–51.

Humphrey, G.L. (1971). Field control of animal rabies. In: *Rabies* (Y. Nagano, F.M. Davenport, eds). pp. 277–342. Baltimore: University Park Press.

Ianshin, IM., Voitanik, L.I., Ospanov, K.S. and Abdullin, B.K. (1979). Therapeutic and preventive vaccination of persons bitten by wolves in the Aktiubinsk region [in Russian]. *Zhurnal mikrobiologii, epidemiologii, i immunobiologii* **9**, 87–89.

Jackson, A.C. (1994). The fatal neurologic illness of the fourth Duke of Richmond in Canada: rabies. *Annals of the Royal College of Physicians and Surgeons of Canada* **27**, 40–41.

Jackson, A.C., Phelan, C.C. and Rossiter, J.P. (2000). Infection of Bergmann glia in the cerebellum of a skunk experimentally infected with street rabies virus. *Canadian Journal of Veterinary Research* **64**, 226–228.

Jay, M.T., Reilly, K.F., Debess, E.E., Haynes, E.H., Bader, D.R. and Barrett, L.R. (1994). Rabies in a vaccinated wolf-dog hybrid. *Journal of the American Veterinary Medical Association* **205**, 1719, 1729–1732.

Johnston, D.H. and Beauregard, M. (1969). Rabies epidemiology in Ontario. *Journal of Wildlife Diseases* **5**, 357–370.

Johnson, N., Black, C., Smith, J. *et al.* (2003) Rabies emergence among foxes in Turkey. *Journal of Wildlife Diseases* **39**, 262–270.

Johnson, R.H. (1995). Rabies vaccination of wolf-dog hybrids. *Journal of the American Veterinary Medical Association* **206**, 426–427.

Jorgenson, R.D. and Gough, P.M. (1976). Experimental rabies in a great horned owl. *Journal of Wildlife Diseases* **12**, 444–447.

Karp, B.E., Ball, N.E., Scott, C.R. and Walcoff, J.B. (1999). Rabies in two privately owned domestic rabbits. *Journal of the American Veterinary Medical Association* **215**, 1824–1827.

Kasempimolporn, S., Saengseesom, W., Lumlertdacha, B. and Sitprija, V. (2000). Detection of rabies virus antigen in dog saliva using a latex agglutination test. *Journal of Clinical Microbiology* **38**, 3098–3099.

Kasempimolporn, S., Saengseesom, W., Tirawatnapong, T., Puempumpanich, S. and Sitprija, V. (2004). Genetic typing of feline rabies virus isolated in greater Bangkok, Thailand. *Microbiology and Immunology* **48**, 307–311.

King, A.A., Meredith, C.D. and Thomson, G.R. (1994). The biology of southern African lyssavirus variants. *Curr Trop Microbiol Immunol* **187**, 267–95.

Knobel, D.L., Cleaveland, S., Coleman, P.G. *et al.* (2005) Re-evaluating the burden of rabies in Africa and Asia. *Bulletin of the World Health Organization* **83**, 360–368.

Krebs, J.W., Mandel, E.J., Swerdlow, D.L. and Rupprecht, C.E. (2005) Rabies surveillance in the United States during 2004. *Journal of the American Veterinary Medical Association* **227**, 1912–1925.

Krebs, J.W., Rupprecht, C.E. and Childs, J.E. (2000). Rabies surveillance in the United States during 1999. *Journal of the American Veterinary Medical Association* **217**, 1779–1811.

Krebs, J.W., Strine, T.W., Smith, J.S., Rupprecht, C.E. and Childs, J.E. (1994). Rabies surveillance in the United States during 1993. *Journal of the American Veterinary Medical Association* **205**, 1695–1709.

Krebs, J.W., Wheeling, J.T. and Childs, J.E. (2003a). Rabies surveillance in the United States during 2002. *Journal of the American Veterinary Medical Association* **223**, 1736–1748.

Krebs, J.W., Williams, S.M., Smith, J.S., Rupprecht, C.E. and Childs, J.E. (2003b). Rabies among infrequently reported mammalian carnivores in the United States, 1960–2000. *Journal of Wildlife Diseases* **39**, 253–261.

Kulonen, K. and Boldina, I. (1996). No rabies detected in voles and field mice in a rabies-endemic area. *Zentralblatt für Veterinärmedizin Reihe B* **43**, 445–447.

Kuzmin, I.V., Botvinkin, A.D., McElhinney, L.M. *et al.* (2004) Molecular epidemiology of terrestrial rabies in the former Soviet Union. *Journal of Wildlife Diseases* **40**, 617–631.

Kuzmin, I.V., Orciari, L.A., Arai, Y.T. *et al.* (2003) Bat lyssaviruses (Aravan and Khujand) from Central Asia: phylogenetic relationships according to N, P and G gene sequences. *Virus Research* **97**, 65–79.

Lafon, M. (1994). Immunobiology of lyssaviruses: The basis for immunoprotection. In: *Lyssaviruses* (C.E. Rupprecht, B. Dietzschold and H. Koprowski, eds). *Current Topics in Microbiology and Immunology*. pp.145–160. Berlin: Springer-Verlag.

Lafon, M. (2005). Modulation of the immune response in the nervous system by rabies virus. *Current Topics in Microbiology and Immunology* **289**, 239–258.

Leffingwell, L.M. and Neill, S.U. (1989). Naturally acquired rabies in an armadillo (Dasypus novemcinctus) in Texas. *Journal of Clinical Microbiology* **27**, 174–175.

Lembo, T., Niezgoda, M., Velasco-Villa, A., Cleaveland, S., Ernest, E. and Rupprecht, C.E. (2006). Evaluation of a direct, rapid immunohistochemical test for rabies diagnosis. *Emerging Infectious Diseases* **12**, 310–313.

Leslie, M.J., Messenger, S., Rohde, R.E. *et al.* (2006). Bat-associated rabies virus in skunks. *Emerging Infectious Diseases* **12**, 1274–1277. Available from http://www.cdc.gov/ncidod/EID/vol12no08/05-1526.htm

Lewis, V.J. and Thacker, W.L. (1974). Limitations of deteriorated tissue for rabies diagnosis. *Health Laboratory Science* **11**, 8–12.

Lodmell, D.L. (1983). Genetic control of resistance to street rabies virus in mice. *Journal of Experimental Medicine* **157**, 451–460.

Lodmell, D.L., Dimcheff, D.E. and Ewalt, L.C. (2006a). Viral RNA in the bloodstream suggests viremia occurs in clinically ill rabies-infected mice. *Virus Research* **116**, 114–118.

Lord, R.D. (1992). Seasonal reproduction of vampire bats and its relation to seasonality of bovine rabies. *Journal of Wildlife Diseases* **28**, 292–294.

Mallory, L.B. (1915). Campaign against rabies in Modoc and Lassen counties. *CA State Board of Health Monthly Bulletin* **11**, 273–277.

Mansfield, K.L., McElhinney, L.M., Hubschle, O. *et al.* (2006a) A molecular epidemiological study of rabies epizootics in kudu (*Tragelaphus strepsiceros*) in Namibia. *BMC Veterinary Research* **2**, 2.

Mansfield, K.L., Racloz, V., McElhinney, L.M. *et al.* (2006b) Molecular epidemiological study of Arctic rabies virus isolates from Greenland and comparison with isolates from throughout the Arctic and Baltic regions. *Virus Research* **116**, 1–10.

Martinez-Burnes, J., Lopez, A., Medellin, J., Haines, D., Loza, E. and Martinez, M. (1997). An outbreak of vampire bat-transmitted rabies in cattle in northeastern Mexico. *Canadian Veterinary Journal* **38**, 175–177.

McLean, R.G. (1975). Raccoon rabies. In: *The Natural History of Rabies* (G.M. Baer, ed.). pp. 53–76. New York: Academic Press.

McQuiston, J.H., Yager, P.A., Smith, J.S. and Rupprecht, C.E. (2001) Epidemiologic characteristics of rabies virus variants in dogs and cats in the United States, 1999. *Journal of the American Veterinary Medical Association* **218**, 1939–1942.

Moro, M.H., Horman, J.T., Fischman, H.R., Grigor, J.K. and Israel, E. (1991). The epidemiology of rodent and lagomorph rabies in Maryland, 1981 to 1986. *Journal of Wildlife Diseases* **27**, 452–456.

Mork, T. and Prestrud, P. (2004) Arctic rabies – a review. *Acta Veterinaria Scandinavica* **45**, 1–9.

Muirhead, T., McClure, J.T., Wichtel, J., McFarlane, D. and Lunn, D.P. (2006) Effect of age on systemic antibody response following rabies and influenza vaccinations in healthy horses. *Journal of Veterinary Internal Medicine* **20**, p. 718. Meeting Abstract American College of Veterinary Internal Medicine, University of Prince Edward Island, Atlantic Veterinary College, Charlottetown, Canada.

Murphy, F.A., Bell, J.F., Bauer, S.P. *et al.* (1980). Experimental chronic rabies in the cat. *Laboratory Investigation* **43**, 231–241.

Nadin-Davis, S.A., Torres, G., Ribas, M.D. *et al.* (2006a) A molecular epidemiological study of rabies in Cuba. *Epidemiology and Infection* **2**, 1–12.

Nadin-Davis, S.A., Muldoon, F. and Wandeler, A.I. (2006b). Persistence of genetic variants of the arctic fox strain of Rabies virus in southern Ontario. *Canadian Journal of Veterinary Research* **70**, 11–19.

Nair, S., Dighe, P.Y. and Nanavati, A.N. (1978). Role of bandicoots in rabies transmission. *Indian Journal of Medical Research* **67**, 347–353.

Nanayakkara, S., Smith, J.S. and Rupprecht, C.E. (2003). Rabies in Sri Lanka: splendid isolation. *Emerging Infectious Diseases* **9**, 368–371.

Nathanson, N. and Gonzalez-Scarano, F. (1991). Immune response to rabies virus. In: *The Natural History of Rabies* (G.M. Baer, ed.). pp. 145–161. Boca Raton: CRC Press.

Nel, L.H., Sabeta, C.T., von Teichman, B., Jaftha, J.B., Rupprecht, C.E. and Bingham, J. (2005). Mongoose rabies in southern Africa: a re-evaluation based on molecular epidemiology. *Virus Research* **109**, 165–173.

Nettles, V.F., Shaddock, J.H., Sikes, R.K. and Reyes, C.R. (1979). Rabies in translocated raccoons. *American Journal of Public Health* **69**, 601–602.

Nichol, S.T., Rowe, J.E. and Fitch, W.M. (1993). Punctuated equilibrium and positive Darwinian evolution in vesicular stomatitis virus. *Proceedings of the National Academy of Sciences USA* **90**, 10 424–10 428.

Niezgoda, M., Diehl, D., Hanlon, C.A. and Rupprecht, C.E. (1991). Pathogenesis of street rabies virus in raccoons. Wildlife Disease Association, 40th Annual Conference. Fort Collins, Colorado, pp. 57–58.

Niezgoda, M., Briggs, D.J., Shaddock, J., Dreesen, D.W. and Rupprecht, C.E. (1997). Pathogenesis of experimentally induced rabies in domestic ferrets. *American Journal of Veterinary Research* **58**, 1327–1331.

Niezgoda, M., Briggs, D.J., Shaddock, J., and Rupprecht, C.E. (1998). Viral excretion in domestic ferrets (*Mustela putorius furo*) inoculated with a raccoon rabies isolate. *American Journal of Veterinary Research* **59**, 1629–1632.

Niezgoda, M., Hanlon, C.A. and Rupprecht, C.E. (2002). Animal rabies. In: *Rabies* (A.C. Jackson and W.H. Wunner, eds). pp. 163–218. San Diego: Academic Press.

Noah, D.L., Smith, M.G., Gotthardt, J.C., Krebs, J.W., Green, D. and Childs, J.E. (1996). Mass human exposure to rabies in New Hampshire: exposures, treatment, and cost. *American Journal of Public Health* **86**, 1149–1151.

Nowak, A. (1991). *Walkers, Mammals of the World*. Baltimore: The Johns Hopkins University Press.

Odegaard, O.A. and Krogsrud, J. (1981). Rabies in Svalbard: infection diagnosed in arctic fox, reindeer and seal. *Veterinary Record* **109**, 141–142.

Orr, P.H., Rubin, M.R. and Aoki, F.Y. (1988). Naturally acquired serum rabies neutralizing antibody in a Canadian Inuit population. *Arctic Medical Research* **47**, 699–700.

Overall, K.L. (2000). Rabies vaccine labeling and the wolf hybrid. *Journal of the American Veterinary Medical Association* **216**, 20.

Páez, A., Saad, C., Nunez, C. and Boshell J. (2005). Molecular epidemiology of rabies in northern Colombia 1994–2003. Evidence for human and fox rabies associated with dogs. *Epidemiology and Infection* **133**, 529–536.

Parker, R.L. and Wilsnack, R.E. (1966). Pathogenesis of skunk rabies virus: quantitation in skunks and foxes. *American Journal of Veterinary Research* **27**, 33–38.

Parkinson, A.J. and Butler, J.C. (2005). Potential impacts of climate change on infectious diseases in the Arctic. *International Journal of Circumpolar Health* **64**, 478–486.

Patabendige, C.G. and Wimalaratne, O. (2003). Rabies in mongooses and domestic rats in the southern province of Sri Lanka. *Ceylon Medical Journal* **48**, 48–50.

Perl, D.P., Bell, J.F. and Moore, G.J. (1977). Chronic recrudescent rabies in a cat. *Proceedings of the Society for Experimental Biology and Medicine* **155**, 540–548.

Randall, D.A., Williams, S.D., Kuzmin, I.V. *et al.* (2004). Rabies in endangered Ethiopian wolves. *Emerging Infectious Diseases* **10**, 2214–2217.

Rausch, R.L. (1975). Rabies in experimentally infected bears, Ursus spp., with epizootiologic notes. *Zentralblatt für Veterinärmedizin Reihe B* **22**, 420–437.

Records, E. (1932). Rabies – its history in Nevada. *California and Western Medicine* **37**, 90–94.

Rhode, R.E., Neill, S.U., Clark, K.A. and Smith, J.S. (1997). Molecular epidemiology of rabies epizootics in Texas. *Clinical and Diagnostic Virology* **8**, 209–217.

Rhodes, C.J., Atkinson, R.P., Anderson, R.M. and Macdonald, D.W. (1998). Rabies in Zimbabwe: reservoir dogs and the implications for disease control. *Philosophical Transactions of the Royal Society of London Series B, Biological Sciences* **353**, 999–1010.

Rosatte, R.C. and Gunson, J.R. (1984). Presence of neutralizing antibodies to rabies virus in striped skunks from areas free of skunk rabies in Alberta. *Journal of Wildlife Diseases* **20**, 171–176.

Rottcher, D. and Sawchuk, A.M. (1978). Wildlife rabies in Zambia. *Journal of Wildlife Diseases* **14**, 513–517.

Rupprecht, C.E., Hamir, A.N., Johnston, D.H. and Koprowski, H. (1988). Efficacy of a vaccinia-rabies glycoprotein recombinant virus vaccine in raccoons (Procyon lotor). *Reviews of Infectious Diseases* **10** (Suppl 4), S803–S809.

Rupprecht, C.E., Gilbert, J., Pitts, R., Marshall, K.R. and Koprowski, H. (1990). Evaluation of an inactivated rabies virus vaccine in domestic ferrets. *Journal of the American Veterinary Medical Association* **196**, 1614–1616.

Rupprecht, C.E., Nesby, S., Fekadu, M. *et al.* (1994). When the 'impossible' happens: Non-bite rabies contamination in a laboratory beagle. *Proceedings of the V annual international meeting, Advances towards rabies control in the Americas*, p. 34. Niagara Falls, Ontario, Canada.

Sabeta, C.T., Bingham, J. and Nel, L.H. (2003) Molecular epidemiology of canid rabies in Zimbabwe and South Africa. *Virus Research* **91**, 203–211.

Sawchuk, A.M. and Rottcher, D. (1978). Mongoose rabies in Zambia. *Journal of Wildlife Diseases* **14**, 54–55.

Schneider, L.G. and Burtscher, H. (1967). Studies on the pathogenesis of rabies in fowls after intracerebral infection. *Zentralblatt für Veterinärmedizin Reihe B* **14**, 598–624.

Seganti, L., Superti, F., Bianchi, S., Orsi, N., Divizia, M. and Pan, A. (1990). Susceptibility of mammalian, avian, fish, and mosquito cell lines to rabies virus infection. *Acta Virologica* **34**, 155–163.

Selimov, M.A., Klyueva, E.V., Aksenova, T.A., Lebedeva, I.R. and Gribencha, L.F. (1978). Treatment of patients bitten by rabid or suspected rabid wolves with inactivated tissue culture rabies vaccine and rabies gammaglobulin. *Development in Biological Standards* **40**, 141–146.

Shah, U. and Jaswal, G.S. (1976). Victims of a rabid wolf in India: effect of severity and location of bites on development of rabies. *Journal of Infectious Diseases* **134**, 25–29.

Shannon, L.M., Poulton, J.L., Emmons, R.W., Woodie, J.D. and Fowler, M.E. (1988). Serological survey for rabies antibodies in raptors from California. *Journal of Wildlife Diseases* **24**, 264–267.

Shope, R. (1975). Rabies virus antigenic relationships. In: *The Natural History of Rabies* (G.M. Baer, ed.). Vol. I, pp. 141–152. New York: Academic Press.

Sidwa, T.J., Wilson, P.J., Moore, G.M. *et al.* (2005) Evaluation of oral rabies vaccination programs for control of rabies epizootics in coyotes and gray foxes: 1995–2003. *Journal of the American Veterinary Medical Association* **227**, 785–792.

Sikes, R.K. and Tierkel, E.S. (1961). Wildlife rabies studies in the southeast. *Proceedings of the 64th Annual Meeting of the US Livestock Sanitation Association,* Charleston, West Virginia, pp. 268–272.

Sikes, R.K. (1962). Pathogenesis of rabies in wildlife. I. Comparative effect of varying doses of rabies virus inoculated into foxes and skunks. *American Journal of Veterinary Research* **23**, 1041–1047.

Sikes, R.K. and Tierkel, E.S. (1966). Wolf, fox and coyote rabies. *Proceedings of the National Rabies Symposium.* pp. 31–33. Atlanta: US Department of Health, Education and Welfare. Public Health Service.

Sillero-Zubiri, C., King, A.A. and Macdonald, D.W. (1996). Rabies and mortality in Ethiopian wolves (Canis simensis). *Journal of Wildlife Diseases* **32**, 80–86.

Smart, N.L. and Charlton, K.M. (1992). The distribution of challenge virus standard rabies virus versus skunk street rabies virus in the brains of experimentally infected rabid skunks. *Acta Neuropathologica* **84**, 501–508.

Smith, J.S. (1996). New aspects of rabies with emphasis on epidemiology, diagnosis, and prevention of the disease in the United States. *Clinical Microbiology Reviews* **9**, 166–176.

Smith, J.S., Sumner, J.W., Roumillat, L.F., Baer, G.M. and Winkler, W.G. (1984). Antigenic characteristics of isolates associated with a new epizootic of raccoon rabies in the United States. *Journal of Infectious Diseases* **149**, 769–774.

Smith J.S. and Seidel H.D. (1993). Rabies: A new look at an old disease. *Progress in Medical Virology* **40**, 82–106.

Somayajulu, M.V. and Reddy. G.V. (1989). Live dogs and dead men. *Journal of the Association of Physicians of India* **37**, 617.

Srinivasan, A., Burton, E.C., Kuehnert, M.J. *et al.* (2005) Transmission of rabies virus from an organ donor to four transplant recipients. *New England Journal of Medicine* **352**, 1103–1111.

Steck, F. (1968). Zoonoses in Britain: some present and potential hazards to man. Rabies, the European situation. *Veterinary Record* **83** (Suppl 15).

Steck, F. and Wandeler, A. (1980). The epidemiology of fox rabies in Europe. *Epidemiologic Reviews* **2**, 71–96.

Steele, J.H. and Fernandez, P.J. (1991). History of rabies and global aspects. In: *The Natural History of Rabies* (G.M. Baer, ed.). pp. 1–24. Boca Raton: CRC Press.

Stoltenow, C.L., Shirely, L.A., Jones, T. and Rupprecht, C.E. (1999). Clinical report – Atypical rabies in a cow. *Bovine Practitioner* **33**, 4–5.

Stoltenow, C.L., Solemsass, K., Niezgoda, M., Yager, P. and Rupprecht, C.E. (2000). Rabies in an American bison from North Dakota. *Journal of Wildlife Diseases* **36**, 169–171.

Swanepoel, R., Barnard, B.J.H., Meredith, C.D. *et al.* (1993). Rabies in southern Africa. *Onderstepoort Journal of Veterinary Research* **60**, 325–346.

Tabel, H., Corner, A.H., Webster, W.A. and Casey, C.A. (1974). History and epizootiology of rabies in Canada. *Canadian Veterinary Journal* **15**, 271–281.

Tang, X., Luo, M., Zhang, S., Fooks, A.R., Hu, R. and Tu, C. (2005). Pivotal role of dogs in rabies transmission, China. *Emerging Infectious Diseases* **11**, 1970–1972.

Taylor, M., Elkin, B., Maier, N. and Bradley, M. (1991). Observation of a polar bear with rabies. *Journal of Wildlife Diseases* **27**, 337–339.

Tepsumethanon, V., Lumlertdacha, B., Mitmoonpitak, C., Sitprija, V., Meslin, F.X. and Wilde, H. (2004). Survival of naturally infected rabid dogs and cats. *Clinical Infectious Diseases* **39**, 278–280.

Tepsumethanon, V., Wilde, H. and Meslin, F.X. (2005). Six criteria for rabies diagnosis in living dogs. *Journal of the Medical Association of Thailand.* **88**, 419–422.

Theberge, J.B., Forbes, G.J., Barker, I.K. and Bollinger, T. (1994). Rabies in wolves of the Great Lakes region. *Journal of Wildlife Diseases* **30**, 563–566.

Thedford, T.R. and Johnson, L.W. (1989). Infectious diseases of New-World camelids (NWC). *Veterinary Clinics of North America. Food Animal Practice* **5**, 145–157.

Tinline, R., Rosatte, R. and MacInnes, C. (2002). Estimating the incubation period of raccoon rabies: a time-space clustering approach. *Preventive Veterinary Medicine* **56**, 89–103.

Toma, B. and Andral, L. (1977). Epidemiology of fox rabies. *Advances in Virus Research* **21**, 1–36.

Tordo, N. (1996). Characteristics and molecular biology of the rabies virus. In: *Laboratory Techniques in Rabies* (F.X. Meslin, M.M. Kaplan and H. Koprowski, eds). pp. 28–51. Geneva: World Health Organization.

Totton, S.C., Tinline, R.R., Rosatte, R.C. and Bigler, L.L. (2002) Contact rates of raccoons (*Procyon lotor*) at a communal feeding site in rural eastern Ontario. *Journal of Wildlife Diseases* **38**,13–19.

Trimarchi, C.V. and Debbie, J. (1977). Naturally occurring rabies virus and neutralizing antibody in two species of insectivorous bats of New York State. *Journal of Wildlife Diseases* **13**, 366–369.

Trimarchi, C.V., Rudd, R.J. and Abelseth, M.K. (1986). Experimentally induced rabies in four cats inoculated with a rabies virus isolated from a bat. *American Journal of Veterinary Research* **47**, 777–780.

Vaughn, J.B., Gerhardt, P. and Paterson, J. (1963). Excretion of street rabies virus in saliva of cats. *Journal of the American Medical Association* **184**, 705.

Vaughn, J.B., Gerhardt, P. and Newell, K.W. (1965). Excretion of street rabies virus in the saliva of dogs. *Journal of the American Medical Association* **193**, 363–368.

Velasco-Villa, A., Orciari, L.A., Souza, V. *et al.* (2005) Molecular epizootiology of rabies associated with terrestrial carnivores in Mexico. *Virus Research* **111**, 13–27.

Vila, C., Amorim, I. R., Leonard, J.A. *et al.* (1999). Mitochondrial DNA phylogeography and population history of the grey wolf *canis lupus. Molecular Ecology* **8**, 2089–2103.

von Teichman, B.F., Thomson, G.R., Meredith, C.D. and Nel, L.H. (1995). Molecular epidemiology of rabies virus in South Africa: evidence for two distinct virus groups. *Journal of General Virology* **76** (Pt 1), 73–82.

Walroth, R., Brown, N., Wandeler, A., Casey, A. and Macinnes, C. (1996). Rabid black bears in Ontario. *Canadian Veterinary Journal* **37**, 492.

Wandeler, A., Muller, J., Wachendorfer, G., Schale, W., Forster, U. and Steck, F. (1974). Rabies in wild carnivores in central Europe. III. Ecology and biology of the fox in relation to control operations. *Zentralblatt für Veterinärmedizin Reihe B* **21**, 765–773.

Wandeler, A.I. (1980). Epidemiology of fox rabies. In: *The Red Fox* (E. Zimen, ed.). pp. 237–249. Hingham: Kluwer Boston Inc.

Wandeler, A.I., Nadin-Davis, S.A., Tinline, R.R. and Rupprecht, C.E. (1994). Rabies epizootiology: an ecological and evolutionary perspective. *Current Topics in Microbiology and Immunology* **186**, 297–324.

Weiler, G.J., Garner, G.W. and Ritter, D.G. (1995). Occurrence of rabies in a wolf population in northeastern Alaska. *Journal of Wildlife Diseases* **31**, 79–82.

Whitby, J.E., Johnstone, P. and Sillero-Zubiri, C. (1997). Rabies virus in the decomposed brain of an Ethiopian wolf detected by nested reverse transcription-polymerase chain reaction. *Journal of Wildlife Diseases* **4**, 912–915.

WHO Expert Consultation on rabies (2005). *World Health Organization Technical Report Series* **931**, 1–88.

Wiktor, T.J. and Koprowski, H. (1978). Monoclonal antibodies against rabies virus produced by somatic cell hybridization: detection of antigenic variants. *Proceedings of the National Academy of Sciences USA* **75**, 3938–3942.

Wilbur, L.A. and Aubert, M.F.A. (1996). The NIH test for potency. In: *Laboratory Techniques in Rabies*, 4th edn (F.-X. Meslin, M.M. Kaplan and H. Koprowski, eds). pp. 360–368. Geneva; World Health Organization.

Wilkinson, L. (1988). Understanding the nature of rabies: an historical perspective. In: *Rabies* (J.B. Campbell and K.M. Charlton, eds). pp. 1–23. Boston: Kluwer Academic Publishers.

Wimalaratne, O. (1997) Is it necessary to give rabies post-exposure treatment after rodent (rats, mice, squirrels and bandicoots) bites? *Ceylon Medical Journal* **42**, 144.

Wimalaratne, O. and Kodikara, D.S. (1999). First reported case of elephant rabies in Sri Lanka. *Veterinary Record* **144**, 98.

Windiyaningsih, C., Wilde, H., Meslin, F.X., Suroso, T. and Widarso, H.S. (2004). The rabies epidemic on Flores Island, Indonesia (1998–2003). *Journal of the Medical Association of Thailand* **87**, 1389–1393.

Winkler, W.G. (1975). Fox rabies. In: *The Natural History of Rabies*, 2nd edn (G.M. Baer, ed.). pp. 3–22. New York: Academic Press.

Winkler, W.G., Baker, E.R. and Hopkins, C.C. (1972). An outbreak of non-bite transmitted rabies in a laboratory animal colony. *American Journal of Epidemiology* **95**, 267–277.

Winkler, W.G., Shaddock, J.S. and Bowman, C. (1985). Rabies virus in salivary glands of raccoons (Procyon lotor). *Journal of Wildlife Diseases* **21**, 297–298.

Winkler, W.G. and Jenkins, S.R. (1991). Raccoon rabies. In: *The Natural History of Rabies* (G.M. Baer, ed.). pp. 325–340. Boca Raton: CRC Press.

Xiang, Z.Q., Knowles, B.B., McCarrick, J.W. and Ertl, H.C.J. (1995). Immune effector mechanisms required for protection to rabies virus. *Virology* **214**, 398–404.

Yamaoka, H. (1962). Rabies virus inoculated into the quail. *Annales Paediatrici Japonici* **8**, 92–96.

6

Bat Rabies

IVAN V. KUZMIN AND CHARLES E. RUPPRECHT

Centers for Disease Control and Prevention, Atlanta, GA 30333, USA

1 INTRODUCTION – GENERAL CONSIDERATIONS ON BAT RABIES

Among mammals, bats (order Chiroptera) are second only to rodents in their number of species (approximately 1000). The order is subdivided into two suborders: flying foxes (Megachiroptera) that inhabit the Old World tropics and the smaller-bodied bats (Microchiroptera) that are distributed globally, except in Antarctica (Nowak, 1999). Rabies is described in representatives of both suborders on all continents. Given the issue of size alone, one can expect a greater absolute number of rabid bats to account for disease prevalence, as well as a greater density of the species occupying any given space, compared with larger-bodied carnivores.

In fact, bats maintain circulation of all known lyssavirus genotypes, or viral species, as recognized by the International Committee on Virus Taxonomy (ICTV) (Tordo *et al.*, 2005) except Mokola virus (MOKV, genotype 3). However, the ecology of MOKV has not been sufficiently studied and it is not possible to determine a principal host range or the circulation properties of the virus. In the New World, bats maintain the circulation of different lineages of the rabies virus (RABV, genotype 1). In the Old World, bats maintain circulation of such non-RABV lyssaviruses as Lagos bat virus (LBV, genotype 2), Duvenhage virus (DUVV, genotype 4), European bat lyssavirus, type 1 (EBLV-1, genotype 5) and type 2 (EBLV-2, genotype 6) and Australian bat lyssavirus (ABLV, genotype 7). Recently discovered lyssaviruses, Aravan (ARAV), Khujand (KHUV), Irkut (IRKV) and West Caucasian bat virus (WCBV) were also isolated from bats (Kuzmin *et al.*, 2003, 2005). Circulation of RABV among bats in the Old World was repeatedly suggested but not confirmed (Kuzmin *et al.*, 2006a).

Considering the variety of divergent lyssaviruses, the order Chiroptera was proposed as being primarily affected during adaptation of plant and arthropod rhabdoviruses to mammalian hosts (Shope *et al.*, 1970; Shope, 1982). Presumably, adaptation occurred somewhere in the Old World, perhaps in Africa or Eurasia, where most divergent non-RABV lyssaviruses are found.

According to the hypothesis of Badrane and Tordo (2001), inferred from phylogenetic reconstructions, there were at least two major switches in lyssavirus history, both from bats to the terrestrial mammals. The first switch occurred on the inter-genotype level and the second one occurred within genotype 1, when RABV switched from bats to carnivores. However, the question remains, in which mammal group and where did the first switch occur? Why are non-RABV lyssaviruses not found in the New World, or RABV in Old World bats? From where have terrestrial animals acquired the RABV in the Old World? Holmes *et al.* (2002) suggested that American RABV lineages are not older than 500 years and they originated from the progenitor terrestrial RABVs introduced with the European colonization. Another hypothesis, based on the Bayesian evolution estimations, was recently suggested for the New World RABV lineages (Hughes *et al.*, 2005). However, none of these suggestions answers the questions mentioned above. Clearly, we are missing some important chains of lyssavirus evolutionary pathways that are extinct or undiscovered to date.

The ecologic advantages offered by flight enhance the potential for invasion of new areas and rapid dispersal of viruses by bats, much more readily than by terrestrial mammals. Furthermore, no region, except Antarctica and a few very isolated islands, can be considered truly rabies-free based on the absence of terrestrial rabies, as in the most recent example, where ABLV was discovered in Australia in 1996 and two human cases have already been documented (Allworth *et al.*, 1996; Hanna *et al.*, 2000). In the UK, EBLV-2 was diagnosed in 1996 and a human rabies case of this virus was diagnosed in 2002 (Fooks *et al.*, 2003a). In the Irkutsk province of Eastern Siberia, which had been considered rabies-free for 35 years, IRKV was isolated in 2002 (Botvinkin *et al.*, 2003).

Bats are now the most prominent source of human rabies in the New World, western Europe and Australia, especially where the disease in carnivores has been controlled. Most of these cases, at least in North America, are cryptic. That is, circumstances of human exposure to bats frequently are peculiar and people do not often pay much attention to rather small lesions caused by bat bites (Messenger *et al.*, 2002). Conjecture and unconfirmed reports repeatedly suggested the existence of a carrier state of rabies among asymptomatic bats (Pawan, 1936; Sulkin *et al.*, 1957; Sulkin, 1962; Echevarria *et al.*, 2001; Serra-Cobo *et al.*, 2002; Wellenberg *et al.*, 2002), although these have not been validated to date. However, another peculiarity, such as aerosol transmission of rabies under field conditions (e.g. in caves inhabited by millions of Mexican free-tailed bats), has been demonstrated experimentally at least from bats to carnivores (Constantine, 1962).

Rabies control methods are different for hematophagous, frugivorous and insectivorous bats. Bats are as numerous as they are different and many bat species are endangered and protected. At the same time, they maintain circulation of lyssaviruses, roost synanthropically and serve as a source of infection for humans, domestic animals and other wildlife. Some experimental trials have demonstrated a potential possibility of oral vaccination in vampires (Aguilar Setien *et al.*, 1998; Almeida *et al.*, 2005), but this approach to other bats has not been studied yet.

In this chapter, we describe major aspects of the bat rabies issue. Phylogenetic, ecological and geographical properties of bat lyssaviruses correlate more or less to each other, so we divide the information geographically, trying to avoid repetition and making cross-references when necessary.

2 BAT RABIES IN THE NEW WORLD

2.1 Vampire bat rabies

Vampire bats constitute a unique group of mammals that evolved to consume the blood of other vertebrates. Vampire bats bite to stay alive and represent a splendid niche for RABV circulation. Three genera and species of vampire bats, belonging to the family Phyllostomidae, subfamily Desmodontinae, include the following species: the common vampire bat (*Desmodus rotundus*), which is the most abundant from Mexico to Argentina and prefers bovine blood; the hairy-legged vampire bat (*Diphylla ecaudata*), which is distributed mainly in cooler areas, up to the south of Texas historically and prefers bovine and equine blood; and the white-winged vampire bat (*Diaemus youngi*), the rarest species, found close to the equator, feeding mainly on birds (Nowak, 1999).

A paralytic disease in cattle and sporadically in humans after vampire bites, was reported from the time of the first Spanish colonists. Initially, the disease was misdiagnosed as 'cattle plague' or botulism and as poliomyelitis or 'acute ascending myelitis' in humans (reviewed by Baer, 1991). Vampires probably maintained RABV circulation for a long time prior to the discovery of America by Europeans. The association between vampire bites and the disease was understood by natives, who cauterized or washed the bites to prevent the disease (Constantine, 1988). However, historical antecedents might be some other progenitor RABV, quite different from those which circulate in bat populations presently. According to recent estimations, based on nucleoprotein gene evolution rates, the vampire lineage is the oldest among modern American bat RABVs. Nevertheless, the compartmentalization was suggested sometime between 1656 and 1892, whereas the common ancestor of all modern American bat RABV lineages was aged to between 1254 and 1782 (Hughes *et al.*, 2005).

Monoclonal antibody (MAb) typing of virus isolates suggests that the most divergent vampire bat RABV variants circulate in Mexico. One antigenic variant (AV3) is distributed throughout the vampire bat area, whereas distribution of another variant (AV11) is limited to certain regions in the northern part of South America and southeastern Mexico. A third variant (AV8) has been detected only along the western coast of Mexico. Rigorous study of phylogenetic relationships between vampire bat RABVs suggests that their common ancestor might have originated in South America rather than in Mexico. The antigenic diversity that has been defined for these viruses may depend on only a few amino acid substitutions, which occurred during the circulation of virus in different vampire bat populations, separated from each other by geographic barriers, such as the mountain ridge of the Sierra Madre Occidental (Velasco-Villa and Rupprecht, 2005).

Vampire bat populations, which subsisted on blood of native animals prior to the arrival of European settlers, were significantly enlarged with the increasing numbers of prey, notably livestock, that they could attack. The diagnosis of rabies in livestock was first made by the identification of Negri bodies in the brain of cattle during the outbreak of a previously undiagnosed disease in Santa Catarina, in southern Brazil in 1911 (Carini, 1911). The disease was reproduced for identification purposes by inoculation of rabbits, and wild animals were suspected as the virus reservoir, because attempts to control rabies in dogs did not reduce the number of rabies cases in cattle. Additionally, the correlation between the disease in cattle and rabies in bats was proposed, since they were observed biting cows.

Rabies diagnosis in cattle as a consequence of virus transmission by vampire bats was also confirmed in Paraguay, Argentina (reviewed by Baer, 1991) and in Trinidad, where rabies was confirmed not only in cattle but in numerous frugivorous and hematophagous bats that had exhibited abnormal behavior (Pawan, 1936). Thus, the relationship between vampire bats and paralytic rabies in cattle was firmly established and soon thereafter reported from most Latin American countries.

Clinically, the disease in cattle, humans and bats is paralytic and virtually never furious. This peculiarity may be dependent on some specific properties of the RABV variant circulating among vampire bats. However, there are no explanations as to which viral elements or functions may be responsible for this clinical presentation. Disease signs in the bovine start with posterior incoordination, followed in later stages with anorexia, bellowing, weakness in the hind legs and other signs of brain damage. Finally, ascending paralysis, apnea and death occur. The morbidity period varies from a few days to a week (Baer, 1991). Similar clinical signs and fatal outcome are seen in humans. In a few documented cases, animals and humans recovered after a paralytic disease caused, as believed, by contact with vampire bats. None of those cases was available for laboratory investigation (reviewed by Constantine, 1988). One of the most recent examples was identified in 1996, during an outbreak of a paralytic disease among humans in Peru. At least nine deaths occurred, but also a few recovered. Two fatal cases were confirmed to be rabies, but no samples were available from suspected survivors (Warner et al., 1999). Often the remoteness of some of the vampire bat rabies outbreaks among indigenous people precludes adequate epidemiological investigation.

The common vampire bat, *D. rotundus*, became one of the most abundant bat species in tropical America due to an extensive food supply and ecological flexibility. According to a number of observations (Lord *et al.*, 1975), rabies virus and the disease it causes spreads through susceptible vampire bat populations as a 'migratory epizootic'. Some bats are killed by the disease, while others survive the exposure and develop immunity. After several years of reproduction, the bat population accumulates enough susceptible animals to be vulnerable to another epizootic. In field studies, the virus prevalence among vampire bats varied from 0% in non-infected areas to as much as 14.3% in the outbreak areas. In extensive random surveys, performed in Trinidad and Brazil, the prevalence was 0.46 to 0.75% (reviewed by Baer, 1991). Antibody dynamics can also be demonstrated. As was reported by Lord *et al.* (1975), before an outbreak, only 3.1% of captured vampire bats demonstrated anti-RABV antibody. Seroprevalence increased up to 6.6% during the outbreak and up to 16.8% after the outbreak.

Vampire bats are gregarious animals. They are in close social contact, including allogrooming and even mutual feeding by regurgitation of ingested blood. They can inoculate each other via infectious saliva delivered by bites, licking, or probably by ingestion of saliva-contaminated regurgitated blood and inhalation of aerosolized saliva (Constantine, 1988).

An idea that vampire bats may be asymptomatic rabies carriers, shedding the virus with their saliva for months, was popular during initial studies of vampire bat rabies (Pawan, 1936). However, a significant limitation of the early studies on vampire bats was that the diagnosis was made by detection of Negri bodies, which are known to be unreliable for proper identification. Further surveillance and experimental work did not confirm the 'carrier' proposal. As was suggested by Constantine (1988), the early workers might have mistaken as rabies one or more other infectious agents that could chronically persist in bat salivary glands.

In a well-documented experimental study of Moreno and Baer (1980), the disease in vampire bats was similar to rabies observed in other infected mammals. The bats that developed signs of disease and excreted the virus via saliva soon died, whereas those that survived the inoculation without clinical signs never excreted the virus or had it in the brain as demonstrated upon euthanasia. Susceptibility of vampire bats to a homologous virus strain, injected intramuscularly, was lower than the susceptibility of terrestrial animals (e.g. foxes) to their homologous strain. Duration of the incubation period depended on the inoculation dose and varied from 14 to 57 days (mainly 2–4 weeks) and duration of the clinical period was 1 to 28 days (usually no longer than 10 days). The virus was detected in saliva 0–8 days before the clinical onset in 80% of bats. In a single case, when the duration of the clinical period was 28 days, virus appeared in saliva only during the last week before death. Hence, it is impossible to confirm that during the previous weeks the clinical signs were caused by rabies.

Economic losses due to paralytic rabies in livestock are tremendous. In the enzootic area, there is an at-risk population of more than 70 million head of cattle. Vampire bats usually bite many animals in a herd. The proportion of animals bitten may vary from 6 to 52% and bites may be multiple. Baer (1991) reported a herd of 42 cows in Oaxaca, Mexico, where each animal had fresh bites and 18 had over five bites. As reviewed by Acha (1967) and Baer (1991), the estimated annual mortality of bovines due to vampire bat rabies in Latin America during the 1960s was more than 500 000 cases with an economic loss of about US$50 million. Clearly, cases of mortality due to vampire bat rabies was an underestimate as cases reported annually varied from 3 to 60% of the actual mortality. Information from rural localities is often missing. During 2000–2002, Latin American countries reported reduced foci of livestock rabies cases compared with previous years, 1869 to 3327 annually (Vigilancia epidemiologica de la rabia en las Americas, 2000–2003), but it is unclear whether this is the result of rabies control campaigns or an underestimation of the real situation.

Constantine (1988) provided information on about 247 human rabies cases of vampire bat origin from 1929 to 1985.Vampire bats do not feed on humans as frequently as upon cattle. For example, 365 people were bitten in Mexico during 1984–1993 and 52 rabies cases attributed to vampire bat bites had been registered (Yanez Valesco, 1994). Seasonal or other factors, such as anthropogenic environmental modifications, e.g. changing former wildlife regions into semi-urban areas, may stimulate displacement of humans, cattle and vampire bats. As a result, more frequent attacks on humans and an increase of epizootic activity may occur. Thus, between January 1 and April 30, 1990, 29 (5%) of the 636 residents of two rural communities in the Amazon Jungle in Peru acquired a fatal illness symptomatically compatible with rabies. Vampire bat attacks were common in the region. An RABV variant identical to a sample isolated from vampire bats was identified in the brain of the only person available for postmortem investigation (Lopez *et al.*, 1992). A significant increase of vampire bat attacks was reported in 1996 from the Amazonian region of Brazil (Schneider *et al.*, 1996). Of 129 people interviewed, 23.33% had been bitten by vampire bats during the last year, with an average of 2.8 bites per attacked person. It appears that bats did not maintain RABV that time, because neither human nor animal rabies was reported. However, the situation changed dramatically at the beginning of 2004, when at least 22 human cases occurred. Among 250 persons interviewed in the state of Para, Brazil, 140 had been bitten by bats during the previous year (ProMed mail, archive 20040520.1349, 20040527.1428), and each sixth or seventh such bite caused rabies. A similar outbreak was shortly registered in Peru, where at least 12 persons died of rabies transmitted by vampire bats (ProMed mail, archive 20050118.0170).

Apparent spillover of RABV from vampire bats into another potential vector is not a common event. Different bat species, which share the roosts with vampires, appear to be infected most easily. Vampire bat RABV variant was diagnosed in frugivorous bats, *Artibeus* spp. (Shoji *et al.*, 2004). However, neither in other bats nor in terrestrial mammals, were secondary transmission events reported. Such spillovers may result in a dead-end infection. A few cases of human rabies caused by exposure to rabid cattle were reported, including a contact with a bovine which presumably was infected by a vampire bat. Vertebrates do consume raw or improperly cooked flesh of rabid livestock. Farmers may be interested in the slaughter and sale of cattle during an epizootic to avoid economic disaster. Of cattle slaughtered for human consumption in Mexico City, 40 of 1000 (4%) were found infected with rabies virus. No apparent human cases caused by the consumption of such carcasses were reported, but dogs, which ate meat from rabid livestock, developed a disease compatible with rabies after a month (reviewed by Constantine, 1988).

The control of vampire bat rabies includes prophylactic vaccination of cattle, post- and pre-exposure vaccination of humans and selective population reduction of vampire bats. One can estimate the benefits obtained with vaccination of cattle, making a comparison of the vaccination cost versus economic losses caused by rabies epizootics in naïve herds. Post-exposure prophylaxis of humans bitten by vampire bats using commercially available rabies vaccines is quite effective. However, vaccination is problematic in remote localities in Latin America due to the necessity for maintaining the cold chain for biologicals, transportation limitations and added expense. Nevertheless, pre-exposure vaccination should be administered to persons of high exposure risk and might be included in childhood vaccination programs in remote areas.

Methods of vampire bat population reduction are performed in roosts or by mist-netting around cattle pastures. Methods include use of anticoagulants (such as warfarin), which are deadly for vampire bats. The warfarin jelly is applied on the backs of captured bats and the animals are released. In the roost, during self- and allogrooming, other vampire bats consume the jelly on those that are treated (up to 20 or more vampire bats may be killed for each one treated). They also smear the jelly on the roost walls and excrete warfarin with urine and feces and, finally, the whole roost becomes deadly for bats. Another approach includes application of warfarin to the fresh vampire bat bite wounds on cattle, because bats often return to bites made the previous night. This strategy may be quite practical for farms with a limited number of livestock, but not for large herds. Warfarin that is injected intramuscularly into cattle will circulate in the blood during the next 3–4 days, killing any vampire bats which may feed on the animal. This approach is preferred by many ranchers, but is dangerous for calves and should never be used for horses (Greenhall, 1993).

Experiments with oral vaccination demonstrated that vampires developed immunity and survived further challenge with homologous RABV (Aguilar Setien et al., 1998; Almeida et al., 2005). However, this approach seems to be not very promising today, because there is no reason to encourage an increase of an immune vampire population. These bats are increasing in number and are quite harmful even without regard of rabies. Bleeding reduces cattle weight and milk production, and multiple wounds provide a portal for bacterial infection. Attacks on humans are also expected to rise with an increase in vampire population. Besides rabies, vampires were also described as potential vectors of surra, Venezuelan equine encephalomyelitis and foot-and-mouth disease (Constantine, 1988).

2.2 Non-hematophagous bat rabies in North and South America

Rabies in non-hematophagous bats was also diagnosed in Latin America during the investigations of vampire bat rabies epizootics (Carini, 1911; Pawan, 1936).

It was not clear, however, whether these were spillover cases from rabid vampire bats or if insectivorous and frugivorous bats could maintain RABV circulation independently.

The first definitive case of rabies in an insectivorous bat, found outside the vampire bat distribution area, occurred in Florida in 1953. A boy was attacked by a yellow bat, *Lasiurus intermedius*, during day time. Relatives of the boy had been informed about vampire bat rabies and the animal was submitted for rabies examination. Negri bodies were found in the bat brain and virus was isolated by mouse inoculation. The boy received post-exposure prophylaxis and survived. Before this, a human case of bat origin was suspected in Texas in 1951 in a woman who was bitten on the arm when she handled a bat. Rabies diagnosis in the patient was confirmed by Negri body detection, the incubation period was compatible with the time passed after the bat bite and no contacts with other mammals prior to disease were established (Sulkin and Greve, 1954).

Since then other cases have been diagnosed in Florida and in other states due to enhanced surveillance. By 1957, bat rabies was reported in Canada and by 1960, 30 states in the USA had reported rabies in at least six different bat species (Figure 6.1). Rabid bats made up a large proportion of the total wildlife rabies cases

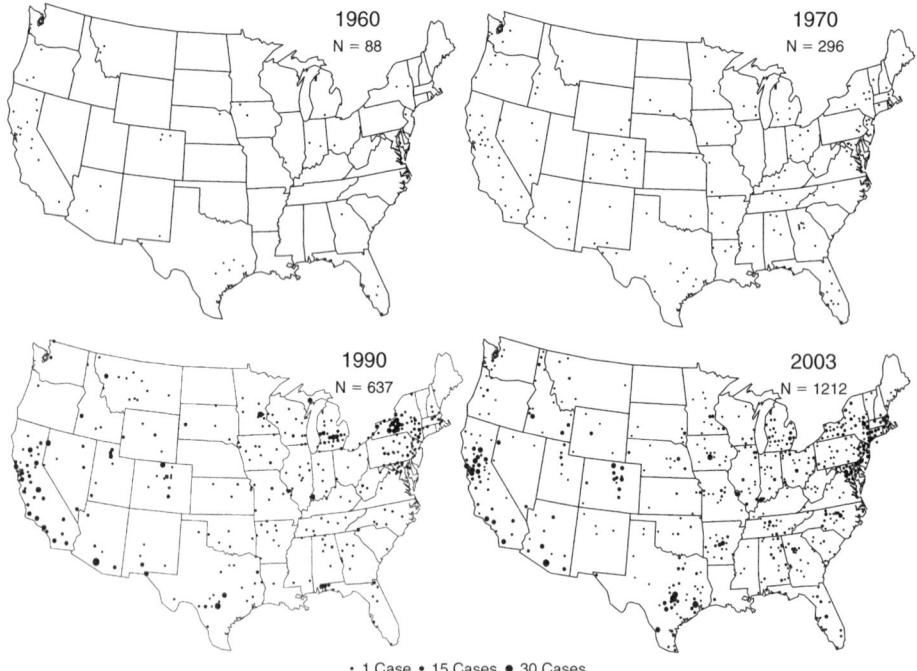

Figure 6.1   Registered cases of rabies in non-hematophagous bats in the USA during selected years (data of the Centers for Disease Control and Prevention, Atlanta, GA).

reported in the USA, increasing from 0.1% in 1953 to more than 10% by 1965. The reason for this increase was not an emergence of rabies in bats, but rather an increase in public health and scientific interest in the problem. For instance, the number of bats reported rabid from New York between 1964 and 1978 increased from 11 to 39, but the number of bats examined increased from less than 200 to over 1000 (Baer and Smith, 1991). More recently, attention to bat rabies has increased since it has been known that these animals have become the main source of indigenous human rabies in the USA and Canada. The maximum annual number of bat rabies cases, 1373, was reported in the USA in 2002, which accounted for 17.2% of all animal rabies cases reported during that year (Krebs *et al.*, 2003). The cases were reported from 47 of the 48 contiguous states (in other years – from all 48 states, Figure 6.1) and some states (Idaho, Illinois, Mississippi, Utah; sometimes Indiana, Nevada and Washington) reported rabies in bats even in the absence of terrestrial rabies. Only a few states, such as Alaska, Hawaii and North Dakota – and also Puerto Rico – did not report bat rabies. Although, in 2003, the number of reported rabid bats slightly decreased from 1373 to 1212 cases, the general pattern of the epizootics has not changed (Krebs *et al.*, 2004). Among the bats reported as rabid during the most recent years, 46% of them were big brown bats, *Eptesicus fuscus*; 27% were Mexican free-tailed bats, *Tadarida brasiliensis*; 6–7% were hoary bats, *Lasiurus cinereus*; 6% were red bats, *Lasiurus borealis*; 3% were little brown bats, *Myotis lucifugus*; 2–3% were silver-haired bats, *Lasionycteris noctivagans*; 1–2% were western pipistrelles, *Pipistrellus hesperus*. Other bat species constituted generally less than 1% of the registered cases (Krebs *et al.*, 2003, 2004).

Rabid bats are reported mostly in geographic areas where they encounter humans often and diagnostic laboratories are available, such as in cities. Some bat species are commonly seen by humans, particularly if they enter human dwellings and use other peri-urban constructions as roosts. Other species are encountered only occasionally. Reliance on enhanced surveillance is key to providing more reliable estimations of rabies prevalence and dynamics compared with the occasional submission of abnormal (sick) bats to diagnostic laboratories. Constantine (1967a) reviewed some surveys of randomly collected bats (Table 6.1) and found in most surveys that the prevalence of rabies in tested bats was less than 1%. One uncommon event was an apparent outbreak registered during 1964–65 in Mississippi, when 103 red bats were diagnosed as rabid (Baer and Smith, 1991).

Seasonal disease patterns were also suggested for some bat species at moderate latitudes. For example, hoary and Mexican free-tailed bats demonstrate two seasonal peaks of disease prevalence. One peak occurs during April–June, after a period of high activity and interactions associated with the return from hibernation. The second peak, which is larger, appears during September–October, which is likely related to a complex set of events. One is that a significant proportion of young bats, never exposed to rabies and lacking immunity, begins to separate from maternity colonies and a second is that the maternity colonies

TABLE 6.1

Results of random active surveillance of bats for rabies

Species	State	Tested/positive	% Positive
Macrotus waterhousii	California	150/1	0.67
M. waterhousii	Arizona	84/0	0.00
Myotis velifer	Texas	100/0	0.00
M. velifer	Arizona	180/2	1.11
Myotis thysanodes, M. volans, M. lucifugus, M. yumanensis, M. californicus	California	100/0	0.00
Myotis thysanodes, M. volans, M. lucifugus, M. yumanensis, M. californicus	Montana	83/0	0.00
M. lucifugus	Illinois	480/0	0.00
M. lucifugus	Massachusetts	83/0	0.00
M. lucifugus	New England	394/2	0.51
Myotis austroriparius	Florida	1998/1	0.05
Myotis grisescens	Florida	281/1	0.36
Pipistrellus subflavus	Florida	327/2	0.61
Eptesicus fuscus	Arizona	88/1	1.14
E. fuscus	Arizona	36/0	0.00
E. fuscus	California, Montana, Illinois, Massachusetts	231/0	0.00
E. fuscus	Ohio	63/1	1.59
E. fuscus	New England	119/3	2.52
Lasiurus cinereus	New Mexico, Arizona	50/1	2.00
Lasiurus borealis	Iowa	44/1	2.27
L. borealis	?	55/0	0.00
Plecotus townsendii (saliva samples only)	New Mexico	97/1	1.03
P. townsendii	New Mexico	12/2	16.67
P. townsendii	California, Arizona	35/0	0.00
Tadarida brasiliensis	California	12/1	8.33
T. brasiliensis	Arizona	86/2	2.33
T. brasiliensis	?	1129/4	0.35

Adapted from Constantine (1967) Bat rabies in the southwestern United States. *Public Health Reports* **82**, 867–888, with permission.

disintegrate. As bats are preparing for seasonal migration, active movements occur between roosts. These activities lead to an increased probability of exposure to rabies. Such enhanced mobility and displacement of bats brings them into contact with humans more frequently than during the maternity nursing period.

Thus, the number of sick bats encountered and submitted to laboratories rises. Since migratory bats are absent in their summer habitations for approximately 4–6 months, few, if any, rabies cases are registered in these species during this time. For non-migratory bats, such as *E. fuscus*, only one major peak is registered, during the second half of the summer. This peak coincides with increased mobility of naïve young individuals and the rearrangement of maternity colonies. For example, among bats of several species tested in Alberta, more than 84% were young animals born that year (Dorward *et al.*, 1977). Only occasionally, *E. fuscus* bats were found rabid in winter (Pybus, 1986).

Bat migrations signify an important pathway of RABV spread. In the Americas, *T. brasiliensis, Lasiurus* spp. and *L. noctivagans* are the most important in this respect. These bats can transmit virus over hundreds and even thousands of kilometers. The reported absence of bat rabies in the Caribbean Islands, as well as in other island territories, which are situated along paths of bat migrations, cannot be confidently believed. As reviewed by Constantine (2003), hoary bats were sometimes found in the Galapagos Islands and one was found in the Orkney Islands, north of Scotland.

As was revealed from the initial investigations, RABV isolates from different bat species are dissimilar to each other in some biological properties. For example, virus isolates from Mexican free-tails demonstrated an extremely short incubation period in mice, whereas in the natural host, the incubation period can be as long as 181 days (Baer *et al.*, 1980). Some viruses killed most of the tested carnivora species injected peripherally, whereas some isolates from red, hoary, silver-haired and big brown bats did not cause rabies in dogs, cats, foxes, skunks and other carnivores injected peripherally (Constantine, 1966a, 1966b, 1966c; Constantine and Woodall, 1966; Constantine *et al.*, 1968). American bat isolates are readily neutralized by commercial anti-rabies immune globulin and consequently are considered as the representatives of 'classical' RABV (serotype 1).

With the development of anti-RABV MAbs, especially those that react with lyssavirus nucleocapsid (anti-N-MAbs), which are used in the indirect immunofluorescent test, subtle antigenic discrimination between bat viruses was possible for taxonomic and epidemiological purposes. For example, it was made clear that certain bat species in the New World maintain circulation of specific virus variants within serotype 1 (Rupprecht *et al.*, 1987; Dietzschold *et al.*, 1988; Smith, 1989; King, 1993). This taxonomic approach was further improved by recruiting with the use of more recent techniques including gene sequencing and phylogenetic analysis. Lyssavirus taxonomy switched from serotypes to genotypes (although the 'species' term is the only one recognized officially by ICTV, the 'genotype' term has still been considered to be appropriate).

All New World bat lyssavirus lineages belong to genotype 1 (formerly serotype 1), which is true also for all New World lyssaviruses isolated to date (Figure 6.2). Bat lineages are segregated together into a cluster which is separated from terrestrial RABVs except for the raccoon branch (and a few particular

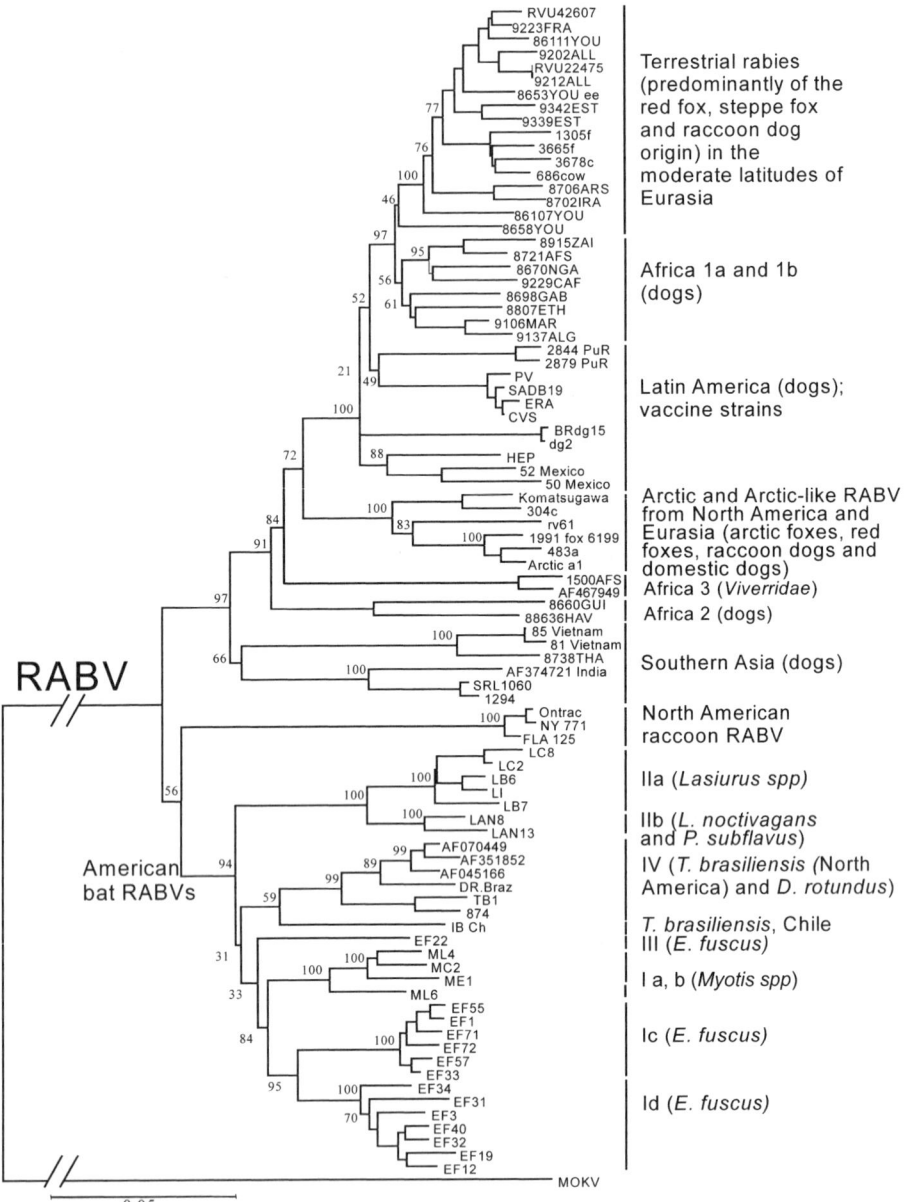

Figure 6.2 Phylogenetic tree of the major RABV (genotype 1) lineages according to the entire N gene sequences, obtained by the neighbor-joining method. African groups 1 (a, b) and 2 are given according to Kissi *et al.* (1995); North American bat lineages I (a, d), II (a, b), III and IV are designated according to Nadin-Davis *et al.* (2001). Principal host species and/or geographic range are designated when possible. MOKV sequence is used as an outgroup. Bootstrap values (% for 1000 replicates) are shown for key nodes and branch lengths are drawn to scale.

skunk clades, related to the raccoon branch but not shown in the figure due to limited sequence availability). Moreover, a number of clades, corresponding to certain species or genera of bats, are distinguished inside this cluster. Phylogenetic groups designated in the figure correspond to those given by Nadin-Davis *et al.* (2001). Group I includes viruses associated with representatives of genus *Myotis* (subgroups Ia and Ib) and *E. fuscus* (Ic and Id). Group II includes viruses circulating among solitary migratory bats of *Lasiurus* genus (IIa) and *Lasionycteris* (IIb) genera. Group III is another lineage associated with *E. fuscus*. Group IV joins *D. rotundus* viruses and isolates from North American *T. brasiliensis*.

Because these same clades are found in the USA, it is apparent that a bat species can maintain circulation of the same virus variant in geographically distant territories, such as *E. fuscus* in Ontario and Colorado (Shankar *et al.*, 2005), or *Lasiurus* species in North and South America (Sheeler-Gordon and Smith, 2001). *T. brasiliensis* also maintains circulation of different virus lineages but, as appears to date, they are separated geographically. The lineage of viruses circulating among North American *T. brasiliensis* is mostly related to the vampire viruses, whereas the lineage from South American *T. brasiliensis* constitutes a different branch. Nevertheless, these branches still segregate together and presumably have a common ancestor (Velasco-Villa and Rupprecht, 2005). The eastern pipistrelle, *Pipistrellus subflavus*, maintains circulation of viruses of group IIb (known as the silver-haired bat rabies virus, SHBRV) along with the primarily established host, *L. noctivagans* (Morimoto *et al.*, 1996; Messenger *et al.*, 2002). However, despite this common aspect of phylogenetic placement, there are conservative amino acid substitutions in the viruses isolated from these two bat species. This suggests that after a primary host shift, RABV circulates in populations of *P. subflavus* and *L. noctivagans* separately. This observation is also made in western North America where *P. subflavus* has not been seen and, similarly, in portions of the southeastern range of *P. subflavus*, where *L. noctivagans* had not been reported. The presence of another monophyletic virus clade was demonstrated in the western pipistrelle, *P. hesperus* (Franka *et al.*, 2006). Further investigations are needed in tropical America, where some additional phylogenetic groups may be discovered (de Mattos *et al.*, 2000).

Special considerations of age and evolution of American bat RABVs were recently emphasized (Hughes *et al.*, 2005). For example, the lineage of the *D. rotundus*/North American *T. brasilensis* virus (group IV in Figure 6.2) is the oldest described to date. This group of viruses, formed sometime between 1656 and 1892, shares common ancestors with a hypothetical progenitor virus, which circulated sometime between 1254 and 1782. This common ancestor likely originated in South America rather than in Mexico and most likely was the ancestor of the vampire bat virus, which subsequently switched to a North American population of *T. brasilensis*, presumably in the territory of Mexico (Velasco-Villa and Rupprecht, 2005). Compartmentalization of these lineages as they became associated with other bat species was estimated to have occurred 18–62 years later and the period for their adaptation was much shorter than might have occurred with the initial adaptation of a progenitor RABV to *D. rotundus* and/or *T. brasiliensis* (Hughes *et al.*, 2005).

Holmes *et al.* (2002) suggested that rabies was introduced into America with terrestrial companion animals during European colonization. According to this analysis, the virus would have switched to bats, perhaps through vampire bats that encountered contacts with rabid terrestrial mammals. Alternatively, terrestrial RABV could have entered the New World independently, via the Bering land bridge. It is possible that a progenitor RABV (as well as some other ancestral lyssaviruses) is the extinct source of the modern bat RABV lineages and circulated in the Americas a long time before that.

Another hypothesis, which emerged from the recently established phylogenetic relationships, was proposed by Badrane and Tordo (2001). They proposed that there were at least two switches in the lyssavirus history, both from bats to the terrestrial mammals. The first switch occurred at the inter-genotype level and the second occurred within genotype 1, when the virus switched from bats to carnivores. These and other hypotheses on the origin, adaptation and radiation of RABV await detailed introspection.

Human rabies of non-hematophagous bat origin has been a significant public health concern in North America. In South America it may be masked by the problem of vampire bat rabies and earlier due to uncontrolled dog rabies. Information about airborne exposure of two humans after a visit to Frio cave in Texas, inhabited by millions of the Mexican free-tailed bats (Constantine, 1962), was considered unreliable. The victims had often visited caves and could have forgotten incidents of bat bites; such often happens with victims of 'cryptic' rabies cases. Interviews with the 'cryptic' patient's relatives or friends usually helped to identify some kind of contact of the victim with a bat that occurred several weeks or months before the onset of symptoms of the disease (Gibbons *et al.*, 2002; Messenger *et al.*, 2002).

Among the 40 indigenous human rabies cases reported in the USA since 1980, 35 (87.5%) were caused by bat virus variants. These include four cases that occurred after transplantation of organs and a vessel from a donor who died of rabies after apparent infection with an RABV variant associated with the Mexican free-tailed bat (Krebs *et al.*, 2004). In 22 of those 35 cases, the virus was attributed not to common house bat species, but was SHBRV (lineage IIb in Figure 6.2), associated with silver-haired and the eastern pipistrelle bats, rarely found around humans or human dwellings. Investigations into SHBRV implied that this virus variant had enhanced pathogenicity. For instance, the SHBRV was better adapted to fibroblasts (BHK-21) and epithelial cells (MA-104) compared with coyote RABV (Morimoto *et al.*, 1996). This trait appeared to correlate with an ability to be more effective in replicating in the dermis at the inoculation site. However, since other bat RABV variants were not tested in this regard and related *in vivo* experiments have not been reported to date, it is not clear whether the reported peculiarities are specific to SHBRV or not. The SHBRV, as well as all RABV lineages, is readily neutralized by commercial anti-rabies immunoglobulin and vaccination with any modern potent vaccine provides adequate protection from the virus (Dietzschold and Hooper, 1998).

Since some bat bites, especially if they were inflicted by small bat species, such as a pipistrelle or *Myotis*, may be ignored because they were not recognized as dangerous by the patient, The Advisory Committee on Immunization Practices (ACIP) has introduced an improved guide for human rabies post-exposure prophylaxis (Human rabies prevention – United States, 1999). According to this document, rabies vaccination should be considered if a bat is found indoors and humans in the same room are unaware that a bite or direct contact might have occurred (e.g. a sleeping person awakened to find a bat in the room, or a bat is found in the room with a previously unattended child, or mentally disabled or intoxicated person).

Because of substantial public health significance, pathogenesis of bat rabies in the Americas (mainly in the USA) has been studied more rigorously than in any other parts of the world. Experimental studies started during the 1950s, shortly after the first case of insectivorous bat rabies was registered in Florida. At that time, scientists were influenced by communications from Latin America, where vampire bats were believed to be asymptomatic carriers of rabies (Pawan, 1936). In a series of experiments aimed at investigating this concern, together with the mechanisms of RABV maintenance during hibernation (Sulkin *et al.*, 1957, 1960; Sulkin, 1962), two significant findings were reported by Sulkin's group. The first finding indicated that it was possible that a virus remained latent in a bat that was injected intramuscularly during hibernation at low temperature. When the inoculated bats were returned to a warm environment and awakened, the viral infection was activated and rabies symptoms appeared after the same incubation period that was documented in the bats that were inoculated and maintained in an active state. The second finding suggested a role for interscapular brown adipose tissue (brown fat) as a depot for the storage of RABV during bat hibernation. *In vivo* experiments performed in two species, Mexican free-tails and little brown bats, demonstrated the presence of RABV in the brown fat of bats, which were serially killed without disease signs. In a few of these bats, virus was not detected in the brain. To correlate these findings *in vivo*, *in vitro* experiments were performed, which demonstrated that RABV could persist in brown fat cell culture for at least 56 days at 37.5°C. Virus replication was suppressed, however, at low temperatures (8°C), but was activated again with rising temperature (Allen *et al.*, 1964a, 1964b). Unfortunately, some of these results appear inconsistent with regard to virus detection and subsequently may lead to misinterpretation. Although conservation of the virus during bat hibernation was confirmed (Sadler and Enright, 1959; Kuzmin *et al.*, 1994; Kuzmin and Botvinkin, 1996), the particular function of the brown fat was not.

In a rigorous experiment performed by Baer and Bales (1967), peripherally inoculated Mexican free-tailed bats had an incubation period of 24 to 125 days, depending in part on the virus dose. Susceptibility to a peripheral inoculation was low: only eight of 46 bats developed rabies. None of the bats inoculated intramuscularly or subcutaneously demonstrated aggressiveness, as seen in the intracerebrally inoculated bats. However, the single bat which developed rabies after intranasal inoculation showed aggressiveness throughout almost the whole period of illness. Distribution and titers of virus in organs is shown in Table 6.2. The virus could not be recovered from any organs in the absence of brain infection and titers in extraneural tissues were low. No evidence of continuous virus excretion via saliva was noted. The longest period during which virus could be isolated from the saliva was 15 days before death (or 12 days before onset of the clinical signs).

Constantine (1967b) provided information on RABV distribution in tissues of 130 naturally infected Mexican free-tailed bats collected in Texas caves. The virus was detected as follows: brain 100%, salivary glands 79%, lungs 30% and kidneys 12%. In further tests on 50 of those 130 bats, brown fat of two (4.0%) contained RABV, but all were negative for virus in liver, spleen, pectoral muscle, intestines and fecal pellets. Later the same author detected RABV in five of 15 (33%) impressions of nasal mucosa from naturally infected *T. brasiliensis*. For two of those five, the virus was also isolated by mouse inoculation (Constantine, 1972).

These findings, coupled with the report about a possible airborne rabies exposure in bat caves, led to a series of transmission experiments performed by D. Constantine. Parts of those experiments were carried out inside the caves inhabited by huge colonies of Mexican free-tailed bats. Several species of Carnivora, including foxes and coyotes, were housed in close-meshed cages under the bat roost. After a variable incubation period, the animals developed furious rabies (Constantine, 1962, 1967b). Similar experiments were also performed in a laboratory environment. Foxes were exposed to an artificial aerosol preparation of the Mexican free-tail RABV. They developed furious disease and were able to infect (presumably by bite) the naïve cage mates. In another experiment, the isolates from Mexican free-tails and from the salivary glands of coyote, which had been infected in the bat cave, were further used for parenteral inoculation of carnivores, including those which could not be infected in the cave. Thereafter, rabies developed in coyotes, foxes, dogs, raccoons and skunks. This was in agreement with the opinion that carnivores, being inoculated from bats in the caves, could be a source of further dissemination of the virus among terrestrial mammals (Constantine, 1966a, 1966b, 1966c). However, as was mentioned above, no outbreaks of the Mexican free-tailed bat RABV variant were registered in terrestrial mammals during decades of surveillance. Thus, even if such spillover events sometimes occur, they apparently lead to a dead-end of the infectious chain.

RABV variants originating from another bat species were used in experimental studies as well. In one experiment, certain carnivores and laboratory rodents were exposed to isolates from leaf-nosed, hoary, silver-haired and big brown bats. Exposures were by bites of infected bats or by intramuscular inoculation of

TABLE 6.2

Titers of RABV in organs of Mexican free-tailed bats dying after inoculation with homologous RABV strain

Inoculation			Virus titers in various organs, $MICLD_{50}$				
Route	Dose, $MICLD_{50}$	Bat #	BR^a	SG	BF	KI	LU
Intracerebral	2000	109	4.62				
		22	4.5	0.63	0.62		
		40	5.16	1.46			1.00
		148	2.16				
		23	4.54	3.46		0.83	1.66
		67	5.5	0.63	0.83		
		35	5.37	1.62	0.62	1.16	
	200	32	5.37	2.93	2.37		
		50	4.68				
		57	4.32	2.93			
		105	5.37	1.83	0.62		
		122					
		158	5.0				
	20	26	4.62	0.67			
		42	4.62	2.68			
		81	3.84				
		85	4.5				
		170	5.5	2.14	1.50		1.33
	2	144	4.0				
	0.2	110	6.17	3.68	2.62		2.16
Intramuscular	2000	5	4.37			0.62	
		11	3.5				
	200	41	4.16				
		134	5.37	2.08	2.83		
Subcutaneous	2000	10	3.74				
		119	4.53		0.83		
	200	52	4.5				
Intranasal	20	129	4.78	2.67	1.47		1.62

Adapted from Baer and Bales (1967), with permission from University of Chicago Press. [a]BR: brain; SG: salivary glands; BF: brown fat; KI: kidneys; LU: lung.

known doses of virus. Only the virus from the leaf-nosed bat produced disease. Hamsters were infected by bat bite and intramuscular inoculation produced rabies in two foxes, a coyote, a ringtail and a raccoon. Viruses from the hoary bat (doses up to $100\,000$ MLD_{50}) and from the big brown bat (doses up to 5400

MLD_{50}) did not cause disease in any animal tested. The virus from the silver-haired bat was not tested by the intramuscular route and bat bites did not cause rabies (Constantine *et al.*, 1968). In another study, a virus originating from the red bat was delivered into rodents and carnivores intramuscularly or via bat bite. Inoculation by bite was successful for suckling mice only. Intramuscular injection of 6000 MLD_{50} produced rabies in ringtail and striped skunks but not in dogs, cats, opossums or rodents. Intramuscular inoculation of red bats with 100 to 500 MLD_{50} caused rabies only in 20% of animals. Virus was detected in brain and salivary glands of all dying bats but not in survivors (Constantine and Woodall, 1966).

In a different study, 35 wild big brown bats were captured and two of them died of rabies within the first month of quarantine. Despite group housing, all remaining bats were healthy over the next 4 months, although one of the rabid bats was seen to bite three cage mates. Five (14.3%) bats had RABV-neutralizing antibodies prior to captivity and two seroconverted during captivity (Shankar *et al.*, 2004). Of relevance, only one of 45 females (2%) screened from the capture site of the first rabid bat seroconverted in summer when bats were collected, but one year later 17 of 73 bats (23%) from this roost were seropositive. High antibody prevalence was also shown in populations of the big brown bat (10%) and the little brown bat (2%) in New York state (Trimarchi and Debbie, 1977) and up to 80% in colonies of Mexican free-tailed bats (reviewed by Baer and Smith, 1991). For the latter species, it was suggested that seropositive females could protect their young by passive transfer of anti-rabies antibodies (Steece and Altenbach, 1989).

Thus, patterns of virus circulation in bats appear to be different from those described for carnivores. Many details of this process have not been studied sufficiently. There must be complex virus–host interactions on different levels. For example, host species susceptibility has to be balanced with pathogenicity of the virus. This balance is directly associated with host population properties (such as population density, mobility, sex and age composition, frequency of social contacts), patterns of behavior and pathogenesis of the disease. These variables should ensure virus transmission to a certain number of susceptibles.

Colonial bat species, e.g. *T. brasiliensis*, *E. fuscus*, *D. rotundus*, demonstrate limited susceptibility to indigenous RABV variants. The disease is rarely furious and predominant clinical signs include general exhaustion, weakness and paralysis. Colonial animals are in tight contact most of their lives. If they transmitted virus more efficiently and actively, the whole colony would die, eliminating the virus from circulation. To the contrary, solitary bat species (*Lasiurus* spp., *Lasionycteris noctivagans*, etc.) develop furious rabies and actively attack bats (or other animals) to ensure virus transmission. When the first case of insectivorous bat rabies was diagnosed in Florida in 1953, it was an unprovoked attack by a *L. intermedius* bat, which suddenly flew out of the bushes and bit a boy on the chest, remaining firmly attached until knocked to the ground (Baer and Smith,

1991). Bell (1980) observed a *L. cinereus* bat, which successively attacked another *L. noctivagans*, *T. brasiliensis* and *E. fuscus* bat during foraging, in which *L. cinereus* chased the victims, caught them and brought them to the ground. Later that night a *L. cinereus* was mist netted approximately 50 m from the site. Presumably, it was the animal that was attacking the bats. It had fresh blood on its face and around its head. However, no injuries were noted. Rabies diagnosis in this bat was confirmed in the laboratory.

It is unclear how often such attacks cause rabies in a 'non-principal' bat species and why 'non-principal' species do not maintain the virus circulation with the same effectiveness as the principal host. According to surveillance data, *Lasiurus* spp. and *L. noctivagans* RABV variants were isolated from the 'non-principal' bat species, *E. fuscus*, in as many as 16–20% of the cases (Shankar *et al.*, 2005). Are these examples of dead-end spillover events, or evidence of concurrent participation of several bat species in the circulation of a single virus variant? Recently, a host shift of the SHBRV was established at least in one additional host, *P. subflavus* (or vice versa) (Messenger *et al.*, 2002; Franka *et al.*, 2006).

Introduction of bat RABV into populations of terrestrial animals occurs as well. Bat viruses were identified in cattle, cats, foxes and other mammals. However, they were never known to cause an outbreak in terrestrial animals, except on a few particular occasions. A rabies epizootic was detected in striped skunks in Arizona during 2001 and the virus was identified as the big brown bat RABV variant (Krebs *et al.*, 2002). Experimental studies demonstrated that skunks were quite susceptible to this virus; they developed furious rabies with high virus titers in the salivary glands and, consequently, were capable of maintaining the epizootic. Big brown bats themselves were much less susceptible to this virus, did not develop furious disease and only limited titers of virus were recovered from their salivary glands (Niezgoda *et al.*, 2003).

3 BAT RABIES IN THE OLD WORLD

3.1 Africa

African non-RABV lyssaviruses were not initially recognized as RABV-related. It was only through a multifaceted collaboration that this finding was established. Boulger and Porterfield (1958) isolated an agent from a pool of brains of six apparently healthy *Eidolon helvum* fruit bats at Lagos Island, Nigeria, in 1956. They called the agent LBV, registered it as a possible arbovirus and distributed it widely to viral taxonomists in the hope of finding out more about its relationships. Since Negri bodies were not detected in the brains of infected mice and the virus was not neutralized by a potent anti-RABV serum, relatedness between LBV and RABV was not established for 14 years. Morphological investigations undertaken during 1969–1970 demonstrated that LBVs, together with recently discovered MOKV, were rhabdoviruses. Further cross-antigenic studies revealed relatedness of their nucleoproteins to each other and to RABV, and the group of rabies-related viruses (genus *Lyssavirus*) was established (Shope *et al.*, 1970; Shope, 1982). LBV was designated as serotype 2 and MOKV as serotype 3 (currently genotypes 2 and 3, respectively).

Later, LBV was isolated from fruit bats, one species of insectivorous bat and domestic cats and dogs in broad territories of Africa, presumably from Egypt to South Africa (Table 6.3).

The other African lyssavirus, MOKV, is the only member of the genus which has never been isolated from bats. However, as well as for LBV, only a few isolates of MOKV are available to date and nothing is known about its distribution, host range and circulation properties. For this reason, and also because MOKV demonstrates a limited similarity to LBV, we are describing this virus here in brief.

The first isolate of MOKV was obtained from a pool of extraneural tissues (lung, liver, spleen and heart) of shrews (*Crosidura* spp.) in the district of Mokola (Ibadan, Nigeria) in 1968, and thereafter from other mammals in different African locations. It was diagnosed also in two children with 'aseptic meningitis' and encephalomyelitis symptoms in Nigeria, one of whom died and the other recovered (Table 6.3).

The distribution ranges of LBV and MOKV cover a large territory of the African continent. A limited number of gene sequences from these viruses are available in GenBank (see also Chapter 2). In all phylogenetic constructions, LBV and MOKV are joined together and separated from other lyssavirus genotypes. Moreover, the extent of their identity to each other is greater than the identity of each of them to other lyssavirus genotypes. They also demonstrate limited serologic cross-reactivity (Badrane *et al.*, 2001; Hanlon *et al.*, 2005), limited pathogenicity for mice by the peripheral route and they were proposed as members of one phylogroup (phylogroup 2), whereas members of genotypes 1 and 4 to 7 were joined into phylogroup 1 (Badrane *et al.*, 2001).

Reduced peripheral pathogenicity of LBV and MOKV is believed to be associated mainly with substitution of arginine or lysine for aspartic acid at position 333 in their glycoprotein ectodomain and with altered structure of the LC8 binding site of the phosphoprotein (Dietzschold *et al.*, 1983; Badrane *et al.*, 2001; Mebatsion, 2001). Mice did not develop rabies after intramuscular, subcutaneous or intraperitoneal inoculation of these viruses. Dogs did not present a productive infection after LBV and MOKV intramuscular administration, even with doses $10^{6.5}$–$10^{7.5}$ MICLD$_{50}$. However, one of six monkeys inoculated intramuscularly with 10^6 MICLD$_{50}$ of LBV, developed bilateral paresis on day 22 but recovered on day 86 and no virus was isolated from the animal after euthanasia (day 108). One of six monkeys inoculated intramuscularly with $10^{7.3}$–$10^{7.5}$ MICLD$_{50}$ of MOKV developed clinical rabies and died. The virus was isolated from the brain of the dead animal (Tignor *et al.*, 1973). In our experience, LBV killed 20–25% of mice and MOKV 50–85% of mice, given per os. Additionally,

TABLE 6.3

Isolation of non-RABV lyssaviruses in Africa

Host species	Year	Location	Reference
LBV			
Fruit bat *Eidolon helvum*	1956	Nigeria	Boulger and Porterfield, 1958
Fruit bat *Micropteropus pussilus*	1974	Central African Republic	Swanepoel, 1994[a]
Fruit bat *Epomophorus wahlbergi*	1980	South Africa	King and Crick, 1988[a]
Domestic cat	1982	South Africa	King and Crick, 1988
Insectivorous bat *Nycteris gambiensis*	1985	Senegal	Swanepoel, 1994
Fruit bat *Eidolon helvum*	1985	Senegal	Swanepoel, 1994
Domestic cat	1986	Zimbabwe	King and Crick, 1988
Domestic dog	1989 or 1990	Ethiopia	Mebatsion *et al.*, 1992
Fruit bat *Epomophorus wahlbergi*	1990	South Africa	Swanepoel, 1994
Fruit bat *Rousettus aegyptiacus*	1999	France ex Togo or Egypt	Rabies Bulletin Europe, 1999
Fruit bat *Epomophorus wahlbergi*	2003	South Africa	Markotter *et al.*, 2006
Fruit bat *Epomophorus wahlbergi*	2004	South Africa	Markotter *et al.*, 2006
Fruit bat *Epomophorus wahlbergi*	2005	South Africa	Markotter *et al.*, 2006
MOKV			
Shrew *Crocidura* spp.	1968	Nigeria	Kemp *et al.*, 1972
Human	1969	Nigeria	Familusi and Moore, 1972
Domestic cat	1970	South Africa	Schneider *et al.*, 1985
Human	1971	Nigeria	Familusi *et al.*, 1972
Shrew *Crocidura* spp.	1974	Cameroon	Swanepoel, 1994
Domestic cat	1981, 1982	Zimbabwe	Foggin, 1983
Domestic dog	1981	Zimbabwe	Foggin, 1983
Rodent *Lophuromys sikapusi*	1983	Central African Republic	Swanepoel, 1994
Domestic cat	1989 or 1990	Ethiopia	Mebatsion *et al.*, 1992
Domestic cats	1995–1998	South Africa	Nel *et al.*, 2000
DUVV			
Human	1970	South Africa	Meredith *et al.*, 1971
Insectivorous bat *Muniopterus* spp.	1981	South Africa	King and Crick, 1988
Insectivorous bat *Nycteris thebaica*	1986	Zimbabwe	King and Crick, 1988

[a] Reference on a review paper is given when the primary literature source was unavailable.

it should not be forgotten that LBV and MOKV are not laboratory strains. They were isolated from a number of naturally infected mammalian species, so there must be some adaptive mechanisms providing their routine circulation. Cats and dogs infected naturally with LBV and MOKV demonstrate signs of brain dysfunction and sometimes signs of furious rabies (Mebatsion *et al.*, 1992; Nel *et al.*, 2000).

LBV gene sequences demonstrate the greatest intragenotypic diversity compared with other lyssavirus genotypes (Badrane *et al.*, 2001; Nadin-Davis *et al.*, 2002). The MOKV sequences also demonstrate heterogeneity, depending mainly on their geographical origin (Nel *et al.*, 2000). Indeed, additional surveillance should be conducted in Africa to address variability, host range and circulation properties of LBV and MOKV.

The third African non-RABV lyssavirus, DUVV, was isolated during 1970 in South Africa from a human (for whom the virus was named), bitten by a bat on a lip while sleeping. He killed the bat and, according to the description given, it was likely a common insectivorous species *Miniopterus schreibersii*. The person did not receive medical treatment and, 5 weeks after the incident, developed rabies (Meredith *et al.*, 1971). The brain impressions were negative when stained by a locally produced anti-rabies FITC-labeled antibody. Since Negri bodies were detected and due to the typical disease symptoms, additional investigations were undertaken, including cross-antigenic and cross-neutralization tests and the isolate was confirmed to be a rabies-related virus. Further isolates of DUVV have also been identified in *Miniopterus* spp. in South Africa and in *Nycteris thebaica* in Zimbabwe (see Table 6.3), but not in terrestrial mammals.

At the beginning of the 1980s, when MAb techniques became readily available for lyssavirus characterization, it was shown that isolates from European bats were related to DUVV. This issue is described in greater detail in Section 3.2, as well as the hypotheses that the *Lyssavirus* genus originated in Africa and that the 'Duvenhage-like' viruses from Africa were introduced into Europe. Further genetic characterization demonstrated that DUVV, similar to its Eurasian relatives, is a member of phylogroup 1 and is more closely related to RABV than to LBV and MOKV (Badrane *et al.*, 2001). In particular, it has R_{333} in the glycoprotein ectodomain, the LC8 binding site of the phosphoprotein is similar to other phylogroup 1 representatives and it is pathogenic for mice by the peripheral route.

One of the consequences of the genetic and antigenic heterogeneity of lyssaviruses is the difficulty to induce ample protection against some members of the genus using standard rabies biologicals. All phylogroup 1 lyssaviruses cross-neutralize each other, but the degree of this effect is different for the various members. Only a limited cross-neutralizing activity is detected between LBV and MOKV within phylogroup 2 and no cross-neutralization was observed between phylogroup 1 and phylogroup 2 representatives (Hanlon *et al.*, 2005). Recently developed anti-glycoprotein MAb CR57 neutralized all phylogroup 1 lyssaviruses and was suggested for inclusion as part of an MAb cocktail aimed at replacing classical polyclonal immunoglobulins (Marissen *et al.*, 2005).

However, commercial immunoglobulins should also be developed specifically to neutralize LBV and MOKV. Human diploid cell vaccine (HDCV) and animal vaccines protect against a challenge with DUVV, but not against LBV and MOKV, whereas animal vaccines Rabisin and Rabiffa protect against LBV but not against MOKV (Fekadu *et al.*, 1988b). Chimeric DNA-vaccines encoding RABV (PV strain) and MOKV glycoproteins were suggested to cover the spectrum of phylogroups 1 and 2 lyssaviruses (Bahloul *et al.*, 1998; Jallet *et al.*, 1999) and another DNA vaccine was suggested to protect against MOKV (Nel *et al.*, 2003).

3.2 Eurasia

In Europe, bat rabies was first diagnosed in Hamburg, Germany in 1954. Although the discovery happened almost simultaneously with the description of rabies in insectivorous bats in North America, not much attention was paid to this issue during the initial few decades. Only 14 cases were diagnosed in different regions of Europe before 1985 (Schneider and Cox, 1994; Table 6.4). These included the first case of human rabies after bat exposure reported from Europe,

TABLE 6.4

European bat lyssavirus isolates, 1954–1985

Year	Location	Bat species	Virus type
1954	1 Germany (Hamburg)	?	?
	3 Yugoslavia	*Nyctalus noctula*	?
1956	1 Turkey	*Rhinolophus ferrumequinum*	?
1963	1 Germany (Jena)	*Eptesicus serotinus*	?
1964	1 Ukraine (Kiev)	*E. serotinus*	?
1968	1 Germany (Hamburg)	?	EBLV-1
1970	1 Germany (Stade)	?	EBLV-1
1972	1 Poland (Krakow)	?	?
1973	1 Germany (Berlin)	*Myotis myotis*	?
1977	1 Ukraine (Voroshilovgrad)	Human of bat origin	?
1982	1 Germany (Bremerhaven)	?	EBLV-1
1983	1 Germany (Aurich)	*E. serotinus*	EBLV-1
1985	10 Denmark	*E. serotinus*	EBLV-1
	3 Germany	*E. serotinus*	EBLV-1
	1 Poland (Gdansk)	*E. serotinus*	EBLV-1
	1 Russia (Belgorod)	Human of bat origin	EBLV-1
	1 Finland (Helsinki)	Human of bat origin	EBLV-2

Adapted from Schneider and Cox (1994), with kind permission of Springer Science and Business Media.

registered in the town of Voroshilovgrad (currently Lugansk) in the Ukraine in 1977. A 15-year-old girl was attacked by a bat during the day and was bitten on her finger. The bat was not examined and the patient did not receive rabies prophylaxis. Approximately 35 days later, the girl developed typical rabies symptoms and the diagnosis was confirmed post-mortem (Scherbak, 1982). The virus was not stored for precise typing.

Further progress of the European bat rabies investigations was tightly associated with the development of MAb technique. The typing of available European bat virus isolates with anti-N-MAbs demonstrated that their antigenic patterns were similar to those of DUVV. For this reason, European bat lyssaviruses were designated initially as serotype 4 viruses (Schneider, 1982).

Another turn in the European bat rabies story happened in 1985. Fourteen bat cases were registered during that year in Denmark, Germany and Poland. It was the beginning of an extensive epizootic which covered certain parts of Europe during the next decade. Two human rabies cases described as having bat origins were reported during 1985 from Russia and Finland.

The Russian case occurred in the town of Belgorod and was quite similar to the earlier one from the Ukraine. An 11-year-old girl was attacked and bitten on her upper lip by a bat during the day. The bat escaped, so neither species identification nor a rabies test was performed. The girl did not seek prophylaxis and she developed the disease about three weeks later. The antigenic profile of the virus, called Yuli, was evaluated with anti-N-MAbs and found to be a representative of the serotype 4 group (Selimov et al., 1989). In Finland, the patient was a 30-year-old zoologist who had been bitten by bats in Malaysia 4.5 years earlier, in Switzerland 1 year previously and in Finland 51 days before his death (Lumio et al., 1986). The virus isolated from that case ('Finman') was primarily found to be similar to DUVV and designated as a serotype 4 virus as well (King and Crick, 1988).

Additional studies with a number of anti-N-MAbs demonstrated that European bat lyssaviruses were similar but not identical to DUVV. Furthermore, they were different from each other (Dietzschold et al., 1988; King et al., 1990; Rupprecht et al., 1991). For a short period of time, the Finman virus and a few other bat isolates originating from *Myotis dasycneme* from the Netherlands, were segregated into a new group, serotype 5, whereas other European bat isolates, obtained through Europe primarily from *Eptesicus serotinus* bats and the Yuli isolate from a human, were still considered as members of serotype 4. Sometimes they were also termed as 'biotypes', to reflect their distinctions from the genuine African DUVV (King et al., 1990; WHO report of the 6th WHO consultation on monoclonal antibodies for rabies diagnosis and research, 1990 – after Schneider and Cox, 1994). EBL-1 included the isolates least different from DUVV and thought to be serotype 4 members, whereas EBL-2 joined the members of serotype 5. This disorder with the classification sometimes led to confusion. When serotype 4 viruses are referenced in communications, one should verify whether the viruses are of African or

European origin, and if a serotype 5 virus was already separated from the serotype 4 members.

At the beginning of the 1990s, with the implementation of genetic methods and phylogenetic analysis, lyssavirus classification switched from serotypes to genotypes (Bourhy *et al.*, 1992, 1993). Both European bat lyssavirus lineages were demonstrated to be different from DUVV and from each other. They were distinguished into two genotypes, e.g. EBLV-1 (genotype 5) and EBLV-2 (genotype 6). Although some authors continued to describe EBLV-1 as DUVV serotype 4 viruses (Schneider and Cox, 1994), it was not helpful since the serotype classification fell out of use and ICTV established the lyssavirus species in concordance with the identified genotypes. This taxonomy appears to have been quite stable during the last decade and is one reason we are using the names EBLV-1 and EBLV-2 in this text.

Extensive phylogenetic evaluation demonstrated that each EBLV can be subdivided into two lineages, 'a' and 'b'. Furthermore, for EBLV-1a, a west–east distribution was suggested, whereas for EBLV-1b, a north–south distribution was proposed (Amengual *et al.*, 1997). When the similarity of these viruses to DUVV had been shown, the hypothesis was proposed that the EBLVs might be introduced from Africa with migratory bats or with bats translocated by ships (especially taking into account that most of the cases were registered along coastal territories) (Schneider, 1982; Shope, 1982).

As was mentioned above, 1985 was the year when significant extension of the rabies epizootic among European bats was detected, with a peak of about 280 cases registered during 1986–1987 (Figure 6.3). At least 99% were EBLV-1

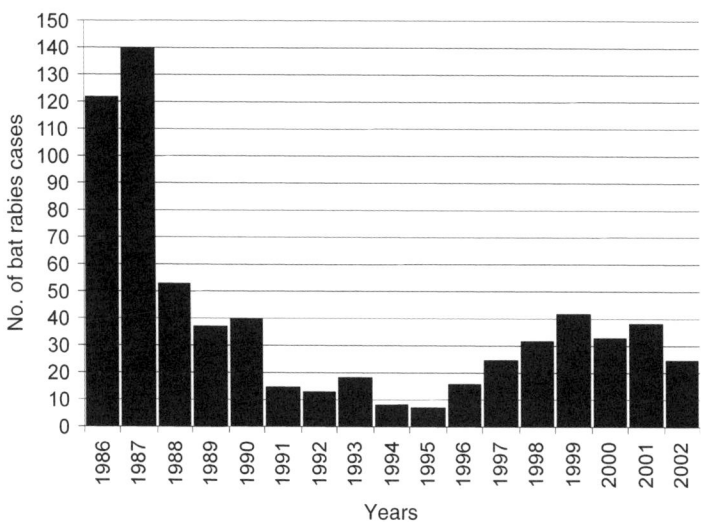

Figure 6.3   Dynamics of bat rabies cases in Europe during 1986–2002 (n=664).

cases, diagnosed throughout Europe, from Spain to the Ukraine and western Russia. The town of Belgorod appears to be the most eastern known point of virus distribution. However, surveillance in the former Soviet Union must be extended to validate this assumption. Most of the cases were reported from the northwestern coastal regions of Germany, Denmark and the Netherlands (Potzsch *et al.*, 2002).

About 95% of EBLV-1 cases have been observed in *E. serotinus* (Schneider and Cox, 1994; Amengual *et al.*, 1997). However, it was reported also in *Nyctalus noctula* and *Vespertilio murinus* (Selimov *et al.*, 1991), and *Myotis myotis, M. dasycneme, M. daubentonii, Pipistrellus pipistrellus, P. nathusii, M. natterreri, Rhinolophus ferrumequinum* and *Miniopterus schreibersii* (Schneider and Cox, 1994; Serra-Cobo *et al.*, 2002; Van der Poel *et al.*, 2005). Antibodies to EBLV-1 were found additionally in *Tadarida teniotis* (Serra-Cobo *et al.*, 2002).

EBLV-2 was diagnosed not as frequently as EBLV-1 (Figure 6.4). This virus was isolated from *M. dasycneme* and *M. daubentonii* with the only exception being when it was detected in *Nyctalus noctula* or *Vespertillio murinus* from the Ukraine. This latter record is confusing since it belongs to an atypical bat species, lies quite far from all other EBLV-2 isolation points and has an inconsistent preliminary history. Two isolates were obtained from bat carcasses collected under a few roosts in the Ukraine during 1987. Both were identified originally as 'Duvenhage'

Figure 6.4 Isolation points of EBLV-2. Adapted from Johnson *et al.* (2003). Bat cases are shown by circles, whereas human cases are shown by squares; the question mark indicates the uncertain case in the Ukraine.

using anti-N-MAbs and their antigenic patterns were identical to each other and to those of the EBLV-1 isolate 'Yuli' (Selimov *et al.*, 1991). Further, one of them was sequenced at the Pasteur Institute (Paris, France) and confirmed to be EBLV-1 (GenBank accession U89465; Amengual *et al.*, 1997), but the second one, which was sequenced in the Veterinary Laboratories Agency (Weybridge, UK), was identified as EBLV-2 (GenBank accession AY212119; Johnson *et al.*, 2003). As was suggested, this latter identification result might be caused by cross-contamination or mislabeling of the samples in the laboratory, because it is unlikely the original isolate could belong to EBLV-2 (Kuzmin *et al.*, 2006a).

Prevalence of EBLVs in the populations of European bats has not been studied sufficiently. All bat species in Europe are protected. Passive surveillance of abnormal or dead bats submitted to veterinary laboratories often gives inappropriate results. Thus, 21% of abnormal or dead *E. serotinus* bats and 4% of *M. dasycneme* bats submitted for diagnosis in the Netherlands during 1984–2003 were EBLV-1 positive (Van der Poel *et al.*, 2005). Active surveillance had been limited mostly to the screening of oral swabs for the presence of viral RNA and to serological tests for antibody. During one survey in Spain, 15 of 71 oral swabs obtained from apparently healthy *E. serotinus* bats were reported positive for EBLV-1 RNA. Additionally, viral RNA was detected in 13 oral swabs but only in five brains of the 34 bats from which simultaneous testing of brains and oral swabs was available (Echevarria *et al.*, 2001). In general, the authors detected viral RNA in 6.7% of harvested brains, but in 20% of harvested oral swabs. These findings suggested that bats might be 'carriers' of EBLV-1 and shed virus in saliva without brain infection. A further study, published by another Spanish team (Serra-Cobo *et al.*, 2002), again reported the presence of EBLV-1 RNA in the esophagous-larynx-pharynx and lung of *Rhinolophus ferrumequinum* bats with negative detection in brain. The same paper reported viral RNA in three of 27 blood pellet samples obtained from apparently healthy *Myotis myotis* bats. These bats were bled and released, and for this reason no further information on the presence of viral RNA and infectious virus in their tissues was available. EBLV-1 RNA was also detected in tissues of apparently healthy zoo bats, *Rousettus aegyptiacus*, which presumably acquired the infection from European insectivorous bats (Wellenberg *et al.*, 2002). However, the possibility of a carrier state has never been confirmed in a number of experimental studies addressed below, as well as presence of viral RNA in blood pellet of experimentally infected bats.

Serologic surveys in Spanish bat populations demonstrated the presence of anti-EBLV-1 antibodies in 7.8% of serum samples. In some colonies, prevalence was as high as 20.8 and 22.7% (Serra-Cobo *et al.*, 2002). Additionally, the authors reported some dynamics in the percentage of seropositive bats in some colonies during a number of successive years. In one location, the primary prevalence in 1995 was 3.3%; in 1996 it became as high as 59.3% and in 1999 again decreased, to 10%. Screening of Scottish bats for antibodies to EBLV-2 demonstrated that 0.05 to 3.8% (95% confidence interval) of *M. daubentonii* bats were seropositive. However, in one location, the prevalence was much higher, 2.9 to 16.3%. Neither viral RNA nor infectious EBLV-2 were detected in 218 oral swabs obtained from Scottish bats (Brookes *et al.*, 2005).

EBLV-2 is the only lyssavirus to have been isolated in the UK, yet the recent finding of antibody to EBLV-1 in *E. serotinus* bats suggests that this virus circulates there as well (ProMed mail, archive 20050606.1576). The last previous indigenous human rabies case was registered in the UK in 1902. In 1993, King and Turner wrote that none of 1250 bats found dead and submitted for rabies diagnosis to the UK's Central Veterinary Laboratory was positive for lyssaviruses. Nevertheless, EBLV-2 was isolated there from *M. daubentonii* bats four times, from 1996 to 2004 (Johnson *et al.*, 2003; Brookes *et al.*, 2005) and also from a human (bat worker) in Scotland in 2002 (Fooks *et al.*, 2003a). During that year, the unvaccinated patient undertook approximately 13 tasks involving bats and was bitten by those animals on some occasions while not wearing gloves.

This was the fourth human rabies case of bat origin in Europe. Thus, there were two well reported human cases of EBLV-2 infection, one in Finland in 1985 and the other in the UK in 2002, and one case of EBLV-1 infection in Russia in 1985. It is important to emphasize that virus from the first case, which occurred in Voroshilovgrad in 1977, was not identified by serotype or genotype. For that reason it is incorrect to describe it as an EBLV (King and Turner, 1993; Ronsholt, 2002; Fooks *et al.*, 2003a, 2003b; Brookes *et al.*, 2005).

A similar case occurred in Lugansk (formerly Voroshilovgrad) province again, this time in 2002. A 34-year-old man died after disease clinically compatible with rabies. He was bitten on the finger when handling a bat, approximately 1.5 months before the disease onset. The patient treated the bite wound himself, using iodine solution, but did not seek rabies prophylaxis. No contacts with other mammals prior to disease were established (Botvinkin *et al.*, 2006). Neither ante-mortem nor post-mortem virological investigation were performed, so there is no way to determine which virus caused the disease. In both these cases the causing virus likely might be an EBLV, but it also might be some other lyssavirus. For example, the location at which the West Caucasian bat virus (WCBV) was discovered in 2002 is quite close to the town of Voroshilovgrad.

Indeed, the isolation of WCBV was surprising. An active bat survey along the Caucasian shore of the Black Sea was undertaken to elucidate a possible EBLV circulation in that region. All 129 bats collected during the survey appeared healthy, and no sick animals or carcasses were found in the roosts. Since bat brains were collected in 50% glycerol, they were not subjected to the routine fluorescent antibody test. The enzyme-linked immunosorbent assay (ELISA) with anti-RABV and anti-EBLV-1 immunoglobulins revealed negative results for all samples. Nevertheless, the mouse inoculation test yielded one neurotropic agent from the brain of a *Miniopterus schreibersii* bat. Brain impressions of the dead mice demonstrated typical lyssavirus fluorescence being stained with commercially available FITC-labeled anti-RABV conjugates. Further analysis with anti-N-MAbs and sequencing of viral genes demonstrated that WCBV is possibly a new member of the *Lyssavirus* genus (Botvinkin *et al.*, 2003).

The description above is given to demonstrate that bat surveillance, performed with the use of PCR-based methods only, may give unreliable negative results (Echevarria *et al.*, 2001; Serra-Cobo *et al.*, 2002). We would have missed WCBV if this kind of surveillance alone were implemented. As was shown, based on the N, P and G gene sequences, this virus is the most distant member of the genus and demonstrates only a limited similarity to phylogroup 2 lyssaviruses (LBV and MOKV) (Kuzmin *et al.*, 2005). The WCBV does not react with many primers designed for EBLVs. Thus, inoculation of laboratory animals and cell cultures still appears to be the most consistent way to carry out a field surveillance when multiple or unknown infectious agents may be encountered.

Only one sample of WCBV has been discovered to date and further active and passive surveys are needed to evaluate the distribution, host range and circulation properties of this virus. The bat *M. schreibersii* is a migratory species distributed in many territories of Eurasia and Africa, roosting mainly in caves. Of interest is the report about the rabid bat, *Rh. ferrumequinum*, from Turkey found in 1956 (Tuncman, 1958). The *M. schreibersii* populations in the Caucasus mountains and in Turkey may have regular contacts and the *Rh. ferrumequinum* bat is a cave roosting species as well.

Very limited information on bat lyssaviruses is available from Asia. One record describes a probable 'rabies virus' isolate in India (Pal *et al.*, 1980) and one in Thailand (Smith *et al.*, 1968), both of Pteropid origin. These have never been confirmed by further identification (evidently, the isolates have not been stored) or other observations. No rabies was found in 1013 bats examined in the Philippines (Beran *et al.*, 1972) or in 478 bats in Malaysia (Tan *et al.*, 1969). One human rabies case after bat bite was suspected in northern China in 2002; however, the diagnosis was based on clinical symptoms only – no virological assay was implemented and post-mortem samples were not stored (Tang *et al.*, 2005).

Three isolates of RABV were reported from Siberian bats. One of them was obtained from a *Myotis daubentonii* bat in the Novosibirsk province of Western Siberia and referred to as 1157 or the Novosibirsk bat (Botvinkin, 1988). Another one, referred to as 1150 or 'Omsk' bat (Botvinkin, 1988; King *et al.*, 1990; Botvinkin *et al.*, 1992; Kuzmin *et al.*, 1994; Kuzmin and Botvinkin, 1996) was obtained in 1984 from a *Vespertilio murinus* bat in the town of Omsk. Gene sequence verification of these isolates after 5–9 mouse brain passages demonstrated the presence of the laboratory RABV strain CVS only. Therefore, it is impossible to establish if these isolations were primary laboratory mistakes or if the original viruses were contaminated by CVS during further passages. The third RABV, referred to as 'Olekma' (Khozinski *et al.*, 1991) was isolated from the brain of an *Eptesicus nilssonii* bat delivered from Yakutia in 1987. The bat samples were processed together with a collection of Arctic fox (*Alopex lagopus*) samples, some of which were found to be positive for RABV (arctic virus variant). In reactions with anti-N-Mabs, the bat isolate demonstrated properties of arctic RABV. This virus was lost very fast and we consider it as a primary laboratory contamination with an Arctic fox sample (Kuzmin *et al.*, 2006a).

An unusual communication on bat rabies in Western Siberia was published in the ProMed mail in 2003 (Archive 20030525.1291). Of 18 *Myotis daubentonii* bats collected in the caves of the Novosibirsk province and the Altay region, six were positive for lyssavirus antigen in the brain by the fluorescent antibody test. The viruses were preliminarily described as genotype 1 representatives. This study was then extended. In total, the authors collected 88 hibernating bats. Brains of 24 (27.3%) were positive for lyssavirus antigen and 17 (19.3%) were positive by RT-PCR performed using internal N gene primers. Mouse inoculation revealed encephalitic signs in a limited number of animals (such as bristled fur, humped backs and incoordination). The authors observed lyssavirus antigen in the mouse brain impressions, but the RT-PCR of mouse brains, performed using the same primers, was negative. During the next intracerebral passages, the number of mice demonstrating encephalitic signs was reduced and, finally, all isolates but one were completely cleared (Zaikovskaia *et al.*, 2004).

Partial N gene sequences were obtained for seven isolates (from positions 319 to 856, according to the Pasteur virus genome). The sequence of at least one of them (BAT-BN1) was compared with sequences of other RABVs, originating from different regions of the world and belonging to different phylogenetic lineages. The bat isolate was clearly placed within the clade of terrestrial RABVs isolated in the same territory of Western Siberia from foxes, a badger, a cat and a human exposed to a fox. This clade subsequently belonged to the generic 'dog-fox' RABV cluster circulating worldwide. The authors proposed that bats in Western Siberia maintain circulation of the same RABVs that are found in terrestrial mammals and that virus transmission between those hosts may be realized along the food chain (Zaikovskaia *et al.*, 2005). Such unusual claims should be corroborated by additional studies, including field sampling and further laboratory surveys to avoid a possibility of cross-contamination of the specimens.

Three novel lyssaviruses were isolated from bats in the Asian territory of the former Soviet Union. Aravan virus (ARAV) was isolated in southern Kyrgyzstan in 1991 from a *Myotis blythi* bat. Anti-N-MAb typing of ARAV demonstrated a unique antigenic pattern and the virus was suggested as a new putative serotype (Kuzmin *et al.*, 1992). A second isolate, Khujand virus (KHUV), was obtained in northern Tajikistan in 2001 from a *Myotis mystacinus* bat. The isolation point of KHUV was separated from the isolation point of ARAV by only 260 km. Surprisingly, the antigenic patterns of KHUV appeared different from other lyssaviruses, including ARAV (Kuzmin *et al.*, 2001). Phylogenetic analysis of viral genes suggested that KHUV was most related to genotype 6, while ARAV was related to KHUV and demonstrated moderate similarity to genotypes 4, 5 and 6 (Kuzmin *et al.*, 2003). One more lyssavirus, named Irkut (IRKV), was isolated from a *Murina leucogaster* bat in eastern Siberia, in the town of Irkutsk, in 2002. Phylogenetic analysis demonstrated that IRKV was a member of the cluster that joined genotypes 4 and 5 (Botvinkin *et al.*, 2003).

As was mentioned before, the ICTV recognizes virus species rather than genotypes. The species definition officially recognized by ICTV since 1991 is: 'A virus species is a polythetic class of viruses that constitute a replicating lineage and occupy a particular ecological niche' (Büchen-Osmond, 2003). Thus, 'species' is quite a subjective term and to decide whether a new virus may be separated into new species or not is entrusted to the collective decision of a group of taxonomy experts. Additionally, according to the given definition, it may be applied for a group of isolates only. As only one isolate of each ARAV, KHUV, IRKV and WCBV is available, these viruses cannot be considered as separate species. However, quantitative separation of lyssavirus by genotypes, which has been appreciated in a number of publications (Bourhy *et al.*, 1993; Kissi *et al.*, 1995; Badrane *et al.*, 2001; Nadin-Davis *et al.*, 2002) allows one to consider them all as four new lyssavirus genotypes (Kuzmin *et al.*, 2005).

In fact, phylogenetic similarity of DUVV, EBLVs, ARAV, KHUV and IRKV is quite obvious (Figure 6.5). They present a monophyletic group of Old World bat lyssaviruses (of African and Eurasian origin) (Kuzmin *et al.*, 2005). Have all of them originated from Africa, as was proposed earlier for the EBLVs (Schneider, 1982; Shope, 1982; Amengual *et al.*, 1997; Serra-Cobo *et al.*, 2002)? This is a possibility, because all putative reservoir species appear to be broadly distributed, due in no small measure to the power of flight. However, only one member of this group, DUVV, is present in Africa. Other lineages are found in Eurasia. Is this observation evidence of independent evolution after an initial introduction of a progenitor virus from Africa? DUVV is no more closely related to LBV and MOKV than it is to EBLVs, ARAV, KHUV and IRKV. Africa was proposed as the continent of primary lyssavirus development and evolution because the two most distant members of the genus, LBV and MOKV, have been found only there. Additionally, MOKV demonstrated antigenic relatedness to African arthropod rhabdoviruses, Obodhiang and Kotonkan, hypothesized as examples for probable ancestors of the lyssaviruses (Shope, 1982). However, further genome sequencing clearly demonstrated that Obodhiang and Kotonkan are ephemeroviruses and not lyssavirus ancestors. WCBV, which is the most divergent member of the genus known to date, demonstrates only a limited phylogenetic relatedness to LBV and MOKV and was isolated in Eurasia. Has this virus derived from Africa as well? Perhaps, because it was found in a migratory bat species broadly distributed in many regions of Africa and Eurasia. However, there are no reasons to reject the alternative hypothesis that ancient lyssaviruses appeared and evolved in the territory of modern Eurasia.

Bat surveys performed in some regions of southern Asia during recent years yielded no lyssavirus isolates but demonstrated presence of anti-lyssavirus

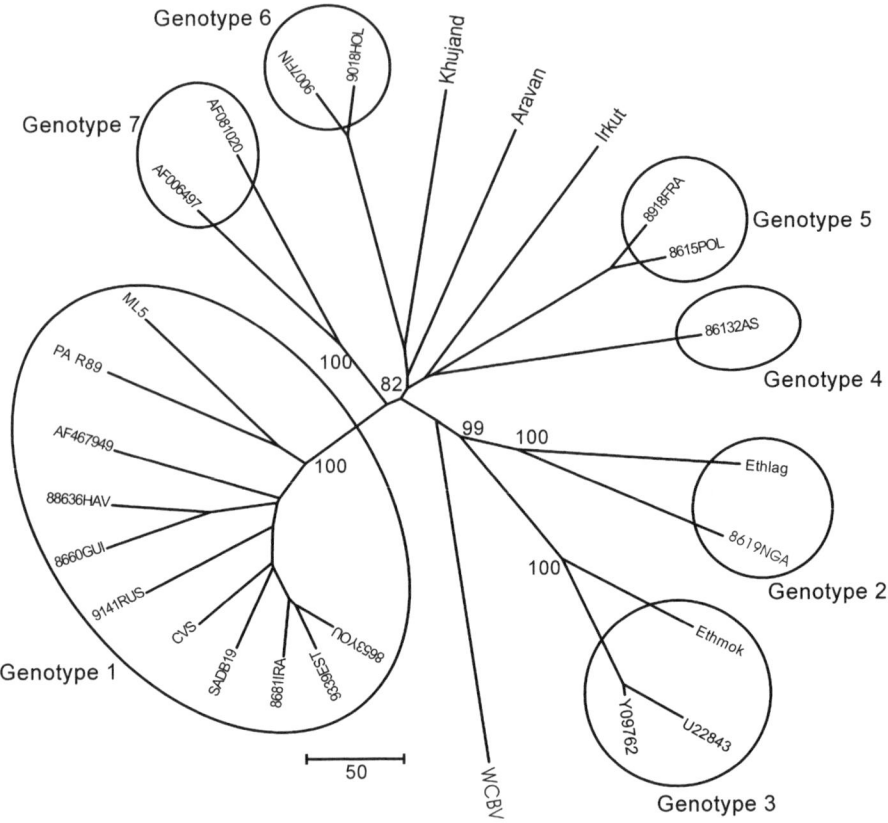

Figure 6.5 Unrooted phylogenetic tree of *Lyssavirus* genus based on the entire N gene sequence, obtained by the neighbor-joining method. Bootstrap values (% for 1000 replicates) are shown for key nodes and branch lengths are drawn to scale.

antibodies in bat sera. Of 231 sera samples collected from the Philippines and screened against RABV and ABLV, 22 (9.5%) demonstrated neutralizing activity against the ABLV, whereas no neutralization of RABV was detected (Arguin *et al.*, 2002). Those bats included six different species from both Megachiroptera and Microchiroptera suborders. At the time of that survey no Asian bat lyssavirus isolates were available for comparative testing. Further studies were performed using ARAV, KHUV and IRKV for comparative challenge *in vitro* and the presence of antibodies neutralizing these viruses was demonstrated in bat sera from Thailand (Lumlertdacha *et al.*, 2005) and Bangladesh (Kuzmin *et al.*, 2006b). In both latter studies, the greatest number of positive findings was made in Pteropid bats. In Thailand, 4.1% of tested *Pteropus lylei* bats were positive and, in Bangladesh, 2.4% of tested *P. giganteus* bats were positive. However, serum samples from small insectivorous bats were often presented in limited small volumes and tested at high dilution. Given this technical limitation, there may be an underestimation of the positive results. In Cambodia, 14.7–16% of bat serum samples (both *Mega-* and *Microchiroptera*) were positive for antibodies against RABV, EBLV-1, ABLV and even LBV, although the authors did not use Asian bat lyssaviruses for a comparison (Reynes *et al.*, 2004). This broad range of reactions may be due to some antigenic cross-reactivity reported among members of the *Lyssavirus* genus (Hanlon *et al.*, 2005). Detectable antibody may cross-react with other related lyssaviruses, as well as viruses yet to be discovered. Definitely, additional surveillance screens in bats are required in the Old World.

Spillover of EBLV-1 into terrestrial animals has occurred infrequently and no spillover information is available for other bat lyssaviruses encountered in Eurasia. Four EBLV-1 cases were registered in sheep in Denmark in 1998 and 2002 (Ronsholt, 2002) and one case in a stone marten in Germany in 2001. This sparsity may depend on a limited susceptibility of some terrestrial mammals to EBLVs. In one experiment, all ferrets inoculated intramuscularly with 10^6 foci-forming units (FFU) of EBLV-1 developed rabies, whereas only 43% of ferrets inoculated with 10^4 FFU of EBLV-1 and none of those which received 10^4 FFU of EBLV-2 developed rabies (Vos *et al.*, 2004). Dogs were not susceptible to intramuscular inoculation with EBLV-1, whereas cats were and they died of rabies within 15 days (Fekadu *et al.*, 1988a). Ferrets survived inoculation with $10^{4.2}$ $MICLD_{50}$ of ARAV (100%) and $10^{4.6}$ $MICLD_{50}$ of KHUV (66.6%), but did not survive inoculation with $10^{5.3}$ $MICLD_{50}$ of IRKV (Hanlon *et al.*, 2005).

One important question addresses peripheral pathogenicity of WCBV. As was mentioned above, D_{333} in the glycoprotein ectodomain of LBV and MOKV was a suggested reason for their reduced peripheral pathogenicity in mice. The LC8 binding site of their phosphoprotein is also different from that of phylogroup 1 lyssaviruses. The WCBV has E_{333} in the glycoprotein ectodomain and the LC8 binding site of its phosphoprotein is different from both phylogroup 1 and 2 lyssaviruses. In our experiments, WCBV was apathogenic for 3-week-old mice and ferrets by intramuscular, subcutaneous, intraperitoneal and oral routes, even when doses of $10^{6.3}$ $MICLD_{50}$ were given. However, 78% of Syrian hamsters, challenged intramuscularly with the same virus dose, developed typical rabies and succumbed. Big brown bats demonstrated limited susceptibility to the same dose of WCBV injected intramuscularly and, after their death, the virus was detected in their brain but not in extraneural tissues. Pathogenesis and circulation patterns of this virus should be studied more extensively.

Clinical signs in bats infected by EBLV-1, ARAV, KHUV, IRKV and WCBV were similar to those observed in bats dying of RABV infection. In some instances, signs of encephalitis, such as tonic-clonic convulsions, ascending paresis and paralysis were observed. Biting behavior was repeatedly reported; however, bats often were too weak, could not actively chase and attack another bat or an artificial irritant, such as forceps (Figure 6.6). In other cases hypersensitivity for

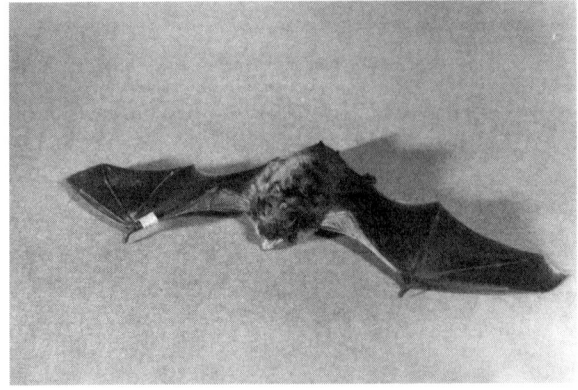

(A)

(B)

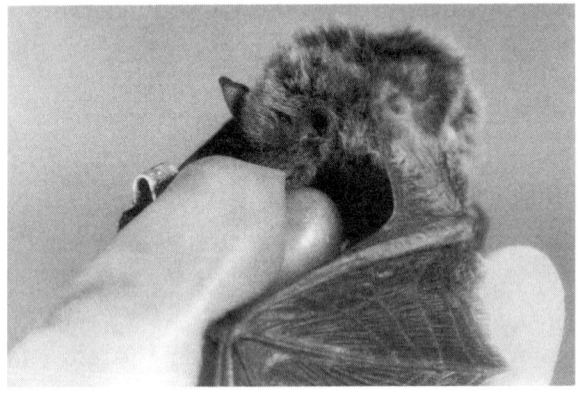

(C)

Figure 6.6 Behavior of bats *Pipistrellus pipistrellus* during clinical onset of experimental ARAV infection. (A) Weak bat, incapable of flying, is stretching the wings and vocalizing, trying to frighten an experimenter; (B) tonic-clonic convulsions in rabid bat; (C) rabid bat biting the experimenter's glove (Photo by I. Kuzmin).

high frequency sounds was reported. Finally, bats often were seen emaciated and exhausted, and unable to fly, without any specific signs of brain dysfunction. They could only utter a prolonged loud vocalization and uncontrolled wing beats when disturbed (Kuzmin and Botvinkin, 1996; Bruijn, 2003; Hughes *et al.*, 2006).

Only a few experimental studies on the pathogenesis of EBLV-1, ARAV, KHUV, IRKV and WCBV infections in bats have been performed to date. In some of these, an additional RABV isolate 1150 was used for bat inoculation (Botvinkin *et al.*, 1992; Kuzmin *et al.*, 1994; Kuzmin and Botvinkin, 1996) but, as we described above, it was identified later as a laboratory strain, CVS.

All experiments demonstrated low to moderate susceptibility of bats to Eurasian bat lyssaviruses, in the same way as it was shown for some varieties of American bat RABV described above. In one experiment, *Pipistrellus pipistrellus* bats were captured at the beginning of hibernation and transported to the laboratory in a cold environment. Half of the animals were awakened, inoculated intramuscularly with EBLV-1 and ARAV and kept in an active state at room temperature. The other half of the bats were inoculated and maintained in hibernation, at 2–5°C, during the first 60 days post challenge. Thereafter, they were awakened and transferred to ambient room conditions as well. The incubation period in the latter group of bats after their awakening was the same as that in bats inoculated and maintained in the active state (Kuzmin *et al.*, 1994; Kuzmin and Botvinkin, 1996). This demonstrates that viruses can be conserved in bats in an inactive state for 60 days of hibernation and possibly longer.

Distribution of virus in bat tissues is presented in Tables 6.5 and 6.6. In general, virus was detected only in bats which succumbed to the disease. In all those animals, the virus was detected in brain and infrequently in some extraneural tissues. Among the extraneural tissues, the salivary glands were the organs where virus was detected most regularly. The function of brown fat for virus conservation was not confirmed. Lungs and kidneys of a few bats were demonstrated to be positive for infectious virus (however, in low titers) and additional investigations of possible respiratory and urinary virus excretion were suggested.

Viral RNA or infectious virus was detected in oral swabs of dying bats 0–2 days before the onset of clinical signs (or 0–4 days before death). All oral swabs available within 3 days of the first positive detection were negative. Therefore, no detectable amounts of viral RNA appeared in the saliva earlier than 5 days before the onset of clinical signs, or within 1 week of death. The animals that survived the inoculation never demonstrated signs of disease during the observation period, and their oral swabs obtained in that period never contained virus – nor did their brains and salivary glands harvested at the end of the experiment (4–6 months post challenge).

Transmission of virus from the inoculated bats to non-inoculated cage mates could never be demonstrated, which was suggested on the basis of different habitation capabilities that occur in natural bat colonies versus laboratory cages.

TABLE 6.5

Isolation of EBLV-1 and ARAV from the tissues of intramuscularly inoculated bats

Bat species	Virus; dose (MICLD$_{50}$); group[a] (if applicable)	Number inoculated	Number dying[b]	Virus isolated from the tissues				
				BR[c]	SG	BF	LU	KI
M. daubentonii and M. brandtii	EBLV-1; $10^{3.6}$	25	23	6	1[d]	0	0	1
	EBLV-1; $10^{5.8}$	33	32	3	1	0	1	0
P. pipistrellus	EBLV-1; $10^{6.2}$; group A	23	23	4	0	0	0	1
	EBLV-1; $10^{6.2}$; group B	23	23	11	4	2	1	0
	ARAV; $10^{5.4}$; group A	24	24	8	3	4	3	1
	ARAV; $10^{5.4}$; group B	24	24	6	4	3	0	0

[a] Group A – bats inoculated and maintained in active state; group B – bats inoculated and maintained in hibernation during the first 60 days post-inoculation and awakened and transferred to the active state thereafter.

[b] Only bats dying more than one week post-inoculation are considered.

[c] BR = brain; SG = salivary glands; BF = brown fat; LU = lung; KI = kidney.

[d] In all cases when virus was detected in an extraneural tissue, the brain was positive as well.

TABLE 6.6

Isolation of IRKV, ARAV, KHUV and WCBV from the tissues of intramuscularly inoculated E. fuscus bats

Virus; dose (MICLD$_{50}$)	Number inoculated	Number dying	Virus isolated from the tissues					
			BR[a]	SG	BF	LU	KI	BL
IRKV; $10^{4.7}$	11	6	6	5[b]	0	2	0	0
ARAV; $10^{3.9}$	4	3	3	0	0	0	0	NT[c]
KHUV; $10^{4.3}$	5	3	3	1	2	1	0	NT
WCBV; $10^{5.7}$	12[d]	3	3	0	0	0	0	0

[a] BR = brain; SG = salivary glands; BF = brown fat; LU = lung; KI = kidney; BL = bladder.

[b] In all cases when virus was detected in an extraneural tissue, the brain was positive as well.

[c] NT – not tested.

[d] Also 6 bats inoculated per os with the same virus dose are not included (all survived, negative).

Experimental inoculation of R. aegyptiacus fruit bats with two EBLV-1 isolates also demonstrated presentation of typical rabies rather than a carrier state. Of 22 bats inoculated, seven (32%) succumbed to the disease. The virus was detected in the brain of each dying animal and in the salivary glands of two of them. The tissues of all survivors were virus-negative (Van der Poel et al., 2000).

TABLE 6.7

Cross-neutralization activity among lyssaviruses (log$_{10}$ differences)

Antibody	Virus[a]						
	CVS	DUVV	EBLV-1	EBLV-2	ARAV	KHUV	IRKV
RABV	0.0	0.7	0.6	0.6	0.7	0.4	0.7
DUVV	1.3	0.0	0.3	0.6	0.8	0.6	1.1
EBLV-1	2.2	1.2	0.0	0.5	0.6	0.3	0.7
EBLV-2	1.1	0.7	0.3	0.0	0.9	0.3	0.8
ARAV	1.1	0.8	0.5	0.1	0.0	0.0	0.6
KHUV	0.9	1.1	0.2	0.7	0.5	0.0	0.8

Antibody	Virus[b]			
	RABV (CVS)	LBV	MOKV	WCBV
RABV	250	<5	<5	<5
LBV	95	3125	17	<11
MOKV	<11	431	989	<11
WCBV	<11	<11	<11	6390

Adapted from Hanlon *et al.* (2005), with permission from Elsevier.

[a] Mouse anti-sera (antibody) were prepared against RABV (challenge virus standard, CVS-11), DUVV, EBLV-1, EBLV-2, ARAV and KHUV using standard techniques and evaluated using the rapid fluorescent focus inhibition test with CVS-11, DUVV, EBLV-1, EBLV-2, ARAV, KHUV and IRKV. Results represent the log titer difference in the amount required for equivalent neutralization.

[b] Mouse anti-sera (antibody) were prepared against RABV, LBV, MOKV and WCBV using standard techniques and evaluated using the rapid fluorescent focus inhibition test with CVS-11, LBV, MOKV and WCBV. Results are reported in reciprocal titers.

Since all commercially available anti-rabies biologicals have been produced using RABV, their capability to protect humans and other animals against non-RABV lyssaviruses is lacking. All phylogroup 1 lyssaviruses cross-neutralize each other, with different degrees of activity and, in most instances, anti-RABV antibodies neutralized non-RABV viruses incompletely (Table 6.7). Recently developed anti-glycoprotein MAb CR57 readily neutralized all phylogroup 1 lyssaviruses but did not neutralize phylogroup 2 lyssaviruses and WCBV (Marissen *et al.*, 2005). The phylogroup 2 lyssaviruses and WCBV are quite different from RABV and from each other, with only a limited cross-neutralization between LBV and MOKV and an absence of cross-neutralization for WCBV (Hanlon *et al.*, 2005). This is another reason why WCBV could not be considered as a phylogroup 2 virus.

Immunization experiments demonstrated similar results. The HDCV and animal vaccines Rabisin and Babiffa protected mice against a challenge with EBLV-1, but the protection against EBLV-2 was only partial (Fekadu *et al.*, 1988b). As was reported, the PV RABV strain induced better protection against EBLV-1 than did the PM strain (100% and 36%, respectively) and both of these strains gave equally incomplete protection against EBLV-2 (80–85%) (Jallet *et al.*, 1999). However, this observation might be interpreted as an effect of different amounts of antigen in the vaccines used rather than the properties of the specific strain, because these differences are not supported by genetic distances between PV and PM versus EBLV-1 and EBLV-2 (Dietzschold *et al.*, 1988).

Pre-exposure and conventional post-exposure prophylaxis gave incomplete protection of hamsters against ARAV, KHUV and IRKV. Neither pre-exposure nor post-exposure courses of both human and veterinary biologicals produced a sufficient protective effect against WCBV (Hanlon *et al.*, 2005). Further biologicals, such as chimeric DNA vaccines and selective MAbs, are needed to provide a consistent protection against this virus as was suggested for LBV and MOKV (Bahloul *et al.*, 1998; Jallet *et al.*, 1999; Nel *et al.*, 2003).

3.3 Australia

Prior to 1996, Australia had been considered free of rabies and rabies-like viruses. An outbreak of rabies involving several dogs occurred in the island state of Tasmania in 1867, but was quickly eradicated. Since then, only a few imported rabies cases have been registered (Fraser *et al.*, 1996).

Following the discovery that flying foxes were a reservoir of Hendra virus, surveillance of these animals was increased, particularly of those which were found sick or injured. In 1996, a young female black flying fox (*Pteropus alecto*) was found under a fig tree, unable to fly, in Ballina, New South Wales. It was euthanized and organs were submitted for virological study, mainly in regard to Hendra virus. The test on this virus was negative, but evidence of severe non-suppurative encephalitis was found in the brain. Typical inclusions of lyssavirus antigen were demonstrated through different brain areas by staining with anti-RABV fluorescein-labeled antibody conjugate. A second case was recognized retrospectively in a juvenile female of the same species from northern Queensland. This bat had been euthanized in 1995 with evidence of unusual aggressiveness.

The only fresh tissues available for isolation from the 1996 case were blood, lung, kidney and spleen. The virus was isolated from a kidney suspension. Signs of disease in mice, and pathomorphological and electron microscopy findings were compatible with rabies. RNA sequencing demonstrated that the isolate, preliminarily named Ballina, was more related to classical RABV than to other non-RABV lyssavirus representatives. Nevertheless, the amount of distinction allowed one to propose that the virus should be considered as a new genotype

(Fraser *et al.*, 1996). Shortly, other isolates were obtained and their affiliation with a monophyletic clade, quite distinguishable from RABV, became apparent. The seventh genotype and species, ABLV, was thus established (Hooper *et al.*, 1997; Gould *et al.*, 1998; Tordo *et al.*, 2005).

There are four flying fox species in continental Australia: *P. alecto, P. poliocephalus, P. scapulatus* and *P. conspicillatus*, and the virus was recognized in each of them, along the eastern coastal territory of the continent (Figure 6.7). Genetically, all of the pteropid isolates were similar or identical to each other, without any correlation to a particular host species or geographic location (Guyatt *et al.*, 2003). Pteropid bats roost in trees in colonies ('camps') that frequently number in the thousands of animals belonging to one or several species. These colonies may fluctuate in size, depending on available food resource and season and animals from one colony can move to another one, especially during periods of migration. This dynamic social structure has been invoked to explain the circulation of similar viruses in these animals (Hooper *et al.*, 1997; Guyatt *et al.*, 2003).

Several samples of ABLV were obtained from the insectivorous bat *Saccolaimus flaviventris*. Their nucleotide sequences formed another monophyletic clade,

Figure 6.7 Location of ABLV isolates within Australia. The bat species and location are given where known: **?**: an unspecified Pteropus bat; X: *P. alecto* ; O: *S. flaviventris*; □: *P. scapulatus*; and ■: *P. poliocephalus*. Adapted from Gould *et al.* (2002), with permission from Elsevier.

clearly distinguishable from the pteropid clade. However, when compared with other lyssavirus genotypes, these two lineages were still confidently segregated into one cluster, demonstrating that their differentiation from each other occurred later than differentiation of ABLV from other lyssavirus genotypes (Gould et al., 2002). It should be additionally clarified whether other species of Australian insectivorous bats participate in ABLV circulation and, if so, do they maintain the same virus strain as S. flaviventris bats or does additional variability of the virus occur?

Two human cases of ABLV infection have been described to date. Both were fatal and the clinical symptoms were compatible with rabies. The first one was reported very shortly after the virus discovery, in 1996. The patient was a 39-year-old female presumably infected by an S. flaviventris bat in her care. The virus that was isolated was compatible with this bat species (Allworth et al., 1996; Gould et al., 2002). The second case occurred in a 37-year-old female who developed rabies in 1998, approximately 27 months after presumable exposure from a bite by an unspecified flying fox. This isolate belonged to the pteropid ABLV variant (Hanna et al., 2000; Warrilow et al., 2002).

Little is known about ABLV circulation patterns. The virus was detected in 6% of sick, injured or orphan flying foxes submitted to diagnostic laboratories. In one study, 9.4% of flying foxes submitted to laboratories because they had bitten or scratched humans, or where testing was considered to be in the interest of public health, were ABLV-positive (Warrilow et al., 2003). Serological surveys of a mixture of sick and apparently healthy bats demonstrated that 16% of the bats were seropositive to ABLV (Hooper et al., 1997). It is interesting that the distribution range of P. alecto bats extends into Papua, New Guinea and the eastern islands of Indonesia (Fraser et al., 1996). There is no reason to expect that distribution of ABLV is limited to continental Australia. For example, the presence of antibodies to this virus was demonstrated in 9.5% of bat serum samples collected in the Philippines (Arguin et al., 2002).

Transmission mechanisms, susceptibility and pathogenesis of ABLV infection in natural hosts have not been sufficiently studied. At least one case of human exposure occurred after a bite. The virus antigen was detected in the salivary glands of one of eight naturally infected flying foxes and all had the antigen in the brain (Hooper et al., 1997). Three of ten P. poliocephalus bats, infected peripherally with 10^5 $TCID_{50}$ of ABLV, developed rabies and were euthanized 15–24 days post challenge. Clinical signs in these animals included weakness, generalized trembling and limb paresis or paralysis (McColl et al., 2002). This incubation period and limited susceptibility is in agreement with one natural observation when a rabid wild P. alecto bat was removed from the top of a wire-mesh container for viewing bats at a zoo, inside of which 23 flying foxes were housed. One month later, one of these captive flying foxes developed signs of abnormal behavior, was euthanized and confirmed ABLV-positive. The isolate was compared with the one originating from the P. alecto bat which had earlier been found on the top of the container. The virus gene sequences were identical. No abnormalities and no antibody response were detected in the other 22 flying foxes following 6 months of quarantine (Warrilow *et al.*, 2003).

Public health strategies to minimize ABLV infection include management to minimize bat–human interactions, extensive awareness to educate the general public and the full range of human rabies prophylaxis (including pre-exposure vaccination for persons of high risk). The main points of public education include: not picking-up sick or injured bats; and preventing bats from gaining roost access in houses (Field *et al.*, 2004). The former was emphasized, as sick and injured bats have been found to have a significantly higher prevalence of infection than do randomly collected bats. The risk posed by sick and injured bats was highlighted by a study which found that volunteer animal handlers accounted for 39% of potential exposures, their family members for 12%, professional animal handlers for 14%, community members who intentionally handled bats for 31% and community members where the contact was initiated by the bat, for 4% (McCall *et al.*, 2000). Modern potent rabies vaccines were demonstrated to be sufficiently protective against ABLV and commercially available anti-RABV immunoglobulins also readily neutralized this virus (Hooper *et al.*, 1997).

REFERENCES

Acha, P.N. (1967). Epidemiology of paralytic bovine rabies and bat rabies. *Bulleten of the Office International des Epizooties* **67**, 343–382.

Aguilar Setien, A., Brochier, B., Tordo, N. *et al.* (1998). Experimental rabies infection and oral vaccination in vampire bats (*Desmodus rotundus*). *Vaccine* **16**, 1122–1126.

Allen, R., Sims, R.A. and Sulkin, S.E. (1964a). Studies with cultured brown adipose tissue. I. Persistence of rabies virus in bat brown fat. *American Journal of Hygiene* **80**, 11–24.

Allen, R., Sims, R.A. and Sulkin, S.E. (1964b). Studies with cultured brown adipose tissue. II. Influence of low temperature on rabies virus infection in bat brown fat. *American Journal of Hygiene* **80**, 25–32.

Allworth, A., Murray, K. and Morgan, J. (1996). A human case of encephalitis due to a lyssavirus, recently identified in fruit bats. *Communicable Diseases Intelligence* **20**, 504.

Almeida, M.F., Martorelli, L.F., Aires, C.C., Sallum, P.C. and Massad, E. (2005). Indirect oral immunization of captive vampires, *Desmodus rotundus*. *Virus Research* **111**, 77–82.

Amengual, B., Whitby, J.E., King, A., Cobo, J.S. and Bourhy, H. (1997). Evolution of European bat lyssaviruses. *Journal of General Virology* **78**, 2319–2328.

Arguin, P.M., Murray-Lillibridge, K., Miranda, M.E., Smith, J.S., Calaor, A.B. and Rupprecht, C.E. (2002). Serologic evidence of lyssavirus infections among bats, the Philippines. *Emerging Infectious Diseases* **8**, 258–262.

Badrane, H. and Tordo, N. (2001). Host switching in *Lyssavirus* history from the *Chiroptera* to the *Carnivora* orders. *Journal of Virology* **75**, 8096–8104.

Badrane, H., Bahloul, C., Perrin, P. and Tordo, N. (2001). Evidence of two lyssavirus phylogroups with distinct pathogenicity and immunogenicity. *Journal of Virology* **75**, 3268–3276.

Baer, G.M. (1991). Vampire bat and bovine paralytic rabies. In: *The Natural History of Rabies*, 2nd edn (G.M. Baer, ed.). pp. 390–406. Boca Raton: CRC Press.

Baer, G.M. and Bales, G.L. (1967). Experimental rabies infection in the Mexican freetail bat. *Journal of Infectious Diseases* **117**, 82–90.

Baer, G.M., Harrison, A.K., Bauer, S.P., Shaddock, J.H. and Murphy F.A. (1980) A bat rabies isolate with an unusually short incubation period. *Experimental Molecular Pathology* **33**, 211–222.

Baer, G.M. and Smith, J.S. (1991). Rabies in nonhematophagous bats. In: *The Natural History of Rabies*, 2nd edn (G.M. Baer, ed.). pp. 341–366. Boca Raton: CRC Press.

Bahloul, C., Jacob, Y., Tordo, N. and Perrin, P. (1998). DNA-based immunization for exploring the enlargement of immunological cross-reactivity against the lyssaviruses. *Vaccine* **16**, 417–425.

Bell, G.P. (1980). A possible case of interspecific transmission of rabies in insectivorous bats. *Journal of Mammalogy* **61**, 528–530.

Beran, G.W., Nocete, A.P., Elvina, O. *et al.* (1972). Epidemiological and control studies on rabies in the Philippines. *Southeast Asian Journal of Medicine and Public Health* **3**, 433–445.

Botvinkin, A.D. (1988). The results of study on the bat role in circulation of rabies virus in the USSR. In: *Chiroptera. Morphology, ecology, echolocation, parasites, protection. Proceedings of the USSR seminar.* pp. 138–141. Kiev: Naukova Dumka.

Botvinkin, A.D., Kuzmin, I.V. and Chernov, S.M. (1992). Experimental inoculation of bats with lyssaviruses serotypes 1 and 4. *Voprosy Virusologii* **37**, 215–218.

Botvinkin, A.D., Poleschuk, E.M., Kuzmin, I.V. *et al.* (2003). Novel lyssavirus isolated from bats in Russia. *Emerging Infectious Diseases* **9**, 1623–1625.

Botvinkin, A.D., Selnikova, O.P., Antonova, L.A., Moiseeva, A.B. and Nesterenko, E. Yu. (2006). New human rabies case caused from a bat bite in Ukraine. *Rabies Bulletin Europe* **3**, 5–7.

Boulger, L.R. and Porterfield, J.S. (1958). Isolation of a virus from Nigerian fruit bats. *Transcriptions of the Royal Society of Tropical Medicine and Hygiene* **52**, 421–424.

Bourhy, H., Kissi, B., Lafon, M., Sacramento, D. and Tordo, N. (1992). Antigenic and molecular characterization of bat rabies virus in Europe. *Journal of Clinical Microbiology* **30**, 2419–2426.

Bourhy, H., Kissi, B. and Tordo, N. (1993). Molecular diversity of the *Lyssavirus* genus. *Virology* **194**, 70–81.

Brookes, S.M., Aegerter, J.N., Smith, G.C. *et al.* (2005). European bat lyssavirus in Scottish bats. *Emerging Infectious Diseases* **11**, 572–578.

Bruijn, Z. (2003). Behavioral observations in some rabid bats. *Rabies Bulletin Europe* **27**, 7–8.

Büchen-Osmond, C. (2003). Taxonomy and classification of viruses. In: *Manual of Clinical Microbiology*, 8th edn, Vol. 2 (P.R. Murray, E.J. Baron, J.H. Jorgensen, M.A. Pfaller and R.H. Yolken, eds). pp. 1217–1226. Washington, DC: ASM Press.

Carini, A. (1911). Sur une grande epizootie de rage. *Annals of the Institute Pasteur* **25**, 843–846.

Constantine, D.G. (1962). Rabies transmission by nonbite route. *Public Health Reports* **77**, 287–289.

Constantine, D.G. (1966a). Transmission experiments with bat rabies isolates: responses of certain carnivora to rabies virus isolated from animals infected by nonbite route. *American Journal of Veterinary Research* **27**, 13–15.

Constantine, D.G. (1966b). Transmission experiments with bat rabies isolates: reaction of certain carnivora, opossum, and bats to intramuscular inoculations of rabies virus isolates from free-tailed bats. *American Journal of Veterinary Research* **27**, 16–19.

Constantine, D.G. (1966c). Transmission experiments with bat rabies isolates: bite transmission of rabies to foxes and coyote by free-tailed bats. *American Journal of Veterinary Research* **27**, 20–23.

Constantine, D.G. (1967a). Bat rabies in the southwestern United States. *Public Health Reports* **82**, 867–888.

Constantine, D.G. (1967b). *Rabies transmission by air in bat caves.* Washington, DC: Public Health Service Publication.

Constantine, D.G. (1972). Rabies virus in nasal mucosa of naturally infected bats. *Science* **175**, 1255–1256.

Constantine, D.G. (1988). Transmission of pathogenic organisms by vampire bats. In: *Natural History of Vampire Bats* (A.M. Greenhall and U. Schmidt, eds). pp.167–189. Boca Raton: CRC Press.

Constantine, D.G. (2003). Geographic translocation of bats: known and potential problems. *Emerging Infectious Diseases* **9**, 17–21.

Constantine, D.G. and Woodall, D.F. (1966). Transmission experiments with bat rabies isolates: reactions of certain carnivora, opossum, rodents, and bats to rabies virus of red bat origin when exposed by bat bite or by intramuscular inoculation. *American Journal of Veterinary Research* **27**, 24–32.

Constantine, D.G., Solomon, G.C. and Woodall, D.F. (1968). Transmission experiments with bat rabies isolates: responses of certain carnivores and rodents to rabies viruses from four species of bats. *American Journal of Veterinary Research* **29**, 181–190.

de Mattos, C.A., Favi, M., Yung, V., Pavletic, C. and de Mattos, C.C. (2000). Bat rabies in urban centers in Chile. *Journal of Wildlife Diseases* **36**, 231–240.

Dietzschold, B., Wunner, W.H., Wiktor, T.J. *et al.* (1983). Characterization of an antigenic determinant of the glycoprotein that correlates with pathogenicity of rabies virus. *Proceedings of the National Academy of Sciences of the United States of America* **80**, 70–74.

Dietzschold, B., Rupprecht, C.E., Tollis, M. *et al.* (1988). Antigenic diversity of the glycoprotein and nucleocapsid proteins of rabies and rabies-related viruses: implications for epidemiology and control of rabies. *Reviews in Infectious Diseases* **10**, S785–S798.

Dietzschold, B. and Hooper, D.C. (1998). Human diploid cell culture rabies vaccine (HDCV) and purified chick embryo cell culture rabies vaccine (PCECV) both confer protective immunity against infection with the silver-haired bat rabies virus. *Vaccine* **16**, 1656–1659.

Dorward, W.J., Schowalter, D.B. and Gunson, J.R. (1977). Preliminary studies of bat rabies in Alberta. *Canadian Veterinary Journal* **18**, 341–348.

Echevarria, J.E., Avellon, A., Juste, J., Vera, M. and Ibanez, C. (2001). Screening of active lyssavirus infection in wild bat populations by viral RNA detection on oropharyngeal swabs. *Journal of Clinical Microbiology* **39**, 3678–3683.

Familusi, J.B. and Moore, D.L. (1972). Isolation of a rabies related virus from the cerebrospinal fluid of a child with 'aseptic meningitis'. *African Journal of Medical Sciences* **3**, 93–96.

Familusi, J.B., Osunkoya, B.O., Moore, D.L., Kemp, G.E. and Fabiyi, A. (1972). A fatal human infection with Mokola virus. *American Journal of Tropical Medicine and Hygiene* **21**, 959–963.

Fekadu, M., Shaddock, J.H., Chandler, F.W. and Sanderlin, D.W. (1988a). Pathogenesis of rabies virus from a Danish bat (*Eptesicus serotinus*): neuronal changes suggestive of spongiosis. *Archives of Virology* **99**, 187–203.

Fekadu, M., Shaddock J.H., Sanderlin D.W. and Smith J.S. (1988b). Efficacy of rabies vaccines against Duvenhage virus isolated from European house bats (*Eptesicus serotinus*), classic rabies and rabies-related viruses. *Vaccine* **6**, 533–539.

Field, H., Mackenzie, J. and Daszak, P. (2004). Novel viral encephalitides associated with bats (*Chiroptera*) – host management strategies. *Archives of Virology* **18** (Suppl.), 113–121.

Foggin, C.M. (1983). Mokola virus infection in cats and a dog in Zimbabwe. *Veterinary Records* **113**, 115.

Fooks, A.R., McElhinney, L.M., Pounder, D.J. *et al.* (2003a). Case report: isolation of a European bat lyssavirus type 2a from a fatal human case of rabies encephalitis. *Journal of Medical Virology* **71**, 281–289.

Fooks, A.R., Brookes, S.M., Johnson, N., McElhinney, L.M. and Hutson, A.M. (2003b). European bat lyssaviruses: an emerging zoonosis. *Epidemiology and Infection* **131**, 1029–1039.

Franka, R., Constantine, D.G., Kuzmin, I. *et al.* (2006). A new phylogenetic lineage of rabies virus associated with western pipistrelle bats (*Pipistrellus hesperus*). *Journal of General Virology* **87**, 2309–2321.

Fraser, G.C., Hooper, P.T., Lunt, R.A. *et al.* (1996). Encephalitis caused by a lyssavirus in fruit bats in Australia. *Emerging Infectious Diseases* **2**, 327–331.

Gibbons, R.V., Holman, R.C., Mosberg, S.R. and Rupprecht, C.E. (2002). Knowledge of bat rabies and human exposure among United States cavers. *Emerging Infectious Diseases* **8**, 532–534.

Gould, A.R., Hyatt, A.D., Lunt, R., Kattenbelt, J.A., Hengstberger, S. and Blacksell, S.D. (1998). Characterisation of a novel lyssavirus isolated from Pteropid bats in Australia. *Virus Research* **54**, 165–187.

Gould, A.R., Kattenbelt, J.A., Gumley, S.G. and Lunt, R.A. (2002). Characterisation of an Australian bat lyssavirus variant isolated from an insectivorous bat. *Virus Research* **89**, 1–28.

Greenhall, A.M. (1993). Ecology and bionomics of vampire bats in Latin America. In: *Bats and Rabies* (A.M. Greenhall, M. Artois and M. Fekadu, eds). pp. 3–57. Lyon: Foundation Marcel Merieux.

Guyatt, K.J., Twin, J., Davis, P. *et al.* (2003). A molecular epidemiological study of Australian bat lyssavirus. *Journal of General Virology* **84** (Pt 2), 485–496.

Hanlon, C.A., Kuzmin, I.V., Blanton, J.D., Weldon, W.C., Manangan, J.S. and Rupprecht, C.E. (2005). Efficacy of rabies biologics against new lyssaviruses from Eurasia. *Virus Research* **111**, 44–54.

Hanna, J.N., Carney, I.K., Smith, G.A. *et al.* (2000). Australian bat lyssavirus infection: a second human case, with a long incubation period. *Medical Journal of Australia* **172**, 597–599.

Holmes, E.C., Woelk, C.H., Kassis, R. and Bourhy, H. (2002). Genetic constraints and the adaptive evolution of rabies virus in nature. *Virology* **292**, 247–257.

Hooper, P.T., Lunt, R.A., Gould, A.R. *et al.* (1997). A new lyssavirus – the first endemic rabies-related virus recognized in Australia. *Bulletin of Institute Pasteur* **95**, 209–218.

Hughes, G.J., Kuzmin, I.V., Schmitz, A. *et al.* (2006). Experimental infection of big brown bats (Eptesicus fuscus) with Eurasian bat lyssaviruses. *Archives of Virology* **151**, 2021–2035.

Hughes, G.J., Orciari, L.A. and Rupprecht, C.E. (2005). Evolutionary timescale of rabies virus adaptation to North American bats inferred from the substitution rate of the nucleoprotein gene. *Journal of General Virology* **86** (Pt 5), 1467–1474.

Human rabies prevention – United States, 1999. (1999). Recommendations of the Advisory Committee on Immunization Practices (ACIP). *Morbidity and Mortality Weekly Report* **48** (RR-1), 1–21.

Jallet, C., Jacob, Y., Bahloul, C. *et al.* (1999). Chimeric lyssavirus glycoproteins with increased immunological potential. *Journal of Virology* **73**, 225–233.

Johnson, N., Selden, D., Parsons, G. *et al.* (2003). Isolation of a European bat lyssavirus type 2 from a Daubenton's bat in the United Kingdom. *Veterinary Records* **152**, 383–387.

Kemp, G.E., Causey, O.R., Moore, D.L., Odeola A. and Fabiyi, A. (1972). Mokola virus. Further studies on IbAn 27377, a new rabies-related etiologic agent of zoonosis in Nigeria. *American Journal of Tropical Medicine and Hygiene* **21**, 356–359.

Khozinski, V.V., Selimov, M.A., Botvinkin, A.D. *et al.* (1991). Rabies virus of bat origin in the USSR. *Rabies Bulletin of Europe* **14**, 10–11.

King, A.A. (1993). Monoclonal antibody studies on rabies-related viruses. *Onderstepoort Journal of Veterinary Research* **60**, 283–287.

King, A. and Crick, J. (1988). Rabies-related viruses. In: *Rabies* (J.B. Campbell and K.M. Charlton, eds). pp. 178–199. Boston: Kluwer Academic Publishers.

King, A., Davis, P. and Lawrie, A. (1990). The rabies viruses of bats. *Veterinary Microbiology* **23**, 165–174.

King, A.A. and Turner, G.S. (1993). Rabies: a review. *Journal of Comparative Pathology* **108**, 1–39.

Kissi, B., Tordo, N. and Bourhy, H. (1995). Genetic polymorphism in the rabies virus nucleoprotein gene. *Virology* **209**, 526–537.

Krebs, J.W., Noll, H.R., Rupprecht, C.E. and Childs, J.E. (2002). Rabies surveillance in the United States during 2001. *Journal of the American Veterinary Medical Association* **221**, 1690–1701.

Krebs, J.W., Wheeling, J.T. and Childs, J.E. (2003). Rabies surveillance in the United States during 2002. *Journal of the American Veterinary Medical Association* **223**, 1736–1748.

Krebs, J.W., Mandel, E.J., Swerdlow, D.L. and Rupprecht, C.E. (2004). Rabies surveillance in the United States during 2003. *Journal of the American Veterinary Medical Association* **225**, 1837–1849.

Kuzmin, I.V., Botvinkin, A.D., Rybin, S.N. and Bayaliev, A.B. (1992). A lyssavirus with an unusual antigenic structure isolated from a bat in southern Kyrgyzstan. *Voprosy Virusologii* **37**, 256–259.

Kuzmin, I.V., Botvinkin, A.D. and Shaimardanov, R.T. (1994). Experimental lyssavirus infection in chiropters. *Voprosy Virusologii* **39**, 17–21.

Kuzmin, I.V. and Botvinkin, A.D. (1996). The behaviour of bats *Pipistrellus pipistrellus* after experimental inoculation with rabies and rabies-like viruses and some aspects of pathogenesis. *Myotis* **34**, 93–99.

Kuzmin, I.V., Botvinkin, A.D. and Khabilov, T.K. (2001). The lyssavirus was isolated from a whiskered bat in Northern Tajikistan. *Plecotus et al.* **4**, 75–81.

Kuzmin, I.V., Orciari, L.A., Arai, Y.T. *et al.* (2003). Bat lyssaviruses (Aravan and Khujand) from Central Asia: phylogenetic relationships according to N, P and G gene sequences. *Virus Research* **97**, 65–79.

Kuzmin, I.V., Hughes, G.J., Botvinkin, A.D., Orciari, L.A. and Rupprecht, C.E. (2005). Phylogenetic relationships of Irkut and West Caucasian bat viruses within the *Lyssavirus* genus and suggested quantitative criteria based on the N gene sequence for lyssavirus genotype definition. *Virus Research* **111**, 28–43.

Kuzmin, I.V., Botvinkin, A.D., Poleschuk, E.M., Orciari, L.A. and Rupprecht, C.E. (2006a). Bat rabies surveillance in the former Soviet Union. In: *First International Conference on Rabies in Europe: Proceedings of a Conference Organized by the World Organisation for Animal Health – OIE, Kiev, Ukraine, 15–18 June, 2005.* pp. 273–282. Basel: S. Karger AG.

Kuzmin, I.V., Niezgoda, M., Carroll, D.S. *et al.* (2006b). Lyssavirus surveillance in bats, Bangladesh. *Emerging Infectious Diseases* **12**, 486–488.

Lopez, A., Miranda, P., Tejada, E. and Fishbein, D.B. (1992). Outbreak of human rabies in the Peruvian jungle. *Lancet* **339**, 408–411.

Lord, R.D., Delpietro, H., Fuenzalida, E., de Diaz, A.M.O. and Lazaro, L. (1975). Presence of rabies neutralizing antibodies in wild carnivores following an outbreak of bovine rabies. *Journal of Wildlife Diseases* **11**, 210–213.

Lumio, J., Hillbom, M., Roine, R. *et al.* (1986). Human rabies of bat origin in Europe. *Lancet* **1**, 378.

Lumlertdacha, B., Boongird, K., Wanghongsa, S. *et al.* (2005). Survey for bat lyssaviruses, Thailand. *Emerging Infectious Diseases* **11**, 232–236.

Marissen, W.E., Kramer, R.A., Rice, A. *et al.* (2005). Novel rabies virus-neutralizing epitope recognized by human monoclonal antibody: fine mapping and escape mutant analysis. *Journal of Virology* **79**, 4672–4678.

Markotter, W., Randles, J., Rupprecht, C.E. *et al.* (2006). Lagos bat virus, South Africa. *Emerging Infectious Diseases* **12**, 504–506.

McCall, B.J., Epstein, J.H., Neill, A.S. *et al.* (2000). Potential exposure to Australian bat lyssavirus, Queensland, 1996–1999. *Emerging Infectious Diseases* **6**, 259–264.

McColl, K.A., Chamberlain, T., Lunt, R.A., Newberry, K.M., Middleton, D. and Westbury, H.A. (2002). Pathogenesis studies with Australian bat lyssavirus in grey-headed flying foxes (*Pteropus poliocephalus*). *Australian Veterinary Journal* **80**, 636–641.

Mebatsion, T. (2001). Extensive attenuation of rabies virus by simultaneously modifying the dynein light chain binding site in the P protein and replacing Arg_{333} in the G protein. *Journal of Virology* **75**, 11 496–11 502.

Mebatsion, T., Cox, J.H. and Frost, J.W. (1992). Isolation and characterization of 115 street rabies virus isolates from Ethiopia by using monoclonal antibodies: identification of 2 isolates as Mokola and Lagos bat viruses. *Journal of Infectious Diseases* **166**, 972–977.

Meredith, C.D., Rossouw, A.P. and van Praag Koch, H. (1971). An unusual case of human rabies thought to be of chiropteran origin. *South African Medical Journal* **45**, 767–769.

Messenger, S.L., Smith, J.S. and Rupprecht, C.E. (2002). Emerging epidemiology of bat-associated cryptic cases of rabies in humans in the United States. *Clinical Infectious Diseases* **35**, 738–747.

Moreno, J.A. and Baer G.M. (1980). Experimental rabies in the vampire bat. *American Journal of Tropical Medicine and Hygiene* **29**, 254–259.

Morimoto, K., Patel, M., Corisdeo, S. *et al.* (1996). Characterization of a unique variant of bat rabies virus responsible for newly emerging human cases in North America. *Proceedings of the National Academy of Sciences of the United States of America* **93**, 5653–5658.

Nadin-Davis, S.A., Huang, W., Armstrong, J. *et al.* (2001). Antigenic and genetic divergence of rabies viruses from bat species indigenous to Canada. *Virus Research* **74**, 139–156.

Nadin-Davis, S.A., Abdel-Malik, M., Armstrong, J. and Wandeler, A. (2002). Lyssavirus P gene characterization provides insight into the phylogeny of the genus and identities structural similarities and diversity within the encoded phosphoprotein. *Virology* **298**, 286–305.

Nel, L., Jacobs, J., Jafta, J., Von Teichman, B. and Bingham, J. (2000). New cases of Mokola virus infection in South Africa: a genotypic comparison of Southern African virus isolates. *Virus Genes* **20**, 103–106.

Nel, L.H., Niezgoda, M., Hanlon, C.A., Morril, P.A., Yager, P.A. and Rupprecht, C.E. (2003). A comparison of DNA vaccines for the rabies-related virus, Mokola. *Vaccine* **21**, 2598–2606.

Niezgoda, M., Hanlon, C.A., Hughes, G.J. *et al.* (2003). Host switching of bat rabies viruses: field and experimental observations of infected skunks. In: *Rabies in the Americas: The XIV International Conference: October 19–24, 2003. Program and abstract book.* p. 65. Philadelphia: Thomas Jefferson University.

Nowak, R.M. (1999). *Walker's Mammals of the World*, 6th edn. Vol. 1. Baltimore: Johns Hopkins University Press.

Pal, S.R., Arora, B.M., Chuttani, P.N., Broer, S. Choudhury, S. and Joshi, R.M. (1980). Rabies virus infection of a Flying fox bat, *Pteropus poliocephalus*, in Chandigarh, Northern India. *Tropical and Geographical Medicine* **32**, 265–267.

Pawan, J.L. (1936). Rabies in the vampire bat of Trinidad, with special reference to the clinical course and the latency of infection. *Annals of Tropical Medicine and Parasitology* **30**, 401–422.

Potzsch, C.J., Muller, Th. and Kramer, M. (2002). Summarizing the rabies situation in Europe 1990–2002 from the Rabies Bulletin Europe. *Rabies Bulletin Europe* **4**, 11–16.

Pybus, M.G. (1986). Rabies in insectivorous bats in Western Canada, 1979 to 1983. *Journal of Wildlife Diseases* **22**, 307–313.

Reynes, J.M., Molia, S., Audry, L. *et al.* (2004). Serologic evidence of lyssavirus infection in bats, Cambodia. *Emerging Infectious Diseases* **10**, 2231–2234.

Ronsholt, L.A. (2002). New case of European bat lyssavirus (EBL) infection in Danish sheep. *Rabies Bulletin Europe* **26**(2), 15.

Rupprecht, C.E., Dietzschold, B., Wunner, W.H. and Koprowski, H. (1991). Antigenic relationships of lyssaviruses. In: *The Natural History of Rabies*, 2nd edn (G.M. Baer, ed.). pp. 69–100. Boca Raton: CRC Press.

Rupprecht, C.E., Glickman, L.T., Spencer P.A. and Wiktor, T.J. (1987). Epidemiology of rabies virus variants. Differentiation using monoclonal antibodies and discriminant analysis. *American Journal of Epidemiology* **126**, 298–309.

Sadler, W.W. and Enright, J.B. (1959). Effect of metabolic level of the host upon the pathogenesis of rabies in the bat. *Journal of Infectious Diseases* **105**, 267–273.

Scherbak, Y. N. (1982). *Epidemiology of natural rabies*. Thesis of dissertation for DSc degree, Kiev, the Ukraine (USSR).

Schneider, L.G. (1982). Antigenic variants of rabies virus. *Comparative Immunology, Microbiology and Infectious Diseases* **5**, 101–107.

Schneider, L.G., Barnard, B.J.H. and Schneider, H.P. (1985). Application of monoclonal antibodies for epidemiological investigations and oral vaccination studies: I. African viruses. In: *Rabies in the Tropics* (E. Kuwert, C. Merieux, H. Koprowski and K. Bogel, eds). pp. 49–53. Berlin: Springer-Verlag.

Schneider, L.G. and Cox, J.H. (1994). Bat lyssaviruses in Europe. *Current Topics in Microbiology and Immunology* **187**, 207–218.

Schneider, M.C., Santos-Burgoa, C., Aron, J., Munoz, B., Ruiz-Velazco, S. and Uieda, W. (1996). Potential force of infection of human rabies transmitted by vampire bats in the Amazonian region of Brazil. *American Journal of Tropical Medicine and Hygiene* **55**, 680–684.

Selimov, M.A., Tatarov, A.G., Botvinkin, A.D., Klueva, E.V., Kulikova, L.G. and Khismatullina, N.A. (1989). Rabies-related Yuli-virus; identification with a panel of monoclonal antibodies. *Acta Virologica* **33**, 542–546.

Selimov, M.A., Smekhov, A.M., Antonova, L.A., Shablovskaya, E.A., King, A.A. and Kulikova, L.G. (1991). New strains of rabies-related viruses isolated from bats in the Ukraine. *Acta Virologica* **35**, 226–231.

Serra-Cobo, J., Amengual, B., Abellan, C. and Bourhy, H. (2002). European bat lyssavirus infection in Spanish bat populations. *Emerging Infectious Diseases* **8**, 413–420.

Shankar, V, Bowen, R.A., Davis, A.D., Rupprecht, C.E. and O'Shea, T.J. (2004). Rabies in a captive colony of big brown bats (*Eptesicus fuscus*). *Journal of Wildlife Diseases* **40**, 403–413.

Shankar, V., Orciari, L.A., De Mattos, C. *et al.* (2005). Genetic divergence of rabies viruses from bat species of Colorado, USA. *Vector-Borne and Zoonotic Diseases* **5**, 330–341.

Sheeler-Gordon, L.L. and Smith, J.S. (2001). Survey of bat populations from Mexico and Paraguay for rabies. *Journal of Wildlife Diseases* **37**, 582–593.

Shoji, Y., Kobayashi, Y., Sato, G. *et al.* (2004). Genetic characterization of rabies viruses isolated from frugivorous bat (*Artibeus spp.*) in Brazil. *Journal of Veterinary Medical Science* **66**, 1271–1273.

Shope, R.E. (1982). Rabies-related viruses. *Yale Journal of Biology and Medicine* **55**, 271–275.

Shope, R.E., Murphy, F.A., Harrison, A.K. *et al.* (1970). Two African viruses serologically and morphologically related to rabies virus. *Journal of Virology* **6**, 690–692.

Smith, J.S. (1989). Rabies virus epitopic variation: use in ecologic studies. *Advances in Virus Research* **36**, 215–253.

Smith, P.C., Lawhaswasdi, K., Vick, W.E. and Stanton, J.S. (1968). Isolation of rabies virus from fruit bats in Thailand. *Nature* **216**, 384.

Steece, R. and Altenbach, J.S. (1989). Prevalence of rabies specific antibodies in the Mexican free-tailed bat (Tadarida brasiliensis mexicana) at Lava Cave, New Mexico. *Journal of Wildlife Diseases* **25**, 490–496.

Sulkin, S.E. (1962). Bat rabies: experimental demonstration of the 'reservoiring mechanism'. *American Journal of Public Health* **52**, 489–498.

Sulkin, S.E. and Greve, M.J. (1954). Human rabies caused by bat bite. *Texas State Journal of Medicine* **50**, 620–621.

Sulkin, S.E., Krutzsch, P.H., Wallis, C. and Allen, R. (1957). Role of brown fat in pathogenesis of rabies in insectivorous bats (*Tadarida b. mexicana*). *Proceedings of the Society for Experimental Biology and Medicine* **96**, 461–464.

Sulkin, S.E., Allen, R., Sims, R., Krutzsch, P.H. and Kim, C. (1960). Studies on the pathogenesis of rabies in insectivorous bats. II. Influence of environmental temperature. *Journal of Experimental Medicine* **112**, 595–617.

Swanepoel, R. (1994). Rabies. In: *Infectious Diseases of Livestock with Special Reference to Southern Africa* (J.A.W. Coetzer, G.R. Thomson and R.C. Tustin, eds). pp. 493–553. Cape Town: Oxford University Press/NECC.

Tan, D.S., Peck, A.J. and Omar, M. (1969). The importance of Malaysian bats in the transmission of oral disease. *Medical Journal of Malaya* **24**, 32–35.

Tang, X., Ming Luo, M., Zhang, S., Fooks, A.R., Hu, R. and Tu, C. (2005). Pivotal role of dogs in rabies transmission, China. *Emerging Infectious Diseases* **11**, 1970–1972.

Tignor, G.H., Shope, R.E., Bhatt, P.N. and Percy, D.H. (1973). Experimental infection of dogs and monkeys with two rabies serogroup viruses, Lagos bat and Mokola (IbAn 27377): clinical, sero-logic, virologic and fluorescent-antibody studies. *Journal of Infectious Diseases* **128**, 471–478.

Tordo, N., Benmansour, A., Calisher, C. *et al.* (2005). Rhabdoviridae. In: *Virus Taxonomy, VIIIth Report of the ICTV* (C.M. Fauquet, M.A. Mayo, J. Maniloff, U. Desselberger and L.A. Ball, eds). pp. 623–644. London: Elsevier/Academic Press.

Trimarchi, C.V. and Debbie, J.G. (1977). Naturally occurring rabies virus and neutralizing antibody in two species of insectivorous bats of New York state. *Journal of Wildlife Diseases* **13**, 366–369.

Tuncman, Z.M. (1958). Research on the rabies virus of bats of Turkey. *Microbiyoloji Dergisi* **11**, 81.

Van der Poel, W.H.M, Van der Heide, R., Van Amerongen, G. *et al.* (2000). Characterisation of a recently isolated lyssavirus in frugivorous zoo bats. *Archives of Virology* **145**, 1919–1931.

Van der Poel, W.H.M., Van der Heide, R., Verstraten, E.R.A.M., Takumi, K., Lina, P.H.C. and Kramps, J.A. (2005). European bat lyssaviruses, the Netherlands. *Emerging Infectious Diseases* **11**, 1854–1859.

Velasco-Villa, A. and Rupprecht, C.E. (2005). Comparative study of the genetic and antigenic diversity of vampire bat rabies in the Americas. In: *The XVI International Conference on Rabies in the Americas. October 16–21, 2005, Ottawa, Ontario, Canada. Program and abstract book*. p. 45. Ottawa: Canadian Food Inspection Agency.

Vigilancia epidemiologica de la rabia en las Americas. Centro Panamericano de Zoonosis. Programa de Salud Publica Veterinaria (OPS/OMS). 2000–2003.

Vos, A., Muller, T., Cox, J., Neubert, L. and Fooks, A.R. (2004). Susceptibility of ferrets (Mustela putorius furo) to experimentally induced rabies with European bat Lyssaviruses (EBLV). *Journal of Veterinary Medicine. B, Infectious Diseases and Veterinary Public Health* **51**, 55–60.

Warner, C.K., Zaki, S.R., Shieh, W.-J. *et al.* (1999). Laboratory investigation of human deaths from vampire bat rabies in Peru. *American Journal of Tropical Medicine and Hygiene* **60**, 502–507.

Warrilow, D., Harrower, B., Smith, I.L. *et al.* (2003). Public health surveillance for Australian bat lyssavirus in Queensland, Australia, 2000–2001. *Emerging Infectious Diseases* **9**, 262–264.

Warrilow, D., Smith, I.L., Harrower, B. and Smith, G.A. (2002). Sequence analysis of an isolate from a fatal human infection of Australian bat lyssavirus. *Virology* **297**, 109–119.

Wellenberg, G.J., Audry, L., Ronsholt, L., Van der Poel, W.H., Bruschke, C.J. and Bourhy H. (2002). Presence of European bat lyssavirus RNAs in apparently healthy *Rousettus aegyptiacus* bats. *Archives of Virology* **147**, 349–361.

WHO (1990). Report of the 6th WHO consultation on monoclonal antibodies for rabies diagnosis and research. Philadelphia (WHO/Rab Res 90.34).

Yanez Valesco, L.B. (1994). Epidemiology of human rabies due to vampire bat exposure. In: *Fifth Annual International Meeting on Rabies in the Americas. November 16–19, 1994, Niagara Falls, Canada. Program and abstracts*. Ottawa: Rabies Unit & Ontario Ministry of Natural Resources.

Zaikovskaia, A.V., Ternovoy, V.A., Tomilenko, A.A., Rassadkin, Yu.V., Aksenov, V.I. and Shestopalov, A.M. (2004). Lyssaviruses of bats inhabiting the south of Western Siberia. *Journal of Infectious Pathology* **11**, 70–73.

Zaikovskaia, A.V., Ternovo, V.A., Aksenov, V.I., Rassadkin, Iu.N. and Shestopalov, A.M. (2005). Molecular genetic characterization of rabies virus, isolated in Western Siberia. *Vestnik Rossiyskoy Akademii Medicinskih Nauk* **2**, 30–35.

7

Human Disease

ALAN C. JACKSON

Departments of Medicine (Neurology) and of Microbiology and Immunology, Queen's University, Kingston, ON K7L 3N6, Canada; currently Departments of Internal Medicine (Neurology) and of Medical Microbiology, University of Manitoba, Winnipeg, MB R3A IR9, Canada

1 INTRODUCTION

Since antiquity, rabies has been one of the most feared diseases. Human rabies remains an important public health problem in many developing countries where dog rabies is endemic. Worldwide there are about 55 000 human deaths each year due to rabies (World Health Organization, 2005). Beginning in the 1990s up to six human cases of rabies were diagnosed per year in the USA and many of these infections were acquired indigenously from unrecognized exposures to insectivorous bats (Noah *et al.*, 1998). A significant number of additional rabies cases probably go unrecognized in the USA and Canada, because undiagnosed acute and fatal neurologic illnesses are common and a history of an animal exposure may not exist.

2 EXPOSURES, INCUBATION PERIOD AND PRODROMAL SYMPTOMS

The infectious cycle of rabies virus is perpetuated mainly through animal bites and the deposition of rabies virus-laden saliva into subcutaneous tissues and muscle. With respect to human rabies, worldwide, dogs are by far the most common and important rabies vector; bats are most important in the USA and Canada, although there are reservoirs in various terrestrial animals. Other types of non-bite exposures, including contamination of an open wound, scratch, abrasion, or mucous membrane by saliva or central nervous system (CNS) tissue from an infected animal, are quite common, although they are rarely responsible for transmission of rabies virus. Handling and skinning of infected carcasses and perhaps consumption of raw infected meat have resulted in transmission of rabies virus (Tariq *et al.*, 1991; Kureishi *et al.*, 1992; Wallerstein, 1999). Rarely, but notably, inhalation of aerosolized rabies virus in caves containing millions of bats (Constantine, 1962) or in laboratories (Winkler *et al.*, 1973; Tillotson *et al.*, 1977b) has resulted in human rabies. At least eight cases of rabies have resulted from

Rabies, second edition. Edited by Alan C. Jackson and William H. Wunner
ISBN 978-012-369366-2. Copyright Elsevier Inc. 2007

TABLE 7.1

Human rabies cases transmitted by corneal transplantation

Location	Year	Age of patient (recipient)	Time to death (days)	Reference
USA	1978	37	50	Houff et al., 1979
France	1979	36	41	Galian et al., 1980
Thailand	1981	41	22	Thongcharoen et al., 1981
Thailand	1981	25	33	Thongcharoen et al., 1981
India	1987	62	15	Gode and Bhide, 1988
India	1988	48	264[a]	Gode and Bhide, 1988
Iran	1994	40	27	Javadi et al., 1996
Iran	1994	35	41	Javadi et al., 1996

[a] Patient received two doses of rabies vaccine about one month after the transplant.

TABLE 7.2

Cases of human rabies associated with organ transplantation in the USA (Srinivasan et al., 2005) and Germany (Johnson et al., 2005)

	Sex/age	Organ transplanted	Onset of clinical rabies post-transplantation (days)
Donor in USA	Male/20	–	–
Recipient 1	Male/53	Liver	21
Recipient 2	Female/50	Kidney	27
Recipient 3	Male/18	Kidney	27
Recipient 4	Female/55	Iliac artery segment	27
Donor in Germany	Female/26	–	–
Recipient 1	Female/45	Lung	7–14 weeks
Recipient 2	Male/70	Kidney	7–14 weeks
Recipient 3	Male/45	Kidney/pancreas	14 weeks

transplantation (human-to-human) of rabies virus-infected corneas (Table 7.1). In other reports, transmission did not occur after corneal transplantation from a donor with rabies in France (Sureau et al., 1981) and from another donor in Germany, in which instance there were two cornea transplant recipients (Johnson et al., 2005). In 2004, organ transplantations in Texas were associated with transmission of rabies virus and the development of fatal rabies in four recipients (Table 7.2) (Srinivasan et al., 2005). The donor in this case presented with gastrointestinal symptoms, throat pain, intermittent periods of confusion and agitation and he had mild fever and ballistic trunk movements (Burton et al., 2005). The initial CT head scan showed a small subarachnoid hemorrhage. In retrospect, it is highly doubtful that this clinical presentation could be explained by a small subarachnoid hemorrhage. Subsequently, there was neurologic deterioration and a repeat CT head scan showed a large subarachnoid hemorrhage with evidence of herniation. He progressed to brain death and his organs (lungs, kidneys and liver) and iliac vessels were harvested (Burton *et al.*, 2005). The four recipients of the liver, kidneys and an iliac artery segment (for a liver transplant) developed clinical rabies within a month and died. The donor had anti-rabies virus antibodies in serum at the time of death and three of the four recipients had antibodies on postoperative days 35 and 36 (Srinivasan *et al.*, 2005). Immunosuppression of the recipients in order to prevent organ rejection results in a favorable environment for viral replication and spread. Only later, it was determined that the donor had been bitten by a bat and antigenic typing indicated that the rabies virus variant was associated with Brazilian (Mexican) free-tailed bats (Krebs *et al.*, 2005). All four transplantation recipients had histopathologic features of encephalitis with cytoplasmic inclusions characteristic of Negri bodies and rabies virus antigen was detected in neurons with immunohistochemical staining from multiple areas of the CNS. Rabies virus antigen was also observed in peripheral nerves of the transplanted kidneys, liver and arterial graft (Figure 7.1), which was the suspected site of infection in the organs/arterial graft responsible for transmission of rabies virus (Jackson, 2004). Transmission occurred again from a donor to organ transplant recipients in Germany that resulted in three fatal cases in 2005 (Johnson *et al.*, 2005) (Table 7.2). It is clear, and it should be emphasized, that tissues or organs should not be transplanted from a donor who dies from an undiagnosed neurologic disease, because the risk of transmitting unsuspected infectious agents, including rabies virus, is too high. Most other reported cases of human-to-human transmission have not been well documented. Two patients with rabies from Ethiopia were described and their only known exposure was contact with family members who died of rabies (Fekadu *et al.*, 1996). In this report, a 41-year-old female died of rabies 33 days after her 5-year-old son died of rabies; he had bitten his mother on her little finger. A 5-year-old boy presented with rabies 36 days after his mother died of rabies; he had repeatedly received kisses from his mother on his mouth during her illness. Sexual transmission of rabies virus has not been documented. Although natural human-to-human transmission of rabies likely occurs very rarely, anyone in direct contact with rabies patients, including family members and health care workers, should employ barrier nursing techniques in order to minimize the risk of transmission of the virus via saliva or other secretions (Remington *et al.*, 1985). Evidence of transplacental transmission of rabies virus exists in a single report from Turkey (Sipahioglu and Alpaut, 1985).

The incubation period for human rabies is usually 20–90 days after exposure, although occasionally disease develops after only a few days (Anderson *et al.*, 1984) and rare cases have occurred a year or more following exposure. Three immigrants from Laos, the Philippines and Mexico developed rabies in the USA

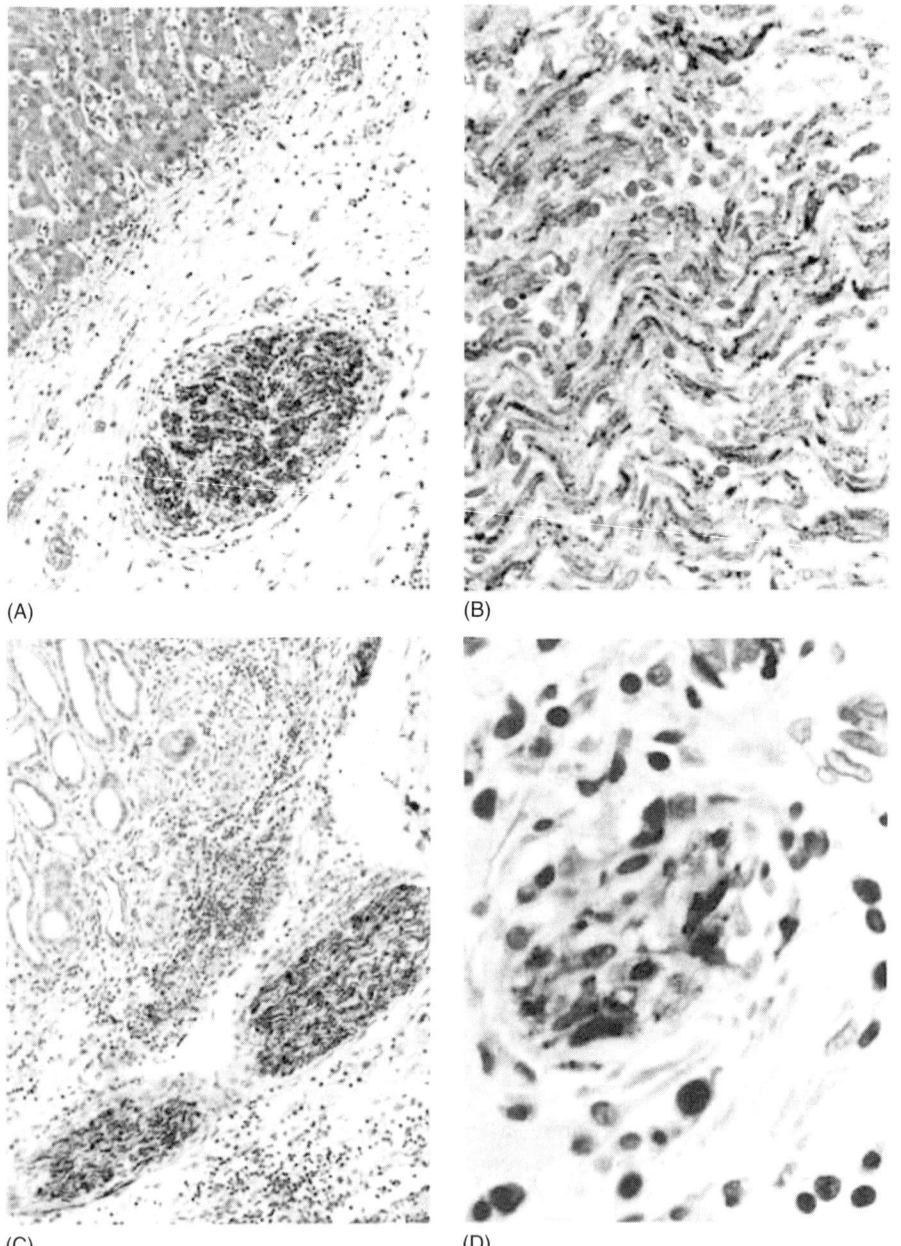

(A)

(B)

(C)

(D)

Figure 7.1 Immunohistochemical staining for rabies virus antigen (red) in peripheral nerves of the liver (A and B), kidney (C) and arterial graft (D) transplants. (Reproduced with permission from Srinivasan *et al.* Transmission of rabies virus from an organ donor to four transplant recipients. *New England Journal of Medicine* **352**, 1103–1111, 2005. Copyright © 2005, Massachusetts Medical Society. All rights reserved.) This figure is reproduced in the color plate section.

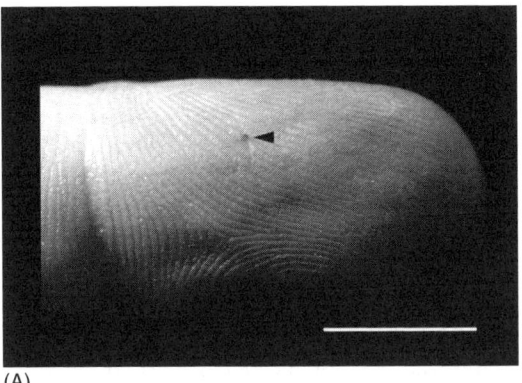

(A)

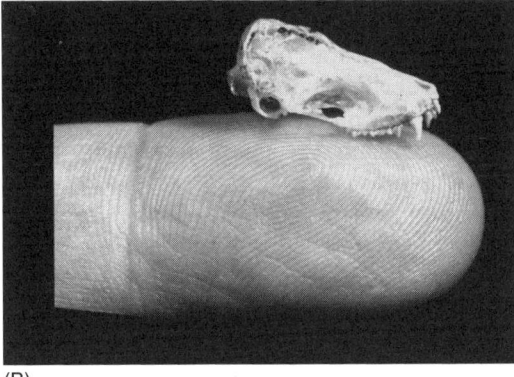

(B)

Figure 7.2   Small puncture wound (arrowhead) involving the right ring finger of a bat biologist (A) caused by a defensive bite from a canine tooth of a silver-haired bat (*Lasionycteris noctivagans*) (Bar = 10 mm). Skull of a silver-haired bat (B) (length 17.1 mm) is resting on a distal phalanx, which demonstrates the small size of the bat and its teeth. (Reproduced from Jackson and Fenton in *Lancet* **357**, 1714, 2001 Copyright © 2001, Elsevier.)

due to rabies virus strains from their countries of origin with incubation periods of at least 11 months and 4 and 6 years, which were based on the time of their immigration (Smith *et al.*, 1991). A case of rabies in a 10-year-old Vietnamese girl in Australia in 1990 was also likely acquired at least 5 years earlier (Bek *et al.*, 1992; McColl *et al.*, 1993). The incubation period (from exposure to onset of disease) in rabies is longer and more variable than for most other infectious diseases, which may cause considerable emotional stress to the patient. Very long incubation periods raise the possibility of another unrecognized or forgotten exposure in rabies endemic areas. Severe multiple bites and facial bites are associated with shorter incubation periods (Warrell and Warrell, 1991), although there is no clear correlation between the site of the bite and the incubation period (Dupont and Earle, 1965). There may be no history of a bite exposure because it was unrecognized, particularly with insectivorous bat bites because they may be very small (Jackson and Fenton, 2001) (Figure 7.2), or because a bite was either forgotten or no inquiry was made while the patient was still lucid. With known bite exposures from rabid animals, the following has been observed in untreated persons who develop rabies: 50–80% occurrence after head bites, 15–40% after hand or arm bites and 3–10% after leg bites. The risk is about 0.1% for contamination of minor wounds with saliva, including scratches (Hattwick, 1974). The biologic bases for these observations are unclear, but a number of factors may be responsible, including the density of rabies virus receptors in affected tissues, the degree of innervation in tissues in different anatomical locations, the quantity of virus inoculated and the properties of the rabies virus variant. Some individuals with rabies virus exposures may have inapparent rabies virus infection and develop naturally acquired immunity (Doege and Northrop, 1974). Low titers of rabies virus neutralizing antibodies (VNAs) have been found in Canadian Inuit hunters (7 of 20) and their wives (2 of 11) (Orr *et al.*, 1988) and also, surprisingly, in 6.6% (15 of 226) of unimmunized students and faculty members of a veterinary medical school at the inception of a rabies vaccine trial (Ruegsegger *et al.*, 1961). Black and Wiktor (1986) also observed low titers of rabies VNA in 17% (5 of 30) of Florida raccoon hunters, but not in a control group of other hunters. These studies suggest the possibility of exposure to rabies virus under natural conditions resulting in low-level immunity, but without the occurrence of clinical disease.

Non-specific prodromal symptoms in rabies, including fever, chills, malaise, fatigue, insomnia, anorexia, headache, anxiety and irritability may last for up to 10 days prior to the onset of neurologic symptoms (Warrell, 1976). About 30–70% of patients develop pain, paresthesias, and/or pruritus at or close to the site of the bite and the bite wound has often healed by the time these symptoms develop (Dupont and Earle, 1965; Hattwick, 1974). These local neurologic symptoms may be more common with bat rabies virus variants than with dog rabies virus variants (Hemachudha, 1997b; Noah *et al.*, 1998) and they are strong clinical clues to a diagnosis of rabies. The pruritus may result in severe excoriations from scratching. Retro-orbital pain also occurred as an early symptom in some patients with transmission by corneal transplantation. Local neurologic symptoms may reflect infection involving local peripheral sensory ganglia (dorsal root or trigeminal ganglia) (Mitrabhakdi *et al.*, 2005). Weakness may also develop in the bitten extremity and this may be more common after transmission of rabies virus from bats rather than from dogs (Hemachudha, 1997b). The initial neurologic symptoms may occasionally occur at a site distant from the bite, although the pathogenic basis for this phenomenon is not clear. Two patients bitten on their toes developed rabies with early severe itching of their ears (Hemachudha, 1994). Tremor has also been described involving the bitten extremity (Warrell, 1976).

3 CLINICAL FORMS OF DISEASE

3.1 Encephalitic rabies

About 80% of patients develop an encephalitic or classical (also called *furious*) form of rabies and about 20% have a paralytic form of disease. In encephalitic rabies, patients have episodes of generalized arousal or hyperexcitability, which are separated by lucid periods (Warrell, 1976) and these features reflect brain involvement with the infection. Intermittent episodes may occur with confusion, hallucinations, agitation and aggressive behavior, which typically last for periods of 1–5 minutes (Hattwick, 1974; Warrell and Warrell, 1991; Hemachudha, 1997a). The episodes may occur spontaneously or be precipitated by a variety of sensory stimuli (tactile, auditory, visual or olfactory). Biting behavior of patients with rabies has been described (Dupont and Earle, 1965; Emmons *et al.*, 1973; Warrell, 1976), but it is unusual. Fever is common and may be quite high (over 42°C/ 107°F) and there may be signs of autonomic dysfunction, including hypersalivation, lacrimation, sweating, piloerection (gooseflesh) and dilated pupils. The autonomic dysfunction may result from the infection directly involving the autonomic nervous system centers or pathways in the hypothalamus, spinal cord and/or autonomic ganglia. Parasympathetic stimulation may increase the production of saliva above the normal volume of about 1 liter per 24 hours. Often patients appear frightened with wide palpebral fissures, dilated pupils and an open mouth (Nicholson, 1994). Movement disorders have been noted (Warrell, 1976). Seizures, including convulsions, may occur, but they are not common and they usually occur late in the illness. Cranial nerve signs may be present, including ophthalmoplegia, facial weakness, impaired swallowing and tongue weakness. There may also be nuchal rigidity, reflecting leptomeningeal inflammation.

About 50–80% of patients develop hydrophobia, which is a characteristic and the most specific manifestation of rabies. Hydrophobia is not a feature of any other diseases. The term *hydrophobia* is derived from the Greek meaning 'fear of water'. Patients may initially experience pain in the throat or difficulty swallowing. On attempts to swallow, they experience contractions of the diaphragm, sternocleidomastoids, scalenes and other accessory muscles of inspiration, which last for about 5 to 15 seconds and may be associated with epigastric pain (Figure 7.3). These symptoms may be followed by contraction of neck muscles, resulting in flexion or extension of the neck and rarely with opisthotonic posturing. There may be associated retching, vomiting, coughing, aspiration into the trachea, grimacing, convulsions and hypoxia (Editorial, 1975). Patients may die during severe spasms with the development of cardiorespiratory arrest if supportive care measures are not initiated (Warrell and Warrell, 1991). During the spasms, there is an associated feeling of terror, often without associated pain. Patients avoid drinking for long periods of time, even despite intense thirst, resulting in dehydration. Subsequently, the sight, sound, or even mention of water (or liquids) may trigger these spasms, indicating that hydrophobia is reinforced by conditioning (Warrell *et al.*, 1976). Hydrophobic spasms may also occur spontaneously, particularly later in the course of the illness. A draft of air on the skin or the breath of an examiner may have the same effect, which has been termed *aerophobia*, and a variety of other stimuli, including water splashed on the skin, attempts by the patient to speak and stimulation from bright lights or loud sounds,

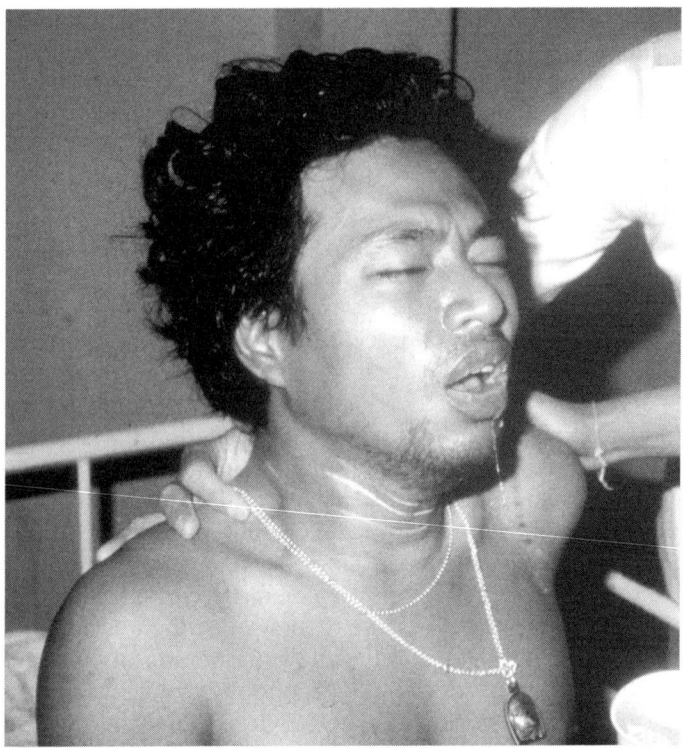

Figure 7.3 Hydrophobic spasm of inspiratory muscles associated with terror in a patient with furious rabies encephalitis attempting to swallow water. (Copyright D.A. Warrell, Oxford, UK). This figure is reproduced in the color plate section.

also may precipitate spasms (Warrell, 1976). Patients may wear heavy clothing in order to avoid drafts. The fan test, elicited by fanning a current of air across the face and observing the patient for spasms of the pharyngeal and neck muscles, has been used as a bedside diagnostic test for the presence of aerophobia (Wilson *et al.*, 1975). Sobbing respiration (like a child who has been crying) with a two-stage (sniff-sniff) inspiration followed by a slow, full expiration has been described (Pearson, 1976). Later these spasms merge with the development of periodic, apneustic or ataxic breathing as the patient's level of consciousness deteriorates (Warrell *et al.*, 1976). The hydrophobia of rabies is likely due to selective infection of neurons that inhibit the inspiratory motor neurons in the region of the nucleus ambiguus in the brainstem (Warrell *et al.*, 1976; Warrell, 1976). This results in exaggeration of defensive reflexes that protect the respiratory tract. Vocal cord weakness may result in a change in the voice and patients may make bark-like sounds. Increased libido, priapism (painful spontaneous erections) and spontaneous ejaculations occasionally occur in rabies and they may be early manifestations of the disease (Gardner, 1970; Talaulicar, 1977; Bhandari and Kumar,

1986; Udwadia *et al.*, 1988; Dutta, 1996). There is often progression to severe flaccid paralysis, coma and multiple organ failure. The paralysis that develops either in association with or after the development of coma should not be confused with paralytic rabies in which the muscle weakness develops, in contrast, early in the course of the illness (see Section 3.2). Rabies is almost always fatal and death often occurs within 14 days of the onset of clinical manifestations, although the time of death may be influenced by critical care measures.

A wide variety of medical complications can develop in patients with rabies. Many of these complications may also occur in critically ill patients with other acute neurological disorders, but some are likely related to the widespread infection in the CNS with systemic (extraneural) organ involvement due to infection of autonomic or sensory neurons (Jackson *et al.*, 1999). Cardiopulmonary complications are the most common and important. Respiratory complications include hyperventilation, hypoxemia, respiratory depression with apnea, atelectasis and aspiration with secondary pneumonia (Hattwick, 1974). Sinus tachycardia is a common cardiac feature and the degree of the tachycardia is often greater than that expected for the degree of fever (Warrell *et al.*, 1976). Cardiac arrhythmias (including wandering atrial/nodal pacemaker, sinus bradycardia and supraventricular or ventricular ectopic beats), hypotension, heart failure and cardiac arrest may occur (Hattwick, 1974; Warrell *et al.*, 1976). Cardiac arrhythmias may account for the sudden death of patients who are alert and do not have advanced neurologic signs of rabies. Cardiac manifestations may reflect infection involving the autonomic nervous system or myocardium (Ross and Armentrout, 1962; Cheetham *et al.*, 1970; Raman *et al.*, 1988; Metze and Feiden, 1991; Jackson *et al.*, 1999). Either hyperthermia or hypothermia may be present, which may reflect hypothalamic involvement of the infection. Gastrointestinal hemorrhage, especially hematemesis, is a common complication (Kureishi *et al.*, 1992). Endocrine complications include both inappropriate secretion of antidiuretic hormone and diabetes insipidus (Hattwick, 1974; Bhatt *et al.*, 1974).

3.2 Paralytic rabies

In paralytic rabies, flaccid muscle weakness develops early in the course of the disease and the weakness is prominent. Patients are frequently misdiagnosed with this clinical form of the disease, especially if a history of an animal bite is not obtained. The earliest description of paralytic rabies was recorded in 1887 (Gamaleia, 1887). Paralytic rabies has also been called *dumb rabies*. Patients may be literally dumb or mute due to laryngeal muscle weakness, but the term *dumb rabies* usually refers to the quieter clinical features and prominent weakness rather than specifically to the presence of anarthria (Editorial, 1978; Mills *et al.*, 1978). The development of paralytic rabies is not related to the anatomical site of the bite (Tirawatnpong *et al.*, 1989) and the incubation period is similar to that in encephalitic rabies. Patients are usually alert with a normal mental status at the onset of this clinical form of rabies. The weakness often begins in the bitten extremity and spreads to involve the other extremities, sometimes in an ascending pattern. Muscle fasciculations may be present (Phuapradit *et al.*, 1985). The facial muscles are frequently weak bilaterally. Associated bilateral deafness has been reported (Phuapradit *et al.*, 1985). Although patients may have local pain, paresthesias or pruritus at the site of the bite, the sensory examination is usually normal in patients with paralytic rabies. The clinical picture may be confused with the Guillain-Barré syndrome, including both the acute inflammatory demyelinating polyradiculopathy and the more severe motor-sensory neuropathy of acute onset with predominant axonal involvement (called the *axonal* Guillain-Barré syndrome) (Feasby *et al.*, 1986; Griffin *et al.*, 1996; Sheikh *et al.*, 2005). Sphincter involvement, especially with urinary incontinence, is common in paralytic rabies, but this is not a feature of the Guillain-Barré syndrome (Asbury and Cornblath, 1990). In addition, pain and sensory disturbances may occur in paralytic rabies. Myoedema has been reported as a sign observed in paralytic rabies, but not in encephalitic rabies (Hemachudha *et al.*, 1987). However, myoedema has not been confirmed as an important sign of paralytic rabies in other reports. In myoedema, percussion of a muscle (e.g. deltoid or thigh muscle) with a tendon hammer results in local mounding of the muscle without propagated contractions and with electrical silence; the mounding disappears over a few seconds. Myoedema is thought to be a normal physiological phenomenon and its presence does not indicate neuromuscular pathology (Hornung and Nix, 1992). Hence, the importance of this sign in rabies needs clarification in the future. Bulbar and respiratory muscles eventually become weak in paralytic rabies, resulting in death. Hydrophobia is more unusual in the paralytic form of the disease, although mild inspiratory spasms are commonly observed (Hemachudha *et al.*, 1988). Survival in paralytic rabies is usually longer (up to 30 days) than in encephalitic rabies (Hemachudha *et al.*, 2005). It is unclear if the hydrophobic spasms *per se* lead to death in the first few days of illness in encephalitic disease (Editorial, 1978) or if they reflect a more life-threatening distribution of the infection in the nervous system.

An unusual human outbreak of rabies affecting over 70 people occurred in Trinidad between 1929 and 1937 with transmission of the virus from vampire bats (Hurst and Pawan, 1931; Pawan, 1939; Waterman, 1959). All patients in this outbreak had the paralytic form of the disease. This led to diagnostic uncertainty and initially poliomyelitis and botulism were suspected. Nine miners died of paralytic rabies transmitted by vampire bats in British Guiana (presently Guyana) in 1953 (Nehaul, 1955). Similarly, seven children died of paralytic rabies in Surinam in 1973–1974 and vampire bats were probably also the responsible vector (Verlinde *et al.*, 1975). However, rabies virus transmitted by vampire bats does not always produce paralytic rabies. A 1990 outbreak of human rabies in Peru with transmission from vampire bats exclusively produced cases of encephalitic rabies (Lopez *et al.*, 1992). Recent outbreaks of human rabies with transmission from vampire bats occurred in the Amazon region of northern Brazil in 2004. Apparently, all of the 21 cases were the paralytic form of rabies, but detailed clinical information or pathological findings have not yet been reported (da Rosa *et al.*, 2006). Furthermore, it has been observed that a dog may bite two individuals and one develops encephalitic rabies and the other develops paralytic rabies (Hemachudha *et al.*, 1988; Wilde and Chutivongse, 1988).

The pathogenetic basis for the two different clinical forms of rabies has not been determined. In a small series, there were no marked differences in the regional distribution of rabies virus antigen or in the inflammatory changes (Tirawatnpong *et al.*, 1989). However, at the time of death, the distribution of the viral infection may be much more widespread and not closely reflect the distribution at the time of the patient's presentation with paralytic rabies. Electrophysiologic studies have indicated that peripheral nerve involvement, including demyelination, likely contributes to the weakness in paralytic rabies (Mitrabhakdi *et al.*, 2005). It is curious that an earlier serum neutralizing antibody response was observed in patients with encephalitic rabies than in those with paralytic rabies (Hemachudha, 1994). There is evidence that patients with paralytic rabies have defects in immune responsiveness, including lack of lymphocyte proliferative responses to rabies virus antigen (Hemachudha *et al.*, 1988) and lower levels of serum cytokines, including interleukin-6 and the soluble interleukin-2 receptor, than in patients with encephalitic rabies (Hemachudha *et al.*, 1993). In contrast, a Chinese case that was misdiagnosed as axonal Guillain-Barré syndrome (Griffin *et al.*, 1996) had pathologic changes that were most marked in the ventral spinal nerve roots without prominent inflammation or motor neuron degeneration (Sheikh *et al.*, 2005). This may have occurred either because axonal degeneration can be an early morphologic consequence of rabies virus-infected motor neurons or because axonal degeneration may be caused by immune injury. In support of the latter hypothesis, a case of encephalitic rabies was treated with high-dose intravenous rabies immune globulin and developed severe paralysis (Hemachudha *et al.*, 2003). However, the lack of anti-rabies virus antibodies in some paralytic rabies cases argues that antibody-mediated injury to nerves is not the only mechanism resulting in paralysis in rabies.

4 INVESTIGATIONS

4.1 Imaging studies

Computed tomographic (CT) studies of the brain usually are normal in rabies (Faoagali *et al.*, 1988; Mrak and Young, 1993; White *et al.*, 1994), although hypodense cortical lesions (Sow *et al.*, 1996) and non-enhancing basal ganglia hypodensities (Awasthi *et al.*, 2001) have been described. There have been reports of magnetic resonance imaging (MRI) studies of the brain with normal findings (Mrak and Young, 1993; Sing and Soo, 1996) and increased signals in gray matter areas (Hantson *et al.*, 1993; Awasthi *et al.*, 2001). Increased signals were observed on T_2-weighted images in the medulla (Figure 7.4) and pons with only minimal gadolinium enhancement in these areas in a patient from California infected by a rabies virus strain associated with Brazilian (Mexican) free-tailed bats (Pleasure and Fischbein, 2000). Gadolinium enhancement of cervical nerve roots was described in a patient with paralytic rabies (Laothamatas *et al.*, 1997). Gadolinium enhancement involving the medulla and hypothalamus was also described in the same report in another patient with paralytic rabies. This indicates imaging evidence of brain infection, which has been shown in histopathologic studies at the time of death (Chopra *et al.*, 1980). MRI findings in both the brain and spinal cord were found to be similar in a small number of patients with encephalitic and paralytic rabies (Laothamatas *et al.*, 2003).

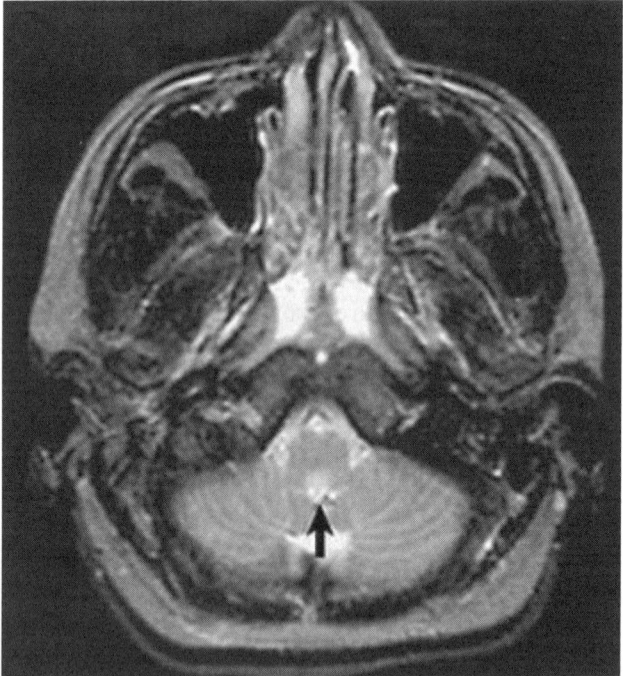

Figure 7.4   An axial T_2-weighted magnetic resonance image through the medulla demonstrating focal increased signal in the dorsal midline (arrow) (General Electric 1.5-T Signa system; TR, 2500 ms, TE, 80 ms). (Reproduced with permission from Pleasure and Fischbein, Correlation of clinical and neuroimaging findings in a case of rabies encephalitis. *Archives of Neurology* **57**, 1765–1769, 2000. Copyright © 2000, American Medical Association. All rights reserved.)

4.2 Laboratory studies

The electroencephalogram may be normal or show non-specific abnormalities in human rabies. Slow wave activity has been observed as well as periodic (Komsuoglu et al., 1981) and epileptiform activity. Electrophysiological studies showed evidence of peripheral nerve and/or anterior horn cell involvement in a small series and features of peripheral nerve demyelination were observed in the paralytic cases (Mitrabhakdi et al., 2005). Electrophysiologic evidence of a primary axonal neuropathy was found in two patients with paralytic rabies in another report (Prier et al., 1979a, 1979b). Hematologic and biochemical tests are usually normal, although hyponatremia may occur secondary to inappropriate secretion of antidiuretic hormone. Cerebrospinal fluid (CSF) analysis often becomes abnormal in human rabies. A CSF pleocytosis (elevated number of white cells) was found in 59% of cases in the first week of illness and in 87% after the first week (Anderson et al., 1984). The white cell count is usually less than 100 cells/μl and the leukocytes are predominantly mononuclear cells. The CSF protein concentration may be mildly elevated and glucose is usually in the normal range, although low CSF glucose levels have occasionally been reported (Roine et al., 1988; Chotmongkol et al., 1991). Serum neutralizing antibodies against rabies virus are not usually present in unimmunized patients until the second week of the illness and patients may die of rabies without developing a detectable serum antibody level (Hattwick, 1974; Anderson et al., 1984; Kasempimolporn et al., 1991). Antibody had not developed in serum by 10 days after the onset of clinical symptoms in five of 18 (28%) patients with rabies in the USA (Noah et al., 1998). One patient, who had received interferon therapy, had not developed antibodies by the time of death 24 days after the onset of symptoms (Sibley et al., 1981). Rabies virus antibodies develop in the CSF later than in the serum and the CSF titer is lower. Very high titers of rabies virus antibodies in the CSF have been interpreted as evidence of rabies encephalitis in vaccinated patients (Hattwick, 1974; Porras et al., 1976; Tillotson et al., 1977b; Alvarez et al., 1994; Madhusudana et al., 2002). Rabies virus may occasionally be isolated from saliva and rarely from the CSF or urine sediment (Anderson et al., 1984). Virus isolation is more likely during early disease before neutralizing antibodies appear, because they produce 'autosterilization' of tissues. Rabies virus antigen may be demonstrated antemortem by using the fluorescent antibody technique in frozen sections from skin biopsies (see Figure 10.1). Biopsies are usually obtained by using a full-thickness punch biopsy (3–7 mm in diameter) from a hairy area at the nape of the neck (Bryceson et al., 1975; Warrell et al., 1988). Many sections should be examined in order to ensure thorough evaluation of several hair follicles. Antigen is found in small sensory nerves adjacent to hair follicles. Antigen detection has also been performed on corneal impression smears, but the sensitivity of the method is low and false positive results may occur (Koch et al., 1975; Anderson et al., 1984; Mathuranayagam and Rao, 1984; Warrell et al., 1988; Noah et al., 1998).

4.2.1 Detection of rabies virus RNA

Small amounts of rabies virus RNA from brain tissue, CSF, or saliva can be amplified using the reverse transcriptase polymerase chain reaction (RT-PCR) and this technique has proven to be a valuable diagnostic tool for rabies. RT-PCR was initially used on CSF specimens and subsequently on saliva to confirm a diagnosis of rabies (Kamolvarin *et al.*, 1993; McColl *et al.*, 1993; Crepin *et al.*, 1998). In a study on both CSF and saliva samples from nine patients with confirmed rabies, the pre-mortem diagnosis of rabies was confirmed by positive RT-PCR in five of nine patients (56%) in saliva and in only two of nine patients (22%) in CSF (Crepin *et al.*, 1998). In comparison, skin biopsies were positive for rabies virus antigen using the fluorescent antibody technique in six of seven patients (86%). These findings led to a recommendation that both skin biopsy and saliva specimens be obtained for testing with immunofluorescence and RT-PCR, respectively. Of 20 human rabies cases diagnosed before death in the USA between 1980 and 1996, rabies virus RNA was detected in saliva from all 10 patients who had the test performed, including three who had negative viral isolation from saliva (Noah *et al.*, 1998).

4.2.2 Brain tissue

The presence of Negri bodies in neurons is a pathologic hallmark of rabies observed on routine histologic staining, but these characteristic inclusion bodies in the cytoplasm of infected neurons (see Chapter 9) may be absent. The diagnosis of rabies in humans using brain biopsies has not been assessed adequately, but rabies virus antigen was detected in brain tissues obtained by biopsy from three of three cases in the USA from 1980 to 1996 (Noah *et al.*, 1998). Post-mortem brain tissue may be obtained by a needle (e.g. Vim-Silverman or trucut needle) aspiration technique through either the orbit or foramen magnum (Sow *et al.*, 1996; Warrell, 1996; Tong *et al.*, 1999) and assessed for viral isolation, rabies virus antigen or rabies virus RNA, although false negative results may occur. Hence, a full autopsy may not be required to confirm a diagnosis when rabies is clinically suspected but unconfirmed ante-mortem. Rabies may not be diagnosed until post-mortem neuropathologic examination of the brain is performed because this diagnosis was not considered by the patient's physicians (King *et al.*, 1978; Munoz *et al.*, 1996; Geyer *et al.*, 1997; Parker *et al.*, 2003; Silverstein *et al.*, 2003). A range of diagnostic investigations may be performed on post-mortem human tissues, including virus isolation, the fluorescent antibody test (on fresh or formalin-fixed, paraffin-embedded [Whitfield *et al.*, 2001] specimens), immunoperoxidase staining for rabies virus antigen or *in situ* hybridization for rabies virus RNA (Jackson and Wunner, 1991) or detection of rabies virus RNA by using reverse transcriptase polymerase chain reaction amplification (see Section 4.2.1).

5 DIFFERENTIAL DIAGNOSIS

For patients and their relatives who are unable to recall an animal exposure, even when questioned directly, it is more difficult to make a diagnosis of rabies. Most cases without a clear history of rabies exposure are due to bat rabies viruses. There may be a history of recent travel in a rabies endemic area. Rabies is most commonly misdiagnosed as either a psychiatric or laryngopharyngeal disorder. The disease may also present with bizarre neuropsychiatric symptoms mimicking conditions such as schizophrenic psychosis or acute mania (Goswami *et al.*, 1984).

Patients often become quite fearful about the possibility of developing rabies after an animal bite or exposure. Rabies hysteria is a conversion disorder in which patients exhibit clinical features similar to rabies with unconscious motivation that involves poorly understood neural networks (Wilson *et al.*, 1975; Ron, 2001), which should not be confused with malingering (feigning) in which there is deception by the patient. Rabies hysteria is probably the most difficult differential diagnosis. In general, it is characterized by a shorter incubation period (often a few hours or a day or two) than rabies, an early onset of inability of the patient to communicate, bizarre spasms, spitting out of water taken in the mouth with no actual attempt at swallowing, barking, biting, aggressive behavior directed toward health care workers, lack of fever and neurologic signs and a long clinical course with recovery. Village practitioners in endemic areas may establish a reputation that they can cure rabies due to recovery of patients with rabies hysteria (Wilson *et al.*, 1975). However, it should be emphasized that the clinical picture may be so bizarre in patients with rabies that they may be misdiagnosed as having hysteria (Bisseru, 1972).

Other viral encephalitides may show behavioral disturbances with fluctuations in the level of consciousness. However, hydrophobic spasms are not observed in these conditions and it is unusual for a conscious patient to have prominent brainstem signs in other encephalitides. Herpes simiae (B virus) encephalomyelitis, which is transmitted by monkey bites, is often associated with a shorter incubation period than in rabies (e.g. 3–5 days); vesicles may be present at the site of the bite (also in the monkey's oral cavity) and recovery may occur (Whitley, 2004). Two recent cases of rabies in the USA were misdiagnosed as Creutzfeldt-Jakob disease (Geyer *et al.*, 1997) and both of these patients had a rapidly progressive neurological illness with prominent myoclonus.

Tetanus, a disease caused by the neurotoxin from the bacterium *Clostridium tetani*, may develop in association with a dirty wound caused by an animal bite. Tetanus has a shorter incubation period (usually 3–21 days) than rabies and, unlike rabies, it is characterized by sustained muscle rigidity involving axial muscles, including paraspinal, abdominal, masseter (trismus), laryngeal and respiratory muscles, with superimposed brief recurrent muscle spasms (Bleck and Brauner, 2004). In tetanus, the mental state is not affected, there is no CSF pleocytosis and the prognosis is much better than in rabies.

Post-vaccinal encephalomyelitis is another important differential diagnosis, particularly in patients who have been immunized with a vaccine derived from neural tissues (e.g. Semple vaccine). Post-vaccinal encephalomyelitis usually develops within 2 weeks of initiation of vaccination, which is helpful in the differential diagnosis. Local sensory symptoms (paresthesias, pain and pruritus), alternating intervals of agitation and lucidity and hydrophobia are clinical features that strongly suggest a diagnosis of rabies rather than post-vaccinal encephalomyelitis.

Paralytic rabies resembles the Guillain-Barré syndrome, including both acute inflammatory demyelinating polyradiculopathy and acute motor-sensory axonal neuropathy. In a recent pathologic series of the latter, one case (case number 1 in the report) was subsequently demonstrated to have paralytic rabies (Griffin *et al.*, 1996; Sheikh *et al.*, 2005). Local symptoms at the site of the bite, piloerection, early or persistent bladder dysfunction and fever are more suggestive of paralytic rabies. The Guillain-Barré syndrome may occasionally occur as a post-vaccinal complication from rabies vaccines derived from neural tissues, particularly the suckling mouse brain vaccine (Toro *et al.*, 1977).

6 THERAPY

Preventative therapy for rabies after exposures is highly effective if current recommendations are followed (Centers for Disease Control and Prevention, 1999). However, even minor deviations from these recommendations may result in the development of rabies. Unfortunately, treatment of rabies has proved to be disappointing. Therapy with human leukocyte interferon in three patients with high-dose intraventricular and systemic (intramuscular) administration was not associated with a beneficial clinical effect, but this therapy was not initiated until between 8 and 14 days after the onset of symptoms (Merigan *et al.*, 1984). Similarly, antiviral therapy with intravenous ribavirin (16 patients given doses of 16–400 mg) was unsuccessful in China (Kureishi *et al.*, 1992). An open trial of therapy with combined intravenous and intrathecal administration of either ribavirin (one patient) or interferon-alfa (three patients) (Warrell *et al.*, 1989) was also unsuccessful. Anti-rabies virus hyperimmune serum of either human or equine origin has been administered intravenously and by the intrathecal route (Emmons *et al.*, 1973; Hattwick *et al.*, 1976; Basgoz and Frosch, 1998; Hemachudha *et al.*, 2003), but there was no clear beneficial effect. In some cases, survival has been prolonged for a few weeks with critical care measures. Therapy is supportive and adequate sedation and analgesia are very important when a palliative approach is taken.

In 2001, a conference was held that included physicians who had experience in the management of human rabies and researchers with expertise in rabies pathogenesis and the opinions of the participants were published in a viewpoint article in *Clinical Infectious Diseases*, including therapeutic options when aggressive therapy is considered desirable (Jackson *et al.*, 2003). Patients in good health with relatively early disease and access to adequate resources and facilities were felt to be potential candidates for an aggressive approach. It was felt that a combination of specific therapies should be considered, including rabies vaccine, rabies immune globulin, monoclonal antibodies (in the future), ribavirin, interferon-α and ketamine (a dissociative anesthetic agent that is a non-competitive antagonist of the *N*-methyl-D-aspartate [NMDA] receptor). Previous studies performed *in vitro* and in an experimental animal model suggested ketamine may be a useful therapeutic agent (Lockhart *et al.*, 1991, 1992). Recovery has occurred in one patient who received therapy with ketamine and other agents (Willoughby *et al.*, 2005) (see below), but therapy was unsuccessful in at least four other cases (Hemachudha *et al.*, 2006). Also, very recent experimental studies in primary neuron cultures and in a mouse model indicate that ketamine therapy is disappointing (Weli *et al.*, 2006).

7 RECOVERY FROM RABIES

Survival from rabies has been well documented in only six patients (Table 7.3) and all but one of these patients received rabies immunization prior to the onset of clinical disease. The first recovery from rabies, which has been the only case without significant neurological sequelae, occurred in 1970 (Hattwick *et al.*, 1972). Matthew Winkler, a 6-year-old boy from Ohio, was bitten on his left thumb by a big brown bat (*Eptesicus fuscus*), which was later shown to be rabid. Vaccination was initiated with duck embryo rabies vaccine beginning 4 days after the bite and shortly after completing the multidose therapy (20 days after the bite) he became ill with fever and meningeal signs. His CSF showed 125 white cells/μl (75% mononuclear cells and 25% polymorphonuclear leukocytes) and the CSF protein was elevated. He developed abnormal behavior and later lapsed into a coma. He had focal neurological signs and seizures and developed cardiac and respiratory complications. He subsequently showed progressive improvement and apparently had a good neurologic recovery. A brain biopsy was consistent with encephalitis. His serum neutralization titer against rabies virus peaked at 1:63 000 at 3 months. This titer was much higher than has been observed secondary to vaccination. He also had very high titers of neutralizing antibodies in the CSF, which have not been observed with vaccination. Rabies virus was not isolated from brain tissue, CSF, or saliva, probably as a result of viral neutralization related to the high antibody levels.

The second case with recovery was a 45-year-old woman who sustained multiple deep bites to her left arm from a dog in Argentina in 1972 (Porras *et al.*, 1976). The dog developed neurologic signs and died 4 days later. The patient received 14 daily doses of suckling mouse brain rabies vaccine beginning 10 days after the bites, which were followed by two booster doses. Twenty-one days after the bites (at the time of her twelfth vaccine dose), she developed left arm paresthesias,

TABLE 7.3

Cases of human rabies with recovery

Location	Year	Age of patient	Transmission	Immunization	Outcome	Reference
USA	1970	6	Bat bite	Duck embryo vaccine	Complete recovery	Baer et al., 1982
Argentina	1972	45	Dog bites	Suckling mouse brain vaccine	Mild sequelae	Porras et al., 1976
USA	1977	32	Laboratory (vaccine strain)	Pre-exposure vaccination	Sequelae	Tillotson et al., 1977b
Mexico	1992	9	Dog bites	Post-exposure vaccination (combination)	Severe sequelae[a]	Alvarez et al., 1994
India	2000	6	Dog bites	Post-exposure vaccination (combination)	Severe sequelae[b]	Madhusudana et al., 2002
USA	2004	15	Bat bite	None	Mild to moderate sequelae	Willoughby et al., 2005

[a] Patient died less than 4 years after developing rabies with marked neurological sequelae (L. Alvarez, personal communication).
[b] Patient died about 2 years after developing rabies with marked neurological sequelae (S. Mahusudana, personal communication).

which subsequently spread and became accompanied by pain; vaccination was continued. She was admitted to hospital with quadriparesis and hyperreflexia 31 days after the bites. She had limb weakness, tremor in her upper extremities (greater on the left), cerebellar signs (asynergia, ataxia, dysmetria and dysdiado-chokinesia), generalized myoclonus and hyperreflexia in her lower extremities. Prominent cerebellar signs are unusual in rabies, despite the characteristic infection of neurons in the cerebellum, including Purkinje cells and deep cerebellar nuclei. Her CSF showed 5 cells/µl and CSF protein was mildly elevated at 0.65 g/l. Her serum neutralization titer against rabies virus peaked at 1:640 000 at about 3 months and she also had very high titers of neutralizing antibodies in the CSF. Rabies virus was not isolated from her saliva or CSF and corneal impression smears were negative for rabies virus antigen. Neurologic deterioration occurred shortly after she received each of the two booster doses of rabies vaccine and included altered mental status, generalized seizures, dysphagia and quadriparesis. She showed neurological improvement over the next few months. Thirteen months after the onset of her symptoms, her recovery was reported as 'nearly complete'. However, there was no description of her residual neurological deficits (Porras et al., 1976). The unusual neurologic features of this patient and the clinical worsening after booster doses of the suckling mouse brain rabies vaccine were administered raise the question of whether encephalomyelitis due to the rabies vaccine played a significant role in this patient's clinical picture.

The third case occurred in a 32-year-old laboratory technician in New York in 1977 who was pre-immunized with duck embryo rabies vaccine (Tillotson et al., 1977a, 1977b). About 5 months prior to his illness, he had a rabies virus neutralizing antibody titer of 1:32. He worked with live rabies virus vaccine strains and he was likely exposed to an aerosol of rabies virus about 2 weeks prior to the onset of his illness. He experienced initial malaise, headache, fever, chills and nausea and then lethargy with intermittent delirium. He was admitted to hospital in Albany, New York 6 days after the onset of his symptoms with expressive aphasia, hyperreflexia and primitive reflexes. CSF showed 230 white cells/µl (95% mononuclear cells) and CSF protein was elevated at 1.17 g/l. The day after admission to hospital, he deteriorated and went into a deep coma. His serum neutralizing antibody titer increased from 1:32 to 1:64 000 and subsequently increased to 1:175 000 over a 10-day period during his illness (Tillotson et al., 1977a). He also developed a high titer of CSF antibodies. Rabies virus antigen was not detectable in a skin biopsy or in corneal impression smears. Four months after the onset of his illness, he was ambulatory, but he had residual aphasia and spasticity (Tillotson et al., 1977a). This was the first report of a case of rabies in a pre-immunized individual and only the fourth well-documented case with transmission due to airborne exposure to the virus.

The fourth case, from 1992, occurred in a 9-year-old boy in Mexico (Alvarez et al., 1994). This boy sustained severe facial bites from a dog and received local wound treatment. On the day after the bites, vaccination was initiated with VERO rabies vaccine, but passive immunization with rabies immune globulin was not given. Nineteen days after the bites, he developed fever and dysphagia. He subsequently had a variety of abnormal neurological signs and convulsions. He never developed hydrophobia or inspiratory spasms. He was admitted to hospital and subsequently became comatose. CSF showed 184 cells/µl (65% mononuclear cells). He required mechanical ventilation for several days. Rabies virus was not isolated from saliva and rabies virus antigen was not found in either a skin biopsy or corneal impression smears. His peak serum neutralizing antibody titer was 1:34 800 (39 days after the bite) and he had a very high CSF antibody titer. He had severe neurologic sequelae, including quadriparesis and visual impairment. Although he recovered for a period, he died almost 4 years later (L. Alvarez, personal communication).

The fifth case was a 6-year-old girl who was bitten on the face and hands by a dog in India and the dog died 4 days later (Madhusudana *et al.*, 2002). She received three doses of rabies purified chick embryo cell vaccine (PCECV) on days 0, 3 and 7, but no local wound treatment was given and rabies immune globulin was not administered. She developed clinical features of rabies 14 days after the bites, which included fever, dysphagia to liquids and visual hallucinations. A rare neurologic complication to the PCECV was considered and she was given methylprednisolone and one dose of rabies human diploid cell vaccine. She subsequently developed hypersalivation and focal motor seizures and she became comatose. An MR scan showed T_2-weighted hyperintense signals in the cerebral cortex, basal ganglia and brainstem. She had a CSF pleocytosis. Her peak serum neutralizing antibody titer was 1:312 000 (7800 IU/ml) after 110 days of illness and she had a CSF antibody titer of 1:182 000 (4550 IU/ml) at this time. Rabies virus was not isolated and both skin biopsies and corneal tests were negative for rabies virus antigen. She had severe neurologic sequelae, including rigidity and involuntary movements of her limbs and she had frequent opisthotonic postures. She died about 2 years later (S. Madhusudana, personal communication).

The sixth and most recent case occurred in Wisconsin in 2004 (Willoughby *et al.*, 2004, 2005). A previously healthy 15-year-old female was bitten by a bat on her left index finger while attending a church service and she subsequently released the bat. The wound was washed with peroxide, but she did not seek medical attention at that time. About one month after the bite she developed numbness and tingling of her left hand and, over the next 3 days, she developed diplopia related to bilateral partial sixth-nerve palsies, unsteadiness and nausea and vomiting. MRI brain was normal. On her fourth day of illness CSF showed 23 white cells/µl (93% lymphocytes) and CSF protein was mildly elevated at 50 mg/dl. She subsequently developed fever (38.8°C), nystagmus, left arm tremor and hypersalivation, and at about that time the history of the bat bite was obtained. The patient was transferred to a tertiary care hospital in Milwaukee 5 days after the onset of neurologic symptoms. A repeat MRI scan was normal. Neutralizing anti-rabies virus antibodies were detected in serum and CSF on the first hospital day (initially 1:102 and 1:47, respectively) and subsequently increased (to 1:1183 and 1:1300, respectively). Nuchal skin biopsies were negative for rabies virus antigen, rabies virus RNA was not detected in the skin biopsies or in saliva by RT-PCR and viral isolation on saliva was negative. The patient was intubated and put into a drug-induced coma, which included the non-competitive NMDA antagonist ketamine at 48 mg/kg/day as a continuous infusion and intravenous midazolam for 7 days. There was a deliberate attempt to maintain a burst-suppression pattern on her electroencephalogram and supplemental phenobarbital was given. She also received intravenous ribavirin and amantadine 200 mg per day administered enterally. She improved and was discharged from hospital with neurologic deficits and she has subsequently shown further progressive neurologic improvement. This is the first documented survivor who did not receive rabies vaccine prior to onset of clinical rabies. As discussed in an accompanying editorial, it is unknown if therapy with one or more specific agents played an important role in the outcome of this case (Jackson, 2005, 2006). Clearly, the efficacy of the drugs used in treating this patient needs to be evaluated further. The induction of coma *per se* has not been shown to be useful in the management of infectious diseases of the nervous system and, to date, there is no evidence supporting this approach in rabies or other forms of viral encephalitis. Hence, this approach should not become routine for the management of rabies at this time. The patient's illness and favorable outcome may have also been due to an attenuated bat rabies virus, possibly that never has been isolated, rather than a result of the therapeutic approach taken. Subsequently, at least four patients received similar therapeutic approaches with fatal outcomes (Hemachudha *et al.*, 2006). In addition, recent studies on ketamine in rabies virus-infected primary neuron cultures and in experimental rabies in mice have cast doubt on the efficacy of ketamine in rabies virus infection (Weli *et al.*, 2006).

Recovery in the preceding patients with rabies has inspired physicians to manage aggressively patients with rabies in critical care units. The hope was that if patients, even when previously unimmunized, could be maintained through the acute phase of their illness and avoid complications, then perhaps they could clear the viral infection and recover. Overall, this approach has been disappointing (Rubin *et al.*, 1970; Emmons *et al.*, 1973; Bhatt *et al.*, 1974; Lopez *et al.*, 1975; Gode *et al.*, 1976; Udwadia *et al.*, 1989), but the recent case from Milwaukee provides optimism that therapy may be much more effective in the future. Further research is needed to identify efficacious therapeutic agents and early clinical diagnosis of rabies with prompt initiation of effective therapy will become important in achieving good clinical outcomes.

8 RABIES DUE TO OTHER *LYSSAVIRUS* GENOTYPES

In addition to rabies virus, which is *Lyssavirus* genotype 1, there are six other *Lyssavirus* genotypes and five have been associated with cases of human rabies: Mokola virus (genotype 3), Duvenhage virus (genotype 4), European bat

TABLE 7.4

Reported human rabies cases due to other *Lyssavirus* genotypes

Virus	Year	Location	Age of patient	Reference
Duvenhage	1970	South Africa	31	Meredith *et al.*, 1971
Duvenhage	2006	South Africa	77	Paweska *et al.*, 2006
Mokola[a]	1968	Nigeria	3.5	Familusi and Moore, 1972
Mokola	1971	Nigeria	6	Familusi *et al.*, 1972
European bat lyssavirus 1	1985	Russia	11	Selimov *et al.*, 1989
European bat lyssavirus 2	1985	Finland	30	Roine *et al.*, 1988
European bat lyssavirus 2	2002	Scotland	55	Nathwani *et al.*, 2003
Australian bat lyssavirus	1996	Australia	39	Samaratunga *et al.*, 1998
Australian bat lyssavirus	1998	Australia	37	Hanna *et al.*, 2000

[a] It is doubtful that this patient's clinical picture was actually caused by Mokola virus infection.

lyssavirus 1 (genotype 5), European bat lyssavirus 2 (genotype 6) and Australian bat lyssavirus (genotype 7) (Table 7.4). They are commonly called *rabies-like* or *rabies-related viruses*. Lagos bat virus (genotype 2), which was first isolated from fruit-eating bats in Nigeria, is the only genotype that has not been associated with human disease. Although not yet designated genotypes, Aravan virus and Khujand virus are lyssaviruses isolated from bats in Central Asia, whereas Irkut virus and West Caucasian bat virus are lyssaviruses isolated from bats in Russia (Kuzmin *et al.*, 2003, 2005 Botvinkin *et al.*, 2003) and none of these four lyssaviruses has been associated with human disease (see Chapter 6).

8.1 Duvenhage virus

In 1970, a 31-year-old man from rural South Africa developed an illness with fever, excessive sweating, hydrophobia and spasms of his face, arms and torso that were precipitated by being touched (Meredith *et al.*, 1971). He also exhibited confusion, irritability and marked aggressiveness. He died after an illness lasting about 5 days. He lived outside the recognized enzootic and epizootic areas for rabies. He had been bitten on the lip by a bat while sleeping about 4 weeks earlier. The virus isolated from his brain was a new virus and was characterized and named *Duvenhage virus* (genotype 4). This patient's clinical illness was indistinguishable from that caused by rabies virus (genotype 1).

A recent report (Paweska *et al.*, 2006) indicates that in February, 2006 a 77-year-old male was scratched on the cheek by an insectivorous bat in North West Province, South Africa. He did not receive post-exposure treatment and he became ill one month later and died on day 14 of his illness. Duvenhage virus was identified by RT-PCR on saliva and brain tissue and confirmed with sequencing of the nucleoprotein amplicons.

8.2 Mokola virus

Mokola virus was first isolated from shrews in Nigeria (Shope *et al.*, 1970). In 1968, a 3½-year-old girl from Nigeria presented with a sudden onset of fever and convulsions (Familusi and Moore, 1972). She rapidly made a complete recovery. There were no cells in her CSF and the CSF protein and glucose were normal. Mokola virus was isolated from her CSF, although the shrew isolate of Mokola virus was handled in the same laboratory during the same time period. Cross-contamination of specimens in the laboratory remains a possible explanation for this viral isolation. The girl's neutralizing antibody titers were very low and disappeared within several months. The febrile convulsion was unlikely related to Mokola virus infection.

A 6-year-old girl died in Nigeria in 1971 after a six-day illness (Familusi *et al.*, 1972). She presented with drowsiness, confusion and weakness involving her extremities and trunk and progressed to coma. Her CSF was normal without a pleocytosis. At autopsy, there were large eosinophilic inclusion bodies in the cytoplasm of neurons and Mokola virus was isolated from her brain. Shrews were known to be plentiful around the house where she lived, although there was no documented evidence that she had actually been bitten. Mokola virus infection was associated with meningoencephalitis in this case without the typical features of brainstem involvement seen in encephalitic rabies.

8.3 European bat lyssavirus 1

In 1985, an 11-year-old girl from Belgorod, Russia, was bitten on the lower lip by an unidentified bat and died with signs of rabies (Selimov *et al.*, 1989). The viral isolate was called *Yuli virus* and classified as European bat lyssavirus type 1 (genotype 5) (Bourhy *et al.*, 1992). There was an earlier fatal case in a 15-year-old female in Voroshilovgrad (now Lugansk), Ukraine in 1977 that developed after a bat bite (Anonymous, 1986). However, no viral isolate is available for molecular characterization and it is uncertain if the infection was actually caused by European bat lyssavirus type 1 or another lyssavirus (see Chapter 6). A similar situation applies in the case of a 34-year-old male who died with clinical rabies (with hypersalivation and hydrophobia) 45 days after a bat bite in the Lugansk province, Ukraine in 2002, which is only 50 km from the site of the 1977 case (Botvinkin *et al.*, 2005).

8.4 European bat lyssavirus 2

In 1985, a 30-year-old zoologist from Finland developed numbness in his right arm and neck with leg weakness (Roine *et al.*, 1988). His CSF was normal without a pleocytosis. Subsequently, he developed myoclonus of his legs, agitation, hyper-excitability, inspiratory spasms, dysarthria, dysphagia and hypersalivation. He had a delirium that progressed to coma. Diabetes insipidus occurred and he died 23 days after the onset of the illness. He had never been vaccinated against rabies and had been bitten by bats in several countries over the prior 5-year period, including an exposure in southern Finland 51 days prior to the onset of his symptoms. A virus was isolated that resembled the enzootic European bat rabies virus isolates and it was classified as European bat lyssavirus type 2 (genotype 6) (Bourhy *et al.*, 1992). This patient also had a clinical illness that was indistinguishable from rabies associated with genotype 1.

In 2002, a 55-year-old bat conservationist presented with hematemesis and a 5-day history of arm paresthesias, left arm and shoulder pain and difficulty swallowing (Fooks *et al.*, 2003; Nathwani *et al.*, 2003). About 19 weeks prior he had been bitten on his left finger by a Daubenton's bat in Angus, Scotland, although he also had a remote history of other bat bites. On admission to hospital in Dundee, Scotland he was febrile and had gaze-evoked nystagmus, dysarthria, truncal, limb and gait ataxia, areflexia in the arms and hyperreflexia in the legs and his behaviour was inappropriately familiar. CT and MRI scans did not show significant abnormalities. CSF showed a normal cell count and a mildly elevated CSF protein (58 mg/dl). He was treated with intravenous immunoglobulin and also subsequently with high-dose methylprednisolone and cyclophosphamide (Fooks *et al.*, 2003). On day 5 of hospitalization he became acutely confused, agitated and aggressive. He was sedated and a repeat CSF examination showed a pleocytosis and CSF protein elevated at 1.09 mg/dl. His mental state subsequently deteriorated and his limbs became flaccid; he died on day 14 of hospitalization. Hemi-nested RT-PCR was positive for lyssavirus RNA in saliva obtained on day 9 of hospitalization with high homology with previous EBLV type 2a isolates obtained from bats in the UK. No rabies virus antibodies were detected in sera or CSF and virus could not be isolated from saliva, skin biopsies or CSF. The lyssavirus was cultured from post-mortem brain tissue. This was the first case of indigenous human rabies in the UK in 100 years and has had important public health implications for bat exposures in the region.

8.5 Australian bat lyssavirus

In 1996, a 39-year-old female from Australia died after a 20-day illness (Samaratunga *et al.*, 1998). She cared for fruit bats and had sustained numerous scratches to her left arm over 4 weeks prior to the onset of her illness and she was likely bitten by a yellow-bellied sheathtail bat, *Saccolaimus flaviventris* (an insectivorous bat), in her care (Hanna *et al.*, 2000). She developed progressive left arm weakness. Her CSF showed 100 white cells/µl (80% mononuclear cells and 20% polymorphonuclear leukocytes). She deteriorated with diplopia, dysarthria, dysphagia and ataxia. She later developed progressive limb and facial weakness with reduced deep tendon reflexes and fluctuations in her level of consciousness prior to her death. Small eosinophilic cytoplasmic inclusions were observed in neurons in gray matter areas. RT-PCR amplification of RNA extracted from brain tissue and CSF indicated that she was infected with a virus identical to Australian bat lyssavirus (genotype 7) that had been identified previously in flying foxes, which are fruit-eating bats. This patient had typical brainstem involvement of rabies, which quickly progressed to diffuse brain involvement.

In 1998, a 37-year-old woman from Mackay, Queensland, was admitted to hospital with a 5-day history of fever, paresthesias around the dorsum of her left hand, pain about the left shoulder girdle and sore throat with difficulty swallowing (Hanna *et al.*, 2000). There were pharyngeal spasms, evidence of autonomic instability and progressive neurologic deterioration. She died 19 days after the onset of the illness. Twenty-seven months prior to the onset of her illness (during 1996) she was bitten at the base of her left little finger by a flying fox (fruit bat) in the course of removing the bat from the back of a young child. She did not receive rabies post-exposure prophylaxis. Hemi-nested PCR analyses on multiple tissues and saliva were positive for the flying-fox (*Pteropus* spp.) variant of Australian bat lyssavirus (Hanna *et al.*, 2000). Although *Lyssavirus* infections of flying foxes have only recently been recognized in Australia, rabies virus infection was recognized in a gray-head flying fox (*Pteropus poliocephalus*) that died in India in 1978 (Pal *et al.*, 1980) and also in two dog-faced fruit bats (*Cyanopterus brachyotis*) from Thailand (Smith *et al.*, 1967). It is unclear exactly when and how Australian bat lyssavirus obtained its foothold in Australian frugivorous and insectivorous bats, but it is clear that the virus poses a threat to human health in this region.

REFERENCES

Alvarez, L., Fajardo, R., Lopez, E. *et al.* (1994). Partial recovery from rabies in a nine-year-old boy. *Pediatric Infectious Disease Journal* **13**, 1154–1155.

Anderson, L.J., Nicholson, K.G., Tauxe, R.V. and Winkler, W.G. (1984). Human rabies in the United States, 1960 to 1979: epidemiology, diagnosis, and prevention. *Annals of Internal Medicine* **100**, 728–735.

Anonymous (1986). Bat rabies in the Union of Soviet Socialist Republics. *Rabies Bulletin Europe* **10**, 12–14.

Asbury, A.K. and Cornblath, D.R. (1990). Assessment of current diagnostic criteria for Guillain-Barré syndrome. *Annals of Neurology* **27** (Suppl), S21–S24.

Awasthi, M., Parmar, H., Patankar, T. and Castillo, M. (2001). Imaging findings in rabies encephalitis. *American Journal of Neuroradiology* **22**, 677–680.

Baer, G.M., Shaddock, J.H., Houff, S.A., Harrison, A.K. and Gardner, J.J. (1982). Human rabies transmitted by corneal transplant. *Archives of Neurology* **39**, 103–107.

Basgoz, N. and Frosch, M.P. (1998). Case records of the Massachusetts General Hospital: a 32-year-old woman with pharyngeal spasms and paresthesias after a dog bite. *New England Journal of Medicine* **339**, 105–112.

Bek, M.D., Smith, W.T., Levy, M.H., Sullivan, E. and Rubin, G.L. (1992). Rabies case in New South Wales, 1990: public health aspects. *Medical Journal of Australia* **156**, 596–600.

Bhandari, M. and Kumar, S. (1986). Penile hyperexcitability as the presenting symptom of rabies. *British Journal of Urology* **58**, 224–233.

Bhatt, D.R., Hattwick, M.A.W., Gerdsen, R., Emmons, R.W. and Johnson, H.N. (1974). Human rabies: diagnosis, complications, and management. *American Journal of Diseases of Children* **127**, 862–869.

Bisseru, B. (1972). Human rabies. In: *Rabies* (B. Bisseru, ed.). pp. 385–453. London: William Heinemann Medical Books Ltd.

Black, D. and Wiktor, T.J. (1986). Survey of raccoon hunters for rabies antibody titers: pilot study. *Journal of the Florida Medical Association* **73**, 517–520.

Bleck, T.P. and Brauner, J.S. (2004). Tetanus. In: *Infections of the Central Nervous System*, 3rd edition (W.M. Scheld, R.J. Whitley and C.M. Marra, eds). pp. 625–648. Philadelphia: Lippincott Williams and Wilkins.

Botvinkin, A.D., Poleschuk, E.M., Kuzmin, I.V. *et al.* (2003). Novel lyssaviruses isolated from bats in Russia. *Emerging Infectious Diseases* **9**, 1623–1625.

Botvinkin, A.D., Selnikova, O.P., Antonova, L.A., Moiseeva, A.B., Nesterenko, E.Y. and Gromashevsky, L.V. (2005). Human rabies case caused from a bat bite in Ukraine. *Rabies Bulletin Europe* **29**, 5–7.

Bourhy, H., Kissi, B., Lafon, M., Sacramento, D. and Tordo, N. (1992). Antigenic and molecular characterization of bat rabies virus in Europe. *Journal of Clinical Microbiology* **30**, 2419–2426.

Bryceson, A.D.M., Greenwood, B.M., Warrell, D.A. *et al.* (1975). Demonstration during life of rabies antigen in humans. *Journal of Infectious Diseases* **131**, 71–74.

Burton, E.C., Burns, D.K., Opatowsky, M.J. *et al.* (2005). Rabies encephalomyelitis: clinical, neuroradiological, and pathological findings in 4 transplant recipients. *Archives of Neurology* **62**, 873–882.

Centers for Disease Control and Prevention (1999). Human rabies prevention – United States, 1999: Recommendations of the Advisory Committee on Immunization Practices (ACIP). *Morbidity and Mortality Weekly Report* **48(No. RR-1)**, 1–21.

Cheetham, H.D., Hart, J., Coghill, N.F. and Fox, B. (1970). Rabies with myocarditis: two cases in England. *Lancet* **1**, 921–922.

Chopra, J.S., Banerjee, A.K., Murthy, J.M.K. and Pal, S.R. (1980). Paralytic rabies: a clinico-pathological study. *Brain* **103**, 789–802.

Chotmongkol, V., Vuttivirojana, A. and Cheepblangchai, M. (1991). Unusual manifestation in paralytic rabies. *Southeast Asian Journal of Tropical Medicine and Public Health* **22**, 279–280.

Constantine, D.G. (1962). Rabies transmission by nonbite route. *Public Health Reports* **77**, 287–289.

Crepin, P., Audry, L., Rotivel, Y., Gacoin, A., Caroff, C. and Bourhy, H. (1998). Intravitam diagnosis of human rabies by PCR using saliva and cerebrospinal fluid. *Journal of Clinical Microbiology* **36**, 1117–1121.

da Rosa, E.S.T., Kotait, I., Barbosa, T.F.S. *et al.* (2006). Bat-transmitted human rabies outbreaks, Brazilian Amazon. *Emerging Infectious Diseases* **12**, 1197–1202.

Doege, T.C. and Northrop, R.L. (1974). Evidence for inapparent rabies infection. *Lancet* **2**, 826–829.

Dupont, J.R. and Earle, K.M. (1965). Human rabies encephalitis. A study of forty-nine fatal cases with a review of the literature. *Neurology* **15**, 1023–1034.

Dutta, J.K. (1996). Excessive libido in a woman with rabies. *Postgraduate Medical Journal* **72**, 554.

Editorial (1975). Diagnosis and management of human rabies. *British Medical Journal* **3**, 721–722.

Editorial (1978). Dumb rabies. *Lancet* **2**, 1031–1032.

Emmons, R.W., Leonard, L.L., DeGenaro, F. Jr *et al.* (1973). A case of human rabies with prolonged survival. *Intervirology* **1**, 60–72.

Familusi, J.B. and Moore, D.L. (1972). Isolation of a rabies related virus from the cerebrospinal fluid of a child with 'aseptic meningitis'. *African Journal of Medical Science* **3**, 93–96.

Familusi, J.B., Osunkoya, B.O., Moore, D.L., Kemp, G.E. and Fabiyi, A. (1972). A fatal human infection with Mokola virus. *American Journal of Tropical Medicine and Hygiene* **21**, 959–963.

Faoagali, J.L., De Buse, P., Strutton, G.M. and Samaratunga, H. (1988). A case of rabies. *Medical Journal of Australia* **149**, 702–707.

Feasby, T.E., Gilbert, J.J., Brown, W.F. *et al.* (1986). An acute axonal form of Guillain-Barré polyneuropathy. *Brain* **109**, 1115–1126.

Fekadu, M., Endeshaw, T., Alemu, W., Bogale, Y., Teshager, T. and Olson, J.G. (1996). Possible human-to-human transmission of rabies in Ethiopia. *Ethiopian Medical Journal* **34**, 123–127.

Fooks, A.R., McElhinney, L.M., Pounder, D.J. *et al.* (2003). Case report: Isolation of a European bat lyssavirus type 2a from a fatal human case of rabies encephalitis. *Journal of Medical Virology* **71**, 281–289.

Galian, A., Guerin, J.M., Lamotte, M. *et al.* (1980). Human-to-human transmission of rabies via a corneal transplant – France. *Morbidity and Mortality Weekly Report* **29**, 25–26.

Gamaleia, N. (1887). Etude sur la rage paralytique chez l'homme. *Annales de L'Institut Pasteur (Paris)* **1**, 63–83.

Gardner, A.M.N. (1970). An unusual case of rabies (Letter). *Lancet* **2**, 523.

Geyer, R., Van Leuven, M., Murphy, J. *et al.* (1997). Human rabies – Montana and Washington, 1997. *Morbidity and Mortality Weekly Report* **46**, 770–774.

Gode, G.R. and Bhide, N.K. (1988). Two rabies deaths after corneal grafts from one donor (Letter). *Lancet* **2**, 791.

Gode, G.R., Raju, A.V., Jayalakshmi, T.S., Kaul, H.L. and Bhide, N.K. (1976). Intensive care in rabies therapy. Clinical observations. *Lancet* **2**, 6–8.

Goswami, U., Shankar, S.K., Channabasavanna, S.M. and Chattopadhyay, A. (1984). Psychiatric presentations in rabies: a clinico-pathologic report from south India with a review of literature. *Tropical and Geographical Medicine* **36**, 77–81.

Griffin, J.W., Li, C.Y., Ho, T.W. *et al.* (1996). Pathology of the motor-sensory axonal Guillain-Barré syndrome. *Annals of Neurology* **39**, 17–28.

Hanna, J.N., Carney, I.K., Smith, G.A. *et al.* (2000). Australian bat lyssavirus infection: a second human case, with a long incubation period. *Medical Journal of Australia* **172**, 597–599.

Hantson, P., Guerit, J.M., de Tourtchaninoff, M. *et al.* (1993). Rabies encephalitis mimicking the electrophysiological pattern of brain death. A case report. *European Neurology* **33**, 212–217.

Hattwick, M.A., Corey, L. and Creech, W.B. (1976). Clinical use of human globulin immune to rabies virus. *Journal of Infectious Diseases* **133** (Suppl), A266–A272.

Hattwick, M.A.W. (1974). Human rabies. *Public Health Review* **3**, 229–274.

Hattwick, M.A.W., Weis, T.T., Stechschulte, C.J., Baer, G.M. and Gregg, M.B. (1972). Recovery from rabies: a case report. *Annals of Internal Medicine* **76**, 931–942.

Hemachudha, T. (1994). Human rabies: clinical aspects, pathogenesis, and potential therapy. In: *Lyssaviruses* (C.E. Rupprecht, B. Dietzschold and H. Koprowski, eds). pp. 121–143. Berlin: Springer-Verlag.

Hemachudha, T. (1997a). Rabies. In: *Central Nervous System Infectious Diseases and Therapy* (K.L. Roos, ed.). pp. 573–600. New York: Marcel Dekker.

Hemachudha, T. (1997b). Rabies. *Current Opinion in Neurology* **10**, 260–267.

Hemachudha, T., Panpanich, T., Phanuphak, P., Manatsathit, S. and Wilde, H. (1993). Immune activation in human rabies. *Transactions of the Royal Society of Tropical Medicine and Hygiene* **87**, 106–108.

Hemachudha, T., Phanthumchinda, K., Phanuphak, P. and Manutsathit, S. (1987). Myoedema as a clinical sign in paralytic rabies (Letter). *Lancet* **1**, 1210.

Hemachudha, T., Phanuphak, P., Sriwanthana, B. *et al.* (1988). Immunologic study of human encephalitic and paralytic rabies: preliminary report of 16 patients. *American Journal of Medicine* **84**, 673–677.

Hemachudha, T., Sunsaneewitayakul, B., Desudchit, T. *et al.* (2006). Failure of therapeutic coma and ketamine for therapy of human rabies. *Journal of Neurovirology* **12**, 407–409.

Hemachudha, T., Sunsaneewitayakul, B., Mitrabhakdi, E. *et al.* (2003). Paralytic complications following intravenous rabies immune globulin treatment in a patient with furious rabies (Letter). *International Journal of Infectious Diseases* **7**, 76–77.

Hemachudha, T., Wacharapluesadee, S., Mitrabhakdi, E., Morimoto, K. and Lewis, R.A. (2005). Pathophysiology of human paralytic rabies. *Journal of Neurovirology* **11**, 93–100.

Hornung, K. and Nix, W.A. (1992). Myoedema. A clinical and electrophysiological evaluation. *European Neurology* **32**, 130–133.

Houff, S.A., Burton, R.C., Wilson, R.W. *et al.* (1979). Human-to-human transmission of rabies virus by corneal transplant. *New England Journal of Medicine* **300**, 603–604.

Hurst, E.W. and Pawan, J.L. (1931). An outbreak of rabies in Trinidad without history of bites, and with the symptoms of acute ascending myelitis. *Lancet* **2**, 622–628.

Jackson, A.C. (2006). Rabies: new insights into pathogenesis and treatment. *Current Opinion in Neurology* **19**, 267–270.

Jackson, A.C. (2004). Screening of organ and tissue donors for rabies (Letter). *Lancet* **364**, 2094–2095.

Jackson, A.C. (2005). Recovery from rabies (Editorial). *New England Journal of Medicine* **352**, 2549–2550.

Jackson, A.C. and Fenton, M.B. (2001). Human rabies and bat bites (Letter). *Lancet* **357**, 1714.

Jackson, A.C., Warrell, M.J., Rupprecht, C.E. *et al.* (2003). Management of rabies in humans. *Clinical Infectious Diseases* **36**, 60–63.

Jackson, A.C. and Wunner, W.H. (1991). Detection of rabies virus genomic RNA and mRNA in mouse and human brains by using in situ hybridization. *Journal of Virology* **65**, 2839–2844.

Jackson, A.C., Ye, H., Phelan, C.C. *et al.* (1999). Extraneural organ involvement in human rabies. *Laboratory Investigation* **79**, 945–951.

Javadi, M.A., Fayaz, A., Mirdehghan, S.A. and Ainollahi, B. (1996). Transmission of rabies by corneal graft. *Cornea* **15**, 431–433.

Johnson, N., Brookes, S.M., Fooks, A.R. and Ross, R.S. (2005). Review of human rabies cases in the UK and in Germany. *Veterinary Record* **157**, 715.

Kamolvarin, N., Tirawatnpong, T., Rattanasiwamoke, R., Tirawatnpong, S., Panpanich, T. and Hemachudha, T. (1993). Diagnosis of rabies by polymerase chain reaction with nested primers. *Journal of Infectious Diseases* **167**, 207–210.

Kasempimolporn, S., Hemachudha, T., Khawplod, P. and Manatsathit, S. (1991). Human immune response to rabies nucleocapsid and glycoprotein antigens. *Clinical and Experimental Immunology* **84**, 195–199.

King, D.B., Sangalang, V.E., Manuel, R., Marrie, T., Pointer, A.E. and Thomson, A.D. (1978). A suspected case of human rabies – Nova Scotia. *Canadian Diseases Weekly Report* **4**, 49–51.

Koch, F.J., Sagartz, J.W., Davidson, D.E. and Lawhaswasdi, K. (1975). Diagnosis of human rabies by the cornea test. *American Journal of Clinical Pathology* **63**, 509–515.

Komsuoglu, S.S., Dora, F. and Kalabay, O. (1981). Periodic EEG activity in human rabies encephalitis (Letter). *Journal of Neurology, Neurosurgery and Psychiatry* **44**, 264–265.

Krebs, J.W., Mandel, E.J., Swerdlow, D.L. and Rupprecht, C.E. (2005). Rabies surveillance in the United States during 2004. *Journal of the American Veterinary Medical Association* **227**, 1912–1925.

Kureishi, A., Xu, L.Z., Wu, H. and Stiver, H.G. (1992). Rabies in China: recommendations for control. *Bulletin of the World Health Organization* **70**, 443–450.

Kuzmin, I.V., Hughes, G.J., Botvinkin, A.D., Orciari, L.A. and Rupprecht, C.E. (2005). Phylogenetic relationships of Irkut and West Caucasian bat viruses within the Lyssavirus genus and suggested quantitative criteria based on the N gene sequence for lyssavirus genotype definition. *Virus Research* **111**, 28–43.

Kuzmin, I.V., Orciari, L.A., Arai, Y.T. *et al.* (2003). Bat lyssaviruses (Aravan and Khujand) from Central Asia: phylogenetic relationships according to N, P and G gene sequences. *Virus Research* **97**, 65–79.

Laothamatas, J., Hemachudha, T., Mitrabhakdi, E., Wannakrairot, P. and Tulayadaechanont, S. (2003). MR imaging in human rabies. *Americal Journal of Neuroradiology* **24**, 1102–1109.

Laothamatas, J., Hemachudha, T., Tulyadechanont, S. and Mitrabhakdi, E. (1997). Neuroimaging in paralytic rabies. *Ramathibodi Medical Journal* **20,** 149–156.

Lockhart, B.P., Tordo, N. and Tsiang, H. (1992). Inhibition of rabies virus transcription in rat cortical neurons with the dissociative anesthetic ketamine. *Antimicrobial Agents and Chemotherapy* **36**, 1750–1755.

Lockhart, B.P., Tsiang, H., Ceccaldi, P.E. and Guillemer, S. (1991). Ketamine-mediated inhibition of rabies virus infection *in vitro* and in rat brain. *Antiviral Chemistry and Chemotherapy* **2**, 9–15.

Lopez, M., Neves, J., Moreira, E.C. *et al.* (1975). Human rabies. I. Intensive treatment. *Revista Do Instituto de Medicina Tropical de Sao Paulo* **17**, 103–110.

Lopez, R.A., Miranda, P.P., Tejada, V.E. and Fishbein, D.B. (1992). Outbreak of human rabies in the Peruvian jungle. *Lancet* **339**, 408–411.

Madhusudana, S.N., Nagaraj, D., Uday, M., Ratnavalli, E. and Kumar, M.V. (2002). Partial recovery from rabies in a six-year-old girl (Letter). *International Journal of Infectious Diseases* **6**, 85–86.

Mathuranayagam, D. and Rao, P.V. (1984). Antemortem diagnosis of human rabies by corneal impression smears using immunofluorescence technique. *Indian Journal of Medical Research* **79**, 463–467.

McColl, K.A., Gould, A.R., Selleck, P.W., Hooper, P.T., Westbury, H.A. and Smith, J.S. (1993). Polymerase chain reaction and other laboratory techniques in the diagnosis of long incubation rabies in Australia. *Australian Veterinary Journal* **70**, 84–89.

Meredith, C.D., Rossouw, A.P. and Koch, H.P. (1971). An unusual case of human rabies thought to be of chiropteran origin. *South African Medical Journal* **45**, 767–769.

Merigan, T.C., Baer, G.M., Winkler, W.G. *et al.* (1984). Human leukocyte interferon administration to patients with symptomatic and suspected rabies. *Annals of Neurology* **16,** 82–87.

Metze, K. and Feiden, W. (1991). Rabies virus ribonucleoprotein in the heart (Letter). *New England Journal of Medicine* **324**, 1814–1815.

Mills, R.P., Swanepoel, R., Hayes, M.M. and Gelfand, M. (1978). Dumb rabies: its development following vaccination in a subject with rabies. *Central African Journal of Medicine* **24**, 115–117.

Mitrabhakdi, E., Shuangshoti, S., Wannakrairot, P. *et al.* (2005). Difference in neuropathogenetic mechanisms in human furious and paralytic rabies. *Journal of the Neurological Sciences* **238**, 3–10.

Mrak, R.E. and Young, L. (1993). Rabies encephalitis in a patient with no history of exposure. *Human Pathology* **24**, 109–110.

Munoz, J.L., Wolff, R., Jain, A. *et al.* (1996). Human rabies – Connecticut, 1995. *Morbidity and Mortality Weekly Report* **45**, 207–209.

Nathwani, D., McIntyre, P.G., White, K. *et al.* (2003). Fatal human rabies caused by European bat lyssavirus type 2a infection in Scotland. *Clinical Infectious Diseases* **37**, 598–601.

Nehaul, B.B.G. (1955). Rabies transmitted by bats in British Guiana. *American Journal of Tropical Medicine and Hygiene* **4**, 550–553.

Nicholson, K.G. (1994). Human rabies. In: *Handbook of Neurovirology* (R.R. McKendall and W.G. Stroop, eds). pp. 463–480. New York: Marcel Dekker.

Noah, D.L., Drenzek, C.L., Smith, J.S. *et al.* (1998). Epidemiology of human rabies in the United States, 1980 to 1996. *Annals of Internal Medicine* **128**, 922–930.

Orr, P.H., Rubin, M.R. and Aoki, F.Y. (1988). Naturally acquired serum rabies neutralizing antibody in a Canadian Inuit population. *Arctic Medical Research* **47** (Suppl 1), 699–700.

Pal, S.R., Arora, B., Chhuttani, P.N. *et al.* (1980). Rabies virus infection of a flying fox bat, *Pteropus poliocephalus* in Chandigarh, Northern India. *Tropical and Geographical Medicine* **32**, 265–267.

Parker, R., McKay, D., Hawes, C. *et al.* (2003). Human rabies, British Columbia – January 2003. *Canada Communicable Disease Report* **29**, 137–138.

Pawan, J.L. (1939). Paralysis as a clinical manifestation in human rabies. *Annals of Tropical Medicine and Parasitology* **33**, 21–29.

Paweska, J.T., Blumberg, L.H., Liebenberg, C. *et al.* (2006). Fatal human infection with rabies-related Duvenhage virus, South Africa. *Emerging Infectious Diseases* **12**, 1965–1967.

Pearson, C.A. (1976). Rabies (Letter). *Lancet* **1**, 206.

Phuapradit, P., Manatsathit, S., Warrell, M.J. and Warrell, D.A. (1985). Paralytic rabies: some unusual clinical presentations. *Journal of the Medical Association of Thailand* **68**, 106–110.

Pleasure, S.J. and Fischbein, N.J. (2000). Correlation of clinical and neuroimaging findings in a case of rabies encephalitis. *Archives of Neurology* **57**, 1765–1769.

Porras, C., Barboza, J.J., Fuenzalida, E., Adaros, H.L., Oviedo, A.M. and Furst, J. (1976). Recovery from rabies in man. *Annals of Internal Medicine* **85**, 44–48.

Prier, S., Gibert, C., Bodros, A. and Krymolieres, F. (1979a). Neurophysiological changes in non-vaccinated rabies patients (Letter). *Lancet* **1**, 620.

Prier, S., Gibert, C., Bodros, A., Vachon, F., Atanasiu, P. and Masson, M. (1979b). Les neuropathies de la rage humaine: etude clinique et electrophysiologique de deux cas [Human rabies neuropathies: clinical and electrophysiological study in two cases]. *Revue Neurologique* **135**, 161–168.

Raman, G.V., Prosser, A., Spreadbury, P.L., Cockcroft, P.M. and Okubadejo, O.A. (1988). Rabies presenting with myocarditis and encephalitis. *Journal of Infection* **17**, 155–158.

Remington, P.L., Shope, T. and Andrews, J. (1985). A recommended approach to the evaluation of human rabies exposure in an acute-care hospital. *Journal of the American Medical Association* **254**, 67–69.

Roine, R.O., Hillbom, M., Valle, M. *et al.* (1988). Fatal encephalitis caused by a bat-borne rabies-related virus: clinical findings. *Brain* **111**, 1505–1516.

Ron, M. (2001). Explaining the unexplained: understanding hysteria (Editorial). *Brain* **124**, 1065–1066.

Ross, E. and Armentrout, S.A. (1962). Myocarditis associated with rabies: report of a case. *New England Journal of Medicine* **266**, 1087–1089.

Rubin, R.H., Sullivan, L., Summers, R., Gregg, M.B. and Sikes, R.K. (1970). A case of human rabies in Kansas: epidemiologic, clinical, and laboratory considerations. *Journal of Infectious Diseases* **122**, 318–322.

Ruegsegger, J.M., Black, J. and Sharpless, G.R. (1961). Primary antirabies immunization of man with HEP Flury virus vaccine. *American Journal of Public Health* **51**, 706–716.

Samaratunga, H., Searle, J.W. and Hudson, N. (1998). Non-rabies lyssavirus human encephalitis from fruit bats: Australian bat lyssavirus (pteropid lyssavirus) infection. *Neuropathology and Applied Neurobiology* **24**, 331–335.

Selimov, M.A., Tatarov, A.G., Botvinkin, A.D., Klueva, E.V., Kulikova, L.G. and Khismatullina, N.A. (1989). Rabies-related Yuli virus; Identification with a panel of monoclonal antibodies. *Acta Virologica* **33**, 542–546.

Sheikh, K.A., Ramos-Alvarez, M., Jackson, A.C., Li, C.Y., Asbury, A.K. and Griffin, J.W. (2005). Overlap of pathology in paralytic rabies and axonal Guillain-Barré syndrome. *Annals of Neurology* **57**, 768–777.

Shope, R.E., Murphy, F.A., Harrison, A.K. *et al.* (1970). Two African viruses serologically and morphologically related to rabies virus. *Journal of Virology* **6**, 690–692.

Sibley, W.A., Ray, C.G., Petersen, E. *et al.* (1981). Human rabies acquired outside the United States from a dog bite. *Morbidity and Mortality Weekly Report* **43**, 537–540.

Silverstein, M.A., Salgado, C.D., Bassin, S. *et al.* (2003). First human death associated with raccoon rabies – Virginia, 2003. *Morbidity and Mortality Weekly Report* **52**, 1102–1103.

Sing, T.M. and Soo, M.Y. (1996). Imaging findings in rabies. *Australasian Radiology* **40**, 338–341.

Sipahioglu, U. and Alpaut, S. (1985). Transplacental rabies in a human [Turkish]. *Mikrobiyoloji Bulteni* **19**, 95–99.

Smith, J.S., Fishbein, D.B., Rupprecht, C.E. and Clark, K. (1991). Unexplained rabies in three immigrants in the United States: a virologic investigation. *New England Journal of Medicine* **324**, 205–211.

Smith, P.C., Lawhaswasdi, K., Vick, W.E. and Stanton, J.S. (1967). Isolation of rabies virus from fruit bats in Thailand. *Nature* **216**, 384.

Sow, P.S., Diop, B.M., Ndour, C.T.Y. *et al.* (1996). Occipital cerebral aspiration ponction: technical procedure to take a brain specimen for postmortem virological diagnosis of human rabies in Dakar. *Medecine et Maladies Infectieuses* **26**, 534–536.

Srinivasan, A., Burton, E.C., Kuehnert, M.J. *et al.* (2005). Transmission of rabies virus from an organ donor to four transplant recipients. *New England Journal of Medicine* **352**, 1103–1111.

Sureau, P., Portnoi, D., Rollin, P., Lapresle, C. and Chaouni-Berbich, A. (1981). [Prevention of inter-human rabies transmission after corneal graft]. [French]. *Comptes Rendus de l'Académie des Sciences – Series III, Sciences de la Vie* **293**, 689–692.

Talaulicar, P.M.S. (1977). Persistent priapism in rabies. *British Journal of Urology* **49**, 462.

Tariq, W.U.Z., Shafi, M.S., Jamal, S. and Ahmad, M. (1991). Rabies in man handling infected calf (Letter). *Lancet* **337**, 1224.

Thongcharoen, P., Wasi, C., Sirikavin, S. *et al.* (1981). Human-to-human transmission of rabies via corneal transplant – Thailand. *Morbidity and Mortality Weekly Report* **30**, 473–474.

Tillotson, J.R., Axelrod, D. and Lyman, D.O. (1977a). Follow-up on rabies – New York. *Morbidity and Mortality Weekly Report* **26**, 249–250.

Tillotson, J.R., Axelrod, D. and Lyman, D.O. (1977b). Rabies in a laboratory worker – New York. *Morbidity and Mortality Weekly Report* **26**, 183–184.

Tirawatnpong, S., Hemachudha, T., Manutsathit, S., Shuangshoti, S., Phanthumchinda, K. and Phanuphak, P. (1989). Regional distribution of rabies viral antigen in central nervous system of human encephalitic and paralytic rabies. *Journal of the Neurological Sciences* **92**, 91–99.

Tong, T.R., Leung, K.M. and Lam, A.W.S. (1999). Trucut needle biopsy through superior orbital fissure for diagnosis of rabies. *Lancet* **354**, 2137–2138.

Toro, G., Vergara, I. and Roman, G. (1977). Neuroparalytic accidents of antirabies vaccination with suckling mouse brain vaccine: clinical and pathologic study of 21 cases. *Archives of Neurology* **34**, 694–700.

Udwadia, Z.F., Udwadia, F.E., Katrak, S.M. *et al.* (1989). Human rabies: clinical features, diagnosis, complications, and management. *Critical Care Medicine* **17**, 834–836.

Udwadia, Z.F., Udwadia, F.E., Rao, P.P. and Kapadia, F. (1988). Penile hyperexcitability with recurrent ejaculations as the presenting manifestation of a case of rabies. *Postgraduate Medical Journal* **64**, 85–86.

Verlinde, J.D., Li-Fo-Sjoe, E., Versteeg, J. and Dekker, S.M. (1975). A local outbreak of paralytic rabies in Surinam children. *Tropical and Geographical Medicine* **27**, 137–142.

Wallerstein, C. (1999). Rabies cases increase in the Philippines. *British Medical Journal* **318**, 1306.

Warrell, D.A. (1976). The clinical picture of rabies in man. *Transactions of the Royal Society of Tropical Medicine and Hygiene* **70**, 188–195.

Warrell, D.A. and Warrell, M.J. (1991). Rabies. In: *Infections of the Central Nervous System* (H.P. Lambert, ed.). pp. 317–328. Philadelphia: B.C. Decker Inc.

Warrell, D.A., Davidson, N.M., Pope, H.M. *et al.* (1976). Pathophysiologic studies in human rabies. *American Journal of Medicine* **60**, 180–190.

Warrell, M.J. (1996). Rabies. In: *Manson's Tropical Diseases* (G.C. Cook, ed.). pp. 700–720. London: W.B. Saunders.

Warrell, M.J., Looareesuwan, S., Manatsathit, S. *et al.* (1988). Rapid diagnosis of rabies and post-vaccinal encephalitides. *Clinical and Experimental Immunology* **71**, 229–234.

Warrell, M.J., White, N.J., Looareesuwan, S. *et al.* (1989). Failure of interferon alfa and tribavirin in rabies encephalitis. *British Medical Journal* **299**, 830–833.

Waterman, J.A. (1959). Acute ascending rabic myelitis. Rabies – transmitted by bats to human beings and animals. *Caribbean Medical Journal* **21**, 46–74.

Weli, S.C., Scott, C.A., Ward, C.A. and Jackson, A.C. (2006). Rabies virus infection of primary neuronal cultures and adult mice: failure to demonstrate evidence of excitotoxicity. *Journal of Virology* **80**, 10270–10273.

White, M., Davis, A., Rawlings, J. *et al.* (1994). Human rabies – Texas and California, 1993. *Morbidity and Mortality Weekly Report* **43**, 93–96.

Whitfield, S.G., Fekadu, M., Shaddock, J.H., Niezgoda, M., Warner, C.K. and Messenger, S.L. (2001). A comparative study of the fluorescent antibody test for rabies diagnosis in fresh and formalin-fixed brain tissue specimens. *Journal of Virological Methods* **95**, 145–151.

Whitley, R.J. (2004). B virus. In: *Infections of the Central Nervous System* (W.M. Scheld, R.J. Whitley and C.M. Marra, eds). pp. 197–203. Philadelphia: Lippincott Williams and Wilkins.

Wilde, H. and Chutivongse, S. (1988). Rabies: current management in Southeast Asia. *Medical Progress* **15**, 14–23.

Willoughby, R.E., Rotar, M.M., Dhonau, H.L. *et al.* (2004). Recovery of a patient from clinical rabies – Wisconsin, 2004. *Morbidity and Mortality Weekly Report* **53**, 1171–1173.

Willoughby, R.E. Jr, Tieves, K.S., Hoffman, G.M. *et al.* (2005). Survival after treatment of rabies with induction of coma. *New England Journal of Medicine* **352**, 2508–2514.

Wilson, J.M., Hettiarachchi, J. and Wijesuriya, L.M. (1975). Presenting features and diagnosis of rabies. *Lancet* **2**, 1139–1140.

Winkler, W.G., Fashinell, T.R., Leffingwell. L., Howard. P. and Conomy. J.P. (1973). Airborne rabies transmission in a laboratory worker. *Journal of the American Medical Association* **226**, 1219–1221.

World Health Organization (2005). *WHO Expert Consultation on Rabies: First Report*. Geneva: WHO.

8

Pathogenesis

ALAN C. JACKSON

Departments of Medicine (Neurology) and of Microbiology and Immunology, Queen's University, Kingston, ON K7L 3N6, Canada; currently Departments of Internal Medicine (Neurology) and of Medical Microbiology, University of Manitoba, Winnipeg, MB R3A IR9, Canada

1 INTRODUCTION

Rabies virus is a highly neurotropic virus that spreads along neural pathways and invades the central nervous system (CNS), where it causes an acute infection. Most of what we know about the events that take place during rabies infection has been learned from experimental models using animals. Fixed laboratory strains of rabies virus and rodent models have commonly been used, because they are easier to handle and less expensive, although the events in these models may not closely mimic the disease under natural conditions either in humans or in rabies vectors. There are a number of sequential steps that occur after peripheral inoculation of rabies virus from an animal bite, which is the most common mechanism of transmission (Figure 8.1). The steps include replication in peripheral tissues, spread along peripheral nerves and the spinal cord to the brain, dissemination within the CNS and centrifugal spread from the CNS along nerves to various organs, including the salivary glands. Each of the pathogenetic steps will be discussed below. In addition, mechanisms of brain dysfunction in rabies will be addressed.

2 EVENTS AT THE SITE OF EXPOSURE

2.1 Earliest events

Early studies in rabies pathogenesis, which were performed in order to establish the pathways and rate of viral spread, involved amputation of the tail or leg of an animal proximal to the site of inoculation with a 'fixed' or 'street' (wild-type) strain of rabies virus. The development of rabies could be prevented with amputation and the timing of the procedure was found to be critical. In later studies, neurectomy of the sciatic nerve was performed instead of amputation and similar results were observed (Baer *et al.*, 1965, 1968). These experiments clearly demonstrated that there was an incubation period in rabies infection during which there was time-dependent movement of virus along peripheral nerves from the site of inoculation

Rabies, second edition. Edited by Alan C. Jackson and William H. Wunner
ISBN 978-012-369366-2.

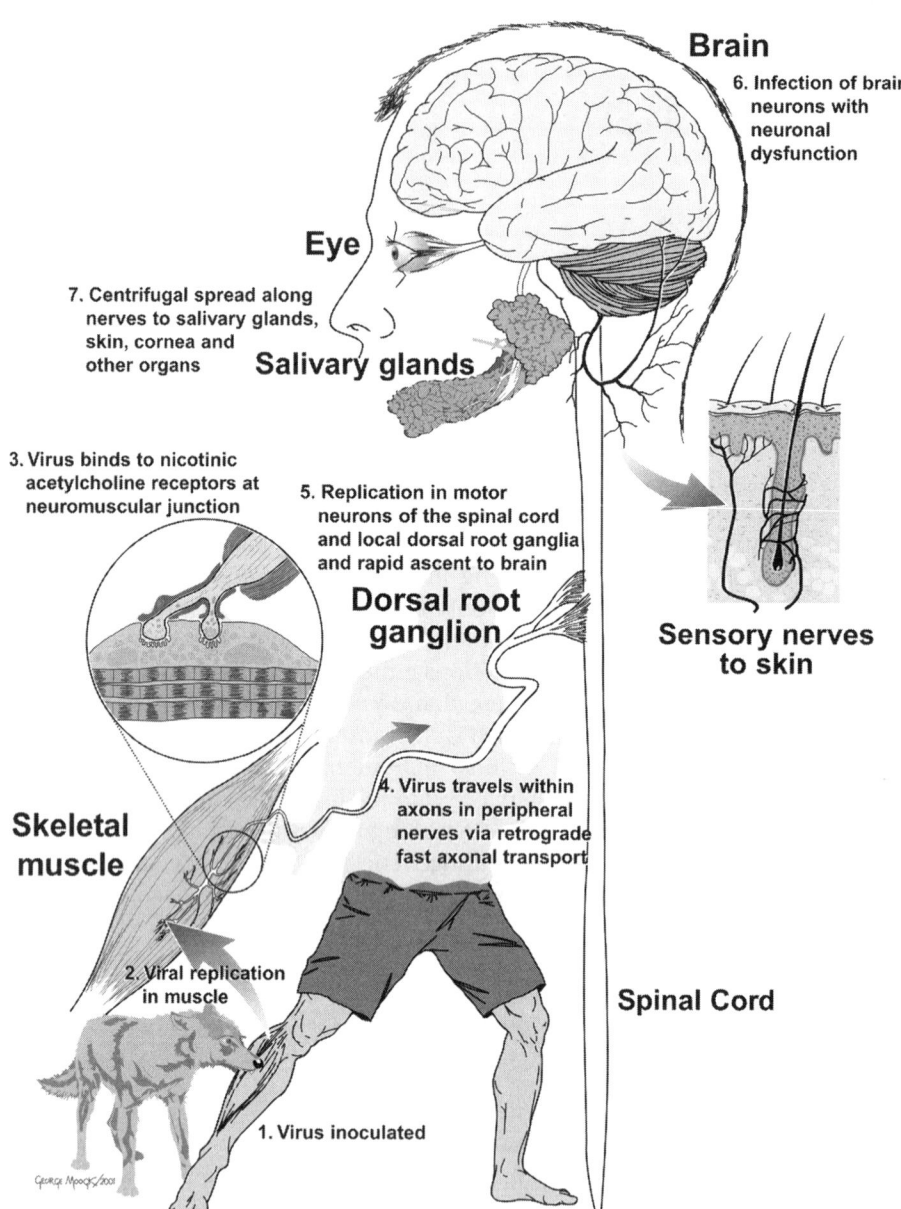

Brain

6. Infection of brain neurons with neuronal dysfunction

Eye

7. Centrifugal spread along nerves to salivary glands, skin, cornea and other organs

Salivary glands

3. Virus binds to nicotinic acetylcholine receptors at neuromuscular junction

5. Replication in motor neurons of the spinal cord and local dorsal root ganglia and rapid ascent to brain

Dorsal root ganglion

Sensory nerves to skin

4. Virus travels within axons in peripheral nerves via retrograde fast axonal transport

Skeletal muscle

2. Viral replication in muscle

1. Virus inoculated

Spinal Cord

Figure 8.1 Schematic diagram showing the sequential steps in the pathogenesis of rabies after an animal bite.

to the CNS and models using street rabies virus supported the idea that the virus remains at or near the site of entry for most of the long incubation period (Baer and Cleary, 1972). However, the time periods in which the procedures were life-saving in rodents infected with fixed rabies virus strains were relatively short (Dean *et al.*, 1963; Baer *et al.*, 1965), suggesting a different mechanism of viral entry for fixed viruses than in natural rabies (due to street viruses).

Under natural conditions, humans and animals may experience long and variable incubation periods following a bite exposure. This may play a role in maintaining enzootic rabies, especially in high density, high contact populations where there is a tendency for the disease to 'burn' itself out by rapidly reducing the number of susceptible animals. In humans, the incubation period is usually between 20 and 90 days, although incubation periods rarely may be as short as a few days or longer than a year (Smith *et al.*, 1991). There is uncertainty about the events that occur during this incubation period. There has been speculation that macrophages may sequester rabies virus *in vivo* because persistent *in vitro* infections of human and murine monocytic cell lines and of primary murine bone marrow macrophages have been demonstrated with different rabies virus strains (Ray *et al.*, 1995). This has not yet been demonstrated in animal models. The best experimental animal studies to date examining the events that take place during the incubation period were performed in striped skunks using a Canadian isolate of street rabies virus obtained from skunk salivary glands (Charlton *et al.*, 1997). These studies, performed using reverse transcriptase polymerase chain reaction (RT-PCR) amplification, showed that viral genomic RNA was frequently present in the inoculated muscle (found in four of nine skunks), but not in either spinal ganglia or the spinal cord when skunks were sacrificed 62–64 days post-inoculation. Immunohistochemical studies performed prior to the development of clinical disease showed evidence of infection of extrafusal muscle fibers and occasional fibrocytes at the site of inoculation. Although it is unclear, the infection of muscle fibers may be a critical pathogenetic step for the virus to gain access to the peripheral nervous system. In a highly susceptible host after intramuscular inoculation, rabies virus-infected suckling hamsters showed early infection of striated muscle cells near the site of inoculation and, shortly afterward, neuromuscular and neurotendinal spindles became infected near the site of inoculation, which was followed by evidence of infection of small nerves within muscles, tendons and adjoining connective tissues (Murphy *et al.*, 1973a). However, these events occurred within a few days of inoculation and do not mimic the situation with long incubation periods seen in natural infections.

In mouse models, early infection of muscle or other extraneural tissues was not observed following inoculation of fixed rabies virus strains (Johnson, 1965; Coulon *et al.*, 1989). Virus-specific RNA was not detected with RT-PCR amplification in the masseter muscle of adult mice between 6 and 30 hours after inoculation of the challenge virus standard (CVS) strain of rabies virus in the muscle, although viral RNA was identified in trigeminal ganglia at 18 hours and in the brainstem at 24 hours after inoculation (Shankar *et al.*, 1991). These studies suggest that rabies virus is capable of direct entry into peripheral nerves without a replicative cycle in extraneural cells during the short incubation period. This is likely the mechanism of viral entry in rodent models using fixed strains of rabies virus, accounting for the short period of time during which amputation or neurectomy is protective after peripheral inoculation of fixed rabies virus (Dean *et al.*, 1963; Baer *et al.*, 1965). Unfortunately, these models provide little information about events that take place during the long incubation period of natural rabies.

2.2 Receptor-mediated entry into nerve endings

The nicotinic acetylcholine receptor (nAChR) was the first identified receptor for rabies virus (Lentz *et al.*, 1982) (see Section 7.1). Rabies virus antigen was detected at sites coincident with the nAChR in infected cultured chick myotubes from chicken embryos and also shortly after immersion of mouse diaphragms in a suspension of rabies virus. It was evident from these studies that the distribution of viral antigen detected by fluorescent antibody staining at sites in neuromuscular junctions corresponded to the distribution of nAChRs. The receptors were stained with the rhodamine-conjugated antagonist α-bungarotoxin. Pretreatment of myotubes with either the irreversible binding nicotinic cholinergic antagonist α-bungarotoxin or the reversible binding *d*-tubocurarine reduced the number of myotubes that became infected with rabies virus. Studies in other laboratories showed that pretreatment of cultured rat myotubes with α-bungarotoxin had an inhibitory effect on infection (Tsiang *et al.*, 1986). Binding of radiolabeled rabies virus to purified *Torpedo* acetylcholine receptor was also inhibited by nicotinic antagonists, but not by atropine (a muscarinic antagonist) (Lentz *et al.*, 1986). Monoclonal antibodies raised against a peptide containing residues 190–203 of the rabies virus glycoprotein also inhibited binding of the rabies virus glycoprotein and α-bungarotoxin to the AChR (Bracci *et al.*, 1988). Both rabies virus and neurotoxins bind to residues 173–204 of the α_1-subunit of the AchR and the highest-affinity virus-binding determinants are located within residues 179–192 (Lentz, 1990). These studies have provided strong evidence that rabies virus binds to nicotinic acetylcholine receptors in neuromuscular junctions.

Snake venom neurotoxins are polypeptides that bind with high affinity to nAChRs and competitively block the depolarizing action of acetylcholine. When the amino acid sequence of the rabies virus glycoprotein was compared with that of snake venom neurotoxins, a significant sequence similarity was found between a segment (residues 151–238) of the rabies virus glycoprotein and the entire long neurotoxin sequence (71–74 residues) (Lentz *et al.*, 1984). The glycoprotein showed identity with residues at the end of loop 2 of the long neurotoxin (the 'toxic loop'), which is a long central loop projecting from the molecule that is highly conserved among all of the neurotoxins (Figure 8.2). This suggests

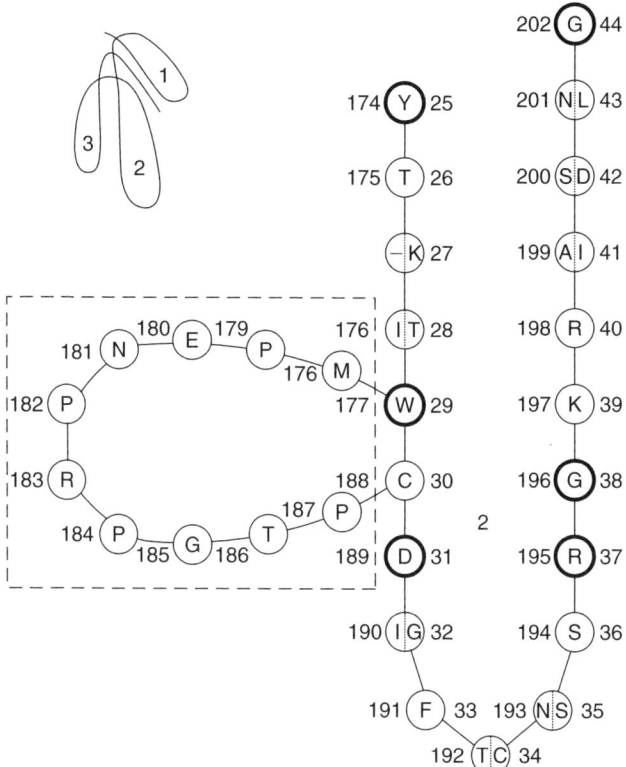

Figure 8.2 A model showing the similarity of the rabies virus glycoprotein with the 'toxic' loop of the neurotoxins. The segment of the glycoprotein (residues 174–202) corresponding to loop 2 of the long neurotoxins (Karlsson positions 25–44) is positioned in relationship to a schematic representation of loop 2. Within circles, residues or gaps in the glycoprotein are shown on the left and those in the neurotoxin on the right. One letter is shown where the glycoprotein and toxin are identical. Bold circles are residues highly conserved or invariant among all of the neurotoxins. A 10-residue insertion in the glycoprotein is enclosed in the box. The rabies virus sequence is of the CVS strain and the neurotoxin sequence is *Ophiophagus hannah*, toxin b. Inset: schematic of neurotoxin structure showing positions of loops 1, 2, and 3. (Reproduced with permission from Lentz *et al.*, Amino acid sequence similarity between rabies virus glycoprotein and snake venom curaremimetic neurotoxins. *Science* **226**, 847–848, 1984. Copyright 1984 by American Association for the Advancement of Science.)

that this region of the rabies virus glycoprotein is likely a recognition site for the acetylcholine receptor (Lentz, 1985).

Lentz and co-workers indicated that binding of rabies virus to AChRs would localize and concentrate the virus on post-synaptic cells, which would facilitate subsequent uptake and transfer of virus to peripheral motor nerves (Lentz *et al.*, 1982). Rabies virus may also bind to another rabies virus receptor, the neural cell adhesion molecule, which is present in presynaptic membranes (Lafon, 2005) (see Section 7.2). Studies performed in chick spinal cord-muscle co-cultures showed that the CVS strain of rabies virus and AChR tracers co-localized at neuro-muscular junctions and nerve terminals, which provided evidence that the neuro-muscular junction is the major site of entry into neurons (Lewis *et al.*, 2000). Subsequently, co-localization with endosome tracers indicated that the virus resides in an early endosome compartment. There is also supporting ultrastructural evidence that rabies virus particles enter nerve terminals by endocytosis (Iwasaki and Clark, 1975; Charlton and Casey, 1979c). The acidic interior of the endosome triggers fusion of the viral membrane with the endosome membrane, which allows the viral nucleocapsid to escape into the cytoplasm. However, it has not yet been resolved whether the viral uncoating actually takes place in nerve terminals or in the cell body (perikaryon) after transport in the axon.

Although rabies virus infection with fixed strains is restricted to a small number of cell types *in vivo*, fixed viruses can infect a much larger variety of cell types *in vitro* (Reagan and Wunner, 1985). There is evidence that carbohydrate moieties, phospholipids, highly sialylated gangliosides and other membrane-associated proteins might contribute to the cellular membrane receptor structure for rabies virus (Superti *et al.*, 1984, 1986; Conti *et al.*, 1986; Broughan and Wunner, 1995). In addition to the nicotinic acetylcholine receptor, the neural cell adhesion molecule and the p75 neurotrophin receptor have also been identified as rabies virus receptors (see Sections 7.2 and 7.3).

Variations in animal susceptibility to rabies virus infection have been recognized for many years. When infected by intramuscular inoculation, foxes are highly sensitive to rabies virus infection, dogs are less sensitive and opossums are highly resistant (Baer *et al.*, 1990a). The difference in susceptibility between the red fox and the opossum could reflect the quantity of acetylcholine receptors in muscle (Baer *et al.*, 1990b). A striking difference in the muscle content, B_{max}, of nicotinic acetylcholine receptors was found with 180.5 fmol/mg protein present in red foxes and only 11.4 fmol/mg protein present in opossums, which was a highly significant difference ($P < 0.001$). No difference was observed in the binding affinity, K_d. In addition, radiolabeled rabies virus bound much better to fox muscles than to opossum muscles. Hence, the susceptibility of different animal species to rabies virus may, at least in part, be related to the quantity of nicotinic acetylcholine receptors in their muscles.

2.3 Superficial and non-bite exposures

The vast majority of human rabies cases that occur without a history of an exposure are thought to be due to unrecognized or forgotten bites. Molecular characterization of the rabies virus strains has indicated that they are most frequently from the strain found in silver-haired bats and eastern pipistrelle bats in the USA

(Noah *et al.*, 1998), which are small bats. Experimental studies on the silver-haired bat virus (SHBV) indicate that the virus replicates well at lower than normal body temperatures (34°C) and is associated with higher infectivity in cell types present in the dermis, including fibroblasts and epithelial cells, than with coyote street virus (Morimoto *et al.*, 1996). Hence, the SHBV may have been selected for efficient local replication in the dermis, which could explain the success of this strain. However, after superficial exposures, it is unclear how or at precisely what sites the virus invades peripheral nerves in the skin or subcutaneous tissues.

Humans have rarely been infected by bat viruses via the airborne route whether in caves, where millions of bats roost (Constantine, 1962), or in laboratory accidents by aerosolized rabies virus (Winkler *et al.*, 1973; Tillotson *et al.*, 1977). Viral entry by the olfactory and oral routes is much less common than by bites. Relatively little experimental work has been done with routes of viral entry other than one simulating a bite exposure (using inoculation techniques). The nasal mucosa has been shown to act as a site of viral entry by suckling guinea pigs that have inhaled street rabies virus (Hronovsky and Benda, 1969). Rabies virus antigen was initially found in nasal mucosa cells 6 days later. Early brain infection was prominent in the olfactory bulbs, suggesting that rabies virus spread into the brain by an olfactory pathway. Similar results were obtained using a variety of rabies virus strains in mice and hamsters (Fischman and Schaeffer, 1971). Rabies virus antigen has been observed in olfactory receptor cells of naturally infected Brazilian free-tailed bats obtained from a cave, suggesting that the nasal mucosa is a portal of entry in natural infection of bats by airborne rabies virus in caves (Constantine *et al.*, 1972). Experimental studies showing transmission of rabies to a variety of species of carnivorous animals caged in a cave containing millions of bats supported infection by the airborne route (Constantine, 1962). However, it is thought that the presence of a very large number (millions) of bats in an unventilated area is necessary for airborne transmission of rabies virus.

Oral transmission of rabies virus might occur naturally by consumption of carcasses of rabid animals by wildlife and may also be important when humans eat raw dog meat (Wallerstein, 1999). Low susceptibility was observed when mice (Charlton and Casey, 1979b) and skunks (Charlton and Casey, 1979a) were given CVS or street rabies virus by either the oral route or intestinal instillation. Mice, hamsters, guinea pigs and rabbits of different ages were infected with CVS either orally or by gastric tube administration (Fischman and Ward, 1968). In CVS-infected weanling mice and hamsters that were infected by this route, rabies virus antigen was not observed in intestinal mucosal cells, but was found in neurons in Auerbach's and Meissner's plexuses of the stomach and intestine (Fischman and Schaeffer, 1971). These findings suggest that viral entry by the oral route likely occurs via breaks in the integrity of the gastrointestinal mucosa. However, the importance of oral transmission in natural rabies of animals remains uncertain.

3 SPREAD TO THE CNS

Centripetal spread of rabies virus to the CNS occurs within motor and perhaps also sensory axons of peripheral nerves. Colchicine, a microtubule-disrupting agent active for tubulin-containing cytoskeletal structures, is an effective inhibitor of fast axonal transport in the sciatic nerve of rats (Tsiang, 1979). When colchicine was applied locally to the sciatic nerve using elastomer cuffs to obtain high local concentrations of the drug, adverse systemic effects were avoided. Propagation of rabies virus was prevented, providing strong evidence that rabies virus spreads from sites of peripheral inoculation to the CNS by retrograde fast axonal transport. Human dorsal root ganglia neurons in a compartmentalized cell culture system were used to show that viral retrograde transport occurs at a rate of between 50 and 100 mm/day (Tsiang *et al.*, 1991b). There is evidence that the rabies virus phosphoprotein, which is a member of the ribonucleocapsid complex (see Chapter 2), interacts with dynein light chain 8 (LC8). Dynein LC8 is a component of both cytoplasmic myosin V and dynein that are involved in actin-based transport (important in early steps of viral entry) and microtubule-based transport (for fast axonal transport) in neurons, respectively (Jacob *et al.*, 2000; Raux *et al.*, 2000). This led to speculation that rabies virus phosphoprotein–dynein interaction may be of fundamental importance in axonal transport of rabies virus. However, studies performed in young mice have shown that deletions of the dynein light chain binding region of recombinant SAD-L16, which contained the genetic sequence of the Street Alabama Dufferin (SAD)-B19 strain, resulted in mutant viruses that demonstrated only minor effects on viral spread after peripheral inoculation and they remained neuroinvasive and neurovirulent (Mebatsion, 2001; Rasalingam *et al.*, 2005b). Mazarakis *et al.* (2001) have demonstrated that rabies virus glycoprotein-pseudotyped lentivirus (equine infectious anemia virus)-based vectors enhance gene transfer to neurons by facilitating retrograde axonal transport. Hence, the rabies virus glycoprotein may play a more important role than the phosphoprotein.

Other investigators used mouse and hamster models to demonstrate early and at least near-simultaneous involvement of motor neurons in the spinal cord and primary sensory neurons in dorsal root ganglia (Johnson, 1965; Murphy *et al.*, 1973a; Coulon *et al.*, 1989; Jackson and Reimer, 1989). After inoculation of mice in the masseter muscle with CVS, early infection was found in trigeminal ganglia (Jackson, 1991b; Shankar *et al.*, 1991). Studies using RT-PCR amplification showed that infection was detectable in trigeminal ganglia (18 hours post-inoculation) before the brainstem (24 hours post-inoculation) (Shankar *et al.*, 1991). However, elegant transneuronal tracer methods using CVS in rats (Tang *et al.*, 1999) and studies in rhesus monkeys (Kelly and Strick, 2000) have not shown early infection of primary sensory neurons. Two days after inoculation of CVS into the bulbospongiosus muscle of rats, the distribution of rabies virus antigen was limited to ipsilateral bulbospongiosus motor neurons in the spinal cord (Tang *et al.*, 1999). One day later (3 days post-inoculation), there was evidence of transfer of antigen to interneurons in the dorsal gray commissure, intermediate zone and sacral parasympathetic nucleus and also to external urethral sphincter motor neurons; at this time there was no labeling of primary sensory neurons in local dorsal root ganglia. This study indicates that a motor pathway rather than a sensory pathway is important in the spread of rabies virus to the CNS. It is unclear if the different results obtained in earlier studies are due to differences in the animal models, including the species of the host and the route of inoculation.

4 SPREAD WITHIN THE CNS

Once CNS neurons (often in the spinal cord) become infected in rodent models, there is rapid dissemination of rabies virus infection along neuroanatomical pathways. Rabies virus also spreads within the CNS, as in the peripheral nervous system, by fast axonal transport. Evidence was provided for axonal transport using stereotaxic brain inoculation in rats (Gillet *et al.*, 1986) and by the administration of colchicine, which inhibited virus transport within the CNS (Ceccaldi *et al.*, 1989, 1990). Studies on cultured rat dorsal root ganglia neurons showed that anterograde fast axonal transport of rabies virus is in the range of 100–400 mm/day (Tsiang *et al.*, 1989). However, the importance of this is unclear because transneuronal tracing studies with CVS in rhesus monkeys have indicated that the spread of rabies virus occurs exclusively by retrograde axonal transport with trans-synaptic transport of rabies virus also occurring exclusively in the retrograde direction (Kelly and Strick, 2000). Studies performed with rabies virus glycoprotein gene-deficient recombinant rabies virus showed limited spread in the brains of mice after intracerebral inoculation (Etessami *et al.*, 2000). After stereotaxic inoculation of the recombinant virus into the rat striatum, infection remained restricted to initially infected neurons and there was no evidence of trans-synaptic spread to secondary neurons. Hence, the rabies virus glycoprotein is necessary for trans-synaptic spread of rabies virus from one neuron to another.

Ultrastructural studies in a skunk model indicated that most viral budding occurs on synaptic or adjacent plasma membranes of dendrites, with less prominent budding from the plasma membrane of the perikaryon (Charlton and Casey, 1979c). Most virions were found partially engulfed by an invaginated membrane of an adjacent axon terminal, indicating transneuronal dendroaxonal transfer of virus. Virions were also occasionally observed budding freely into the intercellular space.

After footpad inoculation of mice with CVS, there was early involvement of neurons in the brainstem tegmentum and deep cerebellar nuclei (Jackson and Reimer, 1989). Subsequently, the infection spread to involve cerebellar Purkinje cells and neurons in the diencephalon, basal ganglia and cerebral cortex. Rabies virus, like Borna disease virus (Carbone *et al.*, 1987), spread to the hippocampus relatively late after peripheral inoculation. Rabies virus predominantly infected pyramidal neurons of the hippocampus, with relative sparing of neurons in the dentate gyrus in adult mice (Jackson and Reimer, 1989). The basis for cell selectivity is uncertain, although Gosztonyi and Ludwig (2001) speculated that if *N*-methyl-D-aspartate (NMDA) NR1 receptors are involved as rabies virus receptors, then cell selectivity can be explained by the fact that rabies virus spreads only by retrograde (not by anterograde) fast axonal transport. Therefore, the virus cannot infect dentate granule cells by the perforant path and mossy fibers from CA3 that predominantly have α-amino-3-hydroxy-5-methyl-4-isoxazole propionate (AMPA) and kainate receptors rather than NMDA receptors. Although rabies virus is highly neuronotropic, skunk rabies virus has been observed to infect Bergmann glia in the cerebellum more prominently than Purkinje cells in experimentally infected skunks (Jackson *et al.*, 2000). In street virus-infected skunks, initial infection was present in the lumbar spinal cord and transit to the brain occurred via a variety of long ascending and descending fiber tracts, including rubrospinal, corticospinal, spinothalamic, spino-olivary, vestibulospinal/ spinovestibular, reticulospinal/spinoreticular, cerebellospinal/spinocerebellar and dorsal column pathways (Charlton *et al.*, 1996).

5 SPREAD FROM THE CNS

Centrifugal spread or viral spread from the CNS to peripheral sites along neuronal routes is essential for transmission of rabies virus to its natural hosts. Salivary gland infection is necessary for the transfer of infectious oral fluids by rabid vectors. The salivary glands receive parasympathetic innervation by the facial (via the submandibular ganglion or Langley's ganglion in some animals) and glossopharyngeal (via the otic ganglion) nerves, sympathetic innervation via the superior (or cranial) cervical ganglion and afferent (sensory) innervation (Emmelin, 1967). Unilateral excision of a portion of the lingual nerve and the cranial cervical ganglion of dogs and foxes resulted in very low viral titers in denervated salivary glands compared with contralateral salivary glands after street rabies virus infection (Dean *et al.*, 1963). Evidence of widespread infection of salivary gland epithelial cells is a result of viral spread along multiple terminal axons rather than spread between epithelial cells (Charlton *et al.*, 1983). Rabies virus antigen was found concentrated in the apical region of mucous acinar cells and ultrastructural studies showed that viral matrices were present in the basal region and there was viral budding on the apical plasma membrane into the acinar lumen and into the intercellular canaliculi and, occasionally, onto membranes of secretory granules (Balachandran and Charlton, 1994). Viral titers in salivary glands may be higher than in CNS tissues (Dierks, 1975).

In addition to salivary gland infection, evidence was found in a suckling hamster model of centrifugal spread involving the central, peripheral and autonomic nervous systems in many peripheral sites (Murphy *et al.*, 1973b). Infection was observed in the ganglion cell layer of the retina and in corneal epithelial cells, which are innervated by sensory afferents via the trigeminal nerve. Epithelial cells in both superficial and deep layers of the cornea were found to be infected (Balachandran and Charlton, 1994). Detection of rabies virus antigen in corneal impression smears has been used as a diagnostic test for human rabies (Koch *et al.*, 1975) and rabies virus has been transmitted by corneal transplantation in humans (see Chapter 7). Infection may be found in free sensory nerve endings of tactile hair in a skin biopsy, which is one of the best diagnostic methods of confirming an ante-mortem diagnosis of rabies in humans (see Chapter 10). Antigen may be demonstrated in small nerves around hair follicles or in epithelial cells of hair follicles in the skin, which is taken from the nape of the neck because it is rich in hair follicles. Widespread infection may be observed in sensory nerve end organs in the oral and nasal cavities, including the olfactory epithelium and taste buds in the tongue.

Studies in both natural and experimental rabies have demonstrated infection involving neurons in a variety of extraneural organs, including the adrenal medulla, cardiac ganglia and plexuses in the luminal gastrointestinal tract, major salivary glands, liver and exocrine pancreas (Debbie and Trimarchi, 1970; Balachandran and Charlton, 1994; Jackson *et al.*, 1999). In addition, there is infection involving a variety of non-neuronal cells, including acini in major salivary glands in rabies vectors, epithelium of the tongue, cardiac and skeletal muscle, hair follicles and even pancreatic islets (Debbie and Trimarchi, 1970; Murphy *et al.*, 1973b; Balachandran and Charlton, 1994; Jackson *et al.*, 1999). There are a few reports of myocarditis in human cases of rabies (Ross and Armentrout, 1962; Cheetham *et al.*, 1970; Araujo *et al.*, 1971).

6 ANIMAL MODELS OF RABIES VIRUS NEUROVIRULENCE

Viral neurovirulence can be defined as the capacity of a virus to cause disease of the nervous system, especially the CNS. Analysis of neurovirulence has frequently been approached in experimental models by comparing infections in a host with closely related viruses (e.g. different rabies virus strains or a parent rabies virus and a variant) (Jackson, 1991a). The ability of a virus to spread to the CNS from a peripheral site, or *neuroinvasiveness*, is an important component of neurovirulence after natural routes of viral entry. The route of inoculation is often very important in evaluating neurovirulence experimentally. Intracerebral inoculation is commonly used for convenience and a number of peripheral sites also have been used in different models, including footpad, intramuscular, intraperitoneal and intraocular inoculation. Species, age and the immune status of the host have also proved to be important factors in neurovirulence (Flamand *et al.*, 1984). Monoclonal antibody-resistant (MAR) variant viruses were selected *in vitro* from CVS and ERA laboratory strains of rabies virus with neutralizing anti-glycoprotein antibodies (Dietzschold *et al.*, 1983; Seif *et al.*, 1985). Mutations involving antigenic site III are located between amino acid residues 330 and 338 of the CVS and ERA glycoprotein. Variants with a single amino acid change at position 333, with loss of either arginine (Dietzschold *et al.*, 1983; Seif *et al.*, 1985) or lysine (Tuffereau *et al.*, 1989), have been found to have diminished virulence in mice after intracerebral inoculation, whereas variants with amino acid changes at other positions remain neurovirulent. Comparisons of avirulent variants with their parent viruses in mouse and rat models using different routes of inoculation have been a useful approach in understanding the biological bases of rabies virus neurovirulence. Both MAR variants RV194-2 (Dietzschold *et al.*, 1983) and Av01 (Coulon *et al.*, 1982) have substitution of a glutamine for the arginine of CVS at position 333 of the glycoprotein.

Avirulent rabies virus variants, but not the parent CVS strain, have been shown to cause infection in extraneural sites close to the site of inoculation in different models. For example, Av01 infected the anterior epithelium of the lens after inoculation into the anterior chamber of the eye in rats (Kucera *et al.*, 1985) (Figure 8.3). Similarly, RV194-2 inoculated into the tongue of mice and rats produced local infection involving epithelial tissues, glandular cells and muscles (Torres-Anjel *et al.*, 1984). In these models, the variant viruses demonstrated less restricted cellular tropism than the highly neuronotropic parental CVS strain.

Two independent studies in mice showed no impairment in neuroinvasiveness after peripheral inoculation of Av01 or RV194-2 (Coulon *et al.*, 1989; Jackson, 1991b). An excellent model was developed for studying the pathways of viral spread to the brain by inoculating rabies virus into the anterior chamber of the eye in rats (Kucera *et al.*, 1985). There are six potential neural pathways for viral spread to occur between the eye and brain (see Figure 8.3). Rabies virus was localized in tissues using immunofluorescent staining. After inoculation of CVS, viral antigen was initially detected at 24 hours in the ipsilateral ciliary ganglion and later in the Edinger-Westphal nucleus of the oculomotor nerve (parasympathetic pathway). At 48 hours, virus also spread to the ipsilateral ganglion of the trigeminal nerve (an afferent sensory pathway) and to neurons of the contralateral area praetectalis medialis, which projects to the retina via pre-opticoretinal fibers. In contrast, Av01 propagated in the trigeminal pathway but not in either parasympathetic or pre-opticoretinal fibers. Neurons in the trigeminal ganglion also were infected at 48 hours, indicating a similar rate of spread. Thus, avirulent Av01 spreads to the brain in this model using more limited pathways than its virulent parent virus.

Intracerebral inoculation is a crude technique in which the inoculum spreads throughout the cerebrospinal fluid (CSF) spaces, including the ventricular system

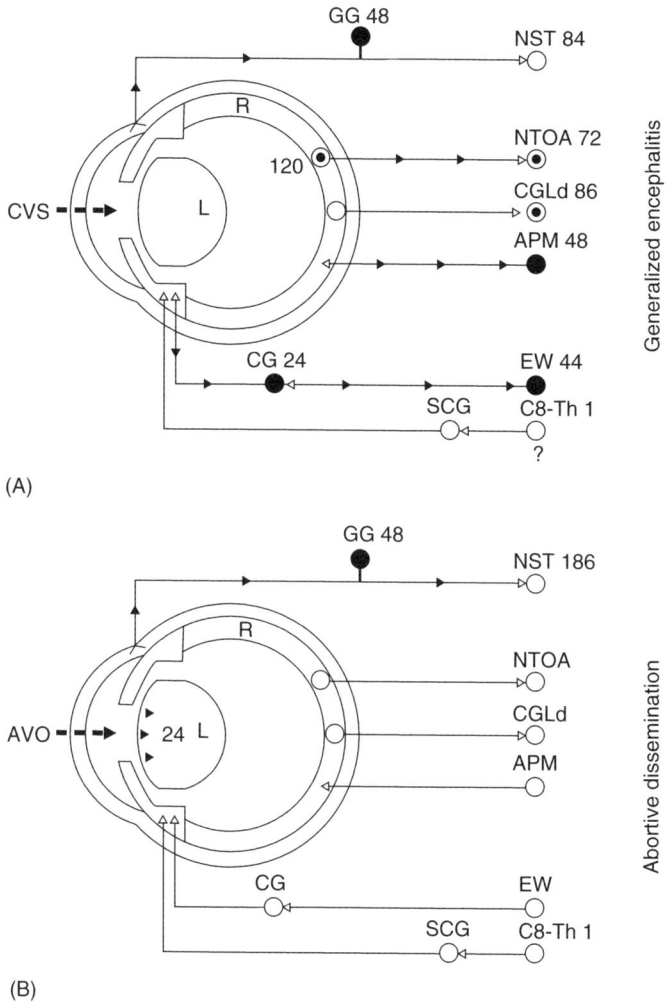

Figure 8.3 Propagation of CVS (A) and avirulent variant rabies virus strain AVO (B) through the trigeminal (top), visual (center) and autonomic (bottom) interconnections between the eye and brain. Symbols: Open arrows, direction of neurotransmission; closed arrows, direction of propagation of the virus; circles, peripheral and central neuronal somata infected primarily (closed), secondarily (dots) and not infected (open) at each interval of time, indicated in hours after inoculation. APM, area praetectalis medialis; C8-Th 1, spinal preganglionic sympathetic neurons; CG, ciliary ganglion; CGLd, lateral geniculate body (dorsal part); EW, Edinger-Westphal nucleus; GG, trigeminal (gasserian) ganglion; L, lens; NST, terminal trigeminal sensory nucleus; NTOA, terminal nuclei of the accessory optic system; R, retina; SCG, superior cervical sympathetic ganglion. (Reproduced with permission from Kucera *et al.*, *Journal of Virology* **55**,158–162, 1985.)

and subarachnoid space (Mims, 1960). A stereotaxic apparatus can deliver an inoculum into a precise location in the brain. Av01 was surprisingly found to be neurovirulent after stereotaxic inoculation into the neostriatum or cerebellum of adult mice (Yang and Jackson, 1992), although Av01 infected fewer neurons and deaths occurred later than after stereotaxic inoculation with CVS (Jackson, 1994). After inoculation of Av01 into the striatum, the infection was widespread in the brain and there were morphologic changes of apoptosis in neurons (A.C. Jackson, unpublished observations) and also infiltration with inflammatory cells. Serum neutralizing antibodies against rabies virus were produced later and at lower levels than after intracerebral inoculation. Av01 is likely neurovirulent after stereotaxic brain inoculation because this route produces both a direct site of viral entry into the CNS and a low level of immune stimulation.

Since centrifugal spread of CVS is limited, comparisons of the spread of CVS and variants from the CNS have not been as useful as comparisons of spread to the CNS and within the CNS. In the model of Kucera *et al.* (1985) using intraocular inoculation of rats, CVS spread from the nuclei of the accessory optic system to ganglionic cells of the retina in both eyes (see Figure 8.3) and Av01 did not show evidence of centrifugal spread in this model.

7 RABIES VIRUS RECEPTORS

In recent years, our knowledge of viral receptors both outside and within the nervous system (Schweighardt and Atwood, 2001) has greatly expanded. These receptors have a role in normal cell function and are hijacked by viruses to gain entry into cells. Viruses must bind to these membrane-associated molecules and their interaction with receptor molecules must facilitate viral entry into cells. The rabies virus glycoprotein is thought to be of prime importance in this process (see Chapter 2). There is evidence of at least three rabies virus receptors and it is very likely that additional ones will be identified in the future.

7.1 Nicotinic acetylcholine receptor

The nAChR was the first rabies virus receptor identified and this receptor is felt to be important for the spread of the virus from the neuromuscular junction at peripheral sites in order to gain access to the CNS along peripheral nerves (Lentz *et al.*, 1982). It has not yet been determined whether the nAChR is also an important rabies virus receptor in the CNS. Binding of rabies virus to nicotinic acetylcholine receptors in the brain could cause neuronal dysfunction. An anti-rabies virus glycoprotein monoclonal antibody was used to generate (by immunization) an anti-idiotypic antibody, B9, that selectively binds to nAChRs (Hanham *et al.*, 1993). Immunostaining of neuronal elements in the brains of rabies virus-infected mice with the B9 antibody was greatly reduced. This suggests that rabies virus binds to nAChRs in the brain, but the pathogenetic significance of this binding in producing neuronal dysfunction in rabies has not yet been established.

7.2 Neural cell adhesion molecule receptor

All cell lines susceptible to rabies virus infection appear to contain the neural cell adhesion molecule (NCAM) receptor, which is a cell adhesion glycoprotein of the immunoglobulin superfamily on their cell surface; NCAM was not found on the surface of resistant cell lines (Thoulouze et al., 1998). Incubation of susceptible cells with rabies virus decreased surface expression of NCAM and had no effect on other integral proteins of the cell membrane, whereas another virus, vaccinia virus, did not affect surface NCAM expression. This is consistent with internalization of rabies virus-NCAM receptor complexes during viral entry by adsorptive endocytosis. Rabies virus infection was also inhibited when NCAM receptor was blocked with heparan sulfate, which is a natural ligand physiologically, and by either polyclonal or monoclonal antibodies directed against NCAM receptor (Thoulouze et al., 1998). Furthermore, soluble NCAM neutralized rabies virus infection, indicating that occupation of the receptor site on virus particles prevented binding to the rabies virus receptors on target cells. When resistant L cells were transfected with NCAM cDNA, the cells became susceptible to rabies virus infection. Hence, there is very strong in vitro evidence that NCAM is a rabies virus receptor.

When primary cortex cultures were prepared from NCAM receptor-deficient ('knockout') and their wild-type littermate mice (Thoulouze et al., 1998) and infected with CVS, a significantly lower mean number of cells became infected in NCAM receptor-deficient cultures (7.8 ± 3.9%) than in wild-type cultures (18.6 ± 8.9%) ($P< 0.005$). In vivo, after inoculation of CVS into the masseter muscle of NCAM receptor-deficient and wild-type mice, significantly less rabies virus antigen was found in the brainstem/cerebellum, diencephalon and cerebral cortex in NCAM receptor-deficient than in wild-type mice, indicating that viral spread was less efficient without NCAM receptor. After inoculation of CVS into hind limb muscles, the mean survival of NCAM receptor-deficient mice was 13.6 days compared to 10.2 days in wild-type mice ($P = 0.002$), indicating that the disease progressed slower without NCAM receptor. The absence of NCAM receptor in vivo only mildly delayed the death of mice. Interestingly, this suggests that there must be other functionally important rabies virus receptors in the CNS in addition to the NCAM receptor. The NCAM receptor is localized in presynaptic membranes and, hence, it is well-positioned for internalization of rabies virus by receptor-mediated endocytosis into vesicles (Lafon, 2005). Subsequently, there is retrograde transport of either these vesicles carrying dissociated rabies virions or of viral nucleocapsids, which are released after uncoating of the virus with fusion of the viral envelope.

7.3 Low-affinity p75 neurotropin receptor

A report that the low-affinity p75 neurotropin receptor ($p75^{NTR}$) is a receptor for street rabies virus further suggests multiple candidates for the rabies virus receptor (Tuffereau *et al.*, 1998). When a random-primed cDNA library from the mRNA of neuroblastoma cells (NG108) was used to transfect COS7 cells, a single plasmid was identified after subcloning which, when transfected into BSR cells, bound soluble rabies virus glycoprotein. The 1.3 kb insert of this plasmid showed high amino acid sequence homology with both rat and human $p75^{NTR}$. Most cell lines of non-neuronal cell origin, including BSR cells, are not permissive for street rabies virus infection. However, the BSR cells with stable expression of $p75^{NTR}$, were able to bind soluble rabies virus glycoprotein. A fox street rabies virus isolate was also able to infect $p75^{NTR}$-expressing BSR cells, but relatively few untransfected control BSR cells. BSR cells expressing $p75^{NTR}$ were only slightly more susceptible to infection with CVS and $p75^{NTR}$-expressing BSR cells were three to 10 times more susceptible to CVS infection than control BSR cells in the presence of 10% serum. Since CVS, like street rabies virus strains, is highly neurotropic *in vivo*, one would expect that CVS would also use the same receptors as street rabies viruses *in vivo*. In addition, evaluation of non-adapted street rabies virus infection of mice would be difficult because, for example, a high incidence of spontaneous recovery with neurologic sequelae has been observed after peripheral inoculation of mice with a fox isolate of street virus (Jackson *et al.*, 1989). When $p75^{NTR}$-deficient mice were infected intracerebrally with CVS, similar clinical features of disease and pathologic changes were observed in the brain as in mice expressing $p75^{NTR}$(Jackson and Park, 1999). $p75^{NTR}$ is not present at the neuromuscular junction and it is mainly present in the dorsal horn of the spinal cord, suggesting that it could be involved in trafficking of rabies virus by a sensory pathway (Lafon, 2005). Ligand–$p75^{NTR}$ complexes are normally internalized by clathrin-coated pits into endosomes (Butowt and Von Bartheld, 2003). Lafon (2005) has speculated that $p75^{NTR}$ may play an important role in the retrograde transport of rabies virus by forming a rabies virus–$p75^{NTR}$ complex that is transported into the cell in endocytic compartments, possibly following caveolae transcytosis.

8 BRAIN DYSFUNCTION IN RABIES

Despite the dramatic and severe clinical neurological signs in rabies, the neuropathological findings are usually quite mild, especially under natural conditions (see Chapter 9). This suggests that neuronal dysfunction occurs in rabies without detectable morphologic changes. Many experimental studies have been performed to gain an understanding of the bases of this neuronal dysfunction. Although no fundamental underlying defect has been identified to explain this dysfunction, major areas of research in this area will be summarized.

8.1 Cellular RNA and protein synthesis

Studies performed *in vitro* have shown that rabies virus has little or no inhibitory effect on cellular RNA and protein synthesis (Madore and England, 1977; Ermine and Flamand, 1977; Tuffereau and Martinet-Edelist, 1985). However, *in vivo* studies using CVS-24-infected rats showed that there was progressive reduction in the expression of the non-inducible housekeeping gene that encodes glyceraldehyde-3-phosphate dehydrogenase and the late response gene that encodes proenkephalin, possibly due to the global suppression of cellular protein synthesis related to extensive synthesis of rabies virus mRNA (Fu *et al.*, 1993). This occurred in association with induction of immediate-early-response genes (*erg-1*, *junB*, and *c-fos*) in the hippocampus and cerebral cortex, where there was colocalization of expression of these genes with viral mRNA expression. In another study, infection of mice with CVS-N2c resulted in downregulation of about 90% of genes in the normal brain at more than fourfold lower levels by using subtraction hybridization (Prosniak *et al.*, 2001). Only about 1.4% of genes became upregulated, including genes involved in regulation of cell metabolism, protein synthesis and growth and differentiation. It is unknown whether in natural rabies there is marked downregulation of cellular RNA and protein synthesis that would explain the observed neuronal dysfunction in the absence of major morphologic changes.

8.2 Defective neurotransmission

8.2.1 Acetylcholine

A hypothesis, that defective cholinergic neurotransmission might be the basis for neuronal dysfunction in rabies, led to the investigation of specific binding to muscarinic acetylcholine receptors in CVS peripherally infected rat brains. ^{3}H-labeled antagonist, quinuclidinyl benzylate (QNB) was used as an indication of defective neurotransmission (Tsiang, 1982). Binding of ^{3}H-labeled QNB to AChRs in infected brain homogenates was decreased by 96 hours after infection compared with controls and the binding was markedly decreased at 120 hours, 10–20 hours before death was expected to occur. The greatest reduction in binding was found in the hippocampus and smaller reductions were observed in the cerebral cortex and in the caudate nucleus.

When cholinergic neurotransmission was examined in mice infected intracerebrally with CVS and compared with mock-infected control mice, the enzymatic activities of choline acetyltransferase and acetylcholinesterase, which are required for the synthesis and degradation of acetylcholine, respectively, were similar in the cerebral cortex and hippocampus of moribund CVS-infected and control mice (Jackson, 1993). In contrast to the findings in infected rats, QNB binding to muscarinic acetylcholine receptors, which was assessed with ^{3}H-labeled QNB using Scatchard plots, was not significantly different in the cerebral cortex or hippocampus of CVS-infected and uninfected control mice. These findings cast doubt on the importance of rabies virus binding to muscarinic acetylcholine receptors in the brain. However, it is possible that differences in the species (mouse versus rat) or in the route of inoculation (peripheral versus intracerebral) account for the differences in the results of the two studies.

In naturally infected rabid dogs, specific binding of ^{3}H-labeled QNB was reduced in the hippocampus (35%) and in the brainstem (27%), but not in other brain regions, compared with uninfected control dogs (Dumrongphol et al., 1996). The results were similar whether the clinical disease was of the furious or dumb form. K_d values were increased, indicating a decrease in receptor affinity and maximum binding B_{max} values, reflecting receptor content, were unchanged in rabid dogs. Curiously, increased K_d values were found to be similar in the hippocampus whether or not rabies virus antigen was detectable at that site. These findings argue against alteration of muscarinic receptor binding as a specific consequence of rabies virus infection of neurons. They suggest an unknown indirect mechanism for altered receptor affinity that is not related to clinical manifestations of disease or the local viral load.

8.2.2 Serotonin

In so far as defective neurotransmission involving other neurotransmitters could be important in the pathogenesis of rabies, the role of serotonin has been examined with great interest. Serotonin has a wide distribution in the brain and it is important in the control of sleep and wakefulness, pain perception, memory and a variety of behaviors (Julius, 1991). Alterations of sleep stages have been recognized in experimental rabies in mice (see Section 8.3). Again, ligand binding to serotonin (5-HT) receptor subtypes was studied in the brains of CVS-infected rats (Ceccaldi et al., 1993). In this case, binding to 5-HT$_1$ receptor sites using [^{3}H] 5-HT was not affected in the hippocampus, but there was a marked decrease in B_{max} in the cerebral cortex 5 days after inoculation of CVS into the masseter muscles. In the presence of drugs that mask 5-HT$_{1A}$, 5-HT$_{1B}$ and 5-HT$_{1C}$ receptors, [^{3}H] 5-HT binding was reduced by 50% in the cerebral cortex 3 days after inoculation, whereas binding of ligands specific for 5-HT$_{1A}$ and 5-HT$_{1B}$ receptor sites was not affected. These results indicate that rabies virus infection must affect other 5-HT receptors in the cerebral cortex. Furthermore, the reduced binding was demonstrated before rabies virus antigen was detected in the cerebral cortex. Hence, the effect of rabies virus on receptor binding is unlikely due to either direct or indirect effects of viral replication in cortical neurons. There are important serotonergic projections from the dorsal raphe nuclei in the brainstem to the cerebral cortex and early infection of the midbrain raphe nuclei in experimental rabies in skunks has been documented (Smart and Charlton, 1992). Is it possible that the reduced binding of serotonin to the 5-HT receptors is an indirect effect of the infection at non-cortical sites by unknown mechanisms or is it part of a physiological response to the stress produced by the infection? In support of impaired serotonergic neurotransmission in rabies, potassium-evoked release of [^{3}H] 5-HT labeled synaptosomes from the cerebral cortex of CVS-infected rats was decreased by 31% compared with controls (Bouzamondo et al., 1993). Hence, there is evidence of both impaired release and impaired binding of serotonin, possibly playing an important role in producing the neuronal dysfunction in rabies.

8.2.3 γ-Amino-n-butyric acid

Impairments of both release and uptake of γ-amino-n-butyric acid (GABA) have been found in CVS-infected primary rat cortical neuronal cultures (Ladogana et al., 1994). A 45% reduction of [^{3}H] GABA uptake was found 3 days after infection, which coincided with the time of peak viral growth in the cultures. Kinetic analysis revealed major reductions in V_{max}, indicating a decrease in the number of fully active GABA transport sites. There were no significant changes in K_m in infected cultures in comparison to controls, reflecting the affinity of the GABA transport system for its substrate. Potassium- and veratridine-induced [^{3}H] GABA release was increased in infected cultures by 98 and 35%, respectively, compared with controls. The importance of these abnormalities in both the uptake and release of GABA in rabies pathogenesis in vivo has yet to be determined.

8.3 Electrophysiological alterations

In addition to effects on neurotransmission, viruses may have important effects on the electrophysiological properties of neurons. Electroencephalographic (EEG) recordings of mice infected with CVS showed that the initial changes were alterations of sleep stages, including the disappearance of rapid-eye-movement (REM) sleep and the development of pseudoperiodic facial myoclonus (Gourmelon et al., 1986). Later, there was a generalized slowing of the EEG recordings (at 2–4 cycles per second) and terminally, there was an extinction of hippocampal slow activity with flattening of cortical activity. Brain electrical activity terminated about 30 minutes before cardiac arrest, indicating that cerebral death in experimental rabies occurs prior to failure of vegetative functions. Street virus-infected mice showed progressive disappearance of all sleep stages with a concomitant increase in the duration of waking stages (indicating insomnia) and these changes occurred before the development of clinical signs of rabies (Gourmelon *et al.*, 1991). There was an absence of EEG abnormalities in street virus-infected mice that lasted through the preagonal phase of the disease. Since pathologic changes are more marked in neurons infected with fixed rabies viruses than with street rabies virus strains, these observations are consistent with the idea that functional impairment of brain neurons is much more important in street rabies virus infection than in infection with fixed rabies virus strains.

8.4 Ion channels

Defective neurotransmission is not the only potential explanation for functional impairment of neurons in rabies. Viral infections might also have important effects on ion channels of neurons. Studies were performed *in vitro* using rabies virus (RC-HL strain) infection of mouse neuroblastoma NA cells and the whole-cell patch clamp technique (Iwata *et al.*, 1999). The infection reduced the functional expression of voltage-dependent sodium channels and inward rectifier potassium channels and there was a decreased resting membrane potential reflecting membrane depolarization. There was no change in the expression of delayed rectifier potassium channels, indicating that non-selective dysfunction of ion channels had not occurred. The reduction in the number of sodium channels and inward rectifier potassium channels could prevent infected neurons from firing action potentials and generating synaptic potentials, resulting in functional impairment.

Rabies virus (RC-HL strain) infection of NG108-15 cells *in vitro* was not found to alter the functional expression of voltage-dependent calcium ion channels (Iwata *et al.*, 2000). NG108-15 cells express both α_2-adrenoreceptors and muscarinic receptors. Induced voltage-dependent calcium ion channel current inhibition with noradrenaline (for α_2-adrenoreceptors) was decreased significantly in rabies virus infection, whereas carbachol (for muscarinic receptors) inhibition remained unchanged. Since α_2-adrenoreceptor-mediated inhibition of voltage-dependent calcium ion current serves as a brake mechanism to keep neurons from releasing their neurotransmitters beyond physiological requirements, the impaired modulation by α_2-adrenoreceptors could possibly contribute to clinical features of rabies, including hyperexcitability and aggressive behavior (Iwata *et al.*, 2000).

8.5 Apoptosis

Apoptosis is a process by which cells undergo physiologic cell death in response to diverse stimuli. It is a normal process in embryonic development, maturation of the immune system and in normal tissue turnover (Buja *et al.*, 1993; Thompson, 1995). Morphologically, apoptosis is characterized by nuclear and cytoplasmic condensation of single parenchymal cells followed by fragmentation of the nuclear chromatin and the subsequent formation of multiple fragments of condensed nuclear material and cytoplasm (Buja *et al.*, 1993). Phagocytosis of this material occurs, although an inflammatory reaction is normally absent. In contrast, cellular death due to necrosis is characterized by preservation of cell outlines and there is variable swelling of the cell and its organelles. Cellular fragmentation occurs as a late event in necrosis. There are derangements in energy and substrate metabolism in necrosis that result in breaks in the plasma membrane and organellar membranes. Apoptosis, on the other hand, is associated with endonuclease-mediated cleavage of the DNA of nuclear chromatin, resulting in DNA fragments with sizes in multiples of a single nucleosome length (180 base pairs). The internucleosomal cleavage of the DNA in apoptosis results in a 'ladder' appearance of the DNA upon agarose gel electrophoresis, whereas in necrosis there is less specific degradation of DNA into a 'smear' containing fragments of various sizes following electrophoresis.

Apoptotic cell death likely plays an important pathogenetic role in a wide variety of viral infections, including those produced by a large number of RNA and DNA viruses and apoptosis occurs in the CNS of humans and experimental animals in many of these infections (Hardwick, 1997; Roulston *et al.*, 1999; Allsopp and Fazakerley, 2000). Strong evidence of apoptotic cell death was found in both cultured cells and neurons in experimental mouse rabies models infected by intracerebral inoculation of fixed rabies virus strains (Jackson and Rossiter, 1997; Jackson and Park, 1998; Jackson, 1999). *In vitro* studies using CVS infected cultured rat prostatic adenocarcinoma (AT3) cells showed striking morphologic changes, revealing apoptosis, at the levels of both light and electron microscopy, whereas AT3 cells transfected with the *bcl*-2 gene (an antiapoptosis gene) did not demonstrate apoptotic changes (Jackson and Rossiter, 1997). Terminal deoxynucleotidyltransferase-mediated dUTP-digoxigenin nick end labeling (TUNEL) staining was also demonstrated in infected AT3 cells, indicating evidence of oligonucleosomal DNA fragmentation typical of apoptosis. In addition, *in vitro* infection of mouse neuroblastoma (N18) cells with CVS was associated with apoptosis (Theerasurakarn and Ubol, 1998). *In vitro* studies have also shown that the ERA strain of fixed rabies virus replicates and induces apoptosis in mouse spleen lymphocytes and the human T-lymphocyte cell line Jurkat (Thoulouze *et al.*, 1997) and that cell death was concomitant with expression of the viral glycoprotein. As CVS induces apoptosis in mouse embryonic hippocampal neurons, the extent of apoptosis and pathogenicity was studied in primary neuron cultures infected with two stable variants of CVS-24, CVS-B2c and CVS-N2c (Morimoto *et al.*, 1999). It was found that the extent of apoptosis in adult mice was actually lower in primary neuron cultures infected with the more pathogenic variant CVS-N2c than with the less pathogenic variant CVS-B2c, indicating an inverse relationship between pathogenicity and apoptosis. Guigoni and Coulon (2002) observed that primary cultures of CVS-infected purified rat spinal motoneurons did not show major evidence of apoptosis over a period of 7 days, while infected purified hippocampal neurons showed apoptosis in over 90% of neurons within 3 days, indicating that different neuronal cell types respond differently to rabies virus infection. CVS and Pasteur virus (PV) strains induce only limited apoptosis, whereas two vaccine strains, Evelyn-Rokitnicki-Abelseth (ERA) and SN-10, induce strong apoptosis in the human neuroblastoma SK-N-SH cell line and in lymphoblastoid Jurkat cells (Thoulouze *et al.*, 1997, 2003; Baloul and Lafon, 2003; Lay *et al.*, 2003; Prehaud *et al.*, 2003). Hence, there are both virus-dependent and cell-dependent mechanisms for induction of apoptosis. Furthermore, rabies virus-induced apoptosis is activated by caspase-dependent and caspase-independent pathways (Thoulouze *et al.*, 2003; Sarmento *et al.*, 2006). There is activation of caspase 8 and caspase 3, but not caspase 9, and poly ADP-ribose polymerase (PARP) is cleaved, confirming activation of downstream caspases and involvement of the extrinsic apoptotic pathway (Ubol *et al.*, 1998; Kassis *et al.*, 2004; Sarmento *et al.*, 2006). Apoptosis-inducing factor is a pro-apoptotic signal transducing molecule that was shown in infection to be upregulated and translocated from the cytoplasm to the nucleus, where it binds to DNA and provokes chromatin condensation, indicating activation of a caspase-independent pathway (Sarmento *et al.*, 2006).

In adult mice infected intracerebrally with CVS, specific morphologic changes associated with apoptosis were observed in neurons, particularly in pyramidal neurons of the hippocampus and cortical neurons and there was positive TUNEL staining in the same regions (Jackson and Rossiter, 1997) (Figure 8.4). Double-labeling studies indicated that infected neurons actually underwent apoptosis. However, not all infected neurons (e.g. Purkinje cells) demonstrated these morphologic features of apoptosis or positive TUNEL staining. Increased expression of the pro-apoptotic Bax protein was observed in neurons in areas where apoptosis was prominent (Jackson and Rossiter, 1997). Studies in *bax*-deficient mice showed that neuronal apoptosis was less marked with similar clinical disease as in wild-type littermates, indicating that the Bax protein plays an important role in modulating rabies virus-induced apoptosis under specific experimental conditions (Jackson, 1999).

Both CVS- and SAD-L16 (a vaccine strain based on SAD-B19)-infected suckling mice show widespread and severe morphologic changes of apoptosis with positive TUNEL staining and activation of caspase 3, a downstream caspase, after intracerebral and peripheral routes of inoculation (Jackson and Park, 1998; Rasalingam *et al.*, 2005a, 2005b) (Figure 8.5). In suckling mice infected with CVS via intracerebral inoculation, uninfected neurons in the external granular layer of the cerebellum also underwent apoptosis (Figure 8.5A) despite the absence of rabies virus antigen, likely due to indirect mechanisms. The role of the adaptive immune response in producing neuronal apoptosis with intracerebral inoculation was evaluated by comparing the infections in adult C57BL/6J mice with nude mice (T cell deficient) and *Rag1* mice (T and B cell deficient) (Rutherford and Jackson,

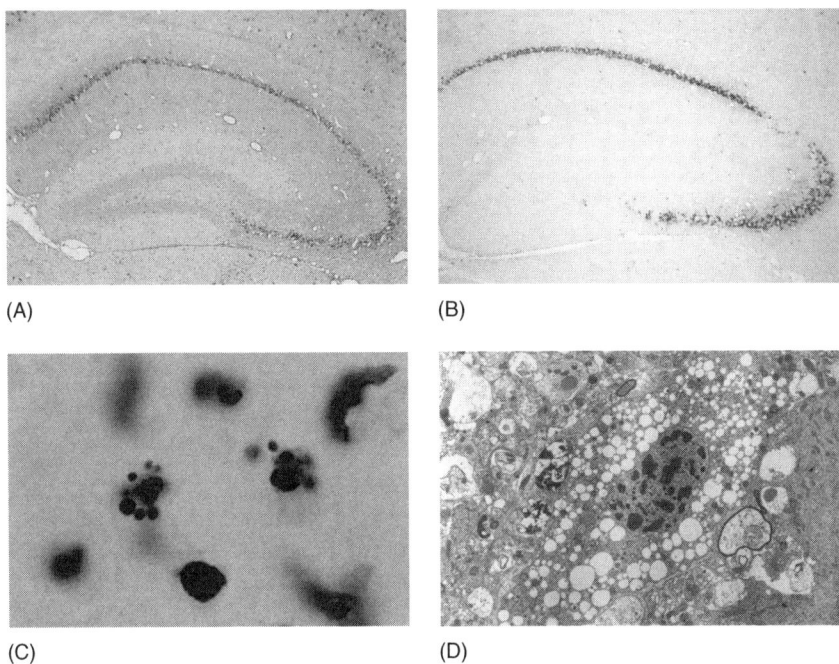

Figure 8.4   (A) Immunostaining for rabies virus antigen in the hippocampus of a mouse 7 days after intracerebral inoculation with CVS showing antigen in pyramidal neurons and in cortical neurons; neurons in the dentate gyrus do not demonstrate staining. (B) TUNEL staining in the hippocampus of a mouse 7 days after intracerebral inoculation with CVS showing marked staining is present in pyramidal neurons but not in neurons of the dentate gyrus. Note the similarity in the distribution of TUNEL staining in (B) and rabies virus antigen in (A). (C) Neurons in the cerebral cortex 8 days after inoculation with CVS showing multiple condensations of nuclear chromatin in two cells. (D) Hippocampal pyramidal neuron showing a pattern of irregular chromatin condensation and marked cytoplasmic vacuolation. (A: immunoperoxidase–hematoxylin; B: TUNEL staining; C: cresyl violet staining; D: transmission electron microscopy; magnifications: A, B, ×27; C, ×1220; D, ×4870. (Adapted with permission from Jackson and Rossiter, *Journal of Virology* **71**, 5603–5607, 1997.)

2004). Both strains of immunodeficient mice showed very similar clinical disease and neuropathological findings, including marked neuronal apoptosis, indicating that the adaptive immune response is unlikely to be of fundamental importance in producing neuronal apoptosis in this model.

Apoptosis in infected cultured cells, including embryonic cells, does not closely correspond to what is observed in infected animals. Animals peripherally inoculated with CVS strains do not show the prominent apoptosis that is observed in neurons after intracerebral inoculation (Reid and Jackson, 2001; Jackson, 2003). After intracerebral inoculation of mice with SHBV, an important bat rabies virus variant, significant neuronal apoptosis was not observed in the brain (Yan *et al.*, 2001; Sarmento *et al.*, 2005), in contrast to observations with fixed (attenuated) strains. Following a low dose of CVS-B2c inoculated intramuscularly into mice,

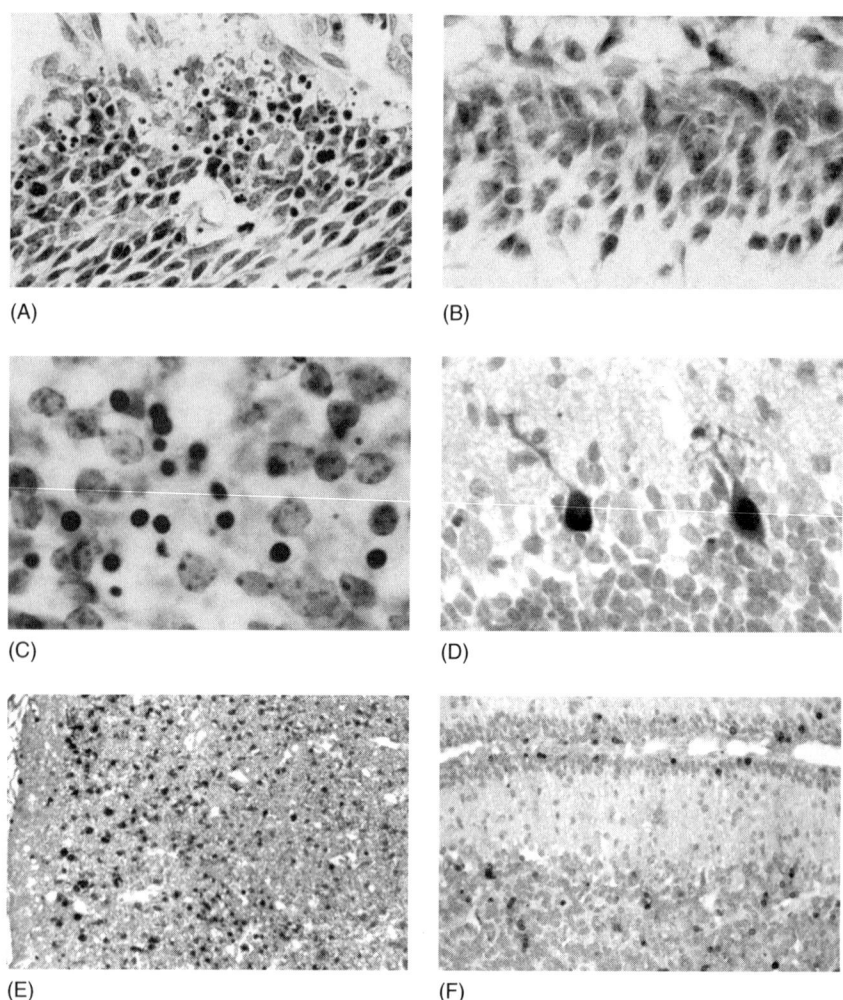

Figure 8.5   Brain sections after intracerebral inoculation of 6-day-old mice with CVS-11 (A–C) and of 7-day-old mice with L16 (D–F). (A) Nuclear chromatin condensations in multiple cells in the external granular layer of the cerebellum in a CVS-infected suckling mouse. (B) External granular layer of the cerebellum of an uninfected mouse of the same age showing the absence of typical apoptotic morphology. (C) Multiple neurons in the dentate gyrus of the hippocampus of a CVS-infected suckling mouse showing chromatin condensations involving entire nuclei. (D) Immunoperoxidase staining for activated caspase-3 and (E, F) TUNEL staining in L16-infected mouse brains. (D) Activated caspase 3 staining is present in Purkinje cells of the cerebellum 4 days p.i. and (E) TUNEL staining is present in many neurons in the cerebral cortex and (F) in neurons in the cerebellar external and internal granular layers 4 and 6 days p.i., respectively. A–C: cresyl violet staining; D: caspase 3 immunostaining; E, F: TUNEL staining – methyl green; magnifications: A, ×400; B, D, ×480; C, ×880; E, ×115; F, ×180. (A–C, Adapted with permission from Jackson and Park, *Acta Neuropathologica* **95**, 159–164, 1998; D–F, Adapted with permission from Rasalingam, Rossiter and Jackson, *Canadian Journal of Veterinary Research* **69**, 100–105, 2005.)

neuronal apoptosis in the spinal cord was associated with failure of the infection to spread to the brain and produce neurological disease, whereas in the infection with the SHBV, apoptosis was not induced in the spinal cord and spread to the brain occurred (Sarmento *et al.*, 2005). Neonatal mice, on the other hand, peripherally inoculated with SAD-L16 virus, were compared with mice infected with the less virulent SAD-D29 virus, which has an attenuating mutation at position 333 of the glycoprotein. The less virulent SAD-D29 virus actually induced more neuronal apoptosis in the brainstem and cerebellum than did SAD-L16 virus (Jackson *et al.*, 2006a), indicating that the inverse relationship between pathogenicity and apoptosis applies *in vivo* in the CNS as well as *in vitro*.

In rabies virus infection there are complex mechanisms involved in the ensuing cell death or survival of neurons, both *in vitro* and in animal models using different viral strains and routes of inoculation. Both *in vitro* and *in vivo* observations demonstrate that apoptosis may be a protective rather than a pathogenic mechanism in rabies virus infections because the less pathogenic viruses induce more apoptosis than the more pathogenic viruses *in vitro* and also *in vivo* using peripheral routes of inoculation (Morimoto *et al.*, 1999; Yan *et al.*, 2001; Prehaud *et al.*, 2003; Sarmento *et al.*, 2005; Jackson *et al.*, 2006a). In addition, it has been recognized that attenuated rabies viruses activate, while pathogenic rabies viruses evade, host innate immune responses in the CNS (Wang *et al.*, 2005). Toll-like receptors (TLRs) alert the host to the presence of microbial pathogens and have been recognized as principal inducers of the innate immune system. TLRs, including TLR 3, are upregulated in the mouse brain in rabies virus infection (McKimmie *et al.*, 2005), in human neurons following rabies virus infection (Prehaud *et al.*, 2005) and also in post-mortem brains of rabies patients (Jackson *et al.*, 2006b). Preservation of the neuronal network by inhibition of apoptosis and limitation of the inflammation and the destruction of T cells that invade the CNS in response to the infection is crucial for successful neuroinvasion of rabies virus and for transmission to another animals (Baloul and Lafon, 2003). Hopefully, a greater understanding of the mechanisms involved in neuronal apoptosis in experimental models will provide insights into the bases for the neuronal dysfunction that occurs in natural rabies. There is a report demonstrating apoptosis in a single human rabies case (Adle-Biassette *et al.*, 1996), but morphologic evidence of neuronal apoptosis has generally not been prominent in natural rabies in humans or animals. Juntrakul *et al.* (2005) reported that TUNEL positive cells were found throughout the neuroaxis in seven cases of human rabies, but this may have been due to non-specific staining. Also, morphologic evidence of neuronal apoptosis was not assessed or illustrated in this report.

8.6 Nitric oxide

Nitric oxide (NO) is a short-lived gaseous radical that acts as a biologic mediator for diverse cell types. It is produced by many different cells and mediates a variety of functions, including vasodilation, neurotransmission, immune cyto-toxicity, production of synaptic plasticity in the brain and neurotoxicity (Nathan, 1992; Lowenstein *et al.*, 1994). NO is released by the enzyme nitric oxide syn-thase (NOS), which also produces other reactive oxides of nitrogen (Nathan, 1992). There are three isoforms of NOS: neuronal NOS (nNOS, also NOS-1), inducible NOS (iNOS, also NOS-2) and endothelial NOS (eNOS, also NOS-3). nNOS is constitutively expressed and inducible by cytokines, including IFN-χ, TNF-α and IL-12, whereas iNOS is inducible with lipopolysaccharides, IFN-χ and TNF-α.

NO plays a variety of roles in different viral infections (Reiss and Komatsu, 1998). In some viral infections (e.g. with Sindbis virus), inhibition of NOS results in increased mortality of infected mice, suggesting that NO plays a pro-tective role in the pathogenesis of the viral infection (Tucker *et al.*, 1996). During infection with vesicular stomatitis virus (a rhabdovirus), NO has been shown to inhibit viral replication and promote viral clearance and recovery of infected mice (Komatsu *et al.*, 1996).

Induction of iNOS mRNA occurred in mice infected experimentally with street rabies virus (Koprowski *et al.*, 1993). iNOS mRNA was detected using RT-PCR amplification in the brains of three of six paralyzed mice, 9–14 days after inocu-lation of rabies virus in the masseter muscle. iNOS mRNA expression was induced rapidly in the brains of the rabid mice. It was speculated that NO and/or other endogenous neurotoxins may mediate the neuronal dysfunction in rabies and other infectious diseases (Koprowski *et al.*, 1993; Zheng *et al.*, 1993). The onset of clinical signs in rabies virus-infected rats and the clinical progression of the disease correlated with increasing quantities of NO in the brain to levels up to 30-fold more than in controls, which was determined using spin trapping of NO and electron paramagnetic resonance spectroscopy (Hooper *et al.*, 1995). iNOS was detected by immunostaining in CVS-infected rats in many cells throughout the brain near blood vessels, which were identified as microglia and macrophages (Van Dam *et al.*, 1995). CVS-24-infected rats developed a reduction in nNOS activity with reductions in nNOS mRNA and nNOS immunoreactivity and an increase in iNOS activity in the brain in a time-dependent manner (Akaike *et al.*, 1995). Choline acetyltransferase activity in the brain remained unchanged, indicating that the decrease in nNOS activity did not reflect general-ized neuronal loss. The NO produced by macrophages may be neurotoxic because its reaction with the superoxide anion O_2^- leads to the formation of per-oxynitrate, which is a reactive oxidizing agent capable of causing tissue damage (Akaike *et al.*, 1995). Ubol *et al.* (2001) found that mice treated with the iNOS inhibitor, aminoguanidine (AG), delayed the death of CVS-11-infected mice by 1.0 to 1.6 days (depending on the dose). A delay in rabies virus replication was observed in the AG-treated mice. The role of NO in rabies pathogenesis clearly needs further study since it exerts both beneficial and detrimental effects and complex mechanisms are likely involved.

8.7 Excitotoxicity

Excitatory amino acids (e.g. glutamate) have been recognized to play a role in neuronal injury in a variety of neurological diseases, including stroke, epilepsy and neurodegenerative disorders. Recently, there is evidence that neurotropic viruses, including human immunodeficiency virus (Nath *et al.*, 2000; Kaul and Lipton, 2004) and Sindbis virus (Nargi-Aizenman and Griffin, 2001; Nargi-Aizenman *et al.*, 2004; Darman *et al.*, 2004), induce neuronal injury through excitotoxic mechanisms. There has been recent speculation that the *N*-methyl-D-aspartate (NMDA) receptor may be one of the rabies virus receptors (Gosztonyi and Ludwig, 2001). Tsiang and co-workers reported that the non-competitive NMDA antagonists ketamine and/or MK-801 inhibited rabies virus infection in primary neuron cultures, inhibited rabies virus genome transcription and restricted viral spread in an experimental model of rabies in rats (Lockhart *et al.*, 1991, 1992; Tsiang *et al.*, 1991a). The *in vitro* doses of ketamine and MK-801 used were much higher than required for stimulation of glutamate receptors, indicating that other mechanisms of actions, including antiviral effects, were likely of primary importance. In more recent studies, Weli *et al.* (2006) observed that CVS-infected cortical and hippocampal mouse embryonic neurons showed loss of trypan blue exclusion, morphologic apoptotic features and activated caspase 3 expression, indicating apoptosis and that the NMDA antagonists, ketamine ($125\,\mu M$) and MK-801 ($60\,\mu M$), had no significant neuroprotective effect. Glutamate-stimulated increases of intracellular calcium were reduced in CVS-infected hippocampal neurons compared with mock-infected neurons. Ketamine ($120\,mg/kg/day$ intraperitoneally) given to adult ICR mice infected with CVS via the hind limb footpad produced no beneficial effects. Hence, there was no supportive evidence that excitotoxicity plays an important role in rabies virus infection or that ketamine is a useful therapeutic agent in this experimental model of rabies.

8.8 Bases for behavioral changes

The neuroanatomical bases for the behavioral changes in animals with rabies have not yet been well characterized. Limbic system infection and dysfunction are suspected to play an important role in the behavioral changes, including alertness, loss of natural timidity, aberrant sexual behavior and aggressiveness (Johnson, 1971). However, experimental rabies studies in these models have not been particularly helpful in giving insights into the neuroanatomical substrate for behavioral changes because these changes are not normally observed in rodent models and hippocampal infection actually occurs relatively late after peripheral routes of inoculation (Jackson and Reimer, 1989). The neural mechanisms of aggressive behavior are not well understood. Aggressive behavior is associated with lesions in a variety of locations in the brain, including the posterior olfactory bulbs, the ventromedial nucleus of the hypothalamus and the septal area (Isaacson, 1989). Offensive aggression, which is often impulsive and seemingly unprovoked, has been associated with low CNS serotonergic activity and also increased testosterone in humans and animal studies (Kalin, 1999). Aggressive behavior is essential in most rabies vectors for horizontal transmission of the virus to other hosts by biting. Early and selective brainstem infection in rabies would allow centrifugal spread of the virus to salivary glands as well as involvement of the serotonergic system in the raphe nuclei, resulting in aggressive behavior of animals with adequate cognitive and motor function in order to execute successful viral transmission by biting. Few studies have been performed in natural models of rabies in which aggressive behavior is exhibited. In the best available study, striped skunks inoculated peripherally with a skunk rabies virus isolate were compared with skunks infected with CVS (Smart and Charlton, 1992). The street virus-infected skunks exhibited aggressive responses to presentation of a stick in their cages, whereas this behavior was not observed in CVS-infected skunks. Heavy accumulations of viral antigen were found in the midbrain raphe nuclei, red nucleus, dorsal motor nucleus of the vagus and hypoglossal nucleus in street virus-infected skunks, but not in CVS-infected skunks. Impaired serotonin neurotransmission from the raphe nuclei in the brainstem, however, may account for the development of aggressive behavior in natural vectors of rabies.

9 RECOVERY FROM RABIES AND CHRONIC RABIES VIRUS INFECTION

Although rabies is usually considered a uniformly fatal disease, it has been recognized that animals sometimes may recover from rabies. Recovery from rabies has also been called *abortive rabies*, which can occur either with or without neurologic sequelae (Bell, 1975). There have been a large number of reports of survival after the development of neurologic illness, particularly in experimental animals (Jackson, 1997). Because of limitations on laboratory diagnostic tests performed during life, a conclusive diagnosis of rabies is only rarely made in natural cases that recover. Animals clinically suspected of having rabies are usually killed and they do not have an opportunity to recover. In a series of five reports from the Pasteur Institute of southern India, the unusual case of a chronically infected dog has been described. A 14-year-old boy died with hydrophobia 48 days after he stepped on a dog and was bitten in November 1965 (Veeraraghavan *et al.*, 1967a, 1967b, 1968, 1969, 1970). The dog was observed at the Pasteur Institute until it died in February 1969 (Veeraraghavan *et al.*, 1970). During that period, rabies virus was isolated from daily saliva samples taken from the dog on 13 occasions between January and May 1966 (Veeraraghavan *et al.*, 1967b) and once in January 1967 after the dog was given a course of prednisolone (Veeraraghavan *et al.*, 1968). Rabies virus was not isolated post-mortem from the dog's brain, spinal cord or salivary glands, although fluorescent antibody staining showed rabies virus antigen in its brain and spinal cord (Veeraraghavan *et al.*, 1970). No anti-rabies virus antibodies were found in the dog's blood at any time (Veeraraghavan *et al.*, 1967b, 1968, 1969, 1970). Although this is an extremely interesting and unusual series of reports, it is unlikely that this seronegative dog excreted a virulent rabies virus that was responsible for the boy's death. The boy may have become infected from an undocumented rabies exposure months or even years earlier (Smith *et al.*, 1991). There was a poor correlation of this laboratory's results with viral isolation and antigen detection in saliva samples and in CNS tissues from the dog. This might be explained by the presence of neutralizing antibodies in tissues, but this dog was seronegative. The viral isolations from saliva samples could be explained by cross-contamination of specimens in the laboratory. Because of a number of inconsistencies in these reports, the validity of this series of reports remains uncertain.

In another report, five dogs in Ethiopia are described that remained healthy for up to 72 months after the first isolation of rabies virus from their saliva (Fekadu, 1972, 1975). However, exposures from these dogs did not result in any human cases of rabies. In a follow-up study, secretion of rabies virus was documented in the saliva of a dog experimentally infected with an Ethiopian strain of dog rabies virus for up to 6 months after its recovery from rabies (Fekadu *et al.*, 1981).

Finally, remarkable cases of experimental rabies in two cats have been reported (Murphy *et al.*, 1980). Cat 1 developed paralysis, most marked in its hind limbs, 17 days after inoculation of a rabies virus strain isolated from a big brown bat. The cat showed slow progressive recovery until 100 weeks after inoculation, when it developed progressive neurologic deterioration with aggressive behavior and weakness and atrophy; it was killed for further study 136 weeks after inoculation. Cat 2 remained well for 120 weeks after inoculation, before it developed progressive neurologic deterioration; it was killed at 136 weeks after inoculation. There were high titers of neutralizing antibody in the serum and CSF of both cats. Rabies virus was not isolated from the saliva or tissue suspensions from these cats, but was isolated from the brain of cat 2 by explant culture techniques. Rabies virus antigen was detected at multiple sites in the CNS and viral inclusions were found in neurons at four sites in the brain of cat 2. Degenerative neuronal changes were noted and there were extensive inflammatory changes in both cats (Figure 8.6). Perl *et al.* (1977) also reported a similar recrudescent form of rabies in a cat experimentally infected with a bat rabies virus isolate. At necropsy, there were features of chronic encephalitis. Rabies virus could not be isolated from CNS tissues, probably because of the presence of neutralizing antibodies. These well-documented extraordinary cases indicate that chronic rabies virus infection may occur rarely, at least under experimental conditions. However, it is unclear if chronic rabies infections have any significance

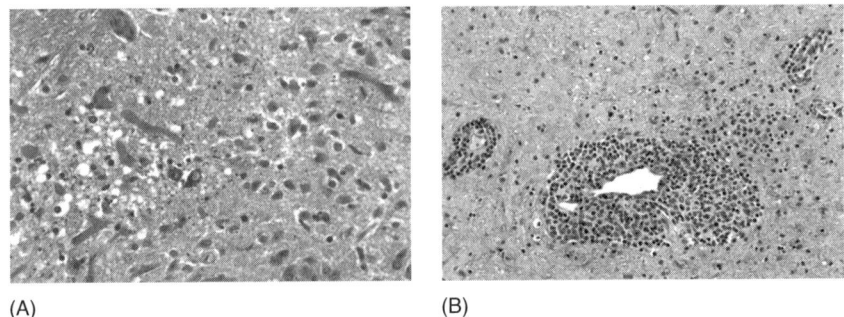

(A)   (B)

Figure 8.6 Medial geniculate body of the thalamus of cat 1, which was killed at 136 weeks post-infection, showing degenerative neuronal changes with vacuolation (A) and massive perivascular lymphocytic and plasmacytic infiltration (B), which were seen throughout the brain. Hematoxylin and eosin; magnifications: A, ×115; B, ×60. (Courtesy of Dr Frederick A. Murphy, University of Texas Medical Branch, Galveston, TX.) This figure is reproduced in the color plate section.

in the natural history of rabies, including a role in perpetuation of rabies in natural reservoirs. If animals with chronic rabies are unable to transmit the virus and are incompetent vectors, then this chronic state may not have any biological importance in nature.

Early studies on rabies pathogenesis in vampire bats, which were performed in Trinidad, suggested that bats might be chronically infected with rabies virus and secrete infectious rabies virus over periods lasting up to several months (Pawan, 1936). These early studies were performed before modern virological methods became available and suffered from inadequate diagnostic evaluations, which was largely limited to examination of tissues for Negri bodies. Infections with a variety of other bat viruses, including Rio Bravo virus, may have been misdiagnosed as rabies virus (Moreno and Baer, 1980; Constantine, 1988). More recent experimental studies have shown that vampire bats have variable incubation periods lasting up to 4 weeks and then develop an acute disease with excretion of virus in the saliva that is not prolonged (Moreno and Baer, 1980). A study of Brazilian free-tailed bats from a dense cave population in New Mexico revealed that 69% of the bats had neutralizing rabies virus antibodies, but only 0.5% had active infection as assessed by direct fluorescent antibody testing of the brain (Steece and Altenbach, 1989). Hence, seroconversion likely occurs in many naturally infected bats, although it is unknown whether any central nervous system involvement normally occurs in this setting and fatal infections may be relatively infrequent. In Spain, serotine bats (*Eptesicus serotinus*), in which EBLV-1 infection has been recognized, were recently studied with RT-PCR amplification of oropharyngeal swabs and simultaneous brain samples. Of 33 bats, a positive RT-PCR result was found in 13 (39%) oropharyngeal swabs and 5 (15%) brains and the positive brains were usually associated with clinical disease (Echevarria *et al.*, 2001). Unless the infection was associated with a previously unrecognized pattern of viral spread in the host, viral RNA was cleared from the brain but not from extraneural tissues in many of these bats. Of course, a positive RT-PCR result does not indicate the presence of infectious virus at a site, rather, it may be a marker of remote infection in some cases.

A recent study of rabies virus infection in spotted hyenas in the Serengeti changes our perspective about naturally occurring variations in rabies pathogenesis (East *et al.*, 2001). In this study, spotted hyenas were monitored in three social groups for periods of 9 to 13 years. Clinical rabies was never observed. On the basis of rabies virus neutralization antibody (VNA) titers, 37% (37 of 100) were found to be seropositive and repeat studies in six of them indicated that half of the seropositive animals became seronegative. High-ranking hyenas had high VNA titers. They also had high oral (open mouths licked by clan members at rates of over twice an hour) and bite contact rates and they lived to a mature age of over 4 years. Although infectious rabies virus was not isolated from saliva, almost half of the seropositive hyenas demonstrated saliva that was positive for rabies virus RNA by RT-PCR. Rabies virus RNA was also detected in three of 23 hyena brain samples in which the hyenas were killed by motor vehicles or other causes. RNA sequence analysis of the isolated viruses showed sequence divergence from strains found in the Serengeti in African wild dogs, bat-eared foxes and the white-tailed mongoose and the sequences in the spotted hyenas more closely resembled those found in dogs in the Middle East and Europe. This interesting report really changes our perspective on the ecology of less virulent viral variants and is an exception to the old dogma that rabies virus kills the great majority of exposed individuals. It is likely that in the future we will learn that under some circumstances the situation is similar in other species, including bats. Much more research is needed in order to gain a better understanding of the full spectrum of the ecology of rabies virus infection.

10 SUMMARY

Rabies is a normally fatal viral infection of the nervous system in humans and animals with characteristic clinical manifestations. Considerable progress has been made in understanding the pathogenesis of rabies. Rabies virus is highly neurotropic. It binds to the nAChR at the neuromuscular junction and it spreads by axonal transport via peripheral nerves to the CNS, where it causes widespread infection in neurons within the CNS. The combination of virus-induced behavioral changes in rabies vectors and centrifugal spread of the virus to salivary glands allows efficient transmission of the infection. An understanding of rabies virus neurovirulence is emerging from basic studies of virus variants in a variety of animal models. A single amino acid change in the rabies virus glycoprotein at position 333 has dramatic effects on the outcome of infection and it affects both the efficiency of viral spread involving different afferent and efferent pathways and cellular tropisms. The precise events at the site of viral entry during the long incubation period of rabies remain poorly understood. The fundamental basis for neuronal dysfunction in rabies has not yet been determined, although there are several hypotheses under active study at the present time. A better understanding of rabies pathogenesis will, hopefully, lead to advances in the treatment of rabies and other viral diseases.

REFERENCES

Adle-Biassette, H., Bourhy, H., Gisselbrecht, M. *et al.* (1996). Rabies encephalitis in a patient with AIDS: a clinicopathological study. *Acta Neuropathologica* **92**, 415–420.

Akaike, T., Weihe, E., Schaefer, M. *et al.* (1995). Effect of neurotropic virus infection on neuronal and inducible nitric oxide synthase activity in rat brain. *Journal of Neurovirology* **1**, 118–125.

Allsopp, T.E. and Fazakerley, J. K. (2000). Altruistic cell suicide and the specialized case of the virus-infected nervous system. *Trends in Neurological Sciences* **23**, 284–290.

Araujo, M.D.F., de Brito, T. and Machado, C.G. (1971). Myocarditis in human rabies. *Revista Do Instituto de Medicina Tropical de Sao Paulo* **13**, 99–102.

Baer, G.M. and Cleary, W.F. (1972). A model in mice for the pathogenesis and treatment of rabies. *Journal of Infectious Diseases* **125**, 520–527.

Baer, G.M., Bellini, W.J. and Fishbein, D.B. (1990a). Rhabdoviruses. In: *Virology: Volume 1* (B.N. Fields, D.M. Knipe, R.M. Chanock *et al.*, eds). pp. 883–930. New York: Raven Press.

Baer, G.M., Shaddock, J.H., Quirion, R., Dam, T.V. and Lentz, T.L. (1990b). Rabies susceptibility and acetylcholine receptor (Letter). *Lancet* **335**, 664–665.

Baer, G.M., Shantha, T.R. and Bourne, G.H. (1968). The pathogenesis of street rabies virus in rats. *Bulletin of the World Health Organization* **38**, 119–125.

Baer, G.M., Shanthaveerappa, T.R. and Bourne, G.H. (1965). Studies on the pathogenesis of fixed rabies virus in rats. *Bulletin of the World Health Organization* **33**, 783–794.

Balachandran, A. and Charlton, K. (1994). Experimental rabies infection of non-nervous tissues in skunks (*Mephitis mephitis*) and foxes (*Vulpes vulpes*). *Veterinary Pathology* **31**, 93–102.

Baloul, L. and Lafon, M. (2003). Apoptosis and rabies virus neuroinvasion. *Biochimie* **85**, 777–788.

Bell, J.F. (1975). Latency and abortive rabies. In: *The Natural History of Rabies* (G.M. Baer, ed.). pp. 331–354. New York: Academic Press.

Bouzamondo, E., Ladogana, A. and Tsiang, H. (1993). Alteration of potassium-evoked 5-HT release from virus-infected rat cortical synaptosomes. *NeuroReport* **4**, 555–558.

Bracci, L., Antoni, G., Cusi, M.G. *et al.* (1988). Antipeptide monoclonal antibodies inhibit the binding of rabies virus glycoprotein and alpha-bungarotoxin to the nicotinic acetylcholine receptor. *Molecular Immunology* **25**, 881–888.

Broughan, J.H. and Wunner, W.H. (1995). Characterization of protein involvement in rabies virus binding to BHK-21 cells. *Archives of Virology* **140**, 75–93.

Buja, L.M., Eigenbrodt, M.L. and Eigenbrodt, E.H. (1993). Apoptosis and necrosis: basic types and mechanisms of cell death. *Archives of Pathology and Laboratory Medicine* **117**, 1208–1214.

Butowt, R. and Von Bartheld, C.S. (2003). Connecting the dots: trafficking of neurotrophins, lectins and diverse pathogens by binding to the neurotrophin receptor p75NTR. *European Journal of Neuroscience* **17**, 673–680.

Carbone, K.M., Duchala, C.S., Griffin, J.W., Kincaid, A.L. and Narayan, O. (1987). Pathogenesis of Borna disease in rats: evidence that intra-axonal spread is the major route for virus dissemination and the determinant for disease incubation. *Journal of Virology* **61**, 3431–3440.

Ceccaldi, P.-E., Ermine, A. and Tsiang, H. (1990). Continuous delivery of colchicine in the rat brain with osmotic pumps for inhibition of rabies virus transport. *Journal of Virological Methods* **28**, 79–84.

Ceccaldi, P.-E., Fillion, M.-P., Ermine, A., Tsiang, H. and Fillion, G. (1993). Rabies virus selectively alters 5-HT$_1$ receptor subtypes in rat brain. *European Journal of Pharmacology* **245**, 129–138.

Ceccaldi, P.E., Gillet, J.P. and Tsiang, H. (1989). Inhibition of the transport of rabies virus in the central nervous system. *Journal of Neuropathology and Experimental Neurology* **48**, 620–630.

Charlton, K.M. and Casey, G.A. (1979a). Experimental rabies in skunks: oral, nasal, tracheal and intestinal exposure. *Canadian Journal of Comparative Medicine* **43**, 168–172.

Charlton, K.M. and Casey, G.A. (1979b). Experimental oral and nasal transmission of rabies virus in mice. *Canadian Journal of Comparative Medicine* **43**, 10–15.

Charlton, K.M. and Casey, G.A. (1979c). Experimental rabies in skunks: immunofluorescence light and electron microscopic studies. *Laboratory Investigation* **41**, 36–44.

Charlton, K.M., Casey, G.A. and Campbell, J.B. (1983). Experimental rabies in skunks: mechanisms of infection of the salivary glands. *Canadian Journal of Comparative Medicine* **47**, 363–369.

Charlton, K.M., Casey, G.A., Wandeler, A.I. and Nadin-Davis, S. (1996). Early events in rabies virus infection of the central nervous system in skunks (*Mephitis mephitis*). *Acta Neuropathologica* **91**, 89–98.

Charlton, K.M., Nadin-Davis, S., Casey, G.A. and Wandeler, A.I. (1997). The long incubation period in rabies: delayed progression of infection in muscle at the site of exposure. *Acta Neuropathologica* **94**, 73–77.

Cheetham, H.D., Hart, J., Coghill, N.F. and Fox, B. (1970). Rabies with myocarditis: two cases in England. *Lancet* **1**, 921–922.

Constantine, D.G. (1962). Rabies transmission by nonbite route. *Public Health Reports* **77**, 287–289.

Constantine, D.G. (1988). Transmission of pathogenic organisms by vampire bats. In: *Natural History of Vampire Bats* (A.M. Greenhall and U. Schmidt, eds). pp. 167–189. Boca Raton: CRC Press.

Constantine, D.G., Emmons, R.W. and Woodie, J.D. (1972). Rabies virus in nasal mucosa of naturally infected bats. *Science* **175**, 1255–1256.

Conti, C., Superti, F. and Tsiang, H. (1986). Membrane carbohydrate requirement for rabies virus binding to chicken embryo related cells. *Intervirology* **26**, 164–168.

Coulon, P., Derbin, C., Kucera, P., Lafay, F., Prehaud, C. and Flamand, A. (1989). Invasion of the peripheral nervous systems of adult mice by the CVS strain of rabies virus and its avirulent derivative AvO1. *Journal of Virology* **63**, 3550–3554.

Coulon, P., Rollin, P., Aubert, M. and Flamand, A. (1982). Molecular basis of rabies virus virulence. I. Selection of avirulent mutants of the CVS strain with anti-G monoclonal antibodies. *Journal of General Virology* **61**, 97–100.

Darman, J., Backovic, S., Dike, S. *et al.* (2004). Viral-induced spinal motor neuron death is non-cell-autonomous and involves glutamate excitotoxicity. *Journal of Neuroscience* **24**, 7566–7575.

Dean, D.J., Evans, W.M. and McClure, R.C. (1963). Pathogenesis of rabies. *Bulletin of the World Health Organization* **29**, 803–811.

Debbie, J.G. and Trimarchi, C.V. (1970). Pantropism of rabies virus in free-ranging rabid red fox *Vulpes fulva*. *Journal of Wildlife Diseases* **6**, 500–506.

Dierks, R.E. (1975). Electron microscopy of extraneural rabies infection. In: *The Natural History of Rabies, volume 1* (G.M. Baer, ed.). pp. 303–318. New York: Academic Press.

Dietzschold, B., Wunner, W.H., Wiktor, T.J. *et al.* (1983). Characterization of an antigenic determinant of the glycoprotein that correlates with pathogenicity of rabies virus. *Proceedings of the National Academy of Sciences of the United States of America* **80**, 70–74.

Dumrongphol, H., Srikiatkhachorn, A., Hemachudha, T., Kotchabhakdi, N. and Govitrapong, P. (1996). Alteration of muscarinic acetylcholine receptors in rabies viral-infected dog brains. *Journal of the Neurological Sciences* **137**, 1–6.

East, M.L., Hofer, H., Cox, J.H., Wulle, U., Wiik, H. and Pitra, C. (2001). Regular exposure to rabies virus and lack of symptomatic disease in Serengeti spotted hyenas. *Proceedings of the National Academy of Sciences of the United States of America* **98**, 15 026–15 031.

Echevarria, J.E., Avellon, A., Juste, J., Vera, M. and Ibanez, C. (2001). Screening of active lyssavirus infection in wild bat populations by viral RNA detection on oropharyngeal swabs. *Journal of Clinical Microbiology* **39**, 3678–3683.

Emmelin, N. (1967). Nervous control of salivary glands. In: *Handbook of Physiology, Section 6, Volume II* (C.F. Code, ed.). pp. 595–632. Washington, DC: American Physiological Society.

Ermine, A. and Flamand, A. (1977). RNA syntheses in BHK_{21} cells infected by rabies virus. *Annals of Microbiology* **128**, 477–488.

Etessami, R., Conzelmann, K.K., Fadai-Ghotbi, B., Natelson, B., Tsiang, H. and Ceccaldi, P.E. (2000). Spread and pathogenic characteristics of a G-deficient rabies virus recombinant: an *in vitro* and *in vivo* study. *Journal of General Virology* **81**, 2147–2153.

Fekadu, M. (1972). Atypical rabies in dogs in Ethiopia. *Ethiopian Medical Journal* **10**, 79–86.

Fekadu, M. (1975). Asymptomatic non-fatal canine rabies (Letter). *Lancet* **1**, 569.

Fekadu, M., Shaddock, J.H. and Baer, G.M. (1981). Intermittent excretion of rabies virus in the saliva of a dog two and six months after it had recovered from experimental rabies. *American Journal of Tropical Medicine and Hygiene* **30**, 1113–1115.

Fischman, H.R. and Schaeffer, M. (1971). Pathogenesis of experimental rabies as revealed by immunofluorescence. *Annals of the New York Academy of Sciences* **177**, 78–97.

Fischman, H.R. and Ward, F.E. (1968). Oral transmission of rabies virus in experimental animals. *American Journal of Epidemiology* **88**, 132–138.

Flamand, A., Coulon, P., Pepin, M., Blancou, J., Rollin, P. and Portnoi, D. (1984). Immunogenic and protective power of avirulent mutants of rabies virus selected with neutralizing monoclonal antibodies. In: *Modern Approaches to Vaccines: Molecular and Chemical Basis of Virus Virulence and Immunogenicity* (R.M. Chanock and R.A. Lerner, eds). pp. 289–294. Cold Spring Harbor: Cold Spring Harbor Laboratory.

Fu, Z.F., Weihe, E., Zheng, Y.M. *et al.* (1993). Differential effects of rabies and Borna disease viruses on immediate, early- and late-response gene expression in brain tissues. *Journal of Virology* **67**, 6674–6681.

Gillet, J.P., Derer, P. and Tsiang, H. (1986). Axonal transport of rabies virus in the central nervous system of the rat. *Journal of Neuropathology and Experimental Neurology* **45**, 619–634.

Gosztonyi, G. and Ludwig, H. (2001). Interactions of viral proteins with neurotransmitter receptors may protect or destroy neurons. *Current Topics in Microbiology and Immunology* **253**, 121–144.

Gourmelon, P., Briet, D., Clarencon, D., Court, L. and Tsiang, H. (1991). Sleep alterations in experimental street rabies virus infection occur in the absence of major EEG abnormalities. *Brain Research* **554**, 159–165.

Gourmelon, P., Briet, D., Court, L. and Tsiang, H. (1986). Electrophysiological and sleep alterations in experimental mouse rabies. *Brain Research* **398**, 128–140.

Guigoni, C. and Coulon, P. (2002). Rabies virus is not cytolytic for rat spinal motoneurons *in vitro*. *Journal of Neurovirology* **8**, 306–317.

Hanham, C.A., Zhao, F. and Tignor, G.H. (1993). Evidence from the anti-idiotypic network that the acetylcholine receptor is a rabies virus receptor. *Journal of Virology* **67**, 530–542.

Hardwick, J.M. (1997). Virus-induced apoptosis. *Advances in Pharmacology* **41**, 295–336.

Hooper, D.C., Ohnishi, S.T., Kean, R., Numagami, Y., Dietzschold, B. and Koprowski, H. (1995). Local nitric oxide production in viral and autoimmune diseases of the central nervous system. *Proceedings of the National Academy of Sciences of the United States of America* **92**, 5312–5316.

Hronovsky, V. and Benda, R. (1969). Development of inhalation rabies infection in suckling guinea pigs. *Acta Virologica* **13**, 198–202.

Isaacson, R.L. (1989). The neural and behavioural mechanisms of aggression and their alteration by rabies and other viral infections. In: *Progress in Rabies Control: Proceedings of the Second International IMVI ESSEN/WHO Symposium on 'New Developments in Rabies Control', Essen, 5–7 July 1988; and, Report of the WHO Consultation on Rabies, Essen, 8 July 1988 WHO Consultation on Rabies* (O. Thraenhart, H. Koprowski, K. Bögel and P. Sureau, eds). pp. 17–23. Royal Tunbridge Wells: Wells Medical.

Iwasaki, Y. and Clark, H.F. (1975). Cell to cell transmission of virus in the central nervous system. II. Experimental rabies in mouse. *Laboratory Investigation* **33**, 391–399.

Iwata, M., Komori, S., Unno, T., Minamoto, N. and Ohashi, H. (1999). Modification of membrane currents in mouse neuroblastoma cells following infection with rabies virus. *British Journal of Pharmacology* **126**, 1691–1698.

Iwata, M., Unno, T., Minamoto, N., Ohashi, H. and Komori, S. (2000). Rabies virus infection prevents the modulation by α_2-adrenoreceptors, but not muscarinic receptors, of Ca^{2+} channels in NG108-15 cells. *European Journal of Pharmacology* **404**, 79–88.

Jackson, A.C. (1991a). Analysis of viral neurovirulence. In: *Molecular Genetic Approaches to Neuropsychiatric Diseases* (J. Brosius and R.T. Fremeau, eds). pp. 259–277. San Diego: Academic Press.

Jackson, A.C. (1991b). Biological basis of rabies virus neurovirulence in mice: comparative pathogenesis study using the immunoperoxidase technique. *Journal of Virology* **65**, 537–540.

Jackson, A.C. (1993). Cholinergic system in experimental rabies in mice. *Acta Virologica* **37**, 502–508.

Jackson, A.C. (1994). Animal models of rabies virus neurovirulence. In: *Current Topics in Microbiology and Immunology, Volume 187: Lyssaviruses* (C.E. Rupprecht, B. Dietzschold and H. Koprowski, eds). pp. 85–93. Berlin: Springer-Verlag.

Jackson, A.C. (1997). Rabies. In: *Viral Pathogenesis* (N. Nathanson, R. Ahmed, F. Gonzalez-Scarano *et al.*, eds). pp. 575–591. Philadelphia: Lippincott-Raven.

Jackson, A.C. (1999). Apoptosis in experimental rabies in *bax*-deficient mice. *Acta Neuropathologica* **98**, 288–294.

Jackson, A.C. (2003). Neuronal apoptosis in experimental rabies: role of the route of viral entry. *Neurology* **60** (Suppl 1), A102.

Jackson, A.C. and Park, H. (1998). Apoptotic cell death in experimental rabies in suckling mice. *Acta Neuropathologica* **95**, 159–164.

Jackson, A.C. and Park, H. (1999). Experimental rabies virus infection of p75 neurotrophin receptor-deficient mice. *Acta Neuropathologica* **98**, 641–644.

Jackson, A.C. and Reimer, D.L. (1989). Pathogenesis of experimental rabies in mice: an immunohistochemical study. *Acta Neuropathologica* **78**, 159–165.

Jackson, A.C. and Rossiter, J.P. (1997). Apoptosis plays an important role in experimental rabies virus infection. *Journal of Virology* **71**, 5603–5607.

Jackson, A.C., Phelan, C.C. and Rossiter, J.P. (2000). Infection of Bergmann glia in the cerebellum of a skunk experimentally infected with street rabies virus. *Canadian Journal of Veterinary Research* **64**, 226–228.

Jackson, A.C., Rasalingam, P. and Weli, S.C. (2006a). Comparative pathogenesis of recombinant rabies vaccine strain SAD-L16 and SAD-D29 with replacement of Arg333 in the glycoprotein after peripheral inoculation of neonatal mice: less neurovirulent strain is a stronger inducer of neuronal apoptosis. *Acta Neuropathologica* **111**, 372–378.

Jackson, A.C., Reimer, D.L. and Ludwin, S.K. (1989). Spontaneous recovery from the encephalomyelitis in mice caused by street rabies virus. *Neuropathology and Applied Neurobiology* **15**, 459–475.

Jackson, A.C., Rossiter, J.P. and Lafon, M. (2006b). Expression of Toll-like receptor 3 in the human cerebellar cortex in rabies, herpes simplex encephalitis, and other neurological diseases in humans. *Journal of Neurovirology* **12**, 229–234.

Jackson, A.C., Ye, H., Phelan, C.C. *et al.* (1999). Extraneural organ involvement in human rabies. *Laboratory Investigation* **79**, 945–951.

Jacob, Y., Badrane, H., Ceccaldi, P.E. and Tordo, N. (2000). Cytoplasmic dynein LC8 interacts with lyssavirus phosphoprotein. *Journal of Virology* **74**, 10 217–10 222.

Johnson, R.T. (1965). Experimental rabies: studies of cellular vulnerability and pathogenesis using fluorescent antibody staining. *Journal of Neuropathology and Experimental Neurology* **24**, 662–674.

Johnson, R.T. (1971). The pathogenesis of experimental rabies. In: *Rabies.* (Y. Nagano and F.M. Davenport, eds). pp. 59–75. Baltimore: University Park Press.

Julius, D. (1991). Molecular biology of serotonin receptors. *Annual Review of Neuroscience* **14**, 335–360.

Juntrakul, S., Ruangvejvorachai, P., Shuangshoti, S., Wacharapluesadee, S. and Hemachudha, T. (2005). Mechanisms of escape phenomenon of spinal cord and brainstem in human rabies. *BMC Infectious Diseases* **5**, 104.

Kalin, N.H. (1999). Primate models to understand human aggression. *Journal of Clinical Psychiatry* **60** (Suppl 15), 29–32.

Kassis, R., Larrous, F., Estaquier, J. and Bourhy, H. (2004). Lyssavirus matrix protein induces apoptosis by a TRAIL-dependent mechanism involving caspase-8 activation. *Journal of Virology* **78**, 6543–6555.

Kaul, M. and Lipton, S.A. (2004). Signaling pathways to neuronal damage and apoptosis in human immunodeficiency virus type 1-associated dementia: Chemokine receptors, excitotoxicity, and beyond. *Journal of Neurovirology* **10** (Suppl 1), 97–101.

Kelly, R.M. and Strick, P.L. (2000). Rabies as a transneuronal tracer of circuits in the central nervous system. *Journal of Neuroscience Methods* **103**, 63–71.

Koch, F.J., Sagartz, J.W., Davidson, D.E. and Lawhaswasdi, K. (1975). Diagnosis of human rabies by the cornea test. *American Journal of Clinical Pathology* **63**, 509–515.

Komatsu, T., Bi, Z. and Reiss, C.S. (1996). Interferon-γ induced type I nitric oxide synthase activity inhibits viral replication in neurons. *Journal of Neuroimmunology* **68**, 101–108.

Koprowski, H., Zheng, Y.M., Heber-Katz, E. *et al.* (1993). *In vivo* expression of inducible nitric oxide synthase in experimentally induced neurologic disease. *Proceedings of the National Academy of Sciences of the United States of America* **90**, 3024–3027.

Kucera, P., Dolivo, M., Coulon, P. and Flamand, A. (1985). Pathways of the early propagation of virulent and avirulent rabies strains from the eye to the brain. *Journal of Virology* **55**, 158–162.

Ladogana, A., Bouzamondo, E., Pocchiari, M. and Tsiang, H. (1994). Modification of tritiated γ-amino-*n*-butyric acid transport in rabies virus-infected primary cortical cultures. *Journal of General Virology* **75**, 623–627.

Lafon, M. (2005). Rabies virus receptors. *Journal of Neurovirology* **11**, 82–87.

Lay, S., Prehaud, C., Dietzschold, B. and Lafon, M. (2003). Glycoprotein of nonpathogenic rabies viruses is a major inducer of apoptosis in human Jurkat T cells. *Annals of the New York Academy of Sciences* **1010**, 577–581.

Lentz, T.L. (1985). Rabies virus receptors. *Trends in Neurological Sciences* **8**, 360–364.

Lentz, T.L. (1990). Rabies virus binding to an acetylcholine receptor α-subunit peptide. *Journal of Molecular Recognition* **3**, 82–88.

Lentz, T.L., Burrage, T.G., Smith, A.L., Crick, J. and Tignor, G.H. (1982). Is the acetylcholine receptor a rabies virus receptor? *Science* **215**, 182–184.

Lentz, T.L., Wilson, P.T., Hawrot, E. and Speicher, D.W. (1984). Amino acid sequence similarity between rabies virus glycoprotein and snake venom curaremimetic neurotoxins. *Science* **226**, 847–848.

Lentz, T.L., Benson, R.J.J., Klimowicz, D., Wilson, P.T. and Hawrot, E. (1986). Binding of rabies virus to purified *Torpedo* acetylcholine receptor. *Molecular Brain Research* **387**, 211–219.

Lewis, P., Fu, Y. and Lentz, T.L. (2000). Rabies virus entry at the neuromuscular junction in nerve-muscle cocultures. *Muscle and Nerve* **23**, 720–730.

Lockhart, B.P., Tordo, N. and Tsiang, H. (1992). Inhibition of rabies virus transcription in rat cortical neurons with the dissociative anesthetic ketamine. *Antimicrobial Agents and Chemotherapy* **36**, 1750–1755.

Lockhart, B.P., Tsiang, H., Ceccaldi, P.E. and Guillemer, S. (1991). Ketamine-mediated inhibition of rabies virus infection *in vitro* and in rat brain. *Antiviral Chemistry and Chemotherapy* **2**, 9–15.

Lowenstein, C.J., Dinerman, J.L. and Snyder, S.H. (1994). Nitric oxide: a physiologic messenger. *Annals of Internal Medicine* **120**, 227–237.

Madore, H.P. and England, J.M. (1977). Rabies virus protein synthesis in infected BHK-21 cells. *Journal of Virology* **22**, 102–112.

Mazarakis, N.D., Azzouz, M., Rohll, J.B. *et al.* (2001). Rabies virus glycoprotein pseudotyping of lentiviral vectors enables retrograde axonal transport and access to the nervous system after peripheral delivery. *Human Molecular Genetics* **10**, 2109–2121.

McKimmie, C.S., Johnson, N., Fooks, A.R. and Fazakerley, J.K. (2005). Viruses selectively upregulate Toll-like receptors in the central nervous system. *Biochemical and Biophysical Research Communications* **336**, 925–933.

Mebatsion, T. (2001). Extensive attenuation of rabies virus by simultaneously modifying the dynein light chain binding site in the P protein and replacing Arg333 in the G protein. *Journal of Virology* **75**, 11 496–11 502.

Mims, C.A. (1960). Intracerebral injections and the growth of viruses in the mouse brain. *British Journal of Experimental Pathology* **41**, 52–59.

Moreno, J.A. and Baer, G.M. (1980). Experimental rabies in the vampire bat. *American Journal of Tropical Medicine and Hygiene* **29**, 254–259.

Morimoto, K., Hooper, D.C., Spitsin, S., Koprowski, H. and Dietzschold, B. (1999). Pathogenicity of different rabies virus variants inversely correlates with apoptosis and rabies virus glycoprotein expression in infected primary neuron cultures. *Journal of Virology* **73**, 510–518.

Morimoto, K., Patel, M., Corisdeo, S. *et al.* (1996). Characterization of a unique variant of bat rabies virus responsible for newly emerging human cases in North America. *Proceedings of the National Academy of Sciences of the United States of America* **93**, 5653–5658.

Murphy, F.A., Bauer, S.P., Harrison, A.K. and Winn, W.C. (1973a). Comparative pathogenesis of rabies and rabies-like viruses: viral infection and transit from inoculation site to the central nervous system. *Laboratory Investigation* **28**, 361–376.

Murphy, F.A., Bell, J.F., Bauer, S P. *et al.* (1980). Experimental chronic rabies in the cat. *Laboratory Investigation* **43**, 231–241.

Murphy, F.A., Harrison, A.K., Winn, W.C. and Bauer, S.P. (1973b). Comparative pathogenesis of rabies and rabies-like viruses: infection of the central nervous system and centrifugal spread of virus to peripheral tissues. *Laboratory Investigation* **29**, 1–16.

Nargi-Aizenman, J.L. and Griffin, D.E. (2001). Sindbis virus-induced neuronal death is both necrotic and apoptotic and is ameliorated by N-methyl-D-aspartate receptor antagonists. *Journal of Virology* **75**, 7114–7121.

Nargi-Aizenman, J.L., Havert, M.B., Zhang, M., Irani, D.N., Rothstein, J.D. and Griffin, D.E. (2004). Glutamate receptor antagonists protect from virus-induced neural degeneration. *Annals of Neurology* **55**, 541–549.

Nath, A., Haughey, N.J., Jones, M., Anderson, C., Bell, J.E. and Geiger, J.D. (2000). Synergistic neurotoxicity by human immunodeficiency virus proteins Tat and gp120: protection by memantine. *Annals of Neurology* **47**, 186–194.

Nathan, C. (1992). Nitric oxide as a secretory product of mammalian cells. *Federation of American Societies for Experimental Biology Journal* **6**, 3051–3064.

Noah, D.L., Drenzek, C.L., Smith, J.S. *et al.* (1998). Epidemiology of human rabies in the United States, 1980 to 1996. *Annals of Internal Medicine* **128**, 922–930.

Pawan, J.L. (1936). Rabies in the vampire bat of Trinidad, with special reference to the clinical course and the latency of infection. *Annals of Tropical Medicine and Parasitology* **30**, 401–422.

Perl, D.P., Bell, J.F., Moore, G.J. and Stewart, S.J. (1977). Chronic recrudescent rabies in a cat. *Proceedings of the Society for Experimental Biology and Medicine* **155**, 540–548.

Prehaud, C., Lay, S., Dietzschold, B. and Lafon, M. (2003). Glycoprotein of nonpathogenic rabies viruses is a key determinant of human cell apoptosis. *Journal of Virology* **77**, 10 537–10 547.

Prehaud, C., Megret, F., Lafage, M. and Lafon, M. (2005). Viral infection switches TLR-3-positive human neurons to become strong producers of beta interferon. *Journal of Virology* **79**, 12 893–12 904.

Prosniak, M., Hooper, D.C., Dietzschold, B. and Koprowski, H. (2001). Effect of rabies virus infection on gene expression in mouse brain. *Proceedings of the National Academy of Sciences of the United States of America* **98**, 2758–2763.

Rasalingam, P., Rossiter, J.P. and Jackson, A.C. (2005a). Recombinant rabies virus vaccine strain SAD-L16 inoculated intracerebrally in young mice produces a severe encephalitis with extensive neuronal apoptosis. *Canadian Journal of Veterinary Research* **69**, 100–105.

Rasalingam, P., Rossiter, J.P., Mebatsion, T. and Jackson, A.C. (2005b). Comparative pathogenesis of the SAD-L16 strain of rabies virus and a mutant modifying the dynein light chain binding site of the rabies virus phosphoprotein in young mice. *Virus Research* **111**, 55–60.

Raux, H., Flamand, A. and Blondel, D. (2000). Interaction of the rabies virus P protein with the LC8 dynein light chain. *Journal of Virology* **74**, 10 212–10 216.

Ray, N.B., Ewalt, L.C. and Lodmell, D.L. (1995). Rabies virus replication in primary murine bone marrow macrophages and in human and murine macrophage-like cell lines: implications for viral persistence. *Journal of Virology* **69**, 764–772.

Reagan, K.J. and Wunner, W.H. (1985). Rabies virus interaction with various cell lines is independent of the acetylcholine receptor: brief report. *Archives of Virology* **84**, 277–282.

Reid, J.E. and Jackson, A.C. (2001). Experimental rabies virus infection in *Artibeus jamaicensis* bats with CVS-24 variants. *Journal of Neurovirology* **7**, 511–517.

Reiss, C.S. and Komatsu, T. (1998). Does nitric oxide play a critical role in viral infections? *Journal of Virology* **72**, 4547–4551.

Ross, E. and Armentrout, S.A. (1962). Myocarditis associated with rabies: report of a case. *New England Journal of Medicine* **266**, 1087–1089.

Roulston, A., Marcellus, R.C. and Branton, P.E. (1999). Viruses and apoptosis. *Annual Review of Microbiology* **53**, 577–628.

Rutherford, M. and Jackson, A.C. (2004). Neuronal apoptosis in immunodeficient mice infected with the challenge virus standard strain of rabies virus by intracerebral inoculation. *Journal of Neurovirology* **10**, 409–413.

Sarmento, L., Li, X., Howerth, E., Jackson, A.C. and Fu, Z.F. (2005). Glycoprotein-mediated induction of apoptosis limits the spread of attenuated rabies viruses in the central nervous system of mice. *Journal of Neurovirology* **11**, 571–581.

Sarmento, L., Tseggai, T., Dhingra, V. and Fu, Z.F. (2006). Rabies virus-induced apoptosis involves caspase-dependent and caspase-independent pathways. *Virus Research* **121**, 144–151.

Schweighardt, B. and Atwood, W.J. (2001). Virus receptors in the human central nervous system. *Journal of Neurovirology* **7**, 187–195.

Seif, I., Coulon, P., Rollin, P.E. and Flamand, A. (1985). Rabies virulence: Effect on pathogenicity and sequence characterization of rabies virus mutations affecting antigenic site III of the glycoprotein. *Journal of Virology* **53**, 926–935.

Shankar, V., Dietzschold, B. and Koprowski, H. (1991). Direct entry of rabies virus into the central nervous system without prior local replication. *Journal of Virology* **65**, 2736–2738.

Smart, N.L. and Charlton, K.M. (1992). The distribution of challenge virus standard rabies virus versus skunk street rabies virus in the brains of experimentally infected rabid skunks. *Acta Neuropathologica* **84**, 501–508.

Smith, J.S., Fishbein, D.B., Rupprecht, C.E. and Clark, K. (1991). Unexplained rabies in three immigrants in the United States: a virologic investigation. *New England Journal of Medicine* **324**, 205–211.

Steece, R. and Altenbach, J.S. (1989). Prevalence of rabies specific antibodies in the Mexican free-tailed bat (*Tadarida brasiliensis mexicana*) at Lava Cave, New Mexico. *Journal of Wildlife Diseases* **25**, 490–496.

Superti, F., Hauttecoeur, B., Morelec, M.J., Goldoni, P., Bizzini, B. and Tsiang, H. (1986). Involvement of gangliosides in rabies virus infection. *Journal of General Virology* **67**, 47–56.

Superti, F., Seganti, L., Tsiang, H. and Orsi, N. (1984). Role of phospholipids in rhabdovirus attachment to CER cells. Brief report. *Archives of Virology* **81**, 321–328.

Tang, Y., Rampin, O., Giuliano, F. and Ugolini, G. (1999). Spinal and brain circuits to motoneurons of the bulbospongiosus muscle: retrograde transneuronal tracing with rabies virus. *Journal of Comparative Neurology* **414**, 167–192.

Theerasurakarn, S. and Ubol, S. (1998). Apoptosis induction in brain during the fixed strain of rabies virus infection correlates with onset and severity of illness. *Journal of Neurovirology* **4**, 407–414.

Thompson, C.B. (1995). Apoptosis in the pathogenesis and treatment of disease. *Science* **267**, 1456–1462.

Thoulouze, M.I., Lafage, M., Montano-Hirose, J.A. and Lafon, M. (1997). Rabies virus infects mouse and human lymphocytes and induces apoptosis. *Journal of Virology* **71**, 7372–7380.

Thoulouze, M.I., Lafage, M., Schachner, M., Hartmann, U., Cremer, H. and Lafon, M. (1998). The neural cell adhesion molecule is a receptor for rabies virus. *Journal of Virology* **72**, 7181–7190.

Thoulouze, M.I., Lafage, M., Yuste, V.J. *et al.* (2003). High level of Bcl-2 counteracts apoptosis mediated by a live rabies virus vaccine strain and induces long-term infection. *Virology* **314**, 549–561.

Tillotson, J.R., Axelrod, D. and Lyman, D.O. (1977). Rabies in a laboratory worker – New York. *Morbidity and Mortality Weekly Report* **26**, 183–184.

Torres-Anjel, M.J., Montano-Hirose, J., Cazabon, E.P.I., Oakman, J.K. and Wiktor, T.J. (1984). A new approach to the pathobiology of rabies virus as aided by immunoperoxidase staining. *American Association of Veterinary Laboratory Diagnosticians* 27th Annual Proceedings, pp. 1–26.

Tsiang, H. (1979). Evidence for an intraaxonal transport of fixed and street rabies virus. *Journal of Neuropathology and Experimental Neurology* **38**, 286–296.

Tsiang, H. (1982). Neuronal function impairment in rabies-infected rat brain. *Journal of General Virology* **61**, 277–281.

Tsiang, H., Ceccaldi, P.-E., Ermine, A., Lockhart, B. and Guillemer, S. (1991a). Inhibition of rabies virus infection in cultured rat cortical neurons by an N-methyl-D-aspartate noncompetitive antagonist, MK-801. *Antimicrobial Agents and Chemotherapy* **35**, 572–574.

Tsiang, H., Ceccaldi, PE. and Lycke, E. (1991b). Rabies virus infection and transport in human sensory dorsal root ganglia neurons. *Journal of General Virology* **72**, 1191–1194.

Tsiang, H., de la Porte, S., Ambroise, D.J., Derer, M. and Koenig, J. (1986). Infection of cultured rat myotubes and neurons from the spinal cord by rabies virus. *Journal of Neuropathology and Experimental Neurology* **45**, 28–42.

Tsiang, H., Lycke, E., Ceccaldi, P.-E., Ermine, A. and Hirardot, X. (1989). The anterograde transport of rabies virus in rat sensory dorsal root ganglia neurons. *Journal of General Virology* **70**, 2075–2085.

Tucker, P.C., Griffin, D.E., Choi, S., Bui, N. and Wesselingh, S. (1996). Inhibition of nitric oxide synthesis increases mortality in Sindbis virus encephalitis. *Journal of Virology* **70**, 3972–3977.

Tuffereau, C. and Martinet-Edelist, C. (1985). Shut-off of cellular RNA after infection with rabies virus. *Comptes Rendus de l'Académie des Sciences – Series III, Sciences de la Vie* **300**, 597–600.

Tuffereau, C., Benejean, J., Blondel, D., Kieffer, B. and Flamand, A. (1998). Low-affinity nerve-growth factor receptor (P75NTR) can serve as a receptor for rabies virus. *European Molecular Biology Organization Journal* **17**, 7250–7259.

Tuffereau, C., Leblois, H., Benejean, J., Coulon, P., Lafay, F. and Flamand, A. (1989). Arginine or lysine in position 333 of ERA and CVS glycoprotein is necessary for rabies virulence in adult mice. *Virology* **172**, 206–212.

Ubol, S., Sukwattanapan, C. and Maneerat, Y. (2001). Inducible nitric oxide synthase inhibition delays death of rabies virus-infected mice. *Journal of Medical Microbiology* **50**, 238–242.

Ubol, S., Sukwattanapan, C. and Utaisincharoen, P. (1998). Rabies virus replication induces Bax-related, caspase dependent apoptosis in mouse neuroblastoma cells. *Virus Research* **56**, 207–215.

Van Dam, A.M., Bauer, J., Manahing, W.K.H., Marquette, C., Tilders, F.J.H. and Berkenbosch, F. (1995). Appearance of inducible nitric oxide synthase in the rat central nervous system after rabies virus infection and during experimental allergic encephalomyelitis but not after peripheral administration of endotoxin. *Journal of Neuroscience Research* **40**, 251–260.

Veeraraghavan, N., Gajanana, A. and Rangasami, R. (1967a). Hydrophobia among persons bitten by apparently healthy animals. In: *The Pasteur Institute of Southern India, Coonoor: Annual Report of the Director 1965 and Scientific Report 1966*. pp. 90–91. Madras: Diocesan Press.

Veeraraghavan, N., Gajanana, A., Rangasami, R. *et al.* (1967b). Studies on the salivary excretion of rabies virus by the dog from Surandai. In: *The Pasteur Institute of Southern India, Coonoor: Annual Report of the Director 1965 and Scientific Report 1966*. pp. 91–97. Madras: Diocesan Press.

Veeraraghavan, N., Gajanana, A., Rangasami, R. *et al.* (1969). Studies on the salivary excretion of rabies virus by the dog from Surandai. In: *The Pasteur Institute of Southern India, Coonoor: Annual Report of the Director 1967 and Scientific Report 1968*. pp. 68–70. Madras: Diocesan Press.

Veeraraghavan, N., Gajanana, A., Rangasami, R. *et al.* (1970). Studies on the salivary excretion of rabies virus by the dog from Surandai. In: *The Pasteur Institute of Southern India, Coonoor: Annual Report of the Director 1968 and Scientific Report 1969*. p. 66. Madras: Diocesan Press.

Veeraraghavan, N., Gajanana, A., Rangasami, R., Saraswathi, K.C., Devaraj, R. and Hallan, K.M. (1968). Studies on the salivary excretion of rabies virus by the dog from Surandai. In: *The Pasteur Institute of Southern Indian, Coonoor: Annual Report of the Director 1966 and Scientific Report 1967*. pp. 71–78. Madras: Diocesan Press.

Wallerstein, C. (1999). Rabies cases increase in the Philippines. *British Medical Journal* **318**, 1306.

Wang, Z.W., Sarmento, L., Wang, Y. *et al.* (2005). Attenuated rabies virus activates, while pathogenic rabies virus evades, the host innate immune responses in the central nervous system. *Journal of Virology* **79**, 12554–12565.

Weli, S.C., Scott, C.A., Ward, C.A. and Jackson, A.C. (2006). Rabies virus infection of primary neuronal cultures and adult mice: failure to demonstrate evidence of excitotoxicity. *Journal of Virology* **80**, 10270–10273.

Winkler, W.G., Fashinell, T.R., Leffingwell, L., Howard, P. and Conomy, J.P. (1973). Airborne rabies transmission in a laboratory worker. *Journal of the American Medical Association* **226**, 1219–1221.

Yan, X., Prosniak, M., Curtis, M.T. *et al.* (2001). Silver-haired bat rabies virus variant does not induce apoptosis in the brain of experimentally infected mice. *Journal of Neurovirology* **7**, 518–527.

Yang, C. and Jackson, A.C. (1992). Basis of neurovirulence of avirulent rabies virus variant Av01 with stereotaxic brain inoculation in mice. *Journal of General Virology* **73**, 895–900.

Zheng, Y.M., Schafer, M.K.-H., Weihe, E. *et al.* (1993). Severity of neurological signs and degree of inflammatory lesions in the brains of rats with Borna disease correlate with the induction of nitric oxide synthase. *Journal of Virology* **67**, 5786–5791.

9

Pathology

JOHN P. ROSSITER[1] AND ALAN C. JACKSON[2]

Department of Pathology and Molecular Medicine, Richardson Laboratory, Queen's University, Kingston, ON K7L 3N6, Canada[1]
Departments of Medicine (Neurology) and of Microbiology and Immunology, Queen's University, Kingston, ON K7L 3N6, Canada; currently Departments of Internal Medicine (Neurology) and of Medical Microbiology, University of Manitoba, Winnipeg, MB R3A IR9, Canada[2]

1 INTRODUCTION

Investigation of the pathological changes in the central and peripheral nervous systems and extraneural organs of human rabies cases, as well as naturally and experimentally infected animals, has provided an important foundation for ongoing study of the pathogenesis of rabies (see Chapter 8). Following an exposure with deposition of rabies virus at or near the site of a bite from a rabid animal, the virus spreads centripetally towards the central nervous system (CNS) by fast axonal transport through the peripheral nervous system, typically to the spinal cord. Rapid neuron-to-neuron trans-synaptic viral dissemination within the spinal cord and brain results in a polioencephalomyelitis (i.e. an inflammatory disease predominantly involving the gray matter of the brain and spinal cord (Love and Wiley, 2002). This is followed by centrifugal spread away from the CNS along peripheral nerve pathways, with resulting infection of salivary glands, skin, heart and other viscera.

Many of the cardinal pathological features of rabies virus infection were first described over about a 30-year period extending from the early 1870s to the early 1900s (Nepveu, 1872; Gowers, 1877; Benedikt, 1878; Pasteur, 1881; Schaffer, 1888; Babes, 1892; Van Gehuchten and Nelis, 1900; Negri, 1903a, 1909; Ramón y Cajal and Garcia, 1904; Abba and Bormans, 1905). However, throughout the subsequent century, especially following the introduction of electron microscopy and immunohistochemistry, pathological studies have continued to provide key insights into our understanding of this dreaded disease.

2 MACROSCOPIC FINDINGS

Macroscopic examination of the brain in rabies victims is frequently unremarkable, or shows a spectrum of relatively mild and non-specific changes (Sukru-Aksel, 1958; Nieberg and Blumberg, 1972; Perl and Good, 1991; Love and Wiley, 2002). There is often mild cerebral edema, but severe cerebral swelling and

383

Rabies, second edition. Edited by Alan C. Jackson and William H. Wunner
ISBN 978-012-369366-2.

associated brain herniation are not features of rabies. There may be congestion of leptomeningeal and parenchymal blood vessels, sometimes associated with multiple petechiae (Lowenberg, 1928; Tangchai et al., 1970). Frank subarachnoid or parenchymal hemorrhage is not a recognized feature of rabies. Thickening of the basal leptomeninges due to prominent inflammatory cell infiltration has been reported in a few cases of rabies in children (Tangchai et al., 1970). Although the brain parenchyma is typically grossly unremarkable, focal changes are seen in some cases, perhaps related to prolongation of survival with critical care measures (Rubin et al., 1970). Multifocal gray and white matter tissue softening and discoloration were found in an immunosuppressed man with prolonged survival (Mackenzie et al., 2003). A variety of macroscopic abnormalities, including discoloration of the cortical mantle, generalized softening of deep gray matter and infarction of the insular cortex, were described in four transplant recipients infected with rabies virus from a common donor (Burton et al., 2005). Softening and congestion of the amygdala and extensive laminar necrosis of the cerebral cortex were described in a case of bat-transmitted human rabies, where the patient received human rabies immune globulin and antiviral therapy and died 33 days after the onset of symptoms. Death was attributed to a direct viral action rather than anoxic brain injury or autolysis of 'respirator brain' (Dolman and Charlton, 1987). Leptomeningeal and vascular congestion may also be seen in the spinal cord and may be intense (Gowers, 1877; Lowenberg, 1928; Tangchai et al., 1970). In their classic study of the 1929–30 Trinidad outbreak of paralytic rabies, Hurst and Pawan (1932) described the victims' spinal cords as having the 'consistency of butter', likely reflecting extensive tissue injury.

3 PATHOLOGY IN THE CENTRAL NERVOUS SYSTEM

3.1 Overview

Despite the catastrophic clinical outcome of rabies virus encephalomyelitis, the histopathological changes observed in the CNS are typically relatively mild, with varying degrees of mononuclear inflammatory cell infiltration of the leptomeninges, perivascular cuffing, microglial activation with formation of 'Babes' nodules' and neuronophagia. Moreover, this combination of features is not unique to rabies and can be seen in a variety of other viral encephalitides (Love and Wiley, 2002). However, Negri bodies, eosinophilic cytoplasmic viral inclusions that are unique to rabies, are found in infected neurons in many cases. The extent of the infection of the CNS by rabies virus is best highlighted by immunostaining for rabies virus antigen.

3.2 Inflammation

Some degree of inflammatory cell infiltration of the leptomeninges is usually seen in human rabies cases, although the extent and intensity can vary greatly (Dupont and Earle, 1965; Tangchai *et al.*, 1970; Perl and Good, 1991). The infiltrates are typically composed predominantly of lymphocytes and monocytes, with smaller numbers of plasma cells, but neutrophils can predominate when inflammation is intense, especially in fulminant childhood cases (Tangchai *et al.*, 1970; Perl and Good, 1991). In the classic study by Dupont and Earle (1965) of 49 cases of human rabies encephalitis, frank leptomeningitis was found in three cases, all of them children. Tangchai *et al.* (1970) observed meningitis in four of 24 cases, again all children, in whom the clinical course was more fulminant by comparison with the other cases. The meninges of the brainstem are frequently involved, especially in cases where leptomeningeal inflammation is sparse overall (Perl and Good, 1991). Paradoxically, in paralytic rabies cases inflammatory cell infiltration of the spinal meninges is typically relatively sparse, despite intense inflammation within the adjacent spinal cord in many of the reported cases (Hurst and Pawan, 1932; Chopra *et al.*, 1980).

Perivascular mononuclear inflammatory cell infiltrates (Figure 9.1), consisting predominantly of lymphocytes and monocytes, are seen in the great majority of

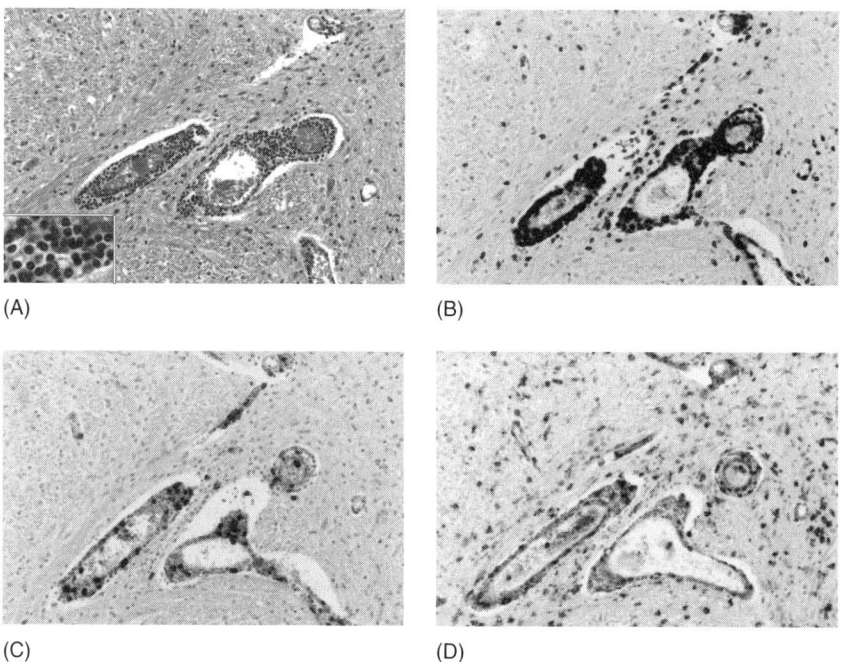

(A)

(B)

(C)

(D)

Figure 9.1 Perivascular mononuclear inflammatory cell infiltrates in the brainstem (medulla oblongata) of a human rabies case. (A) Hematoxylin-phloxine-saffron (HPS) stained section (magnification ×80). Inset shows a higher power view of an infiltrate that is composed predominantly of lymphocytes (magnification ×240). (B, C, D) Adjacent immunoperoxidase-stained sections showing: (B) CD3-positive T lymphocytes, (C) CD20-positive B lymphocytes and (D) CD68-positive monocyte/macrophages in a perivascular distribution and microglia/macrophages in the neuropil of the brainstem.

human rabies cases, but their density and distribution can vary greatly between cases. This perivascular 'cuffing' is seen predominantly in gray matter, especially in the brainstem and spinal cord, with relative sparing of white matter. Dupont and Earle (1965) observed perivascular cuffing in 48 of their 49 cases, in the following locations and approximate proportions of cases: medulla and pons (38%), spinal cord (35%), cerebral cortex (26%), hippocampus (14%), thalamus (29%), basal ganglia (26%) and cerebellum (14%). In the paralytic rabies series of Hurst and Pawan (1932) (3 cases) and Chopra *et al.* (1980) (11 cases), dense perivascular infiltrates of lymphocytes and some neutrophils were seen in the anterior and posterior horn gray matter of most cases. This was especially prominent in the lumbar and lower thoracic segments in the cases of Chopra *et al.* (1980) and was associated with extension of inflammation along perivascular spaces into the adjacent white matter. By contrast, inflammation was considerably less severe in the brain and predominantly involved the medulla (both series) and dorsal half of the pons (Hurst and Pawan, 1932).

In the case of a patient who died 17 days into the clinical course of encephalitic rabies, Iwasaki *et al.* (Iwasaki *et al.*, 1993; Iwasaki and Tobita, 2002) found that 50–70% of perivascular mononuclear cells were CD3 immunopositive T lymphocytes. Approximately one-third of these were CD4-positive helper T cells. Only occasional CD20-positive B cells were observed and the remaining perivascular cells were CD68-positive monocyte/macrophage lineage cells. More than half of the T lymphocytes were found in the CNS parenchyma surrounding the perivascular spaces. However, in a rabies patient who only survived for 9 days, there was virtually no inflammation or tissue injury, despite the finding of numerous neurons containing Negri bodies (see below), which emphasizes an important point that fatal encephalitic rabies may not necessarily be accompanied by significant inflammation (Iwasaki *et al.*, 1993). The degree of inflammation may at least be partially influenced by the strain of rabies virus. In dogs experimentally inoculated with an Ethiopian dog rabies virus strain, widespread inflammation, neuronal degeneration and neuronophagia (see below) were seen, whereas such lesions were generally much less severe in animals infected with a Mexican dog virus strain (Fekadu *et al.*, 1982). An especially florid and widespread encephalitic pattern was reported in a patient presenting with paralytic rabies caused by canine virus transmitted by a fox bite (Suja *et al.*, 2004).

In 1892, Babes described microscopic accumulations of cells surrounding chromatolytic and degenerating neurons in a series of human rabies cases. He called these foci 'les îlots inflammatories pericellulaires de la rage' (pericellular inflammatory islets of rabies) and 'nodules rabiques' (rabidic nodules) (Babes, 1892). Subsequently referred to in the literature as 'Babes' nodules', these microglial nodules (Figure 9.2) are composed predominantly of activated microglia/monocytes and are seen in other viral encephalitides and other infectious disorders (Love and Wiley, 2002). Dupont and Earle (1965) observed activated microglia in many of their cases, especially in the medulla and, in several cases, the microglia had a

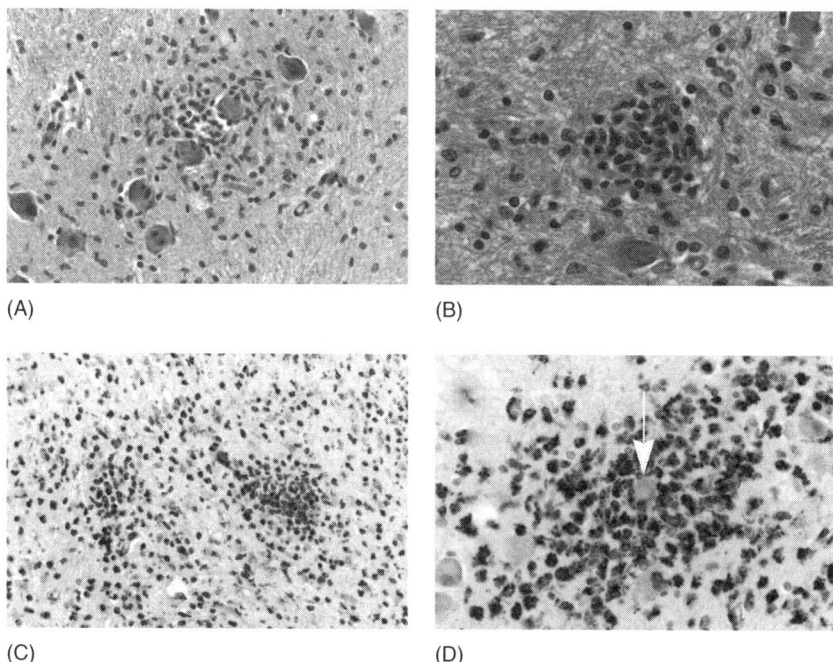

(A)   (B)

(C)   (D)

Figure 9.2   Microglial nodules ('Babes' nodules') in the medulla oblongata of a human rabies case. (A, B) HPS-stained sections (magnifications: A, ×160, B, ×260). (C, D) CD68-immunoperoxidase-stained sections showing (C) diffuse and nodular microglial proliferation (magnification ×85) and (D) a microglial nodule containing a neuron (arrow) undergoing neuronophagia (magnification ×170).

predominantly 'rod cell' morphology. Classic Babes' nodules were seen in 42% of their cases. In paralytic rabies cases, microglial proliferation, both diffuse and nodular, was found throughout the spinal gray matter and focally in the adjacent white matter in most cases. Marked microglial activation was seen in the medulla and dorsal half of the pons in several of these cases (Hurst and Pawan, 1932; Chopra *et al.*, 1980).

3.3 Cell injury and cell death

Neuronophagia (see Figure 9.2D), a microscopic pattern characterized by accumulations of activated microglia/macrophages in the process of phagocytosing degenerating and/or dying neurons, is seen in many rabies cases. Once again, however, the severity and anatomical extent of neuronophagia and resulting neuronal loss can vary greatly between cases. Dupont and Earle (1965) observed neuronophagia in 57% of their cases. The neurons within these foci often have a shrunken appearance, with condensed cytoplasm and pyknotic nuclei (Perl and Good, 1991). Central chromatolysis is a cytological pattern of swelling of the neuronal cell body, disruption and dispersal of Nissl granules from the central part of the perikaryon and peripheral displacement of the nucleus, classically seen in response to axonal injury and which may also be seen in rabies. In some paralytic rabies cases there was extensive neuronal degeneration and loss in the anterior and posterior horns of the spinal cord and to a lesser extent in the medulla (Chopra *et al.*, 1980), whereas in others there was marked neuronal central chromatolysis, but only occasional (Hurst and Pawan, 1932) or no (Sheikh *et al.*, 2005) neuronophagia. Central chromatolysis of spinal motoneurons has also been described in encephalitic rabies cases (Mitrabhakdi *et al.*, 2005). In addition to chromatolysis, vacuolation of neuronal cytoplasm and degenerative changes in nuclear chromatin have been reported (Lowenberg, 1928; Reisman *et al.*, 1933).

In rodents experimentally infected with street virus, neurons typically remain relatively intact, with little alteration in the structure of organelles. However, with fixed virus infection using intracerebral inoculation in adult animals or using any route of inoculation in immature animals, widespread neuronal injury with cytoplasmic condensation, multivesiculation, increase in lysosome content, intercellular edema and cell death are frequently seen (Miyamoto and Matsumoto, 1967; Murphy, 1977). Prominent microvacuolation of the gray matter neuropil, especially cerebral cortex and thalamus, has been documented in experimentally infected skunks and foxes, with the Arctic fox variant, as well as in naturally occurring infection in these species and also in cow, horse and cat (Charlton, 1984; Charlton *et al.*, 1987; Bundza and Charlton, 1988). This spongiform change closely resembled that of traditional spongiform encephalopathies, although the vacuolation was less extensive than that found in skunks experimentally inoculated with scrapie agent (Bundza and Charlton, 1988). However, spongiform change is not a feature of human rabies cases.

The role of apoptotic cell death in the pathogenesis of naturally occurring rabies infection is unclear, but the available data suggest that neuronal apoptosis does not play an essential role in human rabies (Fu and Jackson, 2005). In a series of 10 human rabies cases, Juntrakul *et al.* (2005) have observed cytoplasmic cytochrome-c immunoreactivity in neurons in many regions of the CNS with relative sparing of the spinal cord, despite the presence of abundant rabies virus antigen within the cord. The cytoplasmic cytochrome-c signal was interpreted as evidence of mitochondrial outer membrane permeabilization, an important feature of the mitochondrial pathway of apoptotic cell death and numerous TUNEL-labeled (terminal deoxynucleotidyltransferase-mediated DNA nick-end labeling) cells were observed throughout the neuraxis (Juntrakul *et al.*, 2005). Foci of TUNEL-positive neurons have also been observed in the brainstem and hippocampus of a patient who developed rabies encephalitis on a background of AIDS (Adle-Biassette *et al.*, 1996). However, apoptotic cell death cannot be reliably diagnosed in the absence of morphological features and neither of these two reports showed any morphologic features of apoptosis, such as cell shrinkage, nuclear karyorrhexis and formation of apoptotic bodies. Hence, the role of apoptosis in natural rabies has not yet been confirmed.

In animal models, prominent neuronal apoptosis has been found in the brains of adult and immature mice following intracerebral inoculation with the challenge virus standard (CVS) strain of fixed rabies virus (Jackson and Rossiter, 1997; Jackson and Park, 1998; Theerasurakarn and Ubol, 1998; Yan *et al.*, 2001; Fu and Jackson, 2005). However, following peripheral inoculation of a fruit-eating bat species with rabies CVS strain, apoptosis was not observed (Reid and Jackson, 2001). Extensive TUNEL-positivity has been found in the brains of mice intracerebrally inoculated with street (wild-type) rabies virus strains in one study (Ubol and Kasisith, 2000), but other investigators found little or none with bat rabies virus infections (Yan *et al.*, 2001; Sarmento *et al.*, 2005). A compelling experimentally based model for a 'subversive neuroinvasive strategy of rabies virus' in naturally occurring infection has been advanced. Preservation of the integrity of the neuronal network by avoidance of neuronal apoptosis, together with induction of apoptosis in potentially protective T lymphocytes, permits dissemination of the virus, its excretion in saliva and transmission by bite to another host (Baloul and Lafon, 2003).

3.4 Negri and lyssa bodies

In the early 1900s, Adelchi Negri undertook a series of detailed studies of rabies virus-infected animal brains and described the characteristic neuronal intracytoplasmic inclusions that now bear his name (Negri, 1903a, 1903b). Although he mistakenly interpreted these intracytoplasmic bodies as a protozoan species, he established their detection as being specifically diagnostic for rabies virus infection. Following his premature death from tuberculosis at age 37, his work was summarized by his wife (Negri-Luzzani, 1913). For many decades before the introduction of electron microscopy and immunofluorescence staining of viral material, light microscopic identification of Negri bodies remained the predominant pathological method for diagnosing rabies encephalomyelitis. Furthermore, Negri's work stimulated numerous studies, extending well into the second half of the 20th century, on the nature of Negri bodies and their significance in the pathogenesis of rabies (see Kristensson *et al.*, 1996 and Perl and Good, 1991 for detailed reviews). Although Negri bodies are a characteristic cytological feature in many cases of infection with 'street' rabies virus strains, for poorly understood reasons, they are almost never observed with 'fixed' virus strains (Kristensson *et al.*, 1996).

On hematoxylin and eosin stained sections, Negri bodies appear as dense, well-defined, oval or round, eosinophilic cytoplasmic inclusions (Figure 9.3). They are typically 2–10 μm in diameter, but may range from 0.5 to 27 μm in size (Negri, 1903a; Nieberg and Blumberg, 1972; Perl and Good, 1991). There is

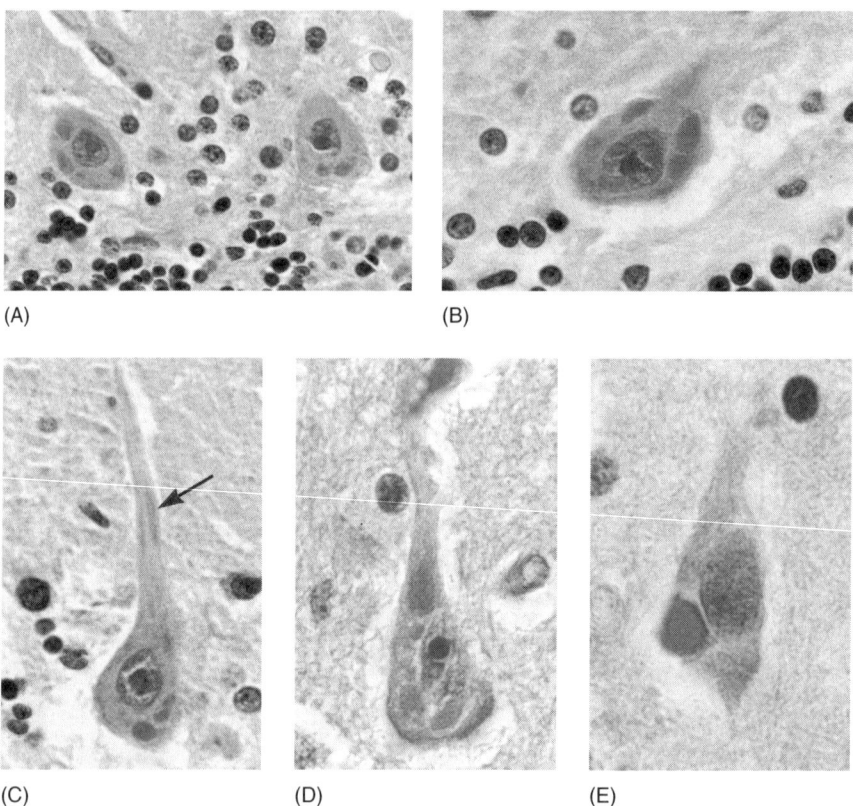

(A)   (B)

(C)   (D)   (E)

Figure 9.3 Hematoxylin and eosin (HE)-stained sections showing Negri bodies in the perikarya of (A–C) cerebellar Purkinje cells and (D, E) pyramidal neurons in the cerebral cortex of human rabies cases. The arrow in (C) indicates a Negri body in an apical dendrite. (Magnifications: A, ×315, B, ×460, C, ×550, D, ×730, E, ×865). This figure is reproduced in the color plate section.

considerable inter-species variation in the average size of Negri bodies, ranging from small in rabbits and raccoons, to large in dogs, guinea pigs and skunks and very large in cows (Perl and Good, 1991). Within an individual neuron, Negri bodies may be single or multiple and are typically located in the perikaryon, but may occasionally be found in dendrites and axons.

Using the Mann methylene blue and eosin staining method (Lepine and Atanasiu, 1996), Negri observed a small, basophilic and granular 'innerkörperchen' (inner body) within the Negri body inclusions (Negri, 1903a, 1909). Negri and subsequent investigators emphasized the presence of this inner body/basophilic granule/basophilic stippling, to the extent that, in 1925, Goodpasture coined the term 'lyssa bodies' to distinguish other inclusions that lacked these staining features of classic Negri bodies (Goodpasture, 1925). The impetus to make this distinction was to recognize the presence of the other small eosinophilic cytoplasmic inclusions that lacked an internal structure, found in healthy neurons in a variety of animal species. The concern was the other inclusion bodies could be misinterpreted as rabies virus inclusions (Goodpasture, 1925; Szlachta and Habel, 1953; Tierkel, 1973). However, this is less of an issue with human tissue as lyssa bodies are typically more numerous than Negri bodies (Sung *et al.*, 1976; Mrak and Young, 1994). They are frequently more irregularly shaped and less clearly demarcated from the surrounding cell cytoplasm than are Negri bodies, although they share ultrastructural features with Negri bodies (Perl and Good, 1991; Mrak and Young, 1994; Iwasaki and Tobita, 2002) and are immunoreactive for rabies virus antigen (see below).

   Negri bodies have been found in 50–90% of street rabies infections in different series of human rabies cases (Negri-Luzzani, 1913; Herzog, 1945; Dupont and Earle, 1965; Jogai *et al.*, 2000), influenced in part by the species, the extent of tissue sampling and possibly by the duration of clinical disease. Dupont and Earle (1965) observed Negri bodies in 71% of their cases and Negri bodies unassociated with inflammation in 15% of cases. Although Negri bodies may be found in virtually any neuronal population in the CNS or peripheral nerve ganglia, they tend to be most numerous and largest in larger neurons (Tangchai *et al.*, 1970), especially in hippocampal pyramidal neurons and cerebellar Purkinje cells. Dupont and Earle (1965) observed Negri bodies in the following locations and approximate proportions of their series of 59 cases: cerebellum (60%), hippocampus (43%), medulla (14%), pontine nuclei (12%), spinal cord (10%), cerebral cortex (7%), midbrain (7%), basal ganglia (5%), thalamus (2%) and peripheral nerve ganglia (5%) and Tangchai *et al.* (1970) in their series of 24 human rabies cases: cerebellum (54%), hippocampus (50%), brainstem (50%), hypothalamus (42%), thalamus and hypothalamus (42%), cerebral cortex (21%) and spinal cord (21%). In their monumental study of over one thousand biologically positive rabies cases from a diverse range of species, Tustin and Smith (1962) found that Negri bodies were present in 71% of cases when the hippocampus alone was examined histologically. Remarkably numerous and diffusely distributed Negri bodies have been reported in two immunocompromised patients, one of whom had AIDS (Adle-Biassette *et al.*, 1996), the other being a renal transplant recipient with prolonged survival (Mackenzie *et al.*, 2003). In a series of experimentally infected dogs, an increasing proportion of neurons contained Negri bodies as the disease progressed (Marinesco and Storesco, 1931). More numerous and widely distributed Negri bodies have also been observed with increasing duration of survival in some human case series (Sandhyamani *et al.*, 1981).

   Negri bodies are more likely to be found in areas of the CNS where there is little inflammation and are less frequently seen in degenerating neurons and/or in association with inflammatory foci (Marinesco and Storesco, 1931; Sukru-Aksel, 1958; Dupont and Earle, 1965; Iwasaki and Tobita, 2002). Although Negri bodies are a specific characteristic of the majority of street rabies virus infections, their pathogenetic significance remains unclear (Kristensson *et al.*,

1996), especially since they are almost never found in infection with 'fixed' virus strains.

The ultrastructure of Negri bodies and rabies virions has been investigated in human rabies cases (Morecki and Zimmerman, 1969; Gonzalez-Angulo et al., 1970; Leech, 1971; de Brito et al., 1973; Sandhyamani et al., 1981; Iwasaki et al., 1985; Manghani et al., 1986; Mrak and Young, 1993; Adle-Biassette et al., 1996) and in experimentally inoculated animals (Hottle et al., 1951; Miyamoto and Matsumoto, 1965; Perl et al., 1972; Murphy et al., 1973a; Iwasaki and Clark, 1975; Iwasaki et al., 1975; Charlton and Casey, 1979; Fekadu et al., 1982). In addition, many aspects of rabies virus replication and virus/host cell interaction have been studied ultrastructurally in vitro (Davies et al., 1963; Hummeler et al., 1967; Matsumoto and Kawai, 1969; Iwasaki et al., 1973; Matsumoto et al., 1974; Iwasaki and Clark, 1977; Iwasaki and Minamoto, 1982; Lewis and Lentz, 1998). Negri bodies show a similar spectrum of ultrastructural features in both human and animal material, being composed of large aggregates of granulo-filamentous matrix material (matrix) and varying numbers of viral particles (Figure 9.4).

The matrix consists of randomly oriented viral nucleocapsids (Hummeler et al., 1968; Schneider et al., 1973). In 'immature' inclusions, it has a filamentous appearance with visible substructural coiling of the nucleocapsid strands, but with maturation and increasing density, the individual strands become increasingly difficult to resolve, such that the matrix has a more granular and electron-dense appearance (Miyamoto and Matsumoto, 1967; Murphy et al., 1973a; Matsumoto et al., 1974). Bullet-shaped or tubular virions are typically associated with the matrix accumulations (for more detailed descriptions, see Perl and Good, 1991; Iwasaki and Tobita, 2002 and Chapter 2), but may be absent in some Negri bodies (Morecki and Zimmerman, 1969; Sandhyamani et al., 1981; Perl and Good, 1991). In a study of experimentally infected rhesus monkey tissue, Perl and colleagues (1972, 1991) described three basic ultrastructural configurations of Negri bodies. In the first type, seen most often in the thalamus and caudate nuclei, many bullet-shaped rabies virions were found around the periphery of the matrix, where they were often attached to dilations in the endoplasmic reticulum. Bullet-shaped virions were also frequently located within deep invaginations of the matrix contour, accounting for the inner bodies (innerkörperchen) seen by light microscopy. However, entrapment of ribosomes and endoplasmic reticulum has been proposed as an alternative explanation (Iwasaki and Tobita, 2002). In the second type of inclusion, found predominantly in brainstem, cerebellar and spinal neurons, tubular virions were dispersed throughout the matrix, while in the third type (in hippocampal neurons), virions were typically not seen within aggregates of matrix. It has been proposed that Negri bodies may represent surplus viral nucleoprotein that, for poorly understood reasons, selectively accumulates in neuronal subpopulations in street rabies virus infection (Kristensson et al., 1996).

Lyssa bodies, without the inner body, exhibit essentially the same ultrastructural features as Negri bodies (Matsumoto, 1963; Miyamoto and Matsumoto, 1965).

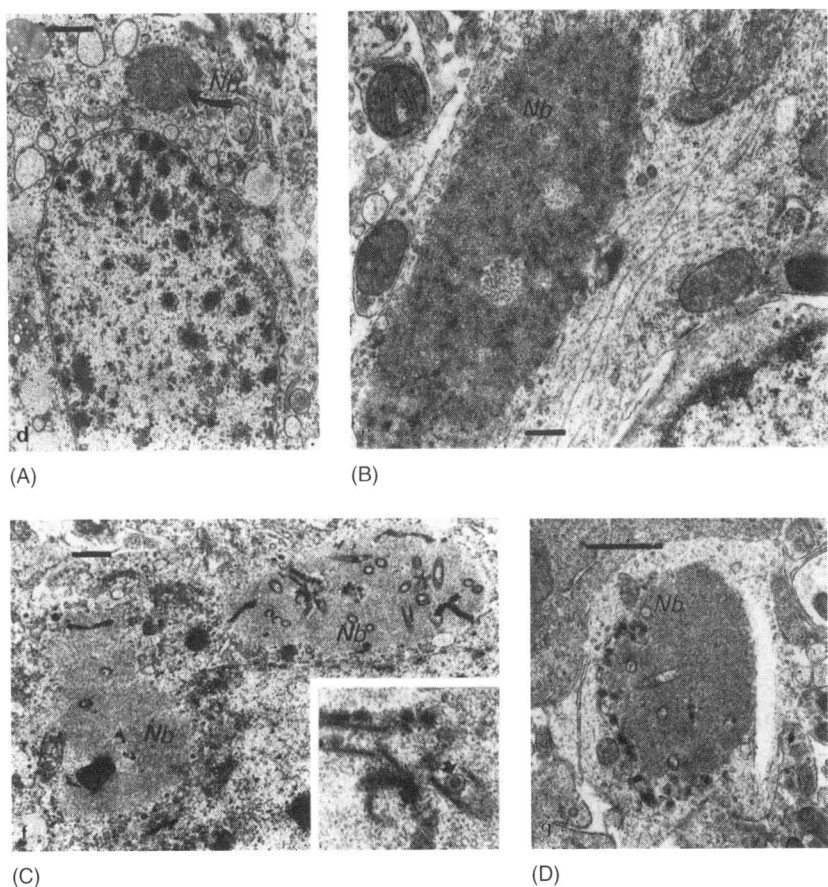

Figure 9.4 Ultrastructural features of Negri bodies. (A) A circular Negri body (Nb) in the perikaryon of a hippocampal neuron from a human patient. (B) A large elongated Negri body in a dendrite of a cortical neuron from a mouse. (C, D) Negri bodies in infected mouse brains containing bullet-shaped virions. The inset in (C) shows a virion with its core. Scale bars correspond in (A, C, D) to 1 µm, in (B) to 0.5 µm. (Adapted with permission from Kristensson *et al.*, *Neuropathology and Applied Neurobiology* **22**, 179–187, 1996.)

Furthermore, a much more extensive distribution of lyssa bodies is observed in viral infection by electron microscopy than can be resolved by routine light microscopy. Lyssa bodies, which form small aggregates of virions and matrix, are undetectable by light microscopy in neuronal perikarya and also in dendrites and axons (Jenson *et al.*, 1967; Murphy *et al.*, 1973a, 1973b; Charlton and Casey, 1979; Fekadu *et al.*, 1982; Iwasaki *et al.*, 1985; Perl and Good, 1991). Small aggregates of matrix and virions have also been observed in astrocytes

(Matsumoto, 1963; Iwasaki and Clark, 1975; Fekadu *et al.*, 1982; Perl and Good, 1991) and oligodendrocytes (Perl and Good, 1991).

In the later stages of rabies virus assembly, the virions acquire a lipid bilayer envelope by budding through host cell plasma membranes into the extracellular space (see Chapter 2). Virions also may bud intracytoplasmically through membranes of the endoplasmic reticulum or, less frequently, Golgi apparatus or outer lamella of the nuclear envelope (Murphy *et al.*, 1973b; Iwasaki and Clark, 1975; Iwasaki *et al.*, 1975, 1985; Charlton and Casey, 1979; Gosztonyi, 1994). Importantly, cell surface viral budding may not be associated with adjacent Negri bodies or nucleocapsid matrix (Iwasaki and Tobita, 2002). Also, neuron-to-neuron transmission of rabies virus, especially at synaptic junctions, has been clearly established ultrastructurally (Iwasaki *et al.*, 1975, 1985; Charlton and Casey, 1979; Burrage *et al.*, 1983).

3.5 Degeneration of neuronal processes

The fact that the pathological abnormalities described above, i.e. perivascular inflammation, microglial activation, neuronophagia and Negri bodies, may be minimal or even absent in some fatal cases (Jogai *et al.*, 2000), indicates that these are not essential neuropathological accompaniments of neurological dysfunction in rabies virus infection. This suggests that morphological correlates of neuronal dysfunction in rabies, should they exist, are likely to be most apparent at a fine structural scale. The neuropil microvacuolation observed by Charlton and colleagues (1984, 1987) in experimentally infected skunks and foxes, consisted of membrane-bound vacuoles in neuronal processes, predominantly dendrites, and has some similarities to excitotoxic amino acid-induced dendritic swelling (Charlton *et al.*, 1987). More recently Li *et al.* (2005) have found severe disorganization and destruction of axons and dendrites, with relative preservation of the neuronal cell bodies, in mice infected with a pathogenic rabies virus strain (N2C virus derived from CVS-24), but not with an attenuated strain (SN-10 derived from the SAD B19 vaccine strain). There was complete loss of neurofilament and MAP-2 (microtubule-associated protein 2) immunoreactivity in neurons infected with the pathogenic strain. These investigators have consequently proposed that pathogenic rabies virus infection may induce degeneration of neuronal processes by disrupting cytoskeletal integrity and that this may form the basis for neuronal dysfunction in rabies (Li *et al.*, 2005).

3.6 Distribution of rabies virus antigen

The development and use of immunofluorescence (Goldwasser and Kissling, 1958; Murphy *et al.*, 1973b; Charlton and Casey, 1979; Johnson *et al.*, 1980;

Bingham and van der Merwe, 2002) and immunoperoxidase techniques (Iwasaki *et al.*, 1985; Fekadu *et al.*, 1988; Jackson and Reimer, 1989; Tirawatnpong *et al.*, 1989; Last *et al.*, 1994) for the detection of rabies virus antigen represented important methodological advances for both the diagnosis and investigation of rabies virus infection. In 1958, Goldwasser and Kissling used immunofluorescence microscopy to demonstrate rabies virus antigen in infected brain tissue and they established that Negri bodies contain viral antigen. The fluorescent antibody test (FAT) subsequently became established as an important routine laboratory method for the rapid diagnosis of rabies (Bingham and van der Merwe, 2002).

Comprehensive immunohistochemical studies of the distribution of rabies virus antigen in human CNS material have been relatively few in number, but have shown a fairly consistent pattern. Polyclonal or monoclonal anti-ribonucleoprotein/nucleocapsid antibodies have typically been used in these studies (Johnson *et al.*, 1980; Iwasaki *et al.*, 1985; Tirawatnpong *et al.*, 1989; Jogai *et al.*, 2000). Rabies virus antigen (RVAg) was found throughout the brain and spinal cord in most cases. It was consistently present in far more neurons than those containing Negri bodies and also in cases in which no Negri bodies were found, despite a meticulous search. In a series of 20 cases with a clinical diagnosis of rabies, a histopathological diagnosis could be made in only 17 cases, whereas all of the cases exhibited positive RVAg-immunohistochemical staining (Jogai *et al.*, 2000). RVAg was typically seen in the cytoplasm of neuronal perikarya and in dendrites and axons and appeared as blob-like masses (10–20 μm) and granules (1–3 μm) (Figure 9.5), with the larger masses corresponding with Negri bodies seen on hematoxylin and eosin-stained sections (Jogai *et al.*, 2000). The intensity of staining may vary from cell to cell and some neurons showed diffuse staining of their cytoplasm and processes (Johnson *et al.*, 1980; Feiden *et al.*, 1985; Iwasaki *et al.*, 1985). RVAg was also found in processes in the neuropil remote from cell bodies as oval or spindle-shaped masses (Iwasaki *et al.*, 1985) and was inconsistently present in some astrocytes and oligodendrocytes (Feiden *et al.*, 1985; Tirawatnpong *et al.*, 1989; Jogai *et al.*, 2000).

In a quantitative study of neuronal infection in a human case, neurons with Negri bodies contained larger mean amounts of RVAg than those without. Of a variety of neuronal populations, cerebellar Purkinje cells and periaqueductal gray matter neurons showed the largest percentage area for both Negri bodies and RVAg signal, whereas neurons in the trochlear nucleus had a much smaller area of Negri bodies, despite a similar RVAg signal (Jackson *et al.*, 2001). Rabies virus genomic RNA and mRNA have been detected by *in situ* hybridization in brain tissue from human rabies cases and in both CVS- and street virus-infected mouse brains. The distribution of virus RNA was similar to that of viral antigen, although the amount of RNA signal was generally lower than that of RVAg, especially in dendrites (Jackson *et al.*, 1989; Jackson and Wunner, 1991; Jackson, 1992).

In a study of three encephalitic rabies cases, Feiden *et al.* (1985) observed RVAg in all brain regions, the highest amounts being found in the hippocampus,

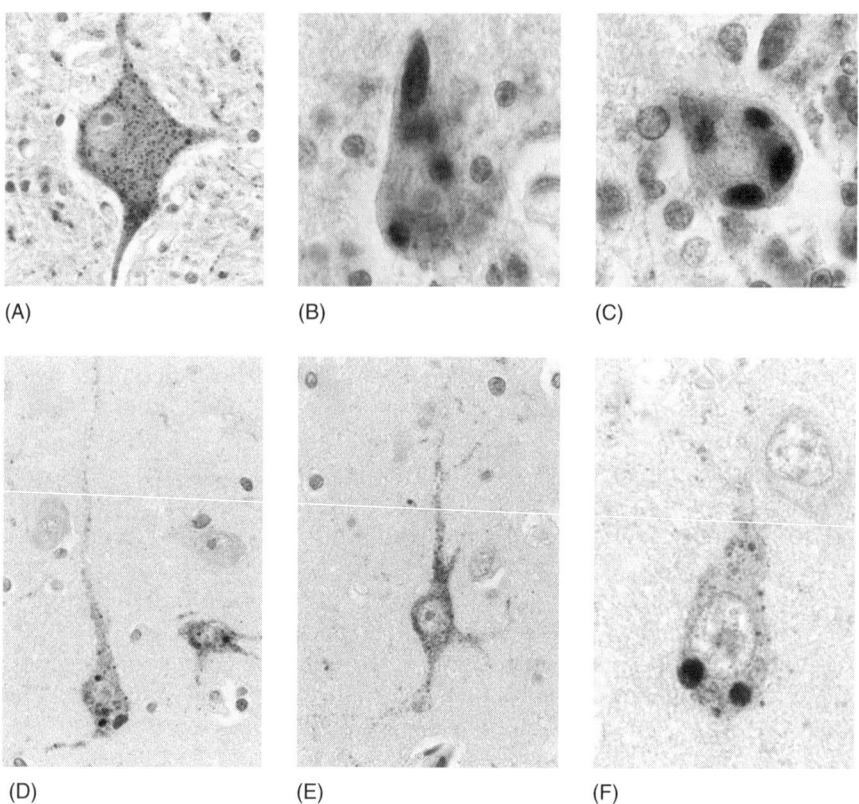

(A)   (B)   (C)

(D)   (E)   (F)

Figure 9.5 Immunoperoxidase staining for rabies virus antigen (mouse monoclonal anti-rabies virus nucleocapsid protein IgG) in human rabies cases. (A) Motoneuron in anterior horn of spinal cord; (B, C) cerebellar Purkinje cells; (D–F) pyramidal neurons in cerebral cortex. The larger immuno-labeled masses correspond with Negri bodies. (Magnifications: A, ×256, B, ×535, C, ×567, D, ×300, E, ×290, F, ×516.) This figure is reproduced in the color plate section.

hypothalamus and tegmental region of the lower brainstem, with many positive neurons also being present in the ventral thalamus and basal portion of the lower brainstem. In the cerebral cortex, basal ganglia and gray matter of the spinal cord, there was a more patchy distribution of fewer virus-containing neurons. In the cerebellar cortex, Purkinje cells, as well as neurons in the molecular and internal granule cell layers, contained numerous immunopositive inclusions (Feiden *et al.*, 1985). Tirawatnpong *et al.* (1989), in a study of four encephalitic and three paralytic rabies cases, did not find any correlation between the distribution of RVAg and the presenting clinical manifestations. In patients who survived 7 days or less, there were a greater number of antigen-positive neurons in the brainstem and spinal cord. In those that survived longer than 7 days, a similar degree of widespread neuronal involvement was seen in the spinal cord and in supratentorial and infratentorial structures. RVAg was found in neurons of the dorsal and ventral horns of the spinal cord, regardless of the clinical pattern and the site of the infecting bite not being associated with a particular antigen distribution. RVAg-positive neurons were typically found in all layers of the cerebral cortex and cortical involvement did not clearly correlate with the degree of disturbance of consciousness (Tirawatnpong et al., 1989). In a study of 12 paralytic and eight encephalitic rabies cases by Jogai et al. (2000), the maximum amount of RVAg was found in the hippocampus, followed by the pons, medulla and cerebellum, whereas antigen was relatively minimal in the cerebral cortex in most cases. In three cases in which Negri bodies were absent, RVAg was present in all regions examined in one case and was restricted to the pons and medulla in the two other cases. Jogai et al. (2000) found a positive correlation between the degree of inflammation and intensity of RVAg-immunopositivity, whereas Tirawatnpong et al. (1989) did not observe a correlation between either the anatomical distribution or amount of RVAg and inflammation in their cases.

4 PATHOLOGY IN THE PERIPHERAL NERVOUS SYSTEM

Spread of rabies virus through the peripheral nervous system (PNS) plays an essential role in the centripetal and centrifugal phases of rabies infection. There are associated inflammatory, reactive and degenerative changes in many of the structural components of the PNS, including neuronal cell bodies in sensory and autonomic ganglia and their capsular/satellite cells, as well as in sensory and motor axons and their enveloping Schwann cells. This results in varying degrees of degeneration and loss of sensory and autonomic neurons (neuronopathy), reactive proliferation of their satellite cells and demyelination and Wallerian degeneration of nerve fibers in spinal nerve roots and peripheral nerves. Historically, there has been particular interest in the diagnostic utility of histological changes in the peripheral nerve ganglia.

4.1 Changes in sensory and autonomic ganglia

The proliferation of capsule (satellite) cells was briefly described in 1872 by Nepveau in the gasserian ganglion (sensory ganglion of the trigeminal nerve) of a human rabies case, but it was Van Gehuchten and Nelis in 1900 who first recognized the importance of ganglionic lesions as a diagnostic marker of rabies. They described marked proliferation of capsular cells surrounding chromatolytic neurons in spinal and cranial nerve ganglia of animal and human rabies cases (Figure 9.6). This capsular cell reaction, together with varying degrees of interstitial lymphocytic infiltration, resulted in grossly apparent enlargement and

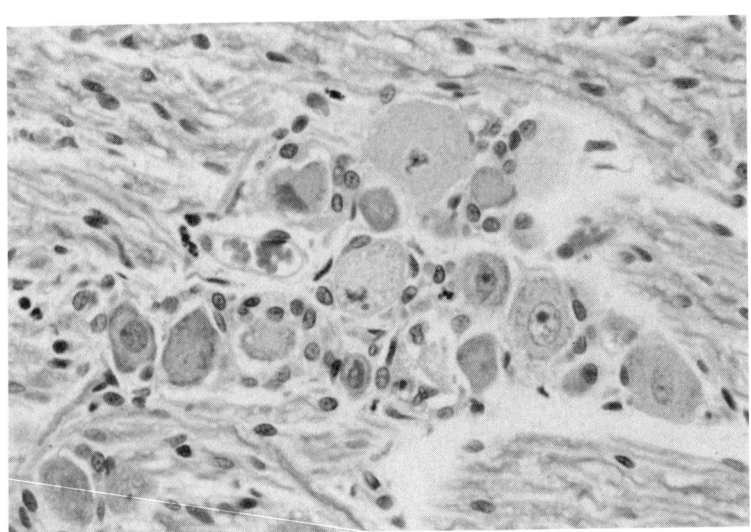

Figure 9.6   Van Gehuchten nodule ('Van Gehuchten and Nelis lesion') in rat trigeminal ganglion following inoculation of street virus into the ipsilateral mental nerve. There is a group of
chromatolytic neurons accompanied by proliferation of satellite cells and a sparse lymphocytic infiltrate. (H&E stain, magnification ×285). (Reproduced with permission from Iwasaki and Tobita, in
Rabies [A.C. Jackson and W.H. Wunner, eds], Academic Press, 2002.)

increased firmness of involved spinal ganglia (Van Gehuchten and Nelis, 1900;
Perl and Good, 1991; Iwasaki and Tobita, 2002). These changes, often referred to
in the literature as 'Van Gehuchten and Nelis lesions' or 'Van Gehuchten nodules'
were reconfirmed in a number of later studies (Hardenbergh, 1916; Marinesco
and Storesco, 1931; Tangchai and Vejjajiva, 1971; Sung *et al.*, 1976; Mitrabhakdi
et al., 2005).

In a series of 52 human rabies cases, Herzog (1945) could not find Negri bodies
in the hippocampus in almost 50% of the cases, whereas he observed capsular cell
proliferation and neuronal degenerative changes in the ganglion nodosum of the
vagus nerve in all 52 cases, with or without accompanying focal or diffuse inflammatory cell infiltration. In a study of nine human rabies cases, Tangchai and
Vejjajiva (1971) reported leukocyte infiltration in three, hypertrophy and proliferation of capsular cells in five, neuronal degeneration in five and extreme vascular
congestion in two cases. There was also increased stromal and epineurial collagen
in the ganglia. Many of the neurons were moderately swollen, with pale finely
vacuolated cytoplasm, and some fragmented neurons undergoing neuronophagia
by histiocytes were seen. Some neurons also contained round acidophilic cytoplasmic inclusions (5–10 μm), although these lacked inner bodies (Tangchai and

Vejjajiva, 1971). Numerous Negri bodies have been found in the gasserian (Garcia-Tamayo *et al.*, 1972) and dorsal root ganglia (Sung *et al.*, 1976) in some cases. Severe lymphocytic inflammation and necrosis have been observed in the inferior cervical sympathetic ganglion of an encephalitic rabies case with prolonged survival (Sandhyamani *et al.*, 1981). Mitrabhakdi *et al.* (2005) reported electrophysiological abnormalities consistent with dorsal root ganglionopathy in one encephalitic rabies patient and two paralytic patients who had severe prodromal paresthesiae. Post-mortem analysis showed severe ganglionitis, with infiltration predominantly by CD3-positive T lymphocytes.

In experimentally infected immature hamsters, Murphy *et al.* (1973a) first observed RVAg immunofluorescence in small numbers of ipsilateral lumbar dorsal root ganglion neurons 60 to 70 hours following hind limb inoculation and in lumbar autonomic ganglia by 72 hours. The RVAg initially had a 'dustlike' distribution, with rapid subsequent progression to involvement of nearly all dorsal root ganglia by brilliant aggregate fluorescence. This was most dense at the peripheral margins of individual ganglion cells, with myelinated axon hillocks being especially heavily infected. Satellite cells did not contain RVAg. Ultrastructurally, large masses of nucleocapsid material were seen within ganglion cells, particularly at their margins, with comparatively small numbers of virus particles budding from intracytoplasmic membranes deeper within the cytoplasm (Murphy *et al.*, 1973a). In skunks, Charlton and Casey (1979) found RVAg immunofluorescence in scattered lumbar dorsal root ganglia 10 days following hind limb inoculation, at which time no fluorescence was seen in peripheral nerve fibers. By 14 days post-inoculation, many dorsal root neurons contained antigen, and linear arrays of granular fluorescent RVAg were seen in axons of hind limb and fore limb peripheral nerves. After 14 days, variable, but increasing degrees of neuronal chromatolysis, neuronophagia and inflammatory cell infiltration were seen in the trigeminal and dorsal root ganglia (Charlton and Casey, 1979).

4.2 Changes in spinal nerve roots and peripheral nerves

Pathological studies of spinal nerve roots and peripheral nerves in human rabies cases have been fewer and have tended to be less comprehensive than those of other parts of the nervous system. Knutti (1929) observed slight focal to extensive necrosis in dorsal nerve roots in a case of paralytic rabies, with only minor changes in the ventral roots. Hurst and Pawan (1932) noted lymphocytic infiltrates in the connective tissue sheath of the sciatic nerve in one of their paralytic cases. In peripheral nerves from nine encephalitic rabies cases, including material from the facial region, and upper and lower limbs, Tangchai and Vejjajiva (1971) observed diffuse perivenous, subepineurial and subperineurial mononuclear inflammatory cell infiltration (three cases), degeneration of nerve fibers (seven cases), proliferation and hypertrophy of Schwann cells (five cases) and subepineurial and subperineurial edema (three cases). Similar changes were seen in dorsal spinal nerve roots, but were milder. Chopra *et al.* (1980) studied a total of four spinal nerves and 17 peripheral nerves in their series of 11 paralytic rabies cases. In all of the spinal nerves there was both Wallerian degeneration and segmental demyelination, but inflammatory cell infiltration was seen in only one of the spinal nerve specimens. The peripheral nerves showed variable degrees of segmental demyelination and remyelination, loss of myelinated fibres and Wallerian degeneration. In nine of 17 nerves, segmental demyelination was the primary lesion. There was no apparent relationship between the degree of spinal or peripheral nerve pathology and the duration of incubation or clinical illness (Chopra *et al.*, 1980). Mitrabhakdi *et al.* (2005) observed heavy lymphocytic infiltration of dorsal and ventral roots in two paralytic cases but, by contrast, only a mild degree of inflammation in dorsal and ventral roots in encephalitic rabies cases.

A predominant pattern of acute motor axonal neuropathy involving ventral spinal roots and peripheral nerves, in the absence of prominent inflammation or motoneuron degeneration, has been found in a paralytic rabies case, which was initially diagnosed and reported as an axonal Guillain-Barré syndrome (Sheikh *et al.*, 2005). RVAg was observed in multiple lumbar anterior horn cells and in their dendrites in this case, as well as in a large proportion of ventral root myelinated axons. Ultrastructurally, mature viral particles were seen in some axons in the ventral root exit zone. Double-label immunostaining showed co-localization of human IgG and C3d complement activation marker with RVAg on ventral root axons, supporting the possibility that the pathogenesis of paralytic rabies may include immune-mediated axonal degeneration (Sheikh *et al.*, 2005).

In experimentally infected hamsters, Murphy (1977) first observed RVAg in peripheral nerves proximal to the inoculation site only concomitant with or later than RVAg detection in ipsilateral spinal ganglia or spinal cord, with individual axons showing a fine linear dustlike pattern of immunofluorescence. With disease progression, the majority of axons in peripheral nerves of the inoculated hind limb, and subsequently throughout the body, showed this pattern. Ultrastructurally, there was concentration of virus particles at nodes of Ranvier, reflecting budding from the high density of membranous organelles at these sites. By contrast, in internodal regions, virus particles only budded individually or in small groups from plasma membranes, resulting in the presence of virions between the axonal and adjacent Schwann cell plasma membranes (Jenson *et al.*, 1969; Murphy *et al.*, 1973a). Murphy *et al.* (1973a) never found viral particles in Schwann cells, whereas Atanasiu and Sisman (1967) did observe Schwann cell infection. In sciatic nerves of mice experimentally inoculated with canine rabies virus, degeneration of approximately 40% of myelinated axons was found, while only occasional degenerating unmyelinated axons were observed (Teixeira *et al.*, 1986). Axonal degeneration and severe demyelination, possibly immunologically mediated, were found in the trigeminal and facial nerves of experimentally infected rats with street rabies virus (Minguetti *et al.*, 1997).

5 PATHOLOGY INVOLVING THE EYE AND EXTRANEURAL ORGANS

Localized replication of rabies virus in extraneural tissue at the inoculation site, including within skeletal muscle, may be an important feature of rabies virus pathogenesis preceding centripetal spread through peripheral nerves to the CNS. Later in the course of infection, centrifugal spread through both somatic sensory and autonomic divisions of the peripheral nervous system results in involvement of a broad range of extraneural tissues and organs, including lacrimal and salivary glands, cornea, skin, heart, gastrointestinal tract and adrenal glands.

5.1 Ocular pathology

Given that the retina is a direct extension of the CNS, it is perhaps surprising how few detailed accounts of ocular pathology exist in the literature. Haltia *et al.* (1989) described the ocular pathology in the case of a 30-year-old man who developed rabies after receiving several bites by a bat. These authors observed lymphocytic and plasma cell infiltrates in the ciliary body and focally in the choroid, focal loss of retinal pigment epithelium, perivascular inflammation in the retinal nerve fiber layer, focal endothelial destruction and occlusion of retinal veins, destruction of many retinal ganglion cells and partial loss of bipolar cells. RVAg was seen in the cytoplasm of many of the surviving ganglion cells. In experimentally infected rabbits, Dejean (1937) observed corneal sensory loss and clouding, retinal venous congestion, choroidal hemorrhages, vitreous clouding and the presence of Negri-like bodies in retinal ganglion cells. Murphy *et al.* (1973b) observed large aggregates of RVAg in the retinal ganglion cell layer of nearly every terminally infected hamster and focal dustlike antigen in the inner and outer retinal nuclear layers and corneal epithelium in some animals.

Centrifugal spread of virus via sensory fibers to corneal epithelium underlies the use of immunofluorescent staining of corneal impressions for ante-mortem diagnosis of rabies (Schneider, 1969) and also explains the rare instances of rabies transmission through infected corneal transplants (Houff *et al.*, 1979; Gode and Bhide, 1988).

5.2 Changes in extraneural organs

Centrifugal viral spread to cutaneous nerve endings surrounding hair follicles (especially in the head region) forms the basis for ante-mortem diagnosis by means of immunostained nuchal skin biopsies, in a large proportion of animal (Blenden *et al.*, 1983) and human rabies cases (Bryceson *et al.*, 1975; Blenden *et al.*, 1986). In some cases, RVAg is also found in epidermal cells (Balachandran and Charlton, 1994; Jackson *et al.*, 1999; Bago *et al.*, 2005).

Centrifugal spread to the major salivary glands, resulting in production of saliva containing high titers of rabies virus, is a central feature of bite transmission of rabies by natural vectors, such as dog (Goldwasser *et al.*, 1959), fox (Dierks *et al.*, 1969; Balachandran and Charlton, 1994) and skunk (Balachandran and Charlton, 1994). Ultrastructurally, budding of numerous virions from the apical membranes of mucogenic cells and their release into intercellular canaliculi has been documented (Dierks *et al.*, 1969; Balachandran and Charlton, 1994). In human rabies cases RVAg was found in acini of minor salivary glands of the tongue, as well as in skeletal muscle fibers of the tongue, but there was no significant involvement of acini in the major salivary glands (Li *et al.*, 1995; Jackson *et al.*, 1999).

Widespread distribution of RVAg has been observed in autonomic nerve plexuses related to multiple organs, including cardiac ganglia and the submucosal plexus of Meissner and myenteric plexus of Auerbach in the gastrointestinal tract, in both animal (Debbie and Trimarchi, 1970; Fischman and Schaeffer, 1971; Murphy *et al.*, 1973b) and human material (Jackson *et al.*, 1999; Jogai *et al.*, 2002). There is typically an associated mild mononuclear cell inflammatory response (Jackson *et al.*, 1999), although more prominent inflammation and degeneration of enteric ganglion cells has been observed in some cases (Love, 1944). The adrenal medulla, as an extension of the sympathetic nervous system, has been found to be frequently involved, with many cells containing RVAg in both animals (Debbie and Trimarchi, 1970; Fischman and Schaeffer, 1971; Murphy *et al.*, 1973b) and humans (Jackson *et al.*, 1999; Jogai *et al.*, 2002) and may be accompanied by moderate to severe inflammation ('medullitis') (Figure 9.7) (Love, 1944; Almeida *et al.*, 1986; Lopez-Corella *et al.*, 1997; Jackson *et al.*, 1999).

Clinical and/or pathological evidence of cardiac involvement is a recognized feature of some human rabies cases (Ross and Armentrout, 1962; Cheetham *et al.*, 1970; Roux *et al.*, 1976; Raman *et al.*, 1988; Burton *et al.*, 2005). Myocarditis,

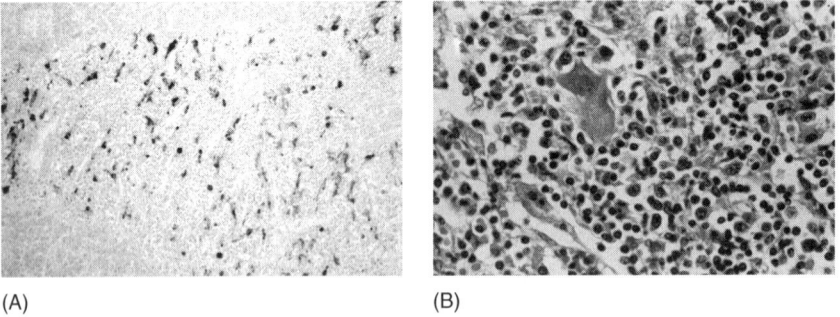

(A)   (B)

Figure 9.7   (A) Adrenal gland showing abundant rabies virus antigen in the adrenal medulla with sparing of the adrenal cortex. (B) Mononuclear inflammatory infiltrate in the adrenal medulla. A, immunoperoxidase–hematoxylin; B, H&E stain. (Magnifications, A, ×80; B, ×140.) (Reproduced with permission from Jackson *et al.*, *Laboratory Investigation* **79**, 945–951, 1999.)

characterized by multifocal muscle fiber degeneration/necrosis, has been observed in some cases (Ross and Armentrout, 1962; Cheetham *et al.*, 1970; Burton *et al.*, 2005). RVAg within cardiac myocytes has also been reported in cases with mild or no associated inflammation (Metze and Feiden, 1991; Jackson *et al.*, 1999).

6 SUMMARY AND CONCLUSIONS

As an almost invariably fatal infection of the CNS, rabies shares a number of histopathological features with other viral encephalitides, such as leptomeningeal and perivascular mononuclear inflammatory cell infiltration, microglial activation and neuronophagia. However, the presence of Negri body viral inclusions in the cytoplasm of neurons is a unique and diagnostic feature in many cases of infection with street rabies virus strains. In paralytic rabies, by comparison with encephalitic cases, there tends to be, at least in some cases, more severe involvement of the spinal cord and brainstem and a greater degree of injury in spinal nerve roots and peripheral nerves. Ultrastructural and immunohistochemical studies of human and animal rabies material have contributed greatly to an understanding of rabies pathogenesis. In some cases, inflammatory changes and neuronal cell death are minimal or even absent, indicating that these are not essential contributors to a fatal outcome in rabies. The role of degenerative changes restricted to neuronal processes needs further study, because these structural changes could potentially explain functional abnormalities in rabies.

REFERENCES

Abba, F. and Bormans, A. (1905). Sur le diagnostic histologique de la rage. *Annales de L'Institut Pasteur (Paris)* **19**, 49.

Adle-Biassette, H., Bourhy, H., Gisselbrecht, M. *et al.* (1996). Rabies encephalitis in a patient with AIDS: a clinicopathological study. *Acta Neuropathologica* **92**, 415–420.

Almeida, H.d.O., Teixeira, V.d.P., de Oliveira, G., Brandao, M.d.C. and Gobbi, H. (1986). Medulite supra-renalica em casos de raiva humana [adrenal medullitis in cases of human rabies]. [Portuguese]. *Memorias do Instituto Oswaldo Cruz* **81**, 439–442.

Atanasiu, P. and Sisman, J. (1967). Morphological aspects of rabies virus [French]. *Bulletin de l'Office International des Epizooties* **67**, 521–533.

Babes, M.V. (1892). Sur certains caractères des lésions histologiques de la rage. *Annales de L'Institut Pasteur* **6**, 209–223.

Bago, Z., Revilla-Fernandez, S., Allerberger, F. and Krause, R. (2005). Value of immunohistochemistry for rapid ante mortem rabies diagnosis. *International Journal of Infectious Diseases* **9**, 351–352.

Balachandran, A. and Charlton, K. (1994). Experimental rabies infection of non-nervous tissues in skunks (*Mephitis mephitis*) and foxes (*Vulpes vulpes*). *Veterinary Pathology* **31**, 93–102.

Baloul, L. and Lafon, M. (2003). Apoptosis and rabies virus neuroinvasion. *Biochimie* **85**, 777–788.

Benedikt, M. (1878). Zur pathologischen Anatomie der Lyssa. *Virchows Archiv fur pathologische Anatomie und Physiologie und fur klinische Medizin* **72**, 425–431.

Bingham, J. and van der Merwe, M. (2002). Distribution of rabies antigen in infected brain material: determining the reliability of different regions of the brain for the rabies fluorescent antibody test. *Journal of Virological Methods* **101**, 85–94.

Blenden, D.C., Bell, J.F., Tsao, A.T. and Umoh, J.U. (1983). Immunofluorescent examination of the skin of rabies-infected animals as a means of early detection of rabies virus antigen. *Journal of Clinical Microbiology* **18**, 631–636.

Blenden, D.C., Creech, W. and Torres-Anjel, M.J. (1986). Use of immunofluorescence examination to detect rabies virus antigen in the skin of humans with clinical encephalitis. *Journal of Infectious Diseases* **154**, 698–701.

Bryceson, A.D.M., Greenwood, B.M., Warrell, D.A. *et al.* (1975). Demonstration during life of rabies antigen in humans. *Journal of Infectious Diseases* **131**, 71–74.

Bundza, A. and Charlton, K.M. (1988). Comparison of spongiform lesions in experimental scrapie and rabies in skunks. *Acta Neuropathologica* **76**, 275–280.

Burrage, T.G., Tignor, G.H. and Smith, A.L. (1983). Immunoelectron microscopic localization of rabies virus antigen in central nervous system and peripheral tissue using low-temperature embedding and protein A-gold. *Journal of Virological Methods* **7**, 337–350.

Burton, E.C., Burns, D.K., Opatowsky, M.J. *et al.* (2005). Rabies encephalomyelitis: clinical, neuroradiological, and pathological findings in 4 transplant recipients. *Archives of Neurology* **62**, 873–882.

Charlton, K.M. (1984). Rabies: spongiform lesions in the brain. *Acta Neuropathologica* **63**, 198–202.

Charlton, K.M. and Casey, G.A. (1979). Experimental rabies in skunks: immunofluorescence light and electron microscopic studies. *Laboratory Investigation* **41**, 36–44.

Charlton, K.M., Casey, G.A., Webster, W.A. and Bundza, A. (1987). Experimental rabies in skunks and foxes: pathogenesis of the spongiform lesions. *Laboratory Investigation* **57**, 634–645.

Cheetham, H.D., Hart, J., Coghill, N.F. and Fox, B. (1970). Rabies with myocarditis: two cases in England. *Lancet* **1**, 921–922.

Chopra, J.S., Banerjee, A.K., Murthy, J.M.K. and Pal, S.R. (1980). Paralytic rabies: a clinicopathological study. *Brain* **103**, 789–802.

Davies, M.C., Englert, M.E., Sharpless, G.R. and Cabasso, V.J. (1963). The electron microscopy of rabies virus in cultures of chicken embryo tissues. *Virology* **21**, 642–651.

de Brito, T., Araujo, M.D.F. and Tiriba, A. (1973). Ultrastructure of the Negri body in human rabies. *Journal of the Neurological Sciences* **20**, 363–372.

Debbie, J.G. and Trimarchi, C.V. (1970). Pantropism of rabies virus in free-ranging rabid red fox *Vulpes fulva*. *Journal of Wildlife Diseases* **6**, 500–506.

Dejean, C. (1937). Les modifications du fond d'oeil dans la rage chez le lapin. *Bulletin des sociétés d'ophtalmologie de France* **50**, 247–254.

Dierks, R.E., Murphy, F.A. and Harrison, A.K. (1969). Extraneural rabies virus infection. Virus development in fox salivary gland. *American Journal of Pathology* **54**, 251–273.

Dolman, C.L. and Charlton, K.M. (1987). Massive necrosis of the brain in rabies. *Canadian Journal of Neurological Sciences* **14**, 162–165.

Dupont, J.R. and Earle, K.M. (1965). Human rabies encephalitis. A study of forty-nine fatal cases with a review of the literature. *Neurology* **15**, 1023–1034.

Feiden, W., Feiden, U., Gerhard, L., Reinhardt, V. and Wandeler, A. (1985). Rabies encephalitis: immunohistochemical investigations. *Clinical Neuropathology* **4**, 156–164.

Fekadu, M., Chandler, F.W. and Harrison, A.K. (1982). Pathogenesis of rabies in dogs inoculated with an Ethiopian rabies virus strain. Immunofluorescence, histologic and ultrastructural studies of the central nervous system. *Archives of Virology* **71**, 109–126.

Fekadu, M., Greer, P.W., Chandler, F.W. and Sanderlin, D.W. (1988). Use of the avidin-biotin perox-idase system to detect rabies antigen in formalin-fixed paraffin-embedded tissues. *Journal of Virological Methods* **19**, 91–96.

Fischman, H.R. and Schaeffer, M. (1971). Pathogenesis of experimental rabies as revealed by immunofluorescence. *Annals of the New York Academy of Sciences* **177**, 78–97.

Fu, Z.F. and Jackson, A.C. (2005). Neuronal dysfunction and death in rabies virus infection. *Journal of Neurovirology* **11**, 101–106.

Garcia-Tamayo, J., Avila-Mayor, A. and Anzola-Perez, E. (1972). Rabies virus neuronitis in humans. *Archives of Pathology* **94**, 11–15.

Gode, G.R. and Bhide, N.K. (1988). Two rabies deaths after corneal grafts from one donor (Letter). *Lancet* **2**, 791.

Goldwasser, R.A. and Kissling, R.E. (1958). Fluorescent antibody staining of street and fixed rabies virus antigens. *Proceedings of the Society for Experimental Biology and Medicine* **98**, 219–223.

Goldwasser, R.A., Kissling, R.E., Carski, T.R. and Hosty, T.S. (1959). Fluorescent antibody staining of rabies virus antigens in the salivary glands of rabid animals. *Bulletin of the World Health Organization* **20**, 579–588.

Gonzalez-Angulo, A., Marquez-Monter, H., Feria-Velasco, A. and Zavala, B.J. (1970). The ultra-structure of Negri bodies in Purkinje neurons in human rabies. *Neurology* **20**, 323–328.

Goodpasture, E.W. (1925). A study of rabies, with reference to a neural transmission of the virus in rab-bits, and the structure and significance of Negri bodies. *American Journal of Pathology* **1**, 547–584.

Gosztonyi, G. (1994). Reproduction of lyssaviruses: ultrastructural composition of lyssavirus and func-tional aspects of pathogenesis. In: *Current Topics in Microbiology and Immunology, Volume 187: Lyssa-viruses* (C.E. Rupprecht, B. Dietzschold and H. Koprowski, eds). pp. 43–68. Berlin: Springer-Verlag.

Gowers, W.R. (1877). The pathological anatomy of hydrophobia. *Transactions of the Pathological Society of London* **28**, 10–23.

Haltia, M., Tarkkanen, A. and Kivela, T. (1989). Rabies: ocular pathology. *British Journal of Ophthal-mology* **73**, 61–67.

Hardenbergh, J.B. (1916). The reliability of cell proliferation changes in the diagnosis of rabies. *Journal of the American Veterinary Medical Association* **49**, 663.

Herzog, E. (1945). Histologic diagnosis of rabies. *Archives of Pathology* **39**, 279–280.

Hottle, G.A., Morgan, G., Peers, J.H. and Wyckoff, R.W.G. (1951). The electron microscopy of rabies inclusion (Negri) bodies. *Proceedings of the Society for Experimental Biology and Medicine* **77**, 721–723.

Houff, S.A., Burton, R.C., Wilson, R.W. *et al.* (1979). Human-to-human transmission of rabies virus by corneal transplant. *New England Journal of Medicine* **300**, 603–604.

Hummeler, K., Koprowski, H. and Wiktor, T. J. (1967). Structure and development of rabies virus in tissue culture. *Journal of Virology* **1**, 152–170.

Hummeler, K., Tomassini, N., Sokol, F., Kuwert, E. and Koprowski, H. (1968). Morphology of the nucleoprotein component of rabies virus. *Journal of Virology* **2**, 1191–1199.

Hurst, E.W. and Pawan, J.L. (1932). A further account of the Trinidad outbreak of acute rabic myelitis: histology of the experimental disease. *Journal of Pathology and Bacteriology* **35**, 301–321.

Iwasaki, Y. and Clark, H.F. (1975). Cell to cell transmission of virus in the central nervous system. II. Experimental rabies in mouse. *Laboratory Investigation* **33**, 391–399.

Iwasaki, Y. and Clark, H.F. (1977). Rabies virus infection in mouse neuroblastoma cells. *Laboratory Investigation* **36**, 578–584.

Iwasaki, Y. and Minamoto, N. (1982). Scanning and freeze-fracture electron microscopy of rabies virus infection in murine neuroblastoma cells. *Comparative Immunology, Microbiology and Infectious Diseases* **5**, 1–8.

Iwasaki, Y. and Tobita, M. (2002). Pathology. In: *Rabies* (A.C. Jackson and W.H. Wunner, eds). pp. 283–306. San Diego: Academic Press.

Iwasaki, Y., Liu, D.S., Yamamoto, T. and Konno, H. (1985). On the replication and spread of rabies virus in the human central nervous system. *Journal of Neuropathology and Experimental Neurology* **44**, 185–195.

Iwasaki, Y., Ohtani, S. and Clark, H.F. (1975). Maturation of rabies virus by budding from neuronal cell membrane in suckling mouse brain. *Journal of Virology* **15**, 1020–1023.

Iwasaki, Y., Sako, K., Tsunoda, I. and Ohara, Y. (1993). Phenotypes of mononuclear cell infiltrates in human central nervous system. *Acta Neuropathologica* **85**, 653–657.

Iwasaki, Y., Wiktor, T.J. and Koprowski, H. (1973). Early events of rabies virus replication in tissue cultures: an electron microscopic study. *Laboratory Investigation* **28**, 142–148.

Jackson, A.C. (1992). Detection of rabies virus mRNA in mouse brain by using *in situ* hybridization with digoxigenin-labelled RNA probes. *Molecular and Cellular Probes* **6**, 131–136.

Jackson, A.C. and Park, H. (1998). Apoptotic cell death in experimental rabies in suckling mice. *Acta Neuropathologica* **95**, 159–164.

Jackson, A.C. and Reimer, D.L. (1989). Pathogenesis of experimental rabies in mice: an immunohistochemical study. *Acta Neuropathologica* **78**, 159–165.

Jackson, A.C. and Rossiter, J.P. (1997). Apoptosis plays an important role in experimental rabies virus infection. *Journal of Virology* **71**, 5603–5607.

Jackson, A.C. and Wunner, W.H. (1991). Detection of rabies virus genomic RNA and mRNA in mouse and human brains by using *in situ* hybridization. *Journal of Virology* **65**, 2839–2844.

Jackson, A.C., Reimer, D.L. and Wunner, W.H. (1989). Detection of rabies virus RNA in the central nervous system of experimentally infected mice using *in situ* hybridization with RNA probes. *Journal of Virological Methods* **25**, 1–11.

Jackson, A.C., Ye, H., Phelan, C.C. *et al.* (1999). Extraneural organ involvement in human rabies. *Laboratory Investigation* **79**, 945–951.

Jackson, A.C., Ye, H., Ridaura-Sanz, C. and Lopez-Corella, E. (2001). Quantitative study of the infection in brain neurons in human rabies. *Journal of Medical Virology* **65**, 614–618.

Jenson, A.B., Rabin, E.R., Bentinck, D.C. and Melnick, J.L. (1969). Rabiesvirus neuronitis. *Journal of Virology* **3**, 265–269.

Jenson, A.B., Rabin, E.R., Wende, R.D. and Melnick, J.L. (1967). A comparative light and electron microscopic study of rabies and Hart Park virus encephalitis. *Experimental and Molecular Pathology* **7**, 1–10.

Jogai, S., Radotra, B.D. and Banerjee, A.K. (2000). Immunohistochemical study of human rabies. *Neuropathology* **20**, 197–203.

Jogai, S., Radotra, B.D. and Banerjee, A.K. (2002). Rabies viral antigen in extracranial organs: a post-mortem study. *Neuropathology and Applied Neurobiology* **28**, 334–338.

Johnson, K.P., Swoveland, P.T. and Emmons, R.W. (1980). Diagnosis of rabies by immunofluorescence in trypsin-treated histologic sections. *Journal of the American Medical Association* **244**, 41–43.

Juntrakul, S., Ruangvejvorachai, P., Shuangshoti, S., Wacharapluesadee, S. and Hemachudha, T. (2005). Mechanisms of escape phenomenon of spinal cord and brainstem in human rabies. *BMC Infectious Diseases* **5**, 104.

Knutti, R.E. (1929). Acute ascending paralysis and myelitis due to the virus of rabies. *Journal of the American Medicial Association* **93**, 754–758.

Kristensson, K., Dastur, D.K., Manghani, D.K., Tsiang, H. and Bentivoglio, M. (1996). Rabies: interactions between neurons and viruses. a review of the history of Negri inclusion bodies. *Neuropathology and Applied Neurobiology* **22**, 179–187.

Last, R.D., Jardine, J.E., Smit, M.M. and van der Lugt, J.J. (1994). Application of immunoperoxidase techniques to formalin-fixed brain tissue for the diagnosis of rabies in southern Africa. *Onderstepoort Journal of Veterinary Research* **61**, 183–187.

Leech, R.W. (1971). Electron-microscopic study of the inclusion body in human rabies. *Neurology* **21**, 91–94.

Lepine, P. and Atanasiu, P. (1996). Histopathological diagnosis. In: *Laboratory Techniques in Rabies* (World Health Organization, F.-X. Meslin, M.M. Kaplan and H. Koprowski, eds). pp. 66–79. Geneva: World Health Organization.

Lewis, P. and Lentz, T.L. (1998). Rabies virus entry into cultured rat hippocampal neurons. *Journal of Neurocytology* **27**, 559–573.

Li, X.Q., Sarmento, L. and Fu, Z.F. (2005). Degeneration of neuronal processes after infection with pathogenic, but not attenuated, rabies viruses. *Journal of Virology* **79**, 10063–10068.

Li, Z., Feng, Z. and Ye, H. (1995). Rabies viral antigen in human tongues and salivary glands. *Journal of Tropical Medicine and Hygiene* **98**, 330–332.

Lopez-Corella, E., Ridaura-Sanz, C. and Samayoa-Palma, J.E. (1997) Human rabies. Systemic pathology in 33 autopsies (Abstract). *Laboratory Investigation* **76**, 140A.

Love, S. and Wiley, C.A. (2002). Viral diseases. In: *Greenfield's Neuropathology*, 7th edition (D.I. Graham and P.L. Lantos, eds). pp. 1–105. London: Arnold.

Love, S.V. (1944). Paralytic rabies: review of the literature and report of a case. *Journal of Pediatrics* **24**, 312–325.

Lowenberg, K. (1928). Rabies in man. Microscopic observations. *Archives of Neurology and Psychiatry* **19**, 638–646.

Mackenzie, I.R., Medvedev, G. and Thiessen, B. (2003) An unusual case of rabies with prolonged survival and extreme neuropathology. *Canadian Journal of Neurological Sciences* **30**, 408 (Abstract).

Manghani, D.K., Dastur, D.K., Nanavaty, A.N. and Patel, R. (1986). Pleomorphism of fine structure of rabies virus in human and experimental brain. *Journal of the Neurological Sciences* **75**, 181–193.

Marinesco, G. and Storesco, G. (1931). Études sur la pathologie de la rage. *Archives Roumaines de Pathologie Experimentales et de Microbiologie* **4**, 243–288.

Matsumoto, S. (1963). Electron microscope studies of rabies virus in mouse brain. *Journal of Cell Biology* **19**, 565–591.

Matsumoto, S. and Kawai, A. (1969). Comparative studies on development of rabies virus in different host cells. *Virology* **39**, 449–459.

Matsumoto, S., Schneider, L.G., Kawai, A. and Yonezawa, T. (1974). Further studies on the replication of rabies and rabies-like viruses in organized cultures of mammalian neural tissues. *Journal of Virology* **14**, 981–996.

Metze, K. and Feiden, W. (1991). Rabies virus ribonucleoprotein in the heart (Letter). *New England Journal of Medicine* **324**, 1814–1815.

Minguetti, G., Hofmeister, R.M., Hayashi, Y. and Montano, J.A. (1997). Ultrastructure of cranial nerves of rats inoculated with rabies virus. *Arquivos de Neuro Psiquiatria* **55**, 680–686.

Mitrabhakdi, E., Shuangshoti, S., Wannakrairot, P. *et al.* (2005). Difference in neuropathogenetic mechanisms in human furious and paralytic rabies. *Journal of the Neurological Sciences* **238**, 3–10.

Miyamoto, K. and Matsumoto, S. (1965). The nature of the Negri body. *Journal of Cell Biology* **27**, 677–682.

Miyamoto, K. and Matsumoto, S. (1967). Comparative studies between pathogenesis of street and fixed rabies infection. *Journal of Experimental Medicine* **125**, 447–474.

Morecki, R. and Zimmerman, H.M. (1969). Human rabies encephalitis. Fine structure study of cytoplasmic inclusions. *Archives of Neurology* **20**, 599–604.

Mrak, R.E. and Young, L. (1993). Rabies encephalitis in a patient with no history of exposure. *Human Pathology* **24**, 109–110.

Mrak, R.E. and Young, L. (1994). Rabies encephalitis in humans: pathology, pathogenesis and pathophysiology. *Journal of Neuropathology and Experimental Neurology* **53**, 1–10.

Murphy, F.A. (1977). Rabies pathogenesis: brief review. *Archives of Virology* **54**, 279–297.

Murphy, F.A., Bauer, S.P., Harrison, A.K. and Winn, W.C. (1973a). Comparative pathogenesis of rabies and rabies-like viruses: viral infection and transit from inoculation site to the central nervous system. *Laboratory Investigation* **28**, 361–376.

Murphy, F.A., Harrison, A.K., Winn, W.C. and Bauer, S.P. (1973b). Comparative pathogenesis of rabies and rabies-like viruses: infection of the central nervous system and centrifugal spread of virus to peripheral tissues. *Laboratory Investigation* **29**, 1–16.

Negri, A. (1903a). Beitrag zum Studium der Aetiologie der Tollwuth. *Zeitschrift fur Hygiene und Infektionskrankheiten* **43**, 507–528.

Negri, A. (1903b). Zur Aetiologie der Tollwuth. Die diagnose der Tollwuth auf Grund der Neuen Befunde. *Zeitschrift fur Hygiene und Infektionskrankheiten* **44**, 519.

Negri, A. (1909). Uber die Morphologie und der Entwicklungszyklus des Parasiten der Tollwut (Neurocytes hydrophobiae Calkins). *Zeitschrift fur Hygiene und Infektionskrankheiten* **63**, 421–440.

Negri-Luzzani, L. (1913). Le diagnostic de la rage par la demonstration du parasite spécifique. Resultats de dix ans d'expériences. *Annales de L'Institut Pasteur (Paris)* **27**, 1039–1064.

Nepveau, M. (1872). Un cas de rage. *Comptes Rendus des Séances et Mémoires de la Société de Biologie* **4**, 133.

Nieberg, K.C. and Blumberg, J.M. (1972). Viral encephalitides. In: *Pathology of the Nervous System* (J. Minckler, ed.). pp. 2266–2323. New York: McGraw-Hill Book Company.

Pasteur, L. (1881). Sur la rage. *Comptes Rendus de l'Académie des Sciences* **92**, 1259–1260.

Perl, D.P. and Good, P.F. (1991). The pathology of rabies in the central nervous system. In: *The Natural History of Rabies* (G.M. Baer, ed.). pp. 163–190. Boca Raton: CRC Press.

Perl, D.P., Callaway, C.S. and Hicklin, M. (1972) An ultrastructural study of Negri bodies in experimental rabies following prolonged incubation (Abstract). *Journal of Neuropathology and Experimental Neurology* **31**, 172.

Raman, G.V., Prosser, A., Spreadbury, P.L., Cockcroft, P.M. and Okubadejo, O.A. (1988). Rabies presenting with myocarditis and encephalitis. *Journal of Infection* **17**, 155–158.

Ramón y Cajal, S. and Garcia, D. (1904). Las lesiones del retículo de las células nerviosas en la rabia. *Trabajos del Laboratorio de Investigaciones biológicas de la Universidad de Madrid* **3**, 213.

Reid, J.E. and Jackson, A.C. (2001). Experimental rabies virus infection in *Artibeus jamaicensis* bats with CVS-24 variants. *Journal of Neurovirology* **7**, 511–517.

Reisman, D., Alpers, B.J. and Cooper, D.A. (1933). Hydrophobia. Report of two fatal cases with pathologic studies in one. *Archives of Internal Medicine* **51**, 643–655.

Ross, E. and Armentrout, S.A. (1962). Myocarditis associated with rabies: report of a case. *New England Journal of Medicine* **266**, 1087–1089.

Roux, F., Bourgeade, A., Salaun, J.J., Bondurand, A., Ette, M. and Bertrand, E. (1976). L'atteinte cardiaque dans la rage humaine [Cardiac involvement in human rabies]. *Coeur et Medecine Interne* **15**, 37–44.

Rubin, R.H., Sullivan, L., Summers, R., Gregg, M.B. and Sikes, R.K. (1970). A case of human rabies in Kansas: epidemiologic, clinical, and laboratory considerations. *Journal of Infectious Diseases* **122**, 318–322.

Sandhyamani, S., Roy, S., Gode, G.R. and Kalla, G.N. (1981). Pathology of rabies: a light- and electron-microscopical study with particular reference to the changes in cases with prolonged survival. *Acta Neuropathologica* **54**, 247–251.

Sarmento, L., Li, X., Howerth, E., Jackson, A.C. and Fu, Z.F. (2005). Glycoprotein-mediated induction of apoptosis limits the spread of attenuated rabies viruses in the central nervous system of mice. *Journal of Neurovirology* **11**, 571–581.

Schaffer, K. (1888). Histologische Untersuchung eines Falles von Lyssa. *Archiv für Psychiatrie und Nervenkrankheiten* **19**, 45–63.

Schneider, L.G. (1969). The cornea test; a new method for the intra-vitam diagnosis of rabies. *Zentralblatt fur Veterinarmedizin – Reihe B* **16**, 24–31.

Schneider, L.G., Dietzschold, B., Dierks, R.E., Matthaeus, W., Enzmann, P.J. and Strohmaier, K. (1973). Rabies group-specific ribonucleoprotein antigen and a test system for grouping and typing of rhabdoviruses. *Journal of Virology* **11**, 748–755.

Sheikh, K.A., Ramos-Alvarez, M., Jackson, A.C., Li, C.Y., Asbury, A.K. and Griffin, J.W. (2005). Overlap of pathology in paralytic rabies and axonal Guillain-Barré syndrome. *Annals of Neurology* **57**, 768–777.

Suja, M.S., Mahadevan, A., Sundaram, C. *et al.* (2004). Rabies encephalitis following fox bite – histological and immunohistochemical evaluation of lesions caused by virus. *Clinical Neuropathology* **23**, 271–276.

Sükrü-Aksel, I. (1958). Pathologische Anatomie der Lyssa. In: *Handbuch Der Speziellen Pathologischen Anatomie Und Histologie* (O. Lubarsch, F. Henke and R. Rossle, eds). pp. 417–435. Berlin: Springer-Verlag.

Sung, J.H., Hayano, M., Mastri, A.R. and Okagaki, T. (1976). A case of human rabies and ultrastructure of the Negri body. *Journal of Neuropathology and Experimental Neurology* **35**, 541–559.

Szlachta, H.L. and Habel, R.E. (1953). Inclusions resembling Negri bodies in the brains of nonrabid cats. *Cornell Veterinarian* **43**, 207–212.

Tangchai, P. and Vejjajiva, A. (1971). Pathology of the peripheral nervous system in human rabies: a study of nine autopsy cases. *Brain* **94**, 299–306.

Tangchai, P., Yenbutr, D. and Vejjajiva, A. (1970). Central nervous system lesions in human rabies: a study of twenty-four cases. *Journal of the Medical Association of Thailand* **53**, 471–488.

Teixeira, F., Aranda, F.J., Castillo, S., Perez, M., Del Peon, L. and Hernandez, O. (1986). Experimental rabies: ultrastructural quantitative analysis of the changes in the sciatic nerve. *Experimental and Molecular Pathology* **45**, 287–293.

Theerasurakarn, S. and Ubol, S. (1998). Apoptosis induction in brain during the fixed strain of rabies virus infection correlates with onset and severity of illness. *Journal of Neurovirology* **4**, 407–414.

Tierkel, E.S. (1973). Rapid microscopic examination for Negri bodies and preparation of specimens for biological test. In: *Laboratory Techniques in Rabies*. (M.M. Kaplan and H. Koprowski, eds). pp. 41–55. Geneva: World Health Organization.

Tirawatnpong, S., Hemachudha, T., Manutsathit, S., Shuangshoti, S., Phanthumchinda, K. and Phanuphak, P. (1989). Regional distribution of rabies viral antigen in central nervous system of human encephalitic and paralytic rabies. *Journal of the Neurological Sciences* **92**, 91–99.

Tustin, R.C. and Smith, J.D. (1962). Rabies in South Africa. An analysis of histological examination. *Journal of the South African Veterinary Medical Association* **33**, 295–310.

Ubol, S. and Kasisith, J. (2000). Reactivation of Nedd-2, a developmentally down-regulated apoptotic gene, in apoptosis induced by a street strain of rabies virus. *Journal of Medical Microbiology* **49**, 1043–1046.

Van Gehuchten, A. and Nelis, C. (1900). Les lésions histologiques de la rage chez les animaux et chez l'homme. *Bulletin de l'Académie royale de médecine de Belgique* **14**, 31–66.

Yan, X., Prosniak, M., Curtis, M.T. *et al.* (2001). Silver-haired bat rabies virus variant does not induce apoptosis in the brain of experimentally infected mice. *Journal of Neurovirology* **7**, 518–527.

10

Diagnostic Evaluation

CHARLES V. TRIMARCHI[1] AND SUSAN A. NADIN-DAVIS[2]

Wadsworth Center, New York State Department of Health, Albany, New York 12202-0509, USA[1];
Centre of Expertise for Rabies, Ottawa Laboratory (Fallowfield), Canadian Food Inspection Agency,
Ottawa, Ontario K2H 8P9, Canada[2]

1 INTRODUCTION

1.1 Role of laboratory diagnostic evaluation

An estimated 7–10 billion laboratory tests of all types are performed each year in the USA alone (Steindel *et al.*, 2000). Very few others of those have a more direct effect on decisions critical to prevent human mortality than do rabies laboratory tests. Rabies diagnostic evaluation is most frequently performed for the post-mortem examination of animals that have bitten a person or have otherwise potentially caused human exposure to the disease. These examinations constitute the most important diagnostic contributions to the control and prevention of rabies. Evidence of rabies virus infection, based on a positive diagnostic test, prompts administration of rabies post-exposure prophylaxis (PEP) to the exposed person, preventing onset of the almost invariably fatal infection. Demonstration of rabies virus infection also initiates proper management of exposed domestic animals, including booster vaccination of previously immunized animals and euthanasia or quarantine of unvaccinated animals. Prompt and reliable negative results, on the other hand, can be used to prevent the initiation of unnecessary PEP in humans, avoiding the small but inherent associated risks and the squandering of expensive and often scarce biologics.

An extremely valuable but less obvious product of these examinations is the surveillance data that they collectively generate revealing the distribution of rabies in domestic and wild mammal populations. These data provide the foundation for good decisions regarding PEP for bites from animals not available for observation or testing (Trimarchi, 2000) and they permit proper targeting of vaccination programs and other rabies-control activities in the region.

Because the diagnosis of rabies in animals can now be completed in a reliable manner in less than one day, the physician's decision whether to initiate or withhold PEP is often based upon the post-mortem examination of the biting animal (Trimarchi and Debbie, 1991). This practice imposes the highest possible standards of sensitivity on the performance of the test, because the consequences of false-negative reports can be expected to include human mortality. Specificity of the test has further implications: false-positive results can lead not only to unnecessary PEP but also to misleading epizootiologic data.

Tests to detect evidence of rabies virus infection are also applied to the ante- and post-mortem diagnosis of rabies in humans afflicted by an encephalitis of unknown etiology. A recent survival of rabies infection following aggressive and novel treatment after onset of disease (Willoughby *et al.*, 2005) and a demonstration of transmission via solid organ transplants (Srinivasan *et al.*, 2005) have emphasized the potential importance of early diagnosis of rabies in human encephalitis cases (Jackson, 2005). Furthermore, rabies diagnostic testing supports studies of rabies pathobiology and the production of vaccines and vaccine potency testing and facilitates surveillance programs that evaluate the success of wildlife vaccination campaigns.

1.2 Range of diagnostic methods

Making a reliable diagnosis of rabies based upon clinical presentation is very difficult, as there are no truly pathognomonic symptoms or signs for this disease. Clinically, rabies in animals can be difficult to distinguish from encephalitic conditions caused by other viral infections, including canine distemper. Although hydrophobia is very indicative of this infection in humans, rabies can be confused with Guillain-Barré syndrome, poliomyelitis and encephalitis of other viral etiology (Plotkin, 2000) (see Chapter 7). Specific histopathologic changes, namely Negri bodies, in the central nervous system (CNS) can provide microscopic evidence of rabies virus infection, but only with a low degree of sensitivity. While the Negri body has been the hallmark of the pathology of rabies for the past 100 years, more definitive evidence of rabies virus infection can now, given recent technological developments, be demonstrated through detection of the entire virion, its proteins, or its genome RNA in tissue samples. This detection can be accomplished by direct visualization of virus particles, demonstration of viral proteins by visualization of reaction with labeled antibodies, cultivation of infectious virus or detection of viral RNA in tissue samples. Methods employed include electron microscopy (EM), direct fluorescent antibody (DFA) and indirect fluorescent antibody (IFA) tests, virus cultivation *in vivo* and *in vitro*, immunohistochemistry, enzyme immunoassay, molecular hybridization with labeled nucleic acid (genetic) probes and reverse transcription polymerase chain reaction (RT-PCR), including conventional, nested and real-time PCR techniques. Rabies virus infection can be inferred from indirect evidence, such as the demonstration of rabies virus-specific neutralizing antibody in the serum of an unvaccinated individual, or antibody in cerebrospinal fluid (CSF).

2 POST-MORTEM DIAGNOSIS OF RABIES IN ANIMALS

Due to the unique pathogenesis of rabies, it is impossible to detect infection with rabies virus during most of the usually long, but variable, incubation period. During this early phase of rabies development, neither the virus nor its antigens or RNA can be reliably identified, since their distribution in the host is unpredictable during this period. Also, there is generally no rise in titer of circulating antibodies until a week or more into the clinical phase (Crepin *et al.*, 1998). Consequently, the diagnosis of rabies can only be achieved with 100% certainty by the post-mortem examination of brain tissue, in which rabies virus replicates to high titer. Fortunately, the viral ascent to the brain is restricted to the nervous system in a strictly retrograde manner, from the peripheral site of exposure to the CNS (Charlton, 1988) (see Chapter 8). Only after its replication in the CNS does the rabies virus spread centrifugally to the salivary glands and other tissues. This pathogenic pattern permits reliable prediction of the ante-mortem presence of virus in the saliva of a biting animal. Because several replication cycles of the virus occur in the brain prior to clinical manifestations (Kaplan, 1985), detection of nascent viral antigen in the brain is possible prior to the onset of signs of the disease. Therefore, when an animal bites a human, the attacking animal can be quickly sacrificed and tested for the potential of rabies virus transmission. It is never necessary to delay euthanasia for the sake of further development of the disease, in order to achieve a reliable diagnosis.

Microscopic examination of brain tissue stained with the DFA has become the standard by which the value of all other rabies diagnostic tests or methods is measured and assessed. The DFA test is applied in a wide variety of other rabies-related laboratory functions, specifically: to indicate residual virus in the mouse inoculation and tissue culture virus isolation tests; for the examination for residual virus in infected cell monolayers in tissue culture neutralization tests for serum antibody; and in research applications such as the evaluation of the non-neural tropisms of rabies virus. We will cover this method in more detail and then discuss confirmatory methods and the advantages and disadvantages of other primary tests.

2.1 Public health laboratory testing

2.1.1 *Indications for testing*

Not all animals that bite or scratch a person need to be killed and tested for rabies. In countries with strong rabies control programs, rabies is relatively uncommon in domestic animals. The early signs and the clinical progression of rabies in companion animals have been described in numerous experimental studies and are easily recognized. Biting incidents involving healthy dogs, cats and ferrets are common, but have never been implicated in North America in a human death from rabies. The public health recommendation drawn from these observations and characteristics of the pathogenesis of rabies is that, in most circumstances, an apparently healthy dog, cat or ferret (Niezgoda *et al.*, 1999; Wandeler and Kumor, 2005) that has bitten a person can be confined and observed daily for 10 days. Unless the animal develops signs suggestive of rabies during this period, it need not be killed and tested for rabies, and no anti-rabies biologics need be administered to the person bitten. In areas where rabies is poorly controlled in domestic animals, anti-rabies treatment is often begun at the time of the bite exposure, but terminated if the biting animal remains healthy during the observation period.

A 10-day confinement is not sufficient or appropriate for animals showing signs of a neurologic illness; it is also not reliable for wild and exotic species, due to inadequate opportunity for observations of virus shedding. In these situations, if rabies is suspected, the animal should immediately be euthanized and promptly tested. For many situations, if testing can be arranged within 48 hours of the potential exposure, initiation of PEP can await laboratory results. However, if delay in testing is unavoidable, and also for high-risk exposures, PEP can be initiated but then discontinued if a reliable laboratory examination excludes a diagnosis of rabies. Delays in sample collection and shipping in some countries may result in an average time of 5 days or longer from death or euthanasia of the animal to available laboratory test results (H. Bourhy, personal communication).

Rabies control efforts can benefit from a specimen acceptance policy that recognizes the value of testing for a wide range of purposes. In addition to those animals that have potentially exposed a human or a domestic animal to the disease, consideration should be given to testing in the absence of reported contacts for:

1 bats captured after close indoor encounters that meet new, more aggressive treatment guidelines pertaining to situations in which a bat bite may be expected to go unrecognized (Debbie and Trimarchi, 1997; Centers for Disease Control and Prevention, 2000)

2 domestic animals that die or are euthanized with signs that would include rabies in the differential diagnosis

3 mammalian species not normally suspected of rabies infection (such as bear, deer, beaver) that present with signs of a neurologic disorder in areas affected by terrestrial rabies cycles

4 symptomatic rabies vector species captured outside the known geographic distribution of terrestrial rabies outbreaks

5 rabies vector species in areas in which enhanced surveillance is being conducted for the evaluation of the efficacy of wildlife rabies control strategies such as oral vaccination.

Combined data, generated by this broad range of rabies surveillance, are used to define the epizootiologic patterns that establish the appropriate animal bite management decisions when the biting animal is not available for observation or testing.

The known natural host range of rabies virus and all other lyssavirus genotypes is limited to mammals. It is therefore not necessary to test species of other classes such as insects, reptiles and birds. In North America, small rodents, including mice, rats, chipmunks, and squirrels, are essentially free of rabies infection, precluding the need for routine testing of these species (Childs *et al.*, 1997) (see Chapters 4 and 5). Larger rodents, such as woodchucks, muskrats and beaver, can be suspect, however, as are smaller rodents that display unusual behavior or are involved in unprovoked bites to humans.

2.1.2 Biosafety

Rabies virus is categorized as a Biosafety Level (BSL)-II pathogen in diagnostic settings in the USA (Centers for Disease Control and Prevention, 1999a) and in many other rabies-endemic countries. In certain research and vaccine-production settings, and for diagnostic samples with the additional suspicion of infection with a BSL-III agent, it may be elevated to BSL-III status. All activities related to the handling of animals and samples for rabies diagnosis should be performed following the appropriate standard guidelines and practices to avoid direct contact with potentially infectious fluids or tissues (Kaplan, 1996). Particular attention should be directed at avoiding percutaneous injuries from contaminated instruments, mucous membrane contamination with infectious fluids and the production of and exposure to aerosols of infectious materials associated with activities such as centrifugation. Use of power saws at necropsy is discouraged, unless restricted to use in a biosafety cabinet and fitted with a high efficiency particulate air (HEPA) filter vacuum attachment. All persons working in rabies diagnostic activities and those capturing, handling, or decapitating rabies-suspect animals should receive rabies pre-exposure immunization with regular serologic assay of antibody titer and booster injections as necessary (Centers for Disease Control and Prevention, 1999b).

2.1.3 Collection, preservation and submission of specimens

An animal can be euthanized for rabies testing by any humane method that does not damage the head, including barbiturate and non-barbiturate injectables or gases. The carcass should be refrigerated immediately following death to retard decomposition and autolysis of the brain. Because the animal species, site of exposure, variant of rabies virus and time and cause of death can all affect the terminal distribution of rabies virus in the brain of an infected animal, multiple areas of two to three regions of the brain must be examined to achieve reliable results. Consequently, the intact head of the animal constitutes the ideal specimen for most species. The entire body of a bat should be submitted to avoid risk of loss

of brain during decapitation of this very small animal and to facilitate identification of the bat species for important epizootiologic considerations. For large livestock, such as cattle and horses, shipping of the entire head to a diagnostic
laboratory/center poses special problems. For these animals, portions of the brainstem and cerebellum can be removed by the veterinary clinician through the
foramen magnum following decapitation at the occipitoaxial juncture (Debbie
and Trimarchi, 1992).

The specimen should be preserved by refrigeration during transport to the laboratory. Glycerol as a transport medium should be avoided because it has been
demonstrated to reduce immunofluorescence intensity even with subsequent
washing (Lennette *et al.*, 1965), particularly when used in conjunction with standard acetone fixation (Andrulonis and Debbie, 1976). In one study (Lembo *et al.*,
2006), glycerol preservation of field-collected specimens for up to 15 months has
been demonstrated to be a satisfactory method when used in conjunction with
the direct rapid immunohistochemical test (DRIT) for rabies. A single freezing
will not deleteriously affect the DFA test or virus isolation, but freezing can delay
diagnosis and exacerbate certain dissection problems. Repeated freeze–thaw
cycles, on the other hand, can seriously affect sensitivity of diagnostic procedures. While decomposition or mutilation of the CNS may affect diagnostic
potential, especially for reliable negative results, laboratory personnel cannot
judge suitability of the specimen based solely on a verbal description of the condition of the carcass. Therefore, unless it is clear that the carcass is in advanced
stages of decomposition or has been mutilated to the extent that there is no intact
CNS, important specimens should be submitted and an evaluation for suitability
will be made at the laboratory (Trimarchi and Briggs, 1999).

To achieve prompt results that will facilitate treatment and to avoid deterioration of tissues, specimen transport to the laboratory should be direct and
immediate. Samples shipped to the laboratory by commercial parcel carriers
(those authorized and participating in hazardous materials transport) must be
properly packaged to avoid exposure hazards and to meet the frequently
updated requirements of many regulators, including the US Department of
Transportation, the United States Postal System and the regulations of individual states or other countries. Unified requirements, following the guidelines of
the International Air Transport Association (47th IATA Dangerous Goods
Regulations, http://www.iata.org/ps/publications/9065.htm), are anticipated.
The present interpretation of these guidelines defines clinical specimens being
shipped to the laboratory for rabies diagnosis as requiring packaging consistent
with 'Biological substance, Category B', but specimens that are known to contain
materials with amplified virus (in animals or cell culture) must be packaged as
'Biological substance, Category A'. Specimens such as entire animals (e.g. bats
and small rodents) or heads of larger animals should be shipped under refrigeration if possible, because freezing may delay the examination or make dissection
and recognition of areas of the CNS more difficult in deteriorated specimens.

Several hard-frozen gel cold packs of appropriate size should be included to refrigerate the contents properly for several days. Wet ice should be avoided because, after thawing, water can leak and potentially cause contamination. If brain tissue is removed by the clinician prior to shipping, it first should be contained in a vial or other firm plastic container to protect it from being crushed or macerated during handling. An envelope containing a fully completed standard rabies specimen history form (if available) should be attached to the outside of the container. If no form is available, all significant information should be provided, including the names, addresses and telephone numbers of the owner, complainant and all humans and animals in contact. Information on the clinical observations, date of death or means of euthanasia, exact location of capture and the person or agency to receive the report should also be provided.

2.1.4 Laboratory reporting practices and emergency examinations

Public health rabies diagnostic laboratories typically operate, for routine examinations, on a traditional 5-day work week schedule. Examination of specimens arriving early on a work day is usually completed the same day and that of those arriving late in the day, or during weekends and holidays, is completed on the next regular work day. Emergency off-hour examinations can generally be arranged by contacting the laboratory or through the local public health agency's epidemiologic unit. Criteria justifying off-hour emergency diagnosis include animals strongly suspected of rabies virus infection that have actually bitten a person and for which a physician is awaiting test results before PEP.

Reports of rabies-positive specimens are generally made immediately to the physician or local public health authority by telephone, either directly from the laboratory or through the epidemiologic branch of the agency. Follow-up written reports of positive findings and routine negative reports are generally issued by mail or facsimile transmission or electronically by laboratory information systems (LIMS). Submitters should never assume negative findings because they do not receive a report; the specimen may not have been received at the laboratory or the specimen may have been unsatisfactory for reliable examination. Because reporting practices vary widely, the submitter should ascertain local practice.

2.2 Immunofluorescence on brain tissue

Immunoglobulin molecules that attach specifically to epitopes on target antigens can be labeled by covalent chemical attachment of a fluorescent tag to produce a fluorescent antibody conjugate. This conjugate is allowed to react with acetone-fixed specimen tissue prepared as smear, impression, section or cultured cell monolayer. During incubation for an appropriate period, the specific antigen–antibody reaction occurs, binding the specific antibody molecule with the attached fluorochrome tag to the desired protein. Washing of the specimen slides removes all conjugate components, except for specifically bound molecules. Examination of the slides is performed with a microscope utilizing illuminating light that is rich in wavelengths that stimulate the fluorochrome to emit radiation in the visible spectrum, permitting visualization and localization of the target protein in the tissue preparation. The DFA test may have no more valuable efficacious application in biomedical science than its utilization for the diagnosis of rabies in post-mortem brain samples of rabies-suspect animals.

2.2.1 Necropsy and dissection

After the intact animal head is received for examination, brain removal and dissection are required. The flesh and muscle are removed from the dorsal aspect of the skull and anterior and lateral cuts are made through the cranium with hammer and chisel, power saw or scissors and forceps, depending on the species, size and age of the specimen and the biocontainment facilities and equipment available. After the meninges are dissected away, the exposed brain can be removed intact. Alternatively, the dissection of areas to be examined can be performed as the exposed brain sits in the base of the skull. Another method has been described (Barrat, 1996; Willis et al., 2000) which employs a soda straw, pipette body or other hollow tube that is forced through the foramen magnum anteriorly through the brain. After extraction of the tube from the head, the core of tissue is extruded from the tube. A rapid, field-applicable method for surveillance in support of wildlife vaccination programs is performed by cutting down to the occipitoaxial junction from the base of the throat, exposing the spinal cord and removing a slice of the cord for examination. Priorities in the necropsy procedure include biocontainment and technician safety, proper accessioning and identification of each specimen and strict avoidance of potential cross-contamination of specimens (Smith et al., 1996b).

Full cross-sections of the brainstem, such as the medulla and pons, provide the most valuable sample for the demonstration of rabies virus infection in most specimens, as confirmed in recent studies (Bingham and van der Merwe, 2002). This is not surprising, given the general model of the virus's route to the CNS from sites of peripheral exposure (see Chapter 7). While it has been suggested that examination of the brainstem alone is adequate (Lee and Becker, 1973; Ito et al., 1985), others have demonstrated the added value of examination of more than one area of the brain (Robinson and DiSalvo, 1980; Maserang and Leffingwell, 1981; Trimarchi et al., 1986). The cerebellum may be ranked next in diagnostic value. The hippocampus, once a prime rabies diagnostic sample area because of its role in the demonstration of Negri bodies in the histologic examination for the disease, may be of limited additional value when brainstem and cerebellum can be examined. Sampling of more than one area of the cerebellum and brainstem may further improve the sensitivity and reliability of the diagnostic method in animals that were killed just before onset of rabies or very early during the clinical phase.

2.2.2   Fluorescent conjugate selection, preparation and evaluation

Since rabies virus nucleoprotein (N) is present in infected cells as microscopically recognizable, discrete intracytoplasmic inclusions, antibody preparations specific for the viral N are used for diagnostic reagent production (Flamand *et al.*, 1980). The fluorochrome most frequently used in rabies immunofluorescence testing is fluorescein isothiocyanate (FITC), selected because it is stable and produces a characteristic bright apple-green fluorescence, which is unlike most autofluorescences produced by animal tissues and cells (Kissling, 1975). The conjugate may also contain a counterstain, such as Evans blue, for background staining of cellular structures (Rudd and Trimarchi, 1997).

The source of the immunoglobulin molecules for immunofluorescence reagents can be sera of hyperimmunized animals or monoclonal antibodies. Hamsters, rabbits or horses hyperimmunized with purified and concentrated 'fixed' laboratory or vaccine strains of rabies virus (Schneider, 1973; Trimarchi and Debbie, 1974) have most often been used for immune serum reagents. Anti-rabies N protein-specific antibodies with the desired characteristics of high specificity, fluorochrome labeling potential and pan-specific reaction with both rabies viruses (genotype 1) and other lyssavirus species (genotypes 2 to 7) (see Chapter 3) have been produced as monoclonal reagents (Wiktor *et al.*, 1980; Bourhy *et al.*, 1992; Fraser *et al.*, 1996). Conjugates produced from cocktails (mixtures) of two or more labeled monoclonal antibodies recognizing different epitopes on the N provide highly specific reagents with uniform staining reactions and can be far more consistent from lot to lot than antisera-based reagents. While immune-serum antibody reagents can contain extraneous antibodies and be potentially cross-reactive, monoclonal antibody reagents risk over-specificity and weak affinity for some virus variants (Smith, 2000). It is beneficial in the interpretation of atypical immunofluorescence patterns to have two diagnostic reagents available for comparison and verification of findings.

The determination of the optimal working dilution of the diagnostic reagent is critical to the sensitivity and specificity of the test as well as to economical conjugate use. If the reagent is too concentrated in routine tests, background fluorescence may be too bright, deleteriously affecting contrast of specific stained inclusions and any non-specific staining problems will be exacerbated. When the conjugate is too dilute, specific staining intensity may be weak, making specific staining difficult to recognize, particularly on suboptimal specimens due to death early in the clinical phase of the disease, tissue decomposition, or repeated freeze–thaw cycles of the sample. The highest dilution that provides very bright green immunofluorescence and the full range of characteristic staining patterns is considered the endpoint dilution. In routine testing, the conjugate is used one dilution more concentrated, to provide a margin of precaution in the direction of optimal sensitivity. Rabies-positive control slides are used to determine a conjugate's working dilution and must be identical in type and preparation to those used in routine specimen testing. Because there is some variability in the affinity of commercial reagents to the many variants of naturally occurring 'street' rabies virus, this important titration should be performed in duplicate on at least two different rabies virus variants.

2.2.3 Immunofluorescence test protocol

While protocols performed in many laboratories around the world may vary to some degree, common practices can be described (see Section 2.3). Touch impressions or slip smears from each area of brain sampled are made on clean glass microscope slides. Multiwell Teflon-coated slides are commonly used to restrict reagents to tissue areas during staining. If other methods are employed, caution must be used to avoid cross-contamination (Wong *et al.*, 2004). For research applications and examination of non-neural tissues, frozen sections are prepared using a cryostat. Duplicate sets of slides are prepared and stored frozen for potential repeat tests on identical slides to evaluate perplexing or uncharacteristic staining patterns. At the same time as the slides are prepared for immunofluorescence examination, a brain suspension can be made in appropriate diluents from the same areas of brain tissue for isolation of virus by *in vivo* or *in vitro* virus cultivation or for RT-PCR for confirmatory testing. Unused tissue samples from diagnostic areas of the brain of each specimen are held at less than $-70°C$ for additional analysis.

After air drying, the slides are fixed for a period ranging from one hour to overnight in acetone at $-20°C$. Fixation enhances permeability of the cells to labeled antibodies and improves tissue adhesion to the slide. A European laboratory has employed a 30-minute fixation period on more than 60 000 specimens and 6000 positives with success (H. Bourhy, personal communication). Very short fixation periods and a microwave fixation procedure, as an alternative to acetone (Davis *et al.*, 1997), have been described but are not generally applied. After removal from acetone, the slides are thoroughly air dried. The diagnostic reagent at working dilution is applied drop-wise to each slide, which is then incubated in a moist chamber at 37°C for 30 minutes. The slides then receive multiple washes in buffered saline and are air dried. Each slide must be carefully identified. Application of the staining reagent, fixation and washing steps all are performed avoiding communal baths and slide-to-slide contact to prevent any potential for contamination between specimens. Coverslips are mounted with a buffered glycerol mountant. Selection of mountant buffer, pH, glycerol concentration and other components is critical to persistence and intensity of immunofluorescence in the preparations (Smith *et al.*, 1986; Rudd *et al.*, 2005). Slides should be read promptly, or kept at 4°C if reading is to be delayed for more than two hours; they should be stored at $-20°C$ or colder if reading will be delayed overnight or longer.

The slides are examined on a standard, incident light fluorescence microscope fitted with a xenon or mercury vapor lamp. Excitation and barrier filter systems are selected for excitation in the blue or ultraviolet spectrum. The performance of illumination-collecting lenses and mirrors, fluorescence objective lenses and filter combinations of the recent-generation fluorescence microscopes dramatically improves the brightness and contrast over the characteristics of earlier models. Specific staining appears as a bright apple-green fluorescence against a dark background. Examination is performed at ×150 to ×450 magnification and about 25 to 50 fields are examined for each slide. Ideally, more than one slide is evaluated for each tissue and all slides from specimens representing possible human exposure to rabies are read by two experienced microscopists.

2.2.4 Interpretation of results and tests for specificity

The recruitment and retention of properly trained, experienced microscopists are essential to the interpretation of slides and for maintenance of the highest standards of sensitivity and specificity of the test (Trimarchi, 2000). Characteristic apple-green FITC color and bright intensity identify foci of antibody attachment. The typical morphologic structures of stained rabies antigen aid in the recognition of specific staining. Intracytoplasmic inclusions (composed of N) typically possess a three-dimensional round or oval shape, with smooth margins and a brighter emission at the periphery (Figure 10.1A). In smears or touch impressions, labeled rabies virus N antigen also appears as fine particles and occasionally in long strands, possibly the result of the slide preparation process. Inclusions are seen in large numbers in the cytoplasm of cell bodies and dendrites of neurons, particularly in the Purkinje cells of the cerebellum and the large neurons of the brainstem. These round or oval inclusions can range in diameter from 2 to 10 μm. Rabies-specific antigen may occasionally be present in glial and other non-neuronal cells of the brain (Jackson et al., 2000). Inclusions and particulate fluorescence are generally widespread throughout the CNS of infected animals, beginning up to several days before the onset of clinical signs of rabies and before infection of salivary glands is evident by the same method. Occasionally, distribution of antigen is limited, appearing sporadically in some areas of the CNS, especially when the animal is euthanized prior to or very early in the clinical phase. This situation occurs more often in large livestock, particularly in horses (Smith, 1995).

Rabies virus antigen-specific staining generally can be easily distinguished from the yellow, white and gold autofluorescences of tissue constituents based upon the very characteristic color of FITC emission. Granular non-specific staining can result from drying of the conjugate on the slide during the staining process or from precipitation of aggregates of free FITC in improperly reconstituted, stored or clarified reagents. Non-specific staining can also be the result of the adherence of labeled IgG elements in the conjugate to Fc receptors on contaminating Gram-positive cocci in the sample or on immune cells infiltrating the brain of the animal

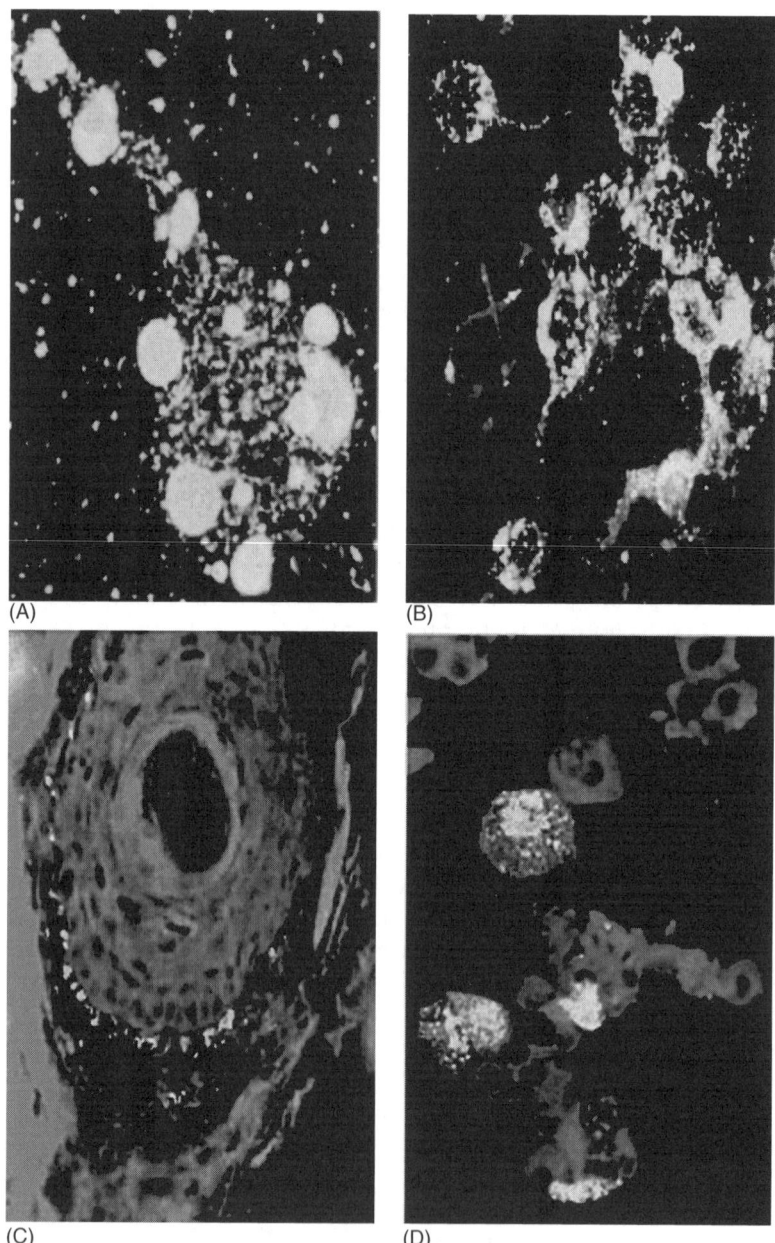

Figure 10.1 (A) Purkinje cell from the cerebellum of a rabies virus-infected bovine showing large intracytoplasmic inclusions and smaller particulate antigen. The DFA method (slip smear; ×540 magnification). (B) Rabies virus-infected monolayer of murine neuroblastoma cells (DFA method with Evans blue counterstain; ×250 magnification). (C) Human hair follicle from nuchal skin biopsy. Nerve cells surrounding follicle are revealed by specific fluorescence associated with presence of rabies antigen (DFA method on frozen section, Evans blue counterstain; ×250 magnification). (D) Human corneal impression with several infected epithelial cells containing inclusions of specifically labeled rabies antigen (DFA method with Evans blue counterstain; ×360 magnification). This figure is reproduced in the color plate section.

in response to an infection other than rabies (e.g. distemper). Diffuse background staining can be the result of an inappropriate conjugate working dilution or an overly high fluorochrome/protein ratio in the reagent.

Unexpected outcomes, unusual antigen distributions and non-characteristic morphology of staining patterns require validation of the specificity of the observed reaction. This can be accomplished when using antisera-based reagents by repeating the staining procedure with an aliquot of the conjugate that has been absorbed with rabies virus to remove all rabies-specific labeled IgG. All fluorescence appearing with this preparation is due to non-specific factors. This procedure is not applicable to conjugates made with monoclonal antibodies, because these reagents contain only antibodies directed to viral proteins. Non-specific staining with monoclonal antibody reagents can be recognized as non-specific if similar staining appears when test slides are stained with a rabies-negative control conjugate made with labeled monoclonal antibodies of the same isotype and at the same concentration as the rabies reagent, but specific for a different agent. The cause of repeated non-specific staining problems should be ascertained as soon as possible and eliminated, since it could mask the few specifically stained areas of a very weakly positive specimen.

A major constraint on the sensitivity of the rabies DFA method is the condition of the brain sample (Lewis and Thacker, 1974). Decomposition and denaturation by heat, repeated freeze–thaw cycles, or exposure to chemicals can contribute to reduced sensitivity. Decomposition will affect the sensitivity of all rabies diagnostic procedures. Immunofluorescence tests may remain positive for a period after virus isolation is no longer possible (Rudd and Trimarchi, 1989). A characteristic positive DFA test result on decomposed or mutilated tissue fragments may support a valid rabies-positive report, confirmed with the appropriate test for specificity or by virus isolation, immunohistochemistry or molecular methods. One of the most difficult decisions confronting the public health rabies laboratory is determining when a reliable negative result can be issued on a specimen received in a partially decomposed condition or otherwise compromised sample, in which no evidence of immunofluorescence is observed. Certainly, once decomposition has advanced to the point that the tissue is in a foul smelling, discolored or liquefied condition due to putrefaction, negative results are invalid. When mutilation or submission of inappropriate samples precludes proper examination of recognizable samples of brainstem and either cerebellum or hippocampus, negative results may not be reliable.

2.3 Quality assurance and quality control in the rabies DFA test

Despite the importance of test results for the medical management of human exposures, post-mortem examination of animals is not categorized as human clinical laboratory testing. Therefore, it has not been subject to the regulatory requirements for stringent adherence to established standard methods by acts such as the United States' Clinical Laboratory Improvement Amendments of 1988. Comparison of methods employed reveals many exceptions to accepted practices (Smith, 1995). Observations of the critical importance of seemingly minor test components, such as composition of coverslip mounting medium to immunofluorescence persistence (Smith *et al.*, 1986), emphasize the value of the recommendation that uniform procedures be adopted (Hanlon *et al.*, 1999). As a result of that recommendation, a committee of reference diagnosticians was established in the USA to develop the 'National Standard Rabies Diagnostic Protocol: A Minimum Standard for Rabies Diagnosis in the United States', which currently includes the Protocol for Post-mortem Diagnosis of Rabies in Animals by Direct Fluorescent Antibody Testing. The protocol is available at the CDC Website (http://www.cdc.gov/ncidod/dvrd/rabies/professional/publications/DFA_diagnosis/DFA_protocol-b.htm). Strict adherence to the protocol assures use of published, properly validated methods. Furthermore, the resulting uniformity of methods permits rapid evaluation of diagnostic problems arising from changes in diagnostic reagents, rabies virus variant introductions or evolutionary genetic drift. For example, a recent observation with one of the commercial rabies-specific conjugates has revealed that high concentrations of glycerol and low pH in coverslip mounting media can contribute to very rapid disappearance of specific fluorescence with certain combinations of reagent and virus variant (Chemicon International, 2000; Rudd *et al.*, 2005).

Voluntary rabies proficiency testing programs were conducted in the USA in 1973 and 1992 and have been run annually since 1994. The performance of diagnostic laboratories enrolled in the recent programs has, mainly, been excellent in the evaluation of positive and negative test slides using the DFA method. Not surprisingly, consensus has been best on strongly positive and negative slides; discrepancies have occurred primarily with very weakly positive specimens (Powell, 1997).

Important factors in the avoidance of false-negative results in rabies diagnosis include strict adherence to a uniform methodology, the use of proper scientific controls for all procedures, diagnostic conjugate quality and proper dilution and optimization of the microscope lamp and instrument performance. False-positive results are avoided through implementation of procedures that reduce the possibility of cross-contamination of negative samples with strong positive samples and through optimization of conjugate diluents and staining and washing conditions to eliminate non-specific fluorescence. Every DFA test on a public health sample requires scientific controls by performance of identical procedures on known rabies-positive and rabies-negative slides. Acceptance of tests is dependent upon positive control slides with consistently bright, characteristically distributed immunofluorescence and the expected range of size and morphology of inclusions. Negative controls must display an absence of fluorescence that mimics or could obscure specific fluorescence. Whenever unexpected outcomes are observed, the findings should be confirmed by the use of the DFA test with another rabies reagent, by another test, or by a corroborative test performed at a regional or national reference laboratory. Unexpected outcomes include sporadic antigen distribution in brain tissues or rabies-positive results on a specimen from an animal unlikely to have been rabid due to species type (small rodent), location of capture (rabies-free area) or history (vaccinated and no clinical signs of rabies). Good quality control of the DFA test requires that unused brain material, taken from internal portions of areas tested, be held in reserve for additional testing (Smith, 2000). The laboratory can use a selected application of tests to confirm specificity of observed fluorescence. With antiserum-based conjugates, paired staining with virus-absorbed and sham-absorbed conjugates still has some application. With the increased dependence on monoclonal antibody-based conjugates, this test for specificity has been replaced by use of rabies-negative control conjugates composed of FITC-labeled IgG conjugates specific for another antigen but containing no rabies-specific antibody. It bears repeating that practice has demonstrated that microscopists who are experienced in examination of rabies DFA slides are essential to the proper interpretation of these tests.

Rabies DFA is comparable in sensitivity to *in vivo* and *in vitro* virus isolation methods. There have been no recorded human rabies cases in North America arising from bites of animals that were diagnosed negative for rabies by DFA. Despite the test's proven reliability, the consequences of a false report for decisions of post-exposure management are so unacceptable that some confirmatory tests are required, both to validate individual negative reports and to maintain confidence in the procedure. This can be accomplished in the public health laboratories where DFA tests are performed, by an alternative back-up test performed on each sample or on selected samples, such as highly suspect human exposure cases. Alternatively, confirmation of DFA results can be arranged by collaboration with a national or regional reference laboratory.

Constant vigilance in the laboratory is required to avoid cross-contamination, especially during rabies epizootics that result in extraordinarily large numbers of rabid animals being processed within the laboratory. An active state health department laboratory identified 2745 rabid animals among a total of 11 896 animals processed and examined during the course of one year (Trimarchi, 1994). Special steps during necropsy must be taken to assure that instruments, examination gloves, work surfaces and all other materials that come into contact with the specimen are either disposable or replaced after each specimen, or are thoroughly disinfected by boiling or autoclaving and washed after each use. Great care must also be taken during the slide fixation, staining, washing and cover-slip mounting processes to avoid contamination from positive controls or from specimen to specimen (Smith *et al.*, 1996a). The standards required to avoid a false-positive outcome, as a result of cross-contamination, are even stricter when molecular methods are employed (see Section 4.1).

2.4 Other methods for detection of viral proteins in brain tissue

Although the speed and reliability of the DFA test have made it the stalwart of rabies diagnosis, other antigen detection methods are also applied and some of these have advantages for certain testing applications. Tests that can be applied to formalin-fixed tissues benefit from freedom from sample preservation concerns and risks associated with transporting and processing samples containing infectious virus. A potential limitation of the procedure to work with formalin-fixed preparations is the inability to cultivate and amplify the virus from an inactivated sample. Immunofluorescence methods applied to formalin-fixed tissues (Figure 10.2) and immunohistochemical methods, such as immunoperoxidase staining of formalin-fixed, paraffin-embedded sections, were previously significantly less sensitive than the DFA method on fresh brain tissue. With recent modifications to achieve better immunofluorescence (Warner *et al.*, 1997) and immunohistochemistry on brain sections (Hamir *et al.*, 1995) and brain impressions (Niezgoda and Rupprecht, 1999), these procedures may now be approaching comparable sensitivity to DFA on fresh tissue. The methods have benefited from four major improvements:

1 digestion of the tissue sections with an enzyme such as proteinase K prior to staining, to expose antigenic sites that formerly were masked by bonds resulting from fixation

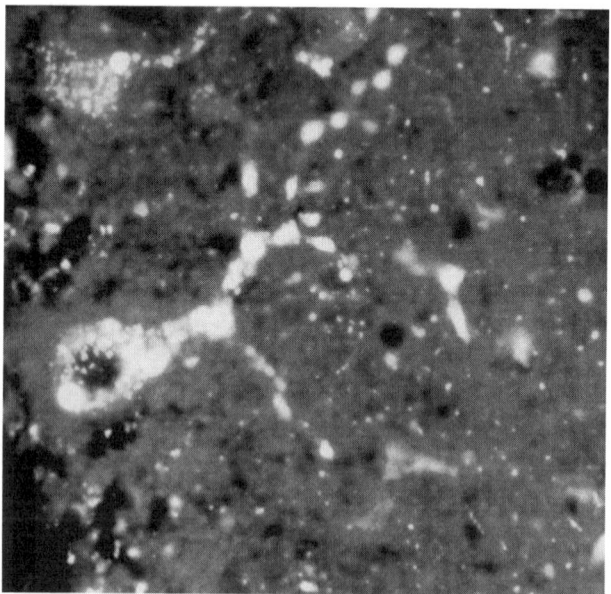

Figure 10.2 Raccoon rabies virus variant infected raccoon cerebellum. Indirect immunofluorescence on formalin-fixed, paraffin embedded section. Two Purkinje cells are shown, one with large intracytoplasmic inclusions, ×540. Primary mouse monoclonal antibody kindly provided by Dr Alex Wandeler (CFIA, Nepean, Ontario, Canada). This figure is reproduced in the color plate section.

2 selection of particularly well-suited monoclonal antibodies as the primary antibody in IFA tests, to improve sensitivity and specificity

3 conversion to direct procedures that use labeled primary antibodies

4 avidin–biotin amplification in the staining process, to increase the signal.

A comparison of the DFA on formalin-fixed tissue with DFA performed on fresh tissue in one laboratory demonstrated agreement approaching 100% (Whitfield *et al.*, 2001). Duplication of this finding by robust evaluation and validation of negative test results is desirable in laboratories considering its use, to form the basis for making decisions to withhold rabies PEP for potentially exposed humans. While these methods can be used for confirmatory testing to back up the DFA test, they are not widely used in public health laboratories for primary diagnosis. Delays in shipping to the few laboratories that perform these tests, and in the processing of the samples, currently limit their application in decisions of post-exposure management. A direct rapid immunohistochemical test (DRIT) (Figure 10.3), employing a short formalin fixation of fresh or glycerol-preserved brain impressions and requiring no specialized equipment, has been demonstrated to be of utility for testing conducted under field conditions and for countries with limited diagnostic resources (Lembo *et al.*, 2006).

Enzyme-linked immunosorbent assays (ELISA) are showing promise of improved sensitivity as a result of avidin–biotin amplification; such assays are employed for rapid and simple diagnosis of other viral infections. A similar method has been developed for the detection of rabies antigen (Bourhy and Perrin, 1996).

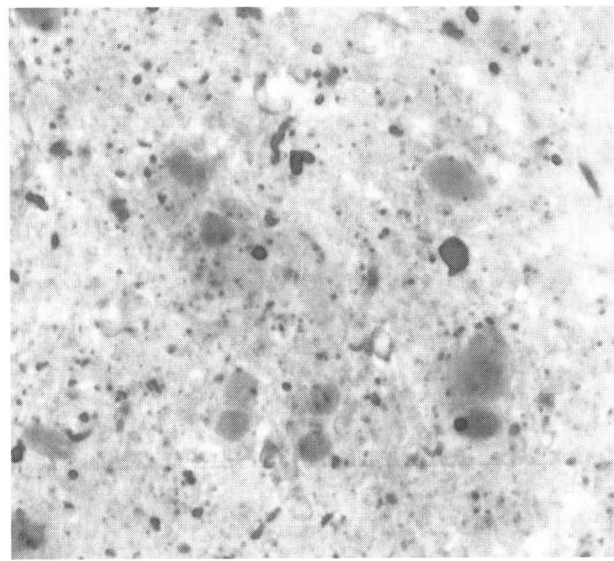

Figure 10.3 Raccoon rabies virus variant infected raccoon brain. Direct Rapid Immunohistochemical Test (DRIT). Gill's hematoxylin formulation 2 counterstain, ×630. Photo kindly provided by Michael Niezgoda, CDC Rabies Unit, Atlanta GA. This figure is reproduced in the color plate section.

The method offers the benefit of sensitivity when applied to poorly preserved specimens and manual or automated reading, which make it well suited for use in field
conditions, but sensitivity on well-preserved specimens may be low in comparison
with other methods (Franka *et al.*, 2004). An avidin–biotin amplified dot–blot
enzyme immunoassay showed good sensitivity and specificity in one study on
brain tissue of rabies-suspect animals, but not on clinical samples such as saliva for
ante-mortem diagnosis of human rabies (Madhusudana *et al.*, 2004).

2.5 Histologic examination

Historically, the microscopic examination of histologic preparations was the primary means of identifying evidence of rabies infection in post-mortem samples
from animals and humans. Fresh brain smears or microtome-cut sections of
formalin-fixed, paraffin-embedded tissue were stained with combinations of basic
fuchsin and methylene blue (Tierkel and Atanasiu, 1996) or with hematoxylin
and eosin (Lepine and Atanasiu, 1996). Histopathologic evidence of encephalitis
includes signs of inflammatory response, such as perivascular cuffing and cellular
infiltrations (see Chapter 8). The presence of acidophilic intracytoplasmic inclusions, called Negri bodies, found prominently in the Purkinje cells of the cerebellum and the pyramidal cells of the hippocampus, is virtually pathognomonic for
the disease. When reported by an experienced pathologist, this provides a reliable
diagnosis of the disease. Negri bodies detected during routine post-mortem examinations of tissue following deaths attributed to encephalitis of unknown etiology,
continue to disclose or raise suspicion of rabies in occasional human cases that
were not suspected ante-mortem or at the time of death (Centers for Disease
Control and Prevention, 1993; Silverstein *et al.*, 2003). The presence, distribution
and size of Negri bodies are related to the species of animal, the variant of the
rabies virus and the duration of the clinical period prior to death. The sensitivity
and reliability of the method are poor, with numerous surveys indicating that 25%
or more of animals have no demonstrable Negri bodies (Perl and Good, 1991).
The method is therefore of limited value for public health purposes. Quantitative
analysis (Jackson *et al.*, 2001), comparing extent of rabies virus infection disclosed
by histologic staining of Negri bodies with amount of antigenic staining by an
immunoperoxidase method, demonstrated that many neurons with demonstrable rabies virus antigen did not have Negri bodies.

2.6 Electron microscopy

Intact rabies virions, viral components and cellular responses can be demonstrated
by electron microscopy (Hummeler and Atanasiu, 1996) in CNS and other tissues
of infected animals and humans, as well as in cell culture. While this technology
has played a key role in studies of the structure and morphogenesis of the virus,

only rarely is this a practical alternative for diagnostic evaluation due to specialized equipment requirements, turnaround time considerations and the enormous number of observations that might be required for a negative determination.

3 VIRUS ISOLATION

The most common confirmatory tests are cultivation of virus by inoculation of animals or cell culture. An asset of either isolation procedure is the availability of cultivated virus for further propagation and characterization by antigenic or genetic analysis (see Chapter 3). The mouse inoculation test (MIT) (Webster and Dawson, 1935) is a sensitive and reliable procedure (Sureau *et al.*, 1991). Small pieces of each area of the brain examined by DFA are combined into one homogenized suspension, made with mortar and pestle or tissue grinders, in a buffered saline diluent containing protein stabilizer and antibiotics. Five weanling laboratory mice are inoculated intracerebrally for each specimen and they are observed for signs of rabies for 30 days. Mice that develop signs of illness are immediately sacrificed and tested by the DFA method. Favorable attributes of the MIT as a back-up test procedure are its applicability to partially decomposed specimens, its high sensitivity in weakly positive specimens and its technical simplicity. Its main drawback, besides the inherent environmental and ethical issues associated with the use of live animals in the laboratory, is the typical 7- to 20-day interval between inoculation and onset of observable signs of infection. The period can be shortened by inoculating families of neonatal mice and sacrificing individual neonates daily, beginning 5 days post-inoculation and examining their brains by DFA. However, this technique requires a larger number of mice per sample and increases the labor-intensive nature of the MIT.

The unpredictable and problematic delay associated with *in vivo* virus isolation (i.e. MIT) can be greatly reduced by inoculation and detection of virus in continuous cell culture. Cell culture medium is used as the diluent for the tissue suspension preparation. After clarification of the suspension preparation by light centrifugation, the sample is inoculated onto cell monolayers or added to cells in suspension. A murine neuroblastoma cell line that is susceptible to rabies virus infection is generally selected (Umoh and Blenden, 1983; Rudd and Trimarchi, 1987). Tissue culture flasks or 96-well plates are seeded with host cells and the cells are incubated for one to several days before they are examined by DFA for evidence of rabies virus infection. Such evidence is the appearance of intracytoplasmic inclusions in clusters of infected cells (fluorescent foci) in the monolayer (see Figure 10.1B). Sensitivity, which can be enhanced by the addition of DEAE-dextran to the cell culture medium (Kaplan *et al.*, 1967), rivals that of the IFA test and MIT (Rudd and Trimarchi, 1989; Webster and Casey, 1996). With results available in a fixed period of only a few days down to as short as 18 hours (Bourhy *et al.*, 1989), cell culture isolation has a much greater practical value than the MIT for prompt initiation of PEP, in the event of a weakly positive specimen that was not detected by the original DFA test (Rudd and Trimarchi, 1980).

A plaque assay based on the cytopathogenic effect of numerous rabies virus variants in monolayers of chick embryo related cell line (CER) was shown in one study to rival the sensitivity of the MIT and tissue culture inoculation test (TCIT) (Cardoso and Pilz, 2003). The application of flow cytometry for detection of vaccine strains of rabies virus, and particularly for study of the kinetics of infection, was reported in one study to be rapid, sensitive and reliable (Bordignon *et al.*, 2002). A fluorometric method employing a cell-enzyme linked immunosorbent assay technique has been used to detect and quantify intracellular rabies virus antigens in medulloblastoma cell culture (Rincon *et al.*, 2005). A rapid avidin–biotin amplified dot–blot enzyme immunoassay has been demonstrated to be sensitive and specific in one study on post-mortem brain samples of humans and animals (Madhusudana *et al.*, 2004).

4 USE OF MOLECULAR METHODS TO DETECT VIRAL RNA

4.1 Advantages and disadvantages of molecular methods

Many diagnostic methods in virology, particularly those developed relatively recently, target the nucleic acids of the infectious agent. These molecular methods either use a nucleic acid probe, an amplification strategy, or a combination of the two approaches, to detect a particular nucleotide sequence target. For many infectious diseases, such molecular methods of diagnosis are rapidly supplanting the more traditional techniques that are directed at detection of a viral protein. Due to the difference in the nature of the target molecule between the two types of assay, differences in their performance need to be considered.

In the case of lyssaviruses, including rabies virus, variation at the level of the nucleic acid genome is significantly greater than at the protein level, due to genetic code redundancy (Bourhy *et al.*, 1993; Kuzmin *et al.*, 2005). Consequently, failure to detect a virus present in a sample (false-negative result) is potentially a larger problem when using molecular methods, which usually rely on the hybridization of relatively short segments of nucleic acid (oligonucleotides) to the RNA target, than when using the antibody–antigen binding strategies that form the basis of serological detection methods, such as the DFA. Since the epitope detected by an antibody is more likely to be conserved than is a particular nucleotide sequence, serological methods have an advantage when a broadly reactive test capable of detecting a wide range of lyssaviruses is required. Moreover, molecular methods are generally more time consuming and costly than the DFA. In their favor, molecular methods have been reported to be highly sensitive, but this is sometimes a two-edged sword due to the potential for false-positive results as a result of the sample cross-contamination that can easily arise when PCR methods are applied without rigorous attention to the operational requirements for such assays (Kwok and Higuchi, 1989). Consequently, routine rabies diagnosis using fresh brain tissue, usually made on animals collected in the terminal stages of disease when levels of viral antigen in the brain are high, still continues to be performed by the DFA, which remains the WHO-recommended gold standard post-mortem diagnostic test for rabies (World Health Organization, 2004). However, there are situations where DFA performance is less than optimal and where molecular methods can, if applied carefully and correctly, provide either a confirmatory or alternative diagnostic capability.

DFA procedure rapidly loses sensitivity when applied to brain tissue that is substantially decomposed: controlled observations have shown that, in such a situation, a molecular method of detection can be greatly more sensitive (Heaton *et al.*, 1997). Also, particularly for the ante-mortem diagnosis in humans (see Section 5), samples appropriate for DFA testing may not be available but nucleic-acid-based methods can detect viral RNA in fluids such as saliva and CSF as well as in skin biopsy material (Crepin *et al.*, 1998; Heaton *et al.*, 1999; Elmgren *et al.*, 2002; Smith *et al.*, 2003; Hemachudha and Wacharapluesadee, 2004). Moreover, in some jurisdictions when an animal has had human contact but is scored as rabies DFA negative, such a result must be confirmed with an alternative test; molecular methods, since they can be completed more quickly than either MIT or TCIT, and sometimes with superior sensitivity (Picard-Meyer *et al.*, 2004), are potentially useful for the routine confirmation of DFA results. Genetic tests can also be used as an adjunct to the DFA, when unexpected or unusual fluorescent staining patterns are observed and confirmation of virus presence is required; sometimes a combination of tests is required to reach a consensus on the disposition of a particular case (McColl *et al.*, 1993). The issue of discordant results between many of these tests can be problematic, when RT-PCR is the only one of the methods that suggests presence of rabies virus. Distinguishing between a false-positive RT-PCR result and a false-negative result for the other assays in such a situation may be a highly complex process, with no clear-cut resolution. The best recourse is to avoid such situations wherever possible, by adherence to the guidelines recommended for the performance of PCR (Cooper and Poinar, 2000). These include use of physically separate areas for performing different parts of the process and use of dedicated pipettors in each of these areas.

4.2 Direct detection of lyssavirus RNA by hybridization with nucleic acid probes

Some of the earliest applications of molecular methods to the detection of viral RNAs involved the use of labeled DNA or RNA probes that were applied to tissue sections and allowed to hybridize *in situ*, in a process that drew extensively from standard immunohistochemical methods. Most of the studies employing these hybridization techniques aimed to elucidate mechanisms of pathogenesis and were not typically used for diagnosis (Jackson and Wunner, 1991). Alternatively, a dot–blot procedure, which employs hybridization methods to detect rabies virus RNA after application to a membrane, was also described (Ermine *et al.*, 1988).

In the development of such assays, consideration must be given to the precise nature of the target sequence; thus probes can be used that are based on either positive- or negative-sense sequence for detection of sequences of the opposite sense, i.e. either negative-sense genomic RNA or positive-sense intermediate RNA/mRNAs, respectively. Although this would not appear to be important from a diagnostic point of view, Nadin-Davis *et al.* (2003) reported an increased sensitivity for *in situ* hybridization when positive-sense sequences were targeted, perhaps due to an accessibility problem when trying to detect genomic RNA, particularly in fixed tissues. While commercial kits for labeling of probes are available, there is no commercial source for rabies virus-specific probes. Accordingly, probes must be developed in-house, with appropriate attention paid to the impact of lyssavirus diversity on the ability of a probe to anneal with its target sequence.

Although not performed routinely, an *in situ* hybridization method that uses a probe to detect the rabies virus N gene transcript has been developed for application to formalin-fixed tissues received by the Rabies Centre of Expertise, CFIA, Canada (Nadin-Davis *et al.*, 2003). This method is occasionally performed to confirm results of immunohistochemical investigations on fixed samples, when direct comparison of the distribution of viral RNAs with viral antigen could facilitate correct interpretation of the observed staining patterns. Refinement of this technique also enables typing of any virus that is detected by using strain-specific probes targeting divergent P gene sequences. Warner *et al.* (1999) also reported on the utility of *in situ* hybridization as a confirmatory test for DFA analysis of fixed tissues. However, both *in situ* and dot–blot hybridization methods are quite labor-intensive, due to the many incubation periods and blocking steps needed; given this limitation, as well as the initial time required for preparation of either tissue sections or the dot–blot itself, these procedures are not suitable for routine application.

4.3 Detection of lyssavirus RNA by reverse transcription polymerase chain reaction

Due to their immense versatility and sensitivity, diagnostic methods based on PCR technology (Saiki, 1989) have been extensively applied to many infectious diseases; lyssaviral diseases are no exception to this trend. The principle of the PCR is that two synthetic oligonucleotides that can hybridize to opposite strands of a dsDNA target are oriented in such a way that when they prime new DNA synthesis, the newly created DNA strands overlap in sequence. The reaction, catalyzed by a thermostable DNA polymerase, requires repeated thermocycling: first, high heat (95°C) to denature the DNA template into single strands; next, a lower temperature (usually 45–60°C), to allow annealing of the oligonucleotides to their target sequences; and finally, an incubation (usually 72°C), to allow the annealed oligonucleotides to prime DNA synthesis using deoxynucleotide triphospate (dNTP) substrates. This cycling is repeated for 25–40 iterations and a successful PCR produces a dsDNA product (amplicon) of specific length, defined at its two ends by the primers used in the reaction. The amplification of this product, usually by more than 100 000-fold, is the basis for the assay's exquisite sensitivity. Since the technique's original description, refinement now allows PCR products of up to several Kb to be generated from good-quality DNA template. However, for this technique to be applied to lyssaviruses, the viral RNA must first be converted to a complementary DNA (cDNA) strand, with the enzyme reverse transcriptase. This reverse transcription (RT) is often the limiting step in the amplification of RNA virus sequences and it usually precludes amplification of segments longer than 2–2.5 Kb in a single reaction. The generation of a significant quantity of DNA corresponding to a particular segment of the viral genome not only provides a highly sensitive diagnostic test, but also readily allows precise nucleotide sequence determination of the product. The analysis and comparison of such data for many lyssaviruses have significantly improved knowledge of the epizootiologic nature of rabies in many regions of the world (see Chapter 3).

Once nucleotide sequence information is available for a particular nucleic acid, primers of appropriate sequence can be readily designed so as to amplify virtually any segment of that nucleic acid. This capability provides tremendous versatility in the use of PCR for a huge variety of diagnostic applications. However, design of PCRs for RNA viruses, such as lyssaviruses, becomes somewhat complicated, due to the extent of the genetic variation between genotypes and strains. To develop a robust, broadly cross-reactive assay, very careful primer design is needed; this will ensure that the primer's targets are sequences that are well conserved throughout the genus.

Early RT-PCR methods for lyssavirus amplification were described by Sacramento et al., (1991) and subsequently more detailed protocols have been reported. These use a combination of commercially available reagents suitable for general RT-PCR applications and oligonucleotides of defined sequence, which can be custom-synthesized by several suppliers (Tordo et al., 1995; Nadin-Davis, 1998). While we do not seek to reiterate here all the information provided in those references, we do provide general comments below about the most important aspects to consider when developing or applying RT-PCR to lyssavirus detection.

4.3.1 Method for RNA extraction

To apply RT-PCR to a sample, total RNA must first be recovered from the tissue to be tested. While a number of methods for total RNA purification from tissues have been described, the most commonly used method utilizes a commercial reagent known as TRIzol®, a phenol/guanidine isothiocyanate solution based on earlier acidic phenol methods of RNA extraction. This reagent rapidly inactivates any nuclease present and quickly dissolves soft tissues such as brain, making it especially suitable for application in rabies diagnosis. After addition of chloroform to the mixture to facilitate a liquid phase separation, RNA recovered in the aqueous phase is readily precipitated by addition of isopropanol. This method is reasonably simple and is amenable to moderate throughput in terms of sample numbers.

Other methods that are sometimes used rely on commercially available kits that avoid the use of noxious chemicals and the requirement for RNA precipitation; Qiagen RNeasy® kits, for example, provide a silica-membrane spin column format for recovery of total tissue RNA from a wide variety of tissue types.

4.3.2 Selection of target strand

Since the rabies-virus life cycle includes production of both full-length negative- and positive-sense copies of its genome and significant amounts of mRNA (see Chapter 2), either positive(messenger)- or negative(genomic)- sense sequences can be targeted for RT-PCR, by use of either negative- or positive-sense primers, respectively, for the generation of the cDNA. Usually, the primer chosen is one of the primers also employed for the subsequent PCR. However, the sensitivity of the assay can be affected by the sense of the sequence initially targeted in the RT step. Many protocols target the negative-sense genomic sequence in consideration of the often less-than-ideal state of the tissue submitted to the laboratory; it is presumed that the encapsulated genomic RNA will be better protected than its corresponding mRNA from degradation occurring due to tissue autolysis. Alternatively, both positive- and negative-sense primers can be used to initiate cDNA synthesis, as described by Wellenberg *et al.* (2002). In some situations in which the amount of sample is limited or the sample is of poor quality, priming of cDNA synthesis is undertaken using random hexamer primers in place of the sequence-specific PCR primer; this approach causes cDNA synthesis to be initiated at several positions within the RNA target. Such an approach has been employed for amplification of a wide range of lyssaviruses (H. Bourhy, personal communication). This strategy also facilitates subsequent performance of multiple PCRs that target distinct viral sequences.

4.3.3 Primer design

A critical factor in determining the specificity of a PCR is the nucleotide sequence of the primer pair used to drive the reaction. For diagnostic purposes, the PCR should ideally be able to amplify successfully all members of the *Lyssavirus* genus; accordingly, highly conserved sequences at the genus level should be targeted. Since the N gene is one of the more conserved regions of the lyssavirus genome

(Le Mercier *et al.*, 1997), this gene has been the target of virtually all efforts to develop a broadly cross-reactive diagnostic test using PCR methodology. Moreover, the N gene was historically favored for analysis, because it permitted direct comparison between the genetic characteristics and the antigenic properties of these viruses, as studied using panels of MAbs directed primarily toward the nucleoprotein product of the N gene.

Tables 10.1 through 10.7 display the sequences of some of the primers most frequently used to amplify the lyssavirus N gene. These primer sequences are compared to the sequences determined for lyssaviruses representative of the known diversity of the genus; many of the same isolates are described in more detail in Chapter 3 of this volume. At least one isolate from each of the seven recognized genotypes (Bourhy *et al.*, 1993; Guyatt *et al.*, 2003), as well as the recently described but yet unclassified bat lyssaviruses recovered from Eurasia (Kuzmin *et al.*, 2005), are included in these comparisons. It should be noted that a small number of base mismatches between the primer and its target sequence will not necessarily prevent the two from annealing. The stringency of primer annealing is dependent on the annealing temperature: the lower the annealing temperature, the less stringent the annealing process and the greater the number of mismatches that can be accommodated. The caveat to this rule of thumb, however, is that as the annealing temperature is lowered, the potential for annealing to poorly related sequences increases and substantial non-specific primer binding arises. Ultimately, this leads to a highly non-specific reaction that is of little value diagnostically. The positions of mismatches also affect the extent to which such mismatches can hinder proper primer annealing. In particular, mismatches at the 3′ terminus of the primer, or within the three 3′-most bases of the primer, are highly detrimental to the PCR, since the 3′ end of the primer must anneal to its template faithfully in order to prime new DNA synthesis. One strategy often employed to overcome variability within the targeted sequence is to use a combination of primers of differing sequence that can each anneal to the same position within different target sequences (see the JW6/JW10 primer combinations in Table 10.3). Alternatively, a single degenerate primer can be employed. Degenerate primers are constructed during synthesis when two or even more choices of base are available for insertion at a certain position; thus, for example, when a residue is denoted as R, 50% of the primer molecules contain an A at this position, while the other 50% contain G; the population as a whole is an equimolar mixture of the two primer sequences. Similarly, Y denotes either T or C at a single position. Alternatively, the base inosine, which has increased flexibility to anneal to all bases, can be inserted into a position in place of any of the four usual bases. The primer 1312Nbdeg (see Table 10.6) is an example of a degenerate primer which has three degenerate positions.

Initial PCRs for lyssaviruses were developed to amplify the complete coding region of the N gene based on the belief that the 5′ and 3′ coding termini were well conserved and also to allow for determination of the complete coding

TABLE 10.1

Primers that target the nucleotide sequence at the start of the N gene for RT-PCR of lyssavirus cDNA

Primer name	Primer sequence	Reference/ GenBank accession
	55 84	
N7	ATGTAACACCTCTACA<u>ATG</u>	Bourhy *et al.*, 1993
JW12	ATGTAACACCYCTACAATG	Heaton *et al.*, 1999
RabN1	AACACCTCTACAATGGATGCCGACAA	Nadin-Davis, 1998
Nseq0	AACACCTCTACAATGGATGCCGAC	Nadin-Davis *et al.*, 1993
10g	CTACAATGGATGCCGAC	Smith *et al.*, 1992

Lyssavirus sequences

RABV:

PV	....	NC_001542
SADB19	............C...................	M31046
NISH	..............................G	AB044824
CVS	............C...................	X55727
RV257FOX	....	AY352464
9147FRA	....	U22474
9221TAN	....	U22645
9107MAR	....	U22852
8708NAM	....	U22632
1500AFS	....	U22628
9218TCH	....	U22644
Pak196p	...........T..	AY352495
HN_TRANSP	............C...................	AY956319
1077SRL	...........T..	AB041967
8738THA	....	U22653
SK5422AZ	......A.......	AY170226
EF2054CO	.........A....	AF394888
MAUS876FL	....	AY039225
LC814CA	....	AF394883
SHBRV-18	............C...................	AY705373
FTB874FL	....	AF394876
pehm3230	....	AF045166
ABLV.HN	............C..........T.T..T..	AF418014
EBLV2.9007	.....C..T.....	U22846
ARAV	......T.T.....	AY262023

TABLE 10.1 (Continued)

Primer name	Primer sequence	Reference/ GenBank accession
KHUV	`............G`	AY262024
EBLV1.8918	`.......TTA...G`	U22845
DUVV.86132	`........T..A.G`	U22848
IRKV	`......T.T....G`	AY333112
LGBV.8619	`......T.A..A.G`	U22842
MOKV	`..........TC.........GT.T.....`	U22843
WCBV	`......T.T..AC.`	AY333113

Genomic sequence for representative lyssaviruses is presented as a DNA positive strand (mRNA sense) and primer sequences are oriented in the same sense for comparison. All primers are used in the positive orientation (mRNA sense). The initiating codon is underlined. Differences in the lyssavirus sequences compared with the primers are indicated; a dot indicates identity and a space indicates that the data are unavailable at those positions. Base numbers refer to the PV reference strain.

TABLE 10.2

Primers that target internal N gene nucleotide sequence for RT-PCR of lyssavirus cDNA

Primer name/ (orientation)	Primer sequence	Reference/ GenBank accession
	135 165	
RabNfor (M)	`TTGTRGAYCAATATGAGTACAA`	Nadin-Davis, 1998
BB6 (M)	`GATCARTATGAGTAYAAATATCC`	Black et al., 2002
N165-146 (G)	`TATGAGTAYAARTACCCTGC`	Wakeley et al., 2005
Lyssavirus sequences		
RABV:		
PV	`.C..G..T.............G..C.....`	NC_001542
SADB19	`.C..G..T.............G..C.....`	M31046
NISH	`.C.CT..T.............G..C.....`	AB044824
CVS	`.C..G..T.............G..C.....`	X55727
RV257FX	`.A..G..T..........T..G..C....`	AY352464
V703IRN	`.C..G..T..G..........G..C..C..`	AY854583
9221TAN	`.C..G..T.............G..C.....`	U22645
867WSKCAN	`.C..G..T.............G........`	AF344306
V590DGMX	`.C..G..C.............G..C.....`	AY854589
1500AFS	`.C..G..T.............G........`	U22628
V461NIG	`.C..G..T................C.....`	AY854600
1578T1ON	`.C..G..T.............G..C.....`	L20673

(*Continued*)

TABLE 10.2 (Continued)

Primer name/ (orientation)	Primer sequence	Reference/ GenBank accession
HN_TRANSP	.C..G..T.............G..C.....	AY956319
1077SRL	....A..T..G...........G..C..G..	AB041967
8738THA	.C..G..T..G........T.....C..A..	U22653
V125RACFL	....A..T.................C..G..	U27220
SK5422AZ	.C..A..T.......A.....G..C..G..	AY170226
EF33CAN	.AA.A..T.......A.....G..C..G..	AF351829
ML06CAN	....A..T..G...........G.....A..	AF351838
SHBRV-18	.C..G..C.............G..C..G..	AY705373
LC01CAN	.C..G..C.............G..C..G..	AF351845
V588IBMX	.C..G..C.............G..C..G..	AY854588
V587VBMX	....G..C.............G.....G..	AY854587
ABLV.HN	.A..A..T...................C..	AF418014
EBLV2.9007	....A..T..G..............C.....	U22846
ARAV	..ACT..T..G........T..G........	AY262023
KHUV	....T..T..G..C...........C..G..	AY262024
EBLV1.8918	..TCT..T..G..............C.....	U22845
DUVV.86132	.CTCT..T..G...................	U22848
IRKV	.CTCT..T..G.....A.......C.....	AY333112
LGBV.8619	.ATCA..T.......A..T........A..	U22842
MOKV	.ATCA..T...........T........C..	U22843
WCBV	.ATCC..C..G.....A..T...........	AY333113

Genomic sequence for representative lyssaviruses and primers are presented as indicated in Table 10.1. The orientation of primer usage is indicated in brackets as messenger sense (M) or genomic sense (G). Base numbers refer to the PV reference strain.

TABLE 10.3

Primers that target internal N gene nucleotide sequence for RT-PCR of lyssavirus cDNA

Primer name	Primer sequence			Reference
	617	636 641	660	
JW10 (P)	GAACACCATACTCTAATGAC			Heaton et al.,
JW10 (ME1)	GAACAYCACACATTGATGAC			(1999)
JW10 (DLE2)	GAGCAYCACACTTTGATGAC			(all primers)
JW6 (DPL)		CACAAAATGTGTGCGAATTG		
JW6 (E)		CACAAGATGTGTGCCAACTG		
JW6 (M)		CATAAGATGTGCGCTAACTG		

TABLE 10.3 (Continued)

Primer name	Primer sequence		Reference
Lyssavirus sequences			
RABV:			
PV	....	..............T.....	
SADB19	....	..............T.....	
NISH	..G...........C.....	..............T.....	
CVS	..........C.........	.....G........T.....	
RV257FOX	..G..........G.....	..............T.....	
V703IRN	.......C...........	..............T.....	
9221TAN	..G.......C........	..............T.....	
867WSKCAN	.....T........G.....	.T............C.....	
V590MX.DG	.....T.....C........	..............A.....	
1500AFS	.....T..C...T.......	..T..G........T.....	
V461NIG	.....T........G.....	T.T.......C..T..C..	
1578T1ON	.....T........G.....	.....G........C..C..	
HN_TRANSP	........C.....G.....	..............C.....	
1077SRL	........C...T.......	.....G........T..C..	
8738THA	..............G.....	.....G.....C..T.....	
V125RACFL	.....T..C...T.......	...........C..T..C..	
SK5422AZ	.....T........G.....	.....G.....C..T..C..	
EF33CAN	.....T.....CT.G.....	..............T..C..	
ML06CAN	.....T........G.....	..............T..C..	
SHBRV-18	..G..T..C...T.G.....	...........C..T..C..	
LC01CAN	..G..T.....GT.G.....	...........C..C..C..	
V588IBMX	.G...T........G.....	...........C..C..	
V587MXVB	.....T.......T.G.....	...........C..T..C..	
ABLV.HN	.....T..C..GT.......	..T..G........A..C..	
EBLV2.9007	..G..T..C...T.G.....	.....G.....C..C..C..	
ARAV	.....T..C..AT.G.....	.....G........A.....	
KHUV	.....T..C..CT.G.....	..............T..C..	
EBLV1.8918	........C..AT.G.....	.....G........C..C..	
DUVV.86132	..G.....C..AT.......	....	
IRKV	.....T..C..CT.G.....	..T..........A.....	
LGBV.8619	..G..T..C...T.G.....	..............A.....	
MOKV	.....T..C..AT.G.....	..T..G.....C..T..C..	
WCBV	..G.....C..A..G.....	.....G........C.....	

Genomic sequence for representative lyssaviruses and primers are presented as indicated in Table 10.1. GenBank accession numbers for all sequences are as provided in Table 10.2. All primers are employed in the negative orientation (genomic sense). Base numbers refer to the PV reference strain.

TABLE 10.4

Primer that targets internal N gene nucleotide sequence for RT-PCR of lyssavirus cDNA

Primer name	Primer sequence	Reference
	876 896	
RabNrev	TAAGAAGAATGTTTGAGCCGG	Nadin-Davis, 1998

Lyssavirus sequences

RABV:

PV	..................A.	
SADB19	..................A.	
NISH	.......G.....C.....A.	
CVS	............C.....A.	
RV257FOX	..................A.	
V703IRN	..................A.	
9221TAN	....	
867WSKCAN	............C.....A.	
V590MX.DG	....G..G.....C.....A.	
1500AFS	.......G.....C.....A.	
V461NIG	............CC.....A.	
1578T1ON	..................A.	
HN_TRANSP	..................A.	
1077SRL	....G.............A.	
8738THA	............C.....A.	
V125RACFL	....	
SK5422AZ	....G.............A.	
EF33CAN	............C.......	
ML06CAN	..................T.	
SHBRV-18	.......G............	
LC01CAN	.......G............	
V588IBMX	..................T.	
V587MXVB	............C.....T.	
ABLV.HN	.T.............A..C.	
EBLV2.9007	.C.AG...............	
ARAV	.C.A.C.G.......A....	
KHUV	.C.AG..G..........A.	
EBLV1.8918	.C.AG.............A.	
DUVV.86132	.C.A...............C.	
IRKV	.C.AG........C.....T.	
LGBV.8619	.T.A...........A..T.	

TABLE 10.4 (Continued)

Primer name	Primer sequence	Reference
MOKV	.T.AG...............	
WCBV	.C..G........CCGT..CA	

Genomic sequence for representative lyssaviruses and the primer are presented as indicated in Table 10.1. GenBank accession numbers for all sequences are as provided in Table 10.2. The primer is employed in the negative orientation (genomic sense). Base numbers refer to the PV reference strain.

TABLE 10.5

Primer that targets internal N gene nucleotide sequence for RT-PCR of lyssavirus cDNA

Primer name	Primer sequence	Reference
	1157 1176	
1087NFdeg	GAGAARGAACTTCARGAATA	McQuiston et al., 2001
Lyssavirus sequences		
RABV:		
PV	.....A........A.....	
SADB19	.....A........A.....	
NISH	.....A........A.....	
CVS	.....A........A.....	
RV257FOX	.....A........A.....	
V703IRN	.....A........A.....	
9221TAN	.....A........A.....	
867WSKCAN	.....A........A.....	
V590DGMX	.....A........A.....	
1500AFS	.....A........A..G..	
V461NIG	..A..A.....C..A.....	
1578T1ON	.....A........A.....	
HN_TRANSP	..A..A........A.....	
1077SRL	.....A.....C..A.....	
8738THA	.....G.....C..G..G..	
V125RACFL	.....G........G..C..	
SK5422AZ	..A..G........G.....	
EF33CAN	.....A........G.....	
ML06CAN	.....A........G.....	
SHBRV-18	.....G........A.....	
LC01CAN	.....A........A.....	

(Continued)

TABLE 10.5 (Continued)

Primer name	Primer sequence	Reference
V588IBMX	..A..G..G.....G.....	
V587VBMX	.....G..G.....G.....	
ABLV.HN	.....A.....G..A..T..	
EBLV2.9007	....GG..G...GCA..GC.	
ARAV	.....G..G..G..G..T..	
KHUV	.....A..G..CA.G..G..	
EBLV1.8918	.....A..GT.A..G..T..	
DUVV.86132	.....A..G..G..A..C..	
IRKV	.....A.....G..G..C..	
LGBV.8619	.....A...A.G..A..T..	
MOKV	..A..A..GA.G..A..T..	
WCBV	..A.GA..GT.G..G..TC.	

Genomic sequence for representative lyssaviruses and the primer are presented as indicated in Table 10.1. GenBank accession numbers for all sequences are as provided in Table 10.2. The primer is employed in the positive orientation (mRNA sense). Base numbers refer to the PV reference strain.

TABLE 10.6

Primers that target the nucleotide sequence at the end of the N gene for RT-PCR of lyssavirus cDNA

Primer name	Primer sequence		Reference
	1382	1425	
1312NBdeg	TTYGCTGARTTTYTAAACAA		McQuiston et al., 2001
RabN2		CAAGACATATTCGAGTGACTCATAAGA	Nadin-Davis et al., 1993

Lyssavirus sequences
RABV:

PV	..C..C..G...C...............................			
SADB19	..C..C..G...C.............................			
NISH	..C..C..G...C.........G.........C.....T....G			
CVS	..C..C..A...T.........G.......A..........G			
RV257FOX	..T.....G...C....T.....G.......AC.....T....G			
V703IRN	..C..C..A...C.......A........A.............G			

(*Continued*)

TABLE 10.6 (Continued)

Primer name	Primer sequence	Reference
9221TAN	..C.....A...T.................T.......	
867WSKCAN	..T.....A...C..........G....................	
V590MX.DG	..T..A..G...C.............C................G	
1500AFS	..T..C..G...C..........C.....T.....T...	
V461NIG	..T.....G..CC......A.......T.....T.....G	
1578T1ON	..T..C..G...C.......C..G..C..T......C.G...	
HN_TRANSP	..T..C..G...C....T..........T.....T........	
1077SRL	..C.....G...C.............C..T.AC.......G.	
8738THA	..T..C..G...C................A......C.G	
V125RACFL	..T.....A...C................T.AC..TC....G	
SK5422AZ	..T.....G...C.C..T..A........A.AC..T.....G..	
EF33CAN	..C.....G...C....T........C..T.....T..G...AG	
ML06CAN	..C.....G...C.............C.......T..G...AG	
SHBRV-18	..C.....G...T.G...........C..A.....T..T...AG	
LC01CAN	..C.....A...T.G...........C..A.....T..T...AG	
V588IBMX	..T.....A...T.....................T...AG	
V587VBMX	..T.....A...T.............C..........G.G..G	
ABLV.HN	..T..G..G...C.C...............A........G...TC	
EBLV2.9007	..T..A..A..CC.G..T.....C...........TC..AG.	
ARAV	..T..C..A..CC.T..T.....C..C...G....TCAGAG.T.	
KHUV	..T.....G...T.G..T..A..C..C..A.....TC..AG.T.	
EBLV1.8918	..T..A..G...C.C.......G.....T.....TC.CAG.	
DUVV.86132	..T..A..G..CC.C.......C..C..C.......CAG.	
IRKV	..T..G..A...C.C..T.....T...G.C....TA..AG.T.	
LGBV.8619	..T..A..A..CC.C......GTG.....AGAGAG....	
MOKV	..T..A..A..CT........GTG...G.AGA..GA......TC	
WCBV	..T..A..G...C....T...GTT......GA....AAT.G..T	

Genomic sequence for representative lyssaviruses and the primers are presented as indicated in Table 10.1. GenBank accession numbers for all sequences are as provided in Table 10.2. The primers are employed in the negative orientation (genomic sense). Base numbers refer to the PV reference strain.

TABLE 10.7

Primers that target the nucleotide sequence at the start of the P gene for RT-PCR of lyssavirus cDNA. Genomic sequence for representative lyssaviruses and the primers are presented as indicated in Table 10.1. The primers are all employed in the negative orientation (genomic sense). Base numbers refer to the PV reference strain.

Primer name	Primer sequence		Reference/ GenBank accession
	1514 *1536* *1568* *1585*		
304	ATGAGCAAGATCTTTGTCAA		Smith, 1995
RabN5	ATGAGCAAGATCTTYGTCAATCC		Nadin-Davis, 1998
N8		GAGATGGCTGAAGAGACT	Bourhy *et al.*, 1993

Lyssavirus sequences

RABV:

PV	....	.............A...	NC_001542
SADB19	....	.............A...	M31046
V034SWZ	..........T..C........	....	AF369272
V661IRL	.......A..T...........	....	AF369281
V027TUN	.............C........	....	AF369322
V040TAN	.............C........	..A..............	AF369275
867WSKCA	.......A.............	....	AF369285
V591MX	....	.......C........	AY998249
V217TX	.....T.....T..........	C	AF369337
V039BWA	.......A...........C..	.......C.......C	AF369304
V461NIG	....	.......G........	AF369326
V121NEP	.......A.............	....	AF369317
1578T1ON	.......A.............	....	AF369265
V113SRL	....	.......C..G.....	AF369320
V125RACFL	..........T........C..	.......A..G.....	AF369294
V854MX	............C.....C..	.......A..G..A...	AY998275
V212TX	...............T..C..	..A.....G........A	AF369288
EF33CA	..........T........C..	.......A..G..A...	AF369338
ML06CA	..........T..C.....C..	.......A..G..A...	AF369344
LAN12CA	....	.......A..G.....	AF369346
LC01CA	..........T..........	........G..G.....	AF369347
V588MX	..........T..C.....C..	...........G.....	AY998246
V229MX	..........T..........	....	AF369362
P10PGY	..........T........C..	....	AF369363
ABLV.V474	....	..A.....A..G.....	AF369369

TABLE 10.7 (Continued)

Primer name	Primer sequence		Reference/ GenBank accession
EBLV2.V286	....................C..	........G........G	AF049121
ARAV	....	..A.....A..G.....G	AY262023
KHUV	..........T..........	........G..G.....A	AY262024
EBLV1.V002	....................C..	...........G.....A	AF049113
DUVV.V008	..........T...A.......	C	AF049115
IRKV	....................C..	........G..G......	AY333112
LBV.V267	.........GGGC.CA.AC....	..A.....A..G......	AF049119
LBV.V006	.........GGA..AA.AC.C..	...........G..A...	AF049114
MOKV.V020	.........GA.C....GC....	..A.....A.....A..C	AF049116
MOKV.V241	.......AGAT..A..AC.C..	..A..............	AF049118
WCBV	..........GTC..A.TC.C..	..A.....A..T......	AY333113

sequence of the gene. As shown in Table 10.1, information on the nucleotide sequence upstream of the initiation codon of the N gene is limited to a few laboratory-adapted strains and a small number of street isolates. The N7 primer, which has a 3' terminus corresponding to the N gene start codon, was successfully used by Bourhy *et al.* (1993) and subsequently by Kissi *et al.* (1995), to amplify a wide range of lyssavirus sequences. The only known mismatch within this primer is at base 65 (numbering based on the PV reference strain), at which some sequences contain a C in place of the T. The JW12 primer (Heaton *et al.*, 1997) is identical to N7, except that it contains a C/T degenerate base at that position in order to address this difference. Other primers, RabN1 and Nseq0 (Nadin-Davis *et al.*, 1993; Nadin-Davis, 1998) and 10g (Smith *et al.*, 1992), which have been used extensively for priming at this target sequence, straddle the initiating codon. Although most of these primers have performed well in the amplification of most genotype 1 rabies virus sequences, they all exhibit significant mismatch with viruses of the other genotypes and appear to be less well suited for a broadly reactive assay. In particular, the mismatch of the 3' A base of primer RabN1 with certain specimens that have a G at this position is problematic; indeed, this primer has been replaced by Nseq0 for routine application (Nadin-Davis, unpublished data). The primers often used to pair with those shown in Table 10.1 target either the sequence around the stop codon of the N gene (see Table 10.6) or, more frequently now, the start codon of the P gene (see Table 10.7). The extensive genetic variation evident among lyssaviruses in the region around the N-gene stop codon (see Table 10.6) explains why primer RabN2 is no longer in routine use (Nadin-Davis, unpublished data); the degeneracy of primer 1312NBdeg makes the latter rather more useful. Primers 304 (Smith, 1995) and its closely related variant RabN5 (Nadin-Davis, 1998), both target the sequence at and down-stream of the P gene start codon; as seen in Table 10.7, these primers (they are employed in the genomic sense) provide excellent concordance with their target sequences for almost all lyssaviruses except the most divergent phylogroup 2 members, Lagos Bat virus (LBV) and Mokola virus (MOKV), and the divergent West Caucasian Bat virus. The primer N8, which targets downstream sequence at bases 1568–1585, appears to be a better match for the divergent lyssaviruses of genotypes 2 and 3 (Bourhy et al., 1993).

Several primers that target internal sequences have also been used extensively. The primer pair RabNfor/RabNrev, which amplifies a fragment of 762 bp, was developed to amplify all known rabies virus strains circulating in Canada, but it has since proven useful for amplification of an even wider range of rabies viruses (Nadin-Davis, 1998). Tables 10.2 and 10.4 illustrate the relatively good match of these primers with many genotype 1 rabies viruses; however, increased numbers of mismatches are evident when these primers are compared with the lyssaviruses of genotypes 2–7, especially for the 3' terminus of primer RabNrev. Modification of these primers by inclusion of some degenerate positions might improve their suitability as broadly cross-reactive lyssavirus primers. Indeed, slightly modified versions of these two primers (SuEli+/−) were able successfully to amplify a range of Mexican rabies virus variants (Loza-Rubio et al., 2005). An overlapping primer, BB6, described by Black et al. (2002), extends into a less conserved sequence and has not found extensive application. More recently, another primer N165-146 targeting this region, but in the opposite orientation (genomic sense), was suggested to be suitable for a wide variety of lyssaviruses (Wakeley et al., 2005). Two sets of primers used for hemi-nested PCR in combination with JW12 are illustrated in Table 10.3. In each case, the viral sequence is targeted by three primers, each of slightly different sequence; the main variability within this region of the gene is covered among the three. Heaton et al. (1997) have reported that these primers are capable of amplifying members of all seven lyssavirus genotypes. However, the data in Table 10.3 suggest that certain rabies viruses will be amplified poorly by these primers. Most notably, the African dog strain, represented by V461NIG, exhibits a C to T substitution at the position corresponding to the 3' end of primer JW6 and certain American insectivorous bat strains, represented by V588IBMX, have an A to G substitution at the position representing the 3' proximal base of primer JW10. While the panel used to evaluate this primer set included 23 rabies viruses of genotype 1 (Heaton et al., 1997), more extensive testing will be needed to establish the universal suitability of these primers for lyssaviruses. Another internal primer, 1087NFdeg (Table 10.5), targets the sequence window at bases 1157–1176 near the 3' terminus of the N gene coding region. Use of degenerate positions allows this primer to anneal broadly to rabies viruses but, again, mismatches at the 3' end may preclude its utility for members of the other genotypes. It is evident from the above that truly universal primers capable of amplifying all lyssaviruses have yet to be developed. However, many of the primers described above are used on a regular basis and are generally satisfactory for the tests being applied. In particular, consideration must be given to the nature of the viruses that are likely to be encountered in a given geographical area. For instance, in the Americas, primers that successfully amplify all known genotype 1 viruses may be sufficiently broad in scope for most situations, since any indigenously acquired lyssavirus infection is presumed to be due to classical rabies virus. In other areas, particularly Africa and Europe, the presence of lyssaviruses of other genotypes dictates the need for broadly cross-reactive assays, except in cases where a specific lyssavirus genotype within a host reservoir is being targeted, as described for a European Bat lyssavirus type 1 (EBLV-1) assay in European bats (Picard-Meyer *et al.*, 2004). Ultimately, most laboratories will maintain several primer combinations, two or more of which may be utilized to evaluate a particular specimen, giving due consideration to the geographical location of the suspect animal and the viruses to which the animal source of the specimen could have been exposed.

As additional nucleotide sequence information becomes available for other portions of the lyssavirus genome, other targets for a broadly cross-reactive PCR assay may be identified. For example, Bourhy *et al.* (2005) were able to design primers that successfully amplified a section of the L (polymerase) gene for a number of rhabdoviruses representing several genera; these primers targeted sequences encoding highly conserved amino acid motifs required for polymerase function. It may thus be possible to develop a lyssavirus-specific PCR based on sequences of the L gene for future diagnostic use.

4.3.4 Assay sensitivity

The success of an RT-PCR is dependent on the integrity of the RNA template and, whereas PCR is less sensitive than is the DFA to the effects of tissue autolysis (Heaton *et al.*, 1997; Whitby *et al.*, 1997b; David *et al.*, 2002), extensive degradation of the sample will ultimately lead to false-negative results. To maximize the sensitivity of RT-PCR and, hence, the chance of detecting any lyssavirus sequence present in a sample, the strategies listed below are helpful.

1 Targeting of a relatively small sequence window: amplification of short DNA strands is more efficient than amplification of complete genes; thus the smaller the fragment amplified by PCR, the more sensitive the assay. Indeed, RNA that is significantly degraded may be far more effectively amplified if only short stretches of intact sequence are required to be present in the sample. Thus, assays that produce products of 200–300 bp (see McQuiston *et al.*, 2001) may be ideal for this purpose.

2 Increasing the number of amplification cycles: in an ideal reaction, the amount of PCR product doubles after each cycle, so that an increase in the number of cycles should significantly increase the product yield. Generally,

25–35 thermocycles of PCR are employed, although this can be increased to 40 or 45. Use of cycle numbers higher than this generally becomes unproductive, due to the gradual loss of enzymatic activity of the DNA polymerase employed in the reaction; such tailing off arises even with the use of thermostable enzymes relatively resistant to high temperature.

3 Using nested PCR: a nested PCR is one in which the product of a PCR is subjected to a second round of amplification that uses primers internal to those employed for the first round (Kamolvarin *et al.*, 1993). A heminested PCR (Heaton *et al.*, 1997; Picard-Meyer *et al.*, 2004) employs one of the first-round primers in combination with an internal primer in the second PCR. Nested strategies increase the sensitivity of the assay enormously, although at the cost of greatly increasing the chance of a false-positive result unless stringent precautions are taken to prevent carry-over contamination of the sample. It has been proposed that the main reason why nested PCRs are sometimes necessary is to compensate for inefficient first-round PCR arising from primer mismatches. The use of well-matched primers for first-round PCR in most circumstances should preclude the need for a nested approach (Trimarchi and Smith, 2002).

To demonstrate the utility of nested PCR, we show in Figure 10.4 the results of an evaluation of several samples from a human case of rabies (Elmgren *et al.*, 2002), by both regular and nested PCR. It is apparent that, for the first round of PCR (Figure 10.4 panel A), apart from the positive control, the only sample to generate a specific band of the correct size is the saliva sample. However, after a second round of PCR (Figure 10.4 panel B), the eye secretion, saliva and skin biopsy samples all generated a specific product of size identical to that of the positive control, while all blank samples, the negative control and the CSF remained negative. Subsequent nucleotide sequencing of these products identified the viral strain responsible for the case as a silver-haired bat strain, thereby confirming the specificity of the rabies diagnosis based on amplicon production.

4 Method of PCR product detection: PCR products are most commonly separated according to size, by electrophoresis through agarose gels, followed by staining with ethidium bromide, a dye that intercalates between DNA bases and is readily visualized under UV light. However, more sensitive methods of detection employ hybridization of the amplicon to rabies-specific oligonucleotide/DNA probes, either after transfer of the product to a membrane (Crepin *et al.*, 1998; Heaton *et al.*, 1999) or else by an ELISA-based method (Whitby *et al.*, 1997a). Traditionally, such probes were labeled with radioisotopes, but they now are more commonly labeled with an easily detectable ligand such as digoxigenin (DIG) or biotin. Probes increase PCR sensitivity by 10- to 100-fold over direct ethidium bromide staining, and they do, moreover, confirm the specific nature of the amplicon (Heaton *et al.*, 1999). In rare instances, false-positives have resulted, due to the production of non-specific bands of a size similar to that of the expected

C E Sa B1 B2 S P N   M   C E Sa B1 B2 S P N   M

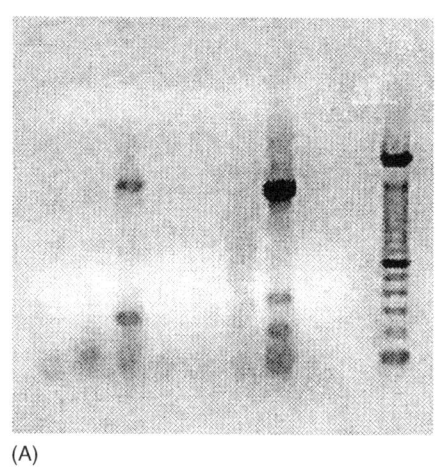
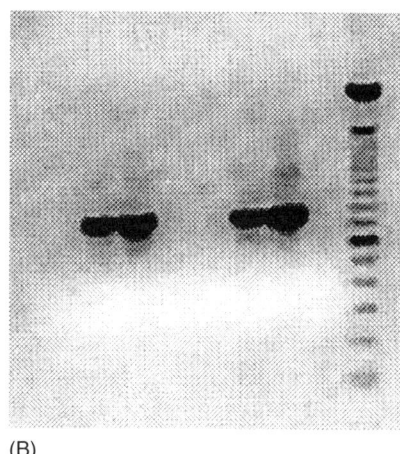

(A)   (B)

Figure 10.4   Analysis by gel electrophoresis of first-round (panel A) and second-round (panel B) PCRs of several samples from a human rabies case. First-round RT-PCR was performed using primer Nseq0 for RT and primers Nseq0/RabN5 for PCR; the expected product has a size of 1478 bp. Re-amplification of an aliquot of each first-round PCR was performed using primers RabNfor/RabNrev that produce an amplicon of 762 bp. The samples tested are as follows: C, CSF; E, eye secretion; Sa, saliva; B1 and B2, water samples extracted and processed in parallel with the tissues; S, skin biopsy. RT-PCR controls included a positive control (P), from a rabies-positive skunk and a water blank as a negative control (N). The marker (M) electrophoresed in parallel with the samples was a 100-bp DNA ladder (Invitrogen). DNA was detected under UV light after ethidium bromide staining of the agarose gel; an inverted image is presented.

product (Trimarchi and Smith, 2002). Only subsequent characterization of the amplicon, either by hybridization to a rabies-specific probe or by nucleotide sequencing, can discriminate between specific and non-specific product in such cases. TaqMan® methods incorporate dye-labeled probes as an integral part of the assay (see below) and thereby provide exquisite sensitivity and specificity.

4.3.5 Use of controls

As in any diagnostic assay, use of appropriate controls is essential to proper interpretation of the results. The controls listed below are strongly recommended.

Mock extraction control

A known rabies-negative sample should be processed for RNA extraction in parallel with the sample under investigation, so as to control for any inadvertent contamination of the specimens through aerosol generation or reagent contamination.

Ideally, the negative sample should comprise rabies-negative brain tissue, although a water sample can also be used if no such tissue is readily available.

Positive and negative samples as PCR controls

Upon RT-PCR set-up, each assay should include at least one water sample and the extracted rabies-negative sample as negative controls and RNA from one or more rabies-positive samples as positive controls. Any reaction run in which either the negative or positive controls fail must be considered invalid and the RT-PCR must be repeated. When large numbers of samples are being examined, use of a number of negative controls interspersed among the samples is recommended.

Control for template integrity

A control to evaluate a sample for its suitability for PCR can be useful if sample integrity is in question or as a means to assess the presence of PCR inhibitors which are sometimes carried through the RNA extraction process from the sample. Smith *et al.* (2000) reported on the use of a ribosomal RNA internal control, used to assay template quality; this control can be incorporated into an RT-PCR for lyssaviruses. The rRNA assay, shown to be suitable for 14 mammalian species, was reported as especially suitable for use as a control, since it exhibited degradation kinetics similar to that of the lyssavirus assay. A slightly modified version of the method, performed separately from the lyssavirus RT-PCR, has proven useful for evaluation of samples at the Rabies Centre of Expertise of the Canadian federal government (Nadin-Davis, unpublished data). Beta-actin mRNA is another target reported to serve as a useful internal control for rabies virus RNA detection by TaqMan® assays (Hughes *et al.*, 2004; Wakeley *et al.*, 2005) (see next section). Use of such controls reduces the chance of false-negative RT-PCR results from samples of tissue infected with a lyssavirus as a consequence of poor sample integrity or interference with the amplification by inhibitors; clearly, any sample that does not support amplification of rRNA or beta-actin mRNA should be declared unfit for testing.

4.4 Real-time PCR

Recent developments in PCR technology have focused on improvements in the quantitative capability of the methodology, as well as on the facilitation of rapid (real time) generation of results through elimination of the need to analyze the results of a PCR by gel electrophoresis. Since this latter aspect of the assay eliminates the need to open the vial during processing, an additional benefit of real-time methods is a significant reduction in the chance for false-positive results arising through sample contamination. The products of a real-time PCR are detected as they are produced, by inclusion of either a DNA-intercalating dye

(SYBR Green), which detects the production of all dsDNA within a reaction, or a dye-labeled oligonucleotide, the emission of which increases as specific amplicon yields rise. One of the most popular chemistries employed by real-time PCR is the TaqMan® format, which uses dual-labeled oligonucleotide probes (DLPs), first described by Lee *et al.* (1993). The DLP is usually labeled at its 5' end by a reporter dye and at its 3' end by a quencher that prevents emission by the reporter dye as long as the two are in close proximity. As the PCR progresses and as specific product is generated, binding of the DLP to its cognate sequence occurs, during the annealing step and just prior to strand extension from one of the PCR primers. During DNA synthesis the probe is cleaved, the reporter and quencher become dissociated from one another and the presence of PCR product is detected as fluorescence emitted by the reporter dye at a defined wavelength. Since this technology performs optimally with PCR products of <200 bp, the small amplicon size and subsequent detection mechanism ensure that this type of one-tube assay is highly sensitive, often comparable with or even more sensitive than nested PCR, but avoiding the contamination potential associated with a multitube assay. However, design of TaqMan® assays is constrained by the requirement for probes having close-to-perfect sequence identity with the intended target; base mismatches between target and probe sequences can substantially reduce assay sensitivity or even preclude detection completely.

To date, relatively few reports of the application of this technology to rabies diagnosis have appeared in the literature. Black *et al.* (2002) reported an ambitious effort to both detect and discriminate among lyssavirus genotypes 1–6 using TaqMan® technology. The assay relied on the previously described JW6 primer set and a newly developed primer BB6, which overlaps in sequence with primer RabNfor, and which generates a product that, at 502 bp, is considered rather long for a TaqMan® assay. In addition, eight different probes, comprising three genotype 1-specific probes and one probe each for genotypes 2–6, were developed; these probes targeted different portions of the amplicon over a 261-bp region. Separate assays were undertaken to evaluate samples for each of the six genotypes. This complex assay design underscores the difficulty in developing a TaqMan® method to detect the range of viral sequences encompassed by the *Lyssavirus* genus. The authors did report the successful detection and assignment of 106 lyssaviruses using the reagents described, with no detectable cross-reactivity with 18 non-lyssavirus isolates. A greatly simplified protocol, which detects lyssaviruses of genotypes 1, 5 and 6, was subsequently described by the same group (Wakeley *et al.*, 2005). The primers employed for that RT-PCR assay were JW12 and N165-146 (a genomic sense primer overlapping in sequence with RabNfor, as shown in Table 10.2), which produced an amplicon of 111 bp, while three distinct probes, each labeled with a different reporter dye for each of the three genotypes and targeting positions between bases 80 and 109, were used for product detection and typing. Of 62 lyssaviruses evaluated by this method, all but one were readily detected and typed; a bat rabies virus from the USA was only weakly detected, possibly due in part to three mismatches between the genotype 1 probe and the target sequence in this isolate. The high efficiency of this modified TaqMan® assay was evident from sensitivity comparisons made between it and the hemi-nested PCR method previously employed; the latter was 100-fold less sensitive than the TaqMan® assay. It is reported that this current version of the TaqMan® assay is in routine use for diagnostic purposes at the Rabies Laboratory of the Veterinary Laboratories Agency (VLA) at Weybridge in the UK.

Specifically to address issues of cross-contamination problems encountered with nested RT-PCR assays for detection of the Australian bat lyssavirus (ABLV), Smith *et al.* (2002) developed a TaqMan® method for this virus. An assay, which employed two forward primers, corresponding to sequences of the flying fox and insectivorous biotypes of ABLV respectively, together with a common reverse primer and two distinct probes, was reported; the assay detected and discriminated between the two viral biotypes. The assay was 10-fold more sensitive than the hemi-nested PCR of Heaton *et al.* (1997). No data on the cross-reactivity of this assay with other lyssaviruses were presented; more extensive evaluation of this test with a larger collection of specimens is needed to confirm its specificity and broad applicability to the ABLV group.

In another, rather more cautious study on the application of TaqMan® technology to rabies virus RNA detection, Hughes *et al.* (2004) developed a series of assays for several rabies virus strains circulating in North American reservoir hosts. Their studies supported the utility of the quantitative aspects of TaqMan® technology but exposed difficulties that might be faced in the design of TaqMan® primers and probes to detect a wide range of rabies virus strains, given the intra-strain genetic diversity. The authors reported that three historical samples (5% of those tested) were not amplified by any of their primer/probe sets. This failure was due to several mismatches between the nucleotide sequences of these reagents and the viral target sequences; in particular, mismatches in the center of the probe target sequence were found to be especially detrimental to the assay. These findings suggest that a simple TaqMan® assay could yield significant numbers of false-negative results and that large numbers of primer/probe combinations would be required to cover adequately the genetic diversity existing within rabies virus populations. Such a requirement for many primer/probe combinations would render TaqMan® technology impractical. The authors also noted, in contrast to observations by others, that the sensitivity of their TaqMan® assay was equivalent to that of a regular PCR, but lower than that achieved by hemi-nested PCR.

While real-time PCR technology offers the potential of a rapid and highly sensitive method for rabies virus RNA detection, further evaluation and refinement of the reagents employed for this assay are needed. The expectations of the assay in terms of the geographical area to be covered may allow this assay to be more readily adopted in certain parts of the world. The speed of TaqMan® assays does make them especially suitable for the diagnosis of suspected human cases for which rapid results are important with respect to patient management and determination of whether PEP of patient contacts is required. Further developments in this area are anxiously awaited.

4.5 Other molecular methods for rabies virus detection

While a number of sensitive amplification methods other than PCR have been developed for detection of specific nucleic acid sequences, few have been applied to rabies virus detection. An exception is nucleic acid sequence-based amplification (NASBA) technology, which has been used to detect rabies virus RNA in saliva and CSF of human patients (Wacharapluesadee and Hemachudha, 2001; Hemachudha and Wacharapluesadee, 2004). NASBA employs three enzymes, avian myeloblastosis virus reverse transcriptase, *E. coli* ribonuclease H and T7 RNA polymerase, for isothermal amplification of an RNA template (Compton, 1991). The method uses a pair of primers that target a specific viral sequence, in this case a window comprising 180 bases within the central region of the rabies virus N gene; the primers incorporate in their sequences the T7 RNA polymerase promoter and a non-related sequence that is recognized by a reporter probe. In terms of sensitivity, this assay appears to compare favorably to standard RT-PCR. Its application, especially in countries where the costs of acquiring the thermocylers required to perform RT-PCR might be prohibitive, should be further investigated.

5 DIAGNOSIS OF RABIES IN HUMANS

Rabies surveillance in humans is of the utmost importance for the management of potential community and health care exposures to rabies from contact with infected individuals, even though human-to-human transmission is rarely reported (see Chapter 4). The lack of reliable and sensitive surveillance for human cases in most of the developing countries does not favor political commitment of resources to rabies control activities.

5.1 Ante-mortem testing

A diagnosis of rabies should be considered in any patient who presents with encephalopathy of unknown cause. The first signs and symptoms of rabies are often non-specific. Without a clear history of animal bite, rabies is often not suspected until late in the clinical illness. A delayed diagnosis obviously increases the number of persons potentially exposed to rabies by contact with the patient.

An early diagnosis can eliminate the expense and discomfort of unnecessary diagnostic tests and medical treatment of the patient. A young patient with rabies, confirmed by laboratory identification of a rapidly increasing rabies virus antibody titer, survived following aggressive treatment that included tertiary support, induced coma and a cocktail of antiviral drugs (Willoughby *et al.*, 2005). This event increases the hope that early diagnosis can facilitate a treatment regimen leading to a desirable clinical outcome. A report of transmission of rabies from an organ donor to four transplant recipients in the USA in 2004 (Srinivasan, *et al.*, 2005) as well as a similar event in Germany in the same year (Johnson *et al.*, 2005), further underscores the potential value of early ante-mortem diagnosis of human rabies infection.

Once considered, a diagnosis of rabies is not always easy to confirm. Ante-mortem diagnosis of rabies is one of the most difficult procedures attempted by the laboratory and should be performed only by experienced laboratories. The risks of performing a brain biopsy make this procedure unacceptable for the routine acquisition of brain tissue for ante-mortem diagnosis of rabies. Methods for intra-vitam rabies diagnosis rely on the demonstration of antibody in serum or CSF, and detection of virus, viral antigen, or viral RNA in tissues. If rabies is suspected, a complete set of such samples should be collected for a battery of testing by all currently used diagnostic procedures. Because of the implications of a positive test, a finding of rabies must be confirmed in more than one tissue or sample. Because antibody generally is produced late in a lyssavirus infection, and because virus may be absent, or present at very low levels in tissues, the analysis of samples taken for ante-mortem diagnosis cannot definitively rule out rabies. If a suspicion of rabies persists despite negative findings, repeated sampling and analysis may be necessary.

5.2 Sample collection

A clinician suspecting rabies in a patient should consult with the state health department or with the Rabies Laboratory at the CDC, Atlanta, GA, the Canadian Food Inspection Agency's Centre of Expertise for Rabies, Ottawa, Canada, or the equivalent in other countries. The course of the illness, additional patient history and laboratory tests for other more common etiologies can help to determine whether samples specific for rabies should be collected. All samples should be considered as potentially infectious. Test tubes and other sample containers must be securely sealed. Tape around the cap will ensure that the containers do not open during transit. If immediate shipment is not possible, samples should be stored frozen at $-20°C$ or below. Samples should be shipped frozen on dry ice by an overnight courier, in water-tight primary containers and leak-proof secondary containers that meet the guidelines of the International Air Transport Association; these guidelines currently categorize rabies virus as a

'Biological Substance, Category B' in clinical specimens and as a 'Biological Substance, Category A' only when the virus has been cultured (International Air Transport Association, 2006). The laboratory should be telephoned at the time of shipment and given information on the mode of shipment, expected arrival time and courier tracking number.

5.2.1 Saliva

Saliva should be collected with a sterile eyedropper pipette and placed in a small sterile container that can be sealed securely. No preservatives or additional material should be added. Laboratory tests to be performed include detection of lyssavirus RNA by RT-PCR of extracted nucleic acids and isolation of infectious virus in cell culture. Tracheal aspirates and sputum are not suitable for rabies tests.

5.2.2 Neck skin biopsy

A full-thickness punch biopsy of skin 5–6 mm in diameter should be taken from the posterior region of the neck at the hairline. The biopsy specimen should contain a minimum of 10 hair follicles and should be of sufficient depth to include the cutaneous nerves at the base of the follicle. The specimen should be placed on a piece of sterile gauze moistened with sterile water or saline in a sealed container. Preservatives or additional fluids should not be added. Neck biopsy samples are tested by immunofluorescence staining for viral antigen in frozen sections of the biopsy and by RT-PCR on extracted RNA. The antigen is generally present in the nerve cells surrounding the bases of hair follicles (see Figure 10.1C).

5.2.3 Serum and cerebrospinal fluid

At least 0.5 ml of serum (not whole blood) is needed to test for antibody by indirect immunofluorescence and virus neutralization. If no vaccine or rabies-immune serum has been administered to the patient, the presence of antibody to the challenge virus standard (CVS) rabies virus in the serum is diagnostic and tests of CSF are unnecessary. If collected, at least 0.5 ml of CSF should be sent for testing by RT-PCR and neutralization tests. Lyssavirus-specific antibody in the CSF, regardless of the immunization history, suggests rabies encephalitis.

5.2.4 Corneal impressions

Corneal epithelium, while it can test positive for lyssavirus antigen in some patients with rabies, is difficult to sample correctly, especially from comatose patients. Because of the risk of permanent damage to the cornea, samples should be taken only by an ophthalmologist, after consultation with the rabies testing laboratory (Zaidman and Billingsley, 1998). The sample is collected by vigorously rubbing a flat surface of a clean microscope slide on each cornea. Corneal impression slides are tested by immunofluorescence staining for viral antigen; they can also be tested by RT-PCR. The antigen appears characteristically as round to oval intracytoplasmic inclusions in corneal epithelial cells (see Figure 10.1D).

5.2.5 Brain biopsy

Brain biopsy is costly and invasive and may be less than 100% sensitive, since the tissue that can be biopsied is generally not taken from ideal diagnostic regions of the brain. The rarity of human rabies and the lack of an as-yet validated effective treatment make routine brain biopsy unwarranted; however, biopsy samples negative in other tests should be tested for evidence of lyssavirus infection. The biopsy material is placed in a sterile gasket-sealed container without preservatives or additional fluids. Laboratory tests to be performed include RT-PCR and immunofluorescence staining for viral antigen in touch impressions or frozen sections.

5.3 Significance of positive and negative findings

In a study of ante-mortem test results for human rabies deaths in the USA between 1980 and 1997 (Noah *et al.*, 1998) that is updated with data from cases through 2000, reported in Table 10.8, RT-PCR of RNA extracted from saliva was the most reliable diagnostic test for early detection. Positive results were obtained for 15 of 15 cases; however, nested PCR was required in almost all cases to compensate for the often extremely limited amount of RNA in ante-mortem samples. In a similar study of a smaller number of cases, Crepin *et al.* (1998) diagnosed rabies in four of nine cases by RT-PCR of saliva, but they did not use nested PCR.

Virus isolation from saliva was positive in nine of the 15 cases in which this diagnostic method was applied (see Table 10.8). Successful virus isolation was related to the antibody status of the patient, suggesting that virus is cleared from salivary glands by the immune response to infection. Virus was isolated from 13 of 15 serial saliva samples from antibody-negative patients. Virus isolation methods were negative in 17 of 17 serial saliva samples from antibody-positive patients.

Frozen sections of skin biopsies were positive by DFA in 15 of 20 cases to which this diagnostic method was applied (see Table 10.8). Observation of 20 or more sections was needed in most cases to reliably detect areas of positive staining. Crepin *et al.* (1998) diagnosed rabies in five of nine patients with this method.

Serum was more likely to be positive when sampled late in the clinical course (see Table 10.8; Crepin *et al.*, 1998), but in one case antibody was absent as late as day 24. Antibody was present in the CSF of only three of 17 patients (see Table 10.8).

TABLE 10.8

Summary of ante-mortem diagnostic test results for 27 human rabies cases in the USA 1981 to 2000

Case	Detection of antigen			Isolation of virus	Detection of RNA	Detection of antibodies	
	Cutaneous nerve	Corneal epithelium	Brain biopsy	Oral secretions		Serum	CSF
81AZ	**d8+**	d8-	ns	d11+ d15+	nt	d8 to d24-	d8 to d24-
83MA	**d6+** d11+	d6- d18-	d8+	d6- d9 to d13+ d16 to d25-	nt	d6 to d14- d16 to d27+	d8 to d19-
83MI	d17- d21-	ns	ns	d17-	nt	**d17+** d22+ d26+	d17- d22- d26+
84TX	ns	ns	d17+	d18-	nt	d11- d18+	**d15+**
84PA	**d7+**	ns	ns	**d7+**	nt	d7-	d7-
90TX	d6-	ns	ns	d6-	nt	d6-	d6-
91TX	d6- d14+	ns	ns	**d6+**	**d6+**	d6-	d6-
92CA	d4- **d8+** d14+	d3-	ns	d8- d12- d14- d17 nt	**d8+** d12+ d14+ d17+	d3- **d8+** d17+	d14-
93TX	ns	ns	ns	ns	ns	?d7-	ns
93CA	**d6+**	d6 to d10-	ns	**d6+**	**d6+**	d6 to d10-	d9-
94WV	**d5+**	ns	d8+	d5-	**d5+**	**d5+**	ns

(Continued)

TABLE 10.8 (Continued)

Case	Detection of antigen			Isolation of virus	Detection of RNA	Detection of antibodies	
	Cutaneous nerve	Corneal epithelium	Brain biopsy	Oral secretions		Serum	CSF
94TN	ns	d15+	ns	ns	ns	ns	ns
94TX	d11+	ns	ns	d11+	d11+	d11-	d11-
95WA	d8+	ns	ns	d8+ d10+	d8+	ns	ns
95CA	d6-	d6-	ns	ns	ns	d6- d12+	d5-
95CT	ns	d14+	ns	d10-	d10+	d12+	d11-
96FL	d9+	ns	ns	d9+	d9+	d8-	ns
96NH	d5- d6-	ns	ns	d5+ d6 nt	d5+ d6 nt	d5- d6+	d5-
96KY	ns	ns	ns	ns	ns	d13+	ns
96MT	d9-	ns	ns	d9 nt	d9+	d9+	ns
97TX	ns	ns	ns	ns	ns	d10-	ns
97NJ	d6+	ns	ns	d6 nt	d6+	d6-	d6-
98VA	d7+	ns	ns	ns	d7+	d8+ d15+	d8+
00CA	d4+	d4+	ns	?	d4+	ns?	ns
00NY	d6+	d6-	ns	ns	d6+	ns	ns
00WI	ns	ns	ns	ns	ns	d6-	ns
00MN	d12+	ns	ns	ns	d12+	ns?	ns
Number of cases with samples submitted for testing	20	9	3	15	15	22	15
Number of diagnoses	15	3	3	9	15	11	3

Corneal impression was the least satisfactory type of diagnostic sample. A positive DFA test was made in only three of nine cases where this sample was taken (see Table 10.8).

Although the nested PCR of saliva was positive in the 15 of 15 cases in which the test was applied, conventional methods are often equally reliable and can produce a diagnosis within a few hours of sample receipt in the laboratory. Positive tests were obtained by serology or DFA on skin biopsy in 20 of 25 cases for which these samples were available (see Table 10.8).

5.4 Post-mortem testing of autopsy samples

In 15 of 42 human rabies cases reported in the USA between 1980 and 2000, samples were not obtained specifically for rabies testing before the patients' deaths. The clinical history in seven cases was sufficiently suggestive of rabies that fresh brain material obtained during autopsy was submitted immediately for DFA testing. In the remaining cases, however, a diagnosis was delayed for 3 weeks to 6 months after autopsy, when findings suggestive of rabies were noted in histologic exams of formalin-fixed brain material. One 2005 case was diagnosed 20 days after onset of illness and 8 days after death, on the basis of a rise in rabies IgG antibody titer that was demonstrated in CSF from 1 in 128 on day 5 after onset to 1 in 8102 on day 10 post-onset (Centers for Disease Control and Prevention, 2006). Formalin-fixed brain material from humans is tested as indicated for animal brains (Section 2.4).

6 RABIES ANTIBODY ASSAYS

Serologic tests for lyssavirus-specific antibody can provide an estimate of vaccine efficacy, an indication of disease prevalence in areas of enzootic rabies and an ante-mortem diagnosis in human rabies cases. The various methods used to measure lyssavirus-specific antibody can be grouped generally as antigen-binding assays, antibody-function assays and antigen-function assays (Smith, 1991).

6.1 Antigen-binding assays

Antigen binding assays measure the extent of attachment of antibody to a lyssavirus or lyssavirus-specific proteins attached to a substrate, typically a slide, microtiter plate or bead. Bound antibody is detected with anti-antibody or Fc-binding protein (staphylococcal protein A or streptococcal protein G) either labeled with an enzyme that is detected in the ELISA or else labeled with FITC, which is detectable in the IFA. The specificity of antigen binding assays is determined by the choice of antigen (whole virion versus purified protein) (Perrin *et al.*, 1986; Grassi *et al.*, 1989; Cliquet *et al.*, 2000) or by use of a labeled control antibody specific for a particular protein in a competitive-binding or blocking format (Sugiyama *et al.*, 1997; Cleaveland *et al.*, 1999). Caution must be exercised in considering these assays reproducible given that the preparation of the antigen may not always follow quality assurance practices (A. Wandeler, Canadian Food inspection Agency, personal communication).

Because vaccine efficacy is closely tied to the generation of virus neutralizing antibodies directed against specific epitopes on the virus glycoprotein, antigen-binding assays involving the viral glycoprotein have not been suitable as a means to estimate immune response to vaccination. Although good correlation has been reported in some studies (Piza *et al.*, 1999), laboratories in the USA are required by the Advisory Committee on Immunization Practices (ACIP) to use virus neutralization tests in assaying antibody titers in persons immunized against rabies (Centers for Disease Control and Prevention, 1999b).

Microtiter plate ELISA methods can be simple and inexpensive and can yield reliable surveys for the presence of rabies antibody in animals in areas of enzootic rabies (Cleaveland *et al.*, 1999; Cliquet *et al.*, 2003). Although species-specific antisera are required for some samples, Fc-binding proteins allow testing of a wide variety of animal sera. Use for the screening of vaccinated companion animals also has been proposed (Cliquet *et al.*, 2004), although one study demonstrated some discrepancy between results from this method and from neutralization assay, in determination of adequate response to vaccination (Bahloul *et al.*, 2005). Although not commonly used in seroconversion surveys, IFA can also be used to test for antibody-positive animals, if FITC-labeled anti-antibodies or Fc-binding proteins are available (Hill *et al.*, 1992). IFA is especially valuable for ante-mortem evaluation of suspected rabies in humans since the test is easily done and results can be obtained quickly. The counter immunoelectrophoresis test, which is simple and comparatively inexpensive, has been evaluated and proposed for use in monitoring efficacy of dog vaccination campaigns (Rani *et al.*, 2002; Da Silva *et al.*, 2002).

6.2 Antibody-function assays

Antibody-function assays are based on detection of a non-virus-related function (e.g. complement fixation or hemagglutination) performed by an antibody, after an interaction with antigen. Antibody-function assays were among the earliest serologic methods to be developed, but are no longer in wide use due to their lack of specificity.

6.3 Antigen-function assays

Antigen-function assays measure the capacity of an antibody to block a specific viral function. Neutralization of virus infectivity is the most widely used antigen-function assay. In all virus neutralization assays, dilutions of heat-inactivated serum are incubated with a constant amount of virus. Although early tests depended on animal inoculation to measure residual virus infectivity after incubation with serum, levels of residual virus are now determined almost exclusively through inoculation of cell cultures.

Most of the variation in virus neutralization tests stems from the precision with which the amount of residual infectious virus is measured. The most precise measurement is made by plaque reduction tests (PRT), in which each infectious unit of virus is counted (Wiktor and Clark, 1973). Because 5–7 days are required for infectious foci (plaques) to reach detectable size, PRT is unsuitable for most diagnostic applications. A more rapid method of measuring antibody present in serum, with many of the advantages of PRT, uses FITC-labeled antibody to detect residual virus. The amount of residual virus is estimated by counting all infectious foci (tissue culture infectious doses, or TCID) within a single well or chamber or by estimating the amount of virus by finding the point at which 50% of inoculated chambers contain virus ($TCID_{50}$) after 3 or 4 days of incubation.

Several laboratories have modified fluorescent methods of calculating residual virus, by increasing the amount of challenge virus and thereby decreasing the length of time required for the virus to reach detectable levels. Although less precise, these methods are suitable as tests of vaccine efficacy, for sero-surveys and ante-mortem serum samples. The rapid fluorescent focus inhibition test (RFFIT) (Smith *et al.*, 1996b) determines a $TCID_{50}$ as the dilution of virus at which 50% of observed microscopic fields contain one or more infected cells after a 20-hour incubation period. Tests are performed either in multichamber slides (one chamber per serum dilution) or in microtiter plates (one or two wells per serum dilution). In both the tissue culture serum neutralization (TCSN) test (Trimarchi *et al.*, 1996) and the fluorescent antibody virus neutralization (FAVN) test (Cliquet *et al.*, 1998), each serum dilution is placed in multiple wells of a microtiter plate and each well is simply scored as having virus present or absent after a 40–48 hour incubation period. In each method, test results are reported in international units, by comparison to a reference standard serum. No statistical difference was found between results obtained by the RFFIT and the FAVN tests (Briggs *et al.*, 1998). The main differences between the two methods are the time to completion (24 versus 40 hours), the volume of infectious material created in a single test dilution (0.4 ml versus 0.8 ml) and the time required to read the test results (evaluation of 20 microscope fields in one chamber versus a thorough scan of four wells). Additionally, manipulation and reading of microtiter plates in the FAVN are easily automated (Hostnik, 2000).

A method that employs a recombinant rabies virus containing a green fluorescent protein (GFP) gene as the challenge virus in the RRFIT permits the direct visualization of the GFP-expressing virus in infected cells, without the use of an FITC-labeled antibody conjugate. Results with this method were shown to correlate well with results from the standard RFFIT (Khawplod *et al.*, 2005). An indirect immunoperoxidase method to demonstrate residual virus, instead of immunofluorescence, in a cell culture neutralization test, has been shown to offer a possible alternative where fluorescence equipment and reagents could be limiting (Cardoso *et al.*, 2004).

ACKNOWLEDGEMENTS

The major contributions of Jean S. Smith (CDC, retired), made as co-author of this chapter in the first edition of *Rabies*, are gratefully acknowledged, particularly to the sections on human ante-mortem detection (including Table 10.8) and antibody assays, which were carried over with only minor updates. We are most grateful to Hervé Bourhy and Alexander Wandeler for their review of this work and for their helpful comments.

REFERENCES

Andrulonis, J.A. and Debbie, J.G. (1976). Effect of acetone fixation on rabies immunofluorescence in glycerin-preserved tissues. *Health Laboratory Science* **47**, 207–209.

Bahloul, C., Tajeb, D., Kaabi, B. *et al.* (2005). Comparative evaluation of specific ELISA and RFFIT antibody assays in the assessment of dog immunity against rabies. *Epidemiology and Infection* **133**, 749–757.

Barrat, J. (1996). Simple technique for the collection and shipment of brain specimens for rabies diagnosis. In: *Laboratory Techniques in Rabies*, 4th edn (F.X. Meslin, M.M. Kaplan and H. Koprowski, eds). pp. 425–432. Geneva: World Health Organization.

Bingham, J. and van der Merwe, M. (2002). Distribution of rabies antigen in infected brain material: determining the reliability of different regions of the brain for the rabies fluorescent antibody test. *Journal of Virological Methods* **101**, 85–94.

Black, E.M., Lowings, J.P., Smith, J., Heaton, P.R. and McElhinney, L.M. (2002). A rapid RT-PCR method to differentiate six established genotypes of rabies and rabies related viruses using TaqMan™ technology. *Journal of Virological Methods* **105**, 25–35.

Bordignon, J., Cordoba Pires Ferreira, S., Maria Medeiros Caporale, G. *et al.* (2002) Flow cytometry assay for intracellular rabies virus detection. *Journal of Virological Methods* **105**, 181–186.

Bourhy, H., and Perrin, P. (1996). Rapid rabies enzyme immunodiagnosis (RREID) for rabies antigen detection. In: *Laboratory Techniques in Rabies*, 4th edn (F.X. Meslin, M.M. Kaplan and H. Koprowski, eds). pp. 105–112. Geneva: World Health Organization.

Bourhy, H., Cowley, J.A., Larrous, F., Holmes, E.C. and Walker, P. J. (2005). Phylogenetic relationships among rhabdoviruses inferred using the L polymerase gene. *Journal of General Virology* **86**, 2849–2858.

Bourhy, H., Kissi, B., Lafon, M., Sacramento, D. and Tordo, N. (1992). Antigenic and molecular characterization of bat rabies virus in Europe. *Journal of Clinical Microbiology* **30**, 2419–2426.

Bourhy, H., Kissi, B. and Tordo, N. (1993). Molecular diversity of the lyssavirus genus. *Virology* **194**, 70–81.

Bourhy, H., Rollin, P.E., Vincent, J. and Sureau, P. (1989). Comparative field evaluation of the fluorescent-antibody test, virus isolation from tissue culture, and enzyme immunodiagnosis for rapid laboratory diagnosis of rabies. *Journal of Clinical Microbiology* **27**, 519–523.

Briggs, D.J., Smith, J.S., Mueller, F.L. *et al.* (1998). A comparison of two serological methods for detecting the immune response after rabies vaccination in dogs and cats being exported to rabies-free areas. *Biologicals* **26**, 347–355.

Cardoso, T. and Pilz, D. (2003). Wild rabies virus detection by plaque assay from naturally infected brains in different species. *Veterinary Microbiology* **103**, 161–167.

Cardoso, T.C., da Silva, L.H., Albas, A., Ferreira, H.L. and Perri, S.H.V. (2004). Rabies neutralizing antibody detection by indirect immunoperoxidase serum neutralization assay performed on chicken embryo related cell line. *Memorias do Instituto Oswaldo Cruz* **99**, 531–534.

Centers for Disease Control and Prevention (1993). Human Rabies – New York, 1993. *Morbidity and Mortality Weekly Report* **42**, 799–806.

Centers for Disease Control and Prevention (1999a). *Biosafety in Microbiological and Biomedical Laboratories*, 4th edn. Washington, DC: US Government Printing Office.

Centers for Disease Control and Prevention (1999b). Human Rabies Prevention – United States, 1999 Recommendations of the Advisory Committee on Immunization Practices (ACIP). *Morbidity and Mortality Weekly Report* **48**, No. RR-1, 1–21.

Centers for Disease Control and Prevention (2000). Human rabies – California, Georgia, Minnesota, New York, and Wisconsin, 2000. *Morbidity and Mortality Weekly Report* **49**, 1111–1115.

Centers for Disease Control and Prevention (2006). Human Rabies – Mississippi, 2005. *Morbidity and Mortality Weekly Report* **55**, 207–208.

Charlton K.M. (1988). The pathogenesis of rabies. In: *Rabies* (J.B. Campbell and K.M. Charlton, eds). pp. 101–150. Norwell: Kluwer Academic Publishers.

Chemicon International (2000). Notification on Light Diagnostics Rabies Reagent. Temecula, California.

Childs J.E., Colby, L., Krebs, J.W. *et al.* (1997). Surveillance and spatiotemporal associations of rabies in rodents and lagomorphs in the United States, 1985–1994. *Journal of Wildlife Diseases* **33**, 20–27.

Cleaveland, S., Barrat, J., Barrat, M.J., Selve, M., Kaare, M. and Esterhuysen, J. (1999). A rabies sero-survey of domestic dogs in rural Tanzania: Results of a rapid fluorescent focus inhibition test (RFFIT) and a liquid-phase blocking ELISA used in parallel. *Epidemiology and Infection* **123**, 157–164.

Cliquet, F., Aubert, M. and Sagne, L. (1998). Development of a fluorescent antibody virus neutralisation test (FAVN test) for the quantitation of rabies-neutralising antibody. *Journal of Immunological Methods* **212**, 79–87.

Cliquet, F., McElhinney, L.M., Servat, A. *et al.* (2004) Development of a qualitative indirect ELISA for the measurement of rabies virus-specific antibodies from vaccinated dogs and cats. *Journal of Virological Methods* **117**, 1–8.

Cliquet, F., Muller, T., Mutinelli, F. *et al.* (2003) Standardization and establishment of a rabies ELISA test in European laboratories for assessing the efficacy of oral fox vaccination campaigns. *Vaccine* **21**, 2986–2993.

Cliquet, F., Sagne, L., Schereffer, J. L. and Aubert, M. F. A. (2000). ELISA test for rabies antibody titration in orally vaccinated foxes sampled in the fields. *Vaccine* **18**, 3272–3279.

Compton, J. (1991). Nucleic acid sequence based amplification. *Nature* **350**, 91–92.

Cooper, A. and Poinar, H.N. (2000) Ancient DNA: Do it right or not at all. *Science* **289**, 1139.

Crepin, P., Audry, L., Rotivel, Y., Gacoin, A., Caroff, C. and Bourhy, H. (1998). Intravitam diagnosis of human rabies by PCR using saliva and cerebrospinal fluid. *Journal of Clinical Microbiology* **36**, 1117–1121.

Da Silva, L.H.Q., Bissoto, C.E., de Carvalho, C., Cardoso, T.C., Pinheiro, D.M. and Perri, S.H.V. (2002). Comparison between the counterimmunoelectrophoresis test and the mouse neutralization test for the detection of antibodies against rabies virus in dog sera. *Memorias do Instituto Oswaldo Cruz* **97**, 259–261.

David, D., Yakobson, B., Rotenberg, D., Dveres, N., Davidson, I. and Stram, Y. (2002). Rabies virus detection by RT-PCR in decomposed naturally infected brains. *Veterinary Microbiology* **87**, 111–118.

Davis, C., Neill, S. and Raj, P. (1997). Microwave fixation of rabies specimens for fluorescent antibody testing. *J. Virol. Methods* **68**, 177–182.

Debbie, J.G. and Trimarchi, C.V. (1992). Rabies. In: *Veterinary Diagnostic Virology* (A.E. Castro and W.P. Heuschele, eds). pp. 116–120. Boston: Mosby Year Book.

Debbie, J.G. and Trimarchi, C.V. (1997). Prophylaxis for suspected exposure to bat rabies. *Lancet* **350**, 1790–1791.

Elmgren, L.D., Nadin-Davis, S.A., Muldoon, F.T. and Wandeler, A.I. (2002). Diagnosis and analysis of a recent case of human rabies in Canada. *Canadian Journal of Infectious Disease* **13**, 129–133.

Ermine, A., Tordo, N. and Tsiang, H. (1988). Rapid diagnosis of rabies infection by means of a dot hybridization assay. *Molecular and Cellular Probes* **2**, 75–82.

Flamand, A., Wiktor, T.J. and Koprowski, H. (1980). Use of monoclonal antibodies in the detection of antigenic differences between rabies virus and rabies-related virus proteins. I. The nucleocapsid protein. *Journal of General Virology* **48**, 105–109.

Franka, R., Svrcek, S., Madar, M. *et al.* (2004). Quantification of the effectiveness of laboratory diagnostics of rabies using classical and molecular-genetic methods. *Veterinary Medicine* **49**, 259–267.

Fraser, G.C., Hooper, P.T., Lunt, R.A. *et al.* (1996). Encephalitis caused by a lyssavirus in fruit bats in Australia. *Emerging Infectious Diseases* **2**, 327–330.

Grassi, M., Wandeler, A. and Peterhans, E. (1989) Enzyme-linked immunosorbent assay for determination of antibodies to the envelope glycoprotein of rabies virus. *Journal of Clinical Microbiology* **27**, 899–902.

Guyatt, K.J., Twin, J., Davis, P. *et al.* (2003). A molecular epidemiological study of Australian bat lyssavirus. *Journal of General Virology* **84**, 485–496.

Hamir, A.N., Moser, Z.F., Fu, F., Dietzschold, B. and Rupprecht, C.E. (1995). Immunohistochemical test for rabies: identification of a diagnostically superior monoclonal antibody. *Veterinary Record* **136**, 295–296.

Hanlon, C., Smith, J., Anderson, G., Trimarchi, C.V. and Schnurr, D. (1999). Laboratory diagnosis of rabies: report of the National Working Group on Prevention and Control of Rabies. *Journal of the American Veterinary Medical Association* **215**, 1444–1446.

Heaton, P.R., Johnstone, P., McElhinney, L.M., Cowley, R., O'Sullivan, E. and Whitby, J.E. (1997). Heminested PCR assay for detection of six genotypes of rabies and rabies-related viruses. *Journal of Clinical Microbiology* **35**, 2762–2766.

Heaton, P.R., McElhinney, L.M. and Lowings, J.P. (1999). Detection and identification of rabies and rabies-related viruses using rapid-cycle PCR. *Journal of Virological Methods* **81**, 63–69.

Hemachudha, T. and Wacharapluesadee, S. (2004). Antemortem diagnosis of human rabies. *Clinical Infectious Diseases* **39**, 1085–1086.

Hill, R.E. Jr, Beran, G.W. and Clark, W.R. (1992). Demonstration of rabies virus-specific antibody in the sera of free-ranging Iowa raccoons (Procyon lotor). *Journal of Wildlife Disease* **28**, 377–385.

Hostnik, P. (2000). The modification of fluorescent antibody virus neutralization (FAVN) test for the detection of antibodies to rabies virus. *Journal of Veterinary Medicine Series B* **47**, 423–427.

Hughes, G.J., Smith, J.S., Hanlon, C.A. and Rupprecht, C.E. (2004). Evaluation of a TaqMan PCR assay to detect rabies virus RNA: influence of sequence variation and application to quantification of viral loads. *Journal of Clinical Microbiology* **42**, 299–306.

Hummeler, K. and Atanasiu, P. (1996) Electron microscopy. In: *Laboratory Techniques in Rabies*, 4th edn (F.X. Meslin, M.M. Kaplan and H. Koprowski, eds). pp. 209–217. Geneva: World Health Organization.

International Air Transport Association (2006). IATA Dangerous Goods Regulations 2006 – 47th edn.

Ito, F.H., Vasconcellos, S.A., Erbolato, E.B., Macruz, R. and Cortes, J.A. (1985). Rabies virus in different segments of brain and spinal cord of naturally and experimentally infected dogs. *International Journal of Zoonoses* **38**, 98–105.

Jackson, A.C. (2005) Recovery from rabies (Editorial). *New England Journal of Medicine* **352**, 2549–2550.

Jackson, A.C. and Wunner, W.H. (1991). Detection of rabies virus genomic RNA and mRNA in mouse and human brains by using in situ hybridization. *Journal of Virology* **65**, 2839–2844.

Jackson, A.C., Phelan, C.C. and Rossiter, J.P. (2000). Infection of Bergmann glia in the cerebellum of a skunk experimentally infected with street rabies virus. *Canadian Journal of Veterinary Research* **64**, 226–228.

Jackson, A.C., Ye, H.T., Ridaura-Sanz, C. and Lopez-Corella, E. (2001). Quantitative study of the infection in brain neurons in human rabies. *Journal of Medical Virology* **65**, 614–618.

Johnson, N., Brookes, S.M., Fooks, A.R. and Ross, R.S. (2005) Review of human rabies cases in the UK and in Germany. *Veterinary Record* **157**, 715.

Kamolvarin, N., Tirawatnpong, T., Rattanasiwamoke, R., Tirawatnpong, S., Panpanich, T. and Hemachudha, T. (1993). Diagnosis of rabies by polymerase chain reaction with nested primers. *Journal of Infectious Diseases* **167**, 207–210.

Kaplan, C. (1985) Rabies: a world wide disease. In: *Population dynamics of rabies in wildlife* (P.J. Bacon, ed.). pp. 1–21. New York: Academic Press.

Kaplan, M.M. (1996). Safety precautions in handling rabies virus. In: *Laboratory Techniques in Rabies*, 4th edn (F.X. Meslin, M.M. Kaplan and H. Koprowski, eds). pp. 105–112. Geneva: World Health Organization.

Kaplan, M.M., Wiktor, T.J., Maes, R.F., Campbell, J.B. and Koprowski, H. (1967). Effect of polyions on the infectivity of rabies virus in tissue culture: construction of a single-cycle growth curve. *Journal of Virology* **1**, 145–151.

Khawplod, P., Inoue, K., Shoji, Y. *et al.* (2005). A novel rapid fluorescent focus inhibition test for rabies virus using a recombinant rabies virus visualizing a green fluorescent protein. *Journal of Virological Methods* **125**, 35–40.

Kissi, B., Tordo, N. and Bourhy, H. (1995). Genetic polymorphism in the rabies virus nucleoprotein gene. *Virology* **209**, 526–537.

Kissling, R.E. (1975). The fluorescent antibody test in rabies. In: *The Natural History of Rabies* (G.M. Baer, ed.). Vol. I, pp. 401–416. Boca Raton: CRC Press.

Kuzmin, I.V., Hughes, G.J., Botvinkin, A.D., Orciari, L.A. and Rupprecht, C.E. (2005). Phylogenetic relationships of Irkut and West Caucasian bat viruses within the Lyssavirus genus and suggested quantitative criteria based on the N gene sequence for lyssavirus genotype definition. *Virus Research* **111**, 28–43.

Kwok, S. and Higuchi, R. (1989). Avoiding false positives with PCR. *Nature* **339**, 237–238.

Lee, L.G., Connell, C.R. and Bloch, W. (1993). Alleleic discrimination by nick-translation PCR with fluorogenic probes. *Nucleic Acids Research* **21**, 3761–3766.

Lee, T.K. and Becker, M.E. (1973). Validity of spinal cord examination as a substitute procedure for routine rabies diagnosis. *Public Health Laboratory* **31**, 149–164.

Le Mercier, P., Jacob, Y. and Tordo, N. (1997). The complete Mokola virus genome sequence: structure of the RNA-dependent RNA polymerase. *Journal of General Virology* **78**, 1571–1576.

Lembo, T., Niezgoda, M., Velasco-Villa, A., Cleaveland, S., Ernest, E. and Rupprecht, C.E. (2006). Evaluation of a direct, rapid immunohistochemical test for rabies diagnosis. *Emerging Infectious Diseases* **12**, 310–313.

Lennette, E.H., Woodie, J.D., Nakamura, K., and Magoffin, R.L. (1965). The diagnosis of rabies by the fluorescent antibody method (FRA) employing immune hamster serum. *Health Laboratory Science* **2**, 24–34.

Lepine P. and Atanasiu, P. (1996). Histopathological diagnosis. In: *Laboratory Techniques in Rabies*, 4th edn (F.X. Meslin, M.M. Kaplan, and H. Koprowski, eds). pp. 66–79. Geneva: World Health Organization.

Lewis, V.J. and Thacker, W.L. (1974). Limitations of deteriorated tissues for rabies diagnosis. *Health Laboratory Science* **11**, 8–12.

Loza-Rubio, E., Rojas-Anaya, E., Banda-Ruiz, V.M., Nadin-Davis, S.A. and Cortez-Garcia, B. (2005). Detection of multiple strains of rabies virus RNA using primers designed to target Mexican vampire bat variants. *Epidemiology and Infection* **133**, 927–934.

Madhusudana, S.N., Paul, J.P., Abhilash, V.K. and Suja, M.S. (2004). Rapid diagnosis of rabies in humans and animals by a dot blot enzyme immunoassay. *International Journal of Infectious Diseases* **8**, 339–345.

Maserang, D.L. and Leffingwell, L. (1981). Single-site localization of rabies virus: impact on laboratory reporting policy. *American Journal of Public Health* **71**, 428–429.

McColl, K.A., Gould, A.R., Selleck, P.W., Hooper, P.T., Westbury, H.A. and Smith, J.S. (1993). Polymerase chain reaction and other laboratory techniques in the diagnosis of long incubation rabies in Australia. *Australian Veterinary Journal* **70**, 84–89.

McQuiston, J.H., Yager, P.A., Smith, J.S. and Rupprecht, C.R. (2001) Epidemiologic characteristics of rabies virus variants in dogs and cats in the United States during 1999. *Journal of the Veterinary Medical Association* **218**, 1939–1942.

Nadin-Davis, S.A. (1998). Polymerase chain reaction protocols for rabies virus discrimination. *Journal of Virological Methods* **75**, 1–8.

Nadin-Davis, S.A., Casey, G.A. and Wandeler, A.I. (1993). Identification of regional variants of the rabies virus within the Canadian province of Ontario. *Journal of General Virology* **74**, 829–837.

Nadin-Davis, S.A., Sheen, M. and Wandeler, A.I. (2003). Use of discriminatory probes for strain typing of formalin-fixed rabies virus-infected tissues by in situ hybridization. *Journal of Clinical Microbiology* **41**, 4343–4352.

Niezgoda, M. and Rupprecht, C.E. (1999). Towards the development of another rabies diagnostic test. 10th Annual Rabies in the Americas Meeting, San Diego.

Niezgoda, M., Briggs, D.J., Shaddock, J. and Rupprecht, C.E. (1999). Viral excretion in domestic ferrets (*Mustela putorius furo*) inoculated with a raccoon rabies isolate. *American Journal of Veterinary Research* **59**, 1629–1632.

Noah, D.L., Drenzek, C.L., Smith, J.S. *et al.* (1998). Epidemiology of human rabies in the United States, 1980 to 1996. *Annals of Internal Medicine* **128**, 922–930.

Perl, D.P. and Good, P.F. (1991). The pathology of rabies in the central nervous system. In: *The Natural History of Rabies*, 2nd edn (G.M. Baer, ed.). pp. 163–190. Boca Raton: CRC Press.

Perrin, P., Versmisse, P., Delagneau, J.F., Lucas, G., Rollin, P.E. and Sureau, P. (1986). The influence of the type of immunosorbent on rabies antibody EIA; advantages of purified glycoprotein over whole virus. *Journal of Biological Standardization* **14**, 95–102.

Picard-Meyer, E., Bruyere, V., Barrat, J., Tissot, E., Barrat, M.J. and Cliquet, F. (2004). Development of a hemi-nested RT-PCR method for the specific determination of European bat lyssavirus 1: Comparison with other rabies diagnostic methods. *Vaccine* **22**, 1921–1929.

Piza, A.S., Santos, J.L., Chaves, L.B. and Zanetti, C.R. (1999). An ELISA suitable for the detection of rabies virus antibodies in serum samples from human vaccinated with either cell culture vaccine or suckling mouse brain vaccine. *Revista do Instituto de Medicina Tropical de Sao Paulo* **41**, 39–43.

Plotkin, S.A. (2000). Rabies. *Clinical Infectious Diseases* **30**, 4–12.

Powell, J. (1997). Proficiency testing in the rabies diagnostic laboratory. Abstracts of the Eighth Annual Rabies in the Americas Conference, November 2–6, 1997, Kingston, Ontario.

Rani, A, Singh, C.K. and Monica, A. (2002). Counter-immuno electrophoresis for detection of rabies antibodies: an experimental study. *Indian Veterinary Journal* **79**, 1127–1128.

Rincon, V., Corredor, A., Martinez-Gutierrez, M. and Castellanos, J.E. (2005). Fluorometric cell-ELISA for quantifying rabies infection and heparin inhibition. *Journal of Virological Methods* **127**, 33–39.

Robinson, S.J. and DiSalvo, A.F. (1980). Rabies in South Carolina: 1969–1979. *Public Health Laboratory* **38**, 315–321.

Rudd, R.J. and Trimarchi, C.V. (1980). Tissue culture technique for routine isolation of street strain rabies virus. *Journal of Clinical Microbiology* **12**, 590–593.

Rudd, R.J. and Trimarchi, C.V. (1987). Comparison of sensitivity of BHK-21 and murine neuroblastoma cells in the isolation of a street strain rabies virus. *Journal of Clinical Microbiology* **25**, 1456–1458.

Rudd, R.J. and Trimarchi, C.V. (1989). The development and evaluation of an *in vitro* virus isolation procedure as a replacement for the mouse inoculation test in rabies diagnosis. *Journal of Clinical Microbiology* **27**, 2522–2528.

Rudd, R.J. and Trimarchi, C.V. (1997). Evans Blue counterstain in the rabies fluorescent antibody test. Abstracts of the Eighth Annual Rabies in the Americas Conference, November 2–6. Kingston, Ontario.

Rudd, R.J., Smith, J.S., Yager, P.A., Orciari, L.A. and Trimarchi, C.V. (2005). A need for standardized rabies-virus diagnostic procedures: Effect of cover-glass mountant on the reliability of antigen detection by the fluorescent antibody test. *Virus Research* **111**, 83–88.

Sacramento, D., Bourhy, H. and Tordo, N. (1991). PCR technique as an alternative method for diagnosis and molecular epidemiology of rabies virus. *Molecular and Cellular Probes* **5**, 229–240.

Saiki, R.K. (1989). The design and optimization of the PCR. In: *PCR Technology: Principles and Applications for DNA Amplification* (H.A. Erlich, ed.). pp. 7–16. New York: Stockton Press.

Schneider, L., (1973). Aluminum phosphate method for rabies virus purification. In: *Laboratory Techniques in Rabies*, 3rd edn. WHO Technical Report Series, No. 661. pp. 181–192. Geneva: World Health Organization.

Silverstein, M.A., Salgado, C.D., Bassin, S. *et al.* (2003). First human death associated with raccoon rabies – Virginia, 2003. *Journal of the American Medical Association* **290**, 2930–2931.

Smith, J.S. (1991). Rabies serology. In: *The Natural History of Rabies*, 2nd edn (G.M. Baer, ed.). pp. 235–252. Boca Raton: CRC Press.

Smith, J.S. (2000). Quality control for fluorescent antibody tests for rabies virus. International Meeting on Research Advances and Rabies Control in the Americas, Lima, Peru.

Smith, J., McElhinney, L.M., Heaton, P.R., Black, E.M. and Lowings, J.P. (2000). Assessment of template quality by the incorporation of an internal control into a RT-PCR for the detection of rabies and rabies-related viruses. *Journal of Virological Methods* **84**, 107–115.

Smith, J., McElhinney, L., Parsons, G. *et al.* (2003). Case report: Rapid ante-mortem diagnosis of a human case of rabies imported into the UK from the Philippines. *Journal of Medical Virology* **69**, 150–155.

Smith, J.S., Orciari, L.A., Yager, P.A., Seidel, H.D. and Warner, C.K. (1992). Epidemiologic and historical relationships among 87 rabies virus isolates as determined by limited sequence analysis. *Journal of Infectious Disease* **166**, 296–307.

Smith J.S., Reid–Sanden, F.L., Roumillat, L.F. *et al.* (1986). Demonstration of antigen variation among rabies virus isolates by using monoclonal antibodies to nucleocapsid proteins. *Journal of Clinical Microbiology* **24**, 573–580.

Smith, J.S., Trimarchi, C.V. and Neill, S.U. (1996a). *Rabies Testing Proficiency Forum: Avoiding Cross Contamination.* Madison: Wisconsin State Laboratory of Hygiene.

Smith, J.S., Yager, P.A. and Baer, G.M. (1996b). A rapid fluorescent focus inhibition test (RFFIT) for determining rabies virus neutralizing antibody. In: *Laboratory Techniques in Rabies*, 4th edn (F.X. Meslin, M.M. Kaplan and H. Koprowski, eds). pp. 181–192. Geneva: World Health Organization.

Smith, I.L., Northill, J.A., Harrower, B.J. and Smith, G.A. (2002). Detection of Australian bat lyssavirus using a fluorogenic probe. *Journal of Clinical Virology* **25**, 285–291.

Srinivasan, A., Burton, E.C., Kuehnert, M.J. *et al.* (2005). Transmission of rabies virus from an organ donor to four transplant recipients. *New England Journal of Medicine* **352**, 1103–1111.

Steindel, S.J., Rauch, W.J., Simon, M.K. and Handsfield, J. (2000). National inventory of clinical laboratory testing services (NICLTS): development and test distribution for 1996. *Archives of Pathology and Laboratory Medicine* **124**, 1201–1208.

Sugiyama, M., Yoshiki, R., Tatsuno, Y. *et al.* (1997). A new competitive enzyme-linked immunosorbent assay demonstrates adequate immune levels to rabies virus in compulsorily vaccinated Japanese domestic dogs. *Clinical and Diagnostic Laboratory Immunology* **4**, 727–730.

Surreau, P., Ravisse, P. and Rollin, P.E. (1991). Rabies diagnosis by animal inoculation, identification of Negri bodies, or ELISA. In: *The Natural History of Rabies*, 2nd edn (G.M. Baer, ed.). pp. 203–217. Boca Raton: CRC Press.

Tierkel E.S. and Atanasiu, P. (1996). Rapid microscopic examination for Negri bodies and preparation of specimens for biological tests. In: *Laboratory Techniques in Rabies*, 4th edn (F.X. Meslin, M.M. Kaplan and H. Koprowski, eds). pp. 55–65. Geneva: World Health Organization.

Tordo, N., Bourhy, H. and Sacramento, D. (1995). Polymerase chain reaction technology for rabies virus. In: *The Polymerase Chain Reaction (PCR) for Human Viral Diagnosis* (J.P. Clewley, ed.). pp. 125–145. Boca Raton: CRC Press.

Trimarchi, C.V. (1994) 1993 Rabies Annual Summary. NYSVMS *Veterinary News,* **59**, 1.

Trimarchi, C.V. (2000). Rabies. In: *Clinical Virology Manual* (S. Specter, R. Hodinka and S. Young, eds). pp. 335–338. Washington, DC: ASM Press.

Trimarchi, C.V. and Briggs, D.J. (1999). The diagnosis of rabies. In: *Rabies: Guidelines for Professionals.* pp. 55–66. Trenton, NJ: Veterinary Learning Systems.

Trimarchi, C.V. and Debbie, J.G. (1974). Production of rabies fluorescent conjugate by immunization of rabbits with purified rabies antigen. *Bulletin of the World Health Organization* **51**, 447–449.

Trimarchi C.V. and Debbie, J.G. (1991). The fluorescent antibody in rabies. In: *The Natural History of Rabies*, 2nd edn (G.M. Baer, ed.). pp. 219–233. Boca Raton: CRC Press.

Trimarchi, C.V. and Smith, J.S. (2002). Diagnostic Evaluation. In: *Rabies* (A.C. Jackson and W.H. Wunner, eds). pp. 307–349. New York: Academic Press.

Trimarchi, C.V., Rudd, R.J. and Abelseth, M.K. (1986). Experimentally induced rabies in four cats inoculated with a rabies virus isolated from a bat. *American Journal of Veterinary Research* **47**, 777–780.

Trimarchi, C.V., Rudd, R.J. and Safford, M. Jr (1996). An in vitro virus neutralization test for rabies antibody. In: *Laboratory Techniques in Rabies*, 4th edn (F.X. Meslin, M.M. Kaplan and H. Koprowski, eds). pp. 193–199. Geneva: World Health Organization.

Umoh, J.V. and Blenden, D.C. (1983). Use of monoclonal antibodies in diagnosis of rabies virus infection and differentiation of rabies and rabies-related viruses. *Journal of Virological Methods* **1**, 33–46.

Wacharapluesadee, S. and Hemachudha, T. (2001). Nucleic-acid sequence based amplification in the rapid diagnosis of rabies. *Lancet* **358**, 892–893.

Wakeley, P.R., Johnson, N., McElhinney, L.M., Marston, D., Sawyer, J. and Fooks, A.R. (2005). Development of a real-time, TaqMan reverse transcription-PCR assay for detection and differentiation of lyssavirus genotypes 1, 5 and 6. *Journal of Clinical Microbiology* **43**, 2786–2792.

Wandeler, A. and Kumor, L. (2005). Canadian Food Inspection Agency's notice to private practition-ers: Biting animals (dogs, cats and ferrets) and public health: Immediate euthanasia for rabies testing vs. the 10-day observation period. *Canadian Veterinary Journal* **46**, 917.

Warner C.K., Whitfield, S.G., Fekadu, M. and Ho, H. (1997). Procedures for reproducible detection of rabies virus antigen, mRNA and genome *in situ* in formalin fixed tissues. *Journal of Virological Methods* **67**, 5–12.

Warner, C.K., Zaki, S.R., Shieh, W. *et al.* (1999). Laboratory investigation of human deaths from vampire bat rabies in Peru. *American Journal of Tropical Medicine and Hygiene* **60**, 502–507.

Webster, L.T. and Dawson, J.R. (1935). Early diagnosis of rabies by mouse inoculation. Measurement of humoral immunity to rabies by mouse protection test. *Proceedings of the Society for Experimental Biology and Medicine* **32**, 570–573.

Webster, W.A. and Casey, G.A. (1996). Virus isolation in neuroblastoma cell culture. In: *Laboratory Techniques in Rabies*, 4th edn (F.X. Meslin, M.M. Kaplan and H. Koprowski, eds). pp. 96–104. Geneva: World Health Organization.

Wellenberg, G.J., Audry, L., Rønsholt, L., van der Poel, W.H.M., Bruschke, C.J.M. and Bourhy, H. (2002). Presence of European bat lyssavirus RNAs in apparently healthy *Rousettus aegyptiacus* bats. *Archives of Virology* **147**, 349–361.

Whitby, J.E., Heaton, P.R., Whitby, H.E., O'Sullivan, E. and Johnstone, P. (1997a). Rapid detection of rabies and rabies-related viruses by RT-PCR and enzyme-linked immunosorbent assay. *Journal of Virological Methods* **69**, 63–72.

Whitby, J.E., Johnstone, P. and Sillero-Zubiri, C. (1997b). Rabies virus in the decomposed brain of an Ethiopian wolf detected by nested reverse transcription-polymerase chain reaction. *Journal of Wildlife Diseases* **33**, 912–915.

Whitfield, S.G., Fekadu, M., Shaddock, J.H., Niezgoda, M., Warner, C.K., Messenger, S.L. and the Rabies Working Group. (2001). A comparative study of the fluorescent antibody test for rabies diagnosis in fresh and formalin-fixed brain tissue specimens. *Journal of Virological Methods* **95**, 145–151.

Wiktor, T.J. and Clark, H.F. (1973). Application of the plaque assay technique to the study of rabies virus-neutralizing antibody interactions. *Annals of Microbiology (Paris)* **124**, 271–282.

Wiktor, T.J., Flamand, A. and Koprowski, H. (1980). Use of monoclonal antibodies in diagnosis of rabies virus infection and differentiation of rabies and rabies-related viruses. *Journal of Virological Methods* **1**, 33–46.

Willis, K., Tims, T., Schweitzer, K., Davis, R. and Briggs, D. (2000). A change in sampling methodol-ogy at the Kansas State University Rabies Diagnostic Laboratory. International Meeting on Research Advances and Rabies Control in the Americas, Lima, Peru.

Willoughby, R.E., Tieves, K.S., Hoffman, G.M. *et al.* (2005). Survival after treatment of rabies with induction of coma. *New England Journal of Medicine* **352**, 2508–2514.

Wong, A.J., Constantine, D.G., Armstrong, O., Wong, W.Y. and Comb, J.C. (2004). A novel tech-nique to eliminate cross-contamination when making wells on slides for rabies diagnosis. *Journal of Virological Methods* **115**, 117–122.

World Health Organization Expert Consultation on Rabies (2004). WHO Technical Report Series No. 931. Geneva: WHO.

Zaidman, G.W. and Billingsley, A. (1998). Corneal impression test for the diagnosis of acute rabies encephalitis. *Ophthalmology* **105**, 249–251.

11

Rabies Serology

SUSAN M. MOORE, CHANDRA R. GORDON, AND DEBORAH J. BRIGGS

Rabies Diagnostic Laboratory, Department of Diagnostic Medicine Pathobiology,
College of Veterinary Medicine, Kansas State University, Manhattan, KS 66506, USA

1 INTRODUCTION

The science of serology involves the detection of immune components in the blood, particularly serum. These components range from immunoglobulins to cytokines. It is the presence of rabies-specific antibodies in serum or cerebrospinal fluid (CSF) that indicates prior exposure to rabies virus antigens. The measurement of rabies virus specific antibodies is a useful tool for evaluating both the level of humoral immunity in vaccinated individuals and for diagnostic purposes in patients suspected of having rabies. Many serological techniques have been developed over the past five decades, which differ not only in their ability to detect the isotype, affinity and specificity of rabies virus antibodies, but also in the ease and practicality with which they are performed. Knowing the specific strengths, weaknesses and limitations of the methods available enables the selection of the appropriate method and aids in the interpretation of the subsequent results. A variety of serological methods have been developed for testing the presence of rabies virus antibodies. Numerous reports indicate that protection against rabies is dependent upon the presence of rabies virus neutralizing antibodies (RVNAs). Thus, assays to detect and quantify RVNAs, such as the rapid fluorescent focus inhibition test and the fluorescent antibody virus neutralization test, are the methods recommended for quantitation purposes in rabies serology. Antigen binding assays have proven to be useful for the detection of specific isotypes of rabies virus antibodies, using either whole virions or specific viral proteins as antigen(s). The sensitivity and specificity along with the accuracy and precision required should be the primary determining factors in the selection of an assay. The laboratory materials, instruments and safety equipment available also must be considered when evaluating the assay to be used. Finally, it is critical to understand exactly which components will be measured in terms of the antibody response itself when choosing an appropriate assay for detection of rabies virus.

Rabies, second edition. Edited by Alan C. Jackson and William H. Wunner
ISBN 978-012-369366-2. Copyright Elsevier Inc. 2007

2 INVESTIGATIVE SEROLOGY

Investigative serology focuses on the detection and measurement of immune components in blood (usually serum) ranging from immunoglobulins of several subclasses directed against specific epitopes to different types of cytokines. Depending on the time point in the course of the humoral immune response after exposure to an antigen, the presence of IgM and/or IgG may or may not be measured. The specificity of the immune response is affected by distinct epitopes present on the rabies viral proteins used to generate the antibodies. Whole rabies virus induces the formation of antibodies potentially against all viral proteins, but predominantly against the rabies virus glycoprotein (G) and nucleoprotein (N). Studies of monoclonal antibodies (MAbs) capable of neutralizing rabies virus indicate that these MAbs are directed against a number epitopes on the G of rabies virus (Tordo, 1996). Mechanisms associated with rabies virus neutralization require a minimum number of antibody molecules per G spike to induce steric hindrance of the virus-receptor-binding activity and other mechanism(s) that may involve conformational changes in the G protein ultimately resulting in the loss of virion receptor-binding ability (Irie and Kawai, 2002). The humoral immune response elicited by an individual after exposure to rabies virus antigen(s) consists of a mixture of polyclonal antibodies that influence a variety of complex neutralization mechanisms.

Antibodies directed against rabies virus antigens can be detected and measured by several different serology methods including precipitation, agglutination, immunoelectrophoresis, radioimmunoassay, enzyme-linked immunosorbent assays (ELISA), Western blots, immunofluorescence, immunoelectron microscopy and virus neutralization assays. All of these assays measure the presence of antibody through an antibody–antigen interaction. Three different types of assays based on antibody function and antibody–antigen interactions have been described: antibody-function assays, antigen-binding assays and antigen-function assays (Smith, 1991). Due to inherent problems associated with antibody-function assays, such as complement fixation (CF), immune adherence hemagglutination (IAHA), complement-mediated cell lysis (CMCL), passive hemagglutination (PHA), mixed hemadsorption (MH) and counter-immunoelectrophoresis, these assays have been largely replaced by assays involving primary binding activity between antibodies and antigens or by virus neutralization assays. Although other components and products of the immune system are involved, protection from clinical rabies after infection relies heavily on the presence of RVNA (Hooper *et al.*, 1998). Therefore, methods to detect and quantify neutralizing antibodies are necessary to quantify the level of immunity after rabies vaccination and subsequently the need for routine booster administration. Rabies virus neutralization assays rely on being able to detect the virus neutralization activity of RVNAs *in vitro* and therefore mimic the protective action of these antibodies *in vivo*. Technical performance of rabies virus neutralization assays requires the use of live virus and is labor intensive and time consuming. There are two rabies virus neutralizing assays recognized by the World Health Organization (WHO) to measure RVNAs the rapid fluorescent focus inhibition test (RFFIT) and the fluorescent antibody virus neutralization test (FAVN).

Antigen binding assays such as ELISAs and indirect immunofluorescence assays (IFAs) are rapid, simple and do not require manipulation of live rabies virus. These assays rely on the interaction of the antibody and antigen, regardless of the ability of the antibody to neutralize rabies virus, and are useful for the detection of rabies virus binding antibodies. The use of whole virions or purified viral proteins as the antigen provides a useful tool to differentiate the specificity of antibodies produced by the humoral immune response and present in sera at the time the blood was withdrawn for testing. In addition, it is possible to identify the presence of specific subclasses of rabies virus antibodies using the appropriate conjugated anti-subclass Ig antibody. Since not all rabies virus binding antibodies neutralize rabies virus, the titers obtained from antigen-binding assays should not be considered to be identical to RVNA titers and therefore extreme caution must be taken when evaluating test results.

Rabies serological testing is routinely used for the confirmation of an immune response in both humans and animals after rabies vaccination and for evaluating the timing of routine booster immunizations after primary immunization. They are used for the diagnosis of rabies in humans and for federal agencies evaluating the immunity afforded by new rabies vaccines or vaccination regimens. Serological testing is also a valuable tool in field trials investigating the immunogenicity of oral rabies vaccines. These vaccines are increasingly being used as a method to eliminate rabies in the wildlife and stray dog populations throughout the world.

Finally, rabies serology continues to play a critical role in the assessment of dogs and cats imported into rabies-free regions of the world. Just a decade ago, long quarantine periods were the only means by which rabies-free countries attempted to prevent the introduction of potentially infected dogs and cats within their borders. Reliable serological testing for the presence of RVNAs, along with vaccination and appropriate documentation, has transformed the quarantine system. Currently, most rabies-free regions of the world rely heavily on a handful of qualified laboratories to conduct rabies serological testing on healthy dogs and cats being imported into their respective countries. Ultimately, it was the availability of reliable rabies virus neutralization tests that dismantled the archaic animal quarantine systems and simplified the process by which people could move to rabies-free countries without either leaving their cherished pets behind or boarding them in unfamiliar, isolated conditions for long periods of time.

The ultimate choice of a serological method will depend on the specific requirements imposed by the investigator, technician or researcher and the purpose of the test. Individual testing requirements may include the acceptable level of sensitivity and specificity as well as the accuracy and precision that are necessary. Consideration of materials, equipment and the safety required for the testing procedure will also influence the decision as to what method is suitable for routine diagnoses. Finally, it is important to understand what component of the antibody response itself is to be measured prior to making a decision as to which assay would be the most appropriate procedure to choose for either research or diagnostic purposes.

In the final analysis, rabies serological testing plays an important role in all well-designed rabies prevention programs, including the registration and licensing of new rabies vaccines, monitoring the immune status of humans, animal vaccination campaigns, surveillance and epidemiology of rabies and, finally, maintaining rabies-free zones. Therefore, the importance of choosing the correct serological assay as well as ensuring that the testing procedure was conducted appropriately with adequate quality assurance procedures in place cannot be overstated. The results obtained will ultimately influence not only the dependability and reputation of the laboratory and its potential clients, but also will influence the evaluation of research results published in scientific journals, international trade agreements, establishment and maintenance of new and existing rabies-free zones and the ever needful development of new vaccines.

3 SEROLOGIC METHODS

3.1 Antigen-binding assays

Antigen-binding assays detect specific antigen–antibody reactions that occur when serum comes in contact with antigen bound to a solid surface, i.e. tube, plate or beads. The binding of the antibody to the antigen is detected by an additional binding step in which either anti-immunoglobulin or Fc-binding protein (staphylococcal protein A or streptococcal protein G) conjugated with an enzyme or a fluorochrome marker is incubated with the antibody–antigen complex (Figure 11.1). The antibody concentration is estimated by comparing the degree of color development or fluorescence with a reference standard. A quantitative evaluation is possible by testing serial dilutions of the serum samples. Additionally, the amount of antibody can be calculated by mathematical extrapolation from the optical density (OD) readings of known positive samples. Adaptations to this basic concept include changing the type of antigen and/or antibody according to the purpose of the test: for example, the selection of a specific viral variant, whole virus or specific viral protein as the antigen in order to detect the presence of specific antibodies in a serum sample. Additionally, the choice of a particular class of antibody used to bind the antigen will allow detection of the presence or absence of specific isotypes of antibodies in the serum sample.

For serological purposes, the IFA technique can be designed to detect the presence of different classes of rabies virus-specific antibodies. In this case, the test serum is incubated with rabies virus-infected cells fixed on slides. Rabies virus antibodies that are present in the serum will bind to rabies virus proteins

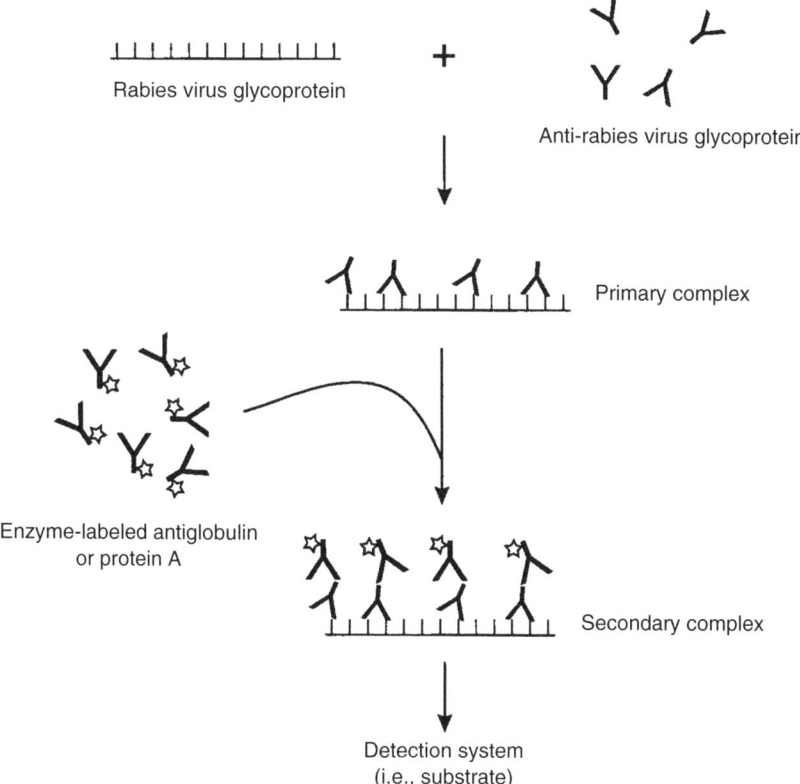

Rabies virus glycoprotein

+

Anti-rabies virus glycoprotein

Primary complex

Enzyme-labeled antiglobulin
or protein A

Secondary complex

Detection system
(i.e., substrate)

Figure 11.1 A variation of a binding assay is the indirect ELISA technique using rabies virus glycoprotein (short straight vertical lines) bound to microtiter wells (bottom horizontal line). The presence of rabies virus glycoprotein binding antibodies (Y-shaped symbols) is detected by an enzyme-labeled antiglobulin or *Staphylococcus aureus* protein A (Y shapes with star attached). A detection system involving the addition of an enzyme substrate leads to a color change proportional to the amount of bound antibody.

in the infected cells and are subsequently detected by FITC-labeled anti-IgG or anti-IgM. Slides are examined under a fluorescence microscope for the presence of the labeled antibodies. The rabies virus-specific antibodies can be quantified by assaying serial dilutions of the serum to determine the endpoint titer. This technique is not time consuming and has an increased sensitivity due to the amplification of fluorescence by multiple labeled antibodies binding to the rabies virus-specific antibody. The antibodies that are detected may include those that bind to all rabies viral proteins present in the infected cell and therefore cannot be used as a measure of RVNA levels.

Nicholson and Prestage (1982) described an ELISA procedure for the detection of anti-rabies virus antibodies that has been used as originally published and also modified for the detection of IgG antibodies to inactivated whole rabies virus. According to the original method, the endpoint titer is calculated from the intercept between the fitted curve of the optical density (OD) values of each serum dilution and the straight line of the mean, plus 2–3 standard deviations of the mean OD values of the negative controls. Alternatively, calculating the value of the unknown serum sample using only one dilution can be accomplished using a standard curve of OD values obtained from the serial dilutions of a positive control of known concentration. Modifications to the ELISA include the use of purified rabies virus G as the antigen enabling the detection of antibodies specific to the G (Grassi *et al.*, 1989). Antibodies to purified G are responsible for rabies virus neutralization and published information indicates that, by using the G as the binding antigen, the correlation between ELISA and RVNA values is improved. The use of species-specific or staphylococcus protein A increases the versatility of the ELISA by enabling the procedure to detect antibodies from other species (Barton and Campbell, 1988; Mebatsion *et al.*, 1989). A sandwich ELISA incorporating a liquid-phase blocking procedure introduces a neutralization step thus eliminating the need for a species-specific conjugated immunoglobulin (Esterhuysen *et al.*, 1995). The inhibition enzyme immunoassay (INH EIA) and the competitive ELISA (cELISA) both utilize competition (the rabies virus-specific antibodies in the unknown serum sample competing with a known amount of enzyme conjugated antibody), thus overcoming some of the disadvantages of the ELISA method described above (Kitala *et al.*, 1990; Elmgren and Wandeler, 1996). The competition for a limited amount of bound antigen between any rabies virus antibodies in the unknown serum sample and the enzyme-labeled 'known' rabies neutralizing antibody allows a more direct measurement of the neutralizing antibody in the unknown sample by the amount of reduction of color intensity. The INH EIA uses a polyclonal goat IgG while the cELISA is a more specific test due to the use of a monoclonal anti-rabies G antibody. The cELISA eliminates the non-specific binding problem of the antibody, therefore making it a more specific test, but the sensitivity may be reduced due to the use of only one monoclonal antibody (MAb). A more subjective ELISA assay, the dot immunobinding assay, utilizes antigen bound to nitrocellulose sheets (Heberling *et al.*, 1987). Serum is applied on paper strips, which are placed over the antigen dots on the sheets and the subsequent conjugated antibody–substrate color development is graded 1+ to 4+.

ELISA methods have several advantages including the fact that they are rapid, require little expertise, do not need high-level biohazard facilities to be performed and several steps of the procedure can be automated (i.e. serial dilution of the sera, addition of reagents and optical density reading). Additionally, the software packages are available to calculate endpoint titers or antibody concentration, thus allowing for objective reading and interpretation. Disadvantages of ELISA methods include the restrictive nature of the conjugated antibody or protein A that limits the isotype of immunoglobulin detected and, also, the use of anti-IgG makes the assay more or less species-specific. Additionally, although the use of protein A increases test utilization by being able to detect IgG from several species, it does not react with all forms of IgG3 and therefore will lead to an underestimation of the level of rabies virus-specific antibodies in serum containing higher proportions of IgG3 antibodies (Carpenter, 1997). It is important to understand that the degree of non-specific binding detected by an ELISA will depend on the purity of the antigen preparation and the efficiency of the coating step because immunoglobulins will non-specifically adhere to glass and plastic and also to contaminating material (i.e. mycoplasma) or cell culture components. Quantification of IgG antibodies that bind to rabies virus will not precisely demonstrate the level of protective virus neutralizing antibodies present in the sera and therefore the ELISA method is not appropriate for measuring the amount of RVNA present. Additionally, reporting of ELISA results in IU/ml is not reflective of this unit of measurement as defined by the WHO, i.e. 1 international unit of neutralizing activity per mg of protein. Therefore, the use of IU/ml to describe an antigen-binding assay for a rabies virus titer result is misleading.

3.2 Neutralizing antibody assays

Globally, the most widely used serologic method to measure the presence of rabies virus antibodies is the virus neutralization assay (Figure 11.2). Although there are other immune mechanisms that protect against the development of clinical rabies, RVNAs are most closely associated with protection (Dietzschold, 1993; Hooper *et al.*, 1998). Therefore, *in vitro* or, in the case of mouse neutralization assays, *in vivo* virus neutralization assay is considered to be the assay that most closely mimics the action of the RVNA after a person is vaccinated or infected with rabies virus. There are many variations of rabies virus neutralization tests but some basic steps are consistent in all of these assays, including the incubation of dilutions of heat-inactivated serum with a fixed amount of live rabies virus for 60–90 minutes at 37°C. Measurement of residual virus infectivity involves inoculation of virus into live animals or cell culture and subsequent calculation of the quantitative titer by using the proportion of deaths that occur or the number of detectable infected cells, respectively. The mouse neutralization test was developed in 1935 and was initially considered to be the gold standard. It is the assay that was used to validate the first *in vitro* neutralization methods (Atanasiu, 1973). The RFFIT, described in 1973 by Smith *et al.* (1973) and the FAVN, developed in 1997 by Cliquet *et al.* (1998), are the only two *in vitro* virus neutralization methods recommended by WHO and are therefore the most widely used. The RFFIT method is conducted in multichamber slides. Serum is serially diluted using fivefold dilutions in each well. Variations of the RFFIT include the use of microtiter plates in place of the slides and the use of twofold or threefold dilutions. The rabies challenge virus should contain 30–100 TCID. After the virus is added to the diluted serum, the slides are incubated at 37°C, after which baby hamster kidney (BHK) or mouse neuroblastoma (MNA) cells are added to each of the wells. Diethylaminoethyl-dextran (DEAE-dextran)

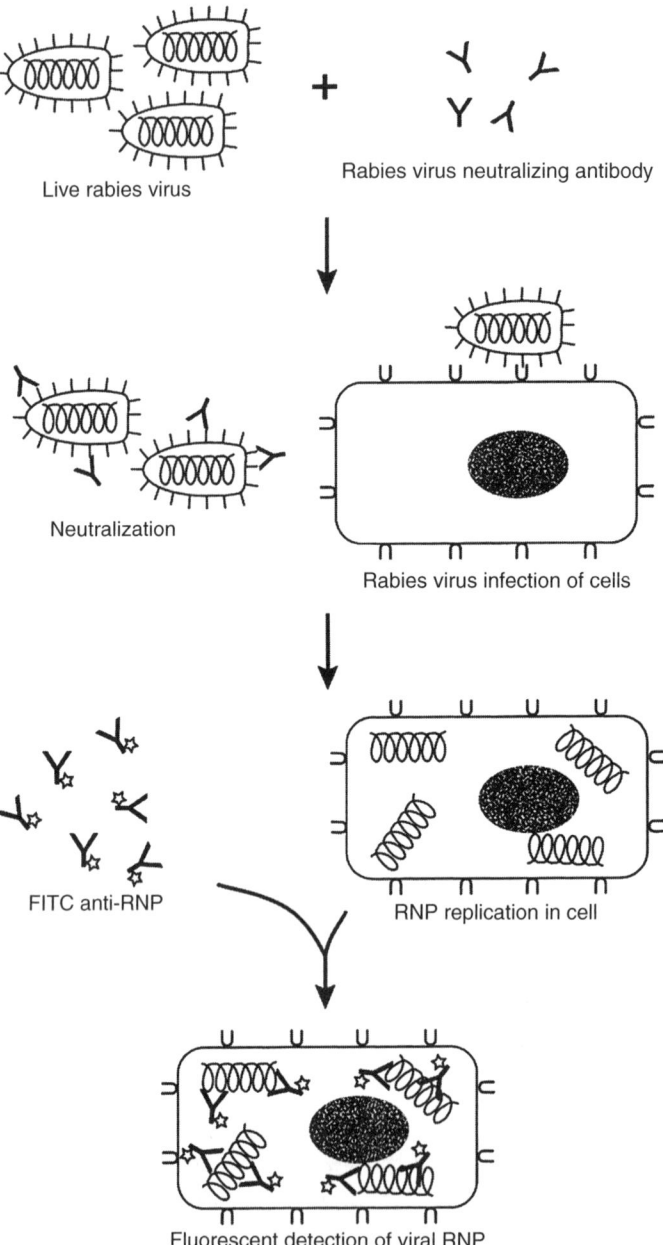

Figure 11.2 Rabies virus neutralization assays involve mixing a fixed amount of live rabies virus (bullet shape with helix in center surrounded by spikes) with serum containing rabies virus neutralizing antibodies (Y shapes) and allowing neutralization to occur through interaction of rabies virus neutralizing antibodies with the rabies virus glycoprotein (spikes). Non-neutralized rabies virus infects cells (oblong with U-shaped rabies virus receptors) added to the mixture. Rabies virus RNP (helix shapes) in the infected cells is detected with fluorescent labeled anti-RNP (Y shape with star attached).

Live rabies virus

Rabies virus neutralizing antibody

Neutralization

Rabies virus infection of cells

FITC anti-RNP

RNP replication in cell

Fluorescent detection of viral RNP

has been used, typically at a 0.01 μg/ml concentration, in some variations of the RFFIT to increase infectivity of the cells (Kaplan *et al.*, 1967). The slides are generally incubated at 37°C in a 5% CO_2 incubator for 20–24 hours, although in some variations of the method conducted in microtiter plates the incubation period is extended to 48 hours. The wells, containing an adherent monolayer of cells, are washed and the cells are fixed with 80% cold acetone. FITC-conjugated anti-rabies antibody directed against the rabies virus N is added in order to detect virus-infected cells. In 8-well chamber slides, 20 fields of each well are examined, using a fluorescence microscope, for the presence of fluorescence in the cells indicating the presence of non-neutralized rabies virus. The titer of RVNA in the serum sample being analyzed is defined as the dilution at which 50% of the observed microscopic fields contain one or more infected cells. Mathematical calculation using the Reed and Muench or Spearman-Karber method will determine the exact quantitative titer of RVNA in the serum sample. Alternatively, the quantitative titer of RVNA can be more simply defined, albeit with less precision, as the highest serum dilution where 100% viral inhibition occurred, thus indicating that there were no infected cells at that dilution and all subsequent higher dilutions exhibit infected cells (Aubert, 1996; Habel, 1996). Transcribing a serum dilution value into a more globally recognized measure of IU/ml is achieved by a simple calculation wherein the value from a serum sample being tested is compared with the serum dilution value of a reference serum standard containing a specific amount of RVNA previously tested and verified to be accurate (WHO, 1985, 1994). The skill and expertise of the technician conducting the test, including the analysis of the microscopic readout, can dramatically affect the precision of the RFFIT. To simplify and reduce the subjectivity of the microscopic counting step, the FAVN method uses four replicates of serum using threefold dilutions in microtiter wells and scores each well as either positive or negative for the presence of rabies virus-infected cells after a 48-hour incubation. The two methods have been demonstrated to be not statistically different upon direct comparison in a laboratory adhering to good quality assurance standards (Briggs *et al.*, 1998). Precision and repeatability of virus neutralization test results can be controlled by strict adherence to the dose and strain of the challenge virus used and the source of the standard reference serum. Early published reports that compared different laboratory RFFIT results found that the use of a high infective dose of challenge virus resulted in reduced sensitivity for testing low titered sera, whereas a low viral dose of challenge virus could result in lower precision when testing high titered sera such as rabies virus immunoglobulin (RIG) preparations (Fitzgerald *et al.*, 1975). In addition, the use of an equine RIG as the reference standard to determine IU/ml values resulted in significantly different titer results than when a human RIG reference standard was used (Lyng *et al.*, 1989). Rabies virus neutralization tests can be used to measure the presence of all classes of immunoglobulin in a sample (both IgM and IgG) and therefore will be able to detect the early production of rabies virus antibody after exposure or vaccination, but may not be as sensitive as the IFA (Smith, 1991). Since the virus neutralization testing method depends on the measurement of residual, or 'non-neutralized' rabies virus infecting the cells, the presence of interference factors in the sera or culture media that adversely affects cell, and ultimately viral, growth will mimic virus neutralization by non-specifically inhibiting viral growth. Therefore, any inhibition of viral growth not directly due to neutralizing antibody will give a false-positive result. As with ELISA techniques, some steps of the virus neutralization test can be automated, especially when performed using microtiter plates, e.g. the addition of media to the plates, serial dilution of the serum samples, addition of virus and anti-rabies virus conjugate, as well as such tedious steps as plate washing. The automated reading of FAVN and RFFIT has been previously described and reduces the work-time required for the microscopic analysis readout and aids in the minimization of errors (Péharpré *et al.*, 1999). The automated reading method was reported to have a lower sensitivity when its results were compared with values recorded from non-automated reading, although no significant difference was noted when comparison was made between the rabies virus antibody levels reported by both methods. The expense of the equipment required to conduct automated reading of the RFFIT or FAVN, the requirement for a consistent cell monolayer and need for a good quality FITC conjugate limits the practicality of this enhancement, especially for laboratories that do not conduct large numbers of tests. Using an alternative to microscopic fluorescence measurement, a microneutralization test (RAMIN), and later the indirect immunoperoxidase virus neutralization (IPVN) technique and the modified FAVN, all use a mouse anti-rabies virus antibody and a peroxidase anti-mouse conjugate which enables automated reading to be conducted using a spectrophotometer (Mannen *et al.*, 1987; Hostnik, 2000; Cardoso *et al.*, 2004). In each of these studies, a good correlation was confirmed between traditional rabies virus neutralization methods and the modifications that were made to each test. A recent publication described an innovative alteration to the RFFIT, wherein a recombinant rabies virus was modified to express green fluorescent protein (Khawplod *et al.*, 2005). This molecular innovation eliminates the requirement for FITC-conjugated anti-rabies virus antibody. By using flow cytometry to detect residual virus that is present after incubation with serum, the testing sensitivity is reportedly increased as each cell is individually assessed for viral infectivity, creating a more precise percentage of viral inhibition (Bordignon *et al.*, 2002).

4 CHOOSING AN ASSAY

4.1 Diagnostic evaluation

Animal control workers, veterinarians and their staff, rabies diagnostic and research laboratory workers and persons who have been exposed to rabies are all concerned with the level of RVNA present in their serum. For routine evaluation of serum samples and to confirm that an immune response has occurred after post-exposure prophylaxis (PEP), the exact level of RVNA present in a patient's serum is of less importance than is the actual detection of RVNA in the serum. In the USA, the Advisory Committee on Immunization Practices (ACIP) currently recommends periodic evaluation of RVNA using the RFFIT and, if the RVNA level falls below complete neutralization at a 1:5 serum dilution (for those at continual risk of exposure), ACIP recommends that a routine booster should be administered (Centers for Disease Control and Prevention, 1999). Additionally, WHO recommends an RVNA level of 0.5 IU/ml as evidence of a specific immune response after rabies vaccination (WHO, 2005). Both of these levels were chosen based on early studies examining the virus neutralization titers of human subjects vaccinated with modern cell culture vaccines at the Centers for Disease Control and Prevention (CDC). These studies reported that non-specific inhibition was not evident at serum dilutions greater than approximately 1:25. The 1:25 level was doubled for safety reasons by WHO: their recommended RVNA value was raised to 1:50 (or 0.5 IU/ml by RFFIT). No specific RVNA level equivalent to protection against rabies virus infection and progression to disease has ever been or is likely to be established due to the unethical nature of conducting efficacy studies in humans, who have various levels of RVNA, and then determining survival rates. Testing serum samples post-vaccination by ELISA methods that only detect IgG may give false low results if the patient's immune system has not switched from producing IgM to IgG at the time that the blood sample used for testing was obtained. Conversely, measurement of rabies virus antibody in serum or CSF from a patient acutely ill with encephalitis would be expected to include a high level of immunoglobulin, which may bind non-specifically to the microtiter plates being used in the ELISA, resulting in a false-positive report.

In summary, the choice of test used to detect the presence of rabies virus antibody can have far-reaching effects. Medical decisions, including treatment choices eminently important for both the physician and the patient, make the choice of the rabies virus antibody test and testing facility crucial.

4.2 Serological surveillance

The purpose of testing the serum of animals being sampled in a surveillance study is to determine whether or not an animal or a group of animals has been exposed to rabies virus antigens, i.e. animals in an oral baiting area or in a rabies endemic area. In this case, monitoring of herd immunity is more important than the quantitative analysis of antibody from individual animals. There is a need for the test to be specific enough to detect antibodies in the target species against epitopes of the immunizing rabies virus strain and yet have a high enough cut-off value to limit the appearance of false positives. Some of the samples obtained in epidemiology surveys may be of poor quality (hemolyzed, contaminated, etc.) and may be in reality body fluids withdrawn from the body cavities of dead carcasses rather than strictly serum due to less than optimal conditions in the field. Poor quality samples can cause cytopathic effect (CPE) in the cells used for rabies virus neutralization tests but may cause less interference in binding assays where cells are not used. An ELISA method used for surveillance purposes in a fox oral baiting survey has been reported to be very satisfactory for this purpose (Cliquet *et al.*, 2000). Use of whole rabies virus (or G subunit) closely related antigenically to the vaccine strain would be expected to increase the specificity and sensitivity of this method. An important point to consider in the use of an ELISA is the ability of the conjugated antibody or protein (antigen) to react effectively enough with the binding antibody to allow identification. Use of a standard test kit that is simple to use and rapid and can be automated improves the ability for multiple research groups directly to compare their results.

4.3 Vaccine evaluation

The focus of serological testing in evaluating the immunogenicity of a human or animal rabies vaccine is to determine the level of RVNA induced. For this purpose, rabies virus neutralization tests are essential. Rabies virus neutralization tests are not restricted to the use of sera of one animal species because they measure residual viral infectivity rather than the presence of immunoglobulins having specific cell markers for one animal species. However, in the evaluation of a rabies vaccine the choice of the challenge virus used in the rabies virus neutralizing assay is extremely important. When the RVNA levels produced against different vaccines are compared, it is the combined effect of the quantity, functionality and specificity of the respective antibody response that is measured. A recent study examining the RVNA response of patients that received vaccines made from different rabies virus seed strains revealed that the RVNA titers obtained will be dependent upon the strain of the challenge virus used in the RVNA testing system (Moore *et al.*, 2005). Thus a homologous testing system (same or similar virus strain used to produce the vaccine is also used as the challenge virus in the rabies virus neutralization test) will report higher RVNA values than a heterologous testing system (different virus strain used to produce the vaccine is used as the challenge virus in the rabies virus neutralization test). This means that the choice of challenge virus strain to use in a rabies virus neutralization assay should accurately evaluate the production of RVNA titers after vaccination, especially since the titer values will be the criteria used for the evaluation of a new or existing vaccine. Clearly, if quantifying the immune response to the vaccine is the objective, then performing the test with a homologous rabies virus strain would most appropriately reflect this goal. It is important to remember that modern cell culture rabies vaccines are highly effective and cross-protection between strains has been demonstrated (Lodmell *et al.*, 1995). The use of RVNA testing systems to measure the immune response to specific rabies antigens and the response to rabies vaccines should not only be accurate and precise, but also meaningful. The most appropriate test method combined with assured quality of the results allows for the highest level of vaccine evaluation and industry standardization.

5 QUALITY ASSURANCE MEASURES

In order to ensure that serological testing results are accurate, globally recognized quality management standards should be developed and adopted by all rabies serological testing laboratories. This would improve the dependability of serological test results and also provide guidelines for rabies researchers and laboratory technicians to understand and evaluate the reliability of the management, operations and outputs of laboratories conducting serological testing for the presence of rabies virus antibodies. As stated in previous sections of this chapter, rabies serological testing can be used for numerous purposes and there are several types of serological methods that can and have been used to measure antibody to rabies virus. The limitations inherent within each rabies serological assay and a determination as to whether a method is 'fit for purpose' are appropriate criteria on which to base selection of a testing method. Additionally, the testing method that will be used must produce valid data that can be used for decision-making. Valid rabies laboratory results, essential for clinical diagnosis, rabies surveillance and animal or human travel, can be achieved by the appropriate use of the methodology, proper testing techniques and basic good laboratory quality assurances and quality control practices. The performance characteristics such as sensitivity, specificity, accuracy, precision and reproducibility for each testing method will aid in determining the best choice to produce valid data. It is important that laboratories are able to produce correct and defensible results using appropriate methods of analysis because the laboratory must be able to defend the validity of its data and opinions. No matter how carefully produced, a test result is meaningless if it cannot be reproduced or if the data do not support the validity of the test method. Quality Assurance/Quality Control does not do much for accuracy of test results if the test is not valid, or if the wrong method was chosen. Once the 'use' has been determined, further steps must be taken to establish and implement the chosen methodology within the testing laboratory. It is the laboratory's responsibility to prove suitability or competency of the test method 'in house' before and during testing. It is possible that a method that functions in one laboratory satisfactorily fails to operate in the same manner in another. It is considered unacceptable for the researcher/laboratorian to use a published 'validated method' without demonstrating their capability in the use of the method (AOAC, 2002). Validation can be accomplished through quality assurance and quality control systems that prove or document the intrinsic quality of the results as well as participation in an inter-laboratory comparison of samples. In this respect, particular emphasis is placed upon the following elements, which are discussed in depth within the International Standard, ISO/IEC 17025:1999, General requirements for the competence of testing and calibration laboratories:

- Trained and technically competent personnel.
- Equipment maintenance and calibration.
- Well documented standard operating procedures.
- Description of or reference to reagents and media preparation and quality control, including criteria for acceptance or rejection of lots to be used in testing.
- Critical specification for certified reference standards and materials (rabies reference standard, virus, etc.).
- Description of the necessary quality control for the test, including evaluations to be made before initiating the test and action to be taken if limits or criteria are not met.
- Proficiency criteria are in place and available to ensure the laboratory is competent to perform the test.
- Method of analysis and presentation of results, including a description of mathematical factors, calculations and statistics, methods of data transformation (e.g. how optical density values are transformed to positive or negative results) and the method of diagnostic interpretation (interpretation of titers).
- Maintenance of data in order to recreate events and validate results.
- Physical and environmental conditions required for safety, including biosecurity level.

There are many testing components that will significantly influence rabies serological results: serum samples, virus titer and strain, cells, incubation time and temperature, equipment issues, interpretation of test results and the calculations used to determine the value of the results. Laboratory staff should be aware of the importance of relevant clinical information when validating test results, especially when cumulative records are available. An unexpected result can highlight the possibility of an error (CLIA, 2003). Lack of technical training, use of uncalibrated, questionable equipment, low quality of reagents and the questionable traceability of a serum control to a recognized standardized international reference serum may present the researcher/laboratorian with a data acceptance, data interpretation and/or publishing dilemma. Obviously, the discovery of questionable data at any point in routine testing or on a research project can be costly. It is always better and less expensive to design quality into a program or research endeavor from the very beginning, thus avoiding the high cost of needless repetition and, perhaps worse, loss of scientific credibility.

When it is necessary or desirable to compare rabies antibody test results between laboratories, a significant amount of standardization is required. The choice of an identical challenge virus strain (such as the recommended CVS-11) and the use of an internationally recognized reference serum standard diluted to comparable levels of IU/ml values are critical. For rabies virus neutralization assays, the titer of the challenge virus standard should not vary more than twofold from the targeted challenge dose in intra- and inter-laboratory comparisons. The acceptable range of titers for the reference serum should be within one standard deviation of a calculated mean titer (Fitzgerald *et al.*, 1975; Smith, 1991). Additionally, as in all testing procedures where biologic function and individual biologic variation are factors, the calculation method should be harmonious. Each laboratory should determine the detection limit of rabies virus antibody deemed to be specific. This level must be validated within the laboratory and, if possible, through comparison testing with laboratories recognized as having expertise in rabies serology. Inter-laboratory testing is a tool that can be used for assessment of comparability of results and the competence of laboratories to perform specific testing methods. Comparison of results between laboratories can also facilitate a critical review of existing standards and limit values by taking into account the current capabilities of the methodologies.

For decision-makers depending on serological test results, it is imperative that they are confident that the rabies serological test results that they receive are correct, appropriate, fit for purpose and comparable. The ultimate aim of the data delivered by testing laboratories is to provide concrete support to the decision-making process. It is therefore of the utmost importance to ensure not only consistency between the information delivered by the testing laboratory and the actual results of the sample, but also consistency between the analytical information delivered by the laboratory and the information needed by those who ultimately make the decisions.

Thorough documentation of all quality assurance activities is essential to every laboratory. This documentation will become an on-going history of the laboratory's activities to be used to improve the reliability, efficiency and quality of the services provided by the laboratory.

Resources for technical support with useful references, guidelines, and/or standards that supplement the general requirements of a quality assurance program are listed below:

CLIA – Clinical Laboratory Improvement Amendment, www.cms.hhs.gov/clia

CLSI – Clinical and Laboratory Standards Institute, www.clsi.org

FDA – Food and Drug Administration, www.fda.gov

OIE – World Organization for Animal Health, www.oie.int

ISO – International Organization for Standardization, www. Iso.org

6 CONCLUSIONS

Rabies serological laboratories serve a critical function in rabies prevention programs. They are vital for providing reliable information required for human diagnoses, vaccine evaluations, animal surveillance and epidemiological studies and for routine testing of professionals working in the field of rabies. Only a handful of designated rabies serological laboratories exist throughout the entire world. As outlined in the chapter, the reason why there are so few creditable serological laboratories in the world is clear – accurate rabies serological testing can be highly complex. The choice of the correct method for testing depends directly on what the intended purpose of the test results will be. Understanding the principle and limitations of the assay chosen and strict adherence to key components of testing will assure appropriate results for decision-making. Finally, the development of a less complicated and less tedious serological testing methods to measure the presence of RVNA activity and increased standardization of available rabies serological methods would greatly improve the quality and utility of rabies virus antibody measurement.

REFERENCES

AOAC Guidelines for Single Laboratory Validation of Chemical Methods for Dietary Supplements and Botanicals: 2002, AOAC International (available at http://www.aoac.org/dietsupp6/Dietary-Supplement-web-site/slv_guidelines.pdf)

Atanasiu, P. (1973). Quantitative assay and potency test of antirabies serum and immunoglobulin. In: *Laboratory Techniques in Rabies*, 3rd edn (M.M. Kaplan and H. Koprowski, eds). pp. 314–318. Geneva: World Health Organization.

Aubert, M.F. (1996). Methods for the calculation of titres. In: *Laboratory Techniques in Rabies*, 4th edn (F.X. Meslin, M.M. Kaplan and H. Koprowski, eds). pp. 445–459. Geneva: World Health Organization.

Barton, L.D. and Campbell, J.B. (1988). Measurement of rabies-specific antibodies in carnivores by an enzyme-linked immunosorbent assay. *Journal of Wildlife Diseases* **2**, 246–258.

Bordignon, J., Comin, F., Ferreira, S.C., Caporale, G.M., Filho, J.H. and Zanetti, C.R. (2002). Calculating rabies virus neutralizing antibodies titres by flow cytometry. *Revista do Instituto de Medicina Tropical de Sao Paulo Sao Paulo* **44**, 151–154.

Briggs, D.J., Smith, J.S., Mueller, F.L. *et al.*, (1998). A comparison of two serological methods for detecting the immune response after rabies vaccination in dogs and cats being exported to rabies-free areas. *Biologicals* **26**, 347–355.

Cardoso, R.C., Queiroz da Silva, L.H., Albas, A., Ferreira, H.L. and Venturoli Perri, S.H. (2004). Rabies neutralizing antibody detection by indirect immunoperoxidase serum neutralization assay performed on chicken embryo related cell line. *Memorias do Instituto Oswaldo Cruz* **99**, 531–534.

Carpenter, A.B. (1997). Enzyme linked immuno-sorbent assay. In: *Manual of Clinical Laboratory Immunology*, 5th edn (N.R. Rose, E.C. De Macario, J. Fahey, H. Fiedman and G.M. Pen, eds). pp. 2–9. Boston: Little Brown.

Centers for Disease Control and Prevention (1999). Human rabies prevention – United States, 1999. Recommendations of the Advisory Committee on Immunization Practices (ACIP). *Mortality and Morbidity Weekly Report* **48** (RR-1), 1–21.

CLIA (2003). US Department of Health and Human Services, Center for Medical and Medicaid Services. Clinical Laboratory Improvement Amendments of 1988; final rule. *Federal Register* (Jan 24).

Cliquet, F., Aubert, M. and Sagné, L. (1998). Development of a fluorescent antibody virus neutralization test (FAVN test) for the quantitation of rabies-neutralising antibody. *Journal of Immunological Methods* **212**, 79–87.

Cliquet, F., Sagné, L., Schereffer, J.L. and Aubert, M.F.A. (2000). ELISA test for rabies antibody titration in orally vaccinated foxes samples in the fields. *Vaccine* **18**, 3272–3279.

Dietzschold, B. (1993). Antibody-mediated clearance of viruses from the mammalian central nervous system. *Trends in Microbiology* **1**, 63–66.

Elmgren, L.D. and Wandeler, A.I. (1996). Competitive ELISA for the detection of rabies virus-neutralizing antibodies. In: *Laboratory Techniques in Rabies*, 4th edn (F.X. Meslin, M.M. Kaplan and H. Koprowski, eds). pp. 200–208. Geneva: World Health Organization.

Esterhuysen, J.J., Prehaud, C. and Thomson, G.R. (1995). A liquid-phase blocking ELISA for the detection of antibodies to rabies virus. *Journal of Virological Methods* **51**, 31–42.

Fitzgerald, E.A., Baer, G.M., Cabassa, V.J. and Vallancourt, R.J. (1975). A collaborative study on the potency testing antirabies globulin. *Journal of Biological Standards* **3**, 273–278.

Grassi, M., Wandeler, A.I. and Peterhans, E. (1989). Enzyme-linked immunosorbent assay for determination of antibodies to the envelope glycoprotein of rabies virus. *Journal of Clinical Microbiology* **5**, 899–902.

Habel, K. (1996). Habel test for potency. In: *Laboratory Techniques in Rabies*, 4th edn (F.X. Meslin, M.M. Kaplan, H. Koprowski, eds). pp. 369–373. Geneva: World Health Organization.

Heberling, R.L., Kalter, S.S., Smith, J.S. and Hildebrand, D.G. (1987). Serodiagnosis of rabies by dot immunobinding assay. *Journal of Clinical Microbiology* **25**, 1262–1264.

Hooper, D.C., Moromoto, K., Bette, M., Weihe, E., Koprowski, H. and Dietzschold, B. (1998). Collaboration of antibody and inflammation in clearance of rabies virus from the central nervous system. *Journal of Virology* **72**, 3711–3719.

Hostnik, P. (2000). The modification of fluorescent antibody virus neutralization (FAVN) test for the detection of antibodies to rabies virus. *Journal of Veterinary Medicine B. Infectious Diseases and Veterinary Public Health* **47**, 423–427.

Irie, T. and Kawai, A. (2002). Studies on the different conditions for rabies virus neutralization by monoclonal antibodies 1-46-12 and 7-1-9. *Journal of General Virology* **83**, 3045–3053.

ISO/IEC 17025:1999 *General requirements for the competence of testing and calibration laboratories*. Geneva: International Organization for Standardization (ISO).

Kaplan, M.M., Wiktor, T.J., Maes, R.F., Campbell, J.B. and Koprowski, H. (1967). Effect of polyions on the infectivity of rabies virus in tissue culture: Construction of a single-cycle growth curve. *Journal of Virology* **1**, 145–151.

Khawplod, P., Inoue, K., Shoji, Y. *et al.* (2005). A novel rapid fluorescent focus inhibition test for rabies virus using a recombinant rabies virus visualizing a green fluorescent protein. *Journal of Virological Methods* **125**, 35–40.

Kitala, P.M., Lindqvist, K.J., Koimett, E. *et al.* (1990). Comparison of human immune responses to purified vero cell and human diploid cell rabies vaccines by using two different antibody titration methods. *Journal of Clinical Microbiology* **28**, 1847–1850.

Lodmell, D.L., Smith, J.S., Esposito, J.J. and Ewalt, L.C. (1995). Cross-protection of mice against a global spectrum of rabies virus variants. *Journal of Virology* **69**, 4957–4962.

Lyng, J., Weis Bentzon, M. and Fitzgerald, E.A. (1989). Potency assay of antibodies against rabies. A report on a collaborative study. *Journal of Biological Standards* **17**, 267–280.

Mannen, K., Mifune, K., Reid-Sanden, F.L. *et al.* (1987). Microneutralization test for rabies virus based on an enzyme immunoassay. *Journal of Clinical Microbiology* **25**, 2440–2442.

Mebatsion, T., Frost, J.W. and Krauss, H. (1989). Enzyme-linked immunosorbent assay (ELISA) using staphylococcal protein A for the measurement of rabies antibody in various species. *Journal of Veterinary Medicine* **B 36**, 532–536.

Moore, S.M., Ricke, T.A., Davis, R.D. and Briggs, D.J. 2005. The influence of homologous vs. heterologous challenge virus strains on the serological test results of rabies virus neutralizing assays. *Biologicals* **33**, 269–276.

Nicholson, K.G. and Prestage, H. (1982). Enzyme-linked immunosorbent assay: A rapid reproducible test for the measurement of rabies antibody. *Journal of Medical Virology* **9**, 43–49.

Péharpré, D., Cliquet, F., Sagné, E., Renders, C., Costy, F. and Aubert, M. (1999). Comparison of visual microscopic and computer-automated fluorescence detection of rabies virus neutralizing antibodies. *Journal of Veterinary Diagnostic Investigations* **11**, 330–333.

Smith, J.S. (1991). Rabies serology. In: *The Natural History of Rabies*, 2nd edn (G.M. Baer, ed.). pp. 235–252. Boca Raton: CRC Press.

Smith, J.S., Yager, P.A. and Baer, G.M. (1973). A rapid reproducible test for determining rabies neutralizing antibody. *Bulletin of the Wildlife Health Organization* **48**, 535–541.

Tordo, N. (1996). Characteristics and molecular biology of the rabies virus. In: *Laboratory Techniques in Rabies*, 4th edn (F.X. Meslin, M.M. Kaplan and H. Koprowski, eds). pp. 28–51. Geneva: World Health Organization.

WHO (1985). WHO Expert Committee on Biological Standardization. Thirty-fifth report, 1985. *WHO Technical Report Series. No 725.* Geneva: World Health Organization.

WHO (1994). WHO Expert Committee on Biological Standardization. Forty-fourth report, 1994. *WHO Technical Report Series. No 848.* Geneva: World Health Organization.

WHO (2005). WHO Expert Consultation on Rabies, first report, 2005. *WHO Technical Report Series. No. 931.* Geneva: World Health Organization.

12

Immunology

MONIQUE LAFON

Department of Virology, Pasteur Institute, 75724 Paris Cedex 15, France

1 INTRODUCTION

Rabies is a unique neurological infection that can be prevented by post-exposure vaccination, at least when the vaccine is administered to patients within a reasonable period of time after a rabies exposure. Thus, vaccine efficiency can be conferred by immunization even after the virus enters the body and reaches the nervous system. Because of the ability of rabies virus (RABV) to impair immune responsiveness, it is intriguing to understand how immunization against rabies can prevent or limit virus progression towards and/or through the nervous system (NS). In this chapter, the mechanisms involved in specific immune responses, as well as the molecular basis of rabies vaccine protection are discussed.

2 MOLECULAR COMPONENTS OF A SPECIFIC IMMUNE RESPONSE

Innate immunity is the first line of defense against invading pathogens. It involves the release of cytokines, including type 1 interferons (IFN-α and -β) and chemokines, the activation of complement and the attraction of macrophages, neutrophils and natural killer (NK) cells into infected tissues. This innate immune response is triggered in the first hours following the entry of pathogens or vaccine antigen and is not pathogen specific. It is opposed to the adaptive immune response, which is tailored to a specific pathogen and requires several days to be set up. Cells are armed by various mechanisms to sense viral components and initiate intracellular signal transduction to respond rapidly to pathogens. Microorganism components are recognized through receptors such as the evolutionarily conserved Toll-like receptors (TLRs). Early signaling events initiated by the recognition of pathogen components lead to the production of type 1 IFN-α and -β, inflammatory cytokines (IL-6, IL-1α, TNF-α) and chemokines (CCL-5, CXCL-10, CCL-3 and CCL-4), which allow the initiation or regulation of the inflammatory and antiviral responses (Sato *et al.*, 1998; Akira *et al.*, 2001; Sharma *et al.*, 2003; Yoneyama *et al.*, 2004). Besides their antiviral properties, type 1 IFNs are important modulators with pleiotropic effects on the innate immune system as well as the adaptive immune

Rabies, second edition. Edited by Alan C. Jackson and William H. Wunner
ISBN 978-012-369366-2.

response. Type I IFNs are effective in promoting survival and activation of dendritic cells (DCs), the cells that present antigens to lymphocytes to build an adaptive immune response. Thus, both efficiency of vaccines and defense against pathogens depend upon the robustness of the innate immune responses, which make the link between innate and adaptive immunity (Le Bon and Tough, 2002) and are also an effective immune adjuvant for antibody production (Le Bon *et al.,* 2001).

After binding to major histocompatibility complex (MHC) molecules that are displayed at the surface of DCs, peptides derived from specific 'foreign' antigens of invading pathogens or vaccines trigger an adaptive immune response against the invading pathogen or vaccine antigens. The CD4$^+$ T lymphocytes recognize foreign and vaccine-specific antigens once they have been processed through the MHC class II exogenous presentation pathway by DCs. DCs, macrophages and B cells can process foreign antigens through the exogenous MHC class II pathway. They sample the extracellular environment for foreign antigens, process them intracellularly and associate the digested antigen with MHC class II molecules. Once presented by the MHC, the peptides of the digested foreign antigen are recognized by T cells bearing the appropriate T cell receptor (TCR) and CD4 cell surface molecule. Signalling via the TCR and CD4 molecule triggers activation of T cells and their differentiation into two functional subsets, the T helper 1 (Th1) and T helper 2 (Th2) cells. The distinction of the two subsets is based on the cytokines they secrete: interferon-gamma (IFN-γ) is the signature cytokine for Th1 cells, whereas interleukin-4 (IL-4) is the signature cytokine for the Th2 cells. Generation of Th1 cells is under the control of IL-12 produced by macrophages and DCs. Th1 cells limit the proliferation of pathogens via IFN-γ production and provide help for antibody production by B lymphocytes.

The CD8$^+$ T cells, in contrast to CD4$^+$ T cells, recognize foreign antigen that has been processed by the endogenous pathway of cells that express MHC class I molecules. Infected cells export pathogen peptides embedded in the groove of MHC class I molecules to the cell surface. The peptide-charged infected cells activate T cells expressing the CD8 accessory surface molecules and the appropriate TCR. Activated CD8$^+$ T lymphocytes produce IFN-γ and kill the infected cells via cytotoxicity by means of perforin and granzyme release and/or Fas-mediated lysis. Maturation of the T cells takes place in the secondary lymphoid organs such as the lymph nodes or the spleen. DCs must migrate to these organs in order to induce an efficient immune response.

Tipping the balance between CD4$^+$ and CD8$^+$ T cell responses to an antigen depends upon the nature of the foreign or invading antigen. Live microorganisms trigger a CD8$^+$ T cell cytotoxic response, whereas recognition of an inert peptide or protein induces CD4$^+$ T cell recruitment. In particular, vaccines composed of live attenuated microorganisms trigger T cell-mediated immunity that may be crucial in mediating protection against pathogens such as viruses that replicate in the cells. DNA vaccines and recombinant virus vaccines, which induce

TABLE 12.1

Comparative analysis of various vaccine formulations

	Live attenuated vaccine	DNA vaccine	Recombinant vaccine	Killed vaccine
B cells	+++	++	++	+++
CD4$^+$ T cells	+/− Th1	+++ Th1	+	+/− Th1
CD8$^+$ T cells	+++	++	+++	−
MHC class	I and II	I and II	I and II	II

production of viral antigens by live cells, belong to the same category as live attenuated vaccines. Both vaccines generate mainly CD8$^+$ T (cytotoxic) cells. In contrast, vaccines prepared with inactivated (killed) virus particles or portions (subunits) of virus particles that are sampled by the MHC class II DCs trigger CD4$^+$ T cell activation and the generation of a humoral (B cell) immune response (Table 12.1).

At the present time, all rabies vaccines for humans and domestic animals are killed virus vaccines containing the full-size intact proteins, G, N, NS/P, M and L, of the virus (see Chapters 13 and 14). They are expected to trigger a CD4$^+$ T cell and a humoral B cell response. The use of live virus vaccines composed of an attenuated RABV strain (Lafay et al., 1994) or recombinant virus such as vaccinia virus that expresses the RABVG (Kieny et al., 1984) is restricted to wild animal immunization. Live vaccine is expected to trigger an additional strong CD8$^+$ T cell response.

3 IMMUNE RESPONSES DURING RABV INFECTION

The native tropism of RABV is to infect the NS. However, at the earliest steps of infection, virus particles are 'injected' into the skin and muscles and before the virus has reached the NS, it can trigger an immune response in the periphery (Irwin et al., 1999; Camelo et al., 2000). It is unlikely that RABV triggers a primary adaptive immune response in the NS once it reaches the NS. This incapacity results from the absence of lymphoid structures and the lack of professional antigen presenting cells in the NS, which make the NS an immunoprivileged site (Medawar, 1948; Barker and Billingham, 1977). Nevertheless, once in the NS, there is evidence that RABV triggers an early innate immune response, characterized by antiviral, chemoattractive and inflammatory responses (Wang et al., 2005) in which infected neurons play an active part (Prehaud et al., 2005; Jackson et al., 2006). This early innate immune response seems to be modulated according to the pathogenicity of RABV strains (Wang et al., 2005). In addition, to enforce the lack of immune efficacy of the NS, the pathogenic strains of RABV have selected immunosubversive strategies to escape the host immune response

(Camelo *et al.*, 2001). As a result, there is a global subversion of the host immune defenses by rabies virus. This can be seen as a successful well-tailored adaptation of RABV to the host. One would expect that the host's natural capacity to fight such well-adapted virus is greatly limited.

3.1 The immune response in the periphery

The immune response triggered by RABV infection has been analyzed by using the mouse as the animal model for infection with RABV. Routes of infection consist of peripheral intramuscular or intranasal injection to mimic the natural exposure by bite or aerosol. Several strains of RABV with different levels of pathogenicity in mice have been selected. After injection in a peripheral site by the intramuscular, subcutaneous, intraplantar or intranasal route, pathogenic virus strains invade the spinal cord and almost all brain regions and cause a fatal acute encephalitis (Camelo *et al.*, 2000; Baloul *et al.*, 2004). In contrast, virus injected by the same routes consisting of less pathogenic virus strains results in a non-fatal abortive disease characterized by a transient and restricted infection of the NS followed by irreversible paralysis of the inoculated limbs (Galelli *et al.*, 2000; Lafon, 2004). Pathogenic virus strains are referred to further in this chapter as 'acute' RABV, whereas less pathogenic virus strains are referred to as 'abortive' RABV strains (Table 12.2).

TABLE 12.2

Rabies virus pathogenicity correlates with the ability of the virus to control the host immune response

	Encephalitic fatal rabies (acute RABV)	Paralytic rabies (abortive RABV)
T lymphocytes are protective	No	Yes
T lymphocyte infiltration	Yes	Yes, mainly CD8+ T cells
Neuronal apoptosis and release of viral antigens	Sporadic	Widespread
CNS associated antibodies	Undetectable	High
Inflammatory cytokines, chemokines	Yes	Yes (no IL-6)
Immunosuppression	Yes	No
Control of the virus infection	No	Yes
Survival of the animal	No	Yes
Virus cycle	Complete (virus reaches the brain)	Abortive (transient CNS infection); CD8+ T cell-mediated immuno-pathology is observed (paralysis)

This table summarizes data from Irwin *et al.*, 1999; Camelo *et al.*, 2000; Galelli *et al.*, 2000; Baloul *et al.*, 2004). Virus inoculation was performed in the hind limbs of the mice.

Injection of both types of viruses (acute or abortive RABV strains) into the hind limbs of mice induces similar local (in lymph nodes of the hind limb) and systemic (in spleen) proliferative and cytotoxic responses in the periphery. Seven days after injection, similar levels of cytokine-secreting T cells can be isolated from peripheral lymph nodes and similar titers of specific RABV antibodies can be detected in the serum of mice (Irwin *et al.*, 1999). Thus, both acute and abortive RABV strains trigger similar early immune responses in the periphery.

3.2 Immune unresponsiveness

In contrast to what is observed in the first days following virus entry, during which early immune responses triggered by acute and abortive RABV strains are identical, after a few days post-infection (p.i., day 7), the host immune response in the spleen is severely impaired in mice injected with an acute RABV strain and not impaired by injection of the abortive RABV strain, indicating that pathogenic RABV strains control the host immune response in the periphery.

Fatal encephalitis induced by pathogenic strains of rabies virus is accompanied by an immunodepression characterized by the impairment of the lymphocyte response marked by a loss of cellular mediated immunity (Wiktor *et al.*, 1977a). This is concomitant with the unresponsiveness of spleen cells to lectin (ConA) stimulation *in vitro* (Hirai *et al.*, 1992; Perrin *et al.*, 1996; Camelo *et al.*, 2001). The decline in the lymphoproliferative response of splenocytes to an *in vitro* ConA stimulation but not to lipopolysaccharide (LPS) from the sixth day p.i. is associated with a decrease in the number of IL-2, IFN-γ and TNF-α secreting splenocytes (Th1 pattern), but not of IL-4 secreting splenocytes (Th2 pattern). No quantitative modifications of the different spleen populations (CD4$^+$and CD8$^+$ T cells, B cells, NK cells) were observed. The decrease of immune responsiveness clearly depends on the pathogenicity of the strain. Abortive strains of RABV do not induce immunosuppression (Perrin *et al.*, 1996; Camelo *et al.*, 2001). The origin of the virus-mediated immune suppression is unknown. So far, it has only been shown (Torres-Anjel *et al.*, 1988) that it is not under the control of the hypothalamo-pituitary-adrenergic (HPA) axis, since adrenalectomy, which interrupts the production of corticosteroid hormones, does not modify the RABV specific cytotoxic response (Wiktor *et al.*, 1985).

3.3 The immune response in the nervous system

3.3.1 *The innate immune response in the central nervous system*

After injection of acute or abortive RABV strains in the hind limbs, progressive infection within the spinal cord and the brain is accompanied by the production of the inflammatory cytokines IL-1, TNF-α and IL-6, as well as chemokines (e.g. CCL-5, CXCL-10) (Marquette *et al.*, 1996; Camelo *et al.*, 2000; Baloul and Lafon, 2003; Wang *et al.*, 2005) (Figure 12.1). IFN-β can also be detected early in the infection in the spinal cord and brain concomitant with virus neuroinvasion. This indicates that the NS can sense RABV entry and mount a reactive innate immune response. This is consistent with the observation that glial cells and neurons express TLR, including TLR-3 (McKimmie and Fazakerley, 2005; McKimmie *et al.*, 2005; Prehaud *et al.*, 2005; Jackson *et al.*, 2006) and that after RABV infection neurons can mount a classical inflammatory, chemoattractive and antiviral response, including expression of antiviral IFN-β, Mx-1 and 2'5'OAS genes, and expression of the inflammatory TNF-α, IL-6 and chemokines CCL-5 and CXCL-10 (Prehaud *et al.*, 2005).

The role of type 1 IFN in rabies has been investigated in animal models. Treatment with the IFN-inducing poly I:C or direct administration of type 1 IFN resulted in various degrees of protection against RABV (Harmon and Janis, 1975; Hilfenhaus *et al.*, 1975). After abortive RABV infection, higher virus titers and lower levels of neutralizing antibodies were reported in mice lacking the IFN receptor than in immunologically intact mice. Virus infection was cleared more slowly in mice lacking type 1 IFN receptor than in control mice, suggesting that, at least for an abortive RABV strain, antiviral IFN activity can control virus progression in the NS. In contrast, it seems that only acute RABV evades the innate immune response in the NS (Wang *et al.*, 2005). It has been proposed that the phosphoprotein, P, of RABV could control type 1 IFN production by interfering with the phosphorylation of IRF-3 (Brzozka *et al.*, 2005; Hengel *et al.*, 2005). The chemokines and inflammatory cytokines probably attract lymphocytes that are activated in the periphery, as demonstrated for other diseases (Ma *et al.*, 2002), since T and B cells can be detected in the infected NS as early as 3 days p.i. (Camelo *et al.*, 2000; Baloul *et al.*, 2004; Lafon, 2004). Thus, there is evidence that the NS can both sense and mount an early immune response in response to RABV attack whatever the pathogenicity of the strain and that only acute RABV escapes, at least partially, the host innate immune response.

3.3.2 The adaptive immune response in the nervous system during fatal rabies infection

Even if acute RABV strains have the capacity to decrease the innate immune response, the control of the innate host response should be only partial since the NS is progressively invaded by T and B lymphocytes. In the case of abortive RABV infection, the brain is protected against virus invasion by the T cell-mediated response, which kills infected neurons. As a consequence, abortive RABV does not reach the brain and mice survive (see Table 12.2). The T cell-dependent destruction of motor neurons induces polio-like paralysis (Galelli *et al.*, 2000). In the absence of T cells, as in nude (*nu* −/−) mice, abortive RABV infection cannot be controlled and mice will die with signs similar to those in acute RABV infection,

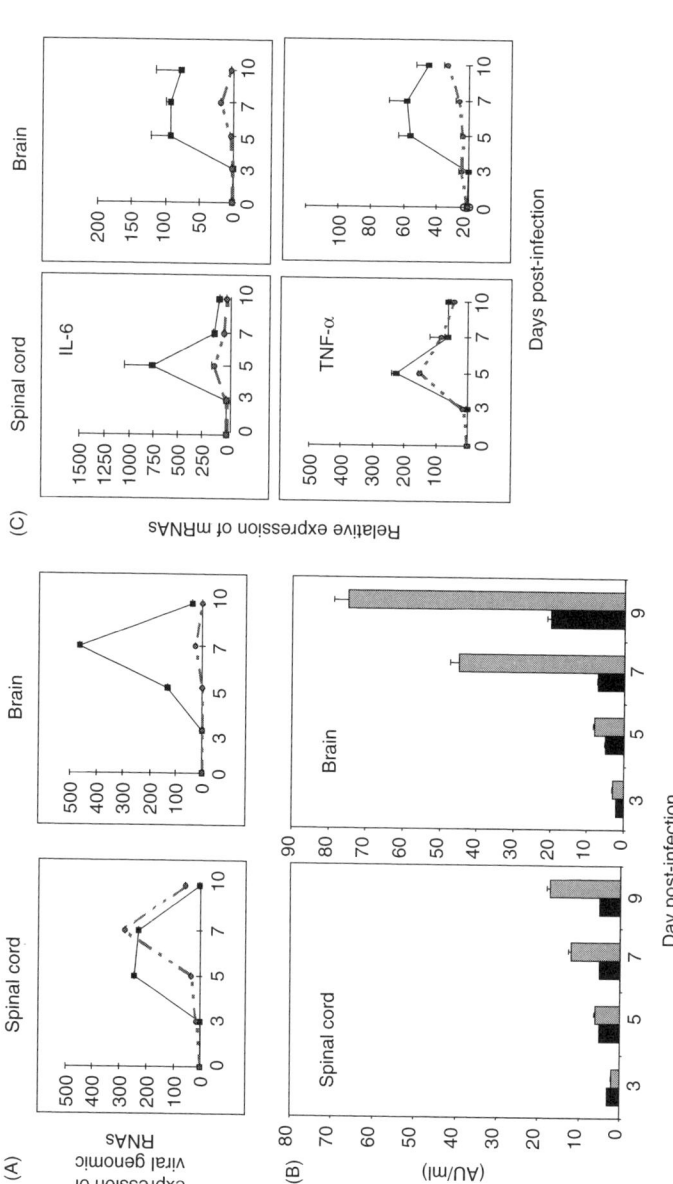

Figure 12.1   Comparative immune reaction in the mouse NS after infection with a highly neuroinvasive rabies virus strain and an abortive rabies virus strain. BALB/c mice received an intramuscular injection of 10^7 infectious particles of acute RABV strain or abortive RABV into both hind legs. (A) Invasion of BALB/c spinal cord (left panel) and brain (right panel) by acute RABV (solid line, black squares) and abortive RABV (dotted line, black circles) was measured by real-time PCR 3, 5, 7 and 10 days post-infection in the spinal cord and brain of acute- and abortive-RABV-infected mice and of non-infected mice (day 0). Each point represents the arithmetic mean of results from three of four mice. The SD was less than 7%. (B) Kinetics of antibody production in the NS of mice infected with acute RABV (black histograms) or abortive RABV (gray histograms) in the spinal cord (left panel) brain (right panel). Rabies virus-specific antibodies were detected by enzyme immunoassay. (C) Kinetics of cytokine mRNA production (IL-6 and TNF-α) of mice infected with acute RABV (line) or abortive RABV (dashed line), in spinal cord (left panel) and brain (right panel) by real time PCR. Results are expressed as relative fold increase compared with non-infected animals (value 1).

indicating that migratory T cells have the capacity to control the viral neuroinvasiveness in abortive rabies.

In striking contrast to abortive RABV, migratory T lymphocytes cannot block
brain access by acute RABV infection (see Table 12.2). This results in the destruction, after a few days, of the T cells that had migrated into the NS (Figure 12.2).
T cells are destroyed by apoptosis caused by the activation of the Fas molecules
they express on their surface, after ligation to FasL. It appears that acute RABV
strains and not abortive RABV strains cause the NS, and infected neurons in particular, to upregulate FasL expression (Figure 12.2). In the absence of FasL, as in
mice lacking a functional FasL molecule (*gld* mice), destruction of T cells is

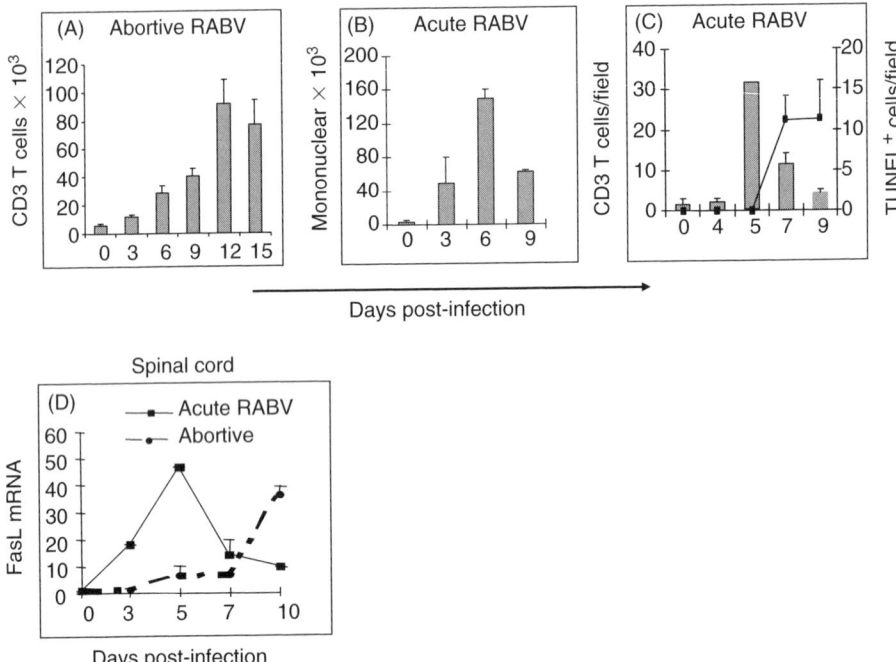

Figure 12.2 Migratory T cells encounter cell death in the NS of mice infected by acute RABV.
C57B16 or BALB/c mice received an intramuscular injection of 10^7 infectious particles of acute
RABV virus strain or abortive RABV into both hind legs. Kinetics of migration of immune cells in
the NS of mice infected by abortive (A) or acute (B) RABV. Cells were isolated by Percoll gradient.
(C) Kinetics of migratory T cells (histogram) and number of apoptotic cells (TUNEL positive
cells, line) per field of acute RABV-infected spinal cord sections. (D) Kinetics of FasL in the NS of
rabies virus-infected C57Bl6 mice. Relative FasL mRNA expression was evaluated at day 0, 3, 5, 7
and 10 post-infection in spinal cord of either acute RABV (solid line, black squares) or abortive
RABV-infected mice (broken line, black circles). Each point represents the arithmetic mean of results
from 3–4 mice.

reduced and RABV disease is less severe (Baloul and Lafon, 2003; Baloul *et al.*, 2004). The early upregulation of FasL by neurons appears then to be one of the mechanisms that contributes to RABV immunosubversion. It cannot be excluded that acute RABV utilizes other immunosubversive molecules to escape T cell immunosurveillance in the NS because, in human neurons, it has been shown that acute RABV upregulates the expression of HLA-G, a non-classical human MHC class I molecule, which may promote tolerance, leading to acceptance of the semi-allogeneic fetus and tumor immune escape (Lafon *et al.*, 2005).

In addition, other factors may contribute to the escape of the rabies virus from the host immune response since, in striking contrast to what was observed in an abortive RABV infection, acute RABV triggers only limited amounts of IFN-γ mRNAs and RABV-specific Ab in the NS (see Figure 12.1) (Baloul and Lafon, 2003; Lafon, 2004, 2005). Thus, RABV has selected a battery of mechanisms to escape host immunosurveillance in the NS.

The sequence of events that occurs during fatal encephalitis is schematically illustrated in Figure 12.3. After the virus enters the NS by terminal nerve endings, neuromuscular junctions and muscle spindles, it travels through the axons of neuronal networks. RABV infects mainly neurons (Murphy *et al.*, 1973; Wunner, 1987). Infection triggers production of chemokines and inflammatory cytokines that attract activated lymphocytes to migrate through the blood–brain barrier

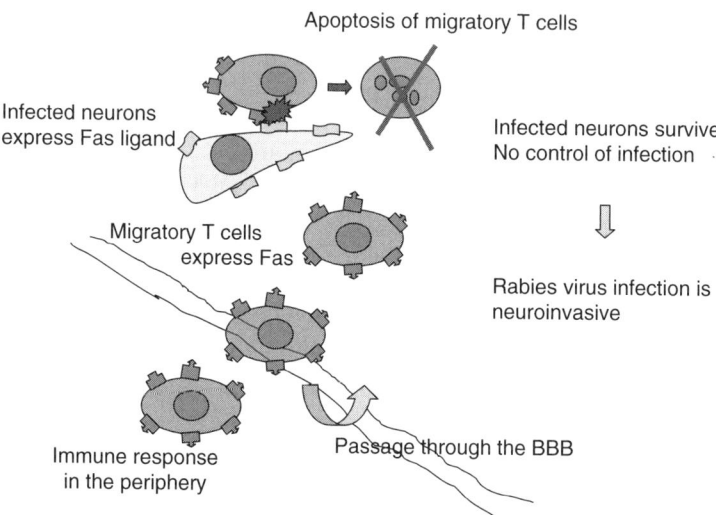

Figure 12.3 Scheme of the mechanisms adopted by acute rabies virus strains to subvert the host defenses. Activated T cells cross the blood–brain barrier (BBB) expressing Fas attracted by cytokines such as TNF-α secreted locally in reaction to the infection. FasL is upregulated by infected neurons. T cells undergo apoptosis after interaction of Fas with FasL. As a consequence, the virus subverts the host immune response and favors its propagation through the NS. This figure is reproduced in the color plate section.

(BBB). RABV strains causing encephalitis produce a non-cytopathogenic infection, which preserves the physical integrity of the neuron and of the neuronal network, favoring neuroinvasiveness through the entire NS, from the site of entry, the terminal nerve endings, up to the salivary glands, the site of exit. Absence of an immune response is strengthened by the capacity of the pathogenic strain of RABV to induce peripheral immunosuppression. As a consequence, the virus escapes the immune response and invades the entire NS.

4 IMMUNOLOGICAL BASIS FOR POST-EXPOSURE VACCINATION EFFICIENCY

Evaluation of the immunopathological events that participate in virus clearance from the NS helps to identify the production of antibodies associated with a T cell response as crucial survival factors. A dual role is assigned for the CD8$^+$ T cells: they participate in the NS clearance by controlling infection together with antibodies and, in contrast, they induce neuronal apoptosis and thus can initiate an immunopathological reaction.

4.1 Protective role of CD4 lymphocytes

T lymphocytes play an important role in the immune defense against rabies virus. Demonstration that RABV is a T cell-dependent antigen was established by experiments in T lymphocyte-deficient mice, immunosuppressed and reconstituted animals. In contrast to immunocompetent control animals, immunosuppressed mice were unable to mount an antibody response after rabies vaccination and resist a challenge of infectious virus (Turner, 1976; Mifune *et al.*, 1981). The use of immunosuppressive agents in mice confirms the protective role of T lymphocytes (Smith, 1981). Depletion of CD4$^+$ T cells did not allow vaccinated mice to resist a peripheral challenge (Celis *et al.*, 1990). Similarly, depletion of the CD4$^+$ subset of T cells in the first 10 days of the infection resulted in fatal infection of mice (Perry and Lodmell, 1991), which were naturally resistant to a street RABV injected by the peripheral route (Lodmell, 1983).

4.2 Protective role of antibodies and B lymphocytes

Humoral immunity has an essential protective function in the course of rabies virus infection. Virus-neutralizing antibodies, under the control of T helper cells, play a critical role in immunoprotection. The glycoprotein (G) of RABV is responsible for the induction of virus-neutralizing antibodies. Its ability to mount a protective immune response depends on its structure. Soluble G, a glycoprotein lacking the 58 carboxy-terminal amino acids, elicited 15 times less neutralizing antibodies than the intact full-length G and failed to protect mice (Dietzschold *et al.*, 1983). Immunoglobulins of the G isotype (IgGs), but not IgM, confer passive protection against RABV (Turner, 1978).

Antibodies against the N protein can be detected in human sera after immunization with inactivated rabies vaccines or after natural infection (Kasempimolporn *et al.*, 1991; Herzog *et al.*, 1992). These antibodies cannot neutralize the virus. Nevertheless, it has been proposed they could confer protection in experimental rabies (Lodmell *et al.*, 1993). Their mode of action remains to be determined and could be linked to their ability to block virus replication inside the cell as observed *in vitro* after antibodies were artificially injected into infected cells (Lafon and Lafage, 1987).

A protective role of B cells as a source of virus-neutralizing antibodies was first suspected in rabies because the resistance of different strains of mice to RABV infection correlated with levels of neutralizing antibodies that they can develop against rabies virus. The high antibody responders were more resistant to peripheral virus challenge, whereas the low antibody responders were more susceptible (Templeton *et al.*, 1986). The second line of evidence that antibodies can protect against RABV infection came from B lymphocyte depletion experiments using either anti-isotype antibody or B cell-deficient mice. Depletion of B cells using anti-isotype antibodies, which compromise the animal's capacity to mount an antibody response while leaving the T cell response intact, demonstrated that B cells play an essential role at least in the clearance of an abortive RABV strain (Miller *et al.*, 1978). Mice lacking B cells, such as J_{hD} knock-out mice, develop a progressive disease when infected intranasally with an abortive RABV and they succumb to infection (Hooper *et al.*, 1998). Vaccinated animals are not protected against a rabies virus challenge in the absence of B cells (μMT cells), whereas cytotoxic T cells and IFN-γ are dispensable for protection (Xiang *et al.*, 1995).

Antibody-mediated clearance of RABV from the NS was demonstrated with a single monoclonal antibody (Dietzschold *et al.*, 1992; Dietzschold, 1993). It was shown that this particular antibody mediates complete clearance of virus from the NS without the help of antibody-dependent cell-mediated cytotoxicity or complement-dependent lysis. This particular function is not shared with other neutralizing monoclonal antibodies. While the mechanisms involved remain unclear and specific transcytosis (entry of antibody into the target cell) cannot be excluded, the types and roles of antibodies involved in clearing neuronal infections need to be delineated. In particular, it will be of interest to characterize precisely the properties of antibodies that can cross the BBB. The BBB is impermeable to most antibodies. However, the blood barrier that surrounds nerves is less impermeable than the BBB (Moalem *et al.*, 1999), giving an opportunity for antibodies to clear the virus in the early steps of nerve infection. In addition, it is likely that a majority of the protective role of antibodies occurs before the virus enters the NS. Hence, antibodies can neutralize RABV particles before they reach the shield of the NS.

4.3 Role of CD8 lymphocytes

Early studies indicate that mice vaccinated with a live attenuated strain of RABV vaccine and, to a lesser extent with inactivated vaccine, produce cytotoxic T cells capable of lysing rabies virus-infected target cells (Wiktor *et al.*, 1977b). In contrast, fully virulent viruses do not induce cytotoxic T cells (Wiktor *et al.*, 1985). Glycoprotein, N and P proteins are inducers of cytotoxic T cells (Reddehase *et al.*, 1984; Wiktor *et al.*, 1984; Celis *et al.*, 1990; Larson *et al.*, 1991). However, a series of experiments designed to identify the immune system components involved in protection against rabies virus infection strongly suggest that cytotoxic T cells alone are not sufficient to confer protection against rabies virus. When mice were vaccinated with a mutated form of recombinant RABVG (expressed by a recombinant vaccinia virus containing a G gene that contains a leucine in position 8 instead of a proline) and compared with mice vaccinated with the non-mutated G, the mice did not develop neutralizing antibodies (Wiktor *et al.*, 1984), whereas they induced a similar CD8$^+$ T cell response (Celis *et al.*, 1990). Mice vaccinated with the mutated G were not protected, indicating the importance of neutralizing antibodies. Moreover, depletion of CD8$^+$ T cells has no effect on the resistance of mice to a street RABV or on the survival rate of vaccinated animals (Perry and Lodmell, 1991). The reconstitution of mice with a cytotoxic T cell clone specific for the G induces protection only in the case of abortive RABV (Kawano *et al.*, 1990). Since this is only observed with abortive RABV strains and not in encephalitic RABV infection, it can be concluded that CD8$^+$ T cells do not play a primary role in immunoprotection.

4.4 How rabies vaccines protect in post-exposure treatment

The most important role of rabies vaccine is in the induction of a sustained antibody response with the help of CD4$^+$ T lymphocyte activation. Rabies is an exception because it is generally thought that cytotoxic T cells are more important to clear virus infection from tissues than antibodies. Moreover, activation of CD8$^+$ T cells induces a pathological reaction, which is clinically associated with paralysis. This information should probably discourage the use of live vaccines, such as DNA vaccines or recombinant virus, as post-exposure vaccines (Lodmell and Ewalt, 2001), because of the tremendous risk of mounting a strong deleterious CD8 response in the NS. Nevertheless, these types of new generation vaccines would still be appropriate for pre-exposure vaccination regimens because of the robustness of live immunization.

The inactivated post-exposure vaccines that induce mainly B cell activation with the help of CD4$^+$ T cells (see Table 12.1) are the most appropriate choice to preserve integrity of the NS. Post-exposure vaccines probably confer protection because they prime an immune response in the periphery in secondary organs. Activated lymphocytes, CD4, antibody-secreting plasmocytes and, possibly, antibodies, can migrate into the NS parenchyma. New investigations are necessary to understand exactly the sequence and nature of events that confer protection.

5   CONCLUSIONS

Thus, RABV has selected a battery of mechanisms to escape host immunosur-veillance, possibly explaining why, in the absence of post-exposure treatment, rabies is one of the very few human infections with a near 100% mortality rate. Despite these well-adapted viral strategies to escape the immune response, RABV infections can be limited if vaccine is injected promptly after exposure suggesting that the viral-mediated paralysis of the host immune response requires some time, which can be exploited for therapy. Today, with an increased understanding of the particular properties of rabies virus infection, we can propose a biological relevance and rationale for heuristic choices elaborated by former vaccine designers. On the other hand, the success of vaccination of patients after exposures (i.e. post-exposure vaccination, or the use of serother-apy) urges us to investigate further the mechanisms that sustain the immune B cell response despite the impermeability of the BBB to most antibodies.

REFERENCES

Akira, S., Takeda, K. and Kaisho, T. (2001). Toll-like receptors: critical proteins linking innate and acquired immunity. *Nature Immunology* **2**, 675–680.

Baloul, L. and Lafon, M. (2003). Apoptosis and rabies virus neuroinvasion. *Biochimie* **85**, 777–788.

Baloul, L., Camelo, S. and Lafon, M. (2004). Up-regulation of Fas ligand (FasL) in the central nervous system: a mechanism of immune evasion by rabies virus. *Journal of Neurovirology* **10**, 372–382.

Barker, C.F. and Billingham, R.E. (1977). Immunologically privileged sites. *Advances in Immunology* **25**, 1–54.

Brzozka, K., Finke, S. and Conzelmann, K.K. (2005). Identification of the rabies virus alpha/beta interferon antagonist: phosphoprotein P interferes with phosphorylation of interferon regula-tory factor 3. *Journal of Virology* **79**, 7673–7681.

Camelo, S., Lafage, M. and Lafon, M. (2000). Absence of the p55 Kd TNF-alpha receptor promotes survival in rabies virus acute encephalitis. *Journal of Neurovirology* **6**, 507–518.

Camelo, S., Lafage, M., Galelli, A. and Lafon, M. (2001). Selective role for the p55 Kd TNF-alpha receptor in immune unresponsiveness induced by an acute viral encephalitis. *Journal of Neuroimmunology* **113**, 95–108.

Celis, E., Rupprecht, C.E. and Plotkin, S.A. (1990). New and improved vaccines against rabies. In: *New Generation Vaccines* (G.C. Woodrow, ed.). pp. 419–437. New York: Dekker.

Dietzschold, B. (1993). Antibody-mediated clearance of viruses from the mammalian central ner-vous system. *Trends in Microbiology* **1**, 63–66.

Dietzschold, B., Kao, M., Zheng, Y.M. *et al.* (1992). Delineation of putative mechanisms involved in antibody-mediated clearance of rabies virus from the central nervous system. *Proceedings of the National Academy of Sciences USA* **89**, 7252–7256.

Dietzschold, B., Wiktor, T.J., Wunner, W.H. and Varrichio, A. (1983). Chemical and immunological analysis of the rabies soluble glycoprotein. *Virology* **124**, 330–337.

Galelli, A., Baloul, L. and Lafon, M. (2000). Abortive rabies virus central nervous infection is controlled by T lymphocyte local recruitment and induction of apoptosis. *Journal of Neurovirology* **6**, 359–372.

Harmon, M.W. and Janis, B. (1975). Therapy of murine rabies after exposure: efficacy of polyriboinosinic-polyribocytidylic acid alone and in combination with three rabies vaccines. *Journal of Infectious Diseases* **132**, 241–249.

Hengel, H., Koszinowski, U.H. and Conzelmann, K.K. (2005). Viruses know it all: new insights into IFN networks. *Trends in Immunology* **26**, 396–401.

Herzog, M., Lafage, M., Montano-Hirose, J.A., Fritzell, C., Scott-Algara, D. and Lafon, M. (1992). Nucleocapsid specific T and B cell responses in humans after rabies vaccination. *Virus Research* **24**, 77–89.

Hilfenhaus, J., Karges, H.E., Weinmann, E. and Barth, R. (1975). Effect of administered human interferon on experimental rabies in monkeys. *Infection and Immunology* **11**, 1156–1158.

Hirai, K., Kawano, H., Mifune, K. *et al.* (1992). Suppression of cell-mediated immunity by street rabies virus infection. *Microbiology and Immunology* **36**, 1277–1290.

Hooper, D.C., Morimoto, K., Bette, M., Weihe, E., Koprowski, H. and Dietzschold, B. (1998). Collaboration of antibody and inflammation in clearance of rabies virus from the central nervous system. *Journal of Virology* **72**, 3711–3719.

Irwin, D.J., Wunner, W.H., Ertl, H.C. and Jackson, A.C. (1999). Basis of rabies virus neurovirulence in mice: expression of major histocompatibility complex class I and class II mRNAs. *Journal of Neurovirology* **5**, 485–494.

Jackson, A.C., Rossiter, E. and Lafon, M. (2006). Expression of Toll Like receptor in the human cerebellar cortex in rabies, herpes simplex encephalitis and other neurological diseases. *Journal of Neurovirology* **12**, 229–234.

Kasempimolporn, S., Hemachudha, T., Khawplod, P. and Manatsathit, S. (1991). Human immune response to rabies nucleocapsid and glycoprotein antigens. *Clinical and Experimental Immunology* **84**, 195–199.

Kawano, H., Mifune, K., Ohuchi, M. *et al.* (1990). Protection against rabies in mice by a cytotoxic T cell clone recognizing the glycoprotein of rabies virus. *Journal of General Virology* **71**, 281–287.

Kieny, M.P., Lathe, R., Drillien, R. *et al.* (1984). Expression of rabies virus glycoprotein from a recombinant vaccinia virus. *Nature* **312**,163–166.

Lafay, F., Benejean, J., Tuffereau, C., Flamand, A. and Coulon, P. (1994). Vaccination against rabies: construction and characterization of SAG2, a double avirulent derivative of SADBern. *Vaccine* **12**, 317–320.

Lafon, M. (2004). Subversive neuroinvasive strategy of rabies virus. *Archives of Virology* **18** (Suppl), 149–159.

Lafon, M. (2005). Modulation of the immune response in the nervous system by rabies virus. *Current Topics in Microbiology and Immunology* **289**, 239–258.

Lafon, M. and Lafage, M. (1987). Antiviral activity of monoclonal antibodies specific for the internal proteins N and NS of rabies virus. *Journal of General Virology* **68**, 3113–3123.

Lafon, M., Megret, F., Lafage, M. *et al.* (2005). Modulation of HLA-G expression in human neural cells after neurotropic viral infections. *Journal of Virology* **79**, 15 226–15 237.

Larson, J.K., Wunner, W.H., Otvos, L. Jr, Ertl, H.C. (1991). Identification of an immunodominant epitope within the phosphoprotein of rabies virus that is recognized by both class I- and class II-restricted T cells. *Journal of Virology* **65**, 5673–5679.

Le Bon, A. and Tough, D.F. (2002). Links between innate and adaptive immunity via type I interferon. *Current Opinion in Immunology* **14**, 432–436.

Le Bon, A., Schiavoni, G., D'Agostino, G., Gresser, I., Belardelli, F. and Tough, D.F. (2001). Type I interferons potently enhance humoral immunity and can promote isotype switching by stimulating dendritic cells in vivo. *Immunity* **14**, 461–470.

Lodmell, D.L. (1983). Genetic control of resistance to street rabies virus in mice. *Journal of Experimental Medicine* **157**, 451–460.

Lodmell, D.L. and Ewalt, L.C. (2001). Post-exposure DNA vaccination protects mice against rabies virus. *Vaccine* **19**, 2468–2473.

Lodmell, D.L., Esposito, J.J. and Ewalt, L.C. (1993). Rabies virus antinucleoprotein antibody protects against rabies virus challenge in vivo and inhibits rabies virus replication in vitro. *Journal of Virology* **67**, 6080–6086.

Ma, M., Wei, T., Boring, L., Charo, I.F., Ransohoff, R.M. and Jakeman, L.B. (2002). Monocyte recruitment and myelin removal are delayed following spinal cord injury in mice with CCR2 chemokine receptor deletion. *Journal of Neuroscience Research* **68**, 691–702.

Marquette, C., Van Dam, A.M., Ceccaldi, P.E., Weber, P., Haour, F. and Tsiang, H. (1996). Induction of immunoreactive interleukin-1 beta and tumor necrosis factor-alpha in the brains of rabies virus infected rats. *Journal of Neuroimmunology* **68**, 45–51.

McKimmie, C.S. and Fazakerley, J.K. (2005). In response to pathogens, glial cells dynamically and differentially regulate Toll-like receptor gene expression. *Journal of Neuroimmunology* **169**, 116–125.

McKimmie, C.S., Johnson, N., Fooks, A.R., Fazakerley, J.K. (2005). Viruses selectively upregulate Toll-like receptors in the central nervous system. *Biochemical Biophysical Research Communication* **336**, 925–933.

Medawar, P.B. (1948). Immunity to homologous grafted skin. III. The fate of skin homografts transplanted to the brain, to subcutaneous tissue, and to the anterior chamber of the eye. *British Journal of Experimental Pathology* **29**, 58–69.

Mifune, K., Takeuchi, E., Napiorkowski, P.A., Yamada, A. and Sakamoto, K. (1981). Essential role of T cells in the postexposure prophylaxis of rabies in mice. *Microbiology and Immunology* **25**, 895–904.

Miller, A., Morse, H.C. 3rd, Winkelstein, J. and Nathanson, N. (1978). The role of antibody in recovery from experimental rabies. I. Effect of depletion of B and T cells. *Journal of Immunology* **121**, 321–326.

Moalem, G., Monsonego, A., Shani, Y., Cohen, I.R. and Schwartz, M. (1999). Differential T cell response in central and peripheral nerve injury: connection with immune privilege. *FASEB Journal* **13**, 1207–1217.

Murphy, F.A., Bauer, S.P., Harrison, A.K. and Winn, W.C. Jr (1973). Comparative pathogenesis of rabies and rabies-like viruses. Viral infection and transit from inoculation site to the central nervous system. *Laboratory Investigation* **28**, 361–376.

Perrin, P., Tino de Franco, M., Jallet, C. *et al.* (1996). The antigen-specific cell-mediated immune response in mice is suppressed by infection with pathogenic lyssaviruses. *Research in Virology* **147**, 289–299.

Perry, L.L. and Lodmell, D.L. (1991). Role of CD4+ and CD8+ T cells in murine resistance to street rabies virus. *Journal of Virology* **65**, 3429–3434.

Prehaud, C., Megret, F., Lafage, M. and Lafon, M. (2005). Virus infection switches TLR-3-positive human neurons to become strong producers of interferon-beta. *Journal of Virology* **79**, 12 893–12 904.

Reddehase, M.J., Cox, J.H. and Koszinowski, U.H. (1984). Frequency analysis of cytolytic T lympho-cyte precursors (CTL-P) generated in vivo during lethal rabies infection of mice. II. Rabies virus genus specificity of CTL-P. *European Journal of Immunology* **14**, 1039–1043.

Sato, M., Tanaka, N., Hata, N., Oda, E. and Taniguchi, T. (1998). Involvement of the IRF family tran-scription factor IRF-3 in virus-induced activation of the IFN-beta gene. *FEBS Letters* **425**, 112–116.

Sharma, S., tenOever, B.R., Grandvaux, N., Zhou, G.P., Lin, R. and Hiscott, J. (2003). Triggering the interferon antiviral response through an IKK-related pathway. *Science* **300**, 1148–1151.

Smith, J.S. (1981). Mouse model for abortive rabies infection of the central nervous system. *Infection and Immunity* **31**, 297–308.

Templeton, J.W., Holmberg, C., Garber, T. and Sharp R.M. (1986). Genetic control of serum neutral-izing-antibody response to rabies vaccination and survival after a rabies challenge infection in mice. *Journal of Virology* **59**, 98–102.

Torres-Anjel, M.J., Volz, D., Torres, M.J., Turk, M. and Tshikuka, J.G. (1988). Failure to thrive, wast-ing syndrome, and immunodeficiency in rabies: a hypophyseal/hypothalamic/thymic axis effect of rabies virus. *Review of Infectious Diseases* **10**, S710–S725.

Turner, G.S. (1976). Thymus dependence of rabies vaccine. *Journal of General Virology* **33**, 535–538.

Turner, G.S. (1978). Immunoglobulin (IgG) and (IgM) antibody responses to rabies vaccine. *Journal of General Virology* **40**, 595–604.

Wang, Z.W., Sarmento, L., Wang, Y. *et al.* (2005). Attenuated rabies virus activates, while pathogenic rabies virus evades, the host innate immune responses in the central nervous system. *Journal of Virology* **79**, 12 554–12 565.

Wiktor, T.J., Doherty, P.C. and Koprowski, H. (1977a). Suppression of cell-mediated immunity by street rabies virus. *Journal of Experimental Medicine* **145**, 1617–1622.

Wiktor, T.J., Doherty, P.C. and Koprowski, H. (1977b). In vitro evidence of cell-mediated immunity after exposure of mice to both live and inactivated rabies virus. *Proceedings of the National Academy of Sciences USA* **74**, 334–338.

Wiktor, T.J., Macfarlan, R.I., Reagan, K.J. *et al.* (1984). Protection from rabies by a vaccinia virus recombinant containing the rabies virus glycoprotein gene. *Proceedings of the National Academy of Sciences USA* **81**, 7194–7198.

Wiktor, T.J., Macfarlan, R.I. and Koprowski, H. (1985). Rabies virus pathogenicity. In: *Rabies in the Tropics* (E. Kuwert, H. Koprowski and K. Bögel, eds). pp. 21–29. New York: Springer-Verlag.

Wunner, W.H. (1987). Rabies viruses – pathogenesis and immunity. In: *The Rhabdoviruses* (R.R. Wagner, ed.). pp. 361–426. New York: Plenum Press.

Xiang, Z.Q., Knowles, B.B., McCarrick, J.W. and Ertl, H.C. (1995). Immune effector mechanisms required for protection to rabies virus. *Virology* **214**, 398–404.

Yoneyama, M., Kikuchi, M., Natsukawa, T. *et al.* (2004). The RNA helicase RIG-I has an essential function in double-stranded RNA-induced innate antiviral responses. *Nature Immunology* **5**, 730–737.

13

Human Rabies Vaccines

DEBORAH J. BRIGGS

Department of Diagnostic Medicine/Pathobiology, College of Veterinary Medicine, Kansas State University, Manhattan, KS 66506, USA

1 INTRODUCTION

Rabies vaccines are virtually 100% efficacious when used according to World Health Organization (WHO) recommendations and they have saved millions of lives since they were first developed in the late 1800s. Modern cell culture rabies vaccines can be administered to protect persons at risk of exposure (pre-exposure) or after a person has been exposed to rabies virus (post-exposure). Industrialized countries utilize cell culture rabies vaccines for both pre- and post-exposure prophylaxis, whereas few medical professionals in developing countries, where canine rabies is endemic and over 50% of rabies victims are children less than 15 years of age, are aware of the value of pre-exposure vaccination. Increasing the use of pre-exposure vaccination in countries where canine rabies is endemic and not well controlled would undoubtedly reduce the death toll in children and other members of the populations at high risk of exposure. Rabies vaccines can be produced from several different types of cell substrates as well as from infected brain tissue. However, outdated nerve tissue rabies vaccines are rapidly being replaced in Asia and Latin America. In North America, there are two types of cell culture rabies vaccines that are currently available, human diploid cell vaccine (HDCV) and purified chick embryo cell rabies vaccine (PCECV). Both of these vaccines are highly efficacious and immunogenic but they are expensive. The cost of cell culture rabies vaccines has been the main deterrent to replacing nerve tissue vaccines in developing countries. In order to reduce the cost of modern cell culture rabies vaccines, the use of intradermal administration of vaccines that meet WHO recommended production standards has been widely accepted in Asian countries with limited budgets. In Thailand, the use of intradermal regimens and increased public awareness concerning rabies prevention has dramatically decreased the number of human fatalities. Other countries, especially India, have been able to convince international pharmaceutical companies to invest in technological transfer of rabies production facilities and thus have reduced the cost of modern cell culture

Rabies, second edition. Edited by Alan C. Jackson and William H. Wunner
ISBN 978-012-369366-2. Copyright Elsevier Inc. 2007

rabies vaccines by producing these vaccines locally rather than relying on imports. In the future, it is worth considering the development of new rabies vaccines that incorporate antigens found on the emerging lyssaviruses that are increasingly being reported from Africa and Asia.

2 VALUE OF VACCINATION

Rabies has the dubious distinction of having the highest case fatality rate of any known infectious disease. Once clinical symptoms are evident, the prognosis is almost certain death. WHO estimates that one person dies of rabies every 15 minutes and during the same time period, 300 people are exposed (WHO, 2005). Unfortunately, over half of these estimated deaths occur in children less than 15 years of age (WHO, 2005) and certainly most, if not all of them, could have been prevented through the administration of either pre- or post-exposure vaccination using one of the most powerful tools available to prevent rabies, modern cell culture rabies vaccine. It is quite evident that the availability and use of cell culture rabies vaccines has dramatically reduced the incidence of human rabies in countries that have implemented wide-scale use of post-exposure prophylaxis (PEP). For example, the dramatic increase in the administration of PEP in Thailand reduced the incidence of human rabies in that country by 80% in 15 years (WHO, 2002). Government officials and medical professionals in Thailand were able to improve the availability of cell culture rabies vaccines and reduce the cost of PEP at government sponsored anti-rabies clinics by using small doses of vaccine intradermally (Harverson and Wasi, 1984; Chutivongse et al., 1990). Other developing countries, including Sri Lanka, the Philippines and India have followed their example and have also implemented and promoted the use of low-dose intradermal regimens using cell culture rabies vaccines (Wilde et al., 1999; APCRI, 2006).

The production of rabies vaccines has changed dramatically since Pasteur, Roux and their colleagues conducted their first experiments and eventually developed the first very crude anti-rabies treatment in 1885 (Centers for Disease Control, 1985). All of the rabies vaccines that were developed immediately after the original Pasteur treatment were produced and applied using the same theory of serial injections of increasingly virulent rabies virus-infected nerve tissue. The virulence of the rabies virus in these crude vaccine suspensions was manipulated to obtain various degrees of infectivity by drying the infected material for increasing amounts of time in a specially designed 'Roux Bottle'. In the early 1900s, Fermi and Semple made two important modifications in the production of rabies vaccines: they introduced the use of phenol chemically to inactivate rabies virus-infected nerve tissue and they used the same inactivated suspension for all injections (Bugyaki et al., 1959). Thus, the chance of infecting a patient with a live virus from the vaccine was significantly reduced. These changes allowed tests of safety and potency to be implemented to produce safer rabies vaccines. In 1973, the WHO recommended that Fermi vaccines, crude rabies vaccines prepared from a variety of animals and partially inactivated with phenol, be discontinued because they contained residual, live, fixed rabies virus (WHO, 1973). However, a recent publication indicates that Ethiopia may still be producing this type of vaccine (Ayele *et al.*, 2001). Up until the last decade, sheep brain tissue was widely used throughout Asia and in parts of Africa but, currently, this vaccine is regarded as being less than adequate and is being replaced by cell culture vaccines. For example, in 2005, India, the largest producer of sheep brain vaccine, stopped producing and distributing sheep brain vaccine and a recent publication from Pakistan indicates that there is increasing pressure to replace the unreliable sheep brain vaccine produced there (Parviz *et al.*, 2004; APCRI, 2006).

Historically, nerve tissue rabies vaccines were produced and distributed at Pasteur Institutes across the world, thus allowing large numbers of exposed patients to be treated at a central location. This concept still exists today in many countries in Asia where the government financially supports PEP for poor patients unable to afford to go to private clinics for treatment. In most of these clinics, rabies immune globulin (RIG), which is recommended in PEP, is not administered on a regular basis (WHO, 2005). A recent publication from Thailand indicates that less than 3% of patients presenting at anti-rabies treatment centers with transdermal wounds are administered RIG (Kamoltham *et al.*, 2003).

Nerve tissue vaccines provided the first hope for survival after an exposure to a rabid animal when they were originally developed in the late 1800s. The gradual improvements that were made to these first crude vaccines over the next century enabled millions of patients to be treated effectively. The side effects and vaccine failures that accompanied their use led to the conclusion that safer and more efficacious vaccines needed to be developed in order to protect humans against rabies. The next major step forward in vaccine development came in 1955, when Fuenzalida and Palacios (1955) developed a less reactogenic vaccine in newborn mice which was quickly established as the standard of care in several Latin American countries.

The first non-nervous tissue culture rabies vaccine to become available in North America was the duck embryo vaccine (DEV). This vaccine was replaced by the first modern cell culture rabies vaccine, the HDCV in 1978 in the USA. HDCV produced significantly higher immunogenicity in vaccinated subjects and lower severe allergic reactions compared with previous vaccines (Plotkin, 1980). After the development of HDCV, two other cell culture rabies vaccines, purified Vero cell rabies vaccine (PVRV) and PCECV, were developed at approximately the same time and are currently distributed globally by international pharmaceutical companies. At the present time, PVRV and PCECV are the most widely used cell culture rabies vaccines in the world and are administered to millions of patients annually. Recently, clinical trials in humans to evaluate a new rabies vaccine, chromatographically purified Vero cell rabies vaccine (CPRV), have been conducted in a few countries including the USA (Quiambao *et al.*, 2000; Jones *et al.*, 2001; Arora *et al.*, 2004). To date this vaccine has not been marketed in any country. Other types of cell culture vaccines have been and are now being developed on a local level for 'in-country' distribution using the following cell substrates: primary baby hamster kidney cells, human diploid cells, Vero cells and embryonated eggs. It is unclear, however, whether all of these cell culture vaccines are being produced according to the recommended production standards and protocols published by the WHO (WHO, 2003, 2004).

3 NERVE TISSUE VACCINES

The nerve tissue vaccines (NTVs) that are presently produced and administered in Africa, Asia and Latin America include vaccines produced from the rabies virus-infected brain tissue of sheep, goats and mice. NTVs produced by infecting young sheep or goats are referred to as 'Semple' vaccine, named after Dr David Semple, who developed this vaccine in 1911 at Central Research Institute in Kasauli, India. When it was developed, almost a century ago, Semple vaccine was an improvement over previous NTVs and several production facilities were established throughout Asia and the rest of the world. Although the production of Semple vaccine was recently stopped in India, it is still produced and used in some Asian and African countries even though safer and more efficacious cell culture vaccines are available (APCRI, 2006). WHO has recommended that the production of all NTVs, including Semple vaccine, be discontinued and that they be replaced by cell culture rabies vaccines (WHO, 2005).

Four decades after Dr David Semple developed sheep brain vaccine in India, another improvement to NTV was made on the other side of the world, in Latin America, by Fuenzalida and Palacios (1955). Fuenzalida-Palacios rabies vaccine, or suckling mouse brain vaccine (SMBV), is produced from rabies virus-infected suckling mouse brain tissue and is still used in some Latin American countries as well as some countries in Asia and Africa. In the very near future, a local production of cell culture rabies vaccine produced from Vero cells may be available within Latin America, which should reduce the cost of cell culture rabies vaccines within the continent and speed the replacement of SMBV.

Due to the fact that the brain tissue of suckling mice is not myelinated, SMBV is less allergenic than Semple vaccine (Meslin and Kaplan, 1996). SMBV is produced from fixed rabies virus strains originally isolated in Chile (strains 51 and 91 of dog and human origin respectively). Mice not older than 1 day are injected intracerebrally and brain tissue is harvested approximately 4 days later (Diaz, 1996). Rabies virus-infected brain tissue is diluted and inactivated with ultraviolet light or β-propiolactone (BPL). The final concentration of brain tissue in SMBV depends on the manufacturer but it is usually around 10%. WHO recommends that all SMBVs have a final potency of 1.3 IU per dose (WHO, 1973). The vaccine is supplied in 1–2 ml vials with a shelf life of 1 year when stored at 4°C (Diaz, 1996). The adverse reactions associated with SMBV are lower (1:8000) than with Semple rabies vaccines (1:142 to 1:7000), but the case mortality rate has been reported to be higher in affected patients (22% for SMBV and 4.8% for Semple vaccines) (Meslin and Kaplan, 1996; Nogueira, 1998).

4 CELL CULTURE VACCINES

4.1 Global distribution

There are many different types of cell substrates that are used for the production of cell culture rabies vaccines including primary cell, diploid cell and continuous cell cultures. In China and the former USSR, primary baby hamster kidney cells are used to produce a rabies vaccine that is widely distributed throughout both regions. Embryonated eggs are a primary cell culture substrate used to produce at least three different types of rabies vaccine in Europe, Japan and India. Additionally, two rabies vaccines are produced from human diploid cells by two different rabies manufacturers, one in Europe and the other in India. Finally, Vero cells are grown from a continuous cell line that is used to produce a number of different quality vaccines in several countries in Asia, Europe and Latin America.

The first widely used cell culture rabies vaccine was developed on the human fetal lung diploid fibroblast cell line WI-38 using the Pittman Moore strain of rabies virus by Wiktor and his colleagues at the Wistar Institute in Philadelphia (Wiktor *et al.*, 1964, 1969). The WI-38 cell line was originally isolated and adapted for viral growth by Hayflick and Moorhead (1961). The accessibility of this 'normal' cell line was a monumental step forward in vaccinology as it provided a cell line free from potentially contaminated foreign and oncogenic particles. Human diploid cells have a higher (but finite) life span than primary cells, whereas continuous cell lines have the capacity to multiply indefinitely *in vitro*. The cells used for vaccine production from continuous cell lines may have differences in karyotype from the original cell line and may be obtained from healthy or tumoral tissues. The Food and Drug Administration (FDA) Center for Biologics Evaluation and Research (CBER) in the USA has specific guidelines for manufacturers using the continuous cell line Vero and they refer to the WHO recommended limit of less than or equal to 10 nanograms of DNA per dose for a parenterally administered vaccine (CBER, 2005).

As mentioned earlier, HDCV was the first cell culture rabies vaccine to be developed and this vaccine dramatically changed the perception of human rabies prevention and the concept of rabies vaccination. In the late 1970s, HDCV gained world attention when it was successfully used in a WHO-sponsored 'field trial' to treat humans exposed to severe bite wounds by rabid wolves in Iran (Bahmanyar *et al.*, 1976). This vaccine provided a safe and immunogenic means by which to expand the use of pre-exposure vaccination, thus giving added protection to persons whose vocation, hobby or living conditions puts them at increased risk of exposure to rabies. HDCV was licensed for use in the USA in 1980 and has been recommended for both pre- and post-exposure since that time. Due to the reduced vaccination regimens and apparent safety of HDCV, the Advisory Committee on Immunization Practices (ACIP) originally recommended routine boosters every 2 years for people at continued risk of exposure to rabies (Centers for Disease Control, 1991). However, due to the fact that HDCV has been associated with increased risk of severe immune-complex reactions after booster vaccination, the ACIP now recommends serological testing in lieu of routine boosters (see Chapter 16) (Centers for Disease Control, 1999). Although HDCV produces high serologic titers, high production costs and low virus yields make this vaccine unaffordable in most developing countries of the world where over 90% of human rabies deaths occur.

After the introduction of HDCV, other cell culture rabies vaccines were developed that are less expensive to produce and cause fewer adverse reactions. These second-generation vaccines include PCECV and PVRV. PCECV was developed by Barth *et al.* (1984) and is propagated in primary chick fibroblast cells using the Flury low egg passage (LEP) strain of rabies virus. Rabies virus is inactivated by BPL and purified by zonal centrifugation. The production of PCECV on chick embryo fibroblasts permits a very high yield of virus as compared with vaccine produced on human diploid cells. Additionally, there are fewer adverse reactions in persons that receive boosters with PCECV after primary vaccination (Bijok, 1985). The first successful transfer of rabies vaccine technology from an international pharmaceutical company to a developing country occurred in the late 1980s when production of PCECV was transferred from Marburg, Germany to Ankleshwar, India. This production facility has enabled the manufacturing costs of the vaccine to be substantially reduced while maintaining the production standards required by the WHO. A second transfer of rabies vaccine production technology, purified duck embryo cell vaccine (PDEV), from Switzerland to India, has also recently occurred. In the future, additional technology transfers may continue to play a role in reducing the cost of cell culture rabies vaccines for developing countries with limited budgets.

PVRV is a second-generation rabies vaccine that is produced on Vero cells (vervet monkey origin). PVRV is cultivated on microcarriers in large-scale biofermentors, thus reducing the production cost of vaccine as compared to HDCV, which is produced on a monolayer cell culture. PVRV is also inactivated with BPL and concentrated by ultracentrifugation. PVRV is used widely throughout the world including Europe and Latin America, but at the current time it is not licensed in North America. Clinical trials have also been conducted in the USA and a few Asian countries on a more purified version of Vero cell vaccine known as chromatographically purified Vero cell rabies vaccine (CPRV) (Quiambao *et al.*, 2000; Jones *et al.*, 2001; Arora *et al.*, 2004). To date, CPRV is not marketed in any country.

Another cell culture rabies vaccine that was originally developed for use in the USA by the Michigan Department of Public Health, but is currently not being produced, is rabies vaccine adsorbed (RVA). RVA is purified by filtration, inactivated with BPL and adjuvanted with aluminum phosphate (Barth and Franke, 1996).

4.2 Production standards

Rabies vaccines produced in cell culture and embryonated eggs have unquestionably been responsible for saving the lives of millions of human victims of animal bites that would have otherwise died of rabies. As with all vaccines, there are specific production standards that need to be followed to ensure safety and reliability. This is especially relevant for the production of rabies vaccines due to the very high case fatality rate of the disease. WHO has recently reviewed and updated their recommendations for the production of human rabies vaccines in order to ensure that quality vaccines are being produced (WHO, 2006). The updated WHO recommendations include, but are not limited to, requirements for potency, inactivation, stability, clinical evaluation and intradermal use of human rabies vaccines.

Regarding potency, WHO recommends that all cell culture rabies vaccines contain at least 2.5 IU per dose for intramuscular administration to humans. Cell culture rabies vaccines used in North America are inactivated by BPL. Stability tests are conducted to guarantee that a vaccine remains potent until the end of its stated shelf life. Clinical evaluation of human rabies vaccines is conducted prior to licensure to ensure that the vaccines are safe and effective for human use. The appropriate federal governmental authorities of each country have specific requirements that must be adhered to prior to the licensing of new rabies vaccines. Additionally, they are involved at all levels in the evaluation process, including inspection of the production site, evaluation of clinical data, monitoring batches of vaccine that will be imported for use and post-marketing surveillance. Initially, the efficacy of cell culture rabies vaccines was evaluated by vaccinating patients exposed to rabid animals and following their progress for a specific amount of time after exposure to ensure that they did not succumb to rabies. The data provided by these initial efficacy clinical trials indicated that the intramuscular use of cell culture rabies vaccines, when used according to the specific recommendations for PEP by WHO, is virtually 100% effective. Immunogenicity data used for evaluating serologic response to vaccines can be complicated and data must be carefully interpreted because the strain of seed virus used in the production of the vaccine will influence the amount of rabies virus neutralizing antibody that is detected (see Chapter 15) (Moore *et al.*, 2005).

4.3 Intradermal administration

Currently, there is no cell culture rabies vaccine that is pre-packaged for use intradermally, either for pre- or post-exposure vaccination. The currently available cell culture rabies vaccines do not have a preservative and therefore the risk of contamination is increased if they are used as a 'multiuse' vial, as is the case for intradermal administration of a vaccine packaged for intramuscular use. Therefore, WHO recommends that when rabies vaccines are administered intradermally, they should be kept at 2–8°C and used within 6–8 hours (WHO, 2005). Intradermal administration of human rabies vaccines for post-exposure vaccination is not currently practiced in North America; however, it is occasionally used for pre-exposure especially when the immunogenic response can be tested at least one month after vaccination (National Advisory Committee on Immunization, 2005). Intradermal administration of rabies vaccines has been used extensively for the last two decades in Asia for PEP and is currently used in some travel clinics in North America and Europe for pre-exposure vaccination. At the present time, WHO only recommends three vaccines for intradermal PEP (Table 13.1). These recommendations are based on previously published efficacy data (Chutivongse *et al.*, 1990; Quiambao *et al.*, 2005). All vaccines recommended for intradermal use are produced by international organizations and meet the published production recommendations of the WHO. Although WHO recommends a potency of 2.5 IU per human dose for the intramuscular administration of rabies vaccines, there is no similar specific potency requirement for an intradermal dose of a cell culture rabies vaccine except that the vaccine adhere to the potency for intramuscular administration. A recent paper reported on the correlation of the immunogenicity of rabies vaccines administered intradermally to the amount of rabies antigen present in each intradermal dose. This report concluded that when subjects received the Thai Red Cross regimen using

TABLE 13.1

Cell culture rabies vaccines that currently meet or exceed WHO production recommendations

	Recommended for		
Vaccine	*Pre-exposure*	*Post-exposure*	*Manufacturer*
Human rabies diploid cell culture rabies vaccine (HDCV)	Yes	Essen regimen (IM) Zagreb regimen (IM) 8-site regimen (ID)	Sanofi Pasteur Swiftwater PA 18370 1-800-VACCINE (822-2463)
Purified chick embryo cell rabies vaccine (PCECV)	Yes	Essen regimen (IM) Zagreb regimen (IM) 8-site regimen (ID) 2-site regimen (ID)	Chiron Vaccines Emeryville, CA 1-800-CHIRON (244-7668)
Purified Vero cell rabies vaccine (PVRV)	Yes	Essen regimen (IM) Zagreb regimen (IM) 2-site regimen (ID)	Sanofi Pasteur Lyon, France

IM: intramuscular; ID: intradermal.

a 0.1 ml intradermal dose from a 1 ml vial of PCECV that contained the minimum WHO recommended potency of 2.5 IU, all subjects responded with acceptable titers by day 14 (Beran *et al.*, 2005). Thus, WHO concluded that there was no need to increase the potency requirements for the intradermal use of cell culture rabies vaccines if these vaccines meet the WHO manufacturing recommendations and contain at least 2.5 IU per intramuscular dose (WHO, 2006). The current WHO vaccination regimens for intradermal administration of rabies vaccines are detailed elsewhere in this book (see Chapter 15).

5 CONCLUSIONS

The currently available cell culture rabies vaccines have been proven to be safe, immunogenic and, most importantly, efficacious when used according to WHO and ACIP recommendations. Their development and use has undoubtedly enabled millions of human lives to be saved from a dreadful disease. However, the under-utilization of cell culture rabies vaccines, lack of awareness of many medical professionals in canine rabies endemic countries as to the usefulness of administering these vaccines for pre-exposure and their limited availability in many regions of the world are all strong indications that cell culture rabies vaccines have not reached their full potential to benefit humankind.

Looking toward the future, the development of new cell culture rabies vaccines may not be on the priority list of very many international pharmaceutical companies. However, due to the recent identification of new lyssaviruses and the likelihood that there are other lyssaviruses circulating in the world that have not as yet been discovered, it is worth considering developing new lyssavirus vaccines with a broader coverage of protection (Hanlon *et al.*, 2005; Nel, 2005). Additionally, recent published data indicate that there are promising data supporting the future prospect that rabies vaccine could be successfully produced in plants or through DNA technology (Yusibov *et al.*, 2002; Girard *et al.*, 2006; Lodmell *et al.*, 2006) (see Chapter 15).

REFERENCES

APCRI (2006). Editorial – Present scenerio of anti-rabies vaccination programme in India. *Association for the Prevention and Control of Rabies in India Journal* **8**, 5–7.

Arora, A., Moeller, L. and Froeschle, J. (2004). Safety and immunogenicity of a new chromatographically purified rabies vaccine in comparison to the human diploid cell vaccine. *Journal of Travel Medicine* **11**, 195–199.

Ayele, W., Fekadu, M., Zewdie, B. *et al.* (2001). Immunogenicity and efficacy of Fermi-type nerve tissue rabies vaccine in mice and in humans undergoing post-exposure prophylaxis for rabies in Ethiopia. *Ethiopian Medical Journal* **39**, 313–321.

Bahmanyar, M., Fayaz, A., Nour-Salehi, S., Mohammadi, M. and Koprowski, H. (1976). Successful protection of humans exposed to rabies infection: Post-exposure treatment with the new human diploid cell rabies vaccine and antirabies serum. *Journal of the American Medical Association* **236**, 2751–2754.

Barth, R. and Franke, V. (1996). Fetal rhesus monkey lung diploid cell vaccine for humans. In: *Laboratory Techniques in Rabies*, 4th edn (F.-X. Meslin, M.M. Kaplan and H. Koprowski, eds). pp. 297–300. Geneva: World Health Organization.

Barth, R., Gruschkau, H., Bijok, U. *et al.* (1984). A new inactivated tissue culture rabies vaccine for use in man: Evaluation of PCEC vaccine by laboratory tests. *Journal of Biological Standards* **12**, 29–46.

Beran, J., Honegr, K., Banzhoff, A. and Malerczyk, C. (2005). Potency requirements of rabies vaccines administered intradermally using the Thai Red Cross regimen: investigation of the immunogenicity of serially diluted purified chick embryo cell rabies vaccine. *Vaccine* **23**, 3902–3907.

Bijok, U. (1985). Purified chick embryo cell (PCEC) rabies vaccine: A review of the clinical development. In: *Improvements in Rabies Postexposure Treatment* (I. Vodoppija, K.G. Nicholson, S. Smerdel and U. Bijok, eds). pp. 103–111. Zagreb Institute of Public Health.

Bugyaki, L., Moons, J.H. and Blockeel, S.R. (1959). Survival of the fixed Pasteur virus in Fermi-Semple type phenicated antirabies vaccine and its relation to the degree of immunity conferred. *Annales de la Societe belge de medecine tropicale* **39**, 275–280.

Chutivongse, S., Wilde, H., Supich, C., Baer, G.M. and Fishbein, D.B. (1990). Postexposure prophylaxis for rabies with antiserum and intradermal vaccination. *Lancet* **335**, 896–898.

CBER (2005). *Food and Drug Administration (FDA) Center for Biologics Evaluation and Research (CBER) Vaccines and Related Biological Products Advisory Committee Meeting (VRBPAC) November 16, 2005*. Bethesda, Maryland, USA. www.fda.gov/cber/advisory/vrbp/vrbpmain.htm

Centers for Disease Control (1985). A centennial celebration: Pasteur and the modern era of immunization. *Morbidity and Mortality Weekly Report* **34**, 389–390.

Centers for Disease Control (1991). Human rabies prevention – United States, 1991. Recommendations of the Advisory Committee on Immunization Practices (ACIP). *Morbidity and Mortality Weekly Report* **40**, (RR-3), 1–14.

Centers for Disease Control (1999). Human rabies prevention – United States, 1999. Recommendations of the Advisory Committee on Immunization Practices (ACIP). *Morbidity and Mortality Weekly Report* **48**, (RR-1), 1–21.

Diaz, A.M. (1996). Suckling-mouse brain vaccine. In: *Laboratory Techniques in Rabies*, 4th edn (F.-X. Meslin, M.M. Kaplan and H. Koprowski, eds). pp. 243–250. Geneva: World Health Organization.

Fuenzalida, E. and Palacios, R. (1955). Un método mejorado en la preparaciūn de la vacuna antirábica. [An improved method for preparation of rabies vaccine.] *Bulletin of the Institute of Bacteriology, Chile* **8**, 3–10.

Girard, L.S., Fabis, M.J., Bastin, M., Courtois, D., Petiard, V. and Koprowski, H. (2006). Expression of a human anti-rabies virus monoclonal antibody in tobacco cell culture. *Biochemical And Biophysical Research Communications* **345**, 602–607.

Hanlon, C.A., Kuzmin, I.V., Blanton, J.D., Weldon, W.C., Manangan, J.S. and Rupprecht. C.E. (2005). Efficacy of rabies biologics against new lyssaviruses from Eurasia. *Virus Research* **111**, 44–54.

Harverson, G. and Wasi, C. (1984). Use of post-exposure intradermal rabies vaccination in a rural mission hospital. *Lancet* **2**, 313–315.

Hayflick, L. and Moorhead, P.S. (1961). The serial cultivation of human diploid cell strains. *Experimental Cell Research* **25**, 585–621.

Jones, R.L., Froeschle, J.E., Atmar, R.L. *et al.* (2001). Immunogenicity, safety and lot consistency in adults of a chromatographically purified Vero-cell rabies vaccine: a randomized double-blind trial with human diploid cell rabies vaccine. *Vaccine* **19**, 4635–4643.

Kamoltham, T., Singhsa, J., Promsaranee, U., Sonthon, P., Mathean, P. and Thinyounyong, W. (2003). Elimination of human rabies in a canine endemic province in Thailand: five-year programme. *Bulletin of the World Health Organization* **81**, 375–381.

Lodmell, D.L., Ewalt, L.C., Parnell, M.J., Rupprecht, C.E. and Hanlon, C.A. (2006). One-time intradermal DNA vaccination in ear pinnae one year prior to infection protects dogs against rabies virus. *Vaccine* **23**, 412–416.

Meslin, F. -X. and Kaplan, M.M. (1996). General considerations in the production and use of brain-tissue and purified chicken-embryo rabies vaccines for human use. In: *Laboratory Techniques in Rabies*, 4th edn (F. -X. Meslin, M.M. Kaplan and H. Koprowski, eds). pp. 204–212. Geneva: World Health Organization.

Moore, S.M., Ricke, T.A., Davis, R.D. and Briggs, D.J. (2005). The influence of homologous vs. heterologous challenge virus strains on the serological test results of rabies virus neutralizing assays. *Biologicals* **33**, 269–276.

National Advisory Committee on Immunization (2005). Canada. June 1. Update on rabies vaccines. *Canada Communicable Disease Report* **31**, (ACS-5), 1–5.

Nel, L.H. (2005). Vaccines for lyssaviruses other than rabies. *Expert Review on Vaccines* **4**, 533–540.

Nogueira, Y.L. (1998). Adverse effect versus quality control of the Fuenzalida-Palacios antirabies vaccine. *Revista do Instituto de Medicina Tropical de São Paulo* **40**, 295–299.

Parviz, S., Chotani, R., McCormick, J., Fisher-Hoch and Luby, S. (2004). Rabies deaths in Pakistan: results of ineffective post-exposure treatment. *International Journal of Infectious Diseases* **8**, 346–352.

Plotkin, S.A. (1980). Rabies vaccine prepared in human cell cultures: progress and perspectives. *Review of Infectious Diseases* **2**, 433–448.

Quiambao, B.P., Dimaano, E.M., Ambas, C., Davis, R., Banzhoff, A. and Malerczyk, C. (2005). Reducing the cost of post-exposure rabies prophylaxis: efficacy of 0.1 ml PCEC rabies vaccine administered intradermally using the Thai Red Cross post-exposure regimen in patients severely exposed to laboratory-confirmed rabid animals. *Vaccine* **23**, 1709–1714.

Quiambao, B.P., Lang, J., Vital, S. *et al.* (2000). Economic issues in postexposure rabies treatment. *Journal of Travel Medicine* **6**, 238–242.

WHO (1973). World Health Organization Expert Committee on Rabies. Sixth report. *WHO Technical Report Series No 523*. Geneva: WHO.

WHO (2002). World Health Organization position paper on rabies vaccines. *Weekly Epidemiological Record* **77**, 109–119.

WHO (2003). World Health Organization. Meeting report: Discussion on WHO requirements for rabies vaccine for human use: potency assay, 20 May 2003. Geneva: WHO.

WHO (2005). World Health Organization. WHO Expert Consultation on Rabies. First Report. *WHO Technical Report Series No 931*, p. 121. Geneva: WHO.

WHO (2006). *New guidelines and recommendations established by the ECBS, 56th meeting, October 2005.* WHO/BS/05.2015. p. 55. Geneva: WHO.

Wiktor, T.J., Fernandes, M.V. and Koprowski, H. (1964). Cultivation of rabies virus in human diploid cell strain WI-38. *Journal of Immunology* **93**, 353–360.

Wiktor, T.J., Sokol, F., Kuwert, E. and Koprowski, H. (1969). Immunogenicity of concentrated and purified rabies vaccine of tissue culture origin. *Proceedings of the Society for Experimental Biology and Medicine* **131**, 799–805.

Wilde, H., Tipkong, P. and Kwahplod, K. (1999). Economic issues in postexposure rabies treatment. *Journal of Travel Medicine* **6**, 238–242.

Yusibov, V., Hooper, D.C., Spitsin, S.V. *et al.* (2002). Expression in plants and immunogenicity of plant virus-based experimental rabies vaccine. *Vaccine* **19**, 3155–3164.

14

Animal Vaccines

DAVID W. DREESEN

Department of Infectious Diseases, College of Veterinary Medicine,
The University of Georgia, Athens, Georgia 30602, USA

1 INTRODUCTION

Rabies in terrestrial animals, primarily carnivores, is caused by the classic genotype 1 rabies virus (RABV) (Nadin-Davis *et al.*, 2002; Wunner, 2002). Even though the widespread vaccination of domestic dogs has been the one most effective factor in the reduction of human rabies, the number of human deaths worldwide is greater than that of the combined deaths from polio, meningococcal meningitis, Japanese encephalitis, yellow fever, severe acute respiratory syndrome (SARS) and avian influenze (bird flu) (Wilde *et al.*, 2005). We have the 'tools' available to us in highly efficacious and safe animal and human vaccines. Multiple factors, discussed elsewhere in this book, can, however, prevent their use effectively in many areas of the world.

2 ANIMAL RABIES VACCINES

2.1 First generation of animal rabies vaccines

In his quest for a means to prevent rabies in humans, Louis Pasteur initiated research into animal rabies vaccine in France in the early 1880s (Bunn, 1991). Virus obtained from a rabid dog was first serially passed in rabbits by intracerebral inoculation at specified time intervals. Dogs were then vaccinated at various time intervals and challenged with rabies virus. Although this method produced acceptable results, Pasteur found that by serial intracerebral inoculation of monkeys with the dog origin virus, the incubation period increased while the virulence of the virus decreased. By using this regimen, Pasteur demonstrated that dogs vaccinated were resistant to subsequent challenge with virulent street (non-laboratory propagated) rabies virus.

In 1885, Pasteur attenuated, or weakened, the virus by desiccation (Bunn, 1991) to improve on the safety of these early attempts to produce a rabies vaccine. In a review by Friedberger and Frohner (1904), it was reported that Hogyes and Protopopoff and others conducted further studies to improve on the safety

Rabies, second edition. Edited by Alan C. Jackson and William H. Wunner
ISBN 978-012-369366-2.

and efficacy of vaccines for dogs and to reduce the number of doses needed. In 1927, the First International Rabies Conference recommended that fixed virus for canine rabies vaccines be completely inactivated or attenuated so that they caused no disease in dogs vaccinated either subcutaneously (SC) or intramuscularly (IM) (Schoenig, 1930). For the next several decades, virtually all rabies nerve tissue origin (NTO) vaccines were inactivated with phenol using the method described by Semple (Bunn, 1991). The NTO vaccines currently in use for mass vaccination campaigns in Africa, Latin America and the Caribbean are primarily produced from rabies virus-infected suckling mouse brains or lamb brains. These vaccines have been shown to be effective in campaigns (WHO, 2004). However, NTO killed vaccines for dogs and other animals have often, in the past, resulted in post-vaccinal nervous system reactions that could result in the death of the vaccinated animals (Bunn, 1991). Better vaccines were needed.

Embryonated chicken eggs were used by Koprowski and Cox (1948) for serial passage of the Flury strain (a human rabies virus isolate). The virus was initially passed 136 times in 1-day-old chicks. Vaccine produced from the 40th to the 50th chicken embryo passage lost its viscerotropic properties but retained some neurotropic properties. This was designated as Flury low-egg passage (LEP). While effective in dogs, the vaccine occasionally caused rabies in young pups, cats and cattle (Bunn, 1991). To increase the safety of the vaccines in these species, Koprowski *et al.* (1954) increased the passages of the Flury strain in embryonated eggs until the virus was found to be non-pathogenic for dogs when inoculated intracerebrally following the 205th passage. This Flury high-egg passage (HEP) vaccine was declared safe for IM use in cats and cattle as well as puppies 3 months of age. However, since cases of vaccine-induced rabies occurred in cats administered IM with the Flury-HEP vaccine, it was later withdrawn from the market (Cabasso *et al.*, 1963; Dean and Guevin, 1963).

2.2 Parenteral modified live virus vaccines

The Flury and Kelev strains of the rabies virus are used to produce chick embryo origin (CEO) modified live virus (MLV) vaccines. Tissue culture (TC) vaccines, such as those derived with the Street Alabama Dufferin (SAD) strain, which was adapted to hamster kidney cells (Fenji, 1960) and the Evelyn-Rokitnicki-Abelseth (ERA) strain, are grown on porcine kidney cells (Abelseth, 1963) and are commonly used to produce MLV vaccines (Reculard, 1996). Several other MLV vaccines have been produced over the years. These MLV vaccines, especially those using the CEO, SAD and ERA strains are still used extensively in Asia and Africa and parts of Europe and have been adapted for oral immunization of carnivores, including domestic dogs and cats (Blancou and Meslin, 1996). The TC MLV vaccines produce fewer allergic reactions than the CEO vaccines. Potency tests for MLV vaccines for animal use consist of measuring the titer of infectious virus in a sample from each filling lot (see Section 2.6.1). If the titer is as high as that proved efficacious in the species of animal for which the vaccine is intended, the vaccine is released for use (Sizaret, 1996).

Even though MLV vaccines have been trustworthy over the years, the use of inactivated (killed) cell culture vaccines is increasing in areas of the world where MLV vaccines are still in use. The WHO does not recommend MLV vaccines for parenteral use in animals (WHO, 2004) and no MLV rabies vaccines are currently licensed for use in the USA.

2.3 Oral modified live vaccines

The concept of oral rabies vaccines (ORV) was first proven to be successful in 1969 (Baer *et al.*, 1971). Using the SAD Berne strain of virus adapted from the ERA strain, several types of MLV ORV vaccines have been produced for use in baits for free-ranging animals that serve as vectors for the maintenance and transmission of the disease in wildlife species (Rupprecht *et al.*, 2004). ORV have been used extensively in Europe since 1977 and in Canada from 1989 with considerable success (Isara *et al.*, 1990; Aubert *et al.*, 1994). Unfortunately, the live-virus SAD vaccines contained some degree of residual pathogenicity for wild rodents (Artois *et al.*, 1992) and resulted in partially impaired immune responses in fox cubs <8 weeks old born from SAD B19-vaccinated vixens, resulting in insufficient protection against rabies (Muller *et al.*, 2001). Since the early to mid-1990s, the SAD strain used in vaccine has been replaced by the SAG-1 and SAG-2 (SAD-Avirulent-Gif) strains in the development of vaccines. The SAG-2 strain, the strain of choice, is a double mutant isolated from the SAD Berne strain after two successive selection steps utilizing anti-glycoprotein monoclonal antibodies. This strain is avirulent following intracerebral inoculation of immunocompetent mice and protects the mice against challenge with challenge virus standard (CVS) (Lafay *et al.*, 1994). The SAG-2 strain of rabies virus, packaged in chicken-head baits, has successfully protected captive African wild dogs against rabies challenge (Knobel *et al.*, 2003). In studies conducted at the Centers for Disease Control and Prevention in Atlanta, Georgia, USA, the SAG-2 vaccine produced no clinical illness in laboratory vaccinates (beagles) and residual SAG-2 virus was isolated from only one of 57 oral swabs from the dogs (Fekadu, *et al.*, 1996; Orciari *et al.*, 2001). No ORV derived from SAD/SAG origin vaccines are currently licensed for use in the USA (Compendium, 2006).

2.4 Oral live vaccinia-rabies virus glycoprotein recombinant vectored vaccine

A recombinant vaccinia virus expressing the rabies virus glycoprotein gene (V-RG) was developed by inserting the cDNA of the glycoprotein gene of the ERA strain into the thymidine kinase gene of the Copenhagen strain of vaccinia virus. Initial studies of this new ORV were conducted to determine whether the V-RG recombinant virus vaccine satisfied the various criteria that had to be met for such a vaccine to be distributed in the wild (Kieny *et al.*, 1984; Wiktor *et al.*, 1984). Criteria included: the vaccine would be effective when delivered by an oral bait; the baits would be readily accepted by target species but would be rabies virus-free; the vaccine in the baits would have reasonably long-term genetic and thermal stability; the vaccine would be biologically contained in the host; oral exposure to baits with the vaccine would produce full protection against rabies virus challenge; that no non-target species would develop rabies if they ingested the baits; and that the baits were clearly identified and safe for contact with humans. In an extensive series of trials carried out in the USA and in France, these criteria were met and the vaccine was licensed in 1995 for raccoons to prevent spread of raccoon rabies (and later to prevent the spread of Mexican dog rabies to Texas coyotes along the south Texas border with Mexico) by the US Department of Agriculture (USDA) (Wiktor *et al.*, 1985; Rupprecht *et al.*, 1986; Brochier *et al.*, 1990, 1991, 1995; Desmettre *et al.*, 1990). The single licensed product is produced only by Merial, Inc., Athens, GA, USA as Raboral V-RG™ for use by governmental (State Public Health) agencies as an ORV for raccoons and coyotes.

In the USA, Raboral V-RG™ is currently delivered to raccoons and coyotes in an extended fishmeal polymer bait, which contains 150 mg of tetracycline hydrochloride as a bone biomarker and a plastic sachet containing 1.8 ml of the vaccine. An extruded poultry-based bait with identical vaccine content has been shown to be more effective for targeting gray foxes (Merial, Inc., Athens, GA, USA).

The successful use of the ORV to achieve containment or elimination of rabies in some terrestrial wildlife animals in the USA and Canada is indicated by the effective containment to near elimination of red fox rabies in southern Ontario (MacInnes *et al.*, 2001), canine rabies in south Texas (Fearneyhough *et al.*, 1998) and raccoon rabies in Ohio (Krebs *et al.*, 2005), southern Ontario (Rosatte *et al.*, 2001) and eastern New Brunswick (Slate *et al.*, 2005). In 2003, over 10 million baits were distributed in 15 states in the USA (Slate *et al.*, 2005). New and potentially more effective oral vectored vaccines and more effective baits, including a fishmeal coated sachet bait, are being developed for ORV (Slate *et al.*, 2005).

2.5 Parenteral live vaccinia-rabies virus glycoprotein recombinant vectored vaccine

A canarypox-rabies glycoprotein recombinant vaccine was developed and found to be as effective as other poxvirus-rabies glycoprotein recombinants (Taylor *et al.*, 1991, 1995). Live canarypox virus that expresses the rabies virus glycoprotein has been licensed in the USA as a parenteral monovalent vaccine for cats and as a combination rabies vaccine for cats with feline panleukopenia virus, feline parvovirus and feline calicivirus vaccines included in the product. A combination canarypox-rabies vaccine with the whole-cell bacterin of *Neorickettsia risticii* included is also licensed for use in the prevention of Potomac fever in horses. These are the only rabies virus glycoprotein vaccines currently licensed in the USA (Compendium, 2006).

A recombinant adenovirus-vectored vaccine expressing rabies virus glycoprotein (Adrab.gp) was shown to be capable of inducing antibody immune responses in greyhound dogs immunized either subcutaneously or intramuscularly. The dogs had been previously vaccinated for rabies but had low or no rabies antibody titers (Tims *et al.*, 2000). This vaccine holds promise as a rabies virus vaccine for dogs.

2.6 Parenteral inactivated (killed) cell culture vaccines

The inactivated vaccines require that the rabies virus be produced in high concentrations. This is initially done by growing the virus strain (primarily CVS-11, Pittman-Moore (PM)-NIL 2 and Pasteur virus (PV)-BHK 21 strains) in the brain tissue of rabbits, baby hamster kidney (BHK) cells, suckling mouse brains (SMB), guinea pig brain cells, chick embryo cells (CEO), Vero cells or other substrates (Precausta and Soulebot, 1991; Reculard, 1996). Neonatal mice can be used as they lack the immunogenic (or allergenic) myelin that caused encephalomyelitis occasionally noted in animals vaccinated with earlier SMB NTO killed vaccines.

The production methods used for the TCO rabies vaccines have allowed less allergenic but more immunologic products (Greene and Rupprecht, 2006). Various methods, which are still valid, have been used to render the virus non-pathogenic or essentially inactivated (killed) as vaccines. These include, but are not limited to, beta propiolactone (BPL), UV light, and acetylethylamine as well as other amines. Phenol and formaldehyde are no longer recommended for virus inactivation (Reculard, 1996). The most commonly used inactivating agent is BPL. Once inactivated, adjuvants are added in order to increase the immune response to the antigen. The most common adjuvants are aluminum hydroxide, aluminum phosphate, saponin (in cattle vaccines) and, rarely, oil adjuvants (Precausta and Soulebot, 1991). Much of the information on cell lines, inactivating methods and adjuvants is proprietary and cannot be reported specifically for any one vaccine. The stability of these inactivated cell culture vaccines has allowed the rabies vaccine to be combined with other vaccines and bacterins such as canine distemper, canine adenovirus type 1, *Leptospira* and parvovirus for canines. For cats, the combination vaccines include feline panleukopenia-virus, feline parvovirus and feline calicivirus. A combined rabies and foot-and-mouth disease vaccine is available for cattle, sheep and goats (WHO, 2004). The potency and safety of the inactivated rabies vaccines have proven to be quite good.

2.6.1 The NIH test

In 1974, the National Institutes of Health (NIH) of the US Department of Health and Human Services (DHHS) adopted a mouse inoculation test to measure the potency of inactivated vaccines (Seligmann, 1973). This was necessitated because of the poor performance of the initial manufactured tissue culture origin vaccines (Bunn, 1991). Although a number of other tests to measure vaccine potency are used throughout the world, the NIH test is considered the 'gold standard' for measuring the ability of an inactivated vaccine to protect a mouse against virus challenge. The NIH test relies on challenge exposure of immunized mice to one virus strain (CVS), a strain thought to be derived from the original Pasteur isolate (Baer, 1997). This test has some inherent bias towards vaccine from the same virus strain origin when comparing vaccine efficacy across the variety of strains (i.e. SAD, Flury strain vaccines) used to prepare vaccines (Barth et al., 1988), but this bias does not occur when non-Pasteur stain vaccines are tested for protective potential against wild virus strains (Baer, 1997; Wunderli et al., 2003a). In addition, the NIH uses two doses of vaccine administered at a one-week interval by an intraperitoneal challenge two weeks later. This vaccination route is quite different from that used for routine administration of rabies vaccine. The second dose prevents an evaluation of the vaccine's primary immunologic potential and the challenge results in a disruption of the blood–brain barrier, allowing neutralizing antibodies in the serum to prevent infection. As a result of these limitations, the WHO has acknowledged that the NIH test needs some improvements or further suggesting that a new rabies potency test may be needed (WHO, 1992, 1994). Two recent reports have proposed an alternative method that avoids these shortcomings (Wunderli et al., 2003a, 2003b).

2.6.2 Post-vaccinal complications

Due to the higher antigenic mass and the use of adjuvants, inactivated rabies vaccines have produced post-vaccinal local and systemic reactions. The most common non-neurologic reactions include soreness, lameness and regional lymphadenopathy in the injected limb. Fever and anaphylaxis have also been reported (Dreesen, 1999; Greene and Rupprecht, 2006). Focal vasculitis and granulomas have been seen 3–6 months after vaccination (Greene and Rupprecht, 2006). Post-vaccinal sarcomas may develop as a result of sustained inflammatory reactions at the site of the vaccination that involve the underlying dermas. Such post-vaccinal sarcomas are often aggressive and invasive, especially in cats, months to years following vaccination (Dubielzig et al., 1993; Kass et al., 1993; Greene and Rupprecht, 2006). A review of 239 cases of fibrosarcomas in cats following single vaccination showed that 37% of the cats with vaccination-site tumors had received rabies vaccine, 33% were administered a non-rabies combination vaccine and 30% received a feline leukemia vaccine (Hendrick et al., 1994). It is not unusual for palpable lesions to occur in cats administered killed vaccine subcutaneously (Schulze *et al.*, 1997). Adverse incidence rates for reactions to rabies vaccination in a retrospective study of 3587 ferrets was 1% when the rabies vaccine was given alone and 0.85% when given in combination with distemper vaccine. The most common adverse events were vomiting and diarrhea (Moore *et al.*, 2005). The new generation of vectored recombinant vaccines now appearing on the market, such as the avipoxvirus vaccine recently licensed for use for cats in the USA (a rabies glycoprotein, live canarypox vectored vaccine) appears to produce few, if any, allergic or neoplastic reactions (Greene and Dreesen, 1998; Greene and Rupprecht, 2006).

2.6.3 WHO Report

Animal rabies vaccines

The WHO's World Survey of Rabies reported that there are at least 23 countries or territories that reported producing animal rabies vaccines during 1999. For the production of animal rabies vaccines, 14 countries use cell culture, seven use neural tissue and six countries use embryonated eggs (WHO, 2002). Four countries produced more than one type of vaccine. Both MLV and inactivated vaccines are produced worldwide.

The 1998 WHO World Survey of Rabies reported that Brazil is the major producer of NTO rabies vaccines for animal use followed by Bangladesh, Romania, Tunisia and El Salvador (WHO, 2000). These five countries account for 99.8% of the 23.5 million doses of NTO vaccine, primarily SMB origin (Fuenzalida strain), reported produced for the year. This same 1998 survey reported that the USA produced approximately 54 million doses of TCO rabies vaccines, 84% of all TCO animal vaccines produced. Vietnam is reportedly the primary source of embryonated egg-origin animal vaccine, producing 88% of this vaccine produced worldwide. It should be noted here that Argentina, France, Germany, India and a number of other countries that presumably produce animal rabies vaccines did not contribute to the 1998 WHO report.

Latin America

During the 2-year period 1998–1999, the availability of rabies vaccines for dogs and cats in Latin America grew by 10.7% and the total doses of vaccine administered for these species rose by 3.1% (REDIPRA, 2001). Vaccine coverage increased from 2.2% in Brazil to 36.7% in the Southern Cone (Argentina, Chile, Paraguay, Uruguay). However, there was a 16.3% decline in the Andean Area (Bolivia, Columbia, Ecuador, Peru, Venezuela) and a 6.8% decline in Central America. This same report denotes that, in the Andean Area, 67% of the canine population was vaccinated, in the Southern Cone 14.7%, Brazil 85%, Central America 38%, Mexico 88% and Latin Caribbean 41%. The WHO recommends that 70% of dogs in a population should be effectively immunized to prevent an epidemic of canine rabies (Coleman and Dye, 1996). There were 3600 laboratory confirmed canine rabies cases in all of Latin America during 1998 and 2500 during 1999. During the same periods, cattle accounted for 3298 and 3225 cases and other domestic animals accounted for 575 and 593 cases respectively.

2.6.4 USA

Vaccine types and licensing requirements

Many types of rabies vaccines are currently marketed in the USA for use in domestic animals. There are 12 inactivated monovalent rabies vaccines licensed for dogs and cats, two for ferrets, four for horses, four for cattle and five for sheep. Two inactivated vaccines are combined with other biologics for use in horses. In 2000, a new generation of vaccines was licensed for use in cats. These are the live canarypox-rabies virus glycoprotein recombinant vectored vaccines, either monovalent (one licensed vaccine) or in combination with feline panleukopenia virus, feline parvovirus and feline calicivirus vaccines (Compendium, 2006). A live vaccinia-rabies virus glycoprotein recombinant vectored vaccine is licensed for restricted use in wildlife raccoons and coyotes. As stated earlier, there are no MLV (attenuated) rabies vaccines licensed for use in the USA. All currently licensed killed rabies vaccines intended for use in carnivores must protect 22 of 25 or 26 of 30 (or a statistically equivalent number) animals from an IM challenge with a rabies virus for 90 days post challenge and 80% of controls must die from the challenge (Code of Federal Regulations, 2004). Alternative challenge requirements have been outlined when the test animals are of a species other than carnivores (Code of Federal Regulations, 2004). The US Department of Agriculture (USDA), Animal and Plant Health Inspection Service (APHIS), Center for Veterinary Biologics has jurisdiction over licensure of rabies vaccines in the USA.

Compendium of Animal Rabies Prevention and Control, 2006

The National Association of State Public Health Veterinarians (NASPHV) publishes annually the Compendium of Animal Rabies Prevention and Control (Compendium, 2006) in the *Journal of the American Veterinary Medical Association* each year. The annual Compendium is also available on the National Centers for Disease Control and Prevention (CDC) website (http://www.cdc.gov/mmwr/). This Compendium is a basis for animal rabies programs and the NASPHV issues it as recommendations. Some states (e.g. Georgia) and various cities and counties adopt the recommendations in the Compendium as regulations for animal rabies control and prevention.

The inactivated TCO vaccines should be used in animals at 3 months of age or older and then again one year later. This minimum age precludes maternal antibody blockage and recognizes the immature immune system's often poor response

(Greene and Dreesen, 1998). Depending on the vaccine, the animal species and, at times, local regulations, the animals should be vaccinated annually or triennially thereafter (Compendium, 2006). Depending on the vaccine type and the species, the vaccine is administered either IM or SC, while some vaccines can be administered either way. The minimum age for animal vaccination is 8 weeks of age for the licensed vectored vaccines. Regardless of the rabies vaccine type, only when the antibody response peaks, at approximately 28 days after primary vaccination, is the animal considered fully immunized, if vaccination has been administered in accordance with the manufacturer's recommendations.

From an epidemiologic viewpoint, the effectiveness of canine rabies prevention and control programs can be measured by comparing reports of rabies in dogs with reports of increases in cat rabies. This was apparent during the recent raccoon rabies epidemic in the Middle Atlantic and northeastern USA (Krebs *et al.*, 1997; Hanlon and Rupprecht, 1998). The increase in rabies cases in cats, while dog rabies cases remained substantially unchanged, reflects the vaccine status of the two populations as well as the number of feral animals in the two populations (Eng *et al.*, 1988; Petronek, 1998; Dreesen, 1999). Of 54 respondents in a survey of state and community health officials by Johnson and Walden (1996), 74% stated that canine rabies vaccination was required by state law while only 52% stated that cat vaccination was state law. The need for cat vaccination and feral population control cannot be overemphasized (Dreesen, 1999). Johnson and Walden's survey (1996) also noted that over-the-counter sales of rabies vaccines was permitted in 22 states and that, at that time, vaccination of wolf-hybrids was permitted in 14 states; however, in all but two of these 14 states the owner must sign a liability statement. Fourteen other states did not address the wolf-hybrid issue at all.

Ferrets and wolf-hybrids

In 1998, after extensive studies at the Centers for Disease Control and Prevention (Niezgoda *et al.*, 1997), a rabies vaccine for ferrets was approved by the USDA, APHIS. The ferret should be treated in a similar manner as a dog or cat in regard to vaccination and post-exposure management (Compendium, 2006).

Vaccination of wolf-hybrids with canine rabies vaccine is still a matter of considerable debate. In a meeting of taxonomists in 1996, it was concluded that rabies vaccines for dogs would *probably* protect wolves and their hybrids as they are genetically virtually indistinct from the domestic dog (Dreesen, 1999). At least one well-documented case of rabies has occurred in a properly vaccinated wolf-hybrid (Jay *et al.*, 1994). This animal was vaccinated with a 3-year vaccine at 4 months of age and received other vaccines and bacterins and an anti-helminthic on the same day. Six months later the animal was found with a dead skunk in its mouth. Within 3 weeks the animal developed signs suggestive of rabies, was euthanized and rabies was confirmed in the laboratory. Currently, there are no licensed rabies vaccines for wolf-hybrids and the 2006 Compendium states that wild animals and hybrids (offspring of wild animals crossbred to domestic animals) should not be kept as pets.

2.6.5 Post-exposure prophylaxis (PEP) for domestic animals

An animal can be considered to be immunized against rabies virus exposure approximately 28 days after the primary rabies vaccination, which is consistent with a peak antibody response (Compendium, 2006). Thus, an animal is considered immunized if the primary vaccination was administered at least 28 days previously and the follow-up vaccinations have been administered as recommended by the package insert and/or the Compendium (2006).

The NASPHV (Compendium, 2006) recommends that unvaccinated dogs, cats and ferrets exposed to a known or suspected rabid animal should be euthanized immediately. If not euthanized, the animal should be placed in strict quarantine for 6 months and vaccinated either upon entry into isolation or one month prior to release. Animals with expired vaccinations should be evaluated on a case-by-case basis. Currently, vaccinated dogs, cats and ferrets should be revaccinated immediately following exposure and kept under control and observation for 45 days. It has been shown that there is some evidence that the use of vaccine alone will not reliably prevent rabies from occurring in an unvaccinated domestic animal (Hanlon et al., 2002). Vaccinated livestock exposed to rabies should be revaccinated and observed for 45 days (Compendium, 2006). If not previously vaccinated, food animals should be slaughtered within 7 days with disposal of tissues in the exposed area. If not slaughtered within this time period, the animal should be closely observed for 6 months.

As previously mentioned, the Compendium (2006) is issued as recommendations only. Some states do not strictly adhere to the recommendations. For example, the Texas Health and Safety Code originally followed the previously noted recommendations for animals exposed to rabies (Clark and Wilson, 1996). However, in 1988, the Code was amended; unvaccinated domestic animals exposed to a rabid animal were to be euthanized or vaccinated immediately after exposure, kept in isolation for 90 days and given booster vaccinations in the third and eighth week of isolation. This regimen was based loosely on recommendations for humans exposed to rabies virus. A retrospective study conducted by Clark and Wilson (1996) found that 99.7% of 713 unvaccinated animals did not develop rabies during the 1979–1987 period during which the recommendations of the NASPHV were followed. Two PEP failures did occur (0.3%). For the period 1988–1994, after the Texas Code was amended to allow PEP for unvaccinated animals exposed to rabies, 629 of 632 animals (99.5%) that received the PEP booster vaccinations did not die of rabies. There was no statistical difference between the two regimens under conditions followed in Texas. In a follow-up study for the years 1995–1998, Wilson and Clark (2001) found only four of 830 (0.5%) domestic animals that received the PEP protocol, as recommended, during the previous 7-year period developed clinical rabies. They concluded that this is an effective PEP protocol and 'has been proven to be effective for the control of rabies in animals'. This alternative method of PEP for unvaccinated domestic animals exposed to rabies, as practiced in Texas, has not been endorsed in the Compendium (2006).

REFERENCES

Abelseth, M.K. (1963). Propagation of rabies virus in pig kidney cell culture. *Canadian Veterinary Journal* **5**, 84–86.

Artois, M, Guittre, C., Thomas, I., Leblois, H., Brochier, B. and Barrat, J. (1992). Partial pathogenicity for rodents of vaccines intended for oral vaccination against rabies: A comparison. *Vaccine* **10**, 524–528.

Aubert, M.F.A., Masson, E., Artois, M. and Barrat, J. (1994) Oral wildlife rabies vaccination field trials in Europe, with recent emphasis on France. *Current Topics in Microbiology and Immunology* **187**, 219–243.

Baer, G.M. (1997). Evaluation of an animal rabies vaccine by use of two types of potency tests. *American Journal of Veterinary Research* **58**, 837–840.

Baer, G.M., Abelseth, M.K. and Debbie, J.G. (1971). Oral vaccination of foxes against rabies. *American Journal of Epidemiology* **93**, 487–490.

Barth, R., Diderrich, G. and Weinmann, E. (1988). NIH test, a problematic method for testing potency of inactivated rabies vaccine. *Vaccine* **6**, 369–377.

Blancou, J. and Meslin, F. -X. (1996). Modified live-virus rabies vaccination for oral immunizations of carnivores. In: *Laboratory Techniques in Rabies*, 4th edn (F. -X. Meslin, M.M. Kaplan and H. Koprowski, eds). pp. 324–337. Geneva: World Health Organization.

Brochier, B., Costy, F. and Pastoret, P.P. (1995). Elimination of fox rabies from Belgium using a recombinant vaccinia-rabies vaccine: An update. *Veterinary Microbiology* **46**, 269–279.

Brochier, B., Kieny, M.P., Costy, F. *et al.* (1991). Large-scale eradication of rabies using recombinant vaccinia-rabies vaccine (Letter). *Nature* **354**, 520–522.

Brochier, B., Thomas, I., Bauduin, B. *et al.* (1990). Use of a vaccinia-rabies recombinant virus for the oral vaccination of foxes against rabies. *Vaccine* **8**, 101–104.

Bunn, T.O. (1991). Canine and feline rabies vaccines, past and present. In: *The Natural History of Rabies*, 2nd edn (G.M. Baer, ed.). pp. 415–425. Boca Raton: CRC Press.

Cabasso, V.J., Sharpless, G.R. and Shor, A.L. (1963). Vaccination of cats against rabies. In: *Proceedings of the 100th Annual Meeting of the American Veterinary Medical Association*. pp. 172–174. Chicago: American Veterinary Medical Association.

Clark, K.A. and Wilson, P.J. (1996). Postexposure rabies prophylaxis and preexposure vaccination failure in domestic animals. *Journal of the American Veterinary Medical Association* **208**, 1827–1830.

Code of Federal Regulations (2004). Title 9, part 113.209. Rabies vaccine, killed virus, pp. 602–604. Washington, DC: US Government Printing Office.

Coleman, P.G. and Dye, C. (1996). Immunization coverage required to prevent outbreaks of dog rabies. *Vaccine* **14**, 185–186.

Compendium (2006). Compendium of Animal Rabies Prevention and Control, 2006. *Journal of the American Veterinary Medical Association* **228**, 858–864.

Dean D.J. and Guevin V.H. (1963). Rabies vaccination of cats. *Journal of the American Veterinary Medical Association* **142**, 367–370.

Desmettre, P., Languet, B., Chappuis, G. *et al.* (1990). Use of vaccinia rabies recombinant for oral vaccination of wildlife. [Review.] *Veterinary Microbiology* **23**, 227–236.

Dreesen, D.W. (1999). Preexposure rabies immunization. In: *Rabies: Guidelines for Medical Professionals.* pp. 36–43. Trenton: Veterinary Learning Systems.

Dubielzig, R.R., Hawkins, K.L. and Miller, P.C. (1993). Mycoplastic sarcoma originating at the site of rabies vaccination in a cat. *Journal of Veterinary Diagnostic Investigations* **5**, 637–638.

Eng, T.R., Hamaker, T.A., Dobbins, J.G., Tong, T.C., Bryson, J.H. and Pinsky, P.F. (1988). Rabies surveillance, United States, 1988. *Morbidity and Mortality Weekly Report Centers for Disease Control Surveillance Summary* **38**, 1–21.

Fearneyhough, M.G., Wilson, P.J., Clark, K.A. *et al.* (1998). Results of an oral vaccination program for coyotes. *Journal of the American Veterinary Medical Association* **212**, 498–502.

Fekadu, M., Nesby, S.L., Shaddock, J.H., Schumacher, C.L., Linhart, S.B. and Sanderlin, D.W. (1996). Immunogenicity, efficacy and safety of an oral rabies vaccine (SAG-2) in dogs. *Vaccine* **14**, 465–468.

Fenji, P. (1960). Propagation of rabies virus in a culture of hamster kidney cells. *Canadian Journal of Microbiology* **6**, 479–483.

Friedberger, E. and Frohner, E. (1904). *Friedberger and Frohner's Veterinary Pathology,* 4th edn. Vol. 1, p. 353. London: Hurst and Blackett.

Greene, C.E. and Dreesen, D.W. (1998). Rabies. In: *Infectious Diseases of the Dog and Cat,* 2nd edn (C.E. Greene, ed.). pp. 114–126. Philadelphia: W.B. Saunders Co.

Greene, C.E. and Rupprecht, C.E. (2006). Rabies and other Lyssavirus infections. In: *Infectious Diseases of the Dog and Cat,* 3rd edn (C.E. Greene, ed.). pp. 167–183. Philadelphia: W.B. Saunders Co.

Hanlon, C.A. and Rupprecht, C.E. (1998). The reemergence of rabies. In: *Emerging Infections* (W.M. Scheld, D. Armstrong and J.M. Hughes, eds). pp. 59–80. Washington, DC: ASM Press.

Hanlon, C.A., Niezgoda, M.N. and Rupprecht, C.E. (2002). Postexposure prophylaxis for prevention of rabies in dogs. *American Journal of Veterinary Research* **63**, 1096–1100.

Hendrick, M.J., Shafer, F.S., Goldschmidt, M.H. *et al.* (1994). Comparison of fibrosarcomas that developed at vaccination sites and at non-vaccination sites in cats: 239 cases (1991–1992). *Journal of the American Veterinary Medical Association* **205**, 1425–1429.

Isara, A., Bressan, G. and Mutinelli, F. (1990). Sylvatic rabies in Italy: Epidemiology. *Journal of Veterinary Medicine* **B37**, 53–63.

Jay, M.T., Reilly, K.F., DeBess, E.E., Haynes, E.H., Bader, D.R. and Barrett, L.R. (1994). Rabies in a vaccinated wolf-dog hybrid. *Journal of the American Veterinary Medical Association* **205**, 1729–1732.

Johnson, W.B. and Walden, M.B. (1996). Results of a national survey of rabies control procedures. *Journal of the American Veterinary Medical Association* **208**, 1667–1672.

Kass, P.H., Barnes, W.G. Jr, Spangler, W.L., Chomel, B.B. and Culbertson M.R. (1993). Epidemiologic evidence for a causal relation between vaccination and fibrosarcoma tumorigenesis in cats. *Journal of the American Veterinary Medical Association* **203**, 396–405.

Kieny, M.-P., Lathe, R., Drillien, R. *et al.* (1984). Expression of rabies virus glycoprotein from a recombinant vaccinia virus. *Nature* **312**, 163–166.

Knobel, D.L., Liebenberg, A. and Du Toit, J.T. (2003). Seroconversion in captive wild dogs (Lycaon pictus) following administration of a chicken head bait/SAG-2 oral rabies combination. *Onderstepoort Journal of Veterinary Research* **70**, 73–77.

Koprowski, H. and Cox, H.R. (1948). Studies on chick embryo adapted rabies virus. I. Culture characteristics and pathogenicity. *Journal of Immunology* **60**, 533–536.

Koprowski, H., Black, J. and Nelson, D.J. (1954). Studies on chick-embryo-adapted rabies virus. VI. Further changes in pathogenic properties following prolonged cultivation in the developing chick embryo. *Journal of Immunology* **72**, 94–97.

Krebs, J.W., Mandel, E.J., Swerdlow, C.E. and Rupprecht, C.E (2005). Rabies surveillance in the United States during 2004. *Journal of the American Veterinary Medical Association* **227**, 1912–1925.

Krebs, J.W., Smith, J.S., Rupprecht, C.E. and Childs, J.E. (1997). Rabies surveillance in the United States during 1996. *Journal of the American Veterinary Medical Association* **211**, 1525–1539.

Lafay, F., Benejean, J., Tuffereau, C., Flamand, A. and Coulon, P. (1994).Vaccination against rabies: construction and characterization of SAG2, a double avirulent derivative of SADBern. *Vaccine* **12**, 317–320.

MacInnes, C.D., Smith, S.M., Tinline, R.R. *et al.* (2001). Elimination of rabies from red foxes in eastern Ontario. *Journal of Wildlife Diseases* **37**, 119–132.

Moore, G.E., Glickman, N.W., Ward, M.P., Engler, K.S., Lewis, H.B. and Glickman, L.T. (2005). Incidence and risk factors for adverse events associated with distemper and rabies vaccine administration in ferrets. *Journal of the American Veterinary Medical Association* **226**, 909–912.

Muller, T.F., Schuster, P., Vos, A.C., Selhorst, T., Wenzel, U.D. and Neubet, A.M. (2001). Effect of maternal immunity on the immature response to oral vaccination against rabies in young foxes. *American Journal of Veterinary Research* **62**, 1154–1158.

Nadin-Davis, S.A., Abdel-Malik, M., Armstrong, J. and Wandeler, A.I. (2002). Lyssavirus P gene characterization provides insights into the phylogeny of the genus and identifies structural similarities and diversity within the encoded phosphoprotein. *Virology* **298**, 286–305.

Niezgoda, M., Briggs, D.J., Shadduck, J.H. and Rupprecht, C.E. (1997). Pathogenesis of experimentally induced rabies in the domestic ferret. *American Journal of Veterinary Research* **58**, 1327–1331.

Orciari, L.A., Niezgoda, M., Hanlon, C.A. *et al.* (2001). Rapid clearance of SAG-2 rabies virus from dogs after oral vaccination. *Vaccine* **19**, 4511–4518.

Petronek, G.J. (1998). Free roaming and feral cats – their impact on wildlife and human beings. *Journal of the American Veterinary Medical Association* **212**, 218–226.

Precausta, P. and Soulebot, J-P. (1991). Vaccines for domestic animals. In: *The Natural History of Rabies*, 2nd edn (G.M. Baer, ed.). pp. 445–459. Boca Raton: CRC Press.

Reculard, P. (1996). Cell-culture vaccines for veterinary use. In: *Laboratory Techniques in Rabies*, 4th edn (F.-X. Meslin, M.M. Kaplan and H. Koprowski, eds). pp. 314–323. Geneva: World Health Organization.

REDIPRA (2001). Report of the VIII Meeting of Directors of National Rabies Control Programs in Latin America, RIMSA 12/14. Washington DC: Pan American Health Organization.

Rosatte, R., Donovan, D., Allan, M. *et al.* (2001). Emergency response to raccoon rabies introduction into Ontario. *Journal of Wildlife Diseases* **37**, 265–279.

Rupprecht, C.E., Hanlon, C.A. and Slate, D. (2004). Oral vaccination of wildlife against rabies: opportunities and challenges in prevention and control. *Developmental Biology (Basel)* **119**, 173–184.

Rupprecht, C.E., Wiktor, T.J., Johnston, D.H. *et al.* (1986). Oral immunization and protection of raccoons (*Procyon lotor*) with a vaccinia-rabies glycoprotein recombinant virus vaccine. *Proceedings of the National Academy of Sciences of the USA* **83**, 7947–7950.

Schoenig, H.W. (1930). Experimental studies with killed canine rabies vaccine. *Journal of the American Veterinary Medical Association* **76**, 25–27.

Schulze, A.E., Frank, L.A. and Hahn, K.A. (1997). Repeated physical and cytologic characterizations of subcutaneous postvaccinal reactions in cats. *American Journal of Veterinary Research* **58**, 718–724.

Seligmann, E.B. Jr (1973). Potency tests requirements of the United States National Institutes of Health (NIH). In: *Laboratory Techniques in Rabies*, 3rd edn (M.M. Kaplan and H. Koprowski, eds), pp. 279–289. Geneva: World Health Organization.

Sizaret, P. (1996). General considerations in testing the safety and potency of rabies vaccines. In: *Laboratory Techniques in Rabies*, 4th edn (F.-X. Meslin, M.M. Kaplan and H. Koprowski, eds). pp. 355–359. Geneva: World Health Organization.

Slate, D., Rupprecht, C.E., Rooney, J.A., Donovan, D., Lein, D.H. and Chipman, R.B. (2005). Status of oral rabies vaccination in wild carnivores in the United States. *Virus Research* **111**, 68–76.

Taylor, J., Meignier, B., Tartaglia, J. *et al.* (1995). Biological and immunogenic properties of a canarypox-rabies recombinant, ALVAC-RG (vCP65) in non-avian species. *Vaccine* **13**, 539–549.

Taylor, J., Trimarchi, C., Weinberg, R. *et al.* (1991). Efficacy studies on a canarypox-rabies recombinant virus. *Vaccine* **9**, 190–193.

Tims, T., Briggs, D.J., Davis, R.D. *et al.* (2000). Adult dogs receiving a rabies booster dose with a recombinant adenovirus expressing rabies virus glycoprotein develop high titers of neutralizing antibodies. *Vaccine* **18**, 2804–2807.

WHO (1992). World Health Organization Expert Consultation on Rabies. *Techical Report Series 824.* Geneva: World Health Organization.

WHO (1994). World Health Organization. Requirements for rabies vaccine for veterinary uses (Requirements for Biological Substances No. 29, Amendment 1992). *Technical Report No. 840.* Geneva: World Health Organization.

WHO (2000). *World survey of rabies no. 34 for the year 1998.* p. 1. Geneva: World Health Organization.

WHO (2002). *World survey for rabies no. 35 for the year 1999.* WHO/CDS/CSR/EPH/2002. p. 1. Geneva: World Health Organization.

WHO (2004). World Health Organization Expert Consultation on Rabies. *Technical Report Series 931.* Geneva: World Health Organization.

Wiktor, T.J., Macfarlane, R.I., Dietzschold, B., Rupprecht, C. and Wunner, W.H. (1985). Immunogenic properties of vaccinia recombinant virus expressing the rabies glycoprotein. *Annales de L'Institute Pasteur Virology* **136E**, 405–411.

Wiktor, T.J., Macfarlane, R.I., Reagan, K.J. *et al.* (1984). Protection from rabies by a vaccinia recombinant containing the rabies virus glycoprotein gene. *Proceedings of the National Academy of Sciences USA* **81**, 7194–7198.

Wilde, H., Khawplod, P., Khamoltham, T. *et al.* (2005). Rabies control in South and Southeast Asia. *Vaccine* **23**, 2284–2289.

Wilson, P.J. and Clark, K.A. (2001). Postexposure rabies prophylaxis protocol for domestic animals and epidemiologic characteristics of rabies vaccination failures in Texas: 1995–1999. *Journal of the American Veterinary Medical Association* **218**, 522–525.

Wunderli, P.S., Dreesen, D.W., Miller, T.J. and Baer, G.M. (2003a). Effects of vaccine route and dosage on protection from rabies after intracerebral challenge in mice. *American Journal of Veterinary Research* **64**, 491–498.

Wunderli, P.S., Dreesen, D.W., Miller, T.J. and Baer, G.M. (2003b). Effect of heterogeneity of rabies virus strain and challenge route and efficacy of inactivated rabies vaccines in mice. *American Journal of Veterinary Research* **64**, 499–505.

Wunner W.H. (2002). Rabies virus. In: *Rabies* (A.C. Jackson and W.H. Wunner, eds). pp. 23–77. San Diego: Academic Press.

15

Next Generation Rabies Vaccines

WILLIAM H. WUNNER

The Wistar Institute, Philadelphia, PA 19104–4268, USA

1 INTRODUCTION

For the past two decades, investigators have been searching for alternative rabies vaccines that have all the advantages of the killed or modified-live (attenuated) rabies virus vaccines for total stimulation of the immune response in animals and humans, without the undesirable side effects sometimes associated with conventional rabies virus vaccines. With the present high-quality cell-culture rabies vaccines for humans that are virtually 100% efficacious when they are used in accordance with the present WHO recommendations for pre- or post-exposure vaccination, one could say that there is little need to develop a more efficacious human rabies vaccine. The principal reason that alternative rabies vaccines are being researched for development is that inexpensive vaccines are desperately needed in developing countries to replace the present nerve tissue origin (NTO) rabies vaccines. It is unrealistic to imagine that the more expensive cell-culture-based vaccines and vaccination protocols currently used in industrialized countries will be widely implemented in many developing countries that have endemic rabies. It is also very impractical in many places to require patients in cases of severe exposure to seek medical treatment four or five times over 21–28 days. Therefore, new rabies vaccines must be developed, which are inexpensive enough, possibly requiring one or two administrations, by oral or parenteral inoculation routes, to vaccinate millions of people at risk following exposure, especially in canine rabies endemic countries. Oral vaccines have certain attractiveness, particularly for humans in developing countries, as they can be administered more rapidly in mass vaccination programs, as was done in the worldwide campaign to eradicate poliovirus.

2 RECOMBINANT VIRUS VACCINES

2.1 Recombinant poxvirus-vectored vaccines

In the early 1980s, soon after the rabies virus glycoprotein (RG) gene was cloned and its sequence was determined, efforts were initiated to produce the

531

Rabies, second edition. Edited by Alan C. Jackson and William H. Wunner
ISBN 978-012-369366-2. Copyright Elsevier Inc. 2007

first heterologous recombinant virus vaccine for use in oral vaccination of wildlife against rabies (Kieny *et al.*, 1984; Wiktor *et al.*, 1984) (see Chapter 14). This occurred at the same time the Street Alabama Dufferin (SAD)-Berne and the modified-live SAD-B19 vaccine strains were being used in baits for immunization of free-ranging foxes in Europe. These vaccine strains, however, were raising serious concerns about their safety for certain wildlife species (Winkler *et al.*, 1976; Artois *et al.*, 1992; Vos *et al.*, 1999). It was essential, from a safety perspective, to develop a vaccine that could be administered orally (in baits), and would be rabies virus-free and still induce full protection against a rabies virus challenge in animals. This was achieved first with vaccinia virus (Copenhagen strain), as a virus vector, in which the RG gene of the Evelyn-Rokitniki-Abelseth (ERA) strain of rabies virus was inserted into the thymidine kinase region of the vaccinia virus genome (Kieny *et al.*, 1984; Wiktor *et al.*, 1984). The vaccinia-rabies glycoprotein (V-RG) recombinant virus induced a rapid virus-neutralizing antibody (VNA) response in mice, both by subcutaneous inoculation and by oral administration and the animals were protected against a lethal rabies virus challenge (Wiktor *et al.*, 1984; Rupprecht *et al.*, 1986). In extensive vaccine trials, the V-RG recombinant virus vaccine was administered to over 40 families of animals, including the primary target species, the raccoon (*Procyon lotor*), to demonstrate its wide capability of inducing high titers of VNA and its ability to protect animals from rabies virus (Rupprecht *et al.*, 1988; Desmettre *et al.*, 1990; Brochier *et al.*, 1990, 1991, 1995) (see Chapter 14). Thus, a 'proof of principle' was established that a recombinant virus, expressing only the heterologous RG gene of rabies virus, would be a safe and efficacious oral vaccine for protection of wildlife against rabies. Similarly, other poxviruses, including the orthopoxvirus of raccoon (Esposito *et al.*, 1988) and avianpox viruses, such as the fowlpox (Taylor *et al.*, 1988) and canarypox (Taylor *et al.*, 1991; Cadoz *et al.*, 1992) viruses, were investigated as possible alternative vectors for expression of rabies virus antigens. Some of these investigations focused on rabies virus nucleoprotein (N) for protective cellular immune responses, as well as RG for the induction of VNA, indicating the importance of rabies virus N in protection (Fekadu *et al.*, 1992; Fujii *et al.*, 1994; Hooper *et al.*, 1994; Lodmell *et al.*, 2004). The concept that a live recombinant virus vector might fulfill the rabies vaccine requirements for the next generation was rapidly gaining credibility.

An example of a recombinant poxvirus that addresses the specific need for a less expensive and safe vaccine for humans is the non-replicating recombinant canarypox virus, which grows readily to high titer in cell culture (Cadoz *et al.*, 1992; Taylor *et al.*, 1995). Live canarypox virus that expresses the RG as the key immunogenic component has already been licensed as a combination-type vaccine for use in cats (Compendium, 2006). Canarypox virus induces a full range of immune responses, including a CD4[+] helper T-cell response and a humoral B-cell response against the RG, resulting in the production of VNA in animals. Importantly also, no adverse signs of infection or disease in animals following inoculation have been observed (Taylor *et al.*, 1991). The concept of using the canarypox-based ALVAC-RG (vCP65) recombinant virus in humans has been tested in Phase I clinical trials (Cadoz *et al.*, 1992; Fries *et al.*, 1996). It is important, also, that the rabies virus-specific immune response was not diminished in monkeys, which had previously received vaccinia virus before inoculation with the ALVAC-RG recombinant virus. This would further indicate that, in humans, prior immunity to vaccinia virus should not limit use of an ALVAC-RG recombinant vaccine (Taylor *et al.*, 1995).

2.2 Recombinant adenovirus-vectored vaccines

To use the adenovirus as an expression vector, the RG gene was inserted into the replication-non-essential E3 locus of the human adenovirus serotype 5 (AdHu5) genome. The E3 gene locus was deleted to downregulate expression of major histocompatibility complex antigens and protect the adenovirus-infected cells from T-cell-mediated destruction (Wold and Gooding, 1991; Yang *et al.*, 1994). When this prototype vaccine was tested by parenteral or oral vaccination in several animal species including mice, dogs, foxes and skunks, it was shown to induce protective immunity against a rabies virus challenge (Prevec *et al.*, 1990; Charlton *et al.*, 1992). However, the potential for *in vivo* replication of this E3-deleted recombinant adenovirus and tendency to cause occasional disease in the vaccinated host presented an undesirable risk from a safety point of view. With a growing interest at this time to use adenoviruses for gene therapy as well as for vaccination, replication-defective adenovirus vector constructs were investigated as possible next generation vaccines for rabies (Xiang *et al.*, 1996; Wang *et al.*, 1997; Vos *et al.*, 2001; Sharpe *et al.*, 2002). AdHu5 and AdHu2 vectors were made replication-defective by the deletion of essential genes from the E1 locus, which are required to initiate viral replication, as well as the E3 gene locus (Ginsberg *et al.*, 1989). Interestingly, the non-replicative heterologous recombinant adenovirus expressing the RG (Ad-RG), given subcutaneously to mice during the neonatal period or at several weeks of age, induced an immune response to rabies virus even in the presence of maternally transferred immunity to rabies virus (Wang *et al.*, 1997). The possibility of vaccinating young animals and humans with the recombinant Ad-RG vaccine during the early postnatal period is significant. Recombinant Ad-RG is also capable of inducing high titers of anti-rabies VNA in dogs previously immunized with conventional rabies vaccine (Tims *et al.*, 2000). The levels of VNA induced in these dogs following the booster immunization with the recombinant adenovirus-RG vaccine are reported to be higher than the levels that can be achieved with the conventional rabies vaccines, suggesting that the recombinant Ad-RG vaccine is capable of inducing a strong anamnestic response in dogs. The important questions are whether the recombinant Ad-RG vaccine will be efficacious in dogs immunized prior to 3 months of age and following a previous exposure to adenovirus (Tims *et al.*, 2000). Xiang *et al.* (2003) showed with the AdHu5 and the chimpanzee adenovirus serotype 68 (AdC68) vectors expressing the RG that induction of RG-specific VNA was not impaired by pre-existing immunity to the vaccine carrier when the vaccine was administered intranasally or orally in newborn mice. Interestingly, while the potential interference from pre-existing neutralizing antibodies in mice previously exposed by intramuscular inoculation to the vaccine carrier (AdHu5) was completely abolished by oral vaccination with the AdHu5 vaccine carrier, the same potential interference was partially reduced by intranasal vaccination and not at all abolished by intramuscular inoculation. These results clearly indicate that recombinant Ad-RG vaccine can be efficacious in the young and pre-exposed individuals when administered orally.

The initiative to develop the AdC68 virus vector as a potential vaccine for rabies arose from the observation that following repeated administration of AdHu5 in gene therapy studies, the expected high-level expression of the RG in animals became transient and diminished *in vivo*. This was due to clearance of adenovirus-infected cells by CD8[+] T cells directed against antigens of the adenovirus vector as well as against the heterologous (RG) gene product (Xiang *et al.*, 2002, 2003). Because AdHu5 is a ubiquitous pathogen that produces circulating serotype-specific neutralizing antibodies in up to 45% of the adult population in the USA, the AdC68 virus vector was developed to circumvent this potential interference and lack of efficacy of the systemically delivered human adenovirus vaccines (Xiang *et al.*, 2002). The advantages of the AdC68-RG vector, as well as the AdHu5-RG vector (as noted above), as oral vaccines are that they:

1 elicit superb B cell and CD8[+] T cell responses
2 achieve good responses with a single moderate subcutaneous or intramuscular dose in the experimental mouse model, which is an appropriate model for human responses
3 provide full, long-lasting protection against severe RV challenge (Xiang *et al.*, 2003).

Moreover, the distribution and persistence of the Ad68-RG vector in the mouse model after oral administration (by inhalation) were broad, this vector having been detected in all tissues examined except brain and able to sustain high rabies virus-specific IgG and IgA titers in sera and mucosal surfaces (Zhou *et al.*, 2006).

2.3 Rabies virus-based recombinant vaccines

It is now possible using reverse genetics also to generate homologous, i.e. rabies virus-based recombinant virus vectors that can express a variety of foreign genes capable of eliciting major immune effectors against rabies virus (Conzelman and

Schnell, 1994; Schnell *et al.*, 1994; Morimoto *et al.*, 2001; Dietzschold and Schnell, 2002; Wang *et al.*, 2005). The first objective in developing such recombinant rabies virus vectors as vaccines was specifically to modify the RG gene to make the virus fully attenuated without diminishing the positive effect of the recombinant virus vector's ability to induce a complete host immune response against rabies virus (Morimoto *et al.*, 2001). The most significant modification that can be made to the RG, a major contributor to the pathogenicity of the virus, is to replace the codon for arginine at position 333 in the RG gene sequence with another amino acid codon. Codons for glutamine (or glutamic acid), glycine, isoleucine, leucine, methionine, cysteine, serine or aspartic acid are all found to convert the virus from a pathogenic to a non-pathogenic phenotype (Dietzschold *et al.*, 1983; Sief *et al.*, 1985; Tuffereau *et al.*, 1989; Mebatsion, 2001; Morimoto *et al.*, 2001). To enhance the immunogenicity of the vaccine, it was reasoned that two identical RG genes placed in tandem in the recombinant virus might be more beneficial (Faber *et al.*, 2002). Indeed, the RG was overexpressed, on average twofold, in neuroblastoma (NA) cells infected with the recombinant virus that has two RG genes (2 × RG virus) compared with the virus that has only one RG gene (1 × RG virus) (Faber *et al.*, 2002). More importantly, overexpression of RG in 2 × RG virus-infected NA cells produces substantially higher antibody titers against the RG and against the rabies virus nucleoprotein (N), than in 1 × RG virus-infected NA cells, enhancing the immunizing potential of the recombinant virus. Over-expression of RG in 2 × RG virus-infected NA cells also causes a significant increase in caspase 3 activity of the apoptosis cascade and a marked decrease in mitochondrial respiration. Marked degeneration of neuronal cell bodies and frag-mentation of neurites within the neuron cultures was also associated with the more rapid induction of apoptosis in the 2 × RG virus-infected versus the 1 × RG virus-infected NA cells. Similarly, when the proapoptotic protein, cytochrome C, was expressed in the rabies virus recombinant (the cytochrome C gene replaced the pseudogene (Ψ) in the rabies virus genome of the recombinant virus, see Chapter 2), mortality in mice infected intranasally with the cytochrome C-con-taining virus was substantially lower than in animals infected intranasally with the attenuated virus that lacked the cytochrome C gene (Pulmanausahakul *et al.*, 2001). If, as suggested, enhanced apoptosis contributes to a decreased pathogenic-ity and also triggers powerful innate and adaptive immune responses (Rustifo, 2000) and the cell injury leads to release of endogenous adjuvants that stimulate cytotoxic T-cell responses (Shi *et al.*, 2000), then the modification and duplication of the RG gene becomes critically important to the attenuation and immunogenic-ity of the engineered recombinant rabies virus vector. Oral vaccination of dogs with these novel live-attenuated rabies virus recombinant vaccines demonstrated their safety and efficacy (Rupprecht *et al.*, 2005).

Since safety has been a primary concern, replication-defective recombinant rabies virus vaccines have also been generated for post-exposure prophylaxis (PEP). These non-replicative recombinant rabies viruses lack one or more of the viral genes that the virus normally requires for replication or assembly into infectious virions. To generate replication-defective virus from a rabies virus genomic cDNA plasmid (reverse genetics) that has a gene deleted, a rescue cell line constitutively expressing the missing viral gene product is required (Morimoto et al., 2005). For example, to produce the phosphoprotein (P) gene-deficient (def-P) live attenuated recombinant rabies virus, the ancillary cell line provides the missing P of rabies virus to produce recombinant def-P virions (Shoji et al., 2004). The cell line becomes a rescue system that effectively supplies P in trans to complement the deficiency of the def-P virus and generates and propagates recombinant def-P virus with high efficiency (Morimoto et al., 2005). In a normal cell infected with the recombinant def-P virus, the def-P virus cannot produce P but can perform the primary RNA transcription of the genome by L, without de novo synthesis of P, and express all the viral and non-viral (foreign) genes of the recombinant def-P virus in the host. Recombinant def-P virus in vivo has been found to be completely non-pathogenic after inoculation into mouse brain (Shoji et al., 2004). The chance of reversion to a pathogenic virus by homologous recombination (reincorporating the P gene into the RNA genome) has been ruled out on the basis that negative-strand RNA viruses rarely undergo homologous recombination (Chare et al., 2003). Also, since rabies virus P has been shown to act as an interferon (IFN) antagonist, inhibiting the IFN-α/β response by interfering with phosphorylation of IFN regulatory factor 3 (Brzózka et al., 2005), absence of P expression in recombinant def-P virus-infected cells would be expected to provide a beneficial effect as a vaccine by increasing the IFN-α/β response in the vaccinated host.

Similarly, a replication-defective recombinant rabies virus has been generated that lacks the matrix protein (M) gene (def-M virus) (Mebatsion et al., 1999). In wild-type virus, the rabies virus M plays an important role in virus assembly (see Chapter 2). Mebatsion et al. (1999) have shown that more than 98% of recombinant def-M virus particles remain cell associated, compared with less than 10% for wild-type virus, suggesting that the assembly intermediates found in def-M virus-infected cells were not competent for envelopment with the plasma membrane and the RG that is inserted in it. In the absence of M, cells infected with replication-defective recombinant def-M virus showed increased cell–cell fusion, which is a function of the RG on the cell surface (Mebatsion et al., 1999). This could be explained by the excessive accumulation of RG on the cell surface as a function of aberrant virus budding, which also would improve the immunogenicity of recombinant def-M virus as vaccine.

2.4 Genetic stability and large-scale growth of recombinant rabies virus-based vaccines

Several of the recombinant rabies virus vaccines (1 × RG, 2 × RG and the cytochrome C viruses) were analyzed by RT-PCR after 5 and 10 serial passages in brains of newborn mice for nucleotide sequence changes in the RG gene, which was modified by substituting glutamic acid at position 333 (Glu333) for arginine in the wild-type RG gene, and the cytochrome C gene. This was done to investigate whether the viruses remained genetically stable after up to 10 serial passages in mouse brain, given the high rate of mutation that RNA viruses are characteristically known for (Dietzschold et al., 2004). By the 10th serial passage, no change was observed in the extra RG gene in the recombinant 2 × RG virus vaccine or in the cytochrome C gene in the recombinant cytochrome C virus vaccine, and no change was detected in the codon Glu333 of the RG gene of all three viruses, which renders the virus non-pathogenic. However, by the fifth passage, a mutation resulting in an asparagine-to-lysine substitution at position 194 emerged in the RG genes of all three viruses. While this mutation was associated with a modest increase in pathogenicity in the 1 × RG virus and the cytochrome C virus vaccines, it did not alter the pathogenicity of the 2 × RG virus vaccine, which contained the change in codon 194 in only one of its two RG genes. Clearly, these viruses will be investigated further to determine whether the strong proapoptotic properties of the second RG gene in 2 × RG virus, which is associated with a decrease in pathogenicity, could have overridden the modest increase in pathogenicity of the 1 × RG virus (Dietzschold et al., 2004). Nevertheless, the relative gene sequence stability is an encouraging finding considering the high frequency of mutations that can occur in RNA viruses.

Experimental evidence indicates the recombinant rabies viruses can be grown in BSR cells in suspension culture (1.3 l of culture medium) using a moderate size (2.2 l-capacity) bioreactor to vastly higher titers (37- to 66-fold higher) than in stationary cultures (Dietzschold et al., 2004). This is also encouraging, as high virus concentrations (up to 10^{10} FFU/ml) will be needed to provide vaccine doses that are high enough to be used for oral immunization, particularly of wildlife (Hanlon et al., 2002).

3 DNA-BASED RABIES VACCINES

Another area of innovative rabies vaccine development is that of DNA-based or plasmid vaccines. DNA-based vaccines offer new approaches and unique strategies for both prophylaxis and post-exposure therapy against rabies. DNA-based vaccines were initially developed as a rather simple (basic plasmid preparation), yet versatile way to induce a broad spectrum of immune responses (both cell-mediated and humoral) when injected directly into the host, compared with conventional vaccines (Donnelly et al., 1997). The most appealing and compelling reason to develop DNA-based rabies vaccines is that plasmids (DNA encoding an RG gene) are easy to construct and can be mass-produced inexpensively. DNA plasmids that carry more than one viral gene component, e.g. RG and N genes of different rabies virus strains, or multiple plasmids encoding single expressible genes combined in a 'plasmid vaccine cocktail', provide a multivalent vaccine approach for simultaneous induction of immunity to more than one viral pathogen. Multiple RG genes, for example, given in one vaccine cocktail theoretically would protect individuals not only against rabies but also 'rabies-like' viruses (non-genotype 1 lyssaviruses) at the same time (Bahloul *et al.*, 1998; Jallet *et al.*, 1999; Nel *et al.*, 2003). Most significantly, it was thought that DNA-based vaccines would provide an efficient way to induce a cell-mediated cytolytic CD8$^+$ T-cell response against a specific antigen in addition to VNA and CD4$^+$ T cells. The CD8$^+$ T cell response is an important aspect of the immune response that often fails to be evoked by recombinant antigens and synthetic peptides (protein antigen subunits), either when used alone or with conventional inactivated vaccines (Germain, 1994; reviewed in Donnelly *et al.*, 1997).

The first DNA vaccine for rabies was developed at The Wistar Institute (Xiang *et al.*, 1994, 1995). DNA vaccination induces long-lasting immunity to rabies virus, although induction of VNA following intramuscular injection is usually slower compared with human diploid cell vaccine (HDCV), which puts the suitability of rabies DNA vaccines for post-exposure vaccination in question. Nevertheless, numerous studies have demonstrated the relative effectiveness of rabies DNA vaccines for eliciting rabies virus-specific VNA (Xiang *et al.*, 1994, 1995; Lodmell *et al.*, 1998a, 2000, 2003; Osorio *et al.*, 1999; Perrin *et al.*, 2000; Biswas *et al.*, 2001; Lodmell and Ewalt, 2001). Various parameters have been investigated including plasmid dosage, route of plasmid vaccine application, whether alone (using the ballistic gene gun to deliver 2.1 μm diameter gold beads coated with the naked plasmid DNA directly through the dermis) as a primary vaccination or as a booster vaccination in combination with rabies HDCV as the primary vaccine, host species and virus challenge following prophylaxis and in a post-exposure scenario. The technology and these parameters have also been studied in non-human primates with the objective of developing a DNA vaccine for humans (Lodmell *et al.*, 1998b, 2002a, 2002b). While the results from experiments using the mouse model and some larger mammals have been encouraging based on the protective effect of DNA vaccines against rabies, both in pre-exposure and post-exposure vaccination, they have not been as successful in totally protecting non-human primates following pre-exposure and post-exposure vaccination (Lodmell *et al.*, 2002a). Since it normally takes longer for DNA vaccination to induce VNA than HDCV vaccine, the critical question is whether DNA vaccination will be able to replace HDCV, plus a one-time treatment with RIG, in protecting humans in a typical post-exposure scenario. In post-exposure DNA vaccination and protection experiments designed to test various immunization protocols using non-human primates and the gene gun versus intradermal (needle) vaccination with DNA vaccine, gene gun vaccination through skin-associated lymphoid tissues and ear pinnae, with booster vaccinations as well (to induce higher VNA titers), was not sufficient to provide total post-exposure protection from developing rabies (Lodmell *et al.*, 2002b).

It appears that more work needs to be done with DNA vaccination technology, to develop it to its full potential, and to provide the desired, fully protective vaccine for animals and humans, at an extraordinarily low cost.

4   ORAL RABIES VACCINES DERIVED FROM PLANTS

Plants have also provided new and promising prospects in the process of developing effective, inexpensive and safe production and delivery systems for the next generation of vaccines for rabies (Yusibov *et al.*, 1999; Koprowski and Yusibov, 2001; Streatfield *et al.*, 2001). Plant viruses, such as tobacco mosaic virus (TMV) and tomato bushy stunt virus, serving as vectors for expression of foreign antigens in plants, have provided a variety of genetically manufactured vaccines from tomatoes and tobacco leaves, respectively (McGarvey *et al.*, 1995; Yusibov *et al.*, 1997). Among the key advantages of plants and plant crops for protein expression, particularly the tobacco plant as a suitable host for plant virus-vectored protein expression, is that they represent a biomass of major proportions for recombinant (foreign) protein production (Yusibov *et al.*, 1999). Having first demonstrated that vaccine antigens can be expressed in plant tissue in a form that is immunogenic in mice immunized intramuscularly (Yusibov *et al.*, 1997), studies later demonstrated that mice immunized orally (by gastric intubation or by feeding on antigen-producing spinach leaves) or parenterally (by intraperitoneal injection) with the plant-derived rabies virus-specific antigen could be protected from a lethal rabies virus challenge (Modelska *et al.*, 1998; Yusibov *et al.*, 2002). These new approaches are promising and they address the concerns of cost, safety and accessibility of vector-produced proteins as vaccines for the future control of rabies. One of the more advanced developments in productive expression of foreign antigens in plants has been the 'strategic gene design' approach to achieve high-level expression in transgenic plants. For rabies, a chimeric gene encoding the RG was designed whereby the native signal peptide in the RG gene was replaced by that of the pathogenesis-related protein, PR-S, of *Nicotiana tabacum*. The PR-S signal sequence, which is known for its efficient transportation of proteins into the endoplasmic reticulum of plant cells, and an endoplasmic reticulum retention signal, was included at the C-terminus of the RG. In addition, a total of 412 out of the 505 native codons in the protein coding sequence of the RG gene were altered to provide plant-preferred codons, while codons ending in CG and TA, putative transcription termination signals, mRNA stability element and potential splice sites, and hairpin loops were either avoided or eliminated (Ashraf *et al.*, 2005). The redesigned recombinant RG was successfully expressed at a higher level (up to 0.38% of total leaf protein) in tobacco leaves compared with expression of the native RG, was stable and showed a molecular mass ($\sim$66 kDa) that is consistent with being a glycosylated RG. The RG from tobacco leaves was immunogenic and protective against intracerebral rabies virus challenge in the mouse. These results provide considerable support for the proposition that oral immunization with plant-derived immunogenic protein, whether by gastric intubation or by feeding, especially as part of food, can generate local and systemic immune responses and protection against rabies virus.

REFERENCES

Artois, M., Guittre, C., Thomas, I., Leblois, H., Brochier, B. and Barrat, J. (1992). Potential pathogenicity for rodents of vaccines intended for oral vaccination against rabies: a comparison. *Vaccine* **10**, 524–528.

Ashraf, S., Singh, P.K., Yadav, D.K. *et al.* (2005). High level expression of surface glycoprotein of rabies virus in tobacco leaves and its immunoreactive activity in mice. *Journal of Biotechnology* **119**, 1–14.

Bahloul, C., Jacob, Y., Tordo, N. and Perrin, P. (1998). DNA-based immunization for exploring the enlargement of immunological cross-reactivity against the lyssaviruses. *Vaccine* **16**, 417–425.

Biswas, S., Reddy, G.S., Srinivasan, V.A. and Rangarajan, P.N. (2001). Preexposure efficacy of a novel combination DNA and inactivated rabies virus vaccine. *Human Gene Therapy* **12**, 1917–1922.

Brochier, B., Costy, F. and Pastoret, P.-P. (1995). Elimination of fox rabies from Belgium using a recombinant vaccinia-rabies glycoprotein vaccine; an update. *Veterinary Microbiology* **46**, 269–279.

Brochier, B., Kieny, M.P., Costy, F. *et al.* (1991). Large-scale eradication of rabies using recombinant vaccinia-rabies vaccine. *Nature* **354**, 520–522.

Brochier, B., Thomas, I., Bauduin, B. *et al.* (1990). Use of vaccinia-rabies recombinant virus for the oral vaccination of foxes against rabies. *Vaccine* **8**, 101–104.

Brzózka, K., Finke, S. and Conzelmann, K.-K. (2005). Identification of the rabies virus alpha/beta interferon antagonist: phosphoprotein P interferes with phosphorylation of interferon regulatory factor 3. *Journal of Virology* **79**, 7673–7681.

Cadoz, M., Strady, A., Meignier, B. *et al.* (1992). Immunisation with canarypox virus expressing rabies glycoprotein. *Lancet* **339**, 1429–1432.

Chare, E.R., Gould, E.A. and Holmes, E.C. (2003). Phylogenetic analysis reveals a low rate of homologous recombination in negative-sense RNA viruses. *Journal of General Virology* **84**, 2691–1703.

Charlton, K.M., Artois, M., Prevec, L. *et al.* (1992). Oral rabies vaccination of skunks and foxes with a recombinant human adenovirus vaccine. *Archives of Virology* **123**, 169–179.

Compendium (2006). Compendium of Animal Rabies Virus Prevention and Control, 2006. *Journal of the American Veterinary Medical Association* **228**, 858–864.

Conzelmann, K.-K. and Schnell, M. (1994). Rescue of synthetic genomic RNA analogs of rabies virus by plasmid-encoded proteins. *Journal of Virology* **68**, 713–719.

Desmettre, P., Languet, B., Chappuis, G. *et al.* (1990). Use of vaccinia rabies recombinant for oral vaccination of wildlife. *Veterinary Microbiology* **23**, 227–232.

Dietzschold, B. and Schnell, M.J. (2002). New approaches to the development of live attenuated rabies vaccines. *Hybrid Hybridomics* **21**, 129–134.

Dietzschold, B., Wunner, W.H., Wiktor, T.J. *et al.* (1983). Characterization of an antigenic determinant of the glycoprotein that correlates with pathogenicity of rabies virus. *Proceedings of the National Academy of Sciences USA* **80**, 70–74.

Dietzschold, M.-L., Faber, M., Mattis, J.A., Pak, K.Y., Schnell, M.J. and Dietzschold, B. (2004). In vitro growth and stability of recombinant rabies viruses designed for vaccination of wildlife. *Vaccine* **23**, 518–524.

Donnelly, J.J., Ulmer, J.B., Shiver, J.W. and Liu, M.A. (1997). DNA vaccines. *Annual Review of Immunology* **15**, 617–648.

Esposito, J.J., Knight, J.C., Shaddock, J.H., Novembre, F.J. and Baer, G.M. (1988). Successful oral rabies vaccination of raccoons with raccoon poxvirus recombinants expressing rabies virus glycoprotein. *Virology* **165**, 313–316.

Faber, M., Pulmanausahakul, R., Hodawadekar, S.S. *et al.* (2002). Overexpression of the rabies virus glycoprotein results in enhancement of apoptosis and antiviral immune response. *Journal of Virology* **76**, 3374–3381.

Fekadu, M., Sumner, J.W., Shaddock, J.H., Sanderlin, D.W. and Baer, G.M. (1992). Sickness and recovery of dogs challenged with a street rabies virus after vaccination with a vaccinia virus recombinant expressing rabies virus N protein. *Journal of Virology* **66**, 2601–2604.

Fries, L.F., Tartaglia, J., Taylor, J. *et al.* (1996). Human safety and immunogenicity of a canarypox-rabies glycoprotein recombinant vaccine: an alternative poxvirus vector system. *Vaccine* **14**, 428–434.

Fujii, H., Takita-Sonoda, Y., Mifune, K., Hirai, K., Nishizono, A. and Mannen, K. (1994). Protective efficacy in mice of post-exposure vaccination with vaccinia virus recombinant expressing either rabies virus glycoprotein or nucleoprotein. *Journal of General Virology* **75**, 1339–1344.

Germain, R.N. (1994). MHC-dependent antigen processing and peptide presentation: providing ligands for lymphocyte activation. *Cell* **76**, 287–299.

Ginsberg, H.S., Lundholm-Beauchamp, U., Horswood, R.L. *et al.* (1989). Role of early region 3 (E3) in pathogenesis of adenovirus disease. *Proceedings of the National Academy of Sciences USA* **86**, 3823–3827.

Hanlon, C.A., Niezgoda, M., Morril, P. and Rupprecht, C.E. (2002). Oral efficacy of an attenuated rabies virus vaccine in skunks and raccoons. *Journal of Wildlife Diseases* **38**, 420–427.

Hooper, D.C., Pierard, I., Modelska, A. *et al.* (1994). Rabies ribonucleocapsid as an oral immunogen and immunological enhancer. *Proceedings of the National Academy of Sciences USA* **91**, 10 908–10 912.

Jallet, C., Jacob, Y., Bahloul, C. *et al.* (1999). Chimeric lyssavirus glycoproteins with increased immunological potential. *Journal of Virology* **73**, 225–233.

Kieny, M.-P. Lathe, R., Drillien, R. *et al.* (1984). Expression of rabies virus glycoprotein from a recombinant vaccinia virus. *Nature* **312**,163–166.

Koprowski, H. and Yusibov, V. (2001). The green revolution: plants as heterologous expression vectors. *Vaccine* **19**, 2735–2741.

Lodmell, D.L. and Ewalt, L.C. (2001). Post-exposure DNA vaccination protects mice against rabies virus. *Vaccine* **19**, 2468–2473.

Lodmell, D.L., Esposito, J.J. and Ewalt, L.C. (2004). Live vaccinia-rabies virus recombinants, but not an inactivated rabies virus cell culture vaccine, protect B-lymphocyte-deficient A/WySnJ mice against rabies: considerations of recombinant defective poxviruses for rabies immunization of immunocompromised individuals. *Vaccine* **22**, 3329–3333.

Lodmell, D.L., Parnell, M.J., Bailey, J.R., Ewalt, L.C. and Hanlon, C.A. (2002a). One-time gene gun or intramuscular rabies DNA vaccination of non-human primates: comparison of neutralizing antibody responses and protection against rabies virus 1 year after vaccination. *Vaccine* **20**, 838–844.

Lodmell, D.L., Parnell, M.J., Bailey, J.R., Ewalt, L.C. and Hanlon, C.A. (2002b). Rabies DNA vaccination of non-human primates: post-exposure studies using gene gun methodology that accelerates induction of neutralizing antibody and enhances neutralizing antibody titers. *Vaccine* **20**, 2221–2228.

Lodmell, D. L., Parnell, M.J., Weyhrich, J.T. and Ewalt, L.C. (2003). Canine rabies DNA vaccination: a single-dose intradermal injection into ear pinnae elicits elevated and persistent levels of neutralizing antibody. *Vaccine* **21**, 3998–4002.

Lodmell, D.L., Ray, N.B. and Ewalt, L.C. (1998a). Gene gun particle-mediated vaccination with plasmid DNA confers protective immunity against rabies virus infection. *Vaccine* **16**, 115–118.

Lodmell, D.L., Ray, N.B., Parnell, M.J. *et al.* (1998b). DNA immunization protects non-human primates against rabies virus. *Nature Medicine* **4**, 949–952.

Lodmell, D.L., Ray, N. B., Ulrich, J.T. and Ewalt, L.C. (2000). DNA vaccination of mice against rabies virus: effects of the route of vaccination and the adjuvant monophosphoryl lipid A (MPL®). *Vaccine* **18**, 1059–1066.

McGarvey P.B., Hammond, J., Denelt, M.M. *et al.* (1995). Expression of the rabies virus glycoprotein in transgenic tomatoes. *Biotechnology* **13**, 1484–1487.

Mebatsion, T. (2001). Extensive attenuation of rabies virus by simultaneously modifying the dynein light chain binding site in the P protein and replacing Arg333 in the G protein. *Journal of Virology* **75**, 11 496–11 502.

Mebatsion, T., Weiland, F. and Conzelmann, K.-K. (1999). Matrix protein of rabies virus is responsible for the assembly and budding of bullet-shaped particles and interacts with the transmembrane spike glycoprotein G. *Journal of Virology* **73**, 242–250.

Modelska, A., Dietzschold, B., Sleysh, N. *et al.* (1998). Immunization against rabies with plant-derived antigen. *Proceedings of the National Academy of Sciences USA* **95**, 2481–2485.

Morimoto, K., McGettigan, J.P., Foley, H.D., Hooper, D.C., Dietzschold, B. and Schnell, M.J. (2001). Genetic engineering of live rabies vaccines. *Vaccine* **19**, 3543–3551.

Morimoto, K., Shoji, Y. and Inoue, S. (2005). Characterization of P gene-deficient rabies virus: propagation, pathogenicity and antigenicity. *Virus Research* **111**, 61–67.

Nel, L.H., Niezgoda, M., Hanlon, C.A., Morril, P.A., Yager, P.A. and Rupprecht, C.E. (2003). A comparison of DNA vaccines for the rabies-related virus, Mokola. *Vaccine* **21**, 2598–2606.

Osorio, J.E., Tomlinson, C.C., Frank, R.S. *et al.* (1999). Immunization of dogs and cats with a DNA vaccine against rabies virus. *Vaccine* **17**, 1109–1116.

Perrin, P., Jacob, Y., Aguilar-Sétien, A. *et al.* (2000). Immunization of dogs with a DNA vaccine induces protection against rabies virus. *Vaccine* **18**, 479–486.

Prevec, L., Campbell, J.B., Christie, B., Belbeck, L. and Graham, F.L. (1990). A recombinant human adenovirus vaccine against rabies. *Journal of Infectious Diseases* **161**, 27–30.

Pulmanausahakul, R., Faber, M. Morimoto, K. *et al.* (2001). Overexpression of cytochrome c by a recombinant rabies virus attenuates pathogenicity and enhances antiviral immunity. *Journal of Virology* **75**, 10 800–10 807.

Rupprecht, C.E., Hamir, A.N., Johnston, D.H. and Koprowski, H. (1988). Efficacy of vaccinia-rabies glycoprotein recombinant virus vaccine in raccoons (Procyon lotor). *Review of Infectious Diseases* **10** (Suppl 4), S803–S809.

Rupprecht, C.E., Hanlon, C.A., Blanton, J. *et al.* (2005). Oral vaccination of dogs with recombinant rabies virus vaccines. *Virus Research* **111**, 101–105.

Rupprecht, C.E., Wiktor, T.J., Johnston, D.H. *et al.* (1986). Oral immunization and protection of raccoons (*Procyon lotor*) with a vaccinia-rabies glycoprotein recombinant virus vaccine. *Proceedings of the National Academy of Sciences USA* **83**, 7947–7950.

Rustifo, N.P. (2000). Building better vaccines: how apoptotic cell death can induce inflammation and active innate and adaptive immunity. *Current Opinions in Immunology* **12**, 597–603.

Schnell, M.J., Mebatsion, T. and Conzelmann, K.K. (1994). Infectious rabies viruses from cloned cDNA. *European Molecular Biology Organization Journal* **13**, 4195–4203.

Sharpe, S., Fooks, A., Lee, J., Hayes, K., Clegg, C. and Cranage, M. (2002). Single oral immunization with replication deficient recombinant adenovirus elicicts long-lived transgene-specific cellular and humoral immune responses. *Virology* **293**, 210–216.

Shi, Y., Zheng, W. and Rock, K.L. (2000). Cell injury releases endogenous adjuvants that stimulate cytotoxic T cells. *Proceedings of the National Academy of Sciences USA* **97**, 14 590–14 595.

Shoji, Y., Inoue, S., Nakamichi, K., Kurane, I., Sakai, T. and Morimoto, K. (2004). Generation and characterization of P gene-deficient rabies virus. *Virology* **318**, 295–305.

Sief, I., Coulon, P., Rollin, P.E. and Flamand, A. (1985). Rabies virus virulence: effect on pathogenicity and sequence characterization of mutations affecting antigenic site III of the glycoprotein. *Journal of Virology* **53**, 926–935.

Streatfield, S.J., Jilka, J.M., Hood, E.E. *et al.* (2001). Plant-based vaccines: unique advantages. *Vaccine* **19**, 2742–2748.

Taylor, J., Meignier, B., Tartaglia, J. *et al.* (1995). Biological and immunogenic properties of a canarypoxrabies recombinant, ALVAC-RG (vCP65) in non-avian species. *Vaccine* **13**, 539–549.

Taylor, J., Trimarchi, C., Weinberg, R. *et al.* (1991). Efficacy studies on a canarypox-rabies recombinant virus. *Vaccine* **9**, 190–193.

Taylor, J., Weinberg, R., Languet, B., Desmettre, P. and Paoletti, E. (1988). Recombinant fowlpox virus inducing protective immunity in non-avian species. *Vaccine* **6**, 497–503.

Tims, T., Briggs, D.J., Davis, R.D. *et al.* (2000). Adult dogs receiving a rabies booster dose with a recombinant adenovirus expressing rabies virus glycoprotein develop high titer of neutralizing antibodies. *Vaccine* **18**, 2804–2807.

Tuffereau, C., Leblois, H., Bennejean, J., Coulon, P., Lafay, F. and Flamand, A. (1989). Arginine or lysine in position 333 of ERA and CVS glycoprotein is necessary for rabies virulence in adult mice. *Virology* **172**, 206–212.

Vos, A., Neubert, A., Aylan, O. *et al.* (1999). An update on safety studies of SAD B19 rabies virus vaccine in target and non-target species. *Epidemiology and Infections* **123**, 165–175.

Vos, A., Neubert, A., Pommerening, E. *et al.* (2001). Immunogenicity of an E1-deleted recombinant human adenovirus against rabies by different routes of administration. *Journal of General Virology* **82**, 2191–2197.

Wang, Y., Xiang, Z., Pasquini S. and Ertl, H.C.J. (1997). The use of an E1-deleted, replicative-defective adenovirus recombinant expressing the rabies virus glycoprotein for early vaccination of mice against rabies virus. *Journal of Virology* **71**, 3677–3683.

Wang, Z.W., Sarmento, L. Wang, Y. *et al.* (2005). Attenuated rabies virus activates, while pathogenic rabies virus evades, the host immune responses in the central nervous system. *Journal of Virology* **79**, 12 554–12 565.

Wiktor, T.J., Macfarlan, R.I., Reagan, K.J. *et al.* (1984). Protection from rabies by a vaccinia recombinant containing the rabies virus glycoprotein gene. *Proceedings of the National Academy of Sciences USA* **81**, 7194–7198.

Winkler, W.G., Shaddock, J.H. and Williams, L.W. (1976). Oral rabies vaccine: evaluation of its infectivity I three species of rodents. *American Journal of Epidemiology* **104**, 294–298.

Wold, W.S.M. and Gooding, L.R. (1991). Region E3 of adenovirus: a cassette of genes involved in host immunosurveillance and virus-cell interactions. *Virology* **184**, 1–8.

Xiang, Z.Q., Gao, G.P., Reyes-Sandoval, A. *et al.* (2002). Novel, chimpanzee serotype 68-based adenoviral vaccine carrier for induction of antibodies to a transgene product. *Journal of Virology* **76**, 2667–2675.

Xiang, Z.Q., Gao, G.P., Reyes-Sandoval, A., Li, Y., Wilson, J.M. and Ertl, H.C.J. (2003). Oral vaccination of mice with adenoviral vectors is not impaired by preexisting immunity to the vaccine carrier. *Journal of Virology* **77**, 10 780–10 789.

Xiang, Z.Q., Spitalnik, S., Cheng, J., Erikson, J., Wojczyk, B. and Ertl, H.C.J. (1995). Immune responses to nucleic acid vaccines to rabies virus. *Virology* **209**, 569–579.

Xiang, Z.Q., Spitalnik, S., Tran, M., Wunner, W.H., Cheng, J. and Ertl, H.C.J. (1994). Vaccination with a plasmid vector carrying the rabies virus glycoprotein gene induces protective immunity against rabies virus. *Virology* **199**, 132–140.

Xiang, Z.Q., Yang, Y., Wilson, J.M. and Ertl, H.C.J. (1996). A replication-defective human adenovirus recombinant serves as a highly efficacious vaccine carrier. *Virology* **219**, 220–227.

Yang, Y., Ertl, H.C.J. and Wilson, J.M. (1994). MHC class I-restricted cytotoxic T lymphocytes to viral antigens destroy hepatocytes in mice infected with E1-deleted recombinant adenoviruses. *Immunity* **1**, 433–442.

Yusibov, V., Hooper, D.C., Spitsin, S.V., Fleysh, N., Kean, R.B. and Mikheeva, T. (2002). Expression in plants and immunogenicity of plant virus-based experimental rabies vaccine. *Vaccine* **20**, 3155–3164.

Yusibov, V., Modelska, A., Steplewski, K. *et al.* (1997). Antigens produced in plants by infection with chimeric plant viruses immunize against rabies virus and HIV-1. *Proceedings of the National Academy of Sciences USA* **94**, 5784–5788.

Yusibov, V., Shivprasad, S., Turpen, T.H., Dawson, W. and Koprowski, H. (1999). Plant viral vectors based on tobamoviruses. *Current Topics in Microbiology and Immunology* **240**, 81–94.

Zhou, D., Cun, A., Li, Y., Xiang, Z.Q. and Ertl, H.C.J. (2006). A chimpanzee-origin adenovirus vector expressing the rabies virus glycoprotein as an oral vaccine against inhalation infection with rabies virus. *Molecular Therapy* **14**, 662–672.

16

Public Health Management of Humans at Risk

DEBORAH J. BRIGGS[1] AND B.J. MAHENDRA[2]

Department of Diagnostic Medicine/Pathobiology, College of Veterinary Medicine, Kansas State University, Manhattan, KS 66506[1]; Department of Community Medicine, Mandya Institute of Medical Sciences, Mandya, Karnataka State 560004, India[2]

1 INTRODUCTION

The public health significance of rabies is often overlooked by national and international organizations. The fact that rabies has the highest case fatality rate of any known infectious disease and about half of the deaths occur in children under the age of 15 is not well publicized, partly due to the fact that the real mortality rate is unknown in most developing countries of the world where canine rabies is endemic. In spite of the reality that all of the tools to prevent rabies have been available for more than a decade, lack of global awareness and political support has prevented their endorsement and utilization on an international scale. The widespread use of canine vaccination and control programs in Latin America has proven that reduction of human rabies can be successfully accomplished through an organized multinational approach. In some Asian countries, increasing the utilization of safe, efficacious cell culture rabies vaccines and implementing low dose vaccination regimens has also reduced the incidence of human rabies cases. However, the unavailability of rabies immune globulins remains a serious problem in many canine rabies endemic countries. For example, in Asia less than 3% of patients presenting with transdermal wounds at hospital or emergency clinics will actually receive rabies immune globulin as part of their post-exposure prophylaxis. Pre-exposure vaccination is also underutilized as a preventative measure for children and other persons living in regions where canine rabies is highly endemic, partly due to the fact that many medical professionals are unaware that cell culture rabies vaccines can be administered as a protective measure prior to exposure. The key to rabies prevention revolves around increasing educational awareness in all levels of society. By increasing educational awareness on a local, national, regional and international level, a majority of the tens of thousands of rabies deaths that occur every year could be prevented.

Animal bite wounds are the primary cause of rabies infection and death in humans (Fevre *et al.*, 2005). They are also, without a doubt, severe traumatic

Rabies, second edition. Edited by Alan C. Jackson and William H. Wunner
ISBN 978-012-369366-2. Copyright Elsevier Inc. 2007

experiences for everyone involved including the victim of the attack, family of the patient, medical personnel involved in caring for and ultimately making decisions about follow-up treatment and even the owner (if there is one) of the attacking animal. For those individuals living in industrialized countries, access to modern cell culture rabies vaccines and biologicals is not perceived to be a serious problem because, in most cases, they can afford to pay for treatment and if they cannot, the cost will be incurred by either a governmental health system or through specific indigent programs set up by vaccine manufacturers. However, even in industrialized countries, easy access to modern cell culture rabies vaccines should not be considered to be guaranteed as unexpected problems can certainly lead to regional, national and even global shortages, as was recently experienced in 2004 when one of the two major manufacturers of rabies vaccine experienced production problems (Centers for Disease Control, 2004).

Although many poor countries are still producing nerve tissue rabies vaccines (NTV) produced from the brain tissue of infected sheep, these outdated vaccines are increasingly being replaced throughout the world by more immunogenic and less reactogenic cell culture rabies vaccines (Figure 16.1). Even in poor countries, cell culture rabies vaccines are now available in most large cities, but they are expensive, limiting their usage to those who can afford them. Ultimately, it is the cost that prevents access to life-saving cell culture rabies vaccines for those at highest risk, the poorest segment of society. Knowing that rabies has the highest case fatality rate of all currently known infectious diseases, it is clear that removing obstacles to these life-saving modern vaccines should be a priority. To reduce

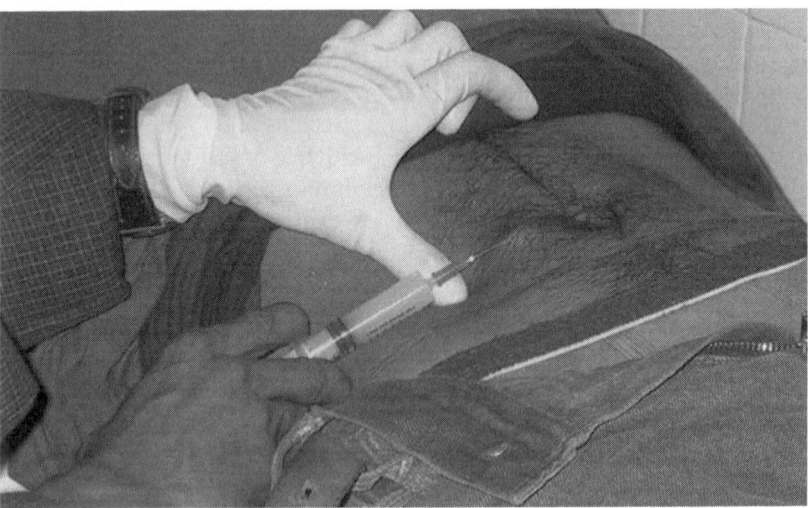

Figure 16.1 The administration of Semple nerve tissue vaccine into the abdominal wall of a patient presenting at an anti-rabies clinic in India with suspected exposure to a rabid dog. (Photo courtesy of Dr Deborah J. Briggs.) This figure is reproduced in the color plate section.

the cost of rabies vaccines, some of the more progressive developing countries have either begun to develop and manufacture their own cell culture rabies vaccines or have been able to acquire the transfer of rabies vaccine technology from abroad to within their own borders (WHO, 1986; Madhusudana *et al.*, 2001). Where technological transfer has occurred, the cost of post-exposure prophylaxis (PEP) has decreased significantly and, in some cases, these national manufacturing facilities have begun to export rabies vaccine to other countries in Asia and Africa. For developing countries that do not have access to technological transfer and cannot afford to import large quantities of cell culture rabies vaccine, the option of using reduced dosage regimens administered intradermally for PEP has proven to be a safe and significantly cheaper alternative.

Even though all of the modern tools required to eliminate the threat of rabies to human health are currently available, this horrific disease continues to infect and kill tens of thousands of humans every year, a large percentage of whom are children (Wilde *et al.*, 2003). In many cases the reasons for not organizing and implementing well-designed national or regional rabies elimination programs center around the lack of political will to initiate and support animal vaccination and control programs. The fact that regional programs of this nature can be very effective is evidenced by the results obtained by the Pan American Health Association (PAHO) in Latin America over the past two decades. Their regional support and organizational efforts to control canine rabies have significantly reduced the incidence of human rabies (Belotto, 2004).

In October of 2004, rabies experts from around the world met at the World Health Organization (WHO) headquarters in Geneva, Switzerland to review the WHO global recommendations for rabies prevention in humans (WHO, 2005). It had been over a decade since the last global WHO Expert Committee Meeting had been convened with the express purpose of updating the entire WHO recommendations for prevention of rabies in humans. New published data presented on intradermal PEP regimens, human rabies treatment and diagnostic testing procedures enabled experts in attendance to implement a number of changes to the WHO recommendations previously published in 1992 (WHO, 1992). In reviewing the current state of rabies prevention programs globally, the WHO experts in attendance agreed that much progress had been made in controlling canine rabies in Latin America, as stated above, but similar progress in Asia had been limited at best. The number of human rabies deaths in both regions had decreased over the last century, but both had used different strategic methodologies. In Latin America, for example, the focus has been on eliminating exposures of humans to rabid dogs by reducing the dog population and by vaccinating those dogs that remain, whereas, in Asia, the focus has been on increasing the availability of modern cell culture rabies vaccines, implementing reduced dosage intradermal vaccination regimens and increasing the awareness levels of rabies prevention techniques in the general public (Wasi *et al.*, 1997; Belotto *et al.*, 2005). The current situation of human rabies in many countries on the

African continent is unknown due to many reasons, including the lack of reliable surveillance systems. Rabies immune globulin (RIG) continues to be in short supply throughout the world (Wilde, 1999; WHO, 2005). In industrialized countries, such as North America and Western Europe, canine rabies has been eliminated, but the threat of exposure and infection to humans still exists due to the presence of rabies in wildlife, often spilling over into domestic pets and also due to the occasional importation of infected dogs from canine rabies endemic countries. Interestingly, rabies vaccination regimens routinely used for PEP in humans may differ slightly between North American and European countries with the USA and Canada recommending RIG for every PEP administered while some European countries only recommend RIG for WHO Category III contacts (Table 16.1) (Centers for Disease Control, 1999; WHO, 2005).

The public health threat of rabies is not only a problem for countries that are aware of its presence in their geographical location. In the few 'rabies-free' regions of the world, new or previously identified rabies virus serotypes can be introduced indirectly through the immigration of infected animals from rabies-endemic countries as occurred recently in Flores Island (Windiyaningsih *et al.*, 2004). Alternately, it can be recognized anew when a human rabies death occurs in an area previously thought to be rabies free as was recently reported in the UK (Nathwani *et al.*, 2003).

In North America, where domestic animal rabies is well controlled, rabies in wildlife species continues to be a problem resulting in many domestic animal and human exposures every year (Krebs *et al.*, 2004). Cooperative international oral rabies vaccination (ORV) programs targeting specific animal species are currently underway in North America and, in the future, may prove to be as successful as the early ORV programs in Europe (Slate *et al.*, 2005).

Until such time as rabies is eliminated in the canine and wildlife populations of the world, public health management of humans at risk of exposure to rabies will continue to depend heavily on rabies immunization, including both pre-exposure and post-exposure prophylaxis (PEP), to prevent infection and death.

2 PRE-EXPOSURE VACCINATION

2.1 Purpose of pre-exposure vaccination

There are several important reasons to utilize pre-exposure vaccination as a strategy to protect persons at risk of exposure to rabies virus (Centers for Disease Control, 1999). First, and perhaps most importantly, pre-exposure can help to protect an individual against an unrecognized exposure to rabies virus that may occur due to one's vocation, hobby, or travel itinerary, or due to the endemicity of rabies in the region in which they live. Secondly, pre-exposure vaccination will simplify PEP in the event of subsequent exposure to rabies virus by reducing the

TABLE 16.1

World Health Organization type of contact and recommended post-exposure prophylaxis[a]

Category	Type of contact with a suspected or confirmed rabid domestic or wild[b] animal, or animal unavailable for observation	Type of exposure	Recommended post-exposure prophylaxis
I	Touching or feeding of animals Licks on intact skin	None	None, if reliable case history is available
II	Nibbling of uncovered skin Minor scratches or abrasions without bleeding. Licks on broken skin	Minor	Administer vaccine immediately.[c] (The ACIP also recommends the administration of RIG.) Stop treatment if animal remains healthy throughout an observation period of 10 days[d] or if animal is proven to be negative for rabies by a reliable laboratory using appropriate diagnostic techniques
III	Single or multiple transdermal bites or scratches, licks on broken skin Contamination of mucous membrane with saliva (i.e. licks)	Severe	Administer rabies immune globulin and vaccine immediately. Stop treatment if animal remains healthy throughout
	Exposures to bats[e]		An observation period of 10 days or if animal is killed humanely and found negative for rabies by a reliable laboratory using appropriate techniques

From: WHO Expert Consultation on Rabies. First report (2005). WHO Technical Report Series *931.* WHO, Geneva Switzerland. Additional information is included regarding the use of RIG from the ACIP (Centers for Disease Control, 1999) and recommendations for exposures to ferrets from the Compendium for Animal Rabies Control (Centers for Disease Control, 2005).

[a] The recommendations from the ACIP include the administration of RIG for both type II and III contacts, whereas WHO recommendations for type II contacts do not include the administration of RIG except for immunosuppressed patients.

[b] Exposure to rodents, rabbits and hares seldom, if ever, requires specific anti-rabies post-exposure prophylaxis.

[c] If an apparently healthy dog or cat (or ferret in North America) from a low-risk area is placed under observation, the situation may warrant delaying initiation of treatment.

[d] This observation period applies only to dogs and cats (and ferrets in North America). Except in the case of threatened or endangered species, other domestic and wild animals suspected as rabid should be humanely killed and their tissues examined for the presence of rabies antigen using appropriate laboratory techniques.

[e] Post-exposure prophylaxis should be considered when contact between a human and a bat has occurred unless the exposed person can rule out a bite or scratch, or exposure to a mucous membrane.

required number of injections of vaccine from five to two doses and by eliminating the need to administer RIG. Additionally, pre-exposure will 'prime' the immune system so that an immediate anamnestic response will occur in a previously vaccinated individual after having received one or two booster doses of vaccine. Considering the fact that rabies has the highest case fatality rate of all currently known diseases, pre-exposure vaccination should be seriously considered as a vaccination strategy for those individuals that may be at increased risk of exposure and infection (Hatz *et al.*, 1995; Sabchareon *et al.*, 1998; WHO, 2005).

2.2 Use of pre-exposure vaccination

It is interesting to note that pre-exposure vaccination is highly recommended and widely used in humans at increased risk of exposure to rabies in industrialized countries where domestic animal vaccination programs are in place and rabies biologicals are generally readily available (Murray and Arguin, 2000). However, pre-exposure vaccination, as a preventative measure, is a definitely under-utilized tool in developing countries where canine rabies is endemic, many human rabies deaths occur, few or no animal control programs exist and the epidemiology of rabies is unknown and definitely under-reported (Centers for Disease Control, 1999; Wilde, 1999; Prasad *et al.*, 2001). Indeed, most veterinarians living in Asia and Africa, where canine rabies is endemic and canine control and vaccination programs are non-existent, have never received rabies pre-exposure vaccination (Ghosh, 2003). Additionally, pediatricians and other medical professionals practicing in canine rabies endemic countries (where from 35 to 55% of human rabies deaths occur in children under the age of 15) are generally unaware of the life-saving value of pre-exposure vaccination (Prasad *et al.*, 2001). While it is true that PEP using modern cell culture rabies vaccines and RIG when given as recommended after an exposure will save lives, it is also true that most children as well as other victims who die of rabies did not receive PEP in a timely manner. There are various reasons why exposed persons do not seek immediate treatment after an exposure to a rabid animal. In the case of children, it is most often because they have not understood the danger of rabies and therefore may not have told their parents or guardians of the incident. For other victims of rabies, the reasons stated for not seeking treatment have also included a misunderstanding of the seriousness of the situation as well as not seeking treatment immediately and, in many cases of humans that die of bat rabies virus variants, the fact that they were unaware that they were exposed (Noah *et al.*, 1998).

2.3 Who should receive pre-exposure vaccination?

Those persons considered to be at increased risk of exposure for rabies should receive pre-exposure vaccination. Specifically, such persons include scientists and

TABLE 16.2

Risk assessment for pre-exposure vaccination[a]

Risk category	Nature of risk	Typical populations	Pre-exposure recommendations
Continuous	Virus present continuously, often in high concentrations. Specific exposures likely to go unrecognized. Bite, non-bite, or aerosol exposure	Rabies research laboratory workers;[b] rabies biologics production workers	Primary course. Serologic testing every 6 months; booster vaccination if antibody titer is below acceptable level[c]
Frequent	Exposure usually episodic, with source recognized, but exposure also might be unrecognized. Bite, non-bite, or aerosol exposure	Rabies diagnostic lab workers,[b] spelunkers, veterinarians and staff and animal-control and wildlife workers in rabies-enzootic areas	Primary course. Serologic testing every 2 years; booster vaccination if antibody titer is below acceptable level[c]
Infrequent (greater than population at large)	Exposure nearly always episodic with source recognized. Bite or non-bite exposure	Veterinarians and animal-control workers in areas with low rabies rates. Veterinary students. Travelers visiting areas where rabies is enzootic and immediate access to appropriate medical care including biologics is limited	Primary course. No serologic testing or booster vaccination
Rare (population at large)	Exposure always episodic with source recognized. Bite or non-bite exposure	US population at large, including persons in rabies-epizootic areas	No vaccination necessary

[a] From Centers for Disease Control (1999). Human rabies prevention – United States, 1999. Recommendations of the Advisory Committee on Immunization Practices (ACIP). *Morbidity and Mortality Weekly Report* **48**, (RR-1), 6.

[b] Judgment of relative risk and extra monitoring of vaccination status of laboratory workers is the responsibility of the laboratory supervisor.

[c] Minimum acceptable antibody level is complete virus neutralization at a 1:5 serum dilution by the rapid fluorescent focus inhibition test. A booster dose should be administered if the titer falls below this level.

technicians working in rabies research and diagnostic laboratories and rabies vaccine production facilities, veterinarians and veterinary assistants, spelunkers, wildlife officers and rehabilitators and animal handlers including animal control officers (Table 16.2) (Centers for Disease Control, 1999; Trevejo, 2000). Pre-exposure vaccination may provide additional protection if a delay in treatment occurs or if incorrect medical advice is given during travel, as has occurred with travelers visiting remote areas (Krause *et al.*, 1999; Arguin *et al.*, 2000). Additionally, WHO recommends that children living in canine rabies endemic countries where animal control and routine animal vaccination programs are not in place should be considered for pre-exposure vaccination (Sabchareon *et al.*, 1998; WHO, 2005). It should be noted that this recommendation does not indicate that all children living in developing countries should receive pre-exposure vaccination. Instead, it is a call for a better evaluation and broader utilization of this life-saving vaccination tool to save the lives of children continually at risk.

2.4 Recommended pre-exposure regimen

Pre-exposure vaccination is administered as a three-dose series. In most instances, pre-exposure is administered intramuscularly as one full dose of vaccine on days 0 and 7 and either day 21 or day 28. In some countries, where cost is a serious consideration and multiple patients are being vaccinated at the same time, a reduced dose of 0.1 ml has been used to replace the full dose using the same days as mentioned above (Dreesen *et al.*, 1989; WHO, 2005). In the event that the pre-exposure schedule is disrupted, a few days variation in the regimen is acceptable. Vaccine should be administered in the upper arm (deltoid) area in the case of adults and in the anterolateral thigh area of young children. Patients who are scheduled to receive anti-malarial treatment should complete their pre-exposure series before the anti-malarial treatment is initiated (Centers for Disease Control, 1983; Bernard *et al.*, 1985; Pappaioanou *et al.*, 1986; Lau, 1999).

There are a few final recommendations to be considered in regard to the administration of pre-exposure vaccination to specific populations as well as booster doses in the event of a subsequent exposure to a confirmed or suspect rabid animal. Regarding the administration of pre-exposure vaccination to persons who are immunosuppressed due to the presence of disease or the concomitant administration of immunosuppressive therapy, these patients should receive the vaccine intramuscularly, rather than intradermally, and have their immune response to vaccination measured by a virus neutralization test such as the rapid fluorescent focus inhibition test (RFFIT) or fluorescent antibody virus neutralization (FAVN) test (WHO, 2005).

2.5 Booster immunizations

Persons who have previously received pre- or post-exposure rabies vaccination with a modern cell culture rabies vaccine and are subsequently exposed to rabies virus should receive two booster doses of vaccine, one given on each of days 0 and 3. Additionally, RIG should not be administered.

Specific questions regarding the frequency with which to administer routine booster vaccinations in previously vaccinated persons has been discussed extensively at several WHO international meetings. Published data indicate that one booster dose of vaccine will induce an adequate immune response in persons who have received a cell culture vaccine 10–14 years earlier (Traenhart *et al.*, 1994; Strady *et al.*, 1998; Malerczyk *et al.*, 2007). Therefore, periodic boosters are only recommended for those persons who are at continual risk of exposure to rabies when their serological titer falls below a serum dilution factor of 1:5 as recommended by the Centers for Disease Control and Prevention (Centers for Disease Control) or below 0.5 international units/ml as recommended by WHO (Centers for Disease Control, 1999; WHO, 2005). Those persons at frequent or continual risk of exposure include rabies researchers, diagnostic laboratory workers where virus is continuously present and where exposures are likely to go unnoticed (see Table 16.2). In order to monitor the titer of rabies virus neutralizing antibody, a serological test every 6 months is recommended for persons with occupations placing them at continuous risk of exposure and, for those at frequent risk, a serological test is recommended every 2 years (Centers for Disease Control, 1999). The monitoring of serological titers of all persons at risk is the responsibility of the administrative authorities in individual rabies laboratories (WHO, 2005).

3 POST-EXPOSURE VACCINATION

3.1 What is an exposure?

The most important decision to be made after a contact with a potentially rabid animal has occurred is whether that interaction constituted a potential exposure to rabies virus. It is critical to make the correct decision at this point because, if an exposure did occur, the patient will need to receive PEP. In the event that an exposure to rabies is questionable, a careful evaluation of the details of the circumstances surrounding the contact should be conducted by a medical professional who can either confirm that PEP is warranted or that it can be avoided, thus eliminating the unnecessary utilization of valuable RIG and rabies vaccine with its associated costs as well as the time required by both the physician and patient to undergo the entire PEP procedure. Factors to consider when evaluating a potential exposure, besides the nature of the contact or injury, include the epidemiology of rabies in the area that the contact occurred, availability of the animal for testing, clinical status of the animal involved and, in some cases, the vaccination and exposure history of the animal (Centers for Disease Control, 2005; WHO, 2005).

The WHO has categorized different types of animal contact according to the severity of the exposure (see Table 16.2). An exposure to rabies occurs when saliva or other potentially infective body tissues from a rabid mammal come into contact with an open wound or mucous membrane of an uninfected mammal.

Human exposures generally occur through a bite wound. However, scratches inflicted from rabid animals should also be considered to be an exposure. An exposure to rabies virus involving a mucous membrane, for example the eyes or mouth, can occasionally occur to veterinarians while they are treating large animals presenting with clinical rabies. For example, cattle infected with rabies often produce excessive amounts of saliva that can come in contact with the mucous membranes of veterinarians and other persons working closely with these animals while caring for them. Regarding aerosol exposure to rabies virus, this method of transmission has been reported to cause human infection on at least two different occasions, but it should be strongly emphasized that these were extremely rare occurrences and involved large amounts of virus in enclosed areas. In one case, the exposure occurred in a cave housing huge colonies of bats and the other case occurred in a laboratory setting (Constantine, 1962; Winkler et al., 1973; Conomy et al., 1977). Standing in the same area and/or touching or petting a rabid animal is not considered to be an exposure. Rabies virus cannot penetrate the skin layer and only enters a person or an animal through a transdermal wound or another body opening as mentioned above. Therefore, rabies virus in infected saliva or other tissues that come in contact with intact skin cannot enter the body and progress to a disease state and therefore this is not considered to be an exposure. Laboratory proven human-to-human transmission of rabies has been reported directly through corneal transplantation and, more recently, through solid organ transplantation (Javadi et al., 1996; Srinivasan et al., 2005). Two unconfirmed cases of human-to-human transmission through saliva have also been reported, one by a kiss and the other by a bite (Fekadu et al., 1996). Medical staff treating suspect human rabies patients should exercise documented safety procedures as recommended for all high-risk infectious diseases, including the use of appropriate protective clothing including gowns, gloves and eye shields (Remington et al., 1985; WHO, 2005).

When determining whether a bite or scratch wound from an apparently healthy dog, cat or ferret constitutes an exposure, health authorities may choose to isolate and observe the animal for a specific period of time, usually 10 days, rather than having it immediately euthanized and tested for the presence of rabies virus. If the dog, cat or ferret remains healthy for the 10-day period, it can be presumed that the animal was not shedding rabies virus at the time of the exposure. The observation period of 10 days is based on historical evidence that a dog or cat infected with a variant of rabies virus from North America would demonstrate clinical signs of rabies infection within 10 days of rabies virus being present in saliva. If the dog, cat or ferret begins to show clinical signs indicative of rabies, it should be immediately and humanely euthanized and the brain tissue examined for the presence of rabies virus. However, because choosing to observe an animal in lieu of euthanizing and testing could increase the risk of infection and disease in the person that was exposed, this decision must be made in agreement with the appropriate health authorities. When a human exposure occurs to other animals that are high risk, for example wild animals or animals for which the shedding period of rabies virus in the saliva prior to the presentation of clinical signs is unknown, the animal should never be observed in lieu of immediate euthanasia and testing (Chomel, 1999; Centers for Disease Control, 2005). If possible, a wild animal that has inflicted a wound should be caught and tested immediately to determine if it is positive for rabies virus infection. If the animal is confirmed positive for rabies or is unavailable for testing, PEP should be initiated as soon as possible for the exposed patient (Trimarchi and Briggs, 1999; Centers for Disease Control, 2005).

In North America, bat-associated variants of rabies virus are confirmed most often to be the cause of human rabies deaths (Noah *et al.*, 1998; Messenger *et al.*, 2002; Mondul *et al.*, 2003). In many cases, the exposed victim did not recall a bite or other type of contact. Guidelines for evaluation of bat rabies exposures have been defined previously (Centers for Disease Control, 1999). Bats that come into contact with humans should be tested immediately for the presence of rabies virus. Direct contact between a bat and a human should be considered to be an exposure unless the exposed person can be sure that a bite or scratch causing a transdermal wound or an exposure to a mucous membrane did not occur, or the bat can be tested and confirmed to be negative for the presence of rabies virus. PEP also should be administered when a bat is found in the same room as a sleeping or incapacitated person or an unattended child and the bat is unavailable for testing. If a person is in the same room or vicinity as a bat and is confident that he or she has not been in physical contact with the bat, then this should not be considered to be an exposure to rabies. If bats are present in an attic or other rooms in a house occupied by humans, measures should be taken to exclude bats from the house and preventative measures taken to prevent their return. It is important to understand that exclusion does not require active removal. For example, bats can be removed without physically coming in contact with them by sealing all entrances into the dwelling after the bats have vacated the premises to forage for food. This can be accomplished most successfully by placing small-mesh netting over the access points into and out of the building. Bats can exit freely and climb down the netting to escape but are unable to find their way back up through the netting into the dwelling again.

Other circumstances that might occur, but do not constitute an exposure, include contact with a domestic pet that has taken food or water from the same bowl as a wild animal; contact with blood, urine or feces; petting the fur of a rabid animal; or an accidental needle stick containing a killed animal vaccine.

3.2   Primary wound care and administration of RIG

After an exposure to a potentially rabid animal has occurred, appropriate primary wound care should be initiated as soon as possible in order to inactivate as much rabies virus as possible that may have infected the wound. Appropriate wound care consists of immediate and thorough washing of the wound for at least 15 minutes using soap, water, detergent, povidone iodine or another antiviral substance. In some developing countries, where soap or detergent is not immediately available, flushing the wound with water should be instituted (WHO, 2005). Experimental evidence indicates that washing a wound helps to reduce rabies virus infection by eliminating or inactivating viral particles that may have been inoculated into the tissue at the time of the exposure (Kaplan et al., 1962; Dean and Baer, 1963; Hatchett, 1991). In addition, tetanus prophylaxis and antibacterial treatment (although antibiotics are not administered routinely, they are almost always indicated for high-risk wounds) should be initiated for all animal bites that cause tissue damage (Fleisher, 1999).

After the wound has been thoroughly cleansed, RIG must be administered. RIG is administered as a passive immune treatment in order to provide immediate access to neutralizing antibodies until the patient's immune system can begin to produce its own antibody. This usually occurs within 7–14 days after the initial dose of rabies vaccine has been administered. There are two types of RIG used globally: human rabies immune globulin (HRIG) is used more frequently in industrialized nations whereas equine rabies immune globulin (ERIG) is more frequently administered to patients in developing countries where federally supplied resources for PEP are limited. The dose for HRIG is 20 international units (IU)/kg of body weight and the dose for ERIG is 40 IU/kg of body weight. Regardless of which RIG is used, as much of this biological product as is anatomically feasible should be infiltrated into and around the wound sites. Any remaining RIG should be administered by deep intramuscular injection at a site that is distant from the vaccination site in order to prevent the RIG from neutralizing the rabies vaccine that was injected. Since animal bite wounds, especially in small children, can be severe and inflicted in multiple sites, the recommended dose of RIG may not be of sufficient volume to inject all wounds. In such a case, the WHO recommends that the dose should be diluted in a sufficient volume of sterile saline solution to enable injection of all of the wounds (Khawplod et al., 1996; Lang et al., 1998; WHO, 2005). However, no more than the recommended dose of RIG per kilogram body weight should be administered.

Lastly, the importance of including RIG in the PEP treatment regimen of victims that have incurred transdermal wounds cannot be overemphasized. This is evidenced by the fact that published data on the investigation of treatment failures (human deaths after PEP was instituted) report that many patients did not receive RIG (Arya, 1999; Sriaroon et al., 2003; Parviz et al., 2004).

3.3 Administration of rabies vaccines

There are two intramuscular and two intradermal post-exposure vaccination regimens recommended for use with cell culture rabies vaccines, but only the intramuscular five-dose Essen regimen is recommended for use in North America (Centers for Disease Control, 1999; WHO, 2005). All four PEP regimens have sufficient published data to insure that when they are used in conjunction with prompt appropriate wound care along with RIG, death from rabies can be averted. All cell culture vaccines should be administered into the upper arm (deltoid) region in adults or into the anterolateral thigh region in young children. Of the two intramuscular regimens, the five-dose Essen regimen is the one most frequently used in the Americas, and Europe and where intradermal administration of rabies vaccines is not yet approved in Asia and Africa. The Essen regimen is administered intramuscularly as one full dose on each of days 0, 3, 7, 14 and 28. The other intramuscular regimen used in some European countries is the 'Zagreb' or '2-1-1' regimen and is administered intramuscularly as two doses, one in each deltoid, on day 0, one additional dose then being administered on each of days 7 and 21 (Vodopija, 1988). This regimen was originally developed to reduce the cost of vaccine and the number of patient visits required to receive PEP. The '2-1-1' regimen reduces the number of doses required from five to four and the number of visits required from five to three.

Intradermal regimens approved for use by the WHO were initially developed as a means to reduce the cost of replacing NTV with cell culture rabies vaccines in Asia. When cell culture rabies vaccines were developed, it was clear that they were more efficacious than NTV and certainly caused significantly fewer adverse reactions, but they were and continue to be expensive (Quiambao *et al.*, 2005). It was also clear that most developing countries with a high incidence of human rabies deaths due to the presence of canine rabies could not afford to treat all of their exposed patients with five or even four doses of cell culture rabies vaccines in government-funded anti-rabies centers. Thailand was the first country to begin using reduced dose regimens in large groups of exposed patients and the results from their initial clinical data proved that cell culture rabies vaccines that met WHO production standards were safe and produced high antibody titers when administered intradermally (Chutivongse *et al.*, 1990). Over the past two decades, millions of patients have received intradermal regimens using cell culture rabies vaccines, thus saving untold numbers of human lives in Asian countries. There are some precautions that should be taken if intradermal vaccines are used. First of all, the vaccines currently recognized by the WHO for intradermal use are lyophilized and do not contain a preservative; therefore, once they are reconstituted, they must be stored at 2–8°C and used within 8 hours or discarded. This is to ensure that a vial of vaccine does not become contaminated and overgrown with bacteria before it is completely used, which could occur due to multiple needle sticks to remove individual intradermal doses of vaccine. Thus, this method of administration should only be used in clinics treating multiple patients on a daily basis. It is not cost effective for a clinic treating only one or two patients every few days to use an intradermal regimen as only a fraction of the vaccine is used and the rest would have to be discarded. In these cases, it is more prudent to use the entire vial intramuscularly. Secondly, intradermal injections of vaccines must be administered by medical personnel experienced in this technique. Vaccine administered correctly will produce a raised 'bleb' in the skin (Figure 16.2). In the event that an intradermal dose was given incorrectly, a new dose should be administered correctly (WHO, 2005).

The two intradermal regimens currently recognized for use by WHO include the updated 'Thai Red Cross' or '2-2-2-0-2' regimen and the 'Eight-site' or '8-0-4-0-1-1' regimen (WHO, 2005). There are only two rabies vaccine manufacturers that currently meet the production standards published by the WHO for intradermal use: Novartis Vaccines (Marburg, Germany and Ankleshwar, India) and Sanofi Pasteur Vaccines (Lyon, France) (WHO, 2005). The 'Thai Red Cross' regimen is the most widely used intradermal regimen and the updated '2-2-2-0-2' regimen is administered as follows: two 0.1 ml doses of vaccine are administered intradermally over the deltoid region on days 0, 3, 7 and 28. The '8-0-4-0-1-1' regimen is administered as follows: on day 0, one 0.1 ml dose of vaccine is administered intradermally at eight different sites (upper arms, lateral thighs, suprascapular region and lower quadrant of abdomen); on day 7, one 0.1 ml dose of vaccine is administered intradermally at four different sites (each upper arm and each lateral thigh); on days 28 and 90, one 0.1 ml dose of vaccine is administered intradermally in the upper arm region.

Caution should be taken to remember that all four of the above PEP regimens require the administration of RIG for all transdermal wounds. No PEP regimens have been developed that eliminate the need to administer passive immunity in the form of RIG.

3.4 Previously vaccinated patients

Some patients who require PEP have received either pre-exposure or PEP previously. If the vaccine that the patient received was a modern cell culture rabies vaccine, they should receive two boosters, one on day 0 and one on day 3 (Centers for Disease Control, 1999). Previously vaccinated persons should not receive RIG because their immune system has already been primed and once they have received an additional booster dose of vaccine, their immune system will initiate an immediate anamnestic immune response (Fishbein et al., 1986; Bernard et al., 1987). In patients who have received a crude preparation of NTV previously, or patients for whom the type and quality of rabies vaccine they received cannot be established, a complete PEP including RIG should be administered (WHO, 2005).

3.5 Immunosuppressed patients

Studies investigating the immune response after rabies vaccination in immunosuppressed patients indicate that patients with very low CD4 counts may not

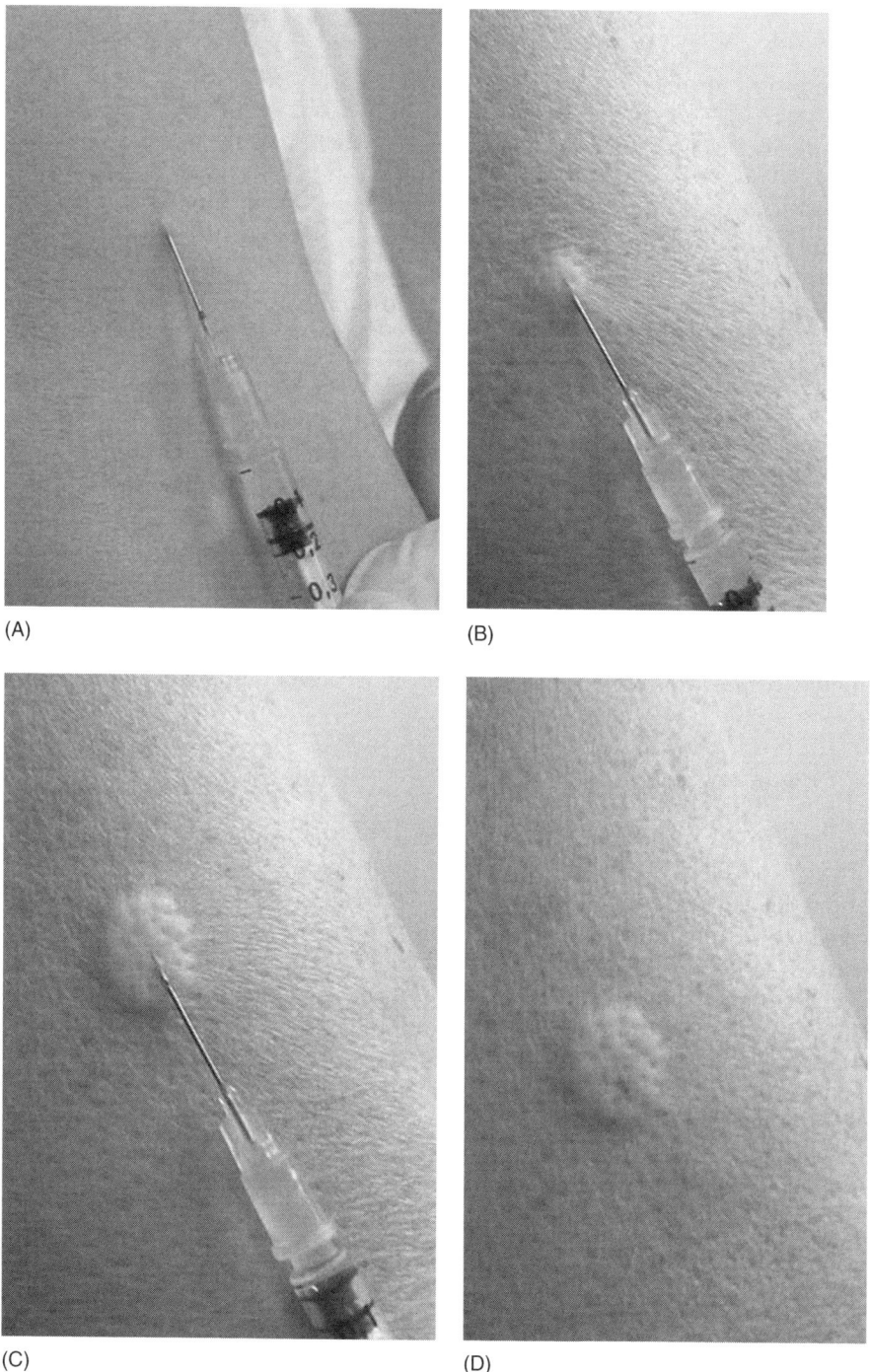

(A)

(B)

(C)

(D)

Figure 16.2 Correct intradermal injection technique of cell culture rabies vaccine. Note the stepwise appearance of the 'bleb' in the skin resembling the surface of an orange peel. (Photo courtesy of Dr Claudius Malerczyk.) This figure is reproduced in the color plate section.

respond well to rabies vaccine (Deshpande and Briggs, 2001; Thisyakorn, 2000).[**16.2] Therefore, when administering rabies vaccine to immunosuppressed patients, an infectious disease specialist with an expertise in rabies prevention should be consulted and all rabies vaccines should be administered intramuscularly (Centers for Disease Control, 1999; WHO, 2005). Corticosteroids and other immunosuppressive agents, including anti-malarials, may interfere with the development of an immune response after vaccination and therefore, if at all possible, persons who are receiving immunosuppressive drugs should postpone pre-exposure vaccination until they are no longer taking these prescription drugs and should also avoid activities that put them at risk of exposure (Bernard et al., 1985; Pappaioanou et al., 1986).

In contrast to North America, the majority of patients exposed to rabies in developing countries do not routinely receive RIG as part of their PEP protocol. Due to the fact that patients with immunosuppressive conditions may have impaired immune systems, they should always receive prompt and adequate wound care as well as passive immune therapy in the form of RIG. Additionally, the WHO has extended the use of RIG in immunosuppressive patients to include both category II and category III exposures (WHO, 2005). Finally, rabies serological testing should be conducted on immunosuppressed patients to monitor their antibody response after vaccination (Centers for Disease Control, 1999).

3.6 Pregnancy

Pregnancy is not a contraindication to PEP if it is confirmed that an exposure to a known or suspect rabid animal has occurred (Varner and McGuinness, 1982; Centers for Disease Control, 1999). In three retrospective studies conducted in Thailand and India, there were no reported fetal abnormalities associated with rabies vaccination (Chutivongse and Wilde, 1989; Chutivongse et al., 1995; Sudarshan et al., 1999). In one of the studies conducted on 202 pregnant women in Thailand bitten by a dog, complete follow-up was available for 190 patients (Chutivongse et al., 1995). Spontaneous abortions occurred in eight (4.5%) of the women, a rate that was reported to be similar to that reported in the general Thai population. Of the eight abortions, three occurred within 24 hours, four between days 3 and 7, and one on day 14 after the dog bite. It was concluded that the eight abortions were caused by trauma from the attack rather than from the rabies vaccine. In another retrospective study conducted on 29 pregnant women in India, 25 were available for follow-up during pregnancy and after birth (Sudarshan et al., 1999). In this report, no congenital deformities occurred and no abortions were reported. Antibody titers were above 0.5 IU/ml in all women tested at 14 days after vaccination was initiated and antibody was detected in six of the babies tested. None of the women available for follow-up at one year after treatment had died of rabies. These data and the pharmacovigilance data collected over decades by international rabies vaccine manufacturers, indicate that PEP in pregnant women is safe and therefore should never be withheld in the event of exposure.

4 ADVERSE REACTIONS TO CELL CULTURE VACCINES

Both local and systemic reactions have been recorded after the administration of cell culture vaccines in clinical trials (Burridge et al., 1982; Dreesen et al., 1986; Jaiiaroensup et al., 1998; Jones et al., 2001). These studies generally reported local reactions, including pain, itchiness, redness and/or swelling at the injection site in between 35 and 45% of the enrolled subjects. Common systemic reactions that were reported in between 10 and 15% of subjects included fever, myalgia, malaise, headaches, dizziness, hives and rash. In one study, localized lymphadenopathy was reported in 10–15% of patients that had received vaccine by the intradermal route (Dreesen et al., 1986).

Systemic allergic reactions in the form of an immune complex-like disease have been reported in between 1 and 10% of subjects that received booster doses of HDCV (Dreesen et al., 1986; Bernard et al., 1987). The disease consisted primarily of urticaria, macular rash, angioedema and arthralgia. All of the subjects in these studies recovered without sequelae. In another study, 3% of subjects who received a booster dose of HDCV developed generalized urticaria or wheezing within one day after receiving a booster and 3% developed urticaria within 6 to 14 days after the booster (Fishbein et al., 1993). The major cause of these immune complex reactions has been identified as being IgE antibodies produced against the beta propiolactone-altered human serum albumin that is present in the manufacturing process of HDCV (Swanson et al., 1987). Beta propiolactone is the inactivating agent and human serum albumin is used as a stabilizer in the manufacturing process of HDCV. These reactions have been successfully treated with antihistamines, epinephrine and steroids (Jones et al., 2001).

5 INTERCHANGEABILITY OF VACCINES

It may be necessary to interchange vaccines during PEP due to the unavailability of one vaccine or to avoid further adverse reactions, should they occur. In the USA, one study indicated that an anamnestic response occurred when persons vaccinated previously with HDCV were subsequently boosted with PCECV, thus indicating that the vaccines were interchangeable (Briggs et al., 2000). Interchangeability of cell culture rabies vaccines has been practiced for many years in countries such as the Philippines, Thailand and Sri Lanka (Hemachudha et al., 1999). Generally, however, it is recommended that the same vaccine be used throughout pre-exposure vaccination and PEP. If this is impossible, the vaccination regimen should be completed with a WHO-recommended vaccine (WHO, 2005).

To date there have been no immunogenicity studies investigating the interchange of intramuscular versus intradermal in one vaccination series, although occasionally it is reported in the literature. Therefore, this practice should be the exception and serologic testing, if possible, is highly recommended for such patients.

6 RABIES IN CHILDREN

Global rabies epidemiological data collected over a number of years reveal that, in many countries where canine rabies is endemic, between 35 and 50% of dog bites and subsequent rabies deaths occur in children less than 15 years of age (Pancharoen *et al.*, 2001; Singh *et al.*, 2001; Chhabra *et al.*, 2004; Parviz *et al.*, 2004; Peigue-Lafeuille *et al.*, 2004; Windiyaningsih *et al.*, 2004). Unfortunately, these statistical facts are relatively unknown in the international health community and at the local level in canine rabies endemic countries. A recent published study from India, the country with the highest reported number of human rabies deaths in the world, indicated that while over 93% of medical health providers knew about cell culture rabies vaccines, only 18% were aware that they could be used for pre-exposure vaccination to protect against rabies (Prasad *et al.*, 2001). Since children make up the highest percentage of rabies deaths throughout Asia, including India, increasing the awareness and understanding among medical professionals regarding the value of pre-exposure vaccination would seem prudent.

Children are most often the victims of dog bites due to their small stature, their curious nature and the fact that they are unaware of the dangers involved in touching a dog. When dog bites do occur in children, they are often sustained in highly innervated areas such as the face and hands and commonly involve multiple bite wounds (Sampath *et al.*, 2005). Severe traumatic exposures in highly innervated areas often reduce the incubation period and critical time for initiating PEP, thus increasing the risk of contracting the disease. Additionally, when minor bites occur, children sometimes do not report the injury to their parents for several reasons, including the fact that they do not understand the risk involved, or they are afraid to tell their parents that they had been playing with an animal that they had been told to avoid. Whatever the reasons, the present death toll in children due to rabies is unacceptable, especially considering the fact that modern cell culture rabies vaccines are available worldwide for both pre-exposure and post-exposure. It has been argued that if PEP works in children, why use pre-exposure? The fact is that if a parent or guardian knew that the child was exposed to a rabid animal, they would certainly seek PEP. The children who are victims of rabies are most often those who do not inform an adult after they have been bitten or scratched and, therefore, the adult is unaware that PEP is necessary. As mentioned above, this situation perfectly corresponds to one of the main purposes of pre-exposure vaccination, i.e. to prevent unknown exposures to rabies.

7 TRAVEL TO CANINE RABIES ENDEMIC COUNTRIES

Many of the human cases of rabies that are reported in industrialized countries result from exposures to rabies in a foreign country where canine rabies is endemic (Shill *et al.*, 1987; Noah *et al.*, 1998; Smith *et al.*, 2003). Generally, these deaths occur because the person exposed was unaware of the danger and did not seek or refused to receive PEP. Increased counseling and educational awareness would help to prevent these unnecessary and tragic deaths. Numerous publications in peer reviewed journals as well as on travel information websites recommend that pre-exposure vaccination should be offered to adventure travelers as well as government, health and religious officials spending extended periods of time in remote areas of developing countries where canine rabies is endemic (Centers for Disease Control, 1999; Fooks *et al.*, 2003; Wilde *et al.*, 2003). Publications regarding pediatric travel are now beginning to recommend pre-exposure vaccination for ambulatory children (Mackell, 2003). The threat of exposure to rabies certainly depends on the endemicity of rabies in the region to be visited. For example, one report indicated that 1.3% of travelers spending an average of 17 days in Thailand, where canine rabies is endemic, experienced a dog bite, 8.9% were licked by a dog and 0.5% required PEP (Phanuphak *et al.*, 1994). In another survey, the estimated number of dog bites that occurred in missionaries and long-term residents was 18.2/1000 persons (Bjorvatn and Gundersen, 1980). Another report has indicated that PEP in Peace Corps volunteers was 550 times higher than in the general population of the USA (Bernard and Fishbein, 1991). In the human rabies cases confirmed in the USA between 1980 and 1996, 12% were associated with a canine variant of rabies found outside of the USA (Noah *et al.*, 1998). None of these patients received a complete PEP series. Rabies deaths have also been reported in victims living in other countries when the correct PEP was not administered, usually after an exposure to a rabid dog (Fescharek *et al.*, 1991). Other human rabies deaths have been documented in spite of reportedly 'correct' PEP treatment but, upon closer evaluation, the WHO/ACIP prescribed recommendations were not followed (Wasi *et al.*, 1997). One such case was reported from South Africa where HRIG and human diploid cell vaccine (HDCV) were administered appropriately, but the HDCV that was administered was injected into the gluteal area rather than the upper deltoid (Shill *et al.*, 1987).

In a study evaluating the management of animal bites among 67 long-term residents in the tropics, 51% of dog bite victims who should have received a complete PEP did not receive any PEP and 19% received insufficient PEP (Hatz *et al.*, 1995). Twenty-eight per cent of the victims were children in whom 85% did not receive PEP or PEP was incomplete. No RIG was administered in 10 cases with category III contacts and 41% did not consider the possibility of rabies exposure when bitten. Additionally, 40% of the bites were unprovoked.

Medical treatment can be complicated and/or frustrating for foreigners to obtain in developing countries (Wilde *et al.*, 2003). Additionally, RIG and/or WHO recommended cell culture rabies vaccines might not be immediately available, depending on where the traveler is visiting. In these circumstances, pre-exposure vaccination would add an additional margin of safety until appropriate medical facilities could be reached. In the event of an exposure, travelers should not omit or delay obtaining the appropriate PEP and should obtain medical advice as soon as possible. Delay or neglecting PEP could result in infection and the development of clinical rabies. At least one case of rabies has been reported in a Peace Corps volunteer who had received intradermal pre-exposure vaccination but did not seek PEP after being bitten by a dog (Bernard *et al.*, 1985). The lack of protection in this case was probably due to the fact that the intradermal route of vaccination was administered concurrently with chloroquine (Pappaioanou *et al.*, 1986).

8 RABIES CONTROL IN DEVELOPING COUNTRIES

The highest number of human rabies cases is reported from developing countries located in the tropics, where canine rabies is endemic (Knobel *et al.*, 2005). Reduction of canine rabies in these countries would cause a significant reduction in the number of human fatalities attributed to exposures to rabid dogs. As stated earlier, this has proven to be the case in Latin America where PAHO has made a concerted effort to vaccinate and control the stray and domestic dog population (Belotto, 2004). The WHO recommends vaccination of 70% of the canine population in order to prevent outbreaks of rabies (WHO, 1987). According to data presented in one report, vaccination of 70% of the canine population would prevent 96.5% of all major outbreaks of rabies (Coleman and Dye, 1996). Regional programs in Asia, following the example in Latin America, would be a tremendous step forward in preventing human rabies. In the future, international and regional support for extensive canine vaccination programs may be possible. However, until that time, human rabies deaths will have to be prevented through the use of increased awareness and use of PEP and pre-exposure vaccination where possible.

As mentioned earlier, NTVs are being replaced in many developing countries. For example, India stopped its production of Semple sheep brain vaccine in 2005. However, if NTVs are replaced without the subsequent implementation of reduced-dose intradermal regimens in governmental anti-rabies clinics treating hundreds of patients each week, the cost of using cell culture rabies vaccines intramuscularly will be daunting. Unfortunately, Semple vaccine is still widely used in Bangladesh, Pakistan and a few other countries in Asia and Africa. Most of the NTV is administered in anti-rabies centers to economically disadvantaged patients who must rely on their government to supply PEP because they cannot afford higher priced cell culture rabies vaccines, much less RIG. Less than 3% of patients in Asian countries receive RIG as part of their PEP regimen (Kamoltham et al., 2003). Due to the pain and neuroparalytic complications associated with administration of Semple vaccine, the complete series of 10 to 14 injections in the abdomen is often not completed (Shankar et al., 2000). Some of the problems associated with rural patients presenting at anti-rabies centers include a delay in the administration of PEP due to the distance of the anti-rabies clinic from the site of the exposure, proper first aid measures not being undertaken immediately following animal bites, suturing of severe bite wounds without prior administration of RIG, unavailability of NTV or cell culture vaccine in rural areas, lack of awareness of the primary treating physician as to proper PEP and non-reporting of human rabies cases (Reddy and Sampath, 2000; Parviz et al., 2004). Additionally, in some regions, patients continue to visit local healers, who often recommend ineffective treatments (APCRI, 2004). Clearly, in these cases, increasing educational awareness as to appropriate PEP methods would be very beneficial.

Increasing the availability of cell culture rabies vaccines in developing countries is only part of the solution required to reduce the incidence of human rabies. In addition, there is a need to increase the availability of highly purified RIG (or a suitable alternative) to fill the demand for passive immune protection as recommended by the WHO and, finally, as mentioned earlier, there is a critical need to eliminate canine rabies through effective vaccination programs.

9 EDUCATIONAL AWARENESS

Rabies prevention is all about educational awareness. Recent reports highlight just how little some regions of the world (where human rabies continues to be a problem) know about preventing the disease (Prasad et al., 2001; Parviz et al., 2004). Often, uninformed patients come to anti-rabies clinics with wounds that are not properly cleaned because they were told that they should not touch the wound. Others have had chili powder or other spices and powders applied directly onto wounds. Still others may be advised to put a specific, non-medicinal item (i.e. a coin or amulet) on the wound (APCRI, 2004). Educational programs would certainly help to alleviate these misconceptions and save lives. Educational awareness is not a single-level entity, it must be approached as a multilevel campaign and include easily understood information for all levels of society. Novel new programs in the Philippines, India and Pakistan are beginning to appear, which include integrating rabies education into existing educational curricula, providing hands-on workshops for medical professionals and even undertaking the education of an entire village on all levels.

Educational programs should include specific information for prevention of rabies in animals as well as humans, including proper animal care and treatment including canine vaccination and control methods, proper wound care and management and the importance of using RIG, as well as what is inappropriate treatment for the prevention of rabies. In the final analysis, education is the real key to rabies prevention at all levels of society. By simply increasing educational awareness on a local, national, regional and international level, most rabies deaths could be prevented.

Finally, although rabies is one of the oldest diseases known to mankind and is virtually 100% preventable, thousands of people continue to die unnecessarily. Although a few human fatalities may be reported due to a deviation from the recommended PEP regimen, most deaths occur due to a failure to seek PEP, or the unavailability of rabies vaccines and RIG in rabies endemic areas, or because an exposed person received a poor quality rabies vaccine. Some of these problems can be overcome when a greater awareness of the impact of human rabies is attained by the international funding organizations and governments of developing countries. It will take the combined efforts of governmental health officials in rabies-endemic countries, the WHO, vaccine manufacturers and rabies researchers themselves significantly to reduce or eliminate the incidence of human rabies in the future.

REFERENCES

APCRI (2004). Assessing the burden of rabies in India. WHO sponsored multi-centric rabies survey 2003, final report. *Association for the Prevention of Rabies in India.* p. 101. www.apcri.org

Arguin, P.M., Krebs, J.W., Mandel, E., Guzi, T. and Childs, J.E. (2000). Survey of rabies pre-exposure and post-exposure among missionary personnel stationed outside the United States. *Journal of Tropical Medicine* 7, 10–14.

Arya, S.C. (1999). Therapeutic failures with rabies vaccine and rabies immunoglobulin. *Clinical Infectious Diseases* 29, 1605.

Belotto, A.J. (2004). The Pan American Health Organization (PAHO) role in the control of rabies in Latin America. *Developmental Biology* 119, 213–216.

Belotto, A.J., Leanes, L.F., Schneider, M.C., Tamayo, H. and Correa, E. (2005). Overview of rabies in the Americas. *Virus Research* 111, 5–12.

Bernard, K.W. and Fishbein, D.B. (1991). Pre-exposure rabies prophylaxis for travellers: are the benefits worth the cost? *Vaccine* 9, 833–836.

Bernard, K.W., Fishbein, D.B., Miller, K.D. *et al.* (1985). Pre-exposure rabies immunization with human diploid cell vaccine: Decreased antibody responses in persons immunized in developing countries. *American Journal of Tropical Medicine and Hygiene* 34, 633–647.

Bernard, K.W., Mallonee, J., Wright, J.C. *et al.* (1987). Preexposure immunization with intradermal human diploid cell rabies vaccine. *American Journal of Tropical Medicine and Hygiene* 257, 1059–1063.

Bjorvatn, B. and Gundersen, S.G. (1980). Rabies exposure among Norwegian missionaries working abroad. *Scandinavian Journal of Infectious Diseases* 12, 257–264.

Briggs, D.J., Dreesen, D.W., Nicolay, U. *et al.* (2000). Purified chick embryo ell culture rabies vaccine: interchangeability with human diploid cell culture rabies vaccine and comparison of a one-dose versus two-dose post-exposure booster regimen for previously immunized persons. *Vaccine* **19**, 1055–1060.

Burridge, M.J., Gaer, G.M., Sumner, J.W. and Sussman, O. (1982). Intradermal immunization with human diploid cell rabies vaccine. *Journal of the American Medical Association* **248**, 1611–1614.

Centers for Disease Control (1983). Human rabies – Kenya. *Morbidity and Mortality Weekly Report* **32**, 494–495.

Centers for Disease Control (1999). Human rabies prevention – United States, 1999. Recommendations of the Advisory Committee on Immunization Practices (ACIP). *Morbidity and Mortality Weekly Report* **48**, 1–21.

Centers for Disease Control (2004). Notice to readers: Manufacturer's recall of human rabies vaccine – April 2, 2004. *Morbidity and Mortality Weekly Report* **53**, 287–289.

Centers for Disease Control (2005). Compendium of animal rabies prevention and control, 2005. *Morbidity and Mortality Weekly Report* **54**, 1–8.

Chhabra, M., Ichhpujani, R.L., Tewari, K.N. and Lal, S. (2004). Human rabies in Delhi. *Indian Journal of Pediatrics* **71**, 217–220.[**16.3]

Chomel, B.B. (1999). Rabies exposure and clinical disease in animals. In *Rabies: Guidelines for Medical Professionals*. pp. 20–26. Trenton: Veterinary Learning Systems.

Chutivongse, S. and Wilde, H. (1989). Postexposure rabies vaccination during pregnancy: experience with 21 patients. *Vaccine* **7**, 546–548.

Chutivongse, S., Wilde, H., Supich, C., Baer, G.M. and Fishbein, D.B. (1990). Postexposure prophylaxis for rabies with antiserum and intradermal vaccination. *Lancet* **335**, 896–898.

Chutivongse, S., Wilde, H., Benjavongkulchai, M., Chomchey, P. and Punthawong, S. (1995). Postexposure rabies vaccination during pregnancy: Effect on 202 women and their infants. *Clinical Infectious Diseases* **20**, 818–820.

Coleman, P.G. and Dye, C. (1996). Immunization coverage required to prevent outbreaks of dog rabies. *Vaccine* **14**, 85–186.

Conomy, J.P., Leibovitz, A., McCombs, W. and Stinson, J. (1977). Airborne rabies encephalitis: demonstration of rabies virus in the human central nervous system. *Neurology* **27**, 67–69.

Constantine, D.G. (1962). Rabies transmission by nonbite route. *Public Health Report* **77**, 287–289.

Dean, D.J. and Baer, G.M. (1963). Studies on the local treatment of rabies-infected wounds. *Bulletin of the World Health Organization* **28**, 477–486.

Deshpande, A. and Briggs, D.J. (2001). Rabies vaccination in immunosuppressed patients. In: *Rabies Control in Asia* (B. Dodet and F.-X. Meslin, eds). pp. 58–60. Paris: John Libby Eurotext.[**16.2]

Dreesen, D.W., Bernard, K.W., Parker, R.A., Deutsch, A.J. and Brown, J. (1986). Immune complex-like disease in 23 persons following a booster dose of rabies human diploid cell vaccine. *Vaccine* **4**, 45–49.

Dreesen, D.W., Fishbein, D.B., Kemp, D.T. and Brown, J. (1989). Two-year comparative trial on the immunogenicity and adverse effects of purified chick embryo cell rabies vaccine for pre-exposure immunization. *Vaccine* **7**, 397–400.

Fekadu, M., Endeshaw, T., Alemu, W., Bogale, Y., Teshager, T. and Olson, J.G. (1996). Possible human-to-human transmission of rabies in Ethiopia. *Ethiopian Medical Journal* **34**, 123–127.

Fescharek, R., Schwarz, S., Quast, U., Gandhi, N. and Karkhanis, S. (1991). Postexposure rabies prophylaxis: when the guidelines are not respected. *Vaccine* **9**, 868–872.

Fevre, E.M., Kayoyo, R.W., Persson, V., Edelsten, M., Coleman, P.G. and Cleaveland, S. (2005). The epidemiology of animal bite injuries in Uganda and projections of the burden of rabies. *Tropical Medicine and International Health* **10**, 790–798.

Fishbein, D.B., Bernard, K.W., Miller, K.D. *et al.* (1986). The early kinetics of the neutralizing antibody response after booster immunizations with human diploid cell rabies vaccine. *American Journal of Tropical Medicine and Hygiene* **35**, 663–670.

Fishbein, D.B., Yenne, K.M., Dreesen, D.W., Teplis, C.F., Mehta, N. and Briggs, D.J. (1993). Risk factors for systemic hypersensitivity reactions after booster vaccinations with human diploid cell rabies vaccine: a nationwide prospective study. *Vaccine* **11**, 1390–1394.

Fleisher, G.R. (1999). The management of bite wounds. *New England Journal of Medicine* **340**, 138.

Fooks, A.R., Johnson, N., Brookes, S.M., Parsons, G. and McElhinney, L.M. (2003). Risk factors associated with travel to rabies endemic countries. *Journal of Applied Microbiology* **94**, 31S–36S.

Ghosh, T.K. (2003). Controversies in rabies vaccination. *Indian Journal of Pediatrics*. **70**, 495–498.

Hatchett, R.P. (1991). Rabies: the disease and the value of intensive care treatment. *Intensive Care Nursing* **7**, 53–60.

Hatz, C.F., Bidaus, J.M., Eichenberger, K., Mikulics, U. and Junghanss, T. (1995). Circumstances and management of 72 animals bites among long-term residents in the tropics. *Vaccine* **13**, 811–815.

Hemachudha, T., Mitrabhakdi, E., Wilde, H., Vejabhuti, A., Siripataravanit, S. and Kingnate, D. (1999). Additional reports of failure to respond to treatment after rabies exposure in Thailand. *Clinical Infectious Diseases* **28**, 143–144.

Jaiiaroensup, W., Lanag, J., Thipkong, P. *et al.* (1998). Safety and efficacy of purified vero cell rabies vaccine given intramuscularly and intradermally. (Results of a prospective randomized trial.) *Vaccine* **16**, 1559–1562.

Javadi, M.A., Fayaz, A., Mirdehghan, S.A. and Ainollahi, B. (1996). Transmission of rabies by corneal graft. *Cornea* **15**, 431–433.

Jones, R.L., Froeschle, J.E., Atmar, R.L. *et al.* (2001). Immunogenicity, safety and lot consistency in adults of a chromatographically purified Vero-cell rabies vaccine: a randomized, double-blind trial with human diploid cell rabies vaccine. *Vaccine* **19**, 4635–4643.

Kamoltham, T., Singhsa, J., Promsaranee, U., Sonthon, P., Mathean, P. and Thinyounyong, W. (2003). Elimination of human rabies in a canine endemic province in Thailand: five-year programme. *Bulletin of the World Health Organization* **81**, 375–381.

Kaplan, M.M., Cohen, D., Koprowski, H., Dean, D. and Ferrigan, L. (1962). Studies on the local treatment of wounds for the prevention of rabies. *Bulletin of the World Health Organization* **26**, 765–775.

Khawplod, P., Wilde, H., Chomchey, P. *et al.* (1996). What is an acceptable delay in rabies immune globulin administration when vaccine alone had been given previously? *Vaccine* **14**, 389–391.

Knobel, D.L., Cleaveland, S., Coleman, P.G. *et al.* (2005). Re-evaluating the burden of rabies in Africa and Asia. *Bulletin of the World Health Organization* **83**, 360–368.

Krause, E., Grundmann, H. and Hatz, C. (1999). Pretravel advice neglects rabies risk for travelers to tropical countries. *Journal of Travel Medicine* **6**, 163–167.

Krebs, J.W., Mandel, E.J. and Rupprecht, C.R. (2004). Rabies surveillance in the United States during 2003. *Journal of the American Veterinary Medical Association* **225**, 1837–1849.

Lang, J., Simanjuntak, G.H., Soerjosembodo, S., Koesharyono, C. and the MAS054 Clinical Investigator Group (1998). Suppressant effect of human or equine rabies immunoglobulins on the immunogenicity of post-exposure rabies vaccination under the 2-1-1 regimen: a field trial in Indonesia. *Bulletin of the World Health Organization* **76**, 491–495.

Lau, J. (1999). Intradermal rabies vaccination and current use of mefloquine. *Journal of Travel Medicine* **6**, 140–141.

Mackell, S.M. (2003). Vaccinations for the pediatric traveler. *Clinical Infectious Diseases* **37**, 1508–1516.

Madhusudana, S.N., Anand, N.P. and Shamsundar, R. (2001). Evaluation of two intradermal vaccination regimens using purified chick embryo cell vaccine for post-exposure prophylaxis of rabies. *National Medical Journal of India* **14**, 145–147.

Malerczyk, C., Briggs, D.J., Dreesen, D.W. and Banzhoff, A. (2007). Duration of immunity: An anamnestic response 14 years after rabies vaccination with Purified Chick Embryo Cell Rabies Vaccine (PCECV). *Journal of Travel Medicine* **14**, 63–64.

Messenger, S.L., Smith, J.S. and Rupprecht, C.E. (2002). Emerging epidemiology of bat-associated cryptic cases of rabies in humans in the United States. *Clinical Infectious Diseases* **35**, 738–747.

Mondul, A.M., Krebs, J.W. and Childs, J.E. (2003). Trends in national surveillance for rabies among bats in the United States (1993–2000). *Journal of the American Veterinary Medical Association* **222**, 633–639.

Murray, K.O. and Arguin, P.M. (2000). Decision-based evaluation of recommendations for preexposure rabies vaccination. *Journal of the American Veterinary Medical Association* **216**, 188–191.

Nathwani, D., McIntyre, P.G., White, K. *et al.* (2003). Fatal human rabies caused by European bat Lyssavirus type 2a infection in Scotland. *Clinical Infectious Diseases* **37**, 598–601.

Noah, D.L., Drenzek, C.L., Smith, J.S. *et al.* (1998). Epidemiology of human rabies in the United States, 1980 to 1996. *Annals of Internal Medicine* **128**, 922–929.

Pancharoen, C., Thisyakorn, U., Lawtongkum, W. and Wilde, H. (2001). Rabies exposures in Thai children. *Wildlife Environmental Medicine* **12**, 239–243.

Pappaioanou, M., Fishbeinn, D.B., Dreesen, D.W. *et al.* (1986). Antibody response to preexposure human diploid-cell rabies vaccine given concurrently with chloroquine. *New England Journal of Medicine* **314**, 280–284.

Parviz, S., Chotani, R., McCormick, J., Fisher-Hoch, S. and Luby, S. (2004). Rabies deaths in Pakistan: results of ineffective post-exposure treatment. *International Journal of Infectious Diseases* **8**, 346–352.

Peigue-Lafeuille, H., Bourhy, H., Abiteboul, D. *et al.* (2004). Human rabies in France in 2004: update and management. *Médécine et maladies infectieuses* **34**, 551–560.

Phanuphak, P., Ubolyam, S. and Sirivichayakul, S. (1994). Should travellers in rabies endemic areas receive pre-exposure rabies immunization? *Annales de Médécine Interne (Paris)* **145**, 167–176.

Prasad, V.S., Duggal, M., Aggarwal, A.K. and Kumar, R. (2001). Animal bite management practices: A survey of health care providers in a community development block of Haryana. *Journal of Communicable Diseases* **44**, 266–273.

Quiambao, B.P., Dimaano, E.M., Ambas, C., Davis, R., Banzhoff, A. and Malerczyk, C. (2005). Reducing the cost of post-exposure rabies prophylaxis: efficacy of 0.1 ml PCEC rabies vaccine administered intradermally using the Thai Red Cross post-exposure regimen in patients severely exposed to laboratory-confirmed rabid animals. *Vaccine* **25**, 1709–1714.

Reddy, A.V. and Sampath, G. (2000). Post exposure treatment – some experiences. *Journal of the Association for Prevention and Control of Rabies in India* **1**, 33–35.

Remington, P.A., Shope, T. and Andrews, J. (1985). A recommended approach to the evaluation of human rabies exposure in an acute-care hospital. *Journal of the American Medical Association* **254**, 67–69.

Sabchareon, A., Chantavanich, P., Pasuralertsakul, S. *et al.* (1998). Persistence of antibodies in children after intradermal or intramuscular administration of preexposure primary and booster immunizations with purified Vero cell rabies vaccine *Pediatric Infectious Diseases Journal* **17**, 1001–1007.

Sampath, G., Parikh, S., Sangram, P. and Briggs, D.J. (2005). Rabies post-exposure prophylaxis in malnourished children exposed to suspect rabid animals. *Vaccine* **23**, 1102–1105.

Shankar, S.K., Madhusudana, S.N. and Satyanarayana, S.H. (2000). Pathological morbidity following Semple type antirabies vaccine in India and why we need to replace it. *Journal of the Association for Prevention and Control of Rabies in India* **1**, 15–20.

Shill, M., Baynes, R.D. and Miller, S.D. (1987). Fatal rabies encephalitis despite appropriate post-exposure prophylaxis. *New England Journal of Medicine* **316**, 1257–1258.

Singh, J., Jain, D.C., Bhatia, R. *et al.* (2001). Epidemiological characteristics of rabies in Delhi and surrounding areas, 1998. *Indian Pediatrics* **38**, 1354–1360.

Slate, D., Rupprecht, C.E., Rooney, J.A., Donovan, D., Lein, D.H. and Chipman, R.B. (2005). Status of oral rabies vaccination in wild carnivores in the United States. *Virus Research* **111**, 68–76.

Smith, J., McElhinney, L., Parsons, G. *et al.* (2003). Case report: rapid ante-mortem diagnosis of a human case of rabies imported from the Philippines. *Journal of Medical Virology* **69**,150–155.

Sriaroon, C., Daviratanasilpa, S., Sansomranjai, P. *et al.* (2003). Rabies in a Thai child treated with the eight-site post-exposure regimen without rabies immune globulin. *Vaccine* **21**, 3525–3526.

Srinivasan, A., Burton, E.C., Kuehnert, M.J. *et al.* (2005). Transmission of rabies virus from an organ donor to four transplant recipients. *New England Journal of Medicine* **352**, 1103–1111.

Strady, A., Lang, J., Lienard, M., Blondeau, C., Jaussaud, R. and Plotkin, S.A. (1998). Antibody persistence following preexposure regimens of cell-culture rabies vaccines: 10-year follow-up and proposal for a new booster policy. *Journal of Infectious Diseases* **177**, 1290–1295.

Sudarshan, M.K., Madhusudana, S.N. and Mahendra, B.J. (1999). Post-exposure prophylaxis with purified Vero cell rabies vaccine during pregnancy-safety and immunogenicity. *Journal of Communicable Diseases* **31**, 229–236.

Swanson, M.C., Rosanoff, E., Gurwith, M., Deitch, M., Schnurrenberger, P. and Reed, C.E. (1987). IgE and IgG antibodies to β-propiolactone and human serum albumin associated with urticarial reactions to rabies vaccine. *Journal of Infectious Diseases* **155**, 909–913.

Thisyakorn, U., Pancharoen, C., Ruxrungtham, K. *et al.* (2000). Safety and immunogenicity of pre-exposure rabies vaccination in children infected with human immunodeficiency virus type 1. *Clinical Infectious Diseases* **30**, 218.

Traenhart, O., Kreuzfelder, E., Hillebrandt, M. *et al.* (1994). Long-term humoral and cellular immunity after vaccination with cell culture rabies vaccines in man. *Clinical Immunology and Immunopathology* **71**, 287–292.

Trevejo, R.T. (2000). Rabies preexposure vaccination among veterinarians and at-risk staff. *Journal of the American Veterinary Medical Association* **217**, 1647–1650.

Trimarchi, C.V. and Briggs, D.J. (1999). The diagnosis of rabies. In: *Rabies: Guidelines for medical professionals.* pp. 55–66. Trenton: Veterinary Learning Systems.

Varner, M.W. and McGuinness, G.A. (1982). Rabies vaccination in pregnancy. *American Journal of Obstetrics and Gynecology* **143**, 717–718.

Vodopija, R., Sureau, P., Smerdel, S. *et al.* (1988). Interaction of rabies vaccine with human rabies immunoglobulin and reliability of a 2-1-1 schedule application for postexposure treatment. *Vaccine* **6**, 283–286.

Wasi, C., Chaiprasithikul, P., Thongcharoen, P., Choomkasien, P. and Sirikawin, S. (1997). Progress and achievement of rabies control in Thailand. *Vaccine* **15**, S7–S11.

WHO (1986). Report of a consultation on transfer of technology for production of rabies vaccine: January 16–17, 1986. *VPH/86.64.* Geneva: WHO.

WHO (1987). Guidelines for dog rabies control. *VPH/83.43 Rev. 1.* Geneva: WHO.

WHO (1992). World Health Organization expert committee on rabies. Eighth report. WHO. *Technical Report Series 824.* Geneva: WHO.

WHO (1996). World Health Organization recommendations on rabies post-exposure treatment and the correct technique of intradermal immunization against rabies. *WHO/EMC/ZOO96.6.* Geneva: WHO.

WHO (2000). Intradermal application of rabies vaccines. Report of a WHO consultation. Bangkok, Thailand 5–6 June, 2000. *WHO/CDS/CDSRAPH/2000.05.* Geneva: WHO.

WHO (2005). *WHO Technical report series 931.* Geneva: WHO.

Wilde, H. (1999). Economic issues in postexposure rabies treatment. *Journal of Travel Medicine* **6**, 238–242.

Wilde, H., Briggs, D.J., Meslin, F.X., Hemachudha, T. and Sitprija, V. (2003). Rabies update for travel medicine advisors. *Clinical Infectious Diseases* **37**, 96–100.

Windiyaningsih, C., Wilde, H., Meslin, F.X., Suroso, T. and Widarso, H.S. (2004). The rabies epidemic on Flores Island, Indonesia (1998–2003). *Journal of the Medical Association of Thailand* **87**, 1–5.

Winkler, W.G., Fashinell, T.R., Leffingwell, L., Howard, P. and Conomy, J.P. (1973). Airborne rabies transmission in a laboratory worker. *Journal of the American Medical Association* **226**, 1219–1222.

17

Dog Rabies and its Control

DARRYN KNOBEL[1]**, MAGAI KAARE**[1]**, ERIC FÈVRE**[2]**, AND SARAH CLEAVELAND**[1]

Centre for Tropical Veterinary Medicine, Royal (Dick) School of Veterinary Studies, University of Edinburgh, Roslin, Midlothian, EH25 9RG, UK[1]; Centre for Infectious Diseases, University of Edinburgh, Ashworth Labs, Edinburgh, EH9 3JF, UK[2]

1 INTRODUCTION

Rabies is caused by a number of genetically closely related viruses belonging to the genus *Lyssavirus*, of which the type species is rabies virus (see Chapter 2). A true generalist pathogen, rabies virus has been isolated from nearly all mammalian orders (Rupprecht *et al.*, 2002) and the disease occurs on all continents except Antarctica (Warrell and Warrell, 2004). Although rabies can infect all mammals, only a few mammalian species are known to act as reservoirs of the disease, with domestic dogs being the major reservoir throughout Africa and Asia (Rupprecht *et al.*, 2002). The association between the bite of a 'mad' dog and rabies has been recognized since antiquity (reviewed by Neville, 2004) and rabid dogs are still responsible for the vast majority (>90%) of human deaths from rabies worldwide (WHO, 1999; http://globalatlas.who.int/globalatlas). This can be true even in some areas where wildlife species are the rabies reservoir, as the proximity between dogs and humans provides a link in transmission between wildlife and people.

2 THE BURDEN OF CANINE RABIES

More than 99% of all human deaths from rabies occur in Africa and Asia (WHO, 1999; http://globalatlas.who.int/globalatlas). Canine rabies is reported in many countries of Africa and Asia and is likely to be maintained endemically in areas where the dog density exceeds the threshold for persistence, considered to be about 4.5 dogs/km^2 (Brooks, 1990; Bishop, 1995; Cleaveland and Dye, 1995; Kitala *et al.*, 2002) (Figure 17.1). In these areas, the disease is responsible for some 55 000 human deaths each year (Knobel *et al.*, 2005). Globally, dog rabies kills more people than yellow fever, dengue fever or Japanese encephalitis and more than 7 million people are potentially exposed to the virus annually (Coleman *et al.*, 2004; Knobel *et al.*, 2005), leading to a high demand for expensive post-exposure prophylaxis (PEP).

Rabies, second edition. Edited by Alan C. Jackson and William H. Wunner
ISBN 978-012-369366-2. Copyright Elsevier Inc. 2007

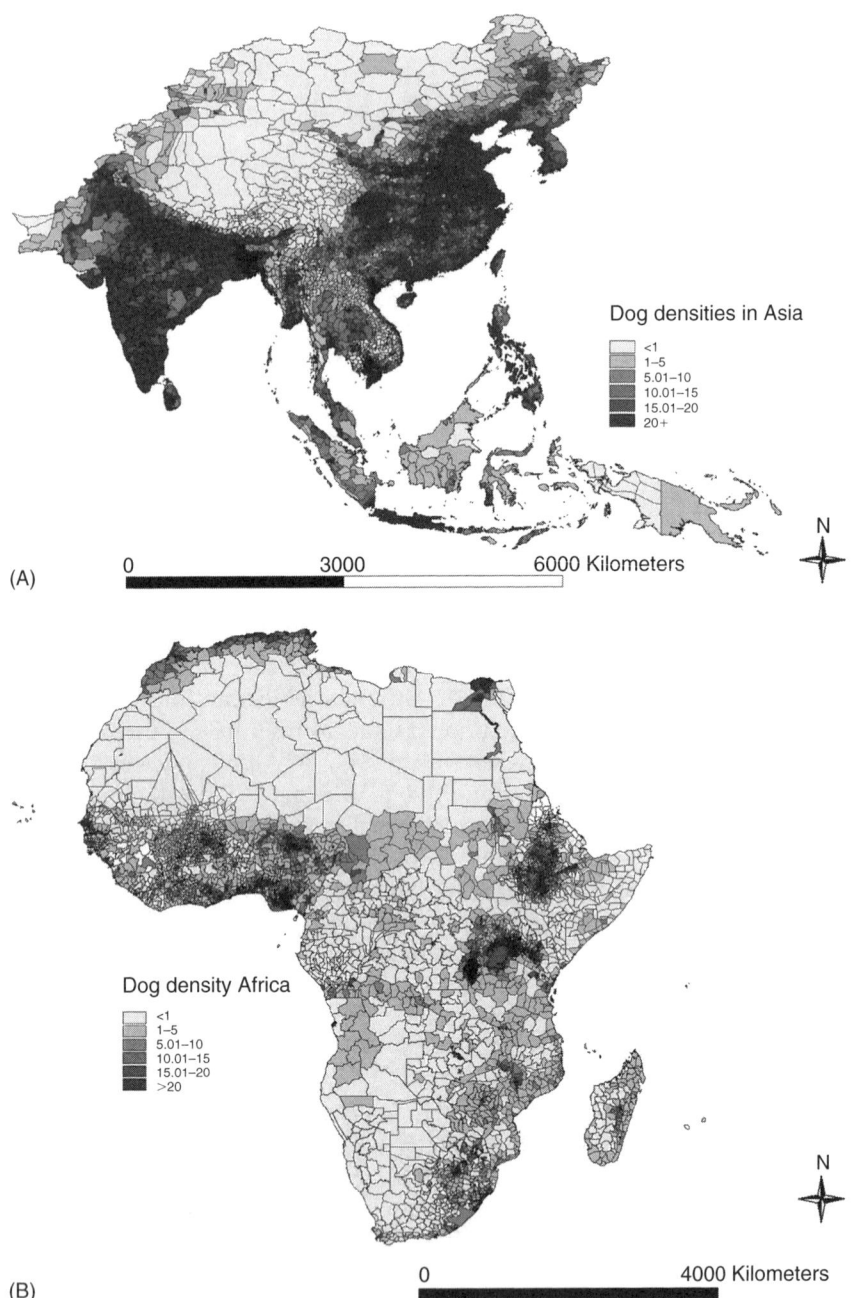

Figure 17.1 (A) The density of domestic dogs in Asia, using human densities derived from an Asia population density dataset and a human:dog ratio of 7.5 for urban settings and 14.3 for rural settings. (B) The density of domestic dogs in Africa, using human densities derived from an Africa population density dataset and a human:dog ratio of 21.2 for urban settings and 7.4 for rural settings (for further details see Knobel *et al.*, 2005). This figure is reproduced in the color plate section.

The high demand for PEP resulting from dog rabies exerts a substantial economic burden in Africa and Asia, not only as a result of the high costs of the human vaccine and immunoglobulin that are required for prevention, but also because of considerable indirect (patient) costs associated with travel and income loss (Knobel *et al.*, 2005). The total (direct and indirect) cost of PEP is equivalent to 5.8% of annual per capita gross national income in Africa ($40 per treatment) and 3.9% ($49 per treatment) in Asia (Meltzer and Rupprecht, 1998; Knobel *et al.*, 2005). Additional economic losses associated with dog rabies accrue through livestock deaths which, although poorly quantified, may be significant. Socio-economic factors influence the distribution of rabies cases, with people who are poor, less well educated and those living far from treatment centres least able to access prompt and appropriate PEP. Children of less than 16 years of age are the major victims of dog rabies, as they are more often bitten by dogs than adults and, when bitten, are frequently bitten on the head and neck, sites which carry a much higher risk than bites to other parts of the body (Pancharoen *et al.*, 2001; Cleaveland *et al.*, 2002; Fèvre *et al.*, 2005; Knobel *et al.*, 2005). Although safe and highly effective cell culture vaccines are available for PEP (WHO, 2004), the high cost of treatment is resulting in the continued use of cheaper nerve-tissue origin vaccines in parts of Asia, with the attendant risks of severe neurological side effects.

An additional disease burden relating to canine rabies is the morbidity associated with bites from rabid dogs, which can often result in severe injury. The incidence of bite injuries from suspected rabid dogs in various communities in Africa and the Middle East has been estimated as 40–288 bite victims/100 000 people (Zeynali *et al.*, 1999; Kitala *et al.*, 2000; Cleaveland *et al.*, 2002; Fèvre *et al.*, 2005). First-aid care, such as washing the bite wound with soap and water, can significantly reduce the risk of rabies infection; however, many bite victims do not receive even this simple treatment (Parviz *et al.*, 1998). Aside from the physical trauma, the psychological impacts of rabid dog bite injuries are not trivial. In many parts of Africa and Asia, the post-exposure prophylaxis given to animal bite victims is deficient by all World Health Organization (WHO) standards (Parviz *et al.*, 1998; Fèvre *et al.*, 2005). Access to human rabies PEP is often undependable, rabies immunoglobulin is unavailable and the provision of a full course of vaccination is not only extremely costly but also logistically demanding. As a result, PEP is often unreliable and the bite from a suspect rabid dog can cause great distress, with months of anxiety as the individual awaits an uncertain outcome. Preliminary work carried out in Tanzania revealed that people are more concerned about the potential threat from rabies than from malaria, despite the higher prevalence of the latter (Kaare, 2007). When human cases do occur, the distressing clinical signs and invariably fatal outcome result in considerable psychological trauma for families, communities and health care professionals involved with the victim (Warrell and Warrell, 2004).

Further consequences of dog rabies relate to animal welfare and conservation. While the effects of canine rabies on public attitudes and treatment of dogs have been poorly investigated, there is no doubt that in areas where canine rabies is endemic, fear of the disease has important implications for animal welfare, with suspected rabid dogs and unknown, stray dogs often killed inhumanely in an attempt to control rabies and human exposure. Infectious disease is also increasingly recognized as a threat to endangered carnivore species (Funk *et al.*, 2001; Woodroffe *et al.*, 2004). Population viability analyses indicate that canine rabies poses an important extinction risk for both Ethiopian wolves *Canis simensis* (Haydon *et al.*, 2002) and African wild dogs *Lycaon pictus* (Ginsberg and Woodroffe, 1997; Vucetich and Creel, 1999), particularly in small- and medium-sized populations. Both these endangered species have suffered high mortality during previous rabies outbreaks (Gascoyne *et al.*, 1993; Kat *et al.*, 1995; Sillero-Zubiri *et al.*, 1996; Randall *et al.*, 2004).

Despite the multitude of problems associated with dog rabies and the considerable disease burden, the disease remains widely neglected and uncontrolled throughout much of Africa and Asia. In many regions of sub-Saharan Africa, the past three decades have seen a dramatic increase in the number of reported cases of the disease and in the range of species in which canine rabies has been recorded (Perry, 1995; Cleaveland, 1998). Many factors are likely to have contributed to this, including the enormous increase in domestic dog populations, which are typically growing at a rate of 5–10% per annum (Kitala and McDermott, 1995; Cleaveland, 1996; Laurenson *et al.*, 1997b) and the high mobility of human (and dog) populations. In many areas, these factors are compounded by deteriorating infrastructure available to veterinary services for effective disease control, allowing rabies outbreaks to spread unchecked. The escalation in rabies cases in Africa and Asia poses considerable challenges and, in this chapter, discussion of issues relating to dog rabies control will therefore focus primarily on these regions of the world.

3 HISTORICAL PERSPECTIVES ON DOG RABIES CONTROL

Successful canine rabies control or even elimination programmes have been carried out since as early as the latter half of the 19th century. In 1831, a bill was drafted in the UK 'to prevent the spreading of canine madness' (Fooks *et al.*, 2004) and enforcement of this legislation, which included the muzzling of dogs, restriction of their movement and the destruction of stray and rabid dogs, along with a strict import control policy, led to the elimination of terrestrial rabies from that country in 1902 and again in 1922, after its reintroduction in 1918 (Muir and Roome, 2005). These measures – movement and contact restrictions, notification and observation of cases, tracing of contacts and the killing of rabid, suspect-rabid and free-roaming dogs not in compliance with legislation – constitute the so-called 'classical' methods of canine rabies control. An additional and powerful control tool came in the 1920s, with the advent of the first effective veterinary vaccines against rabies (Umeno and Doi, 1921). Japan, in 1921, was the first country to apply mass vaccination of dogs, but it was not until 1957 that rabies was eventually eliminated from within its borders (WHO, 1966). The first field trials to demonstrate that canine rabies could be eliminated through a combination of mass vaccination and classical control measures were in fact carried out in Hungary in 1937. This country went on to conduct the first successful national rabies control campaign, from 1939 to 1944 (Manninger, 1968). In the 1950s, several other countries followed this example, with canine rabies being eliminated from Hong Kong in 1956 (Hung Cheuk, 1969) and Taiwan and Portugal in 1961 (WHO, 1966). Early success in Malaysia, where the disease was quickly brought under control following the initiation in 1952 of compulsory vaccination of dogs and destruction of strays (Wells, 1954), indicates that it is not only in the more developed countries that dog rabies control is possible. In Africa, both Zimbabwe (Shone, 1962) and Uganda (Arvo and Kaumba, 1971) reported dramatic declines in canine rabies cases over the 10 years prior to 1961.

Although dog population control through the killing of strays, either by shooting or poisoning, commonly forms part of national rabies control strategies, current consensus suggests that the persistence of this approach is due to a number of erroneous assumptions and poor insight into the epidemiology of rabies and the ecology of dog populations (WH0, 1992a). A dearth of successful canine rabies control programmes during the 1970s and a growing awareness of the lack of information on the dynamics of dog populations (particularly in comparison with the increasing body of knowledge on fox ecology in relation to the rise of sylvatic rabies in Europe at that time), prompted the WHO in 1984 to instigate three research projects into the ecology and population dynamics of dogs in developing countries. The sites chosen were in Sri Lanka, Ecuador and Tunisia, and the results were to change significantly the approach to canine rabies control at a national and international level (WHO, 1988). It was shown that, in order to attain a lasting reduction in dog population size, lethal control programmes would need to remove 50–80% of the population each year. In Sri Lanka, where sustained dog elimination campaigns had been conducted since 1977, researchers found that, despite removing between 35 000 and 50 000 dogs annually, these programmes were only reaching 5% or less of the targeted dog population. A concerted effort in Guayaquil, Ecuador, which resulted in the removal of 24% of the dog population over a period of 12 months, was found to have no lasting impact on the size of the dog population, or on the incidence of canine rabies. In fact, the community response to dog elimination campaigns in Guayaquil and elsewhere has been to buy new puppies or to adopt free-roaming dogs that move into the area to fill the vacant niche. Because these dogs are mostly unvaccinated, and because the elimination campaigns themselves often remove vaccinated dogs (4% of killed dogs in the Sri Lankan study were found to have demonstrable serum rabies virus neutralizing antibody titres), the net result of dog elimination campaigns is often to reduce the proportion of immunized individuals in a population, and thus decrease the level of 'herd immunity'. The WHO report (1988) concluded that not only did dog elimination programmes have no significant long-term effect on dog population sizes, but they incited animosity towards rabies control personnel in local communities, resulting in decreased cooperation during mass vaccination campaigns. This viewpoint is further supported by results of recent campaigns to control dog rabies in Flores, an island in Indonesia that had previously been rabies-free, and in which 295 565 dogs were killed over a 4-year period as a means to tackle a rabies epidemic. While this impacted on rabies in some areas of the island, it was not successful at preventing the further spread of rabies among dogs and transmission to humans (Windiyaningsih et al., 2004), in large part due to the unpopularity of the culling scheme, as most dogs on Flores are owned (Bingham, 2001). By May 2004, 48% of the island's dog population had been destroyed, but rabies transmission continued.

The results of the WHO (1988) and other studies led the WHO Expert Committee on Rabies to the following conclusion in 1992 (WHO, 1992a):

> ... the Committee recommended drastic changes in rabies control policies as compared with those previously adopted and practised by most national authorities and communities. There is no evidence that removal of dogs has ever had a significant impact on dog population densities or the spread of rabies. The population turnover of dogs may be so high that even the highest recorded removal rates (about 15% of the dog population) are easily compensated for by increased survival rates. In addition, dog removal may be unacceptable to local communities. Therefore, this approach should not be used in large-scale control programmes unless ecological and sociocultural studies show it to be feasible.

4 DOG ACCESSIBILITY

The WHO-instigated studies also overturned a number of other assumptions relating to the accessibility of dogs to vaccination. Previously it had been believed that, in any given human social or economic setting, 30–60% of dogs were 'stray' animals, which could not be vaccinated (Bögel and Joshi, 1990). Researchers, however, found that in each of the three study areas most dogs were associated with either one or several households. In Sri Lanka, less than 10% of the total observed dog population was unowned, while in Tunisia, only 7% of dogs were described as 'feral'. No dogs unassociated with any household could be identified in the study in Guayaquil. The authors of the final report stated that 'whether owned or unowned, dogs which are not catchable by at least one person are rare and represent generally less that 15% of the dog population'. Similarly, Bögel and Joshi (1990) concluded that 86–97% of dogs in four areas of the Kathmandu Valley, Nepal, were accessible to parenteral vaccination.

A new and important concept in the control of canine rabies thus emerged: that of dog accessibility (Bögel, 2002). Before this concept could be more fully explored however, it was necessary to revisit and clarify the definitions applied to different segments of the dog population, in light of the recent findings. Previous terms such as 'pet', 'owned', 'feral' or 'stray' were thought either context-specific or ecologically irrelevant. New definitions were proposed, based on two quantifiable parameters: the level of a dog's dependency on humans and the level of its restriction by humans (WHO, 1988). The first of these refers to the *intentional* provision of those needs, such as food, shelter and care, necessary for the survival and well-being of the dog. Although it was recognized that this dependency of dogs on humans was a gradient ranging from total dependency to none, the following three categories of dependency were proposed:

- Full dependency – the dog is given all of its essential needs intentionally by humans
- Semi-dependency – the dog is given some of its essential needs intentionally by humans
- No dependency – the dog is given none of its essential needs intentionally by humans.

The second parameter, restriction by humans, refers to the control of any contact, association and communication with other dogs and people, either through physical restriction (e.g. confinement to premises, leashing) or through direct supervision and control when outside the premises. Again, three categories of restriction are recognized:

- Full restriction – fully restricted or supervised
- Semi-restricted – movements and associations only partially restricted
- No restriction – not subject to any restriction whatsoever.

The use of these two parameters gives rise to a matrix classification system (Table 17.1), which reveals four broad categories of dogs:

- Restricted dog – fully dependent on their owners and fully restricted or supervised
- Family dogs – fully dependent on their owners, but movement and contacts only partially restricted
- Neighbourhood dogs – partially dependent on the intentional fulfilment of basic needs by humans, but subject to only partial or no restriction
- Feral dogs – independent of intentional human provisioning and unrestricted.

TABLE 17.1

Dog categorization matrix and index of accessibility[a] to parenteral vaccination

	Fully restricted	Semi-restricted	Unrestricted
Full dependency	Restricted dog + + +	Family dog + +	
Semi-dependency		Neighbourhood dog +	Unrestricted dog + / −
No dependency			Feral dog −

[a] Index of accessibility is an assessment of the typical ease of accessibility of dogs within each category to parenteral vaccination, ranging from very easily accessible (+ + +) to inaccessible (−).

Missing from this classification are the terms 'owned' or 'stray' dog. Although long employed in rabies control campaigns, reference to 'stray' dogs was considered misleading, as under the new classification system a stray dog could be a feral dog, an unrestricted neighbourhood dog, or a free-roaming family dog. Dogs in the latter two categories are accessible through one or more reference households. The WHO recommended that the term be reserved for application to dogs not in compliance with local rabies control ordinance, for example those not vaccinated, on a leash, muzzled or confined. Similarly, the term 'owned dog' was deemed ecologically irrelevant, appropriate only in connection with certain administrative measures applied to dogs through their owners, such as licensing, taxation or vaccination.

5 VACCINATION COVERAGE

But has this improved understanding of dog accessibility translated into improved control of canine rabies through mass vaccination coverage and, importantly, a reduction in the number of human exposures to rabies and deaths from the disease? The results of an array of studies conducted in different socio-economic and cultural situations suggest that it has. In Nepal, Bögel and Joshi (1990) demonstrated that 75–80% of dogs in a suburban area near Kathmandu could be vaccinated during a month-long campaign incorporating fixed 'central-point' vaccination stations and follow-up house-to-house visits. Using similar techniques, and specifically engaging the help of children over the school-holiday period to bring their dogs to the vaccination stations and to help in identifying dog-owning households, Perry et al. (1995) achieved a vaccination coverage of 68–75% in a densely populated suburb of Nairobi, Kenya. Comparable coverage levels (64–87%) were attained in three study areas of N'Djaména, Chad with the use of central-point stations kept open for 1–2 days (Kayali et al., 2003). The number of inaccessible dogs was considered to be less than 8% of the total dog population in that city. In Thailand, which established a national rabies control programme in

1995 with the goal of being rabies-free within 5 years, mass parenteral vaccination reached 64–78% of the approximately 100 000 dogs in the canine-rabies endemic province of Petchabun (Kamoltham *et al.,* 2003). Notable success in reducing human and dog rabies has been achieved in Central and South America by the programme for the elimination of canine rabies based on mass dog vaccination initiated and coordinated by the PAHO/WHO Regional Office for the Americas (WHO, 2004; Belotto *et al.,* 2005). A key factor in these successes has been the leading role played by the various Ministries of Health or their equivalents, emphasizing the importance of intersectoral collaboration for effective rabies control. In addition, efforts were coordinated through the designation of a national 'rabies day', during which community resources are mobilized locally to allow synchronization of dog vaccination campaigns nationwide (Belotto *et al.,* 2005).

A vaccination coverage of between 60 and 80% has been shown to result in significant decreases in dog rabies incidence and human exposures in rural Tanzania, where central-point dog vaccination campaigns were conducted annually between October 1996 and February 2001 reaching 65, 61, 71 and 74% of the dog population over the 4 years. As a result, the incidence of dog rabies declined by 97% after the second campaign (Cleaveland *et al.,* 2003) with a concomitant decrease in the incidence of human bite injuries from suspected rabid dogs and thus a significantly reduced demand for human post-exposure rabies treatment. In conjunction with rabies awareness campaigns and the increased accessibility of low-cost human intradermal post-exposure treatment, mass dog vaccination in Petchabun province of Thailand, begun in 1996, reduced human rabies deaths to zero between 1999 and 2001 (Kamoltham *et al.,* 2003).

Although parenteral vaccination has been the mainstay of dog rabies control worldwide, the accessibility of dogs to parenteral vaccination is not uniformly high in all areas. Nearly half of all unrestricted but owned dogs in some rural areas of Turkey could not be caught by their owners (WHO, 1992b). During a house-to-house vaccination campaign in the Philippines, teams found that only 10% of eligible dogs could be caught and restrained without difficulty (Estrada *et al.,* 2001). A similar phenomenon has been observed in pastoralist communities in East Africa and in rural areas of Ethiopia, where dogs are kept to raise the alarm in the event of livestock raids by hyenas and other predators. Central-point vaccination campaigns habitually result in low turnout, while during house-to-house visits owners are often reluctant or unable to handle household dogs (Laurenson *et al.,* 1997a; Coleman, 1999). In Ethiopia, low accessibility appears to be the product of a human–dog relationship characterized by a low level of care and a lack of perceived value in an individual dog (although dogs as a whole may have generic worth). Poor awareness of rabies risks and control measures and socio-cultural factors such as religion may also contribute. In pastoral communities, accessibility is also limited by the remote and highly dispersed nature of households. In Tanzanian and Kenyan pastoral (Maasai) communities, for example, a central-point parenteral strategy achieved less than 15% coverage

(Cleaveland, 1996; Coleman, 1999). Given the limited ability of many governments to deliver sustainable veterinary services to remote pastoral locations, the community-based animal health worker (CAHW) approach has emerged as an alternative veterinary service model. CAHWs have been successfully used for rabies vaccination in remote pastoral communities in Tanzania which, in combination with a centralized strategy, achieved coverage levels of up 86% (Kaare, 2007). The combined CAHW-central point approach was also shown to be more cost effective than other conventional approaches, such as house-to-house, central-point and their combinations.

6 THE EPIDEMIOLOGICAL THEORY OF DOG RABIES CONTROL

Failure of dog vaccination to control rabies outbreaks has been noted in some areas where relatively high coverage was achieved (e.g. Mexico [Eng *et al.*, 1993]). A brief examination of the theoretical basis for the design of vaccination programmes may clarify some reasons for programme failure and provide managers with guidelines for successful implementation. Prevention of rabies outbreaks through vaccination is achieved by reducing the density of susceptible (i.e. non-immune) dogs sufficiently so that the basic reproduction number (R_0) in the population falls below one. This parameter is defined as the number of secondary cases arising from a single primary case introduced into a completely susceptible population (Anderson and May, 1991). If $R_0 > 1$, each primary case will, on average, produce more than one secondary case and the infection will spread exponentially through the population, leading to an epidemic. Conversely, when $R_0 < 1$, each primary case will, on average, produce less than one secondary case and, although some secondary cases may occur, the infection will tend to die out without a major epidemic (Woolhouse *et al.*, 1997a). Several studies of canine rabies outbreaks in dogs have shown this figure to be between about 1.5 and 2.5 (cited in Coleman and Dye, 1996). Using these estimates, rabies epidemiologists predict that 70% vaccination coverage will reduce R_0 below the critical threshold of one (sufficient to prevent major outbreaks) approximately 95% of the time (Coleman and Dye, 1996). This figure of 70% was in fact first recommended by the WHO as the critical percentage of dogs which need to be vaccinated to prevent or control an outbreak of rabies, based on a consensus reached among veterinary practitioners in New York State in the 1940s – a neat example of the convergence of empirical and theoretical outcomes.

The theoretical underpinnings of this result are, however, themselves based on a number of assumptions, divergence from which could account for some cases of programme failure at high coverage levels. These assumptions can be broadly divided into epidemiological factors and those pertaining to the consequences of vaccination (Woolhouse *et al.*, 1997a). Regarding vaccination, it has been shown that the level of 'herd immunity' necessary for the prevention of outbreaks is dependent not only on the basic reproduction number R_0, but also on the impact of vaccination on the population (Anderson and May, 1982; McLean and Blower, 1993). Vaccine impact, in its turn, depends on the proportion of vaccinated hosts which actually become protected (vaccine 'take') and on the duration of this protection relative to host life-expectancy (Woolhouse *et al.*, 1997a). In general, dog rabies vaccines have a high relative duration of protection, effective for up to 3 years in host populations whose average life expectancy is often shorter (cited in Coleman, 1999). However, an extension of life expectancy (particularly in populations in which recruitment remains high), perhaps through combined dog vaccination or health-care programmes which reduce mortality from causes such as canine distemper, parvovirus or helminthiasis, may reduce the impact of rabies vaccination. Further study into factors, anthropogenic or otherwise, regulating dog population growth and turnover will be necessary; what is currently obvious, however, is that aspects of host populations relating to vaccination coverage are not uniform across populations. Vaccine 'take' may also vary between programmes, although this is more likely due to factors affecting vaccine efficacy, such as the deficiencies in the cold-chain commonly encountered in the developing world. Differences in the genetic makeup of host populations in regions of the genome, such as the major histocompatibility complex (MHC) involved in the regulation of immune response (Kennedy *et al.*, 2002), may also influence host response to vaccination and account for some variation in vaccine impact between populations.

Host population characteristics may also affect the outcomes of vaccination campaigns in other ways. The figure of 70% coverage assumes that all dogs within a population are equally exposed to infection, in other words that infected and susceptible individuals are perfectly mixed (Anderson and May, 1991). Although this is rarely the case in practice, little work has been done to assess the effects of heterogeneity within dog populations on the impact of rabies vaccination campaigns. Fully restricted dogs, such as those confined to a kennel or chained in a closed yard, do not mix freely with the remaining dog population and may be better considered a separate subpopulation with little danger of infection, yet these are the animals easily accessible to vaccination. Conversely, feral and unrestricted dogs that are most at risk of infection from other dogs within this category and from spillover of infection from wildlife reservoirs are less likely to be vaccinated. Within these subpopulations, age and sex differences in behaviour and home ranges and clustering of dogs in high-density areas may exacerbate inequalities in probability of exposure. The existence of any significant heterogeneities within given dog populations will require higher levels of overall vaccination coverage than those predicted by models working on an assumption of homogeneity (Anderson and May, 1991). It is also important under these conditions that vaccination be focused on the dogs most at risk of infection, a strategy which, in some areas, may necessitate the use of ancillary techniques such as oral vaccination or the promotion of good dog 'husbandry' practices to improve confinement. Identification of high-risk groups may allow for the specific targeting of vaccination effort (Woolhouse et al., 1997b), which could potentially improve the cost-effectiveness of dog vaccination as a rabies control method even further. Until such time as such studies are conducted, however, the core of any rabies control campaign should remain the parenteral vaccination of dogs with the objective of achieving a 70% or higher coverage.

7 ORAL VACCINATION OF DOGS

Since the successful eradication of sylvatic rabies in western Europe through the use of oral rabies vaccines (Müller, 2000), attention has turned to the application of this technology to domestic dog populations, particularly those characterized by a high proportion of dogs inaccessible to traditional mass parenteral vaccination methods (WHO, 1998). Several candidate vaccines are available, broadly categorized as either modified live or live recombinant vaccines (see Chapter 14). These can then be used in combination with a bait matrix formulated to appeal to domestic dogs and delivered to the target population using several possible strategies. Baits generally consist either of commercially produced artificial polymers with an appropriate aroma and flavour, or of locally available, 'home-made' substances such as chicken heads or minced meat. The appropriate bait for a given setting will depend on a number of factors, including its acceptability to the target dog population, the number of baits required and the chosen delivery method. For example, researchers in Sri Lanka found that commercially available baits were too hard and lost their aroma after a long period of storage. These problems were overcome by flavouring the bait with canned fish and leaving it at room temperature for a few hours before delivery (Harischandra, 2001).

Bait delivery methods themselves will vary according to the objectives of the oral vaccination programme and the ecology of the local dog population. Strategies that can be considered include the distribution of baits to dog owners at a central point, handout of baits to dogs encountered in the street by vaccination personnel, presentation of baits to owned dogs through household visits, or placement of baits at locations known to be visited by feral or poorly restricted dogs. By delivering baits to dog owners at a central point in Tunisia, Ben Youssef et al. (1998) showed that 85–90% of owned dogs could potentially be vaccinated in such a way. In Sri Lanka, delivery of oral vaccines to unvaccinated dogs during house-to-house visits following a parenteral vaccination campaign increased vaccination coverage from 45 to 76% (Harischandra, 2001), while in South Africa, 68% of owned dogs at least partially consumed a bait after house-to-house delivery by the vaccination team (Bishop, 2001). In Turkey, 30% of a sample of free-roaming dogs was found to have ingested baits placed out overnight in selected sites in an urban area of Istanbul (WHO, 1998).

Oral vaccination, either alone or more likely in combination with parenteral vaccination, may thus allow for improvement in dog vaccination coverage or for the targeting of high-risk, inaccessible segments of the population; however, further field studies to evaluate economy, efficiency and effectiveness and to demonstrate safety are required before this method can be recommended for broader application. A continuing obstacle to the widespread deployment of oral rabies vaccines for dogs is human health concerns. Although the danger posed to human health by either recombinant or modified live rabies vaccines is extremely slight (particularly when weighed against the threat of canine rabies in endemic areas), it is not negligible, given the extremely close association of dogs with human populations. In the developing world at least, the human population may contain a high number of immuno suppressed individuals. The potential advantages of oral rabies vaccines need to be weighed against the attendant risks by national authorities contemplating their use. It must be said, however, that in specific circumstances, such as in rabies-endemic areas with high dog densities and large numbers of feral or poorly supervised animals, the benefits are likely to be substantial if appropriate strategies are used. Further research will hopefully allow these benefits to be realized safely and effectively.

8 AGE AT FIRST VACCINATION

Current vaccination practices exclude an important and easily accessible segment of the dog population from immunization – puppies below the age of 3 months. This stems from vaccine manufacturers' recommendations that for dogs born of immunized bitches the minimum age of vaccination should be 3 months, so that vaccine-induced active immunity is not affected by the presence of maternally derived antibodies. Conflicting evidence for this position is to be found in the published literature. One laboratory study (Précausta et al., 1985) indeed found that antibody production in puppies born to immunized bitches was inhibited by the persistence of maternal antibodies when the puppies were administered an inactivated rabies vaccine at one month. A further study, however, showed that, although vaccination with an inactivated vaccine of 14-day old puppies with high levels of maternal antibodies did not raise those animals' antibody titres, all puppies survived when challenged 6 months later with a dose of field rabies virus that killed all unvaccinated controls (Chappuis, 1998). A possible explanation could be that maternal antibodies do not appear to inhibit specific T cell responses, even while completely impeding antibody production (Siegrist et al., 1998). T cells, thus primed, may offer protection through accelerated antibody response following subsequent exposure to antigen. In a field trial in Tunisia, Seghaier et al. (1999) found that puppies' serological response to rabies vaccination was similar, whether in the presence or absence of maternal antibodies. Taken together, these findings strongly suggest that vaccination of puppies as young as 2 weeks of age will confer protection against rabies, whatever the pre-existing immune status of the animal.

Given that few adult dogs and fewer puppies have detectable antibody titres in many settings in the developing world anyway and that young dogs are important sources of human rabies cases despite being easily accessible to vaccination (Mitmoonpitak *et al.*, 1998), vaccination of dogs younger than 3 months should no longer be precluded on the grounds of poor immune response. A valid consideration, and one whose effect may vary between populations, is the high mortality rate suffered by this segment of the population (up to 50% – Coleman, 1999), which would negatively impact the cost-effectiveness of any campaign that targeted it. Mitigating this is the 'pulse' nature of typical dog rabies vaccination campaigns, in which mass immunization of the host population is conducted over a short period of time at fixed (though often infrequent) intervals. Puppies considered too young for vaccination may be one or even two years old, and constitute up to a third of the population (Coleman, 1999), before a second opportunity for immunization arises. A cost-effectiveness model incorporating age-specific mortality rates will be useful in assessing the optimal strategy.

9 DOG RABIES CONTROL IN WILDLIFE CONSERVATION

Several approaches have been suggested as a means to minimize the threat of rabies to endangered wildlife (reviewed by Laurenson *et al.*, 1997b). The issue rose to prominence in the wake of a prolonged debate regarding the risk and benefits of directly vaccinating African wild dogs in the Serengeti-Mara ecosystem of Tanzania and Kenya against rabies in an attempt to prevent further outbreaks of disease. The debate focused on a hypothesis that the stress of rabies vaccination was responsible for causing mortality and, ultimately, the local extinction of the wild dog population, through reactivation of 'latent' rabies infection (Burrows *et al.*, 1994, 1995). Although this hypothesis has been widely refuted (e.g. Macdonald *et al.*, 1992; Creel *et al.*, 1997; Woodroffe, 2001), the debate had widespread repercussions, with wildlife managers in many parts of the world reluctant to permit vaccination of wildlife. As a result, vaccination of domestic dog reservoirs has been adopted as the main approach for protecting endangered wildlife against rabies, with large-scale programmes implemented around the Serengeti National Park (to protect African wild dogs) and the Bale Mountains National Park (to protect Ethiopian wolves).

Population viability analyses indicate that for both Ethiopian wolves (Haydon *et al.*, 2002) and African wild dogs (Vial *et al.*, 2006), it may be sufficient to vaccinate only a core of 20–40% of individuals, as the objective (from the conservation perspective) need not be the elimination of rabies, but rather the avoidance of the largest outbreaks that could result in extinction. Given the challenges of maintaining vaccination coverage in large and rapidly growing domestic dog reservoir populations, a policy of core vaccination of target species may prove to be a more feasible and cost-effective approach for conservation management. For example, despite enormous efforts in Ethiopia to vaccinate domestic dogs surrounding wolf habitat in the Bale Mountains National Park, the incursion of a single rabid dog triggered a major new epidemic in wolves in 2003 (Randall *et al.*, 2004). Nonetheless, the added public health and economic benefits of dog vaccination to communities may justify a dual approach.

10 ECONOMICS OF DOG VACCINATION FOR RABIES CONTROL

The primary objective for controlling canine rabies is the elimination of disease in the human population. The disease can be completely eliminated in humans through two major approaches:

1 greater availability of PEP

2 elimination of canine rabies by controlling the disease in the principal reservoir host.

Economic studies indicate that controlling the disease in the canine reservoir is the most cost-effective approach for preventing rabies in humans (Bögel and Meslin, 1990). Available country-level estimates on the macro-economic impact of rabies suggest that rabies impinges greatly on national economies and control of the disease should result in significant savings in national health budgets. In the Philippines, for example, it is estimated that rabies control would result in a net economic benefit of up to $2.5 million annually (Fishbein *et al.*, 1991). In Tanzania, a country which is reported to spend at least $400 000 of its health budget on PEP alone (Meslin, 1994), control of dog rabies could result in a net saving of up to $12 060 per 100 000 people per year (Kaare, 2007). In both cases, the saving represents a significant proportion of respective country health budgets.

Rabies also affects individual household economies. A large proportion of the rural population in Africa and Asia subsists below the poverty line, yet it is in these rural areas that rabies has its most profound economic impact. Knobel *et al.* (2005) predict that five times more deaths due to rabies occur in rural areas than in urban areas (it is ironic, in fact, that the term 'urban rabies' is so often used synonymously with canine rabies). The greatest problem at the household level is the expense associated with PEP. The human rabies vaccine is not only expensive relative to rural household incomes but is also rarely available in the vicinity of most rural populations who, more often than not, are compelled to travel long distances to obtain PEP, paying for both travel and boarding costs. In Tanzania, studies indicate that only 33% of households with dog-bite victims were able to meet the costs of PEP from their own family savings (Kaare, 2007). The remaining proportion had to raise funds from other means including borrowing money, selling household properties and livestock and mortgaging land. The same experience has been noted in India, where a labourer was compelled to sell one third of his land to raise funds for his child's PEP (Dutta, 1996), although the child eventually died. Household economies are also affected through livestock losses. Household losses up to a mean of $7.5 per year have been reported in Ethiopia (Laurenson *et al.,* 1997a), a considerable loss in a country where many households earn less than $1 a day. Although studies on livestock losses due to rabies are few, anecdotal evidence from some parts in north-western Tanzania suggests that the problem is likely to be greater that originally thought.

The economic benefits of ensuring the survival of endangered wildlife populations through dog rabies control remain to be quantified. However, wild dogs are reported to be one of the major attractions for tourists visiting South African national parks (Lindsey *et al.* 2005) and, given the importance of wildlife tourism in the economies of many countries in sub-Saharan Africa, these impacts should not be ignored.

11 DOG POPULATION MANAGEMENT

As mentioned previously, there is no evidence that removal of dogs alone has ever had a significant impact on dog population densities or the spread of rabies. However, dog population management through movement restriction, habitat control, reproductive control and, in certain select circumstances, humane killing of dogs, may be used to supplement mass dog vaccination campaigns in national rabies control programmes. This could be achieved through increased confinement of owned dogs to premises by either kennelling or collaring, securing potential food resources such as waste dumps, slaughter slabs or latrines against access by feral dogs, or by reducing the need for dogs through improvement in husbandry methods to protect livestock against predators. The importance of resource availability in rabies epidemiology is illustrated by the reported rise in dog rabies cases in the wake of the dramatic declines in vulture populations across the Indian subcontinent (Prakash *et al.,* 2003). One of the most notable effects of the absence of vultures was the accumulation of carcasses that permitted a 20-fold increase in domestic dog numbers (Pain *et al.,* 2003).

The effectiveness of any population control measures will depend on a thorough understanding of local conditions pertaining to dog ecology and human socio-cultural attitudes towards dogs and dog-keeping. Reproductive sterilization of dogs by castration, ovariohysterectomy or delivery of hormonal contraceptives has been proposed as means of reducing population turnover and (in combination with vaccination) creating a stable, immunized dog population. Such ABC (animal birth control) programs have been launched in a number of countries with reportedly encouraging reductions in unsupervised dog numbers, human bite injuries and rabies cases (WHO, 2004; Reece and Chawla, 2006). However, independent evaluation of these projects in terms of impact and cost-effectiveness is needed before general recommendations on reproductive management as a rabies control method can be made.

12 CONCLUSION

In this chapter, we have shown that the tools for dog rabies control, particularly those targeting vaccination of domestic dogs, are well developed and effective where they have been deployed appropriately. Successful dog rabies control programs have had a substantial impact on human public health by reducing rabies-related mortality. The tools for canine rabies control are available. However, the neglected status of rabies and the lack of sufficient information on the public health burden of the disease result in its low prioritization against competing health interests; for example, in the Global Burden of Disease studies undertaken by the WHO and the World Bank (1993), rabies was not even mentioned, even though other disease problems such as dengue (Rigau-Pérez *et al.*, 1998), which have a lower global disease burden, were well documented (Coleman *et al.*, 2004). At national levels, there is a lack of motivation, commitment and particularly, resources, to gather data to promote a rabies control agenda. If rabies were to be better prioritized, efficient delivery systems, public education campaigns and resources to apply these technologies would soon follow. It is encouraging that, in an existing case study, the demand for PEP dropped significantly when rabies was well controlled in the dog population (Cleaveland *et al.*, 2003).

Canine rabies control must, however, be locally adapted and culturally sensitive. The relationship between humans and dogs and the roles the animals play differs between societies and these differences will drastically impact on the acceptability of one or other control activity – e.g. dog accessibility and the choice of vaccination strategy. Similarly, dog ecology differs regionally; in some areas, dogs may roam over great distances, while in others, individual dogs may be more restricted with limited contact with other dogs. Rabies transmission between dogs and from dogs to humans is affected by these variations. A fuller understanding of dog ecology in different settings is thus required to optimize strategies targeting rabies transmission in canines.

Zoonotic disease control requires strengthening of partnerships in both the medical and veterinary fields, both in the public and private sectors (King *et al.*, 2004). The nature of this partnership might simply be in the exchange of information (occurrence of outbreaks, results of laboratory testing) or more formally arranged, such as in jointly managed – and funded – control projects. This would be the application of the 'One Medicine' concept (Schwabe, 1984), in which human and animal diseases are considered under a single paradigm. Rabies control focusing on the canine reservoir is an ideal example of where such veterinary and medical collaborations may have great benefits. Institutions in affected countries must widen their perspectives and see beyond their specialized budgets;

if both veterinary and medical budgets include rabies control, the winner in the end will be society as a whole and the overall economic and disease burden in both humans and animals will be diminished.

ACKNOWLEDGEMENTS

DK and SC were supported in this work by the Wellcome Trust, EF by the Department for International Development Animal Health Programme and MK by the National Science Foundation under Grant No. 0225453. Any opinions, findings and conclusions or recommendations expressed in this material are those of the authors and do not necessarily reflect the views of the National Science Foundation. Support for work on re-evaluating the burden of rabies in Africa and Asia was provided by the World Health Organization.

REFERENCES

Anderson, R.M. and May, R.M. (1982). Directly transmitted infectious diseases: control by vaccination. *Science* **215**, 1053–1060.

Anderson, R.M. and May, R.M. (1991). *Infectious Diseases of Humans*. Oxford: Oxford Scientific Press.

Arvo, S.K. and Kaumba, A. (1971). *Rabies Control in Uganda*. Ministry of Animal Industry, Game and Fisheries, Kampala, Uganda.

Belotto, A., Leanes, L.F., Schneider, M.C., Tamayo, H. and Correa, E. (2005). Overview of rabies in the Americas. *Virus Research* **111**, 5–12.

Ben Youssef, S., Matter, H.C., Schumacher, C.L. *et al.* (1998). Field evaluation of a dog owner, participation-based, bait delivery system for the oral immunization of dogs against rabies in Tunisia. *American Journal of Tropical Medicine and Hygiene* **58**, 835–845.

Bingham, J. (2001). Rabies on Flores Island, Indonesia: is eradication possible in the near future? In: *Proceedings of the Fourth International Symposium on Rabies Control in Asia, Hanoi, 5–9 March 2001*. Montrouge: John Libbey Eurotext, pp. 148–155.

Bishop, G.C. (1995). Canine rabies in South Africa. In: *Proceedings of the Third International Conference of the Southern and Eastern African Rabies Group, Harare, March 1995*. pp. 104–111. Harare: Veterinary Research Group, pp. 104–111.

Bishop, G.C. (2001). Increasing dog vaccination coverage in South Africa: is oral vaccination the answer? In: *Proceedings of the Fourth International Symposium on Rabies Control in Asia, Hanoi, 5–9 March 2001*. Montrouge: John Libbey Eurotext, pp. 105–109.

Bögel, K. (2002). Control of dog rabies. In: *Rabies* (A. Jackson and W.H. Wunner, eds). pp. 429–443. San Diego: Elsevier Science.

Bögel, K. and Joshi, D.D. (1990). Accessibility of dog populations for rabies control in Kathmandu valley, Nepal. *Bulletin of the World Health Organisation* **68**, 611–617.

Bögel, K. and Meslin, F.-X. (1990). Economics of human and canine rabies elimination – guidelines for program orientation. *Bulletin of the World Health Organization* **68**, 281–291.

Brooks, R. (1990). Survey of the dog population of Zimbabwe and its level of rabies vaccination. *Veterinary Record* **127**, 592–593.

Burrows, R., Hofer, H. and East, M.L. (1994). Demography, extinction and intervention in a small population – the case of the Serengeti wild dogs. *Proceedings of the Royal Society of London Series B* **256**, 281–292.

Burrows, R., Hofer, H. and East, M.L. (1995). Population dynamics, intervention and survival in African wild dogs (*Lycaon pictus*). *Proceedings of the Royal Society of London Series B* **262**, 235–245.

Chappuis, G. (1998). Neonatal immunity and immunisation in early age: lessons from veterinary medicine. *Vaccine* **16**, 1468–1472.

Cleaveland, S. (1996). *The epidemiology of rabies and canine distemper in the Serengeti, Tanzania*. PhD thesis, University of London.

Cleaveland, S. (1998). Epidemiology and control of rabies. The growing problem of rabies in Africa. *Transactions of the Royal Society of Tropical Medicine and Hygiene* **92**, 131–134.

Cleaveland, S. and Dye, C. (1995). Maintenance of a microparasite infecting several host species: rabies in the Serengeti. *Parasitology* **111**, S37–47.

Cleaveland, S., Fèvre, E., Kaare, M. and Coleman, P.G. (2002). Estimating human rabies mortality in Tanzania from dog bite injuries. *Bulletin of the World Health Organization* **80**, 304–310.

Cleaveland, S., Kaare, M., Tiringa, P., Mlengeya, T. and Barrat, J. (2003). A dog rabies vaccination campaign in rural Africa: impact on the incidence of dog rabies and human dog-bite injuries. *Vaccine* **21**, 1965–1973.

Coleman, P.G. (1999). *The epidemiology and control of domestic dog rabies*. PhD thesis, University of London.

Coleman, P.G. and Dye, C. (1996). Immunization coverage required to prevent outbreaks of dog rabies. *Vaccine* **14**, 185–186.

Coleman, P.G., Fèvre, E.M. and Cleaveland, S. (2004). Estimating the public health burden of rabies. *Emerging Infectious Diseases* **10**, 140–142.

Creel, S., Creel, N.M. and Monfort, S.L. (1997). Radiocollaring and stress hormones in African wild dogs. *Conservation Biology* **11**, 544–548.

Dutta, J.K. (1996). Rabies prevention: cost to an Indian laborer. *Journal of the American Medical Association* **276**, 32.

Eng, T.R., Fishbein, D.B., Talamante, H.E. *et al.* (1993). Urban epizootic of rabies in Mexico: epidemiology and impact of animal bite injuries. *Bulletin of the World Health Organization* **71**, 615–624.

Estrada, R., Vos, A., De Leon, R. and Mueller, T. (2001). Field trials with oral vaccination of dogs against rabies in the Philippines. *BMC Infectious Diseases* **1**, 23.

Fèvre, E., Kaboyo R.W., Persson, V., Edelsten, M., Coleman, P. and Cleaveland, S. (2005). The epidemiology of animal bite injuries in Uganda and projections of the burden of rabies. *Tropical Medicine and International Health* **10**, 790–798.

Fishbein, D.B., Miranda, N.J., Merrill, P. *et al.* (1991). Rabies control in the Republic of the Philippines: benefits and costs of elimination. *Vaccine* **9**, 581–587.

Fooks, A.R., Roberts, D.H., Lynch, M., Hersteinsson, P. and Runolfsson, H. (2004). Rabies in the United Kingdom, Ireland and Iceland. In: *Historical perspective of rabies in Europe and the Mediterranean Basin* (A.A. King, A.R. Fooks, M. Aubert and A.I. Wandeler, eds). pp. 25–32. Paris: OIE, pp. 25–32.

Funk, S.M., Fiorella, C.V., Cleaveland, S. and Gompper, M.E. (2001). The role of disease in carnivore conservation. In: *Carnivore Conservation* (J.L. Gittleman, S.M. Funk, D. Macdonald and R.K. Wayne, eds). pp. 443–466. Cambridge: Cambridge University Press, pp. 443–466.

Gascoyne, S.C., Laurenson, M.K., Lelo, S. and Borner, M. (1993). Rabies in African wild dogs (*Lycaon pictus*) in the Serengeti region, Tanzania. *Journal of Wildlife Diseases* **29**, 396–402.

Ginsberg, J.G. and Woodroffe, R. (1997). Extinction risks faced by remaining wild dog populations. In: *The African wild dog: status survey and conservation action plan* (R. Woodroffe, J.R. Ginsberg, and D.W. Macdonald, eds). Gland: IUCN, pp. 75–87.

Harischandra, P.A.L. (2001). Dog vaccination coverage and oral rabies vaccination in Sri Lanka, an update. In: *Proceedings of the Fourth International Symposium on Rabies Control in Asia, Hanoi, 5–9 March 2001*. Montrouge: John Libbey Eurotext, pp. 97–100.

Haydon, D.T., Laurenson, M.K. and Sillero-Zubiri, C. (2002). Integrating epidemiology into population viability analysis: managing the risk posed by rabies and canine distemper to the Ethiopian wolf. *Conservation Biology* **16**, 1372–1385.

Hung Cheuk (1969). A review of the history and control of rabies in Hong Kong. *Agricultural Science, Hong Kong* **1**, 141–147.

Kaare, M.T. (2007). Rabies control in Tanzania: optimising the design and implementation of domestic dog mass vaccination programmes. PhD Thesis, University of Edinburgh, UK.

Kamoltham, T., Singhsa, J., Promsaranee, U., Sonthon, P., Mathean, P. and Thinyounyong, W. (2003). Elimination of human rabies in a canine endemic province in Thailand: five-year programme. *Bulletin of the World Health Organization* **81**, 375–381.

Kat, P.W., Alexander, K.A., Smith, J.S. and Munson, L. (1995). Rabies and African wild dogs in Kenya. *Proceedings of the Royal Society of London Series B* **262**, 229–233.

Kayali, U., Mindekem, R., Yémadji, N. *et al.* (2003). Coverage of pilot parenteral vaccination campaign against canine rabies in N'Djaména, Chad. *Bulletin of the World Health Organization* **81**, 739–743.

Kennedy, L.J., Barnes, A., Happ, G.M. *et al.* (2002). Extensive interbreed, but minimal intrabreed, variation of DLA class II alleles and haplotypes in dogs. *Tissue Antigens* **59**, 194–204.

King, L.J., Marano, N. and Hughes, J.M. (2004). New partnerships between animal health services and public health agencies. *Revue Scientifique et Technique de l'Office International des Epizooties* **23**, 717–725.

Kitala, P.M. and McDermott, J.J. (1995). Population dynamics of dogs in Machakos District, Kenya: implications for vaccination strategy. In: *Proceedings of the Third International Conference of the Southern and Eastern African Rabies Group, Harare, March 1995* (J. Bingham, G.C. Bishop and A. King, eds). Harare: SEARG, pp. 95–103.

Kitala, P.M., McDermott, J.J., Kyule, M.N. and Gathuma, J.M. (2000). Community-based active surveillance for rabies in Machakos District, Kenya. *Preventive Veterinary Medicine* **44**, 73–85.

Kitala, P.M., McDermott. J.J., Coleman, P.G. and Dye, C. (2002) Comparison of vaccination strategies for the control of dog rabies in Machakos District, Kenya. *Epidemiology and Infection* **129**, 215–222.

Knobel, D.L., Cleaveland, S., Coleman, P.G. *et al.* (2005). Re-evaluating the burden of rabies in Africa and Asia. *Bulletin of the World Health Organization* **83**, 360–368.

Laurenson K., Shiferaw, F. and Sillero-Zubiri C. (1997a). Rabies as a threat to the Ethiopian wolf (*Canis simensis*). In: *Proceedings of the Southern and Eastern African Rabies Group (SEARG) Meeting, Nairobi, 4–6 March 1997* (P. Kitala, B. Perry, J. Barrat and A. King, eds). Lyon: Foundation Merieux, pp. 94–100.

Laurenson, M.K, Shiferaw, F. and Sillero-Zubiri, D. (1997b). Disease, domestic dogs and the Ethiopian wolf: the current situation. In: *The Ethiopian wolf. Status survey and conservation action plan* (C. Sillero-Zubiri and D.W. Macdonald, eds). Gland: IUCN, pp. 32–40.

Lindsey, P.A., Alexander, R., Du Toit, J.T. and Mills, M.G.L. (2005) The cost efficiency of wild dog conservation in South Africa, *Conservation Biology* **19**, 1205.

Macdonald, D.W., Artois, M., Aubert, M. *et al.* (1992). Cause of wild dog deaths. *Nature* **360**, 633–634.

Manninger, R. (1968). Rabies in Hungary during the past forty years. *Magyar Allatorvosok Lapja* **23**, 5–13.

McLean, A.R. and Blower, S.M. (1993). Imperfect vaccines and herd immunity to HIV. *Proceedings of the Royal Society of London Series B* **253**, 9–13.

Meltzer, M.I. and Rupprecht, C.E. (1998). A review of the economics of the prevention and control of rabies. Part 1: Global impact and rabies in humans. *Pharmacoeconomics* **14**, 365–383.

Meslin, F.-X. (1994). The economic burden of rabies in developing countries: costs and benefits from dog rabies elimination. *Kenyan Veterinarian*, **18**, 523–527.

Mitmoonpitak, C., Tepsumethanon, V. and Wilde, H. (1998). Rabies in Thailand. *Epidemiology and Infection* **120**, 165–169.

Muir, P. and Roome, A. (2005). Indigenous rabies in the U.K. *Lancet* **365**, 2175.

Müller, W.W. (2000). Review of rabies case data in Europe to the WHO Collaborating Centre Tübingen from 1977 to 2000. *Rabies Bulletin Europe* **4**, 11–19.

Neville, J. (2004). Rabies in the ancient world. In: *Historical perspective of rabies in Europe and the Mediterranean Basin* (A.A. King, A.R. Fooks, M. Aubert and A.I. Wandeler, eds). Paris: OIE, pp. 1–14.

Pain, D.J., Cunningham, A.A., Donald, P.F. *et al.* (2003). Causes and effects of tempero-spatial declines of *Gyps* vultures in Asia. *Conservation Biology* **17**, 661–671.

Pancharoen, C., Thisyakorn, U., Lawtongkum, W. and Wilde, H. (2001). Rabies exposures in Thai children. *Wilderness and Environmental Medicine* **12**, 239–243.

Parviz, S., Luby, S. and Wilde, H. (1998). Postexposure treatment of rabies in Pakistan. *Clinical Infectious Diseases* **27**, 751–756.

Perry, B.D. (1995). Rabies control in the developing world: can further research help? *Veterinary Record* **137**, 527–522.

Perry, B.D., Kyendo, T.M., Mbugua, S.W., Price, J.E. and Varma, S. (1995). Increasing rabies vaccination coverage in urban dog populations of high human population density suburbs: a case study in Nairobi, Kenya. *Preventive Veterinary Medicine* **22**, 137–142.

Prakash, V., Pain, D.J., Cunningham, A.A. *et al.* (2003). Catastrophic collapse of Indian white-backed *Gyps bengalensis* and long-billed *Gyps indicus* vulture populations. *Biological Conservation* **109**, 381–390.

Précausta, P., Soulebot, J.P., Chappuis, G., Brun, A., Bugand, M. and Petermann, H.G. (1985). Nil cell inactivated tissue culture vaccine against rabies: immunization of carnivores. In: *Rabies in the Tropics* (E. Kuwert, C. Mérieux, H. Koprowski and K. Bögel, eds). Berlin: Springer Verlag, pp. 227–240.

Randall, D.A., Williams, S.D., Kuzmin, I.V. *et al.* (2004). Rabies in endangered Ethiopian wolves. *Emerging Infectious Diseases* **10**, 2214–2217.

Reece, J.F. and Chawla, S.K. (2006) Control of rabies in Jaipur India, by the sterilisation and vaccination of neighbourhood dogs. *Veterinary Record* **159**, 379–383.

Rigau-Pérez, J.G., Clark, G.G., Gubler, D.J., Reiter, P., Sanders, E.J. and Vorndam, A.V. (1998). Dengue and dengue haemorrhagic fever. *Lancet* **352**, 971–977.

Rupprecht, C.E., Hanlon, C.A. and Hemachudha, T. (2002). Rabies re-examined. *Lancet Infectious Diseases* **2**, 327–343.

Schwabe, C.W. (1984). *Veterinary Medicine and Human Health*, 3rd edn. Baltimore: Williams and Wilkins.

Seghaier, C., Cliquet, F., Hammami, S., Aouina, T., Tlatli, A. and Aubert, M. (1999). Rabies mass vaccination campaigns in Tunisia: are vaccinated dogs correctly immunized? *American Journal of Tropical Medicine and Hygiene* **61**, 879–884.

Shone, D.K. (1962). Rabies in Southern Rhodesia 1900–1961. *Journal of the South African Veterinary Medical Association* **33**, 567–571.

Siegrist, C.-A., Córdova, M., Brandt, C. *et al.* (1998). Determinants of infant responses to vaccines in presence of maternal antibodies. *Vaccine* **16**, 1409–1414.

Sillero-Zubiri, C., King, A.A. and Macdonald, D.W. (1996). Rabies and mortality in Ethiopian wolves (*Canis simensis*). *Journal of Wildlife Diseases* **32**, 80–86.

Umeno, S. and Doi, Y. (1921). A study in the antirabic inoculation of dogs. *Kitasato Archives of Experimental Medicine* **4**, 89.

Vial, F., Cleaveland, S., Rasmussen, G. and Haydon, D.T. (2006). Development of vaccination strategies for the management of rabies in African wild dogs. *Biological Conservation* **131**, 180–192.

Vucetich, J.A. and Creel, S. (1999). Ecological interactions, social organization and extinction risk in African wild dogs. *Conservation Biology* **13**, 1172–1182.

Warrell, M.J. and Warrell, D.A. (2004). Rabies and other lyssavirus diseases. *Lancet* **363**, 959–969.

Wells, C.W. (1954). The control of rabies in Malaya through compulsory mass vaccination of dogs. *Bulletin of the World Health Organization* **10**, 731–742.

WHO (1966). World Survey of Rabies VI for 1964. Rabies/Inf./66.19 Corr 67.1 and Add 67.1. Geneva: WHO.

WHO (1988). Report of a WHO Consultation on Dog Ecology Studies Related to Rabies Control. WHO/Rabies Research/88.25. Genova: WHO.

WHO (1992a). WHO Expert Committee on Rabies, 8th report. *Technical Report Series no. 824.* Geneva: WHO.

WHO (1992b). Report of the 3rd Consultation on Oral Immunization of Dogs Against Rabies. WHO/Rabies Research/92.38. Geneva: WHO.

WHO (1998). *Field Application of Oral Rabies Vaccines for Dogs:* report of a WHO consultation. WHO/EMC/ZDI/98.15. Geneva: WHO.

WHO (1999). World survey of rabies No. 34 for the year 1998. WHO/CDS/CSR/APH/99.6. Geneva: WHO.

WHO (2004). WHO Expert Consultation on Rabies, 1st report. *Technical Report Series no. 931.* Geneva: WHO.

Windiyaningsih, C., Wilde, H., Meslin, F.-X., Suroso, T. and Widarso, H.S. (2004). The rabies epidemic on Flores island, Indonesia (1998–2003). *Journal of the Medical Association of Thailand* **87**, 1389–1393.

Woodroffe, R. (2001). Assessing the risks of intervention: immobilization, radio-collaring and vaccination of African wild dogs. *Oryx* **35**, 234–244.

Woodroffe, R., Cleaveland, S., Courtenay, O., Laurenson, M.K. and Artois, M. (2004). Infectious disease in the management and conservation of wild canids. In: *The Biology and Conservation of Wild Canids* (D.W. Macdonald and C. Sillero-Zubiri, eds). Oxford: Oxford University Press, pp. 123–142.

Woolhouse, M.E.J., Haydon, D.T. and Bundy, D.A.P. (1997a). The design of veterinary vaccination programmes. *Veterinary Journal* **153**, 41–47.

Woolhouse, M.E.J., Dye, C., Etard, J.-F. *et al.* (1997b). Heterogeneities in the transmission of infectious agents: Implications for the design of control programs. *Proceedings of the National Academy of Sciences USA* **94**, 338–342.

World Bank (1993). *World development report 1993: Investing in health.* New York: Oxford University Press.

Zeynali, M., Fayaz, A. and Nadim, A. (1999). Animal bites and rabies: situation in Iran. *Archives of Iranian Medicine* 2. Available at http://www.ams.ac.ir/AIM/9923/contents9923.html.

18

Rabies Control in Wild Carnivores

**RICHARD C. ROSATTE[1], ROWLAND R. TINLINE[2], AND
DAVID H. JOHNSTON[3]**

Ontario Ministry of Natural Resources, Rabies Research and Development Unit
Trent University, DNA Complex, Peterborough, Ontario K9J 7B8, Canada[1];
Queen's GIS Laboratory, Queen's University, Kingston, Ontario K7L 3N6, Canada[2];
Johnston Biotech, Sarnia, Ontario N7V 3B5, Canada[3]

1 INTRODUCTION

Wildlife rabies control is a complex business where the outcome is impacted
by many things including the animal habitat, biology and species of the disease-
carrying vector. Add to this the rabies virus variant transmitted by the vector,
public attitudes toward control tactics, the effectiveness of the wildlife rabies control strategy and human practices such as wildlife relocation, and the complexity
increases enormously (Rosatte and MacInnes, 1989; Hanlon *et al.*, 1999). Historically, wildlife rabies control has evolved from strictly a culling operation to a more
comprehensive approach involving the use of several tactics such as point infection
control and oral immunization with vaccine baits (Rosatte *et al.*, 2001; Johnston
and Tinline, 2002). By far, the most successful rabies control programs worldwide
are those that have taken a multiagency approach (Brochier *et al.*, 1996, 2001).
In this chapter, we will examine the traditional and contemporary approaches to
wildlife rabies control.

2 HISTORICAL ASPECTS OF RABIES CONTROL IN WILDLIFE

2.1 North America

During the 1940s to the 1960s, in some areas of North America, trapping of foxes,
coyotes (*Canis latrans*) and/or poisoning of skunks were methods used in attempts
to control rabies (Ballantyne and O'Donoghue, 1954; Schnurrenberger *et al.*, 1964;
Rosatte *et al.*, 1986). Many of those operations were unsuccessful (Linhart, 1960)
and research was initiated in the late 1960s to develop an oral rabies vaccine for
foxes (Black and Lawson 1970; Baer *et al.*, 1971; Winkler *et al.*, 1975). In Ontario,
Canada, oral vaccination using baits containing the attenuated Evelyn-Rokitnicki-
Abelseth (ERA) strain of rabies virus was later proved to be effective for the control

Rabies, second edition. Edited by Alan C. Jackson and William H. Wunner
ISBN 978-012-369366-2. Copyright Elsevier Inc. 2007

of rabies in foxes in rural and urban habitats (Rosatte *et al.*, 1992, 1993; MacInnes *et al.*, 2001). Following this, the vaccinia-rabies glycoprotein (V-RG) recombinant vaccine in baits was shown to be successful in Texas for controlling rabies in coyotes and gray foxes (*Urocyon cinereoargenteus*) (Farry *et al.*, 1998; Fearneyhough *et al.*, 1998; Steelman *et al.*, 1998; Krebs *et al.*, 2003; Sidwa *et al.*, 2005).

Currently, raccoons (*Procyon lotor*) and skunks are the primary terrestrial reservoirs of rabies in the USA. Baits containing the V-RG vaccine have been used in several eastern states from Maine to Florida to control the disease in raccoons (Rupprecht *et al.*, 1988; Roscoe *et al.*, 1998; Olson *et al.*, 2000; Russell *et al.*, 2005). Unfortunately, V-RG has not been as effective in skunks, using similar doses as used for raccoons (Briggs *et al.*, 1988; Hanlon *et al.*, 2002). However, the technique of trap-vaccinate-release (TVR) or vaccination by injection with conventional animal vaccine(s) has been used successfully to control rabies in skunks in urban and rural areas of Ontario, Canada (Rosatte *et al.* 1990, 1992, 1993, 2001). TVR was also used successfully to control an outbreak of rabies in skunks in Arizona (Engeman *et al.*, 2003). Street Alabama Dufferin (SAD) B19, a modified live rabies virus vaccine has also been tested orally in skunks (Vos *et al.*, 2002) and the attenuated SAG-2 rabies virus vaccine (a variant of the SAD vaccine strain) also proved to be effective when given orally to skunks as well as raccoons and Arctic foxes, showing promise as a candidate vaccine for wildlife rabies control in the USA (Hanlon *et al.*, 2002; Follmann *et al.*, 2002, 2004).

The Point Infection Control (PIC) strategy (see Section 8), utilizing population reduction in combination with TVR and oral rabies vaccination (ORV) with V-RG baits has proven effective for the containment and control of raccoon rabies in Ontario (Rosatte *et al.*, 2001). For 2006, Ontario received approval from the Canadian Food Inspection Agency to distribute baits containing a human adenovirus type 5 rabies recombinant oral rabies vaccine (AdRG1.3) (Lutze-Wallace *et al.*, 1995; Yarosh *et al.*, 1996) for the control of rabies in skunks in Ontario (Rosatte *et al.*, unpublished).

2.2 Europe

Historically, thousands of foxes were culled in an effort to control rabies throughout Europe (Matouch and Polak, 1982). Fox depopulation generally only resulted in a transient lull in the prevalence of rabies (Aubert, 1999). Because of this, parenteral vaccination programs targeting foxes were initiated in Switzerland and Germany (Aubert *et al.*, 1994). However, this method was viewed as impractical since too few foxes were captured in a limited area (Aubert *et al.*, 1994). Greater advances were made in 1977, when the first field trial, using baits containing oral rabies vaccine, was initiated in Switzerland (Wandeler *et al.* 1982; Wandeler, 1991). In 1978, Steck *et al.* (1982a, 1982b) attempted to stop an advancing fox rabies epizootic in the Rhone Valley area of Switzerland using chicken-head baits containing

SAD-Berne rabies vaccine. By 1985, the disease was under control in Switzerland (Steck *et al.*, 1982a,b; Kappeler *et al.*, 1988; Aubert *et al.*, 1994).

Later, field trials with baits containing ORV for the control of rabies in foxes were initiated in West Germany in 1983 (Wachendorfer *et al.*, 1986; Schneider and Cox, 1988). By 1987, complete elimination of fox rabies was achieved over large areas of West Germany (Schneider *et al.*, 1988; Wilhelm and Schneider, 1990). In the mid to late 1980s and early 1990s, fox rabies control programs were initiated using oral baits containing SAD-B19 vaccine in Austria, Belgium, Czechoslovakia, East Germany, France, Hungary, Italy, Luxembourg, the Netherlands and Slovenia (Irsara *et al.*, 1990; Gerletti *et al.*, 1991; Curk and Carpenter, 1994; Pastoret *et al.*, 1997, 2004; Brochier *et al.*, 2001). Baits containing SAD-B19 vaccine were also utilized beginning in 1988 in Finland to control rabies in raccoon dogs (*Nyctereutes procyonoides*) (Westerling, 1991; Nyberg *et al.*, 1992; Pastoret *et al.*, 2004).

The SAD-B19 and SAD-Berne vaccines, however, were found to be pathogenic for a variety of rodents and other species and research was initiated to develop more attenuated and safer vaccines (Wandeler, 1988). The V-RG recombinant vaccine that was shown to be effective in foxes (Kieny *et al.*, 1984; Wiktor *et al.*, 1984) was successfully used to control rabies in foxes in Belgium and France beginning in 1989 (Pastoret and Brochier, 1999; Brochier *et al.*, 2001). In 1991, the SAD-Berne vaccine was replaced by an SAG-1 vaccine for use in controlling rabies in foxes in Switzerland (Aubert *et al.*, 1994). SAG-1 was also used in France beginning in 1990 (Coulon *et al.*, 1992; Aubert *et al.*, 1994; Artois *et al.*, 1997). A SAD mutant (SAG$_2$) has also been developed for rabies control in Europe (Lambot *et al.*, 2001).

Oral rabies vaccination of foxes and raccoon dogs has reduced rabies cases throughout Europe (Pastoret and Brochier, 1999; Zanoni *et al*, 2000; Brochier *et al.*, 2001; Artois, 2003; Matouch and Vitasek, 2005). However, re-infections of fox rabies have occurred in France, Belgium and Germany due to a lack of geographic isolation, cross-country contamination, increasing fox populations and budgetary restrictions (Aubert *et al.*, 1994; Brochier *et al.*, 2001; Chautan *et al.*, 2000). Oral rabies vaccination campaigns were reinitiated in those countries during 2005 to bring the disease under control. Rabies is currently on the increase in other European countries such as Lithuania, Latvia, Estonia, Poland and Belarus, primarily due to an increase in fox populations (Zienius *et al.*, 2003). As a result, rabies control strategies have had to be re-designed to achieve more effective results (Aubert, 1999; Artois, 2003).

2.3 Asia

Canine rabies (see Chapter 16) is endemic in most of south and south-east Asia and wildlife rabies plays a very minor role (Kureishi *et al.*, 1992; Wilde *et al.*, 2005). The primary vectors of rabies in areas of northern Asia, such as the former Soviet

Union, are Arctic foxes (*Alopex lagopus*), red foxes (*Vulpes vulpes*) and raccoon dogs (Kuzmin, 1999; Kuzmin *et al.*, 2004). Candidate vaccines under consideration for arctic regions include an SAG$_2$ lyophilized product (Kovalev *et al.*, 1992; Follmann *et al.*, 2004). However, intensive control programs are generally lacking in those northern areas.

Since the 1970s, sylvatic rabies, primarily in foxes and jackals (*Canis* spp.), accounted for nearly 50% of the reported rabies cases in Israel with foxes being the primary reservoir (Yakobson *et al.*, 1998). Linhart *et al.* (1997a) evaluated the use of vaccine-laden baits for controlling rabies in red foxes and golden jackals (*Canis aureus*) in Israel. The results indicated that ORV with baits for the control of rabies in the Middle East was a feasible option. Yakobson *et al.* (1998) recommended the implementation of an oral rabies vaccination program (ORVP) for wildlife, which included extension of the program to include Israel, Egypt, Jordan and Palestine (Yakobson *et al.*, 1998; Devers *et al.*, 2001). Turkey has experienced a recent increase in fox rabies cases (Johnson *et al.*, 2003). In view of this, an oral immunization proposal for the control of fox rabies in Turkey was recently designed and submitted for approval (Office of International Epizootics, 2003).

2.4 Africa

Dog rabies is prevalent throughout much of Africa (see Chapter 16), however, wildlife rabies is becoming increasingly important due to the threat of endangered species extinction and the emergence of new wildlife hosts (Cleaveland, 1998). Sylvatic rabies has been reported in a number of African countries in African wild cats (*Felis libyca*) in Botswana, in jackals (*Canis adustus*) and hyena (*Hyaena brunnea*) in Zimbabwe, and in yellow mongoose (*Cynictus penicillata*) and black-backed jackal (*Canis mesomelas*) in South Africa, to name a few (Keightley *et al.*, 1987; Bingham *et al.*, 1999a, 1999b). Rabies is a threat to the wild dog population in parts of Namibia and thousands of kudu have also recently succumbed to the disease (Laurenson *et al.*, 1997; Office of International Epizootics, 2003). Unfortunately, wildlife rabies control (as well as dog rabies control) is not a priority in Africa due to a lack of human and monetary resources (Cleaveland, 1998). Some research has been initiated with respect to oral rabies vaccination of species such as jackals in Zimbabwe (Bingham *et al.*, 1997; Perry *et al.*, 1988). In addition, endangered African wild dogs (*Lycaon pictus*) were administered inactivated rabies vaccine by dart or via parenteral vaccination during the early 1990s in the Serengeti region of Tanzania following an outbreak of rabies that threatened to eliminate the population (Gascoyne *et al.*, 1993). Recently, an oral vaccination system for free-ranging wild dogs has been developed (Knobel *et al.*, 2002; Knobel and Toit, 2003). In Ethiopia, an outbreak of rabies occurred in the highly endangered Ethiopian wolf (*Canis simensis*) during 2003. Control was attempted by live-capture and parenteral vaccination and the use of ORV is being considered (Office of International Epizootics, 2003). However, despite the diversity of wildlife species in Africa, some of which are capable of being rabies vectors, there is no coordinated effort to organize wildlife rabies control programs, either locally or nationally.

2.5 Central and South America including the West Indies

In Latin America, the domestic dog accounted for more than 80% of the reported cases during the 1990s (de Mattos *et al.*, 1999) (see Chapter 16). Rabies in vampire bats (see Chapter 6) is also prevalent in Latin America and rabies in mongooses is a public health concern on some Caribbean islands such as Cuba, Grenada and Puerto Rico (Diaz *et al.*, 1994; Belotto 2001; 2004). A field trial was initiated to evaluate baits for delivery of oral rabies vaccine to the Indian mongoose (*Herpestes auropunctatus*) on the island of Antigua, West Indies (Linhart *et al.*, 1993). In that study, baiting stations showed promise for vaccinating mongooses (Linhart *et al.*, 1993). Sylvatic rabies is maintained in certain geographic areas of Mexico (Diaz *et al.*, 1994; de Mattos *et al.*, 1999) and wildlife rabies control measures in Latin American countries such as Brazil and Mexico were recommended by Belotto (2001), including identification of the primary wildlife reservoirs of rabies and the implementation of oral rabies vaccination programs for wildlife (Belotto, 2001). However, the extent of rabies infection in wildlife species is not well documented (Almeida *et al.*, 2001; Sato *et al.*, 2004) and wildlife rabies control programs are still non-existent (other than vampire bat control – see Chapter 6).

3 THE CONCEPT OF CONTROLLING RABIES IN WILDLIFE

In many areas of the world, culling of wildlife was not a successful rabies control tactic and, in other situations, it was not an acceptable practice (Debbie, 1991). Therefore, researchers focused on developing alternate methods for rabies control. In 1966, this led the World Health Organization (1966) Expert Committee on Rabies to call for research into the possible rabies vaccination of free-ranging wild animals. Oral vaccination had long been the goal of pharmaceutical enterprise, but to carry this approach to wildlife seemed insurmountable. One of the first researchers, George Baer, tried to vaccinate foxes orally with the existing injectable rabies vaccines Flury LEP and HEP (see Chapter 13) and he showed that the concept of oral rabies vaccination was possible (Baer, 1988). During the 1980s, Rosatte *et al.* (1990) demonstrated that it was possible to immunize free-ranging skunks and raccoons against rabies in Ontario by injection of inactivated rabies vaccine in trapped animals.

Baits laced with poisons had long been used to kill predator animals. Would it be possible to incorporate a vaccine in place of poisons within baits? Many prototype baits were first field tested by ground and aerial distribution in small-scale field trials without an oral vaccine while those vaccines were being developed (Black and Lawson, 1970; Baer *et al.*, 1971; Johnston, 1975; Linhart *et al.*, 1997a, 1997b). Instead of using an actual vaccine, chemical fluorochrome biomarkers were put into the baits to test the feasibility of reaching the high percentage of animals necessary to provide herd immunity for a large, wild population of rabies vectors (Ellenton and Johnston, 1979; Wandeler *et al.*, 1982; Linhart *et al.*, 1997a, 1997b). This early work showed that the concept of mass baiting of a wild species, in this case the red fox, was feasible and would lead to the first release in the wild of an oral vaccine in Switzerland in 1977 (Steck *et al.*, 1982a, 1982b). These early beginnings provided a glimmer of hope that a vaccination system for wildlife could be developed, but also showed the need for the integration of a wide range of technical resources to implement large-scale ORVPs. The scientific and technical evolution that followed this impetus by the 1966 WHO Expert Committee on Rabies has been one of the broadest-based endeavors in modern medicine. Since then, over 30 specialities have contributed to ORV development (Figure 18.1).

4 INITIATION OF WILDLIFE RABIES CONTROL PROGRAMS

The initiation of wildlife rabies control programs has been stimulated by impending epizootics but also precipitated by unexpected happenings or events. In Ontario, it is interesting that the site of the first recorded human death from wildlife rabies in 1819 (Jackson, 1994) precipitated the Canadian wildlife rabies control program. In 1967, at the same location, 4-year-old Donna Featherstone died of rabies contracted from the bite of a stray cat. Her death led to public demand for the Province of Ontario to mount a major control program against fox rabies, which had been epizootic in the Province since 1954 (Johnston and Beauregard, 1969). That program continues to have strong provincial support and has been successful in nearly eliminating Arctic fox-strain rabies from the province (MacInnes *et al.*, 2001). A similar series of events involving two human deaths from a coyote-associated strain of rabies in Texas stimulated the initiation of the Texas ORVP (Finley, 1998).

The initiation of wildlife rabies control, as experience has shown, requires the careful integration of a complex series of events and disciplines. To be effective, industry, government and science must function together as a coherent system (McGuill *et al.*, 1997). Modification or omission of any component of the control system may alter the efficiency and outcome of the whole control program, sometimes deleteriously (Stöhr and Meslin, 1996). Timing is important with regard to both the initiation and continuation of a control program to achieve and maintain a rabies-free status in an area. In the past, rabies epizootics have flourished in the absence of timely control efforts or appropriate technology. We can only surmise that the northward spread of raccoon rabies in the ridge and valley

Rabies Control in Wild Carnivores

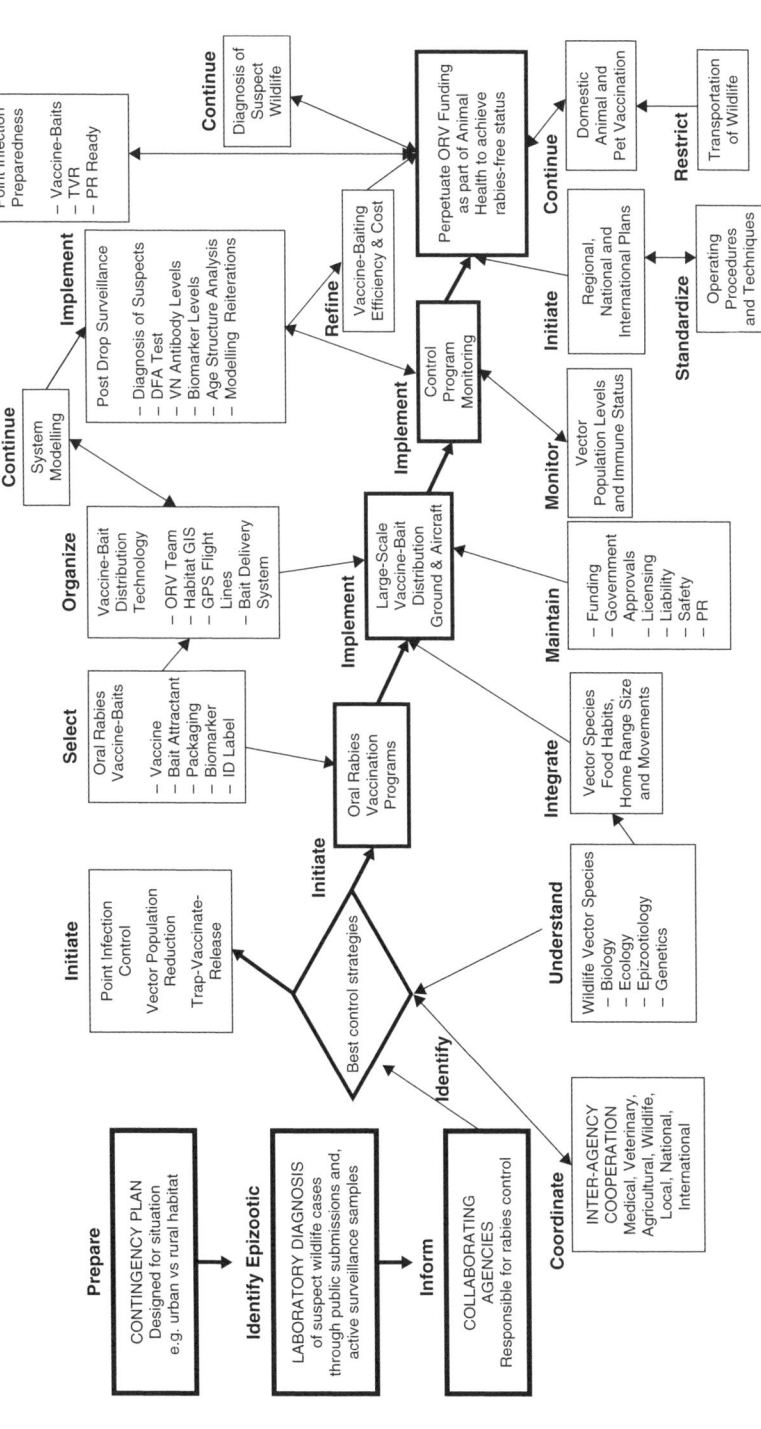

Figure 18.1 Flow chart for wildlife rabies control. TVR = Trap-vaccinate-release; PR = Population reduction; ORV = Oral rabies vaccine; VN = Virus neutralizing; DFA = Fluorescent antibody test.

topography of Pennsylvania in the late 1980s and subsequently into New York and New England could have been stopped with vaccine barriers in the valleys had newly developed vaccines and the oral baiting technology, tested and proven in Ontario in the early 1990s, been utilized earlier (Johnston *et al.*, 1988; Rupprecht *et al.*, 1986). Unfortunately, even when appropriate control technologies are at hand, lack of political will or financial resources have often wasted precious time, allowing epizootics to gain hold and advance into virgin animal populations. Indeed, even with support, budget restraints force trade-offs between investment in prevention weighed against the risk of invasion and subsequent higher costs. Programs that are interrupted due to budgetary or other reasons can exacerbate existing ecologic and epidemiologic conditions and prolong epizootics (Müller, 1997). Conversely, timely, efficient programs have proved effective in halting advancing epizootic fronts in Vermont. The key to being prepared for rabies epizootic control is a well prepared contingency plan that denotes control tactics for a given epizootic scenario and having rapid access to the resources to implement the plan quickly (Rosatte *et al.*, 1997; Whitney *et al.*, 2005).

5 DIAGNOSIS OF SUSPECT WILDLIFE

Diagnosis of rabies is the basis for the control of the disease in wildlife (Rupprecht *et al.*, 1995). Paramount to the initiation of any wildlife rabies control program is a delineation of the problem in wildlife. The diagnosis of suspect cases (see Chapter 10) and identification of the strains of rabies virus that are isolated (see Chapter 3), the species involved (see Chapter 5) and where in the local ecology and epizootiology these species fit in relation to humans, pets and domestic animals, are important to determine. A comprehensive and continuing diagnostic system that focuses on secondary as well as primary vector species surveillance is imperative if wildlife rabies is to be controlled and eventually eradicated. Furthermore, it is critical that a sufficient sample of the vector population be obtained to portray an appropriate picture of the epidemiological situation.

6 VECTOR SPECIES BIOLOGY IN RELATION TO RABIES
   EPIDEMIOLOGY

In order to manage wildlife rabies scientifically, a firm understanding of the behavioral ecology of wild mammals that are the vectors in rabies epizootics is critical (Childs *et al.*, 2000; Guerra *et al.*, 2003). Ecological studies of the primary and secondary vectors of rabies are required to understand how the vector ecology impacts the epizootiology of the disease (Voigt *et al.*, 1985; Artois and Aubert, 1991; Broadfoot *et al.*, 2001; Totton *et al.*, 2002, 2004; Rosatte and Lariviere, 2003; Rosatte *et al.*, 2006). To control rabies in wildlife, it is important to understand the relationship among the potential wildlife hosts, including vector metapopulation structure, the genetic composition of vector species, and human demographic and environmental features, particularly if oral vaccination is proposed as a control method (Ellenton and Basrur 1982; Wandeler *et al.*, 1994; Jones *et al.*, 2003).

7 TRANSPORTATION OF WILDLIFE

Globally, the translocation of native and exotic wildlife species is increasing annually for purposes ranging from restoration of extirpated species to endangered species restoration (Woodford and Rossiter, 1993). Often, species relocations occur without proper disease risk assessments being completed and it is becoming clear that species translocations by humans often are the origin of infected wildlife (Rosatte and MacInnes, 1989; Rupprecht *et al.*, 1995; Rosatte, 2001; Messenger *et al.*, 2002; Morner *et al.*, 2002). For example, raccoons are known to have arrived on the island of Newfoundland, where they are not native, via transport trucks ferried from Nova Scotia on the mainland (H. Whitney, personal communication). In New York, raccoons are a common occurrence at dumpster and trash transfer sites and have often been observed tumbling from newly arriving trash trucks (B. Laniewicz, personal communication). In addition, relocation of wildlife can result in extraordinary movements leading to the spread of infectious diseases over large areas (Rosatte and MacInnes, 1989; Roscoe *et al.*, 1998). Therefore, jurisdictions are encouraged to implement stronger legislation regulating the importation, distribution and relocation of wildlife (Centers for Disease Control and Prevention, 2001a). It is also imperative that a proper risk assessment be undertaken before wild animal translocations occur.

8 POINT INFECTION CONTROL: THE FIRST CALL
FOR CONTROL

The objective with population reduction is to reduce the density of the animal vector population below that which is required for successful transmission of rabies to a susceptible individual. The aim is also to remove incubating and clinical animals from the population as vaccination will not work on those animals (Rosatte *et al.*, 2001). Unfortunately, birth rates in primary vectors, such as foxes and raccoons, are high, resulting in rapid population growth and recovery following control (Bogel *et al.*, 1974; Rosatte, 2000). This means that vector removal restrictions will have to be applied annually until the disease is eliminated. In Denmark, population reduction (gassing and shooting) was used to control fox rabies during the 1960s (Müller, 1967) and euthanasia of foxes was used extensively in Czechoslovakia for the control of rabies in the 1970s and 1980s (Matouch and Polak, 1982). An Arctic fox population was also reduced in an area of Alaska

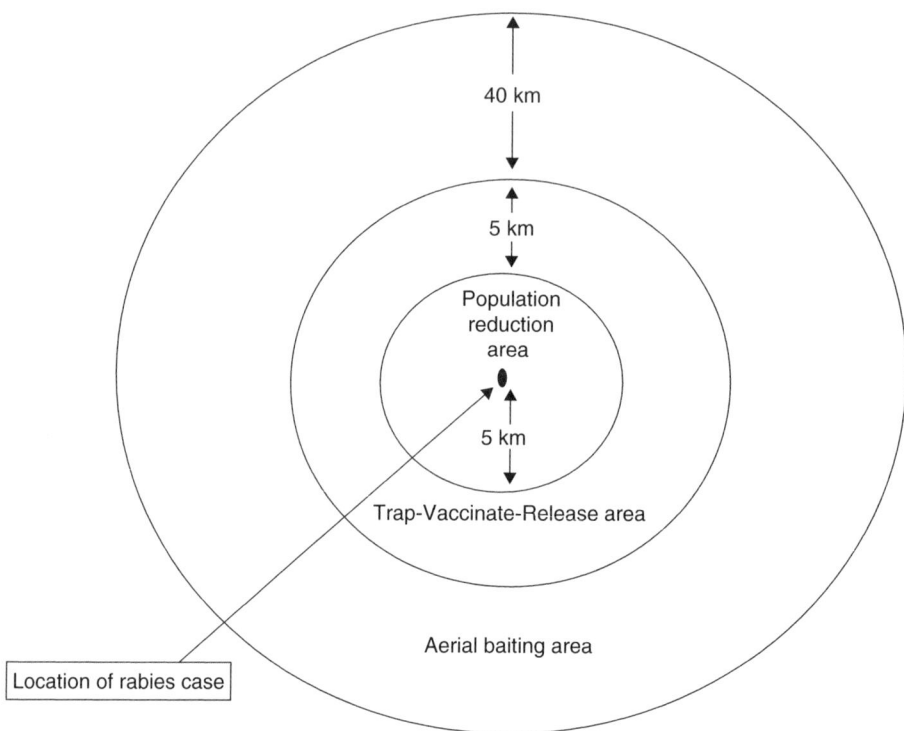

Figure 18.2 Point infection control diagram.

during 1994 in response to a rabid Arctic fox (*Alopex lagopus*) attacking two humans (Ballard *et al.*, 2001).

Population reduction may have its place as a component of control technology when applied to a newly established point of infection that springs up in a naïve population. However, vaccination strategies (both parenteral and oral), in addition to population reduction, have proven to be a more effective control strategy and have been termed point infection control (PIC) (Rosatte *et al.*, 2001) (Figure 18.2). In this situation, incubating and clinically rabid animals are removed from the immediate rabies outbreak area by population reduction. Vaccination of vector species beyond this zone provides a buffer to contain any dispersing infected animals that may have been missed during culling operations. PIC has been successful in limiting the spread of a new focus of raccoon-strain rabies that appeared in Ontario in 1999 after moving across the St Lawrence River from adjacent New York State (Rosatte *et al.*, 1997, 2001). In addition, in Ontario, a 700 km² TVR area has been in place in the Niagara Falls area since 1994 (Rosatte *et al.*, 1997). That TVR program has been so successful that the area has been

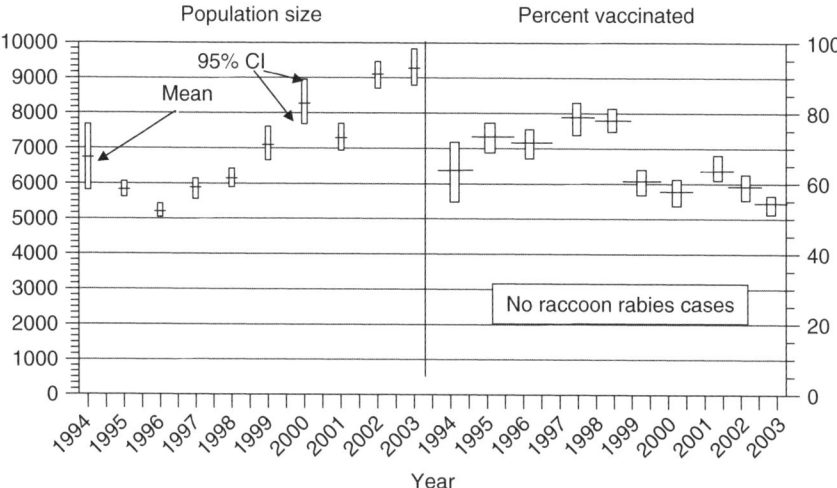

Figure 18.3   Raccoon population size and percentage of the raccoon population that was vaccinated against rabies during a trap-vaccinate-release (TVR) program in Niagara Falls, Ontario during 1994–2003. The TVR area was about 700 km². No rabies cases were detected in the area during 1994–2005 despite rabies being present on the New York side of the Niagara River.

free of reported cases of raccoon rabies despite the disease being enzootic in nearby New York State (Figure 18.3).

Although PIC is effective in some situations, such as a new focus of wildlife rabies, generally, the population reduction and TVR portions of the strategy are only feasible for areas smaller than about 2000 km² as the logistics of these types of operation becomes difficult for larger areas. In an enzootic situation, where several thousands of square kilometers of area are infected, the most feasible rabies control tactic is the aerial distribution of ORV baits. However, PIC may still be used to contain 'hot spots' within large enzootic areas, especially in situations where secondary vectors do not respond to the vaccine used to immunize the primary vector (e.g. skunks infected with raccoon-strain in Massachusetts and Ontario; Robbins *et al.*, 1998; Rosatte *et al.*, 2001).

9 BAIT DEVELOPMENT FOR DELIVERY OF ORAL RABIES VACCINE

Virtually any material thought attractive to a target species has been tested as a potential bait to carry the oral rabies vaccine (Linhart *et al.*, 1997b). In addition to the safety of the rabies vaccine itself, safety of the entire bait material matrix is important. The release of a prion-bearing component could have far-reaching complications in a large-scale free wild release. As a physical object, the bait's shape and substance can cause deleterious consequences. Heavy, hard baits may become lethal projectiles when dropped from the air; sharp-cornered vaccine packages may become lodged in the throat of ingesting animals if gulped. In addition, there is the possibility of human contact with rabies vaccine baits that must be considered (McGuill *et al.*, 1998; Rupprecht *et al.*, 2001). The mass-production capability and cost of baits are also critical determinants when developing baits for the delivery of rabies vaccine to wildlife (MacInnes, 1988).

Three main types of rabies vaccine baits are currently in use in large-scale ORV programs in North America. The first type is the Raboral V-RG® Bait manufactured by Merial, Inc., Athens, Georgia. It contains the V-RG vaccine produced by Merial which is inserted into a fishmeal polymer cube (Rupprecht *et al.*, 1986, 2004; Slate *et al.*, 2005) manufactured by Bait-Tek, Inc., Orange, Texas. The second type of bait is the Ontario Bait manufactured by Artemis Technologies, Inc., Guelph, Ontario, Canada. It contains either the attenuated ERA vaccine (Lawson and Bachmann, 2001) or the same vaccine (Merial V-RG) as used in the Raboral V-RG Bait (Bachmann *et al.*, 1990; MacInnes *et al.*, 2001; Rosatte and Lawson, 2001; Rosatte *et al.*, 2001). The third bait type is a matrix-coated sachet containing the Merial V-RG. These baits (except the Merial sachet) contain tetracycline-HCl as a biomarker. There are many other recent bait developments in North America for the various carnivore species (Robbins *et al.*, 1998; Rosatte *et al.*, 1998; Masson *et al.*, 1999; Bruyere *et al.*, 2000; Olson *et al.*, 2000; Linhart *et al.*, 2002; Follmann *et al.*, 2004). In Europe, several different types of baits have been used for delivery of ORV to targeted rabies vectors. These include the chicken head and fishmeal/ fish-oil baits which contain the V-RG vaccine for fox rabies control in Belgium (Brochier *et al.*, 1990) and Tubingen baits containing SAD B19 vaccine for fox rabies control in Germany and for the control of rabies in raccoon dogs as well as foxes in Finland (Schneider and Cox, 1988; Westerling, 1991). Linhart *et al.* (1997a) found that fishmeal baits were feasible for delivery of ORV to foxes and jackals in Israel.

10 ORV INITIATION/CONSIDERATIONS

Oral rabies vaccination is a total system and has been defined as an attempt at zoonoses control intended to protect human health and prevent economic losses (Wandeler, 2000). If the success of the ORV program is to be evaluated after treatment, specimens of the target species should be tested before treatment to establish levels of existing virus neutralizing antibody (VNA) (Rosatte and Gunson, 1984; Hanlon *et al.*, 1989) and biomarker levels (Hanlon *et al.*, 1993; Nunan *et al.*, 1994; Fearneyhough *et al.*, 1998; Rosatte and Lawson, 2001; Sidwa *et al.*, 2005) in the target species population. If a recombinant vaccine is to be used, a survey of related viruses present in the target population should be initiated before and after bait distribution to determine the potential for recombination between the vaccine virus and naturally occurring virus present in the target population (e.g. animal poxviruses present in an area to be baited with V-RG) (Boulanger *et al.*, 1996). Also, the safety of the vaccine virus should be tested in non-target species that may be present in the target area (Follmann *et al.*, 2002; Rupprecht *et al.*, 2004). In addition, legal and liability issues need to be examined before releasing a biologic into the environment. Other considerations include the following.

10.1 Biomarkers in vaccine baits

An ideal biomarker should be homogeneous with the vaccine but not impair vaccine efficacy and should be detectable for months or years following bait ingestion by an animal. To date an ideal marker that is compatible with vaccine, long-lasting, inexpensive and safe has not been developed. Tetracyclines form a long-term mark in bones and teeth (Linhart and Kennelly, 1967). They have proven useful in verifying not only bait contact, but also the actual time-specific bait ingestion regimens using counts of fluorescent lines in tooth dentin and cementum (Figure 18.4). The age of the animal can also be determined from a count of the cementum growth zones in the tooth. When yearly biomarker lines in cementum from multiyear baitings are correlated with VNA levels, they can indicate the duration of detectable levels of VNA and any anamnestic rise in VNA from yearly boosters (Figure 18.4 D and Figure 18.5) (Vrzal and Matouch, 1996; Johnston and Tinline, 2002). When an area is baited, all animals that eat baits will be 'tagged' with the biomarker and this has proven useful for tracking animal dispersal from a baited area. However, tetracyclines can be found in wildlife at low levels derived from agricultural sources, usually less than 5% (Nunan *et al.*, 1994). This has been a concern in some programs (Linhart *et al.*, 1997a, 1997b) but, if specimens of the target species are examined for ambient levels prior to baiting, this can be taken into account (Fearneyhough *et al.*, 1998; Johnston *et al.*, 1999; Sidwa *et al.*, 2005: Whitney *et al.*, 2005). There is, however, a need for unique, long-lasting marker materials that can be used in combination with tetracycline to produce unique time-specific life-long 'biomarks' in teeth. Such biomarkers will further help to monitor the proportion of a population that has been exposed to vaccine and, in the absence of declining VNA evidence, to determine when to rebait the area (Sidwa *et al.*, 2005). Specimens, (heads), submitted for routine rabies diagnosis, can also be used to establish the sex–age structure of the infected vector populations (Johnston and Beauregard, 1969) and are of value to establish pre-baiting levels of tetracycline and for post-ORV monitoring of biomarkers in animals that have dispersed from the baited area.

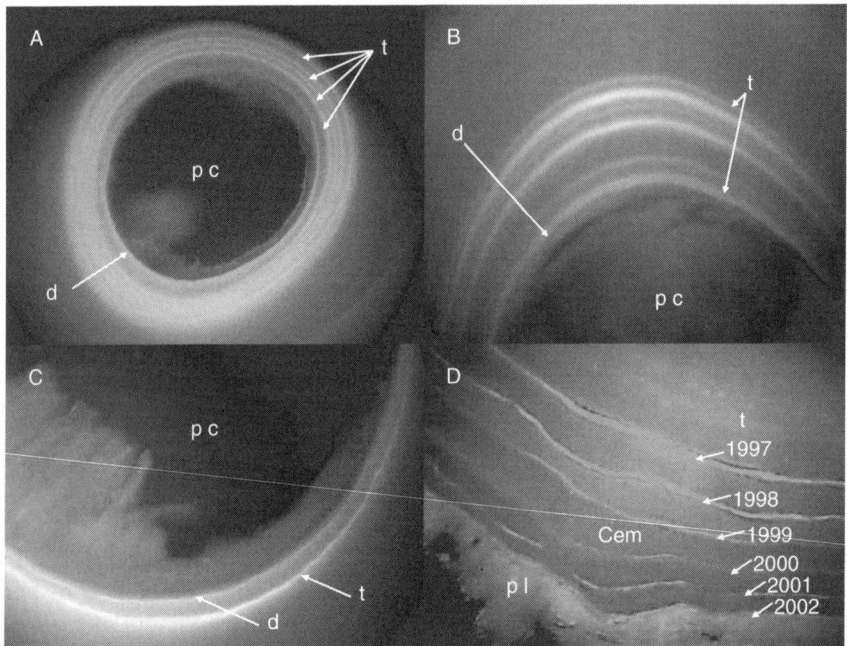

Figure 18.4   Sections of canine teeth showing yellow fluorescent lines from the ingestion of rabies vaccine baits containing 150 mg of tetracycline as a biomarker (undecalcified cross-sections; ultraviolet fluorescence; ×100). (A) Captive raccoon with four tetracycline lines from RABORAL V-RG® baits fed 44, 35, 29 and 22 days before death. 'd', growing edge of dentin at time of death; 'p c', pulp cavity; 't', tetracycline lines. (B) Wild juvenile coyote with a regimen of eight or more tetracycline lines from vaccine baits distributed by aircraft. (C) Wild juvenile coyote with a single tetracycline line from a bait eaten just prior to death. (D) Wild, 6-year-old coyote with one tetracycline line per year in six annual cementum growth zones. Cem, cementum; pl, periodontal ligament. (Figure 18.4 A(® Merial Ltd, LLC.); Figures 18.4 B, C, D (Texas Department of State Health Services, Zoonosis Control; Sidwa *et al.*, 2005). This figure is reproduced in the color plate section.

10.2 Bait density

Control ultimately depends on establishing herd immunity through animals taking baits and sero-converting. Hence, bait density must correlate with animal density in some positive fashion (Rosatte and Lawson, 2001). This includes the density of all bait-consuming species, not just the target vector. Non-target species, such as opossum (*Didelphis virginiana*), may consume a considerable proportion of baits intended for raccoons. In urban habitats, raccoon density may be extremely high, e.g. in Washington DC raccoon density was 67–333/km^2 (Riley *et al.*, 1998). High bait densities will be required to reach a substantial portion of the population. In Scarborough, Ontario, where raccoon density ranged from

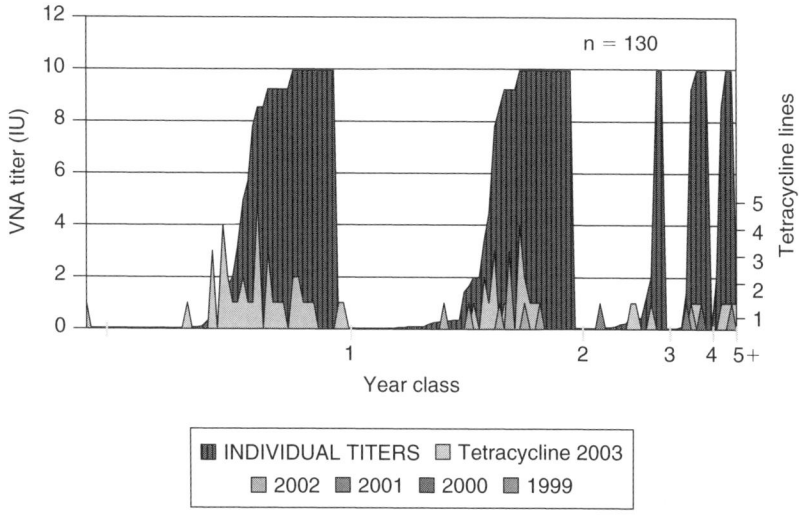

Figure 18.5 2003 West Texas Oral Rabies Vaccination Program. Individual gray fox Virus Neutralizing Antibody titers correlated with tetracycline biomarker line profiles by year class (Department of Defense, Food and Diagnostic Laboratory, San Antonio, TX) and (Texas Department of State Health Services, Zoonosis Control; Sidwa *et al.*, 2005). This figure is reproduced in the color plate section.

37 to 94/km^2 (Broadfoot *et al.*, 2001), raccoon acceptance of baits was 74% when bait density was 200/km^2 (Rosatte and Lawson, 2001). Roscoe *et al.* (1998) used fishmeal polymer V-RG baits at a density 64/km^2 in the Cape May area of New Jersey to control raccoon rabies during 1992–1994. Tetracycline was detected in 73% of the sampled raccoons and 61% of the raccoons tested seroconverted (Roscoe *et al.*, 1998). In Newfoundland, under severe winter snowconditions, Whitney *et al.* (2005) eradicated an invading epizootic of arctic-strain rabies in red fox using a density of 35 baits/km^2 and, in France, Vuillaume *et al.* (1998) distributed vaccine baits at fox dens at a density of 11.4 baits/den. Given the wide variation in results in a small number of trials, it is necessary in planning ORVPs to consider other factors, such as the methods of vaccine placement, timing of vaccination, spatial variations in animal density and bait design (Table 18.1) (Bachmann *et al.*, 1990; Robbins *et al.*, 1998; Olson *et al.*, 2000; Rupprecht *et al.*, 2004).

10.3 Method of vaccine bait placement

In urban areas of Ontario, hand placement of vaccine baits was used for the control of fox and raccoon rabies (Rosatte *et al.*, 1992, 2001, 2007). In Germany, hand-placement of vaccine baits by hunters was also used as a tactic to control

TABLE 18.1

Baiting success guidelines derived from counts of tetracycline biomarker lines in teeth of carnivore vector species

Species (location)[a]	Sample (n) (year class)	Total tetracycline biomarker (%+)	Median (range), tetracycline biomarker lines/(+) animal	Baiting success, guideline (Type)
Red fox (NL)[b]	25 (1st year class)	66	6.3 (1–11)	BSG, Type 1
Coyote (TX)[c]	117 (1st year)	82	3.3 (1–8)	BSG, Type 1
Red fox (ON)[d]	210 (all years)	70	2.8 (1–28)	BSG, Type 2
Red fox (DE-Rügen)[e]	28 (all years)	75	2.5 (1–7)	BSG, Type 2
Red fox (CZ)[f]	19 210 (all years)	78	2.0	BSG, Type 2
Raccoon (NY)[g]	207 (all years)	73	2.0 (1–9)	BSG, Type 2
Raccoon (OH)[h]	22 (1st year)	76	1.5 (1–6)	BSG, Type 2
Gray fox (TX)[c]	141 (1st year)	36	2.1 (1–5)	BSG, Type 3

[a]NL = Newfoundland, TX = Texas, ON = Ontario, DE = Deutschland, CZ = Czech Republic, OH = Ohio; [b]Whitney *et al.* (2005); [c]Sidwa *et al.* (2005); [d]Johnston and Voigt (1982); [e]Müller *et al.* (unpublished); [f]Matouch and Vitasek (2005); [g]Bigler and Lein (2001); [h]Nohrenberg, USDA(unpublished).

rabies in foxes (Wilhelm and Schneider, 1990). Vuillaume *et al.* (1998) targeted fox dens via hand-placement of vaccine baits and also used helicopters for aerial placement of baits in France. Roscoe *et al.* (1998) and Olson *et al.* (2000) used helicopters and motor vehicles to distribute baits for raccoon rabies control in New Jersey and Florida, respectively. In the New Jersey program, raccoon habitats or ecotones were targeted rather than random dispersion of baits. In Ontario, Bachmann *et al.* (1990) found that aerial distribution of baits was more cost effective than ground distribution. In Ontario, a variety of aircraft including Cessna, Turbo Beaver, Bell helicopter and Twin Otter have been used to distribute baits for the control of rabies in wildlife (Bachmann *et al.*, 1990; Rosatte *et al.*, 1993, 2001, 2007; MacInnes *et al.*, 2001; Rosatte and Lawson, 2001). Aircraft were also used for a massive ORVP in Texas for the control of the disease in coyotes (Fearneyhough *et al.*, 1998; Sidwa *et al.*, 2005). In addition, hand placement, airplanes and helicopters were used to distribute V-RG baits for the control of rabies in foxes in Belgium (Brochier *et al.*, 1990).

Control programs using aircraft typically distribute baits along flight lines with fixed spacing. Optimum spacing must ultimately match the movement behavior/territoriality of the target species and the pattern of distribution on the ground. As Johnston and Tinline (2002) have argued, at least one bait must land in the territory traversed by a target species. Tinline and Ball (2003) have demonstrated that the deviations in navigating a fixed wing aircraft along a designated flight line can lead to placement of baits spaced up to 250 meters from the line suggesting that the minimum spacing between flight lines should be in the order of 500 meters. Thus, in practice, flight line spacing has ranged from 500 meters in North America in areas where raccoons were the target species to 2 km where the red fox was the target species in Ontario. Recently, new bait distribution technology has made it possible to analyze the X,Y locations of baits released from an aircraft (see Section 12). Patchy patterns of bait distribution can thus reflect navigator decisions on when to stop/start baiting and inherent errors in navigation. We do not know how these patterns affect the viability of vaccine barriers because we have little knowledge at this scale of how they impact animals finding baits.

10.4 Time of year for vaccination campaigns

There has been much experimentation with respect to the best time of year to distribute rabies vaccine baits in order to maximize the percentage of the target population that is vaccinated against rabies. Fall ORV baiting campaigns resulted in the control of the Arctic fox strain of rabies in Ontario (MacInnes *et al.*, 2001). However, for the control of raccoon rabies in Ontario, PIC and TVR programs were successful when deployed during the summer as capture success was higher during the summer than during the spring or fall (Rosatte *et al.*, 2001). ORV campaigns using V-RG to control raccoon rabies were also initiated during the summer in Ontario to allow time for assessment of vaccine bait uptake for raccoons prior to winter denning (Rosatte *et al.*, 2006). Both summer and fall baiting campaigns have been used in many US states to combat raccoon rabies (United States Department of Agriculture, 2004). For example, Roscoe *et al.* (1998) deployed V-RG baits during the spring and fall in New Jersey to control raccoon rabies.

In most areas of Europe, ORV campaigns for the successful control of fox rabies occurred during the spring (March to May) and autumn (September to October) (Vos, 2003). In France, spring and fall ORV campaigns have been successful for the control of fox rabies, but summer campaigns met with lower success (Masson *et al.*, 1999). In Germany, Vos *et al.* (2001) concluded that if the objective is only to vaccinate adult foxes, then baits should be distributed during the first half of March. If juveniles are to be vaccinated, then baits should not be deployed before the end of May in previously vaccinated areas as many young foxes would have maternally transferred immunity, which could interfere with vaccination via baits (Bruyere *et al.*, 2000; Muller *et al.*, 2001a, 2001b; Vos *et al.*, 2001). Selhorst *et al.* (2001) concluded that June is the recommended time for vaccine-bait distribution for the control of fox rabies in Europe through modeling. However, as foxes are territorial, Vos (2003) concluded that the optimal time for bait distribution in Europe to target territorial foxes should be late autumn (November) or early winter (December). Regardless of the time of year for baiting, consideration needs to be given to the effect of ambient temperature on the vaccine as well as the effects of storage on the titer of the vaccine (Lawson and Bachmann, 2001; Selhorst *et al.*, 2001). In Arctic regions, extreme temperatures may require the use of lyophilized oral vaccines for immunizing vector species such as Arctic foxes (Follmann *et al.*, 2004; Kuzmin *et al.*, 2004). However, in Newfoundland, under severe winter snow conditions, Whitney *et al.* (2005) used frozen or semi-frozen ERA vaccine in Ontario baits to reach 66% of red foxes and eradicate an invading epizootic of arctic-strain rabies.

11 IMPORTANCE OF VECTOR HOME RANGE AND DENSITY FOR VACCINE BAIT DISTRIBUTION

In the wild, vaccine baits must be placed within the home range of the target animal. If we could rely on each member of a wild species only taking one bait and vector population density was constant, there would be no problem. Unfortunately, this is not the case. For example, Blackwell *et al.* (2004) found that 3.3 baits/raccoon were consumed on average (bait density of $75/km^2$) with a 23% non-target uptake where raccoon density was about 25 raccoons/km^2. Unfortunately, there are many non-target species which consume baits and target vector density is usually variable (Blackwell *et al.*, 2004). The spatial distribution of raccoons changes over time, making it difficult to design baiting strategies with a single animal density in mind. At present, the only way to decide on the parameters of bait density and distribution pattern for a particular area is by empirical experimentation. Fortunately, there are a growing number of ORVPs from which to draw experience, but each geographic area and vector species possess unique factors that can modify ORV results (see Table 18.1) (Tinline and MacInnes, 2004).

There must be a link between the current home range distribution of the animals and the bait-available period. If the baits are not placed within that bait-available home range, the individual will not encounter a bait and have the opportunity for vaccination. Further, enough baits must be distributed to allow for bait-eating competitors. Conversely, if too many baits are dropped, the baiting system is not efficient. Another related factor is the variability in home-range size between sex and age classes in the population. Depending on the time of year, adults usually have larger home ranges than young animals and yet the young are usually the largest cohort in a carnivore population (Rosatte, 2000; Rosatte *et al.*, 2001). The bait distribution parameters therefore should be tailored to reach this cohort provided the bait drop is carried out after the young are immunocompetent (Lawson *et al.*, 1997; Olson *et al.*, 2000; Müller *et al.*, 2001a, 2001b).

Most bait uptake studies have shown that more than 50% of baits are consumed by 1 week and more than 80% by 1–3 weeks (Johnston and Voigt, 1982; Bachmann *et al.*, 1990; Linhart *et al.*, 1997a; Blackwell *et al.*, 2004). Therefore, the extent of the target animal's movements inside its home range is critical to bait uptake during this short period. If the home range is small and movement is limited, e.g. the movement of female raccoons during the spring perinatal period, distribution patterns with widely spaced lines may miss many individuals. Home-range size and movement information from telemetry studies during the bait-available period can be incorporated into simulation models and geographic information system (GIS) flight-planning programs further to enhance bait distribution success (Tinline *et al.*, 1999; Hauschildt *et al.*, 2001).

12 LARGE-SCALE VACCINE BAIT DISTRIBUTION TECHNOLOGY

During the past 30 years, development of vaccine bait distribution technology has progressed from hand-dropped test baits to systems in many countries capable of air dropping thousands of baits per day (Johnston, 1975; Johnston and Voigt, 1982; MacInnes *et al.*, 1992; Müller *et al.*, 1993; Fearneyhough *et al.*, 1998; Robbins *et al.*, 1998; Rosatte and Lawson, 2001; Hauschildt *et al.*, 2001; Sidwa *et al.*, 2005). Currently, one of the larger scale operations is the aerial distribution of V-RG baits in the eastern USA to control variants of the rabies virus unique to raccoons, gray foxes and coyotes (United States Department of Agriculture, 2004; Sidwa *et al.*, 2005; Slate *et al.*, 2005). During 2003, an approximate 180 000 km^2 area was treated with more than 10 million rabies-vaccine baits (Slate *et al.*, 2005) (Table 18.2). Over the period between 1989 and 2004, Ontario dropped almost 57 million baits (Table 18.2). In programs such as this there are three principal aspects for consideration with respect to delivery of rabies vaccine baits:

1 ground versus airborne distribution
2 automation of airborne bait delivery systems
3 organization of the bait delivery team.

Ground distribution includes placing baits by hand or throwing baits along roadways from moving vehicles. Airborne distribution ranges from hand baiting by helicopter to automated dropping systems with fixed-wing aircraft. The decision to use a particular method depends largely on the scale of the operation, budget considerations, target species and the ability to select habitats. In practice, hand baiting has been useful for urban areas and specific habitat types. Airborne delivery by helicopter or fixed-wing aircraft is best to cover large areas uniformly.

The use of global positioning systems (GPS) in navigation technology in the 1990s has greatly enhanced aircraft baiting operations. GPS allows aircraft to follow pre-programmed flight lines and, since ground speed can be calculated, the speed of the baiting machine can be continually adjusted to match the prescribed target drop rate per kilometer. Furthermore, the X,Y location of baits at the drop point can be recorded and compared with the distribution plan. In practice,

TABLE 18.2

Number of ORV baits distributed in North America[a]

Year	Ontario	Quebec	New York	Texas	VT, NH, MA	OH, PA, WV	South	Total
1989	285 010							285 010
1990	744 088							744 088
1991	674 126							674 126
1992	687 001							687 001
1993	699 385							699 385
1994	1 485 000							1 485 000
1995	129 092	148 000	159 379	830 000				2 766 471
1996	1 280 289		343 507	2 565 253				4 189 049
1997	1 230 840	67 075	347 032	2 549 040	71 640			4 265 627
1998	965 796	118 404	396 312	2 663 000	114 800	576 720		4 835 032
1999	1 009 096	211 392	615 383	2 755 000	186 239	1 215 768		5 992 878
2000	1 600 460	90 720	866 670	1 970 000	195 258	863 280		5 586 388
2001	1 500 487	140 220	900 000	1 671 000	234 721	1 777 086		6 223 514
2002	1 317 225	0	1 281 720	0	332 033	3 242 615	504 865	6 678 458
2003	1 193 705	0	117 325	0	605 685	2 987 348	586 012	6 546 075
2004	956 680	0	1 179 900	0	465 690	1 649 865	846 571	5 098 706
Total	17 258 280	775 811	7 263 228	15 003 293	2 206 066	12 312 682	1 937 448	56 756 808

[a] ORV = oral rabies vaccination, VT = Vermont, NH = New Hampshire, MA = Massachusetts, OH = Ohio, PA = Pennsylvania, WV = West Virginia.

Ontario has found that the system permits accurate accounting (± 1–2%) on a bait drop and prevents costly errors in distribution especially when millions of baits are being dropped. In addition to using GPS-based flight control, Ontario has developed software (FPLAN) to pre-plan flight routes and upload waypoints to the GPS navigation system. This software has been used successfully in a number of large-scale ORVPs in North America (Hauschildt *et al.*, 2001). In addition, Texas, the USDA and Ontario have cooperated to add crew-management database software to coordinate a team of 30–60 personnel.

13 SURVEILLANCE PRIOR TO, DURING AND AFTER A RABIES CONTROL PROGRAM

Adequate surveillance and direct fluorescent antibody (DFA) diagnosis are the critical tools that can be utilized to determine the magnitude of a rabies problem

(Meslin *et al.*, 1994). These are key tactics that should be employed as rabies approaches an area, and during rabies control operations, as well as following the control or elimination of rabies in an area. The level of surveillance should be determined by the intensity of the epizootic/enzootic situation. For example, in Ontario during a raccoon rabies outbreak in 1999–2004, less than 1% (21/8483) of raccoons (live-trapped) submitted by the Ontario Ministry of Natural Resources (OMNR) for rabies diagnosis were rabid (Rosatte *et al.*, unpublished; Rosatte *et al.*, 2001, 2006). This implied that very intensive rabies surveillance programs had to be implemented in order to maintain an early warning system for the rapid implementation of control programs. However, in other areas, the prevalence of raccoon rabies was quite high, as the epizootic was very intense (Winkler and Jenkins, 1991). In those situations, fewer animals had to be submitted to confirm that rabies was present in an area. Unfortunately, intensive rabies surveillance programs are lacking in most countries due to economic reasons. Some countries only test animals for rabies that have bitten humans. Furthermore, many animals that are involved in human biting incidents escape and are not tested (Meslin *et al.*, 1994). Modeling showed that the worst case scenario of rabies resurgence occurs when rabies has persisted at low levels despite control efforts and has remained undetected by surveillance (Thulke *et al.*, 2000). In addition, as prolonged vaccination programs eventually make continuance uneconomic, surveillance during the termination stage of the control program is imperative. Unfortunately, many wildlife rabies control programs relax surveillance once the prevalence of the disease decreases. However, surveillance should be pursued aggressively during field control operations (Hanlon *et al.*, 1999) as well as following rabies control to ensure the disease does not become re-established. This will allow for early detection of the disease and a rapid control response (Rosatte *et al.*, 1997).

Following an ORV bait drop, there are many questions that arise as to the success and efficiency of the project. How many target vectors contacted the baits and how many were immunized by the vaccine? Under controlled laboratory conditions, efficacy trials can show that a particular vaccine-bait combination will immunize a high percentage of a vector species. However, once released into the wild, there is no control over this vaccine baiting system. To evaluate success and to provide information for improving the design of subsequent control efforts, our experience has demonstrated that the following aspects of surveillance are important.

13.1 Vaccine bait distribution parameters

The controllable parameters of vaccine bait distribution include the maintenance of the cold chain storage prior to dropping, the values of bait density and dispersal pattern over the target habitat and the time of year of placement.

13.2 Post-drop surveillance sampling techniques

While not controllable because of the variation in field sampling procedures, these variables can be standardized for all projects, and should include:

1 A brain tissue sample for DFA diagnosis (Smith, 1995) of a sample of normal animals as well as any suspect rabid animals from the treated area
2 A serum sample for the detection and titer of VNA (Smith, 1995)
3 A maxillary bone sample including the canine tooth to determine the animal's age and the presence or absence of the bait biomarker (Johnston et al., 1997, 1999; Johnston and Tinline, 2002).

13.3 Post-drop surveillance sampling – what size of sample is required for significant results?

It is imperative to collect an adequate sample of target and non-target animals from the baited area before as well as after the bait drop. A statistically adequate sample of animals can be determined through statistical modeling, but questions unique to field samples remain. In laboratory experiments under controlled conditions with uniform test animals, an optimal sample size may be 25 individuals or less. However, results from the Texas ORVP have verified that at least 100 specimens are required to give a confident evaluation of coyote and gray fox populations from the wild (Sidwa et al., 2005).

13.4 When to begin post-drop surveillance specimen collection?

If feasible, post-drop surveillance specimens should not be collected until 6 weeks following the last day of a bait drop. This allows at least 4 weeks for bait contact and vaccine ingestion and at least 2 weeks for the formation of detectable VNA. Past experience indicates that 50% of baits are contacted by 7 days, 80% by 2 weeks and up to 100% by 3–4 weeks (Bachmann et al., 1990). Collections taken too early, i.e. less than 6 weeks post-drop, will include specimens that have contacted a bait, and are tetracycline-positive but VNA negative due to lack of time to develop detectable antibody, which can take 2–3 weeks (see Figure 18.4C and Figure 18.5).

13.5 Monitoring vaccine efficacy

Once out of the production plant, rabies vaccine baits are open to the vagaries of storage temperatures and field distribution. When on the ground, each vaccine bait is subject to local microclimatic changes, which may affect the attractiveness of the bait to the target vector species and the ability of the vaccine to immunize successfully. All of these influences will be different for each individual vaccine and ORVP. In order to verify the efficacy of the vaccine for the duration of the 'bait-available period' of approximately 4 weeks, a protocol can be established to harvest vaccine baits from the field at various intervals to monitor vaccine titer.

13.6 Rabies virus neutralizing antibody and biomarker interconnection

Serum VNA is normally used in laboratory experiments to verify vaccine efficacy, but these titers often fail to indicate the true immune status in a wild population. If the serum sample is taken too soon after baiting, seroconversion may not have yet occurred; if taken months or years later, the antibody titer may have declined below detectable levels. For these reasons, incorporation of a time-specific bio-marker such as tetracycline into the bait along with the vaccine can give additional evidence of vaccine bait contact and the potential duration of immunity (Linhart et al., 1997b; Fearneyhough et al., 1998; Johnston et al., 1999; Olson et al., 2000; Rosatte and Lawson, 2001; Sidwa et al., 2005). Tetracycline is a calciphilic marker which persists for the life of the animal in hard tissues such as teeth and bone. With multiyear baitings, as animals grow older there is a build up of tetra-cycline in the population while VNA levels in these older animals decline unless they are rebooted. This can result in the false conclusion that vaccine efficacy is declining in the face of apparent increase in bait uptake as indicated by total tetracycline positivity. To overcome this phenomenon, analysis of yearly tetracy-cline line profiles in canine teeth in comparison with the annual dentin and cementum growth zones (age lines) can verify the true tetracycline-positive to VNA-positive ratio (see Figure 18.4C and see Figures 2 and 3 in Johnston and Tinline, 2002).

13.7 Standardization of post-drop surveillance sampling techniques

The VNA tests for rabies antigen include the rapid fluorescent focus inhibition test (RFFIT), the fluorescent antibody virus neutralization test (FAVN) and the enzyme-linked immunosorbent assay (ELISA). These have undergone various degrees of standardization over time (Smith, 1995). Concerning VNA titer levels in ORVPs, there is the need to establish a standard level in international units (IUs) that will be taken as evidence of ORV vaccination. This is needed particularly for comparative purposes among different ORV systems. Various studies report VNA levels ranging from 0.012 IU or 0.05 IU, up to a presumed human WHO standard protective level of 0.5 IU. In field specimens, correlation between VNA levels and challenge protection are hard to pin down due to variation in the sampling time after vaccine contact. Therefore, a VNA minimum titer standard needs to be established. To date there has been virtually no standardization of the tetracycline biomarker technique from either the tissue sampling or interpretation points of view. The tetracycline line count has only been employed to advantage in a few studies (see Table 18.1). Here again, there is room for standardization of methods among laboratories so that all field results will be comparable among programs.

14 VACCINE BAITING COSTS/BENEFITS AND COSTS OF RABIES CONTROL

Rabies vaccine baiting over large areas is expensive (Meltzer, 1996; Kreindel *et al.*, 1998; Selhorst *et al.*, 2000, 2001). An example of baiting large geographic areas occurred in 1995–1996 when Texas initiated two ORVPs to control epizootics in coyotes and gray foxes (Fearneyhough *et al.*, 1998). Implementation of these two programs averaged $3.8M per year. This is in contrast to the initial projected cumulative cost of $63M from 1994 to 2004 for resulting post exposure prophylaxis (PEP) alone. Other than PEP biologicals, this estimate did not include any additional medical care, lost productivity, human suffering, or impact on agriculture and tourism. By 2000, the number of cases in South Texas had declined to zero. Given this success and the cost-effectiveness of these programs, the South Texas Coyote ORVP has continued in maintenance mode from 2000 to 2005 at a cost of $1.2M annually and the Gray Fox ORVP will be continued in order to close in on the focus of that epizootic in west Texas (Sidwa *et al.*, 2005).

The containment of raccoon-strain rabies in eastern Ontario has been successful using a PIC strategy (Rosatte *et al.*, 2001). Total costs to distribute V-RG baits for raccoon rabies control were about $200Cdn/km^2 and costs for population reduction (PR) and TVR programs were about $500Cdn/km^2; however, areas treated by PR and TVR were much smaller than those treated by ORV making overall costs for the former two strategies less costly than with ORV (Rosatte *et al.*, 2001). Rabies only moved 50 km in 6 years in Ontario due to control efforts. Without control, modeling suggests the disease would have progressed at least 180–240 km and would have cost the government an additional $8M–$12M/year in rabies-associated costs (Rosatte *et al.*, 2001, 2006).

In North America, when fox rabies was controlled in Ontario, human postexposure treatments and rabies cases declined, resulting in significant economic benefits (MacInnes *et al.*, 2001). The same scenario can be assumed when raccoon rabies was controlled in Ontario using point infection control strategies (Rosatte *et al.*, 2001). However, vaccinating raccoons with vaccine baits requires four to eight times as many baits per unit area as does the control of fox rabies (MacInnes *et al.*, 2001; Rosatte and Lawson, 2001; Rosatte *et al.*, 2001). Meltzer and

Rupprecht (1998) have questioned the benefits of using oral rabies vaccination to eliminate raccoon rabies where the disease is enzootic. However Kemere *et al.* (2002), through modeling, have suggested that the net benefits to be gained would be substantial ($48M–$496M US) if raccoon rabies were eliminated using ORV. The question of the economic feasibility of wildlife rabies control is controversial to say the least and is an area that warrants further study.

15 VACCINE BAITING EFFICIENCY AND BAITING SUCCESS GUIDELINES

Use of ORV is a relatively inefficient method of vaccinating an animal rabies vector, i.e. many vaccine doses are required per animal vaccinated versus one dose per animal by parenteral injection. Thulke *et al.* (2000) suggested there is no overall optimal strategy for deploying ORV over large areas and increasing bait density does not necessarily improve acceptance by the target species. Vuillaume *et al.* (1998) found that the cost of hand placement of baits for fox rabies control in France was 3.5 times greater than for aerial vaccination and took 63.5 times longer. Therefore, with limited budgets for rabies control, efficient use of vaccine baits is imperative for the operating agency (Stöhr and Meslin, 1996). Variables affecting baiting success are many and include the geographic area to be baited, target species and target species density, bait-eating competitor density, habitat variability and year-to-year climate and phenological changes. Results to date indicate that for a vaccine baiting system to be successful in limiting rabies spread and at the same time to be cost efficient, the following criteria must be met:

1 To become immune, an individual animal of the target species needs to consume only one bait containing an efficacious vaccine.

2 For an individual animal to eat only one bait, the bait must be dropped within its active home range during the bait-available period. Also, enough baits must be dropped within the home range so that at least one bait is found by the target individual before all the baits are eaten by other competitors.

3 The target vector population is able to resist continuing a rabies epizootic if approximately 50–70% of the population finds a vaccine bait, becomes immune and maintains that immunity for an extended period (Voigt *et al.*, 1985; Tischendorf *et al.*, 1998).

4 A cost-efficient program will achieve the 50–70% population bait consumption level using only an average of one bait per individual of the target species (Johnston and Tinline, 2002).

A potential advantage of producing a more effective bait is that lower bait densities can be used in oral baiting and costs will be reduced. Two other potential methods of reducing oral baiting costs are to target habitat and to partition large areas into bait and no-bait zones. Targeting habitat is problematic for species like raccoons because they are generalists and can survive in a wide variety of areas at highly variable densities. An alternative strategy would be to take advantage of natural barriers (mountains, large rivers and lakes) to delineate areas for intensive baiting and areas for 'emergency' baiting depending on the presence or absence of rabies. In an epizootic, natural barriers can be used selectively to complement oral vaccine barriers and reduce or eliminate bait distribution behind or within the barriers. In an enzootic situation, two other effects offer promise for partitioning strategies. First, Tinline and MacInnes (2004) have demonstrated successful rabies control in Ontario, for red foxes, by targeting 'Rabies Units' in geographical sequence, breaking the cycle in one unit before moving on to another. Second, at least for rabies in foxes, there appears to be a minimum area of about 3000 km^2 below which rabies will not persist (MacInnes, 1988). If this area/population limitation applies to other species, as it appears to do so from preliminary simulation model studies (Tinline and Sheriff, 2004), then enzootic rabies in small areas will die out on its own provided natural or man-made barriers isolate those areas.

Guidelines for baiting success have been developed and refined using three variables from post-drop surveillance results:

1   the age of the animal
2   the VNA titer level
3   the count of tetracycline biomarker lines in canine teeth (see Figure 18.4 and Table 18.1).

The presence of tetracycline fluorescence in bones or teeth indicates an animal has contacted a bait, but the proportion of a population that is tetracycline-positive does not by itself indicate baiting efficiency if over-baiting is occurring, i.e. there has been more than one bait contact per animal. However, a count of tetracycline biomarker lines per animal, determined by ultraviolet-fluorescence microscopy (Johnston *et al.*, 1999), can show if baits are being distributed uniformly, or whether some animals are getting too few or too many baits (see Figure 18.4B, C). Specimens up to 1 year of age give the best count of tetracycline biomarker lines because rapid tooth growth separates daily ingestions of single baits more clearly than in older animals where growth is slow. In older animals, multiple ingestions often appear clumped together into a single line or band of fluorescence (see Figure 18.4D).

Comparing surveillance results with the following baiting success guidelines (BSG) (see Table 18.1) should help improve bait distribution success and efficiency. When post-drop surveillance results indicate:

• Type 1, i.e. tetracycline (+) is greater than 80% and/or median tetracycline biomarker line count is greater than 2.0, then too many baits are being dropped and bait flight lines are too far apart to spread the baits uniformly. Therefore, reduce bait density slightly and decrease bait flight-line interval.

- Type 2, i.e. tetracycline (+) is less than 80% but median tetracycline biomarker line count is greater than 2.0, then enough baits are being dropped but flight line spacing is too wide. Therefore, maintain bait density, and decrease bait flight-line interval.

- Type 3, i.e. tetracycline (+) is less than 70% and median tetracycline biomarker line count is less than 2.0, then too few baits are being dropped and line spacing is too wide. Therefore, increase bait density and decrease bait flight-line interval.

Slight adjustments in bait density and flight-line interval can accomplish better distribution with little effect on budget. The Texas ORVP program (see Table 18.1) is an example where over-baiting of a coyote population may be occurring. Conversely, using virtually the same baiting parameters, gray foxes were underbaited, probably due to their smaller home ranges compared with coyotes (see Figure 18.5) (Windberg, 1988; Sidwa *et al.*, 2005).

16 CONTINGENCY PLANNING

Contingency planning is critical for the rapid control of disease outbreaks. Ontario had a contingency plan for raccoon rabies 6 years before the disease was reported in the province (Rosatte *et al.*, 1997). This proactive approach allowed staff to implement a control tactic 24 hours after the first case of raccoon rabies was confirmed (Rosatte *et al.*, 2001). Although Australia is free of reported wildlife rabies, experiments were initiated regarding oral rabies vaccination of foxes. This tactic is being considered should rabies occur in the wildlife population (Marks and Bloomfield, 1999). In the UK, contingency plans are in place that involve the distribution of poison baits as well as ORV should an outbreak occur in foxes and/or badgers (*Meles meles*) (Harris *et al.*, 1990; Smith, 2002). In Ohio, the deployment of ORV was successful for controlling raccoon rabies from 1998 to 2003; however, during 2004 there was a significant outbreak in eastern Ohio and the disease now threatens to expand westward (Russell *et al.*, 2005). Neighboring states are currently discussing contingency plans (Slate *et al.*, 2005).

17 MODELING

Mathematical models have played a number of roles in the development of vaccination programs. Early deterministic models of the dynamics of fox rabies in Europe (Anderson *et al.*, 1981; Smith and Harris, 1991) looked at the proportion of the target population that must be vaccinated to achieve herd immunity and, therefore, stop the spread of rabies. In the USA, Coyne *et al.* (1989) investigated the relative merits of culling and/or vaccination and concluded that vaccination or vaccination plus culling would be required to control rabies if it became established in the host population. Deterministic models are, however, highly simplified representations of reality and do not examine how spatial behaviors such as territoriality, dispersal and daily interaction would affect the patterns of spread and the persistence of the disease. Other groups began to develop spatial and stochastic models of spread (Voigt *et al.*, 1985; Smith and Harris, 1991). In stochastic models, animal behaviors at any given time, e.g. dispersal, litter size, breeding success and incubation period, are determined from a random draw from probability distributions representing those behaviors. Thus, unlike deterministic models, every run of a stochastic model will produce slightly different results and many runs are required to establish output variance. While more computation is involved, documenting this variance means that stochastic models are powerful tools for investigating the impact that small changes in input variables have on output and, therefore, provide better understanding of the variables that are important in controlling rabies. For example, field and laboratory experiments in Ontario with air-dropped oral vaccine bait delivery systems for red foxes (Lawson and Bachmann, 2001) have indicated that uptake (74%) combined with seroconversion (80%) meant that only 50–60% of animals would have immune titers. This is a level just on the border between eradication and persistence (Anderson *et al.*, 1981). Using the Ontario fox rabies model (Voigt *et al.*, 1985), one of us (R.T.) demonstrated that immunity levels of 50–60% had the highest probability of eradicating fox rabies in eastern Ontario if the vaccination program began after a peak in rabies incidence. In this way, rabies defeats itself by culling many susceptible animals from the population thereby allowing baiting campaigns to produce herd immunity in the remaining susceptible animals. The success of the campaign in eastern Ontario has been described by MacInnes *et al.* (2001).

During the past decade, many others have explored the power of simulation models to study various aspects of the spread of infectious disease. Smith *et al.* (2002) developed a best-fit stochastic spatial model for the spread of raccoon rabies through Connecticut demonstrating that major rivers acted as semi-permeable barriers to rabies spread. Building on this model, Russell *et al.* (2005), illustrated how the river system in Ohio would affect the spread of raccoon rabies following a breach in an ORV barrier in 2004. Tinline (2005) argued that natural barriers should be an integral part of rabies control planning. Several others have examined the question of the appropriate response should rabies reappear or breach a barrier (Thulke *et al.*, 2000; Broadfoot *et al.*, 2001; Tinline and Sheriff, 2004). Smith and Wilkinson (2003) used a model to evaluate culling, oral vaccination and fertility control as strategies for the control of rabies in foxes. They suggested that the best strategy to control a point source wildlife rabies outbreak would include an area of culling in the center of the disease focus followed with an outer ring of vaccination or an outer ring of vaccination and fertility control. Eisinger *et al.* (2005) also evaluated the concept of point infection control by ring vaccination for the control of sylvatic rabies. Prior to those two exercises, Rosatte *et al.* (2001) controlled an outbreak of raccoon rabies in Ontario, Canada with similar strategies. Optimizing delivery strategies has also been examined (Tinline *et al.*, 1999; Selhorst *et al.*, 2000, 2001). Recently, there has been a renewal of interest in the impact of other control methods such as contraception (Kreeger *et al.*, 1997; Suppo *et al.*, 2000) and comparing combinations of strategies. Smith and Cheeseman (2002) demonstrated that culling or lethal control can be more effective than vaccination for diseases such as rabies in isolated populations. They also suggested that permanent contraception would increase the chances of disease eradication. Barlow (1996) suggested that culling was likely to be more effective than vaccination but equally as effective as sterilization. Tinline and Ball (2003) investigated the impact of animal density on the survival of rabies in a contained raccoon population finding persistence at medium densities (5–10 animals per km²) and die out or burn out at lower and higher densities respectively. Haydon *et al.* (2002) suggested that, in the endangered Ethiopian wolf population, a population that is dispersed and isolated, herd immunity could be achieved if only 20–40% of the population were vaccinated. Gordon *et al.* (2004) have developed an epidemiological model to estimate the risk of infection of secondary species in areas epizootic/enzootic for raccoon rabies which has proved to be a valuable tool in studying the evolution of wildlife disease epizootics.

In comparison with field experiments, simulation models are inexpensive, fast, pose no physical danger and, most important, allow controlled experimentation. The major disadvantage of these models is that they are based on limited data from the field supplemented by the opinions of field biologists and trappers. Paradoxically, this disadvantage is also a major advantage. Simulation models force researchers to make explicit their assumptions and operating rules and, in so doing, force them to state clearly what is known and not known about the many variables in vector ecology. This, in turn, lays out clear directions for further field research and may, in many cases, represent the major gain from the exercise of modeling.

18 CONCLUSION

In the short term, the success of wildlife rabies control programs has helped secure public funds for additional control efforts. For the medium and long term, however, the danger is that there is no perceived need for control and therefore no need for funds to perpetuate a state of preparedness. Furthermore, it is imperative to understand that oral rabies vaccination with baits is not a panacea for wildlife rabies control in all situations worldwide. In some situations, managers have made decisions to continue with tactics that have failed instead of modifying the tactic or implementing new untried tactics. Managers as well as researchers must take an adaptive management approach to meet the needs of wildlife rabies control in any given area. For instance, the use of an ORV is currently the most feasible tactic for wildlife rabies control over large areas. In other situations, e.g. V-RG is not effective in skunks, alternate strategies will have to be employed such as population reduction, TVR and/or high density baiting, until alternate effective oral rabies vaccines are ready for field application. Diagnostic and assessment techniques should be standardized to allow comparisons between various control programs. Adaptive management also means that rabies control programs must be treated as ongoing experiments to enable a fair assessment of the success/failure of the program. This approach also implies the need for continuing surveillance programs, both during and after rabies control program operations, with appropriate sample sizes to allow for a valid assessment of the situation. If rabies is not completely eliminated from an area, the disease may perpetuate at low levels and form a reservoir of rabies to fuel future outbreaks. Adaptive management also requires inter-agency cooperation and a well-designed communication plan to ensure continued cooperation and public understanding and support of a rabies control program.

REFERENCES

Almeida, M., Massas, E., Aguiar, E., Martorelli, L. and Joppert, A. (2001). Neutralizing antirabies antibodies in urban terrestrial wildlife in Brazil. *Journal of Wildlife Diseases* **37**, 394–398.

Anderson, R.M., Jackson, H.C., May, R.M. and Smith, M. (1981). Population dynamics of fox rabies in Europe. *Nature* **289**, 765–771.

Artois, M. (2003). Wildlife infectious disease control in Europe. *Journal of Mountain Ecology* **7**, 89–97.

Artois, M. and Aubert, M. (1991). Foxes and rabies in Lorraine: a behavioural-ecology approach. *Hystrix* **3**, 149–158.

Artois, M., Cliquet, F., Barrat, J. and Schumacher, C. (1997). Effectiveness of SAG1 Oral vaccine for long-term protection of red foxes (*Vulpes vulpes*) against rabies. *Veterinary Record* **140**, 57–59.

Aubert, M. (1999). Costs and benefits of rabies control in wildlife in France. *Revue Scientific et Technique de L'Office International des Epizootics* **18**, 533–543.

Aubert, M., Masson, E., Artois, M. and Barrat, J. (1994). Oral wildlife rabies vaccination field trials in Europe with recent emphasis on France. In: *Lyssaviruses* (C. Rupprecht, B. Dietzschold and H. Koprowski, eds). pp. 219–243. Berlin: Springer-Verlag.

Bachmann, P., Bramwell, R.N., Fraser, S.J. *et al.* (1990). Wild carnivore acceptance of baits for delivery of liquid rabies vaccine. *Journal of Wildlife Diseases* **26**, 486–501.

Baer, G.M. (1988). Oral rabies vaccination: An overview. *Reviews of Infectious Diseases* **10** (Suppl. 4), S644–S648.

Baer, G.M., Abelseth, M.K. and Debbie, J.G. (1971). Oral vaccination of foxes against rabies. *American Journal of Epidemiology* **93**, 487–490.

Ballantyne, E. and O'Donoghue, J. (1954). Rabies control in Alberta. *Journal of the American Veterinary Medical Association* **125**, 316–326.

Ballard, W.B., Follmann, E., Ritter, J., Robards, M. and Cronin, M. (2001). Rabies and canine distemper in an arctic fox population in Alaska. *Journal of Wildlife Diseases* **37**, 133–137.

Barlow, N. (1996). The ecology of wildlife disease control: simple models revisited. *Journal of Applied Ecology* **33**, 303–314.

Belotto, A. (2001). Vaccination against dog rabies in urban centers of Latin America: experience in Brazil and Mexico. In: *Rabies control in Asia* (B. Dodet and F.-X. Meslin, eds). pp. 73–76. London: John Libby.

Belotto, A. (2004). The Pan American Health Organization (PAHO) role in the control of rabies in Latin America. In: *Control of Infectious Animal Diseases by Vaccination* (A. Schudel and A. Lombard, eds). *Developmental Biology* **119**, 213–216.

Bigler, L.L. and Lein, D.H. (2001). Wildlife rabies vaccination program – St Lawrence region. USAHA Conference, Harrington, Delaware.

Bingham, J., Schumacher, C., Aubert, M., Hill, F. and Aubert, A. (1997). Innocuity studies of SAG-2 oral rabies vaccine in various Zimbabwean wild non-target species. *Vaccine* **15**, 937–943.

Bingham, J., Foggin, C., Wandeler, A. and Hill, F. (1999a). The epidemiology of rabies in Zimbabwe. 2. Rabies in jackals (*Canis adustus* and *Canis mesolelas*). *Onderstepoort Journal of Veterinary Research* **66**, 11–23.

Bingham, J., Schumacher, C.L., Hill, F.W. and Aubert, A. (1999b). Efficacy of SAG-2 oral rabies vaccine in two species of jackal (*Canis adustus* and *Canis mesomelas*). *Vaccine* **17**, 551–558.

Black, J.G. and Lawson, K.F. (1970). Sylvatic rabies studies in the silver fox (*Vulpes vulpes*): Susceptibility and immune response. *Canadian Journal of Comparative Medicine* **34**, 309–311.

Blackwell, B., Seamans, T., White, R., Patton, Z., Bush, R. and Cepek, J. (2004). Exposure time of oral rabies vaccine baits relative to baiting density and raccoon population density. *Journal of Wildlife Diseases* **40**, 222–229.

Bogel, K., Arata, A., Moegle, H. and Knorpp, F. (1974). Recovery of reduced fox populations in rabies control. *Zentralblatt für Veterinarmedizin* **21**, 401–412.

Boulanger, D., Crouch, A., Brochier, B. *et al.* (1996). Serological survey for orthopoxvirus infection of wild mammals in areas where a recombinant rabies virus is used to vaccinate foxes. *Veterinary Record* **138**, 247–249.

Briggs, D., Briggs, J. and Howard, D. (1988). Prevalence of rabies in the striped skunk (*Mephitis mephitis*) in Kansas from 1966 to 1986. *Transactions of the Kansas Academy of Science* **91**, 123–131.

Broadfoot, J.D., Rosatte, R.C. and O'Leary, D.T. (2001). Raccoon and skunk population models for urban disease control planning in Ontario, Canada. *Ecological Applications* **11**, 295–303.

Brochier, B., Thomas, I., Baudin, B. *et al.* (1990). Use of a vaccinia-rabies recombinant virus for the oral vaccination of foxes against rabies. *Vaccine* **8**, 101–104.

Brochier, B., Aubert, M.F., Pastoret, P. P. *et al.* (1996). Field use of a vaccinia–rabies recombinant vaccine for the control of sylvatic rabies in Europe and North America. *Revue Scientifique et Technique de L'Office International des Epizootics* **15**, 947–970.

Brochier, B., Deschamps, P., Costy, F. *et al.* (2001). Elimination of sylvatic rabies in Belgium by oral vaccination of the red fox (*Vulpes vulpes*). *Annales de Medecine Veterinaire* **145**, 293–305.

Bruyère, V., Vuillaume, P., Cliquet, F. and Aubert, M. (2000). Oral rabies vaccination of foxes with one or two delayed distributions of SAG2 baits during the spring. *Veterinary Research* **31**, 339–345.

Centers for Disease Control and Prevention (2001a). Compendium of animal rabies prevention and control, 2001. National Association of State Public Health Veterinarians. *Morbidity Mortality Weekly Reports* **50 (RR-08)**, 1–9.

Centers for Disease Control (2001b). Vaccinia (smallpox) vaccine recommendations of the Advisory Committee on Immunization Practices. *Morbidity Mortality Weekly Reports* **50 (RR-10)**, l–25.

Chautan, M., Pontier, D. and Artois, M. (2000). Role of rabies in recent demographic changes in red fox (*Vulpes vulpes*) populations in Europe. *Mammalia* **64**, 391–410.

Childs, J.E., Curns, A.T., Dey, M.E. *et al.* (2000). Predicting the local dynamics of epizootic rabies among raccoons in the United States. *Proceedings of the National Academy of Sciences USA* **97**, 13 666–13 671.

Cleaveland, S. (1998). Epidemiology and control of rabies: the growing problem of rabies in Africa. *Transactions of the Royal Society of Tropical Medicine and Hygiene* **92**, 131–134.

Coulon, P., Lafay, F., LeBlois, H. *et al.* (1992). The SAG1, a new attenuated oral rabies vaccine. In: *Wildlife Rabies Control* (K. Bogel, F.-X. Meslin, and M. Kaplan, eds). pp. 105–111. Royal Tunbridge Wells: Wells Medical.

Coyne, M.J., Smith, G. and McAllister, F.E. (1989). Mathematical model for the population biology of rabies in raccoons in the mid-Atlantic states. *American Journal of Veterinary Research* **12**, 2148–2154.

Curk, A. and Carpenter, T. (1994). Efficacy of the first oral vaccination against fox rabies in Slovenia. *Revue Scientific et Technique de L'Office International des Epizootics* **13**, 763–775.

Debbie, J. (1991). Rabies control in terrestrial wildlife by population reduction. In: *The Natural History of Rabies*, 2nd edn (G. Baer, ed.). pp. 477–484. Boca Raton: CRC Press.

de Mattos, C., Mattos, C., Loza-Rubio, E., Aguilar-Setien, A., Orciari, L. and Smith, J. (1999). Molecular characterization of rabies virus isolates from Mexico: implications for transmission dynamics and human risk. *American Journal of Tropical Medicine and Hygiene* **61**, 587–597.

Devers, D., Rotenberg, D. and Yakobson, B. (2001). New rabies variant in Israel. *Israel Journal of Veterinary Medicine* **56** (2), 1.

Diaz, A., Papo, S., Rodriguez, A. and Smith, J. (1994). Antigenic analysis of rabies-virus isolates from Latin America and the Caribbean. *Journal of Veterinary Medicine* **41**, 153–160.

Eisinger, D., Thulke, H., Selhorst, T. and Muller, T. (2005). Emergency vaccination of rabies under limited resources – combating or containing? *BMC Infectious Diseases* **5**, 10–26.

Ellenton, J.A. and Basrur, P.K. (1982). Microchromosomes of the Ontario red fox (*Vulpes vulpes*): Distribution of chromosome numbers and relationship with physical characteristics. *Genetica* **57**, 13–19.

Ellenton, J.A. and Johnston, D.H. (1979). Oral biomarkers of calciferous tissues in carnivores. In: *Transactions of the 1975 Eastern Coyote Workshop*. pp. 60–67. New Haven, Connecticut.

Engeman, R., Christensen, K., Pipas, M. and Bergman, D. (2003). Population monitoring in support of a rabies vaccination program for skunks in Arizona. *Journal of Wildlife Diseases* **39**, 746–750.

Farry, S.C., Henke, S.E., Beasom, S.L. and Fearneyhough, M.G. (1998). Efficacy of bait distributional strategies to deliver canine rabies vaccines to coyotes in southern Texas. *Journal of Wildlife Diseases* **34**, 23–32.

Fearneyhough, M.G., Wilson, P.J., Clark, K.A. *et al.* (1998). Results of an oral rabies vaccination program for coyotes. *Journal of the American Veterinary Medical Association* **212**, 498–502.

Finley, D. (1998). *Mad Dogs: New Rabies Plague*. College Station: Texas A&M University Press.

Follmann, E., Ritter, D. and Hartbauer, D. (2002). Safety of lyophilized SAG2 oral rabies vaccine in collared lemmings. *Journal of Wildlife Diseases* **38**, 216–218.

Follmann, E., Ritter, D. and Hartbauer, D. (2004). Oral vaccination of captive arctic foxes with lyophilized SAG2 rabies vaccine. *Journal of Wildlife Diseases* **40**, 328–334.

Gascoyne, S., King, A., Laurenson, M., Borner, M., Schildger, B. and Barrat, J. (1993). Aspects of rabies infection and control in the conservation of the African wild dog (*Lycaon pictus*) in the Serengeti region, Tanzania. *Onderstepoort Journal of Veterinary Research* **60**, 415–420.

Gerletti, G., Guidali, F., Scherini, G. and Tosi, G. (1991). Management of the fox (*Vulpes vulpes*) in Lombardy region (Northern Italy) in relation to rabies. *Hystrix* **3**, 191–195.

Gordon, E., Curns, A., Krebs, J., Rupprecht, C., Real, L. and Childs, J. (2004). Temporal dynamics of rabies in a wildlife host and the risk of cross-species transmission. *Epidemiology and Infection* **132**, 515–524.

Guerra, M., Curns, A., Rupprecht, C., Hanlon, C., Krebs, J. and Childs, J. (2003). Skunk and raccoon rabies in the eastern United States: temporal and spatial analysis. *Emerging Infectious Diseases* **9**, 1143–1150.

Hanlon, C.L., Hayes, D.E., Hamir, A.N. *et al.* (1989). Proposed field evaluation of a rabies recombinant vaccine for raccoons (*Procyon lotor*): Site selection, target species characteristics, and placebo baiting trials. *Journal of Wildlife Diseases* **25**, 555–567.

Hanlon, C.A., Buchanan, J.R., Nelson, E., Niu, H.S., Diehl, D. and Rupprecht, C.E. (1993). A vaccinia-vectored rabies vaccine field trial: Ante- and post-mortem biomarkers. *Revue Scientific et Technique de L'Office International des Epizootics* **12**, 99–107.

Hanlon, C.A., Childs, J.E., Nettles, V.F. and the National Working Group on Rabies Prevention and Control (1999). Recommendations of a national working group on prevention and control of rabies in the United States: III. Rabies in wildlife. *Journal of the American Veterinary Medical Association* **215**, 1612–1619.

Hanlon, C., Niezgoda, M., Morrill, P. and Rupprecht, C. (2002). Oral efficacy of an attenuated rabies vaccine in skunks and raccoons. *Journal of Wildlife Diseases* **38**, 420–427.

Harris, S., Cheeseman, C., Smith, G. and Trewhella, W. (1990). Rabies contingency planning in Britain. In *Wildlife Rabies Contingency Planning in Australia* (P. O'Brien and G. Berry, eds). pp. 63–77. Canberra: Australian Government Publishing Service.

Hauschildt, P., Tinline, R.R. and Ball, D.G.A. (2001). Outfoxing rabies in Ontario. *GPS World*, May 2001, 34–39.

Haydon, D., Laurenson, M. and Sillero-Zubiri, C. (2002). Integrating epidemiology into population viability analysis: managing the risk posed by rabies and canine distemper to the Ethiopian wolf. *Conservation Biology* **16**, 1372–1385.

Irsara, A., Bressan, G. and Mutinelli, F. (1990). Sylvatic rabies in Italy: epidemiology. *Journal of Veterinary Medicine* **B37**, 53–63.

Jackson, A.C. (1994). The fatal illness of the Fourth Duke of Richmond in Canada: Rabies. *Annals of the Royal College of Physicians and Surgeons Canada* **27**, 40–41.

Johnson, N., Black, C., Smith, J. *et al.* (2003). Rabies emergence among foxes in Turkey. *Journal of Wildlife Diseases* **39**, 262–270.

Johnston, D.H. (1975). The Principles of Wild Carnivore Baiting with Oral Rabies Vaccines. *WHO Consultation on Oral Vaccination of Foxes*, Frankfurt am Main, Item 5, Document 7. Geneva: WHO.

Johnston, D.H. and Beauregard, M. (1969). Rabies epidemiology in Ontario. *Bulletin of the Wildlife Disease Association* **5**, 357–370.

Johnston, D.H. and Tinline, R.R. (2002). Rabies control in Wildlife. In *Rabies* (A. Jackson and W. Wunner, eds). pp. 445–471. San Diego, California: Academic Press.

Johnston, D.H. and Voigt, D.R. (1982). A baiting system for the oral rabies vaccination of wild foxes and skunks. *Comparative Immunology Microbiology and Infectious Diseases* **5**, 185–186.

Johnston, D.H., Fearneyhough, M.G., Hicks, B.N. and Moore, G.M. (1997). L'influence de l'âge de la population sur le succès apparent des essais de vaccination orale antirabique utilisant de la tétracycline comme marqueur. *Colloque International sur la Rage*. pp. 5–12. Paris: Institut Pasteur.

Johnston, D.H., Joachim, D.G., Bachmann, P. *et al.* (1999). Aging furbearers using tooth structure and biomarkers. In: *Wild Furbearer Management and Conservation in North America*, CD edn (M. Novak, J.A. Baker, M.E. Obbard and B. Malloch, eds). pp. 228–243. Sault Ste Marie, Ontario: Ontario Fur Managers Federation.

Johnston, D.H., Voigt, D.R., MacInnes, C.D., Bachmann, P., Lawson, K.F. and Rupprecht, C.E. (1988) An aerial baiting system for the distribution of attenuated or recombinant rabies vaccines for foxes, raccoons and skunks. *Reviews of Infectious Diseases* **10**, S660–664.

Jones, M., Curns, A., Krebs, J. and Childs, J. (2003). Environmental and human demographic features associated with epizootic raccoon rabies in Maryland, Pennsylvania, and Virginia. *Journal of Wildlife Diseases* **39**, 23–32.

Kappeler, A., Wandeler, A. and Capt, S. (1988). Ten years of rabies control by oral vaccination of foxes. In: *Vaccination to control rabies in foxes* (P. Pastoret, B. Brochier, I. Thomas and J. Blancou, eds). pp. 55–60. Luxembourg: Commission of the European Communities.

Keightley, A., Struthers, J., Johnson, S. and Barnard, B. (1987). Rabies in South Africa: 1980–1984. *South African Journal of Science* **83**, 466–472.

Kemere, P., Liddel, M., Evangelou, P., Slate, D. and Osmek, S. (2002). Economic analysis of a large scale oral vaccination program to control raccoon rabies. In *Human Conflicts with Wildlife Economic Considerations*. (L. Clark, ed.). pp. 109–116. Fort Collins: National Wildlife Research Center.

Kieny, M., Lathe, R., Drillien, R. *et al.* (1984). Expression of rabies virus glycoprotein from a recombinant vaccinia virus. *Nature* **312**, 163–166.

Knobel, D. and Toit, J. (2003). The influence of pack social structure on oral rabies vaccination coverage in captive African wild dogs (*Lycaon pictus*). *Applied Animal Behavior Science* **80**, 61–70.

Knobel, D., Toit, J. and Bingham, J. (2002). Development of a bait and baiting system for delivery of oral rabies vaccine to free-ranging African wild dogs (*Lycaon pictus*). *Journal of Wildlife Diseases* **38**, 352–362.

Kovalev, N., Sedov, V., Shashenko, A., Osidse, D. and Ivanovsky E. (1992). An attenuated oral vaccine for wild carnivores in the USSR. In: *Wildlife Rabies Control* (K. Bogel, F.-X. Meslin and M. Kaplan, eds). pp. 112–114. Royal Tunbridge Wells: Wells Medical.

Krebs, J., Wheeling, J. and Childs, J. (2003). Rabies surveillance in the United States during 2002. *Journal of the American Veterinary Medical Association* **223**, 1736–1750.

Kreeger, T.J. (1997). Contraception in Wildlife Management. *Technical Bulletin No. 1853*, US Department of Agriculture, Animal and Plant Health Inspection Service, Washington.

Kreindel, S.M., McGuill, M., Meltzer, M., Rupprecht, C. and DeMaria, A. (1998). The cost of rabies post-exposure prophylaxis: One state's experience. *Public Health Reports* **113**, 247–251.

Kureishi, A., Xu, L.Z. and Stiver, H.G. (1992). Rabies in China: recommendations for control. *Bulletin of the World Health Organization* **70**, 443–450.

Kuzmin, I. (1999). An arctic fox rabies virus strain as the cause of human rabies in Russian Siberia. *Archives of Virology* **144**, 627–629.

Kuzmin, I., Botvinkin, A., McElhinney, M. *et al.* (2004). Molecular epidemiology of terrestrial rabies in the former Soviet Union. *Journal of Wildlife Diseases* **40,** 617–631.

Lambot, M., Blasco, E., Barrat, J. *et al.* (2001). Humoral and cell-mediated immune responses of foxes (*Vulpes vulpes*) after experimental primary and secondary oral vaccination using SAG_2 and V-RG vaccines. *Vaccine* **19**, 1827–1835.

Laurenson, K., Van Heerden, J., Stander, P. and Van Vuuren, M.J. (1997). Seroepidemiological survey of sympatric domestic and wild dogs (*Lycaon pictus*) in Tsumkwe District, north-eastern Namibia. *Onderstepoort Journal of Veterinary Research* **64**, 313–316.

Lawson, K.F. and Bachmann, P. (2001). Stability of attenuated live virus rabies vaccine in baits targeted to wild foxes under operational conditions. *Canadian Veterinary Journal* **42**, 368–374.

Lawson, K.F., Chiu, H., Crosgrey, S.J., Matson, M., Casey, G.A. and Campbell, J.B. (1997). Duration of immunity in foxes vaccinated orally with ERA vaccine in a bait. *Canadian Journal of Veterinary Research* **61**, 39–42.

Linhart, S. (1960). Rabies in wildlife and control methods in New York State. *New York Fish & Game Journal* **7**, 1–13.

Figure 1.1 A votive giving thanks to the holy virgin for saving the life of a boy from rabies: 'We give you thanks, little virgin, for keeping my son from dying after he was bitten by a dog with rabies while playing with him, so when we saw that tragedy we called on you to perform the miracle of saving his life, and that's the way it was.' Soledad District, May 22, 1927.

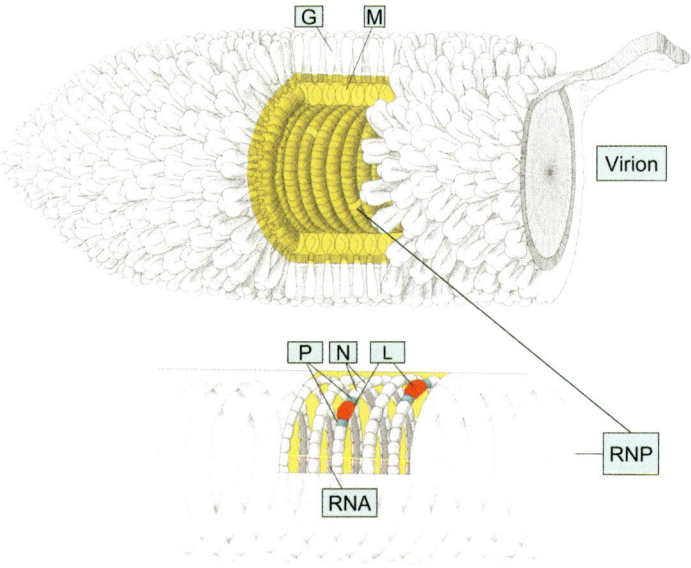

Figure 2.1 Schematic representation of the rabies virion. The drawing shows the internal ribonucleoprotein (RNP) core consisting of the single-strand, negative-sense genome RNA encapsidated with nucleocapsid protein (N), the virion-associated RNA polymerase (L) and polymerase cofactor phosphoprotein (P). The RNP core in association with the matrix protein (M) is condensed into the typical bullet-shape particle that is characteristic of rhabdoviruses. A lipid bilayer envelope (or membrane) in which the surface trimeric glycoprotein (G) spikes are anchored surrounds the RNP-M structure. The membrane 'tail' depicted in the drawing represents the trailing piece of envelope that is frequently observed in the electron microscope attached to the virus as it buds from the plasma membrane of the infected cell. (Reproduced from Wunner, W. H., Larson, J. K., Dietzschold, B. and Smith, C. L. *Review of Infectious Diseases* **10**, Supplement 4, S771–S784, 1988, with permission.)

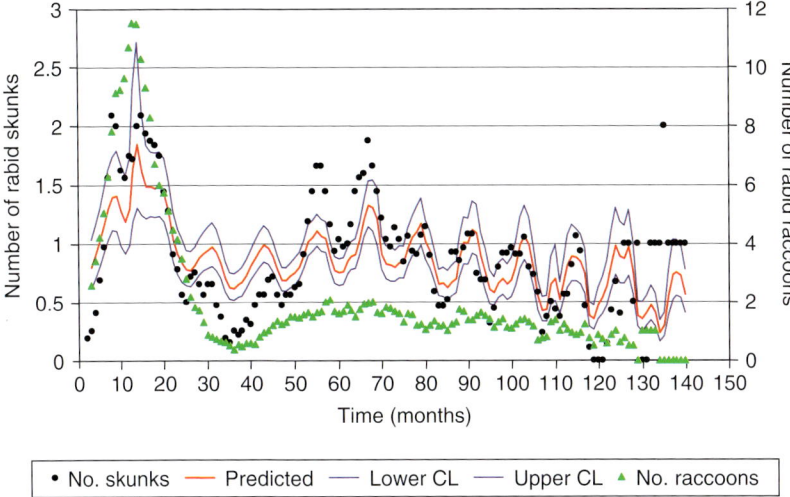

Figure 4.12 The temporal dynamics of the raccoon-associated variant of rabies virus among raccoons and spillover of this variant to skunks are indicated relative to the first month of the first epidemic of raccoon rabies, averaged from individual determinations from 32 counties in 11 eastern states. The green triangles (rabid raccoons) and black points (rabid skunks) are actual average values obtained from individual counties from 11 eastern states. The fitted lines predict the temporal dynamics of rabies in skunks and is best modeled with a lag of one month following the onset of an epidemic among raccoons. The epidemic pattern of rabies among skunks maintains a distinct bimodal annual pattern that is also present among rabid skunks in the midwestern USA where skunks are the primary reservoir of skunk-adapted rabies virus variants (see Figure 4.2 and text for more details). Figure from Guerra *et al.* (2003).

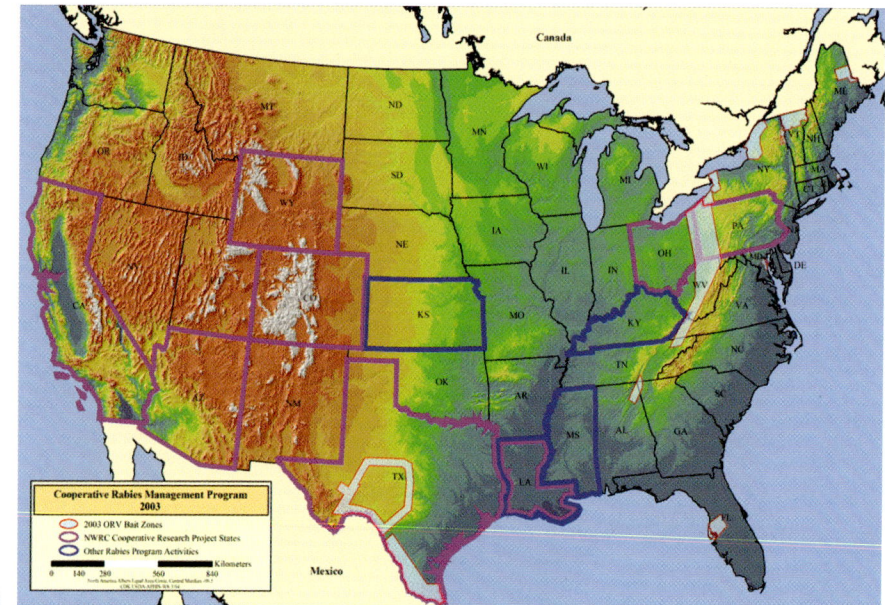

(A)

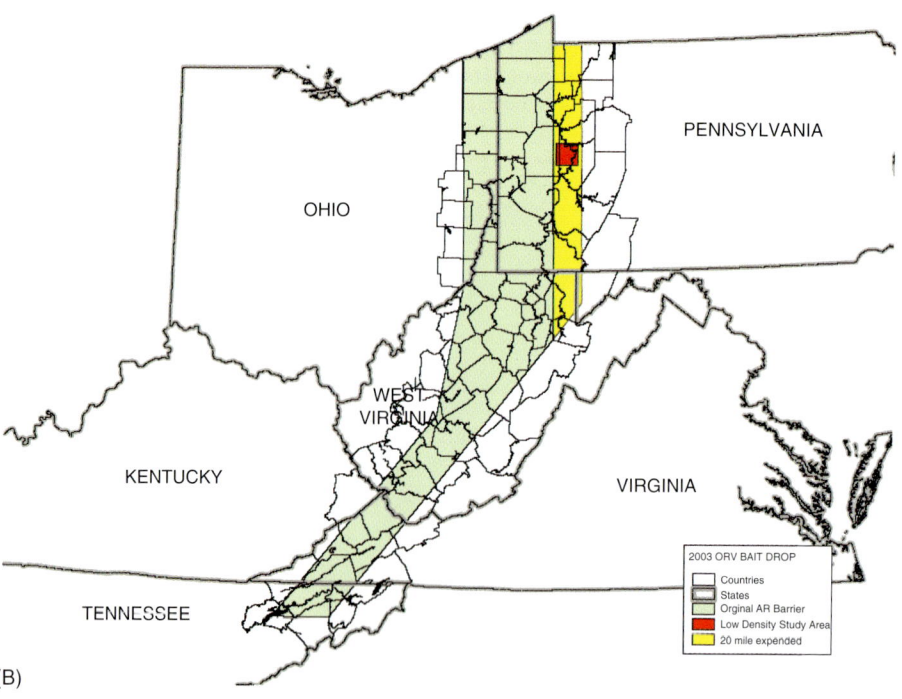

(B)

Figure 4.16 (A) The extent of a national effort to control the westward expansion of the raccoon variant of rabies virus into the Ohio Valley and into western Alabama coordinated by the USDA with individual states. (B) Shows a detail from (A), illustrating the extent and breadth of the ORV barrier established in Ohio, Pennsylvania, West Virginia and Virginia.

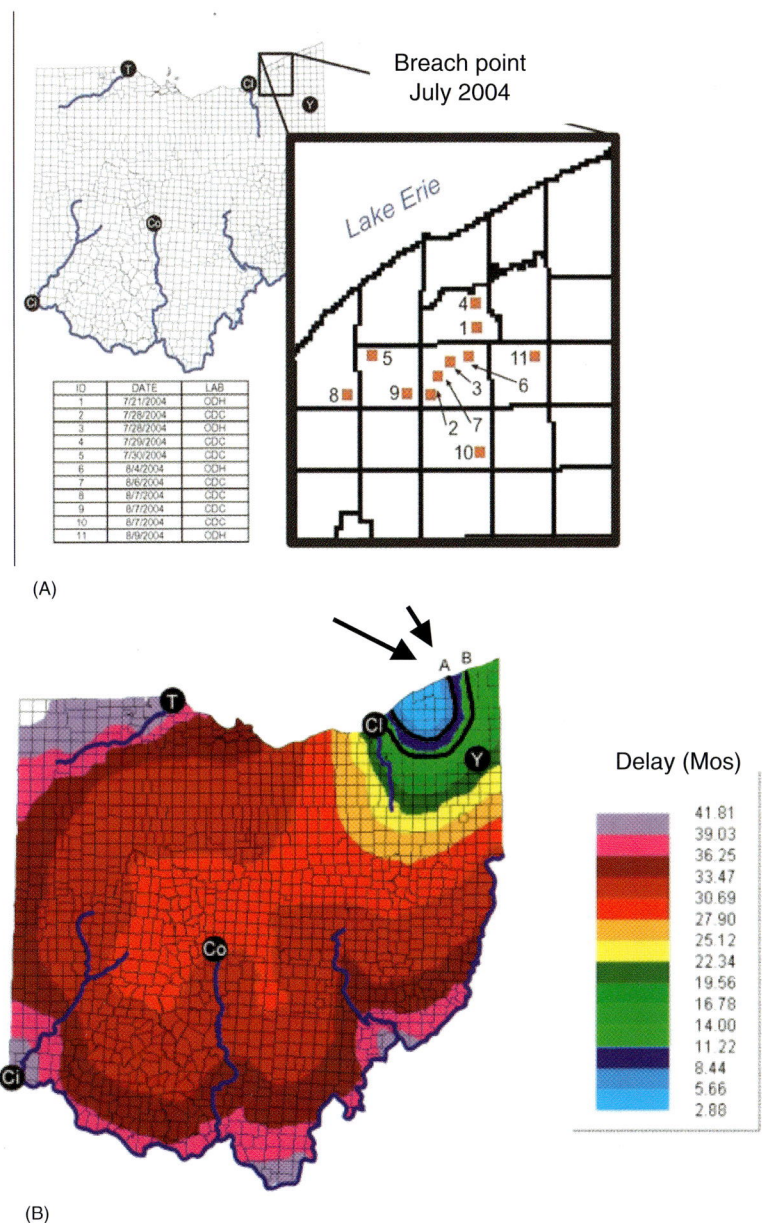

ID	DATE	LAB
1	7/21/2004	ODH
2	7/28/2004	CDC
3	7/28/2004	ODH
4	7/29/2004	CDC
5	7/30/2004	CDC
6	8/4/2004	ODH
7	8/6/2004	CDC
8	8/7/2004	CDC
9	8/7/2004	CDC
10	8/7/2004	CDC
11	8/9/2004	ODH

Breach point
July 2004

Lake Erie

(A)

Delay (Mos)

41.81
39.03
36.25
33.47
30.69
27.90
25.12
22.34
19.56
16.78
14.00
11.22
8.44
5.66
2.88

(B)

Figure 4.18   (A) A potential breach in the ORV barrier established by the USDA in cooperation with the State of Ohio occurred in Leroy Township in July 2004 (Anonymous, 2004). (B) The use of a stochastic simulator model to predict the potential trajectory of raccoon rabies spread from the nascent foci permitted identification of two boundaries for remedial ORV control by ring vaccination to contain raccoon rabies spread (Russell *et al.*, 2005). Arrows indicate the inner boundary in (A) and outer boundary in (B). The boundary lines were estimated by delays anticipated in passive surveillance based on detection of raccoon rabies and the estimated incubation period for rabies among raccoons (Tinline *et al.*, 2002). Ci = Cincinatti; Co = Columbus; Cl = Cleveland; T = Toledo; Y = Youngstown, all part of Ohio.

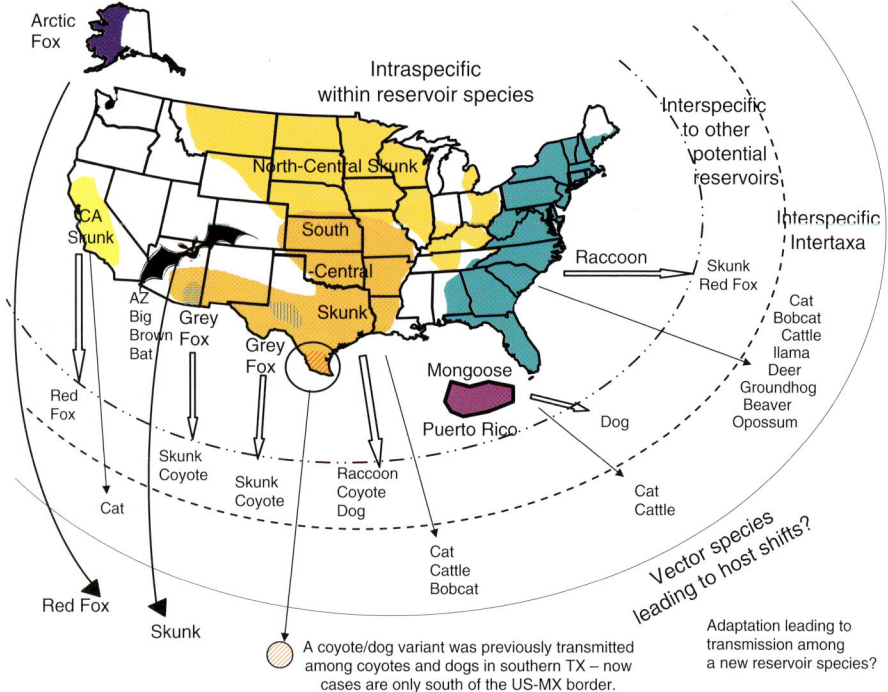

Figure 5.2 Transmission patterns of rabies in the USA.

Arctic
Fox

Intraspecific
within reservoir species

Interspecific
to other
potential
reservoirs

Interspecific
Intertaxa

North-Central Skunk

CA
Skunk

South
-Central

Raccoon

Skunk
Red Fox

Cat
Bobcat
Cattle
llama
Deer
Groundhog
Beaver
Opossum

AZ
Big
Brown
Bat

Grey
Fox

Skunk

Grey
Fox

Mongoose

Red
Fox

Puerto Rico

Dog

Skunk
Coyote

Skunk
Coyote

Raccoon
Coyote
Dog

Cat
Cattle

Cat

Vector species
leading to host shifts?

Red Fox

Cat
Cattle
Bobcat

Adaptation leading to
transmission among
a new reservoir species?

Skunk

A coyote/dog variant was previously transmitted
among coyotes and dogs in southern TX – now
cases are only south of the US-MX border.

Figure 5.3 A rabid fox with multiple embedded porcupine quills as evidence of abnormal behavior. Photograph courtesy of New York State Department of Health, Rabies Laboratory.

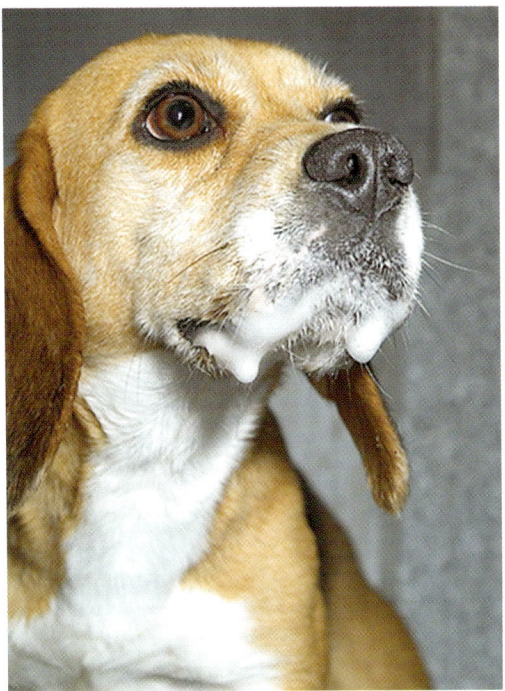

Figure 5.4 Salivation in a clinically rabid dog. Photograph by Ivan Kuzmin.

Figure 5.6 Enhanced rabies surveillance using the direct rapid immunohistochemical test (DRIT) in Tanzania. Arrows: identification of rabies virus antigen on a brain impression using a cocktail of monoclonal antibodies directed against the virus nucleoprotein. (Animal photographs courtesy of the Serengeti Carnivore Disease Project (Tiziana Lembo); leopard photograph provided by Sarah Durant; DRIT photograph by Michael Niezgoda.)

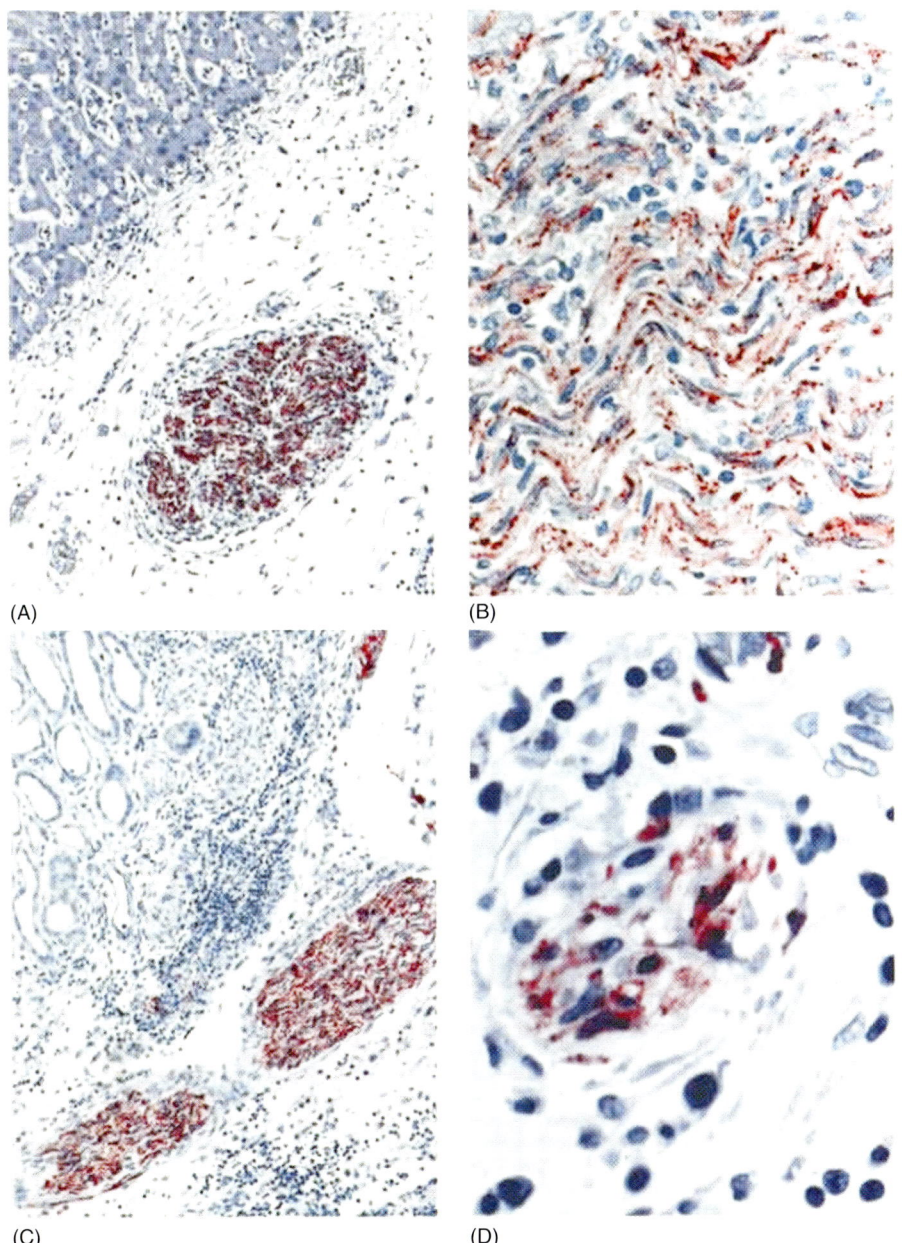

Figure 7.1 Immunohistochemical staining for rabies virus antigen (red) in peripheral nerves of the liver (A and B), kidney (C) and arterial graft (D) transplants. (Reproduced with permission from Srinivasan *et al.* Transmission of rabies virus from an organ donor to four transplant recipients. *New England Journal of Medicine* **352**, 1103–1111, 2005 Copyright © 2005, Massachusetts Medical Society. All rights reserved.)

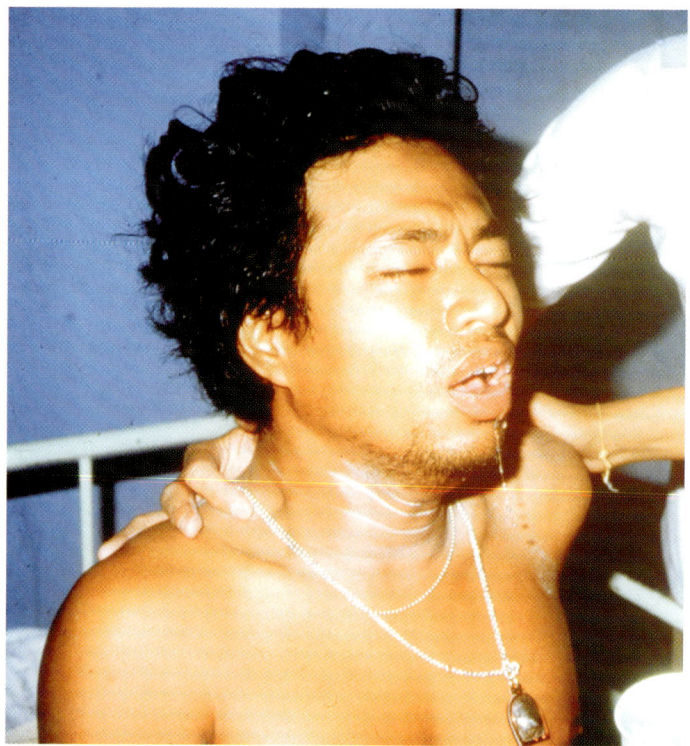

Figure 7.3 Hydrophobic spasm of inspiratory muscles associated with terror in a patient with furious rabies encephalitis attempting to swallow water. (Copyright D.A. Warrell, Oxford, UK).

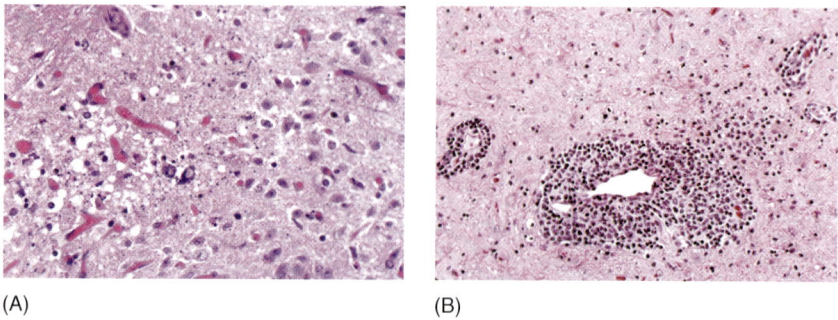

(A) (B)

Figure 8.6 Medial geniculate body of the thalamus of cat 1, which was killed at 136 weeks post-infection, showing degenerative neuronal changes with vacuolation (A) and massive perivascular lymphocytic and plasmacytic infiltration (B), which were seen throughout the brain. Hematoxylin and eosin; magnifications: A, ×115; B, ×60. (Courtesy of Dr Frederick A. Murphy, University of Texas Medical Branch, Galveston, TX.)

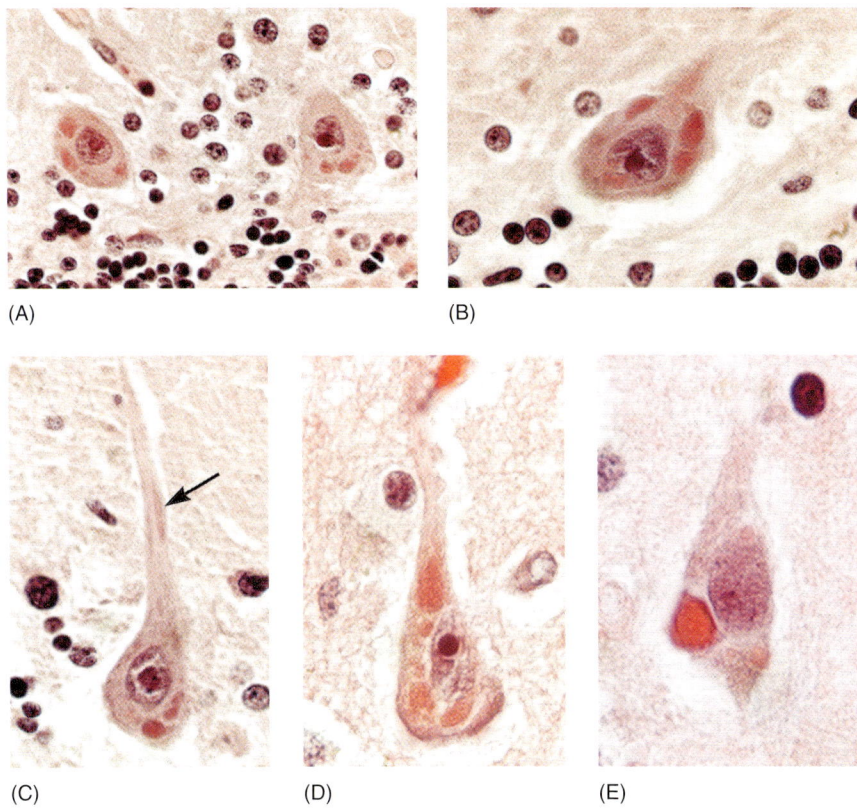

Figure 9.3 Hematoxylin and eosin (HE)-stained sections showing Negri bodies in the perikarya of (A–C) cerebellar Purkinje cells and (D, E) pyramidal neurons in the cerebral cortex of human rabies cases. The arrow in (C) indicates a Negri body in an apical dendrite. (Magnifications: A, ×315, B, ×460, C, ×550, D, ×730, E, ×865).

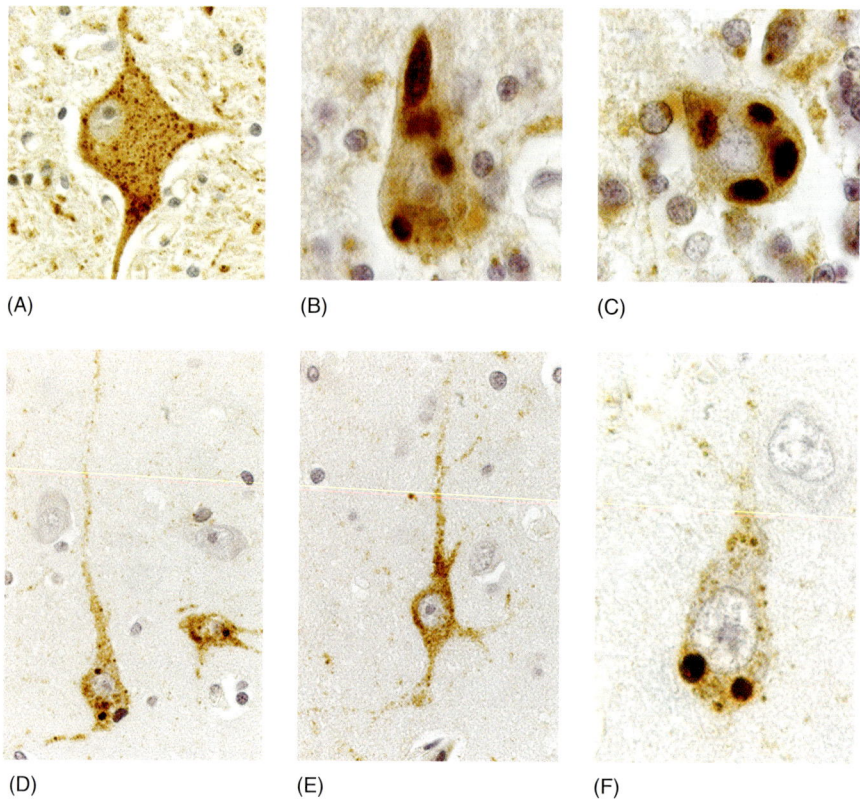

Figure 9.5 Immunoperoxidase staining for rabies virus antigen (mouse monoclonal anti-rabies virus nucleocapsid protein IgG) in human rabies cases. (A) Motoneuron in anterior horn of spinal cord; (B, C) cerebellar Purkinje cells; (D–F) pyramidal neurons in cerebral cortex. The larger immuno-labeled masses correspond with Negri bodies. (Magnifications: A, ×256, B, ×535, C, ×567, D, ×300, E, ×290, F, ×516.)

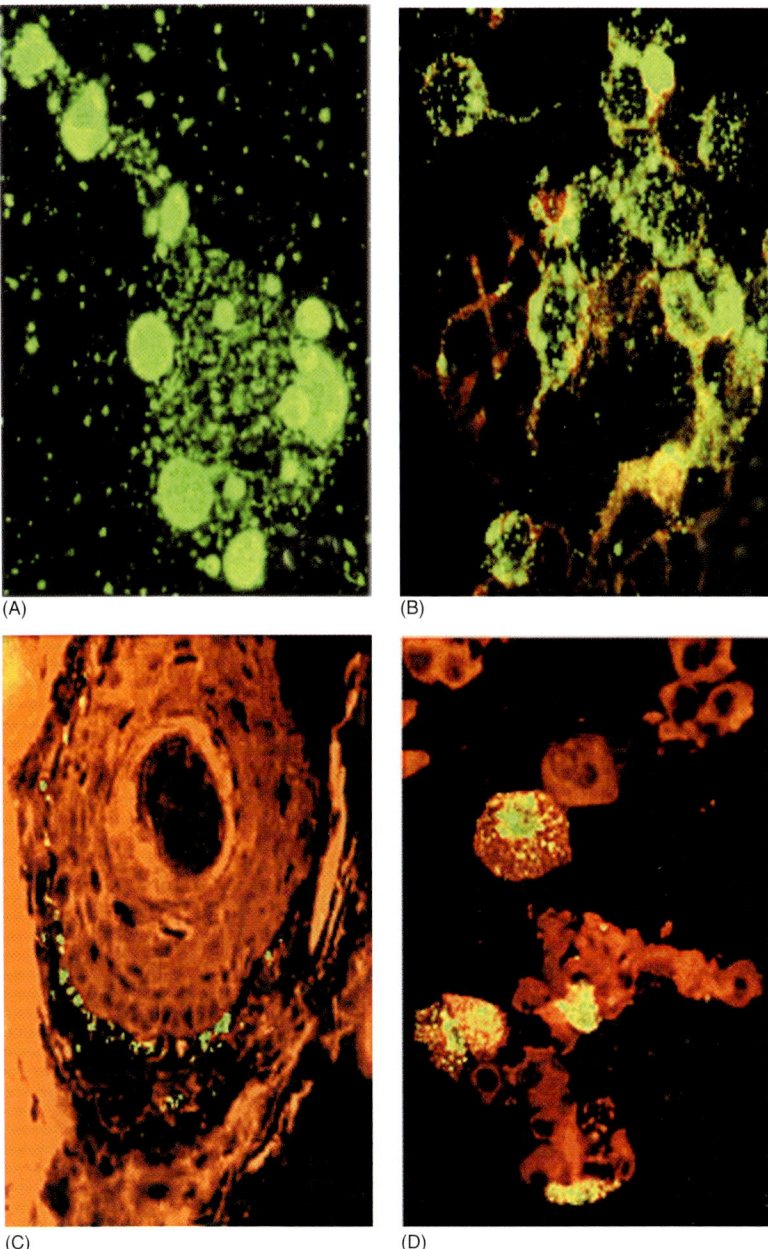

Figure 10.1   (A) Purkinje cell from the cerebellum of a rabies virus-infected bovine showing large intracytoplasmic inclusions and smaller particulate antigen. The DFA method (slip smear; ×540 magnification). (B) Rabies virus-infected monolayer of murine neuroblastoma cells (DFA method with Evans blue counterstain; ×250 magnification). (C) Human hair follicle from nuchal skin biopsy. Nerve cells surrounding follicle are revealed by specific fluorescence associated with presence of rabies antigen (DFA method on frozen section, Evans blue counterstain; ×250 magnification). (D) Human corneal impression with several infected epithelial cells containing inclusions of specifically labeled rabies antigen (DFA method with Evans blue counterstain; ×360 magnification).

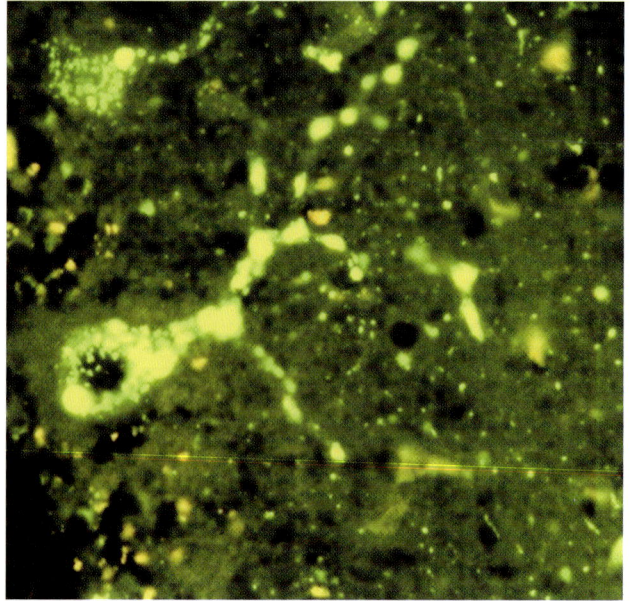

Figure 10.2 Raccoon rabies virus variant infected raccoon cerebellum. Indirect immunofluorescence on formalin-fixed, paraffin embedded section. Two Purkinje cells are shown, one with large intracytoplasmic inclusions, ×540. Primary mouse monoclonal antibody kindly provided by Dr Alex Wandeler (CFIA, Nepean, Ontario, Canada).

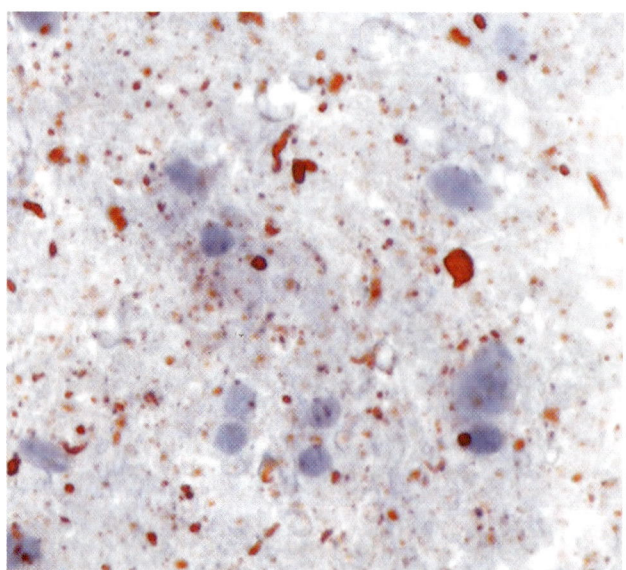

Figure 10.3 Raccoon rabies virus variant infected raccoon brain. Direct Rapid Immunohistochemical Test (DRIT). Gill's hematoxylin formulation 2 counterstain, ×630. Photo kindly provided by Michael Niezgoda, CDC Rabies Unit, Atlanta GA.

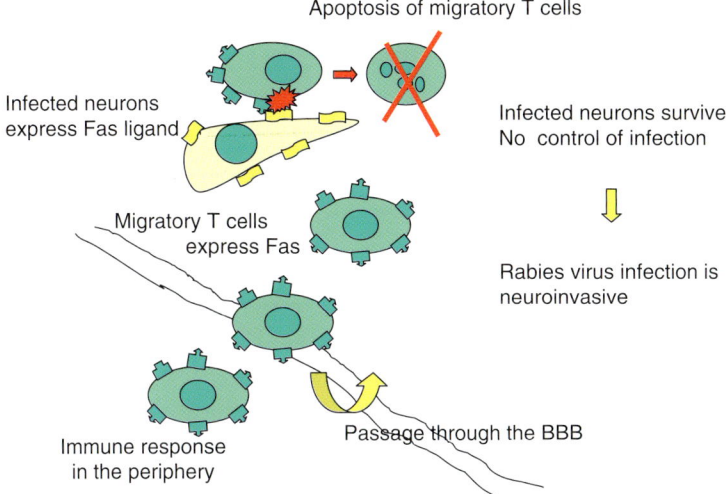

Figure 12.3 Scheme of the mechanisms adopted by acute rabies virus strains to subvert the host defenses. Activated T cells cross the blood–brain barrier (BBB) expressing Fas attracted by cytokines such as TNF-α secreted locally in reaction to the infection. FasL is upregulated by infected neurons. T cells undergo apoptosis after interaction of Fas with FasL. As a consequence, the virus subverts the host immune response and favors its propagation through the NS.

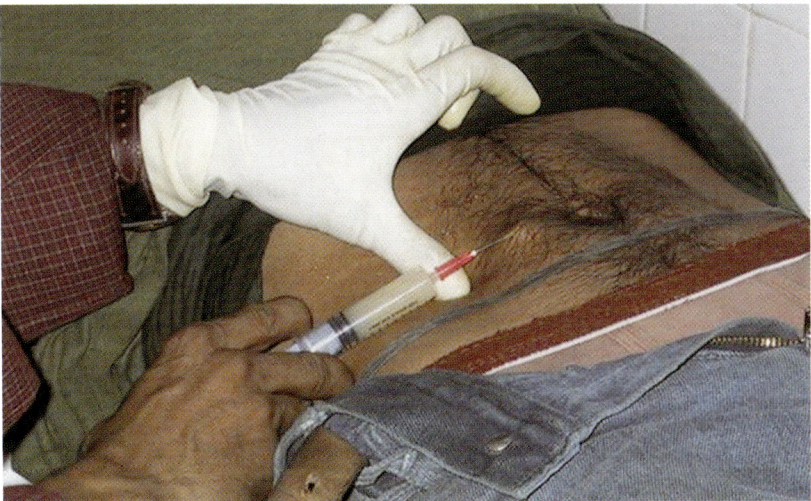

Figure 16.1 The administration of Semple nerve tissue vaccine into the abdominal wall of a patient presenting at an anti-rabies clinic in India with suspected exposure to a rabid dog. (Photo courtesy of Dr Deborah J. Briggs.)

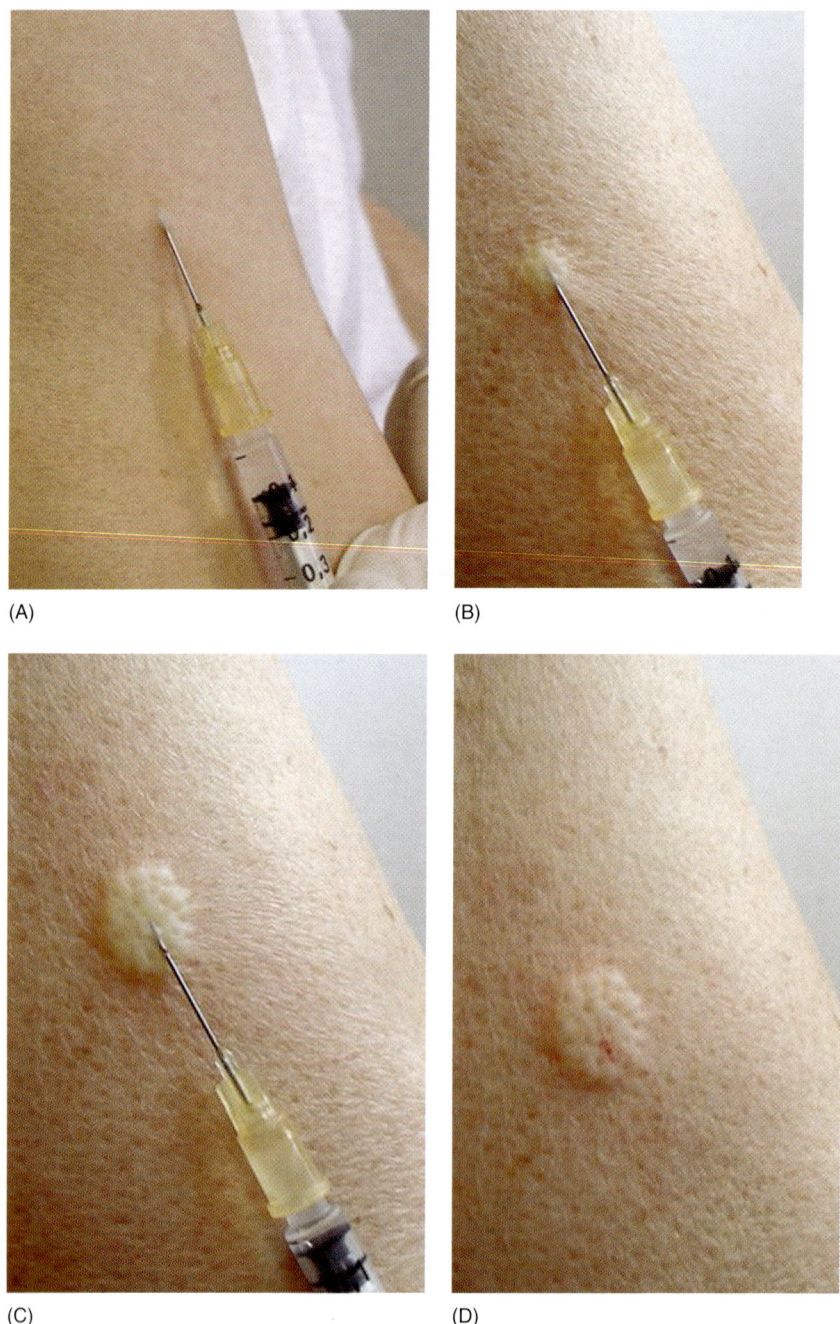

(A)   (B)

(C)   (D)

Figure 16.2 Correct intradermal injection technique of cell culture rabies vaccine. Note the step-wise appearance of the 'bleb' in the skin resembling the surface of an orange peel. (Photo courtesy of Dr Claudius Malerczyk.)

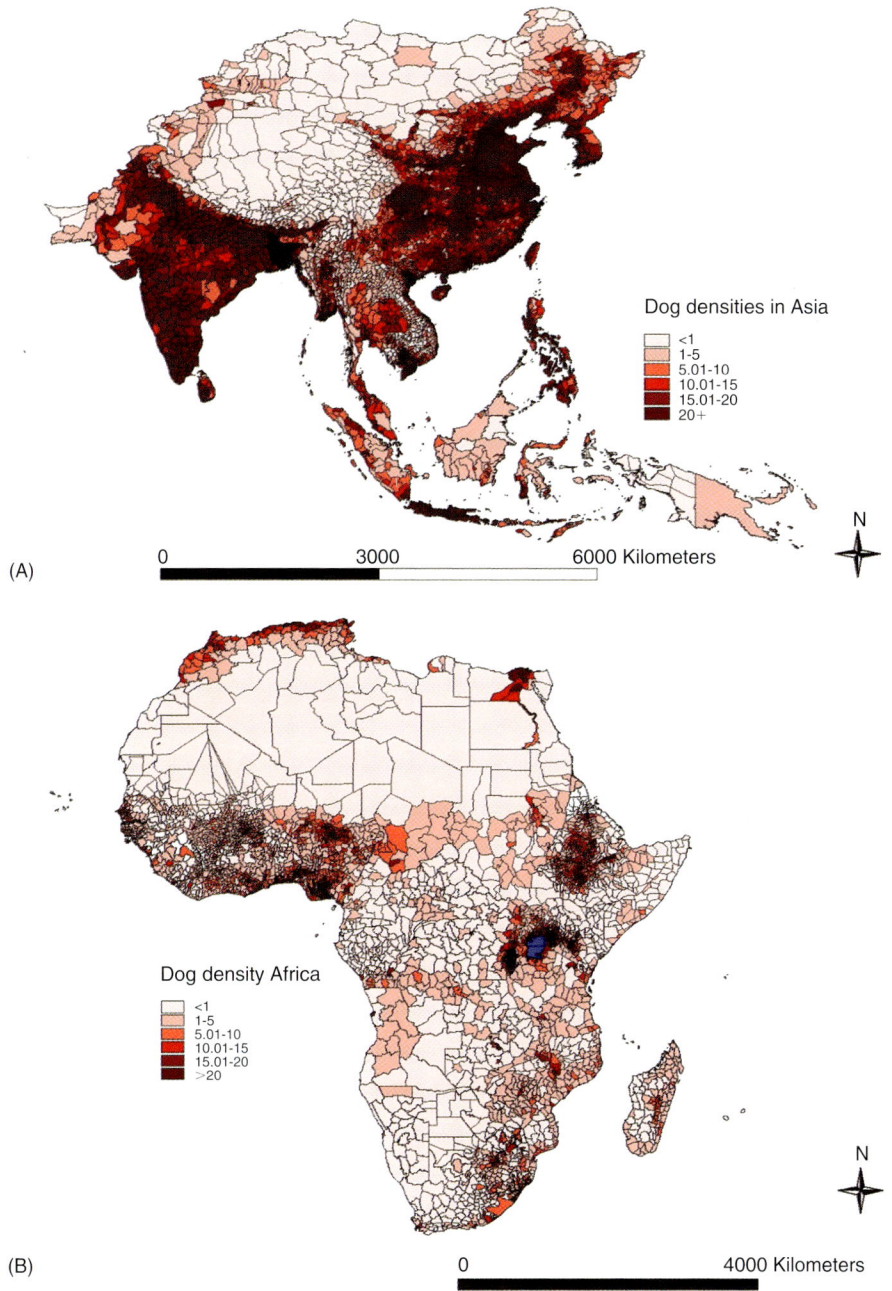

Figure 17.1 (A) The density of domestic dogs in Asia, using human densities derived from an Asia population density dataset and a human:dog ratio of 7.5 for urban settings and 14.3 for rural settings. (B) The density of domestic dogs in Africa, using human densities derived from an Africa population density dataset and a human:dog ratio of 21.2 for urban settings and 7.4 for rural settings (for further details see Knobel *et al.*, 2005).

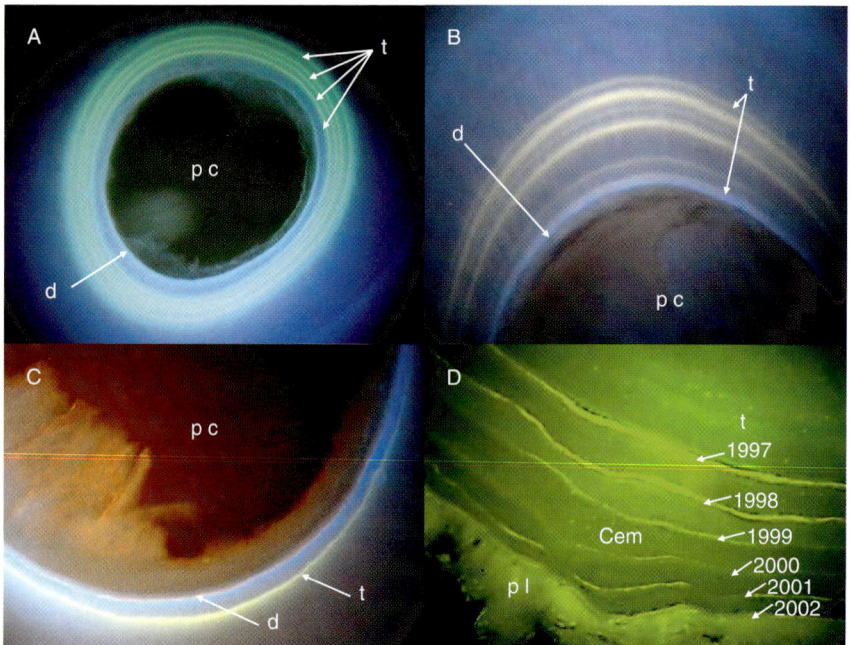

Figure 18.4 Sections of canine teeth showing yellow fluorescent lines from the ingestion of rabies vaccine baits containing 150 mg of tetracycline as a biomarker (undecalcified cross-sections; ultraviolet fluorescence; ×100). (A) Captive raccoon with four tetracycline lines from RABORAL V-RG® baits fed 44, 35, 29 and 22 days before death. 'd', growing edge of dentin at time of death; 'p c', pulp cavity; 't', tetracycline lines. (B) Wild juvenile coyote with a regimen of eight or more tetracycline lines from vaccine baits distributed by aircraft. (C) Wild juvenile coyote with a single tetracycline line from a bait eaten just prior to death. (D) Wild, 6-year-old coyote with one tetracycline line per year in six annual cementum growth zones. Cem, cementum; pl, periodontal ligament. (Figure 18.4 A (® Merial Ltd, LLC.); Figures 18.4 B, C, D (Texas Department of State Health Services, Zoonosis Control; Sidwa *et al.*, 2005).

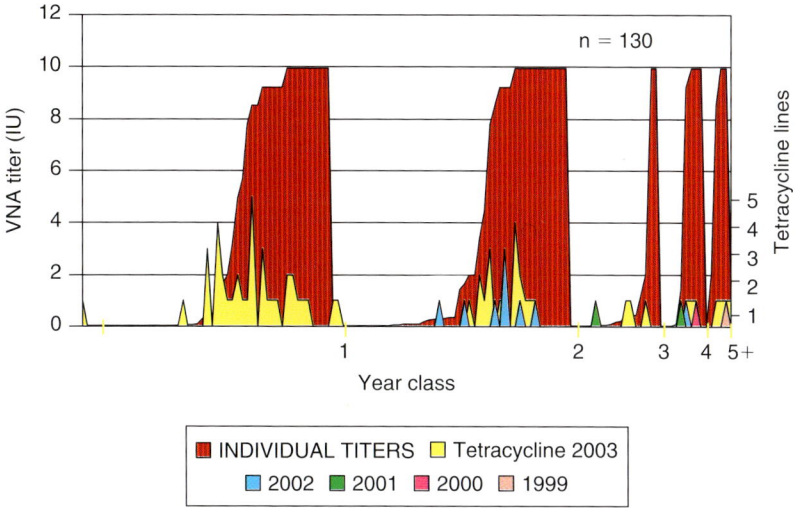

Figure 18.5 2003 West Texas Oral Rabies Vaccination Program. Individual gray fox Virus Neutralizing Antibody titers correlated with tetracycline biomarker line profiles by year class (Department of Defense, Food and Diagnostic Laboratory, San Antonio, TX) and (Texas Department of State Health Services, Zoonosis Control; Sidwa *et al.*, 2005).

Linhart, S.B. and Kennelly, J.J. (1967). Fluorescent bone labelling of coyotes with demethylchlorte-tracycline. *Journal of Wildlife Management* **31**, 317–321.

Linhart, S., Creekmore, T., Corn, J., Whitney, M., Snyder, B and Nettles, V. (1993). Evaluation of baits for oral rabies vaccination of mongooses: pilot field trials in Antigua, West Indies. *Journal of Wildlife Diseases* **29**, 290–294.

Linhart, S.B., Kappeler, A. and Windberg, L. (1997b). A review of baits and bait delivery systems for free-ranging carnivores and ungulates. In: *Contraception in Wildlife Management* (T.J. Kreeger, ed.). pp. 69–132. Technical Bulletin No. 1853. Washington: US Department of Agriculture, Animal and Plant Health Inspection Service.

Linhart, S., King, R., Zamir, S., Naveh, U., Davidson, M. and Perl, S. (1997a). Oral rabies vaccination of red foxes and golden jackals in Israel; preliminary bait evaluation. *Revue Scientific et Technique de L'Office International des Epizootics* **16**, 874–880.

Linhart, S., Wlodkowski, J., Kavanaugh, D. *et al.* (2002). A new flavour-coated sachet bait for deliv-ering oral rabies vaccine to raccoons and coyotes. *Journal of Wildlife Diseases* **38**, 363–377.

Lutze-Wallace, C., Sapp, T., Sidhu, M. and Wandeler, A. (1995). In vitro assessments of the genetic sta-bility of a live recombinant human adenovirus vaccine against rabies. *Canadian Journal of Veterinary Research* **59**, 157–160.

MacInnes, C. (1988). Control of wildlife rabies: the Americas. In: *Rabies* (J. Campbell and K. Charlton, eds). pp. 381–405. Boston: Kluwer Academic.

MacInnes, C.D., Johnston, D.H., Bachmann, P. *et al.* (1992). Design considerations for large-scale aerial distribution of rabies vaccine-baits in Ontario. In: *Wildlife Rabies Control* (K. Bögel, F.-X. Meslin and M. Kaplan, eds). pp.160–167. Royal Tunbridge Wells: Wells Medical.

MacInnes, C.D., Smith, S.M., Tinline, R.R. *et al.* (2001). Elimination of rabies from red foxes in east-ern Ontario. *Journal of Wildlife Diseases* **37**, 119–132.

Marks, C. and Bloomfield, T. (1999). Bait uptake by foxes (*Vulpes vulpes*) in urban Melbourne: the potential of oral vaccination for rabies control. *Wildlife Research* **26**, 777–787.

Masson, E., Bruyère-Masson, V., Vuillaume, P., Lemoyne, S. and Aubert, M. (1999). Rabies oral vac-cination of foxes during the summer with the VRG vaccine bait. *Veterinary Research* **30**, 595–605.

Matouch, O. and Polak, L. (1982). Rabies epizootiology and control in Czechoslovakia. *Comparative Immunology Microbiology and Infectious Diseases* **5**, 303–307.

Matouch, O. and Vitasek, J. (2005). Elimination of rabies in the Czech Republic by oral vaccination of foxes. *WHO Rabies Bulletin Europe* **29** (1), 10–15.

McGuill, M.W., Kreindel, S.M., DeMaria, A. and Rupprecht, C.E. (1997). Knowledge and attitudes of residents in two areas of Massachusetts about rabies and an oral vaccination program in wildlife. *Journal of the American Veterinary Medical Association* **211**, 305–309.

McGuill, M.W., Kreindel, S.M., DeMaria, A. *et al.* (1998). Human contact with bait containing vaccine for control of rabies in wildlife. *Journal of the American Veterinary Medical Association* **213**, 1413–1417.

Meltzer, M.I. (1996). Assessing the cost and benefits of an oral vaccine for raccoon rabies: A possi-ble model. *Emerging Infectious Diseases* **2**, 343–349.

Meltzer, M. and Rupprecht, C. (1998). Economics of the prevention and control of rabies. Part 2: rabies in dogs, livestock and wildlife. *Pharmacoeconomics* **14**, 481–498.

Meslin, F., Fishbein, D. and Matter, H. (1994). Rationale and prospects for rabies elimination in devel-oping countries. In: *Lyssaviruses* (C. Rupprecht, B. Dietzschold and H. Koprowski, eds). pp. 1–26. Berlin: Springer-Verlag.

Messenger, S.L., Rupprecht, C.E. and Smith, J.S. (2002). Bats, emerging virus infections, and the rabies paradigm. In: *Bat Ecology* (T.H. Kunz and M.B. Fenton, eds). Chicago: University of Chicago Press.

Morner, T., Obendorf, D., Artois, M. and Woodford, M. (2002). Surveillance and monitoring of wildlife diseases. *Revue Scientific et Technique de L'Office International des Epizootics* **21**, 67–76.

Müller, J. (1967). The present rabies situation in Denmark. *Bulletin of the Office of International Epizootics* **67**, 3–4.

Müller, T.F., Stöhr, K., Teuffert, J. and Stöhr, P. (1993). Erfahrungen mit der Flügzeugausbringung von Ködern zur oralen Immunisierung der Füchse gegen Tollwut in Ostdeutschland. *Deutsche Tierarztl Wochenschrift* **100**, 203–207.

Müller, T.F., Schuster, P., Vos, A.C., Selhorst, T., Wenzel, U.D. and Neubert, A.M. (2001a). Effect of maternal immunity on the immune response of young foxes to oral vaccination with SAD B19. *American Journal of Veterinary Research* **62**, 1154–1158.

Muller, T., Vos, A., Selhorst, T. *et al.* (2001b). Is it possible to orally vaccinate juvenile red foxes against rabies in spring campaigns? *Journal of Wildlife Diseases* **37**, 791–797.

Müller, W.W. (1997). Where do we stand with oral vaccination of foxes against rabies in Europe? *Archives of Virology* **13**, 83–94.

Nunan, C.P., MacInnes, C.D., Bachmann, P., Johnston, D.H. and Watt, I.D. (1994). Background prevalence of tetracycline-like fluorescence in teeth of free ranging red foxes (*Vulpes vulpes*), striped skunks (*Mephitis mephitis*) and raccoons (*Procyon lotor*) in Ontario, Canada. *Journal of Wildlife Diseases* **30**, 112–114.

Nyberg, M., Kulonen, K., Neuvonen, E., Ek-Kommonen, C., Nuorgam, M. and Westerling, B. (1992). An epidemic of sylvatic rabies in Finland – Descriptive epidemiology and results of oral vaccination. *Acta Veterinaria Scandinavica* **33**, 43–57.

Office of International Epizootics. (2003). Comments on selected list B diseases. *World Animal Health* **1**, 1–27. http://www.oie.int/ (accessed October 2005).

Olson, C. and Werner, P. (1999). Oral rabies vaccine contact by raccoons and nontarget species in a field trial in Florida. *Journal of Wildlife Diseases* **35**, 687–695.

Olson, C.A., Mitchell, K.D. and Werner, P.A. (2000). Bait ingestion by free-ranging raccoons and nontarget species in an oral rabies vaccine field trial in Florida. *Journal of Wildlife Diseases* **36**, 734–743.

Pastoret, P. and Brochier, B. (1999). Epidemiology and control of fox rabies in Europe. *Vaccine* **17**, 1750–1754.

Pastoret, P., Frisch, R., Blancou, J., Wolff, F., Brochier, B. and Schneider, L. (1997). Campagne internationale de vaccination antirabique du renard par voie orale menee au grans-duche de Luxembourg, en Belgique et en France. *Annales de Medecine Veterinaire* **131**, 441–447.

Pastoret, P., Kappeler, A. and Aubert, M. (2004). European rabies control and its history. In: *Historical perspective of rabies in Europe and the Mediterranean basin* (A. King, A. Fooks, M. Aubert and A. Wandeler, eds). pp. 337–350. Paris: OIE. World Organization for Animal Health.

Perry, B.D., Brooks, R., Foggin, C.M., Bleakley, J., Johnston, D.H. and Hill, F.W.G. (1988). A baiting system suitable for the delivery of oral rabies vaccine to dog populations in Zimbabwe. *Veterinary Record* **123**, 76–79.

Riley, S., Hadidian, J. and Manski, D. (1998). Population density, survival, and rabies in raccoons in an urban national park. *Canadian Journal of Zoology* **76**, 1153–1164.

Robbins, A.H., Borden, M.D., Windmiller, B.S. *et al.* (1998). Prevention of the spread of rabies to wildlife by oral vaccination of raccoons in Massachusetts. *Journal of the American Veterinary Medical Association* **213**, 1407–1412.

Rosatte, R. (2000). Management of raccoons (Procyon lotor) in Ontario, Canada: do human intervention and disease have significant impact on raccoon populations? *Mammalia* **64**, 369–390.

Rosatte, R.C. (2001). Long distance movement by a coyote and a red fox in Ontario: Implications for disease-spread. *Canadian Field Naturalist* **116**, 129–131.

Rosatte, R.C. and Gunson, J.R. (1984). Presence of neutralizing antibodies to rabies virus in striped skunks from areas free of skunk rabies in Alberta. *Journal of Wildlife Diseases* **20**, 171–176.

Rosatte, R. and Lariviere, S. (2003). Skunks. In: *Wild Mammals of North America; biology, management and conservation*, 2nd edn (G. Feldhamer, B. Thompson and J. Chapman, eds). pp. 692–707. Baltimore: Johns Hopkins University Press.

Rosatte, R. and Lawson, K. (2001). Acceptance of baits for delivery of oral rabies vaccine to raccoons. *Journal of Wildlife Diseases* **37**, 730–739.

Rosatte, R.C. and MacInnes, C.D. (1989). Relocation of city raccoons. In: *9th Great Plains Wildlife Damage Control Workshop Proceedings* (A.J. Bjugstad, D.W. Uresk and R.H. Hamre, eds). pp. 87–92. USDA Forest Service General Technical Report RM-171. Fort Collins, Colorado: Great Plains Agricultural Council Publication 127.

Rosatte, R., Allan, M., Warren, R. *et al.* (2005). Movements of two rabid raccoons (*Procyon lotor*) in eastern Ontario, Canada. *Canadian Field Naturalist* **119**, 453–454.

Rosatte, R.C., Donovan, D., Allan, M. *et al.* (2001). Emergency response to raccoon rabies introduction in Ontario. *Journal of Wildlife Diseases* **37**, 265–279.

Rosatte, R., Howard, D., Campbell, J. and MacInnes, C. (1990). Intramuscular vaccination of skunks and raccoons against rabies. *Journal of Wildlife Diseases* **26**, 225–230.

Rosatte, R.C., Lawson, K. and MacInnes, C. (1998). Development of baits to deliver oral rabies vaccine to raccoons in Ontario. *Journal of Wildlife Diseases* **34**, 647–652.

Rosatte, R., MacInnes, C., Power, M. *et al.* (1993). Tactics for the control of wildlife rabies in Ontario, Canada. *Revue Scientific et Technique de L'Office International des Epizootics* **12**, 95–98.

Rosatte, R., MacInnes, C., Williams, R. and Williams, O. (1997). A proactive prevention strategy for raccoon rabies in Ontario, Canada. *Wildlife Society Bulletin* **25**, 110–116.

Rosatte, R., Power, M. and MacInnes, C. (1991). Ecology of urban skunks, raccoons, and foxes in Metropolitan Toronto. In: *Wildlife conservation in Metropolitan environments* (L. Adams and D. Leedy, eds). pp. 31–38. Columbia: National Institute for Urban Wildlife.

Rosatte, R.C., Power, M.J. and MacInnes, C.D. (1992). Trap-vaccinate-release and oral vaccination techniques for rabies control in urban skunks, raccoons and foxes. *Journal of Wildlife Diseases* **28**, 562–571.

Rosatte, R.C., Power, M., Donovan, D. *et al.* (2007). Elimination of arctic variant rabies in red foxes, Metropolitan Toronto. *Emerging Infectious Diseases*, **13**, 25–27.

Rosatte, R., Pybus, M. and Gunson, J. (1986). Population reduction as a factor in the control of skunk rabies in Alberta. *Journal of Wildlife Diseases* **22**, 459–467.

Rosatte, R., Sobey, K., Donovan, D. *et al.* (2006) Behavior, movements, and demographics of rabid raccoons in Ontario, Canada: management implications. *Journal of Wildlife Diseases* **42**, 589–605.

Roscoe, D.E., Holste, W.C., Sorhage, F.E. *et al.* (1998). Efficacy of an oral vaccinia–rabies glycoprotein recombinant vaccine in controlling epidemic raccoon rabies in New Jersey. *Journal of Wildlife Diseases* **34**, 752–763.

Rupprecht, C., Blass, L., Smith, K. *et al.* (2001). Human infection due to recombinant vaccinia-rabies glycoprotein virus. *New England Journal of Medicine* **345**, 582–586.

Rupprecht, C., Hamir, A., Johnston, D. and Koprowski, H. (1988). Efficacy of vaccinia-rabies glycoprotein recombinant virus vaccine in raccoons (*Procyon lotor*). *Reviews of Infectious Diseases* **10** (Suppl. 4), S803–S809.

Rupprecht, C., Hanlon, C. and Slate, D. (2004). Oral vaccination of wildlife against rabies: Opportunities and challenges in prevention and control. In: *Control of Infectious Animal Diseases by Vaccination* (A. Schudel and A. Lombard, eds). *Developmental Biology* **119**, 173–184.

Rupprecht, C.E., Smith, J.S., Fekadu, M. and Childs, J.E. (1995). The ascension of wildlife rabies: A cause for public health concern or intervention? *Emerging Infectious Diseases* **1**, 107–114.

Rupprecht, C.E., Wiktor, T.J., Johnston, D.H. *et al.* (1986). Oral immunization and protection of raccoons (*Procyon lotor*) with a vaccinia–rabies glycoprotein recombinant virus vaccine. *Proceedings of the National Academy of Sciences USA* **83**, 7947–7950.

Russell, C., Smith, D., Childs, J. and Real, L. (2005). Predictive spatial dynamics and strategic planning for raccoon rabies emergence in Ohio. *PLoS Biology* **3**, 382–388.

Sato, G., Itou, T., Shoji, Y. *et al.* (2004). Genetic and phylogenetic analysis of glycoprotein of rabies virus isolated from several species in Brazil. *Journal of Veterinary Medical Science* **66** (7), 747–753.

Schneider, L. and Cox, J. (1988). Eradications of rabies through oral vaccination. The German field trial. In: *Vaccination to Control Rabies in Foxes* (P. Pastoret, B. Brochier, I. Thomas and J. Blancou, eds). pp. 22–38. Luxembourg: Commission of the European Communities.

Schneider, L., Cox, J., Muller, W. and Hohnsbeen, K. (1988). Current oral rabies vaccination in Europe, an interim balance. *Reviews of Infectious Diseases* **10**, S654–S659.

Schnurrenberger, P., Beck, J. and Peden, D. (1964). Skunk rabies in Ohio. *Public Health Reports* **79**, 161–166.

Selhorst, T., Thulke, H.H. and Muller, T. (2000). Threshold analysis of cost-efficient oral vaccination strategies against rabies in fox populations. *Proceedings of the Society of Veterinary Epidemiology Preventative Medicine 2000*, 71–84.

Selhorst, T., Thulke, H. and Muller, T. (2001). Cost-efficient vaccination of foxes (*Vulpes vulpes*) against rabies and the need for a new baiting strategy. *Preventive Veterinary Medicine* **51**, 95–109.

Sidwa, T., Wilson, P., Moore, G. *et al.* (2005). Evaluation of oral rabies vaccination programs for control of rabies epizootics in coyotes and gray foxes: 1995–2003. *Journal of the American Veterinary Medical Association* **227**, 785–792.

Slate, D., Rupprecht, C., Rooney, J., Donovan, D., Lein, D. and Chipman, R. (2005). Status of oral rabies vaccination in wild carnivores in the United States. *Virus Research* **111**, 68–76.

Smith, D.L., Brendan, L., Waller, L., Childs, J. and Real, L. (2002). Predicting the spatial dynamics of rabies epidemics on heterogeneous landscapes. *PNAS* **99**, 3668–3672.

Smith, G. (2002). The role of the badger (*Meles meles*) in rabies epizootiology and the implications for Great Britain. *Mammal Review* **32**, 12–25.

Smith, G. and Cheeseman, C. (2002). A mathematical model for the control of diseases in wildlife populations: culling, vaccination and fertility control. *Ecological Modeling* **150**, 45–53.

Smith, G.C. and Harris, S. (1991). Rabies in urban foxes in Britain: The use of a spatial stochastic simulation model to examine the pattern of spread and evaluate the efficacy of different control regimes. *Philosophical Transactions of the Royal Society of London B* **334**, 459–479.

Smith, G. and Wilkinson, W. (2003). Modeling control of rabies outbreaks in red fox populations to evaluate culling, vaccination, and vaccination combined with fertility control. *Journal of Wildlife Diseases* **39**, 278–286.

Smith, J.S. (1995). Rabies virus. In: *Manual of Clinical Microbiology* (P. Murray, E. Baron, M. Pfaller, F. Tenover and R. Yolken, eds). pp. 997–1003. Washington, DC: American Society for Microbiology.

Steck, F., Wandeler, A., Bichsel, P., Capt, S. and Schneider, L. (1982a). Oral immunisation of foxes against rabies. A field study. *Zentralblatt für Veterinärmedizin* **29**, 372–396.

Steck, F., Wandeler, A., Bichsel, P., Capt, S., Hafliger, U. and Schneider, L. (1982b). Oral immunisation of foxes against rabies. Laboratory and field studies. *Comparative Immunology Microbiology and Infectious Diseases* **5**, 165–171.

Steelman, H., Henke, S. and Moore, G. (1998). Gray fox response to baits and attractants for oral rabies vaccination. *Journal of Wildlife Diseases* **34**, 764–770.

Stöhr, K. and Meslin, F.-X. (1996). Progress and setbacks in the oral immunisation of foxes against rabies in Europe. *Veterinary Record* **13**, 32–35.

Suppo, C., Nauline, J., Langlais, M. and Artois, M. (2000). A modeling approach to vaccination and contraception programmes for rabies control in fox populations. *Proceedings of the Royal Society of London B* **267**, 1575–1582.

Thulke, H., Tischendorf, L., Staubach, C. *et al.* (2000). The spatiotemporal dynamics of a post-vaccination resurgence of rabies in foxes and emergency vaccination planning. *Preventative Veterinary Medicine* **47**, 1–21.

Tinline, R. (2005). Natural barriers and rabies control in the North East. Presentation to *North East United States Animal Health Association Annual Meeting*, Connecticut, April.

Tinline, R. and Ball, D. (2003). Why does rabies persist or die out in a region? Presentation to the *14th Annual Rabies in the Americas Meeting*, Philadelphia, October.

Tinline, R. and MacInnes, C. (2004). Ecogeographic patterns of rabies in southern Ontario based on time series analysis. *Journal of Wildlife Diseases* **40**, 212–221.

Tinline, R. and Sheriff, C. (2004). Breaching the barrier: decisions to be made on when to vaccinate. Presentation to *North East United States Animal Health Association Annual Meeting*, Vermont, April.

Tinline, R.R., Ball, D.G.A., Nunan, C.P. and Venodrai, T. (1999). The impact of flight line spacing and baiting density on the uptake of oral rabies vaccine via aerial distribution (Abstract). *10th Annual Rabies in the Americas Meeting*, San Diego.

Tischendorf, L., Thulke, H., Staubach, C. *et al.* (1998). Chance and risk of controlling rabies in large-scale and long-term immunized fox populations. *Proceedings of the Royal Society of London B* **265**, 839–846.

Totton, S., Rosatte, R., Tinline, R. and Bigler, L. (2004). Seasonal home ranges of raccoons, *Procyon lotor*, using a common feeding site in rural eastern Ontario: rabies management implications. *Canadian Field-Naturalist* **118**, 65–71.

Totton, S., Tinline, R., Rosatte, R. and Bigler, L. (2002). Contact rates of raccoons (*Procyon lotor*) at a communal feeding site in rural eastern Ontario. *Journal of Wildlife Diseases* **38**, 313–319.

United States Department of Agriculture (2004). Cooperative rabies management program, National report 2004. United States Department of Agriculture, Animal and Plant Health Inspection Service, Wildlife Services, Compact Disk (CD). Concord, New Hampshire.

Voigt, D.R., Tinline, R.R. and Broekhoven, L.H. (1985). Spatial simulation model for rabies control. In: *Population Dynamics of Rabies in Wildlife* (P.J. Bacon, ed.). pp. 311–349. London: Academic Press.

Vos, A. (2003). Oral vaccination against rabies and the behavioural ecology of the red fox (*Vulpes vulpes*). *Journal of Veterinary Medicine* **50**, 477–483.

Vos, A., Muller, T., Selhorst, T., Schuster, P., Neubert, A. and Schluter, H. (2001). Optimising spring oral vaccination campaigns against rabies. *Deutsche Tierarztliche Wochenschrift* **108**, 55–59.

Vos, A., Pommerening, E., Neubert, L., Kachel, S. and Neubert, A. (2002). Safety studies of the oral rabies vaccine SAD B19 in striped skunk (*Mephitis mephitis*). *Journal of Wildlife Diseases* **38,** 428–431.

Vrzal, V. and Matouch, O. (1996). Annual testing of immunity in foxes after oral rabies immunization. *Veterinary Medicine (Praha)* **4**, 107–111.

Vuillaume, P., Bruyere, V. and Aubert, M. (1998). Comparison of the effectiveness of two protocols of antirabies bait distribution for foxes (*Vulpes vulpes*). *Veterinary Research* **29**, 537–546.

Wachendorfer, G., Frost, J., Gutmann, B. *et al.* (1986). W. Erfahrungen mit der oralen immunisierung von Fuchsen gegen Tollwut in Hessen. *Tierarztliche Praxis* **14**, 185–196.

Wandeler, A. (1988). Control of wildlife rabies: Europe. In: *Rabies* (J. Campbell and K. Charlton, eds). pp. 365–380. Boston: Kluwer Academic.

Wandeler, A. (1991). Oral immunization of wildlife. In: *The Natural History of Rabies*, 2nd edn (G. Baer, ed.). pp. 485–503. Boca Raton: CRC Press.

Wandeler, A. (2000). Oral immunization against rabies: afterthoughts and foresight. *Schweizer Archiv fur Tierheilkunde* **142**, 455–462.

Wandeler, A., Bauder, W., Prochaska, S. and Steck, F. (1982). Small mammal studies in a SAD baiting area. *Comparative Immunology Microbiology and Infectious Diseases* **5**, 173–176.

Wandeler, A.I., Nadin-Davis, S.A., Tinline, R.R. and Rupprecht, C.E. (1994). Rabies epizootiology: An ecological and evolutionary perspective. *Current Topics in Microbiology and Immunology* **186**, 297–324.

Westerling, B. (1991). Rabies in Finland and its control 1988–90. *Suomen Riista* **37**, 93–100.

Whitney, H., Johnston, D.H., Wandeler, A., Nadin-Davis, S. and Muldoon, F. (2005). Elimination of rabies from the island of Newfoundland: 2002–2004. Rabies in the Americas Conference abstract, Ottawa.

Wiktor, T., Macfarlan, R., Reagan, K. *et al.* (1984). Protection from rabies by a vaccinia virus recombinant containing the rabies virus glycoprotein gene. *Proceedings of the National Academy of Sciences USA* **81**, 7194–7198.

Wilde, H., Khawplod, P., Khamoltham, T. *et al.* (2005). Rabies control in South and Southeast Asia. *Vaccine* **23**, 2284–2289.

Wilhelm, U. and Schneider, L. (1990). Oral immunization of foxes against rabies: practical experiences of a field trial in the Federal Republic of Germany. *Bulletin of the World Health Organization* **68**, 87–92.

Windberg, L.A. (1988). Management implications of coyote spacing patterns in southern Texas *Journal of Wildlife Management* **52**, 632–640.

Winkler, W. and Jenkins, S. (1991). Raccoon rabies. In: *The Natural History of Rabies*, 2nd edn (G. Baer, ed.). pp. 325–340. Boca Raton: CRC Press.

Winkler, W., McLean, R. and Cowart, J. (1975). Vaccination of foxes against rabies using ingested baits. *Journal of Wildlife Diseases* **11**, 382–388.

Woodford, M. and Rossiter, P. (1993). Disease risks associated with wildlife translocation projects. *Revue Scientific et Technique de L'Office International des Epizooties* **12**, 115–135.

World Health Organization (1966). WHO Expert Committee on Rabies. Fifth Report. *WHO Technical Report Series 321*. Geneva: WHO.

Yakobson, B., Manalo, D., Bader, K., Perl, S. and Haber, A. (1998). An epidemiological retrospective study of rabies diagnosis and control in Israel, 1948–1997. *Israel Journal of Veterinary Medicine* **53**, 114–126.

Yarosh, O., Wandeler, A., Graham, F., Campbell, J. and Prevec, L. (1996). Human adenovirus type 5 vectors expressing rabies glycoprotein. *Vaccine* **14**, 1257–1264.

Zanoni, R., Kappeler, A., Muller, U., Wandeler, A. and Breitenmoser, U. (2000). Rabies-free status of Switzerland after 30 years of fox rabies. *Schweizer Archiv fur Tierheilkunde* **142**, 423–429.

Zienius, D., Bagdonas, J. and Dranseika, A. (2003). Epidemiological situation of rabies in Lithuania from 1990–2000. *Veterinary Microbiology* **93**, 91–100.

19

Future Developments and Challenges

ALAN C. JACKSON[1] AND WILLIAM H. WUNNER[2]

Departments of Medicine (Neurology) and of Microbiology and Immunology, Queen's University, Kingston, ON K7L 3N6, Canada; currently Departments of Internal Medicine (Neurology) and of Medical Microbiology, University of Manitoba, Winnipeg, MB R3A IR9, Canada[1]
The Wistar Institute, Philadelphia, PA 19104-4268, USA[2]

1 INTRODUCTION

Enormous strides have been made, most marked over the past three decades and also over the past five years, since the first edition of 'Rabies' (Jackson and Wunner, 2002), which described the current status of rabies and attempts to conquer this deadly disease. What then is likely to occur in the future to allow for a better understanding of rabies and for improved control of rabies? Notwithstanding the 55 000 human deaths per year that occur due to rabies worldwide (World Health Organization, 2005) and the millions of persons who require vaccination after exposure to rabid domestic and wild animals, recent improvements in rabies control and reductions in human cases in Latin America (Belotto *et al.*, 2005) and Thailand (Denduangboripant *et al.*, 2005) are very impressive. Economic factors and a lack of political will, probably partially related to many competing health problems, have impeded the control of rabies in much of Asia and Africa. Although rabies is an important disease of children, this aspect of the disease has not received much attention and, consequently, rabies has not received the necessary resources for control throughout the world. This final chapter identifies some of the critical areas that will likely be the focus of ongoing research and development of strategies and approaches to help overcome the disease and threat of animal rabies to public health worldwide.

2 PATHOGENESIS

Our understanding of many aspects of the pathogenesis of rabies is incomplete and many questions remain unanswered. For example, the precise steps that occur during the sometimes long incubation period in rabies have not been defined in terms of what cell types are infected and when the virus enters its life cycle. Studies on the pathogenesis of rabies have focused predominantly on rodent models using laboratory 'fixed' rabies virus strains and rarely 'street' strains and, at times, using

Rabies, second edition. Edited by Alan C. Jackson and William H. Wunner
ISBN 978-012-369366-2. Copyright Elsevier Inc. 2007

unnatural routes of inoculation. Although these experimental animal models are convenient and less expensive than more natural models, it is time that more appropriate models are used such as bats, dogs and even non-human primates. The development of these models will be an important area of research for the future. The fundamental mechanisms underlying brain dysfunction in rabies remain elusive, although structural abnormalities affecting neuronal processes remain a possibility. Profiles of host gene expression through microarray analysis using real-time polymerase chain reaction (PCR) to confirm microarray data is a technology (Wang *et al.*, 2005) that needs to be applied to understand better the virus–host relationship and pathogenetic mechanisms in rabies. Coupling microarray technology and real-time PCR will allow almost all genes involved in the pathogenesis of rabies virus to be monitored, taking the 'systems biology' approach, and evaluated as to whether they are upregulated, downregulated or unchanged in relation to CNS inflammation, neuronal dysfunction or neuronal cell death by apoptosis. It is hoped that a better understanding of how rabies virus infection produces fatal clinical disease will lead to the development of novel, effective therapies for human rabies patients.

3 EPIDEMIOLOGY

As a result of successful rabies control measures in Latin America (Belotto *et al.*, 2005), it has become apparent that rabies also occurs in a variety of wildlife species, including insectivorous bats, foxes, raccoons, mongooses, skunks and monkeys, as well as the longstanding problem of rabies in vampire bats that affects humans and cattle. When dog rabies is prevalent in a country, rabies in wildlife species is often overlooked. This is likely the case in Asia and Africa, where other lyssaviruses are present in wildlife, some having been recently identified in bats, which have the potential to cause disease in humans and animals. Additional studies are necessary to investigate the epizootiology of these new lyssaviruses and their potential impact on human and domestic animal health in the regions where the bat vector species from which these isolates were discovered reside.

Human and canine rabies in Asia and Africa remain a significant problem, although quantitative information is limited (Knobel *et al.*, 2005). Recently, the number of cases in China has increased to over 2500 fatal cases per year, with rural areas in the southern part of the country predominantly affected (Zhang *et al.*, 2006). Increasing dog populations are likely a factor. Thailand has instituted human preventive measures and the number of cases has dropped significantly to relatively low levels (Denduangboripant *et al.*, 2005), indicating that human disease can be prevented in Asian countries when there is a strong commitment. Hopefully, this commitment will come soon from many other Asian countries.

During 2004 and 2005, there were many human rabies cases due to transmission from vampire bats to a native population in the Amazon region of northern

Brazil (da Rosa *et al.*, 2006). Interestingly, these cases have been paralytic rabies, similar to what occurred in Trinidad in the 1920s and 1930s, indicating that specific bat rabies virus variants may be responsible for the strong association with paralytic disease. Appropriate implementation of rabies education and preventive measures is important in this remote population before the death toll grows higher.

4 PREVENTION OF HUMAN RABIES

Since about half of rabies victims are children, rabies can be recognized as a more important infectious disease if the years of life lost are factored into evaluating the burden of disease. A disability-adjusted life year (DALY) score is an attempt to evaluate both disability and years of life lost from a disease and the years of life lived with a disability in order to assess disease impact; when evaluated in this way, rabies becomes the seventh most important infectious disease in the world (Coleman, 2004). Indirect effects of rabies, including the burden related to mortality and disability due to the use of nerve-tissue rabies vaccines, can also be expressed as a DALY score (Knobel *et al.*, 2005).

Strategies for the prevention of human rabies include those measures that can be directed at humans and those that can be directed at the main animal vectors. Clearly, the availability and use of modern cell culture-derived rabies vaccines has dramatically reduced the incidence of human rabies in countries that have implemented wide-scale use for post-exposure prophylaxis (PEP). While vaccines from cell cultures are undoubtedly the vaccines of current choice, it must be recognized that they are currently inaccessible to a significant part of the world. Yet, if vaccines were delivered in smaller doses intradermally in government sponsored anti-rabies clinics dramatically to increase PEP, as was done in Thailand, for example, where the incidence of human rabies was reduced by 80% in 15 years (World Health Organization, 2002), many more countries would also be able to reduce the number of human rabies cases. The key to improving the availability of cell culture rabies vaccines, while at the same time reducing the cost of PEP at government sponsored anti-rabies clinics, is to use small doses of vaccine intradermally (Harverson and Wasi, 1984; Chutivongse *et al.*, 1990; Warrell and Warrell, 2000; APCRI, 2006). The use of rabies vaccines of nerve tissue origin should be eliminated worldwide as soon as possible because of the unacceptable risk of neurologic complications. The challenge is to replace these less expensive vaccines with effective alternative strategies for using cell culture-derived rabies vaccines, such as by lowering the dose, or to produce altogether different (next generation) rabies vaccines that can be manufactured much more cheaply and made accessible to a much greater segment of the world's population. While it may not be in the economic interest of pharmaceutical companies to develop new cell culture-derived rabies vaccines for humans, which is what developing countries need, there are new lyssavirus vaccines on the horizon, now in development, that in 5–10 years might replace the traditional, more expensive cell culture vaccines and provide a broader protective coverage (Hanlon *et al.*, 2005; Nel, 2005) (see Chapter 15). Other challenges to be faced include whether widespread pre-exposure rabies immunization of children should be part of a 'risk-avoidance' strategy in developing countries where canine rabies is endemic and rabies immunoglobulin (RIG), which is recommended as a passive immune treatment for those at risk after exposure, is either too expensive or unavailable. Another is the large number of patients with human immunodeficiency virus (HIV) infection and low $CD4^+$ T lymphocyte counts worldwide (Jaijaroensup *et al.*, 1999; Tantawichien *et al.*, 2001) and this is a particularly important problem in Asia and Africa.

The world production of human rabies immunoglobulin (HRIG) and equine rabies immunoglobulin (ERIG) for administration in combination with vaccine in PEP is currently limited and economic factors significantly restrict the use of these products (Haupt, 1999). As a replacement of HRIG and ERIG, a single monoclonal antibody (MAb) or cocktail of two or more MAbs against the viral glycoprotein (G) that specifically neutralize challenge virus are needed and are being developed (Champion *et al.*, 2000; Bakker *et al.*, 2005; Hanlon *et al.*, 2005). Administration of MAbs instead of HRIG or ERIG with vaccine for the treatment of patients at risk after exposures, in addition to providing passive immunization at the site of virus inoculation, could potentially promote clearance of the infection from the CNS in cases where rabies virus breaks through the blood–brain barrier (Dietzschold *et al.*, 1992; Prosniak *et al.*, 2003). Human MAbs or humanized mouse MAbs would be preferable to conventional mouse MAbs. With the emergence of new bat lyssaviruses, such as the four novel lyssaviruses recently isolated from bats in Eurasia, Aravan (ARAV), Khujand (KHUV), Irkut (IRKV) and West Caucasian bat (WCBV) virus, human rabies PEP may ultimately require additional targeted antiviral biologicals as well as new vaccines that provide virus-selective or broader coverage and protection (Hanlon *et al.*, 2005).

5 DIAGNOSIS AND THERAPY OF HUMAN RABIES

An improved understanding of the pathogenesis of rabies may lead to novel approaches to the therapy of the human disease. New antiviral agents and/or immunotherapies may be developed that have efficacy in established rabies. Clearly, anti-rabies virus-specific drugs are needed for successful therapy. Combinations of therapies may be more effective than single agents, similar to the current situation in other diseases such as cancer and in other viral infections, including human immunodeficiency virus infection and chronic hepatitis C infection (Jackson *et al.*, 2003). Novel methods for antiviral peptide discovery (Real *et al.*, 2004) as well as novel methods to identify antiviral targets for drug discovery (Wunner *et al.*, 2004) and to silence virus-specific gene expression with small interfering RNA molecules that prevent virus replication (Carmichael, 2002; Gitlin *et al.*, 2002; Plasterk, 2002) are needed and already are being considered. A better understanding of the role and importance of rabies virus receptors may be useful in the design of new therapies for rabies. Novel neuroprotective therapies may also be developed that will be applicable to many viral infections of the nervous system as well as non-infectious neurologic diseases.

Despite future therapeutic advances that may occur, it is unlikely that neurologic outcomes will improve or there will be good outcomes if the diagnosis is not made until late in the course of the disease. Rabies diagnosis is particularly difficult when there is no history of an animal exposure. Astute clinical practitioners always will be essential, even if there are improved diagnostic investigations (e.g. imaging and sensitive assays on body fluids, including saliva and cerebrospinal fluid) that allow confirmation of a diagnosis of rabies much earlier than is presently possible.

6 CONTROL OF ANIMAL RABIES

About half of the world's population lives in areas where canine rabies remains endemic and poses a continuing threat of exposure to humans (Meslin *et al.*, 1994; Knobel *et al.*, 2005). This is because approaches that have been successful in controlling canine rabies in developed countries have not been applied successfully in the developing countries for a variety of reasons. Among these are the high costs, lack of adequate infrastructure for management of dog rabies, widespread prevalence of feral dogs and cultural and religious objections to various animal control measures. Appropriate education of the population and health care professionals is also essential; many human deaths occur in developing countries because victims of dog bites do not seek medical treatment. One has only to realize what can be done by examining the strategies, methods and results of those countries in Latin America that have made great advances to date in the control and prevention of rabies transmitted by dogs (Belotto, 2004; Belotto *et al.*, 2005). They acquired the necessary knowledge for implementing rabies surveillance, control and prevention where the disease was transmitted. The World Health Organization recommends vaccination of 70% of the canine reservoir population in order to prevent outbreaks of rabies as the most economic and effective approach for preventing rabies in humans (Bögel and Meslin, 1990). If regional programs would support vaccination of 70% of the canine population and prevent over 90% of the major human rabies cases as predicted by Coleman and Dye (1996), a tremendous step forward would be taken to prevent human rabies. In the future, international and regional support for extensive canine vaccination programs in developing countries globally must be forthcoming (see Chapter 17).

Controlling rabies in other domestic and wildlife animal populations takes a multipronged approach. Modified live (attenuated) and inactivated rabies virus vaccines of cell or tissue culture origin are generally used as animal vaccines for domestic species (Blancou and Meslin, 1996; Reculard, 1996; Sizaret, 1996). Vaccination of wildlife, using oral rabies vaccines (ORVs), has been more challenging and, at the same time, very successful over the past 30 years, particularly for control of wildlife animal rabies in western Europe, southern Ontario and Canada and in Texas and the eastern USA (Aubert *et al.*, 1994; Stohr and Meslin, 1996; Fearneyhough *et al.*, 1998; MacInnes *et al.*, 2001; Krebs *et al.*, 2002; Rupprecht *et al.*, 2004; Slate *et al.*, 2005). Key priorities such as coordination of surveillance and rabies control programs within and between adjoining states and/or countries and effective communication and legal strategies must be addressed and adhered to in order to ensure that ORV strategies and programmatic goals for eliminating specific rabies virus variants from a dominant reservoir are met to produce optimal results (Slate *et al.*, 2005). Important lessons and guidelines can be drawn from the recent history, since 1998, of the challenges that the US Department of Agriculture, Animal and Plant Health Inspection Service, Wildlife Services (APHIS-WS) faced when it received its first federal appropriation to cooperate in existing ORV projects, to expand ORV programs to states of strategic importance in preventing the spread of specific terrestrial variants of the rabies virus and to assist in coordinating cooperative interstate ORV projects (Slate *et al.*, 2005). The National Rabies Management Team that was formed, with a vision for the National ORV Program to eliminate rabies in terrestrial carnivores, provided the guidance and recommendations to carry out the vision. The immediate goals were to prevent specific variants of rabies virus in the raccoon and gray fox (a strain unique to Texas) from spreading to new, uninfected areas (Krebs *et al.*, 2002; Slate *et al.*, 2002). One example to show how this has worked is through the effective implementation of 'point-infection control' strategies following a recent introduction of raccoon rabies into regions of Ontario, Canada. As a result of this emergency response plan and implementation, both Ontario and New Brunswick have controlled the spread of raccoon rabies (Rosatte *et al.*, 2001); nevertheless, programs like this are young in their development and total success cannot be fully evaluated without continued surveillance to monitor project effectiveness (Slate *et al.*, 2005). These programs are costly and they must be repeated on a regular basis until disease is eradicated. Hopefully, there will be greater emphasis in the future on assessing the reason why particular programs either succeed or fail. Local population dynamics and spatial spread of rabies and other infectious diseases must be understood and superimposed on efforts to assess the long-term success of expensive interventions (Childs *et al.*, 2000; Dobson, 2000). In the future, we can expect that geographic information systems and epidemiologic modeling will be used more widely to develop spatially explicit models in order to help refine and guide ORV distribution strategies as well as to aid in the development of contingency plans for targeting breaks in vaccine zones. Even current use of anticoagulants for the prevention of rabies in cattle transmitted by vampire bats (Thompson *et al.*, 1972) may be replaced by such novel strategies as using ORVs (Almeida *et al.*,

2005). In Latin America, the hematophagous bat (*Desmodus rotundus*) is one of the main reservoirs of rabies virus causing heavy economic losses in livestock annually. Bats in Brazil have been a frequently identified vector related to human rabies cases. Consequently, the recent demonstration that an ORV administered directly into the mouth or applied through paste onto the backs of *Desmodus rotundus* bats is immunogenic and protective against rabies virus challenge represents an important 'proof of concept' that oral vaccination may be a feasible method of controlling the disease in bats (Almeida *et al.*, 2005).

7 SUMMARY

An improved understanding of the pathogenesis of rabies may lead to novel therapeutic approaches to the management of human disease. There are many challenges in order to control and eliminate rabies in animals and prevent human disease and strong commitment from the governments of many countries is essential for this to be accomplished. Although costs are high, a long-term economic benefit can be anticipated as well as a reduction in human suffering with a large impact on children in developing countries.

REFERENCES

Almeida, M.F., Martorelli, L.F., Aires, C.C., Sallum, P.C. and Massad E. (2005). Indirect oral immunization of captive vampires, *Desmodus rotundus*. *Virus Research* **111**, 77–82.

APCRI (2006). Present scenerio of anti-rabies vaccination programme in India (Editorial). *Association for the Prevention and Control of Rabies in India Journal* **8**, 5–7.

Aubert, M.F.A., Masson, E., Artois, M. and Barrat, J. (1994). Oral wildlife vaccination field trials in Europe, with recent emphasis on France. In: *Lyssaviruses* (C.E. Rupprecht, B. Dietzschold and H. Koprowski, eds). pp. 219–243. Berlin: Springer-Verlag.

Bakker, A.B., Marissen, W.E., Kramer, R.A. *et al.* (2005). Novel human monoclonal antibody combination effectively neutralizing natural rabies virus variants and individual in vitro escape mutants. *Journal of Virology* **79**, 9062–9068.

Belotto, A.J. (2004). The Pan American Health Organization (PAHO) role in the control of rabies in Latin America. *Developmental Biology* **119**, 213–216.

Belotto, A., Leanes, L.F., Schneider, M.C., Tamayo, H. and Correa, E. (2005). Overview of rabies in the Americas. *Virus Research* **111**, 5–12.

Blancou, J. and Meslin, F.-X. (1996). Modified live-virus rabies vaccines for oral immunization of carnivores. In: *Laboratory Techniques in Rabies* (F.-X. Meslin, M.M. Kaplan and H. Koprowski, eds). pp. 324–337. Geneva: World Health Organization.

Bögel, K. and Meslin, F.-X. (1990). Economics of human and canine rabies elimination: Guidelines for programme orientation. *Bulletin of the World Health Organization* **68**, 281–291.

Carmichael, G.G. (2002). Medicine: silencing viruses with RNA. *Nature* **418**, 379–380.

Champion, J.M., Kean, R.B., Rupprecht, C.E. *et al.* (2000). The development of monoclonal human rabies virus-neutralizing antibodies as a substitute for pooled human immune globulin in the prophylactic treatment of rabies virus exposure. *Journal of Immunological Methods* **235**, 81–90.

Childs, J.E., Curns, A.T., Dey, M.E. *et al.* (2000). Predicting the local dynamics of epizootic rabies among raccoons in the United States. *Proceedings of the National Academy of Sciences of the United States of America* **97**, 13666–13671.

Chutivongse, S., Wilde, H., Supich, C., Baer, G.M. and Fishbein, D.B. (1990). Postexposure prophylaxis for rabies with antiserum and intradermal vaccination [see comments]. *Lancet* **335**, 896–898.

Coleman, P.G. (2004). Estimating the public health impact of rabies. *Emerging Infectious Diseases* **10**, 140–142.

Coleman, P.G. and Dye, C. (1996). Immunization coverage required to prevent outbreaks of dog rabies. *Vaccine* **14**, 185–186.

da Rosa E.S.T., Kotait, I., Barbosa, T.F.S. *et al.* (2006). Bat-transmitted human rabies outbreaks, Brazilian Amazon. *Emerging Infectious Diseases* **12**, 1197–1202.

Denduangboripant, J., Wacharapluesadee, S., Lumlertdacha, B. *et al.* (2005). Transmission dynamics of rabies virus in Thailand: implications for disease control. *BioMed Central Infectious Diseases* **5**, 52.

Dietzschold, B., Kao, M., Zheng, Y.M. *et al.* (1992). Delineation of putative mechanisms involved in antibody-mediated clearance of rabies virus from the central nervous system. *Proceedings of the National Academy of Sciences of the United States of America* **89**, 7252–7256.

Dobson, A. (2000). Raccoon rabies in space and time. *Proceedings of the National Academy of Sciences of the United States of America* **97**, 14041–14043.

Fearneyhough, M.G., Wilson, P.J., Clark, K.A. *et al.* (1998). Results of an oral rabies vaccination program for coyotes. *Journal of the American Veterinary Medical Association* **212**, 498–502.

Gitlin, L., Karelsky, S. and Andino, R. (2002). Short interfering RNA confers intracellular antiviral immunity in human cells (Letter). *Nature* **418**, 430–434.

Hanlon, C.A., Kuzmin, I.V., Blanton, J.D., Weldon, W.C., Manangan, J.S. and Rupprecht, C.E. (2005). Efficacy of rabies biologics against new lyssaviruses from Eurasia. *Virus Research* **111**, 44–54.

Harverson, G. and Wasi, C. (1984). Use of post-exposure intradermal rabies vaccination in a rural mission hospital. *Lancet* **2**, 313–315.

Haupt, W. (1999). Rabies – risk of exposure and current trends in prevention of human cases. *Vaccine* **17**, 1742–1749.

Jackson, A.C. and Wunner, W.H. (2002). *Rabies*. San Diego: Academic Press.

Jackson, A.C., Warrell, M.J., Rupprecht, C.E. *et al.* (2003). Management of rabies in humans. *Clinical Infectious Diseases* **36**, 60–63.

Jaijaroensup, W., Tantawichien, T., Khawplod, P., Tepsumethanon, S. and Wilde, H. (1999). Postexposure rabies vaccination in patients infected with human immunodeficiency virus. *Clinical Infectious Diseases* **28**, 913–914.

Knobel, D.L., Cleaveland, S., Coleman, P.G. *et al.* (2005). Re-evaluating the burden of rabies in Africa and Asia. *Bulletin of the World Health Organization* **83**, 360–368.

Krebs, J.W., Noll, H.R., Rupprecht, C.E. and Childs, J.E. (2002). Rabies surveillance in the United States during 2001. *Journal of the American Veterinary Medical Association* **221**, 1690–1701.

MacInnes, C.D., Smith, S.M., Tinline, R.R. *et al.* (2001). Elimination of rabies from red foxes in eastern Ontario. *Journal of Wildlife Diseases* **37**, 119–132.

Meslin, F.-X., Fishbein, D.B. and Matter, H.C. (1994). Rationale and prospects for rabies elimination in developing countries. Lyssaviruses. In: *Current Topics in Microbiology and Immunology* (C.E. Rupprecht, B. Dietzschold and H. Koprowski, eds). Vol. 187, pp. 1–26. Berlin: Springer-Verlag.

Nel, L.H. (2005). Vaccines for lyssaviruses other than rabies. *Expert Review of Vaccines* **4**, 533–540.

Plasterk, R.H. (2002). RNA silencing: the genome's immune system. *Science* **296**, 1263–1265.

Prosniak, M., Faber, M., Hanlon, C.A., Rupprecht, C.E., Hooper, D.C. and Dietzschold, B. (2003). Development of a cocktail of recombinant-expressed human rabies virus-neutralizing monoclonal antibodies for postexposure prophylaxis of rabies. *Journal of Infectious Diseases* **188**, 53–56.

Real, E., Rain, J.C., Battaglia, V. *et al.* (2004). Antiviral drug discovery strategy using combinatorial libraries of structurally constrained peptides. *Journal of Virology* **78**, 7410–7417.

Reculard, P. (1996). Cell-culture vaccines for veterinary use. In: *Laboratory Techniques in Rabies* (F.-X. Meslin, M.M. Kaplan and H. Koprowski, eds). pp. 314–323. Geneva: World Health Organization.

Rosatte, R., Donovan, D., Allan, M. *et al.* (2001). Emergency response to raccoon rabies introduction into Ontario. *Journal of Wildlife Diseases* **37**, 265–279.

Rupprecht, C.E., Hanlon, C.A. and Slate, D. (2004). Oral vaccination of wildlife against rabies: opportunities and challenges in prevention and control. *Developmental Biology (Basel)* **119**, 173–184.

Sizaret, P. (1996). General considerations in testing the safety and potency of rabies vaccines. In: *Laboratory Techniques in Rabies* (F.-X. Meslin, M.M. Kaplan and H. Koprowski, eds). pp. 355–359. Geneva: World Health Organization.

Slate, D., Chipman, R.B., Rupprecht, C.E. and DeLiberto, T. (2002). Oral rabies vaccination: a national perspective on program development and implementation. In: *Proceedings of the 20th Vertebrate Pest Conference*. pp. 232–240. Davis: University of California.

Slate, D., Rupprecht, C.E., Rooney, J.A., Donovan, D., Lein, D.H. and Chipman, R.B. (2005). Status of oral rabies vaccination in wild carnivores in the United States. *Virus Research* **111**, 68–76.

Stohr, K. and Meslin, F.M. (1996). Progress and setbacks in the oral immunisation of foxes against rabies in Europe. *Veterinary Record* **139**, 32–35.

Tantawichien, T., Jaijaroensup, W., Khawplod, P. and Sitprija, V. (2001). Failure of multiple-site intradermal postexposure rabies vaccination in patients with human immunodeficiency virus with low CD4+ T lymphocyte counts. *Clinical Infectious Diseases* **33**, E122–E124.

Thompson, R.D., Mitchell, G.C. and Burns, R.J. (1972). Vampire bat control by systemic treatment of livestock with an anticoagulant. *Science* **177**, 806–808.

Wang, Z.W., Sarmento, L., Wang, Y. *et al.* (2005). Attenuated rabies virus activates, while pathogenic rabies virus evades, the host innate immune responses in the central nervous system. *Journal of Virology* **79**, 12554–12565.

Warrell, M.J. and Warrell, D.A. (2000). Intradermal postexposure rabies vaccine regimens (Letter). *Clinical Infectious Diseases* **31**, 844–845.

World Health Organization (2002). World Health Organization position paper on rabies vaccines. *Weekly Epidemiological Record* **77**, 109–119.

World Health Organization (2005). World Health Organization Expert Consultation on Rabies: First Report. Geneva: World Health Organization.

Wunner, W.H., Pallatroni, C. and Curtis, P.J. (2004). Selection of genetic inhibitors of rabies virus. *Archives of Virology* **149**, 1653–1662.

Zhang, Y.Z., Xiong, C.L., Zou, Y. *et al.* (2006). Molecular characterization of rabies virus isolates in China during 2004. *Virus Research* **121**, 179–188.

Index